all eoprdi

tint 13 brag

$$\frac{20}{10} = \frac{x}{2}$$

$5k$ j $3k$

$z = 5.83k \angle 30.96°$

$R_T = 72.11$

0.00000002.65

26.53nF

Fundamentals of Electronics: DC/AC Circuits

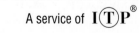

Fundamentals of Electronics: DC/AC Circuits

David L. Terrell

Delmar Publishers

an International Thomson Publishing company I(T)P®

Albany • Bonn • Boston • Cincinnati • Detroit • London • Madrid
Melbourne • Mexico City • New York • Pacific Grove • Paris • San Francisco
Singapore • Tokyo • Toronto • Washington

NOTICE TO THE READER

Cover Design: Nicole Reamer

Delmar Staff

Publisher: Michael A. McDermott
Acquisitions Editor: Gregory L. Clayton
Developmental Editor: Michelle Ruelos Cannistraci
Senior Project Editor: Christopher Chien
Production Manager: Larry Main

Art Director: Nicole Reamer
Marketing Manager: Kitty Kelly
Marketing Coordinator: Paula Collins
Editorial Assistant: Amy E. Tucker

COPYRIGHT © 2000
By Delmar Publishers
a division of International Thomson Publishing Inc.

The ITP logo is a trademark under license.
Printed in the United States of America

Dedication

To Linda, my best friend, whose talents, wit, and charm are without end.

For more information, contact:

Delmar Publishers
3 Columbia Circle, Box 15015
Albany, New York 12212-5015

International Thomson Publishing Europe
Berkshire House
168-173 High Holborn
London, WC1V7AA
United Kingdom

Thomas Nelson Australia
102 Dodds Street
South Melbourne,
Victoria, 3205 Australia

Nelson Canada
1120 Birchmont Road
Scarborough, Ontario
M1K 5G4, Canada

ITE Spain/Paraninfo
Calle Magallanes, 25
28015-Madrid, Espana

International Thompson Editores
Seneca 53
Colonia Polanco
11560 Mexico D. F. Mexico

ITP Southern Africa
Building 18, Constantia Square
138 Sixteenth Road
P.O. Box 2459
Halfway House
1685 South Africa

International Thomson Publishing Asia
60 Albert Street
#15-01 Albert Complex
Singapore 189969

International Thomson Publishing Japan
Hirakawa-cho Kyowa Building, 3F
2-2-1 Hirakawa-cho, Chiyoda-ku,
Tokyo 102, Japan

1 2 3 4 5 6 7 8 9 10 XXX 05 04 03 02 01 00 99

Library of Congress Cataloging-in-Publication Data
Terrell, David L.
        Fundamentals of electronics (DC/AC circuits) / David Terrell.
            p.   cm.
        Includes index.
        ISBN 0-8273-5340-5
        1. Electronics.   I. Title.
TK7816.T42   1998   98-44678
621.381—dc21   CIP

CONTENTS

chapter 5

chapter 6

chapter 7

chapter 8

chapter 9

chapter 10

Magnetism and Electromagnetism   **473**

chapter 11

Alternating Voltage and Current   **551**

chapter 12

Inductance and Inductive Reactance   **611**

chapter 13

Resistive-Inductive Circuit Analysis   **665**

preface

Introduction

This fundamental electronics text covers dc and ac circuits, passive components, circuit analysis, and troubleshooting. It is designed as a core text for electronics courses in community colleges, technical institutes, and vocational/technical schools. The material can be presented in a two-semester or two-quarter sequence covering dc and ac circuits.

In today's job market, it is no longer adequate for a technician to be "exposed" to an array of topics in electronics. Rather, an entry-level technician is expected to demonstrate a practical knowledge of important skills and concepts. One primary objective of this book is to present the material in such a way that the student learns the true foundations of electronics (e.g., Ohm's and Kirchhoff's Laws) early in the sequence of topics. Then, subsequent concepts such as numerical circuit analysis are not only easier to master because they are based on familiar tools, but the coverage of the advanced material inherently reviews those previously-learned concepts that form the key building blocks for effective learning. Unlike many other texts, this book is designed specifically to be used by a student. The conversational writing style used throughout the book conveys technical concepts in terms that can be readily visualized and understood by students. For example, when introducing the various units of measure for magnetic quantities, and how these quantities interact, the author contrasts the analysis of magnetic circuits with the previously learned analysis of electrical circuits. Contrasts, similarities, and analogies are used frequently in the book to tie new material to more familiar material that is already known to the student. For example, some aspects of passive filter circuits (Chapter 18) are viewed as voltage divider circuits, which have been used extensively in earlier chapters.

Teaching Philosophy

One very important belief that permeates this text is that if the knowledge and skills being presented are to be truly practical and ultimately valuable to the student, they must be acquired as an aggregate of electronics knowledge. That is, each new principle, technique, or fact must fit naturally and logically into the student's growing knowledge base. If subjects are presented as independent, standalone topics, then memorization tends to replace learning. While either method may produce adequate grades in an academic environment, only applied, integrated learning (as opposed to rote memorization) can be expected to produce a graduate with skills that will support a lifetime of continued learning. Additionally, a graduate who has integrated knowledge understands the bigger picture with reference to electronics. This understanding can provide a significant edge for an entry-level technician during interviews with potential employers.

One example of how this belief is manifest in the text is the fact that Ohm's and Kirchhoff's Laws are presented early in the text and provide the basis for nearly all subsequent analyses. Where practical, each new circuit type or analysis technique is introduced in terms of previously mastered material, so it fits naturally into the student's knowledge base.

Like no other text available, *Fundamentals of Electronics: DC/AC Circuits* will help students learn and understand faster and better. An important teaching principle practiced within the text is formally called the multipass, multilevel learning system. During the first pass, each new type of circuit is presented from a qualitative, nonmathematical viewpoint. This initial pass develops vocabulary and provides perspective on how the new skills compare with previously mastered subjects. One or more subsequent passes provide the numerical analysis aspects of the circuit. However, since the overall behavior of the circuit has already been mastered, the numerical analysis is much less confusing. This enables students to achieve greater depths of learning in the same or an even shorter time period than conventional methods would require.

Following are some brief examples to illustrate this teaching philosophy:

- Positive identification of series, parallel, series-parallel, and complex circuits is mastered before even the simplest series circuit is analyzed numerically.
- When numerical analyses for series and for parallel circuits are introduced, the overall voltage and current characteristics of the circuit are presented first. Second, circuit analysis based on Ohm's and Kirchhoff's Laws is discussed in a completely specified circuit (i.e., all component values are known). A subsequent pass then presents the numerical analysis of partially specified series and parallel circuits.
- In a similar manner, the topic of series-parallel circuits relies heavily on the concepts and computational strategies discussed with reference to series and parallel circuits.

This same underlying teaching strategy extends to include discussions on magnetism and electromagnetics. The student first learns qualitative concepts (e.g., what factors affect the value of voltage induced into a wire as it passes through a magnetic field, and what relationship–direct or inverse–these factors have). Subsequently, the numerical analysis of magnetic circuits is presented, but the student already has the vision needed to conceptualize the overall problem and the knowledge needed to judge the "reasonableness" of an answer. Without this important sequence of learning, students often compute absurd answers to problems, but are unable to detect the error. Instructors frequently hear the defensive comment, "But that's what the calculator said!"

Overall Organization

Fundamentals of Electronics: DC/AC Circuits contains 18 chapters. The book begins with a brief overview of electronics, an introduction to atomic structure, and a discussion of technical notation. The student is then introduced to various electronic components and electrical quantities that form the basis for further studies. Vocabulary building is an important part of a student's initial efforts.

Identification of all circuit types is then presented *before* a detailed mathematical analysis is made of any circuit type. Thus, a student is able to readily classify a circuit as series, parallel, series-parallel, or complex before learning to compute circuit values in even a

preface

basic series circuit. This simple step goes a long way toward reducing student confusion and increasing student success when more formal circuit analyses are presented.

In-depth analysis of series, parallel, series-parallel, and complex circuits then follows. The more intense numerical procedures are preceded by introductory discussions to clarify principles and to help students grasp the important concepts before becoming focused on intricate calculations. The introductory discussions provide the student with an intuitive or qualitative understanding of circuit or device operation; subsequent numerical analyses provide the technical depth required to support subsequent studies in advanced electronics courses. This sequence increases the student's ability to judge the reasonableness of mathematical results, and greatly strengthens his or her practical knowledge of the subject.

Kirchhoff's Voltage Law is presented as an integral part of series circuit analysis, and Kirchhoff's Current Law is presented as an integral part of parallel circuit analysis. Both of these laws and Ohm's Law are then used as the basis for most subsequent analyses. It is believed that if a student has a genuine command of these basic laws as they apply to circuit analysis, that student is armed with tools that will streamline learning throughout his or her career.

The DC portion of the text (Chapters 1 through 10) concludes with discussions on power sources, wire and cable, DC test equipment, and magnetism and electromagnetism. The magnetism topics include both qualitative and numerical analyses of magnetic circuits. Even here, however, the numerical computations are strongly rooted in previously studied topics such as Ohm's Law and series circuit analysis.

The AC portion of the book (Chapters 11 through 18) begins with the generation of AC and defines the various characteristics, values, and relationships associated with sine waves. Inductance and *RL* circuits are presented before capacitance and *RC* circuits. There is considerable disagreement (even within the same school) about the "best" sequence for these important topics. These two general subject areas consisting of Chapters 12 through 15 can be presented in a different sequence if the instructor so chooses. Thus, the expected sequence of Chapters 12, 13, 14, and 15 could be altered to the sequence of Chapters 14, 15, 12, and 13 if the instructor prefers to introduce capacitance and *RC* circuits prior to inductance and *RL* circuits. Depending on the intended depth of coverage, it might be necessary to present complex-number operations separately, since this subject is currently introduced as part of the *RL* circuits chapter. In either case, *RLC* circuits (Chapter 16) follows naturally. Both intuitive characteristics and numerical analyses of resonant and nonresonant *RLC* circuits are presented.

Transformers and mutual inductance are presented in Chapter 17. Again, the important sequence of intuitive introduction based on previous learning followed by more rigorous numerical computations is employed.

The book concludes with a discussion of passive filter circuits. This may be taught as a standalone chapter. Alternatively, an instructor may successfully teach the various filter circuits presented in Chapter 18 as isolated applications used to illustrate the principles discussed in the *RL, RC,* and *RLC* circuit chapters.

Chapter Format

Each chapter consists of the following sections:

- **Objectives** start off each chapter and identify the main areas to be discussed and the main skills to be acquired in the chapter. The general format of behavioral objectives is used, but with the stringent requirements on verbiage relaxed to simplify their expressions. The chapter objectives can be a powerful tool to guide the student's progress through the chapter, and to clarify the answer to the otherwise prevalent question in the student's mind, "What do I really need to know?".
- **Key Terms** are listed at the opening of each chapter. They are boldfaced and are defined with their first use in the text.
- **Examples and Solutions** guide students through step-by-step procedures to solve problems. Each discussion of a new analytical procedure is followed by one or more detailed examples that illustrate the steps in the solution.

Key Terms are listed at the opening of each chapter. They are boldfaced and defined with their first use in the text.

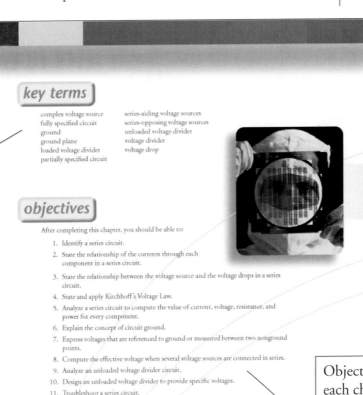

key terms

complex voltage source	series-aiding voltage sources
fully specified circuit	series-opposing voltage sources
ground	unloaded voltage divider
ground plane	voltage divider
loaded voltage divider	voltage drop
partially specified circuit	

objectives

After completing this chapter, you should be able to:

1. Identify a series circuit.
2. State the relationship of the currents through each component in a series circuit.
3. State the relationship between the voltage source and the voltage drops in a series circuit.
4. State and apply Kirchhoff's Voltage Law.
5. Analyze a series circuit to compute the value of current, voltage, resistance, and power for every component.
6. Explain the concept of circuit ground.
7. Express voltages that are referenced to ground or measured between two nonground points.
8. Compute the effective voltage when several voltage sources are connected in series.
9. Analyze an unloaded voltage divider circuit.
10. Design an unloaded voltage divider to provide specific voltages.
11. Troubleshoot a series circuit.

Objectives start off each chapter; they identify the main areas to be discussed and the main skills to be acquired in the chapter.

preface

- **Practice Problems** allow students to solve problems on their own and to reinforce important concepts. Answers to practice problems follow for immediate feedback. The combination of discussion, example, and practice exposes the student to a "Here's how you do it"–"Now watch me do it"–"OK, now you do it" learning sequence.

Examples and Solutions guide students with step-by-step procedures to solve problems

Art consists of four-color illustrations, schematics, and photos.

Practice Problems allow students to solve problems on their own and reinforce important concepts.

Key Points are highlighted in the margin to reinforce essential concepts.

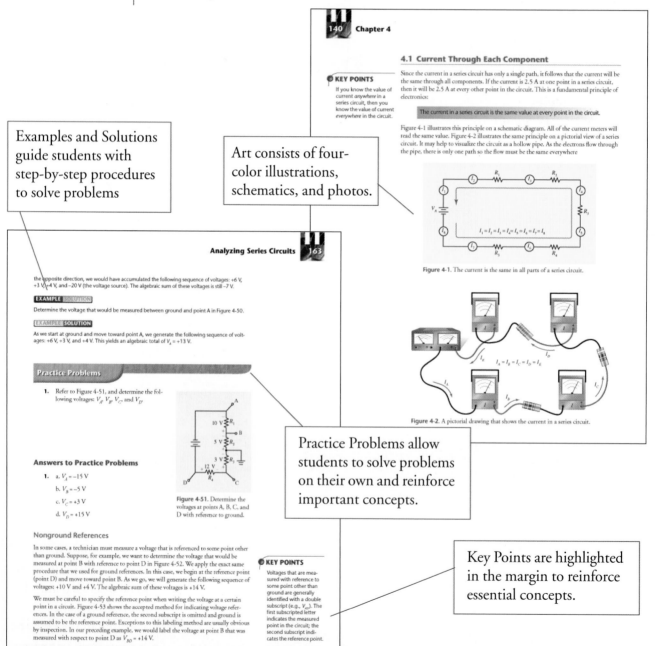

140 Chapter 4

4.1 Current Through Each Component

KEY POINTS

If you know the value of current *anywhere* in a series circuit, then you know the value of current *everywhere* in the circuit.

Since the current in a series circuit has only a single path, it follows that the current will be the same through all components. If the current is 2.5 A at one point in a series circuit, then it will be 2.5 A at every other point in the circuit. This is a fundamental principle of electronics:

The current in a series circuit is the same value at every point in the circuit.

Figure 4-1 illustrates this principle on a schematic diagram. All of the current meters will read the same value. Figure 4-2 illustrates the same principle on a pictorial view of a series circuit. It may help to visualize the circuit as a hollow pipe. As the electrons flow through the pipe, there is only one path so the flow must be the same everywhere

Figure 4-1. The current is the same in all parts of a series circuit.

Figure 4-2. A pictorial drawing that shows the current in a series circuit.

Analyzing Series Circuits 163

the opposite direction, we would have accumulated the following sequence of voltages: +6 V, +3 V, +4 V, and –20 V (the voltage source). The algebraic sum of these voltages is still –7 V.

EXAMPLE SOLUTION

Determine the voltage that would be measured between ground and point A in Figure 4-50.

EXAMPLE SOLUTION

As we start at ground and move toward point A, we generate the following sequence of voltages: +6 V, +3 V, and +4 V. This yields an algebraic total of $V_A = +13$ V.

Practice Problems

1. Refer to Figure 4-51, and determine the following voltages: V_A, V_B, V_C, and V_D.

Answers to Practice Problems

1. a. $V_A = -15$ V
   b. $V_B = -5$ V
   c. $V_C = +3$ V
   d. $V_D = +15$ V

Figure 4-51. Determine the voltages at points A, B, C, and D with reference to ground.

Nonground References

In some cases, a technician must measure a voltage that is referenced to some point other than ground. Suppose, for example, we want to determine the voltage that would be measured at point B with reference to point D in Figure 4-52. We apply the exact same procedure that we used for ground references. In this case, we begin at the reference point (point D) and move toward point B. As we go, we will generate the following sequence of voltages: +10 V and +4 V. The algebraic sum of these voltages is +14 V.

We must be careful to specify the reference point when writing the voltage at a certain point in a circuit. Figure 4-53 shows the accepted method for indicating voltage references. In the case of a ground reference, the second subscript is omitted and ground is assumed to be the reference point. Exceptions to this labeling method are usually obvious by inspection. In our preceding example, we would label the voltage at point B that was measured with respect to point D as $V_{BD} = +14$ V.

KEY POINTS

Voltages that are measured with reference to some point other than ground are generally identified with a double subscript (e.g., V_{45}). The first subscripted letter indicates the measured point in the circuit; the second subscript indicates the reference point.

xiv

preface

- **Key Points** are highlighted in the margin to reinforce essential concepts so the learner has immediate capsulation of the most important points in a particular section. These also serve as an excellent review tool, since they effectively summarize the more critical topics. If a particular key point remains unclear, the student can refer to the adjacent discussion for further details and clarification.

Troubleshooting and Circuit Analysis Tables help students organize circuit analysis problems.

Exercise Problems provide self-check and review questions at the end of each section.

Exercise Problems 4.6

1. If three voltage sources having voltages of 5 V, 25 V, and 50 V are connected in a series-aiding configuration, what is the effective circuit voltage?
2. What is the net applied voltage in the circuit shown in Figure 4-69?

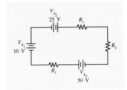

Figure 4-69. Find the net applied voltage in this circuit.

3. Refer to Figure 4-69. Which way does current flow in the circuit?
4. If a 12-V battery and a 24-V battery were connected in a series-opposing configuration, what would be the value of the net circuit voltage? Describe the polarity of the effective voltage.
5. Determine the effective circuit voltage in the circuit shown in Figure 4-70.

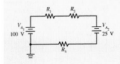

Figure 4-70. What is the effective circuit voltage?

6. In which direction does current flow in the circuit shown in Figure 4-70?
7. Classify the voltage sources in Figure 4-70 (series-aiding, series-opposing, or complex).
8. Classify the voltage sources in Figure 4-71 (series-aiding, series-opposing, or complex).
9. Calculate the effective voltage in the circuit shown in Figure 4-71.
10. In which direction does current flow in Figure 4-71?

5. Troubleshoot the circuit shown in Figure 4-87 using the Pshooter data provided in Table 4-6.

TESTPOINT	VOLTAGE		RESISTANCE	
	NORMAL	ACTUAL*	NORMAL	ACTUAL*
TP1	−25 V	[38]	94 kΩ	[113]
TP2	−17.8 V	[41]	67 kΩ	[131]
TP3	−15.1 V	[88]	57 kΩ	[93]
TP4	−10.4 V	[151]	39 kΩ	[73]
R_1	7.18 V	[51]	27 kΩ	[58]
R_2	2.66 V	[27]	10 kΩ	[24]
R_3	4.79 V	[144]	18 kΩ	[135]
R_4	10.4 V	[188]	39 kΩ	[89]
V_A	25 V	[70]	—	—

* Numbers in brackets refer to value numbers in Appendix A.

Table 4-6. Pshooting Data for the Circuit in Figure 4-87

Chapter Summary

- A series circuit is characterized by having one and only one path for current flow. The current is the same through every component in every part of the circuit. Each resistor in a series circuit drops a portion of the applied voltage. The portion of voltage dropped is in direct proportion to the resistance of the resistor as compared to the total resistance in the circuit. Total resistance in a series circuit is simply the sum of the individual resistances. Kirchhoff's Voltage Law states that the algebraic sum of the voltages (drops and sources) in any closed loop is equal to zero. This means that the sum of the voltage drops across the circuit resistance must always be equal to the value of applied voltage.

- Each resistor in a series circuit dissipates a certain amount of power, which is supplied by the applied voltage source. Total power consumption from the voltage source is equal to the sum of the individual resistor power dissipations.

- Fully specified series circuits can generally be analyzed by direct substitution of component values into an applicable equation. Analysis of partially specified circuits requires creative application of Ohm's Law, Kirchhoff's Voltage Law, power equations, and basic series circuit theory. In order to compute a circuit quantity, you must select an equation with only one unknown value. It is essential that properly subscripted values be used in all calculations to avoid accidental mixing of quantity types.

- Ground is a point in the circuit from which voltage measurements are referenced. Ground generally takes the physical form of a metal chassis, a wide PCB trace, or an entire plane in a multilayer PCB.

Chapter Summaries highlight the most important topics in the chapter.

preface

Equation List

(4-1) $V_A = V_1 + V_2 + V_3 + \ldots + V_N$

(4-2) $R_T = R_1 + R_2 + R_3 + \ldots + R_N$

(4-3) $I_T = I_1 = I_2 = I_3 = \ldots = I_N$

(4-4) $P_T = P_1 + P_2 + P_3 + \ldots + P_N$

(4-5) $V_T = V_{A_1} + V_{A_2} + \ldots + V_{A_n}$

(4-6) $V_T = V_{A_1} - V_{A_2}$

(4-7) $V_T = \pm V_{A_1} \pm V_{A_2} \pm \ldots \pm V_{A_n}$

(4-8) $V_X = \left(\dfrac{R_X}{R_T}\right) V_T$

Equation Lists provide a summary of all formulas presented in the chapter.

---

 196 Chapter 4

As a computer repair technician, you are troubleshooting a graphics display monitor. You have determined that the brightness control circuit is defective. Unfortunately, the required components cannot be purchased because the manufacturer has gone out of business. The portion of the circuit that is defective is essentially an unloaded voltage divider. It uses a potentiometer as a movable tap to vary the available voltage. You know that the current through the divider circuit was originally designed to be 20 mA. You also know that the divider is supplied by a 25-V source. In order to control the monitor properly, the potentiometer must be able to adjust the tapped voltage between the limits of +5 V and +10 V.

You have decided to design your own voltage divider circuit to use in the defective monitor. Figure 4-102 shows the schematic that you plan to use. Determine the resistance required for the potentiometer and for the two fixed resistors. Be sure to specify the minimum power rating for all three resistances.

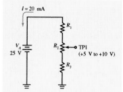

Figure 4-102. A voltage divider design problem using a potentiometer.

Technician Challenges provide a way to challenge the learner and exercise acquired knowledge in the solution of unfamiliar problems.

---

188 Chapter 4

• A series circuit may have more than one voltage source. If the multiple voltage sources all have the same polarity, the sources are said to be series-aiding. The effective voltage of series-aiding sources is the sum of the individual sources. When two voltage sources are connected with opposite polarities, they are said to be series-opposing. The net voltage of two series-opposing voltage sources is the difference between the individual source voltages. The polarity of the resulting voltage is the same as the larger of the two individual sources. If a series circuit has both series-aiding and series-opposing voltage sources, then the circuit has a complex voltage source. The effective voltage of a complex voltage source is equal to the algebraic sum of the individual source voltages.

• An unloaded voltage divider is an application of a series circuit. Voltage dividers are used to provide one or more lower voltages from a single higher source voltage. Unloaded voltage dividers cannot supply current to other circuits or devices. Because they are series circuits, unloaded voltage dividers can be analyzed and designed using the same equations and techniques used with other series circuits.

• Logical, systematic troubleshooting of series circuits is an important technician skill since it provides the foundation for troubleshooting of more complex circuits. The possible defects in a series circuit can be loosely categorized into two classes: opened (or increased resistance) and shorted (decreased resistance). When a resistor increases in value in a series circuit, all voltages in the circuit will measure nearer to the value of the supply terminal on the same side of the open. When a resistor decreases in value, all voltages in the circuit will measure nearer to the value of the supply terminal on the opposite side of the defect.

Review Questions

Section 4.1: Current Through Each Component

1. Current is the same in all parts of a series circuit. (True or False)

2. If a certain series circuit has 1.2 A leaving the voltage source, what percentage of this current will flow through the first resistor in the series chain? The last resistor in the series chain?

Section 4.2: Voltage Across Each Resistor

3. The voltage across a resistor is (*directly, indirectly*) proportional to its resistance value.

4. If one resistor in a series circuit is increased, what happens to the amount of voltage dropped across it?

5. If one resistor in a series circuit is increased, what happens to the amount of voltage dropped across the remaining resistors?

6. The largest resistor in a series circuit will always drop the (*least, most*) voltage.

7. The (*smallest, largest*) resistor in a series circuit will drop the least voltage.

8. Equal-valued resistors in a series circuit will have equal voltage drops. Explain why this is true.

End-of-Chapter Problems are grouped by section and provide testing.

- **Summaries** at the end of each chapter highlight the most important topics in the chapter. This can be a valuable study resource for a student.
- **End-of-Chapter Problems** are grouped by section to enable quick referencing of related discussions in the body of the chapter. The large number of available problems permits considerable flexibility based on the needs of the learner.
- **Equation Lists** provide a collection of all of the equations formally introduced within a given chapter. They provide a quick reference tool during homework assignments or when referencing material in a previous chapter.

Important Features

There are numerous unique features incorporated throughout the book to help stimulate critical thinking, problem solving, troubleshooting and circuit analysis skills.

- **Technician Challenge**. This feature provides a way to challenge the learner and to exercise acquired knowledge in the solution of unfamiliar problems. All of the Technician Challenge problems can be solved by applying the principles and methods presented in the particular chapter or preceding chapters. However, in many cases, they are tough enough to interest even the most accomplished students.
- **Use of Mathematics.** A brief look through the text will quickly reveal that mathematics is an integral part of this text, just as it is an integral part of electronics. However, the mathematics is not used to introduce a subject, nor is it generally used to explain circuit operation. Rather, an effort is made to give the student a solid intuitive understanding of a circuit or component's operation or behavior *before* the introduction of rigorous calculations. The mathematics simply provides a tool for the next level of analysis.

  There is debate as to whether calculations involving polar and rectangular coordinates should be included in an introductory electronics text. In this text, the use of complex numbers to represent circuit quantities and the associated conversions is presented in the text; however, the inclusion appears in key areas only (e.g., numerical analysis of series-parallel *RL* and *RC* circuits). These sections can be skipped in the classroom, if desired.

  Network theorems is another area of debate. This text presents all of the relevant theorems in a single chapter, which may be omitted if desired. However, since each of eight theorems is presented as a separate section within the chapter, instructors may select those areas that are most valuable to their particular classroom. The step-by-step procedures and examples provided in the chapter also make self-study a viable alternative.
- **Circuit Analysis Tables.** Solution matrices or tables are an integral part of detailed circuit analysis procedures for all basic circuit configurations. This is a powerful teaching tool that helps students organize circuit analysis problems and quickly contrast known quantities with unknown quantities for a given problem.

preface

- **Applied Technology.** Applied Technology sections illustrate how textbook concepts relate to practical applications outside the classroom. In each case, one or more applications are described that utilize the components, circuits, or principles discussed in the chapter. This helps the student appreciate the value of the material being studied by seeing the relationship to real applications.

- **Troubleshooting Exercises.** A great number of electronics students initially come to school with the belief that they are going to learn to "fix" anything that has a wire coming out of the back. In principle, they do, but in practice students rarely feel or believe that they can really troubleshoot a circuit. A majority of the chapters in this text include a section on troubleshooting.

The troubleshooting sections address three important areas: application of relevant test equipment, the development of a logical and systematic procedure, and troubleshooting practice. Of course, there is no substitute for the troubleshooting of actual defective equipment, but the PShooter exercises presented in this text require the student to utilize many of the same skills that would be required in an actual troubleshooting environment. These skills include test equipment selection, testpoint selection, measurement interpretation, schematic interpretation, and minimization of the number of circuit checks.

PShooter exercises consist of a schematic diagram with numbered testpoints. Each problem is accompanied by a table that lists the normal values (i.e., values with no defect) for each component and each testpoint in the circuit. Also associated with each measurable point are a series of numerical codes indicating the actual values of circuit quantities such as voltage, current, and resistance that are present with the defect in the circuit. The student chooses a testpoint and a measurement type. The associated code number is then used to look up the actual measured value in a table in the appendix of the book. A student should strive to make as few measurements as possible to develop a systematic troubleshooting procedure in preference to a simple sequence of guesses. While either method may ultimately locate the defect, the student is shown why only a systematic procedure is conducive to long-term success in the field.

The author also conveys important techniques that represent practicalities in a business environment. Examples include minimizing the soldering/desoldering of surface-mount components on an expensive circuit board, the discarding and subsequent replacement of suspected surface-mounted capacitors even if they are found to be good, and the value of in-circuit ohmmeter tests.

The Learning Package

The complete ancillary package was developed to achieve two goals:

1. To assist students in learning essential information needed to succeed to in the exciting field of electronics.
2. To assist instructors in planning and implementing their instructional programs for the most efficient use of time and other resources.

For The Student

Lab Manual–A lab manual is available to support *Fundamentals of Electronics: DC/AC Circuits.* The 54 experiments are closely related to the material presented in the text. Emphasis is placed on practical applications of theory and development of practical technician skills. The experiments are designed to promote student thinking. Cognitive skills are developed as the student progresses through the projects. Step-by-step procedures are minimized in preference to equivalent procedures, which provide flexibility, allow room for creativity, and demand authentic application of the principles presented in the text. The projects are also designed to provide the instructor with quick but effective means for determining learning success in each key area.

Order #: 0-8273-5342-1

Troubleshooting DC/AC Circuits with Electronics Workbench (with enclosed Circuits Data Disk)–This workbook teaches students how to troubleshoot circuits with the help of Electronics Workbench®. Students learn how to make measurements just as they would in an actual electronics lab. They replace components, test results, and observe effects on circuit operation, via a powerful computer-based learning tool.

Order #: 0-7668-1133-6

For The Instructor

Instructor's Guide –This comprehensive Instructor's Guide provides answers from the text and lab manual along with instructional strategies, sample course schedules, and transparency masters.

Order #: 0-8273-5341-3

Instructor's Teaching System (ITS)–An innovative Teaching System composed of tools that are designed to correlate, reinforce each other, and provide synergy, but can also be used individually. This binder of educational tools includes:

- **Instructor's Guide**
- **e.resource™**–Available all on one CD-ROM, this custom-designed teaching resource acts as an organizer and launch pad for all of the following applications:

  Computerized Testbank–Over 1,500 questions that provide mix-and-match capabilities for designing test for any phase of training.

  Electronic Gradebook–Tracks student performance, prints student progress reports, organizes assignments, and more, to simplify administrative tasks.

  On-line Testing–This is a feature of the computerized testbank that allows exams to be administered on-line via a school network or standalone PC.

PowerPoint Lecture Outline–Provides customizable teaching outlines for every chapter in the textbook. Graphics from the Image Library or your own can be imported to create individualized classroom presentations.

Image Library–Selected color illustrations from the textbook provide the instructor with another means of promoting student understanding. The Image Library allows the instructor to display or print images for classroom presentations.

• **Circuits Data Disk**–Diskette includes circuits from the textbook created in Electronics Workbench® to be used in conjunction with the accompanying troubleshooting workbook. Instructors may copy and distribute to students free of charge.

User Documentation–Additional print material provides information, instruction, and teaching hints on how to use all of the tools together for maximum benefit.

ITS Order #: 0-7668-0655-3

 Electronics into the Future: Circuit Fundamentals CD–*Electronics into the Future* is a new CD-Based Interactive Learning product that is the perfect technology accompaniment to your DC/AC Circuits text or as a standalone learning tool. This full multimedia product includes animation, video, and troubleshooting. Interactive, interesting, innovative–it is a great way to visualize and understand electronics concepts!

Network version Order #: 0-7668-0657-X
Student version Order #: 0-7668-0659-6

Online Companion™–This text has a companion website, which will have high appeal to both educators and students. Features include:

• Technology Updates
• Internet Activities
• Periodic Discussion Forums
• Ask the Author: Frequently Asked Questions
• Comprehensive listing of links to electronics industry and educational sites
• Instructors Resources
• Periodic RealAudio® broadcasts from author and contributors on new technology and educational strategies in electronics.
• Free Internet-based online syllabus through Thomson's World Class Course.
• Come visit our site at **www.electronictech.com**

Technology

Electronics Workbench

New to this edition are optional Electronics Workbench® projects. In the accompanying *Troubleshooting DC/AC Circuits with Electronics Workbench®* workbook, circuits are selected from the textbook and are built using Electronics Workbench®. Students learn how to make measurements as they would in the lab. Students replace components and then test the results to see how the changes affect the operation on the circuits. This is all done on a computer.

Using Electronics Workbench® to learn electronics still permits the learner to solve a problem in student-sized steps, yet gives a deeper understanding of the subject. The teaching of electronics has changed because today's students have much more to learn in the same or less time. Application of a powerful yet intuitive learning tool like Electronics Workbench® is essential so students can master all the important concepts and skills in the time available.

This provides a new and exciting dimension to learning electronics by providing 'live' circuits. The troubleshooting strategies used in the computer simulation are the same as those used with physical circuits in the lab. Circuits are provided on a diskette included in this package. The workbook coupled with Electronics Workbench insures that learning how to troubleshoot is swift and effective.

Electronics into the Future: Circuit Fundamentals CD

Electronics into the Future is a new interactive CD-ROM-based product from Delmar that is the perfect technology accompaniment to your DC/AC circuits text or as a stand-alone learning tool.

Electronics into the Future offers your students, and you as an instructor, a wide range of multimedia presentations and interactive simulations designed to develop and clarify important concepts in electronics.

Electronics into the Future puts the power of nonlinear learning technology in your and your students' hands. As a classroom learning tool, *Electronics into the Future* gives you the power to use technology to illustrate difficult concepts. As a standalone learning tool, *Electronics into the Future* enables students to learn theory and troubleshooting from interactive practical applications.

FEATURES:

- Available in either student or network version
- Contains six interactive modules featuring elements such as: presentation video (with practical application video opening and exposition circuit modeling of a practical example), Interactive Conceptualization, Troubleshooting, and much more.

  The content modules will cover the following topics: fundamental quantities, Ohm's law, power, series circuits, parallel circuits, series-parallel circuits, network theorems (Kirchhoff's, superposition, Thevenin's), and magnetism and electromagnetism.
- Mathematics for Electronics content module. This model can be accessed as a standalone module and is linked logically into each content module.

preface

- Recognizing that Troubleshooting is one of the key learning objectives for students in DC/AC Circuits, special emphasis is placed on Troubleshooting. *Electronics into the Future* has circuits for troubleshooting that were created in Electronics Workbench®.
- *Electronics into the Future* can launch the student edition of Electronics Workbench®. The network version allows instructors to add links to their own Electronics Workbench® files.

  Runs in an easy-to-navigate browser environment.

  Includes free copy of Netscape Navigator® and Microsoft Internet Explorer®.

- Please visit our web site at www.electronictech.com for more details and a demonstration. Please contact your sales representative for a full demonstration and for pricing details.

Acknowledgments

The Author and Delmar Publishers would like to extend their appreciation to the following reviewers who evaluated the textbook at different stages of development and provided valuable comments to publishing the best quality textbook:

John Avakian, College of San Mateo

R. Gary Bennett, Bryant & Stratton

Edward L. Bowling, Johnston Community College

Mike Brandt, Western Dakota Vo-Tech

Mel Bratley, Skagit Valley Community College

Dick Bridgeman, DeVry Institute of Technology

Kelley M. Brumley

Bruce Bush, Albuquerque Technical–Vocational Institute

William Campbell, Bay Area Vocational Technical School

John H. Carpenter, Sandhills Community College

Ron Craig, DeVry Institute of Technology

Donnin Custer, Western Iowa Tech

Joe Etminan, Rock Valley College

David Fridenmaker, ITT Technical Institute

John Giancola, DeVry Institute of Technology

Thomas Giasomo, Bryant & Stratton

Randy Goldsmith, Southeast Community College

Jim Hallam, Spokane Community College

John Edward Hart, DeVry Institute of Technology

Dr. Gaby Hawat, Valencia Community College

Timothy Haynes, Haywood Community College

David Heiserman, SweetHaven Publishing Services

William Hill, DeVry Institute of Technology

Robert S. Hockman, Texas State Technical College

Mark Hughes, Orangeburg-Calhoun Technical College

Wade Jung, ITT Technical Institute

Marc Kalis, Winona Technical College

Peter Kerckhoff, DeVry Institute of Technology

Stephen Kuchler, Ivy Tech

James R. Mallory, Rochester Institute of Technology

Brent Meyers, Oakland Community College

Dave Miller, New Hampshire Technical Institute

Shayan Mirabi, ITT Technical Institute

Donald Montgomery, ITT Technical Institute

Patrick O'Connor, DeVry Institute of Technology

Jeff Rankinen, Pennsylvania College of Technology

Jim Rhodes, Trident Technical College

Gus Rummel, Central Texas College

Gary Saylor, Crowder College

Allan Souder, Seneca College

George Sweiss, DeVry Institute of Technology

Ken Teel, ITT Technical Institute

Dale Willett, Lansing Community College

preface

The Author would like to thank many people and companies who provided photographs, physical components, and technical literature to help make this text representative of the most current technologies. The following individuals deserve special appreciation for their efforts on this project:

Denise Anderson, NEC Technologies (Golin-Harris Communications, Inc.), Chicago, IL

Robert K. Beachler, Altera Corporation, San Jose, CA

Derek Brooke, PREM Magnetics Incorporated, Johnsburg, IL

Linda Capcara, Motorola, Inc., Phoenix, AZ

Baldev Chaudhari, SMEC (Surface Mountable Electronic Components), Austin, TX

Linda Chocholka, Technipower, Danbury, CT

DeVry Institutes of Technology

Trisha Di Diego, LeCroy Corporate Headquarters, Chestnut Ridge, NY

Chris Dunlap, Murata Erie North America, Inc., Smyrna, GA

Greg Elmore, Maxtec International Corporation, Chicago, IL

Claude Forter, Fair-Rite Products Corporation, Wallkill, NY

Karl Grubb, Potter and Brumfield, Princeton, NJ

Kim Hall, Duracell Inc., Bethel, CT

Angie Hatfield, Motorola, Inc., Phoenix, AZ

Robyn L. Hensel, Meunier Electronic Supply, Indianapolis, IN

Roslyn Howard, IRC, Inc., Boone, NC

Johansen Manufacturing Corporation (John E. Deimel Company), Northfield, IL

Bryan Johnson, GE Medical Systems, Milwaukee, WI

Terri Johnson, Mallory (North American Capacitor Company), Indianapolis, IN

Dr. Kenneth Keenan, The Keenan Corporation, Pinellas Park, FL

Mary Levitt, AVX Corporation, New York, NY

Matt Maready, Ledtronics, Torrance, CA

Jane Marks, Deltrol Controls, Milwaukee, WI

Mike Morone, Meunier Electronic Supply, Indianapolis, IN

Julie Nusom, Tektronix, Inc., Wilsonville, OR

Darwa Renshaw, IDEC Corporation, Sunnyvale, CA

Janet Roberts, MicroSim Corporation, Irvine, CA

Will Rodgers, Florida Power Corporation, St. Petersburg, FL

Joan Roy, Leader Instruments Corporation, Hauppauge, NY

Loren Santow, Chicago, IL

Keith Schopp, Eveready Battery Company, Inc., St. Louis, MO

Craig Skarpiak, Andrew Corporation, Orland Park, IL

Ted Suever, Triplett Corporation, Bluffton, OH

Joan Sykes, Hewlett-Packard Company, Atlanta, GA

Robert S. Waselewski, Tampa Electric Company, Tampa, FL

Chris Willson, Cornell-Dubilier, Pickens, SC

atom

compound

electron

electron current flow

element

free electron

ionization

matter

mixture

molecule

negative ion

neutron

positive ion

powers of ten

proton

valence electrons

objectives

After completing this chapter, you should be able to:

1. List several major electronics applications.
2. Describe several career opportunities in electronics.
3. Recognize several electrical and electronic components.
4. Describe the composition of matter.
5. Explain the nature of current flow.
6. List several electrical quantities, their symbols, and their units.
7. Perform calculations in scientific and engineering notation.
8. Utilize metric prefixes in the expression of multiple and submultiple quantities.

Introduction to Electricity and Electronics

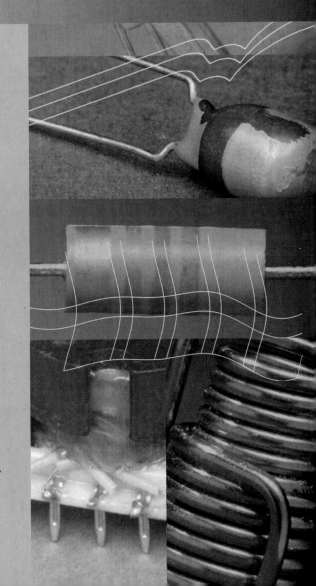

The field of electricity and electronics is exciting and rewarding, and it has a major impact on the lives of nearly every person in the world. This chapter will highlight a few applications of electronics and describe several of the endless career opportunities in electronics. You will learn about the composition of matter, including particles that are so small they cannot be seen—even with a microscope. This chapter will also introduce you to many of the units used to measure electrical quantities. Since the measurement of electrical quantities requires the use of both very small and very large numbers, you will also learn some new methods for manipulating and expressing numbers.

1.1 Electronics in Today's World

It is not practical to list all the applications of electronics devices and circuits. First, the list would be unbelievably long. Second, new applications are being added so quickly, you could never reach the end of the list. The following section illustrates the wide range of existing applications by describing a few of the major fields that utilize electronics.

Applications

COMPUTERS

All forms of modern computers are electronic devices. Computers are used to control such diverse systems as airplane navigation and flight, business record keeping, automotive emissions monitoring, elevator operation and scheduling, satellite communications, and environmental control. Figure 1-1 shows a computer system being used to simplify the measurement and display of complex electronic measurements.

COMMUNICATIONS

Communications takes a wide range of forms, including telephone systems, fiber-optic links, microwave transmissions, private and commercial radio networks, cellular phone systems, television, satellite links, and underwater communications for submarines.

AGRICULTURE

Electronics is used for many agricultural applications, including livestock monitoring and feeding, automatic equipment controls, livestock environmental control, tractor engine control, and dairy production control and monitoring.

Figure 1-1. Electronic computers are used in many fields.
(Courtesy of National Instruments)

CONSUMER ELECTRONICS

Consumer electronics includes such common items as televisions, video cassette recorders (VCRs), radios, stereo systems, laser disk players, home security systems, and many electronic convenience items.

AUTOMOTIVE ELECTRONICS

This area includes automotive sound systems, emissions control, engine control, radar sensors, two-way radios, mobile telephones, garage door openers, and complete onboard monitoring systems.

MARINE ELECTRONICS

Marine electronics includes sonar equipment, navigational systems, communications equipment, emergency beacons, marine telephones, and many engine control and monitoring devices.

AEROSPACE APPLICATIONS

There are incredible numbers of electronic systems aboard commercial and military aircraft. These include systems for navigation, weapons control, communications, autopilot, automatic electronic identification, environmental control, and electronic countermeasures for interfering with enemy electronics. Of course, aerospace applications extend to include the various forms of spacecraft and satellites, which have an extensive array of electronic systems. Figure 1-2 shows a spacecraft application of aerospace electronics.

INDUSTRIAL ELECTRONICS/ELECTRICITY

Industrial electronics is a broad classification that includes such things as automated factory equipment, computerized monitoring and control for chemical processes, automated warehouses, and many forms of process control systems for monitoring and controlling manufacturing processes.

ROBOTICS

Robots are essentially a subset of industrial electronics. They are computer-controlled machines that can perform repetitive tasks with exceptional precision, Figure 1-3 (See page 5). Some robots are mobile and resemble a vehicle. Others consist of a mechanical arm and hand that can be positioned very accurately.

SPORTS

There are numerous electronics applications in sports such as timers, scoreboards, electronic training equipment, speed-measuring equipment, communications systems, and video equipment.

Figure 1-2. A satellite tethered to the space shuttle to gather data on a variety of plasma physics and electrodynamics experiments. (Courtesy of NASA)

MEDICAL ELECTRONICS

Many medical procedures rely on sophisticated electronics devices. A few of these include x-ray machines, laser surgery, ultrasound imaging, magnetic resonance imaging (MRI), heart-pacing equipment, artificial kidneys, electrocardiograms, and many other diagnostic tools.

CAD/CAM

Computer-aided design (CAD) and computer-aided manufacturing (CAM), collectively called CAD/CAM, describe an integrated manufacturing environment where all phases of a product's development are done with the aid of a computer. Once a designer has finalized the design of a new product with an engineering computer, the engineering computer then communicates directly with other computers to order and distribute materials, and to actually build the new product.

Although the preceding applications only begin to scratch the surface of all existing applications for electronics devices, they do illustrate the extreme range of electronics applications. An interesting experiment is to try to name fields that have absolutely no relationship with electronics. It's very hard to do. For example, fields such as fishing, hog

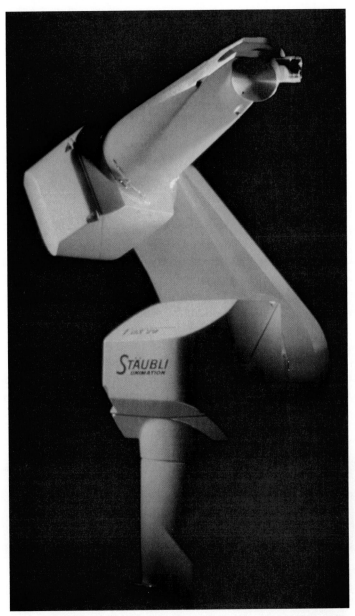

Figure 1-3. A modern industrial robot. (Courtesy of Stäubli)

raising, model building, spelunking, and scuba diving, which seem to be unrelated to electronics, all benefit from applied electronics.

Career Opportunities

Because there is such an endless list of electronics applications, there is a similar list of possible career opportunities in the field of electronics. The following career positions were selected to show the range of opportunities. The job titles are generic. Many companies refer to similar positions by different job titles.

BENCH TECHNICIAN

A bench technician troubleshoots and repairs defective electronic equipment that has been brought to the shop, Figure 1-4. The equipment may be any type of electronic device. Some examples are consumer products, computers, telephones, and medical devices.

FIELD SERVICE TECHNICIAN

The field service technician goes to the customer's site to perform equipment calibration, troubleshooting, and repair. Some equipment that is serviced in this manner includes computers, radar installations, security systems, environmental control systems, and point-of-sale (POS) terminals.

INDUSTRIAL TECHNICIAN/ELECTRICIAN

This technician performs the tasks of both bench and field service technicians, but is generally limited to a single site. For example, an industrial technician working for a chemical company would install, calibrate, and repair many types of industrial electronics equipment within her employer's facility—both in the plant and in the repair shop.

ENGINEERING TECHNICIAN

An engineering technician—also called a research and development (R&D) technician— generally works in a laboratory environment. He builds prototype systems, installs and

Figure 1-4. Bench tech at work. (Courtesy of National Instruments)

tests modifications to existing equipment, and plays a key role in the development cycle for new products. This technician often serves as a coordinator between the engineering and manufacturing departments during the early production phase of a product.

APPLICATIONS ENGINEER

Many technicians become applications engineers once they have become experts on their employer's products. An applications engineer meets with customers, analyzes their applications, and shows the customer how to effectively utilize the products sold by her company. This position requires strong technical skills and well-developed interpersonal skills.

TECHNICAL WRITER

Most companies produce documentation to accompany their products. Some of the documentation is proprietary and is used only within the company. Other documentation, such as the operator's manual, is delivered to the customer. A technical writer develops this documentation.

KEY POINTS

In addition to technical skills, the successful technician also needs to develop interpersonal skills and communication skills.

If you want to succeed in career positions similar to those cited here, then you will want to master the concepts and techniques presented in this text. This text is about basic electronics. An understanding of basic electronics is essential for success in any branch of electronics.

Important Milestones

This section is included to provide you with a brief historical perspective. Figure 1-5 shows a time line extending from the seventeenth century with several key milestones along the way. Each milestone marker lists a major invention or discovery that helped shape the field of electronics. The time line represents over 330 years of electronics developments.

It is particularly interesting to note that the rate of significant discoveries is continuing to increase. For example, the first nine discoveries listed in Figure 1-5 account for 241 years of the time line. The next six developments occurred over a 52-year period. The last three milestones shown in Figure 1-5 cover only 18 years. A complete chart of all significant discoveries in electronics would be enormous, but it would provide even more evidence of the tremendous increase in the rate of important discoveries.

KEY POINTS

The development of new devices and new technologies is continuing at an incredible, and increasingly rapid, rate.

Electronic Components

As you study the material in this and subsequent texts, you will learn the electrical characteristics of many electrical and electronic components. This section is meant to introduce you to the physical appearance of some of the components you will be studying. Similar devices can often have dramatically different appearances. As a technician, you must learn to recognize a component from its physical appearance. You will learn the operation of some of these components later in this text.

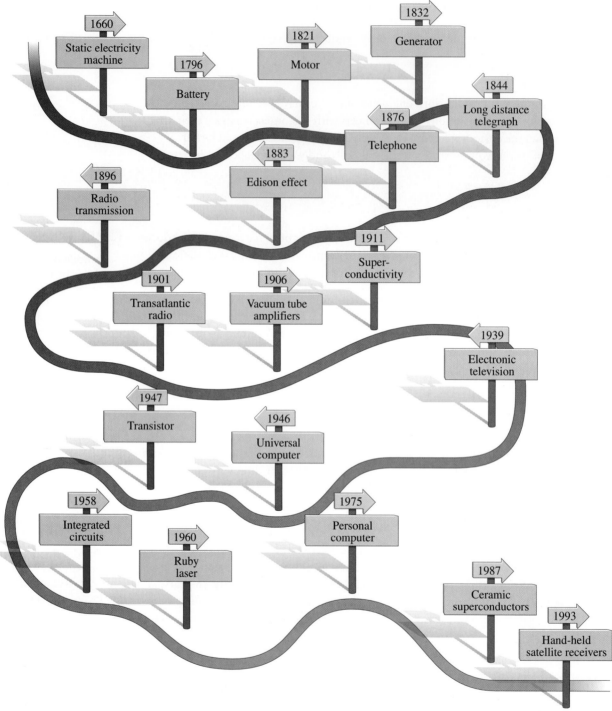

Figure 1-5. Today's electronics is a result of many historic discoveries.

RESISTORS

Resistors are used to limit the amount of current flow in an electrical circuit. Resistors vary in size from less than 0.04 inch to well over 12 inches in length. They are generally coded—either with numbers or colored bands—to indicate their value. Figure 1-6 shows examples of several resistor types.

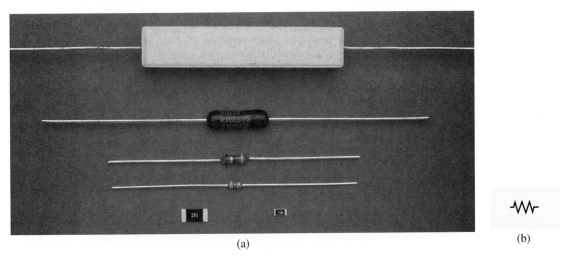

(a)

(b)

Figure 1-6. (a) Resistors vary in value and physical appearance. Their value is indicated with a printed number or with color-coded bands. (b) The schematic symbol for a resistor.

CELLS AND BATTERIES

You are already familiar with several types of cells and batteries that are used as sources of voltage: flashlight battery (D cell), 9-volt transistor radio battery, AAA cell, 12-volt car battery, watch battery, and so on. Figure 1-7 shows several typical cells and batteries. Some batteries are small enough to fit inside a watch or hearing aid. Others have a volume of several cubic feet and weigh hundreds of pounds.

(a)

(b)

(c)

Figure 1-7. (a) Batteries vary in voltage, current capacity, chargeability, and physical appearance. (Courtesy of Duracell) (b) Schematic symbol for a cell. (c) Schematic symbol for a battery.

INDUCTORS

This is a fundamental electrical component—also called a coil—that consists of a spiraled or coiled wire. The inductor is used to store electrical energy in the form of an electromagnetic field. Figure 1-8 shows several representative inductors.

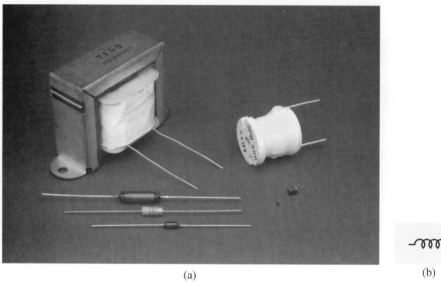

(a)   (b)

Figure 1-8. (a) Representative inductors. (b) Schematic symbol for an inductor.

CAPACITORS

A capacitor is another basic electrical component that is used to store electrical energy in the form of an electrostatic field. Capacitors vary widely in value, appearance, and physical size. Figure 1-9 shows several typical capacitors.

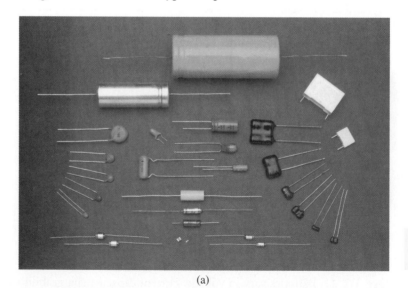

(a)   (b)

Figure 1-9. (a) Typical capacitors. (b) Schematic symbol for a capacitor.

TRANSISTORS AND DIODES

These are called solid-state components or semiconductors. They are used in many circuits, including amplifiers. Figure 1-10 shows several transistors and diodes.

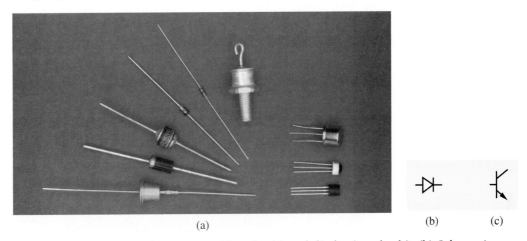

(a) (b) (c)

Figure 1-10. (a) Assorted transistors (three leads) and diodes (two leads). (b) Schematic symbol for a diode. (c) Schematic symbol for a transistor.

INTEGRATED CIRCUITS

These are incredibly complex devices that are actually miniature versions of complete circuits. Some integrated circuits have over a million equivalent components (resistors, transistors, and so on). Figure 1-11 shows several types of integrated circuits.

Figure 1-11. Several types of integrated circuits. (Courtesy of Altera Corporation)

TRANSFORMERS

A transformer is basically two or more inductors whose electromagnetic fields interact. Transformers are used to transform (i.e., increase or decrease) alternating voltages like the 120-volt, 60-hertz voltage that is supplied to your house by the power company. Figure 1-12 shows some sample transformers. Transformers range in size from less than 0.2 inch on a side to more than 10 feet on a side.

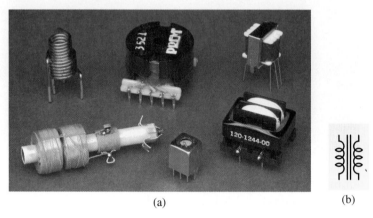

(a)   (b)

Figure 1-12. (a) Typical transformers. (b) Schematic symbol for a transformer.

SWITCHES

Switches are used to open or close an electrical circuit. They are used in your home to control the lights, in your auto to operate the horn, on your radio to apply power, and on your vacuum cleaner to turn it on and off. Switches vary in complexity, but always serve the basic purpose of opening or closing a circuit. Figure 1-13 shows a representative assortment of switches.

(a)   (b)

Figure 1-13. (a) A switch assortment. (b) Schematic symbol for a switch.

FUSES AND CIRCUIT BREAKERS

Figure 1-14 shows a variety of fuses and circuit breakers. These devices are used to protect a circuit from excessive current flow.

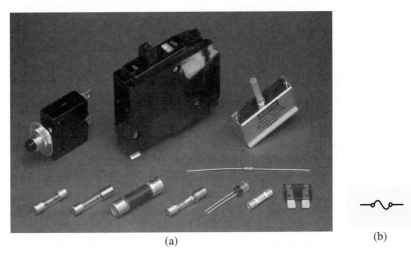

(a)

(b)

Figure 1-14. (a) An assortment of fuses and circuit breakers. (b) Schematic symbol for a fuse.

MOTORS

A motor is an electromechanical device that converts electrical energy into mechanical energy. Motors vary in size from microscopic (inside an integrated circuit) to large motors that are several feet high. Figure 1-15 shows some representative motors.

CATHODE-RAY TUBES

The screen of your television is an example of a cathode-ray tube (CRT). These devices are also used for radar displays and in some electronic test equipment. Figure 1-16 shows a representative CRT.

INDICATORS

There are many types of indicating devices that emit light. Some are light sources such as a common lightbulb. Others are more sophisticated and provide alphanumeric information.

Exercise Problems 1.1

1. Name at least three major areas that utilize electronics.
2. There are many career opportunities for an electronics technician. (True or False)
3. Name two kinds of electronics components that are made of coiled wire.

Figure 1-15. Motors change electrical energy into mechanical energy. (Courtesy of Barber-Colman Company)

Figure 1-16.
A cathode-ray tube.
(Courtesy of NEC
Technologies, Inc.)

4. Name two electronics components that are used to store energy.
5. Name a device that converts electrical energy into mechanical energy.

1.2 Atomic Structure

It is important for an electronics technician to have a good mental picture of the atomic structure of matter. An intuitive appreciation for atomic structure will enhance your understanding of many important phenomena in electronics. This basic knowledge is also needed to understand the behavior of the electronic components introduced in the previous section.

States of Matter

⦿ **KEY POINTS**

All physical things consist of matter.

All of the things in the world that we can see, smell, or touch are various forms of **matter.** Matter can have three physical states. Some matter is in the solid state at all times: trees, rocks, furniture, basketballs. Some matter is usually in the liquid state: gasoline, milk, orange juice. Still other types of matter are naturally in a gaseous state: air, carbon dioxide, ozone. Finally, some kinds of matter can be found in more than one physical state: water, oxygen, nitrogen. You are already familiar with the three states of water: solid (ice), liquid (water), and gaseous (steam). Materials used to construct electronic components are taken from each of the three classes or states of matter.

Composition of Matter

To appreciate the composition of matter, let's closely examine a container of salt water. We shall look closer and closer at the salt water to discover its physical makeup. Figure 1-17 shows our original container of salt water.

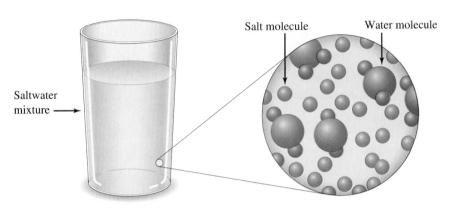

Figure 1-17. Salt water is a mixture made up of salt and water.

MIXTURE

If we remove a drop of salt water from the container, the drop is still composed of salt water. If we divide the drop into progressively smaller quantities, we can eventually reach the point shown in the inset of Figure 1-17. As you can see from Figure 1-17, our magnified sample consists of *discrete* quantities of salt and water. They are not chemically altered, they are simply in close proximity. We call this type of material a **mixture.** Other mixtures could be made by combining sand and sugar, salt and pepper, water and oil, or gold dust and sawdust.

COMPOUND

If we continue to examine the salt water more closely, we will find that each of the materials that form the salt water are composed of smaller building blocks. If we sort the various building blocks such that only one type is present, then we will have pure salt or pure water. Each of these substances is called a **compound.** This is illustrated in Figure 1-18.

MOLECULE

Now let us examine either of the compounds more closely. If we further magnify the building blocks that form the water in our original saltwater mixture, we will find that each of the building blocks is similar. Further, each of the blocks exhibits all of the characteristics of pure water. We call these building blocks molecules. A **molecule** is the smallest quantity of a compound that retains the characteristics of the compound. Figure 1-19 illustrates this idea.

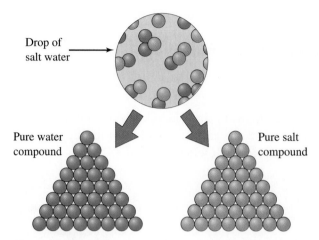

Figure 1-18. Salt water can be separated into pure salt and pure water.

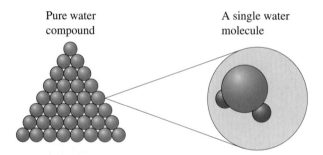

Figure 1-19. A molecule is the smallest quantity of a compound that still has the properties of the original compound.

ELEMENT

Now let's select a single water molecule and examine it more closely. Figure 1-20 reveals that each molecule is composed of smaller building blocks. These smaller blocks are called elements. A water molecule is composed of two different elements: hydrogen and oxygen. If we had examined the salt portion of the original saltwater mixture, we would have found that the salt molecules were composed of the elements sodium and chlorine.

An **element** is a fundamental building block that can combine with other elements to form compounds. In the case of water, two parts of hydrogen combine with one part of oxygen to form water. The rules for allowable combinations are well defined and are described in the study of chemistry.

ATOM

If we now examine a given element more closely, we will again see that it is made up of still smaller building blocks. As we have seen with the larger blocks, each of these

KEY POINTS

An element is composed of atoms.

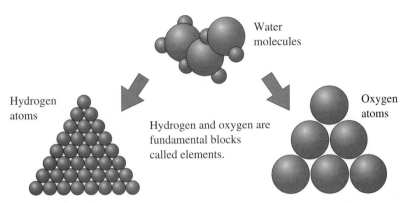

Water
molecules

Hydrogen
atoms

Hydrogen and oxygen are
fundamental blocks
called elements.

Oxygen
atoms

Figure 1-20. Water molecules are composed of the elements hydrogen and oxygen.

smaller building blocks is identical to one another. These smaller blocks are called **atoms,** and are the smallest quantity of an element that can exist and still retain the properties of the element.

SUBATOMIC PARTICLES

We shall now take a single atom and zoom in even closer. Figure 1-21 shows that a single atom is composed of a miniature planetary system similar in concept to our own solar system. Each atom has a dense nucleus, which contains two major subatomic particles: **protons** and **neutrons**. Protons are positively charged particles that have a mass of 0.000000000000000000000001672 grams. Later in this chapter, we will learn to express numbers like this in a more convenient form. Neutrons have a mass similar to that of protons, but neutrons have no charge. For the purposes of our present discussion, the mass of a particle can be thought of as being equivalent to the weight of the particle.

Electrons are negatively charged particles that travel in orbits around the nucleus of an atom. The mass of a proton (or neutron) is about 1,836 times more than the mass of an electron. The electric charges of a proton (+) and an electron (–), however, are similar in magnitude.

Since the nucleus is composed of positive (proton) particles and neutral (neutron) particles, it has an overall positive charge. The positive charge of the nucleus attracts the electrons, which are negatively charged, and helps keep them in orbit. The complete explanation for how and why electrons remain in stable orbits is not yet known. But, for our purposes, it is adequate to visualize that the positive attraction of the nucleus to the negative electrons is counterbalanced by the centrifugal force of the orbiting electrons. This is similar to the way the earth's gravity keeps a satellite in orbit. The atom shown in Figure 1-21 is a representative hydrogen atom. Most hydrogen atoms are even simpler and have no neutrons in their nuclei. Atoms of the other elements are more complex and have more protons, neutrons, and orbiting electrons.

KEY POINTS

Atoms consist of sub-atomic particles called protons, electrons, and neutrons.

KEY POINTS

Protons and electrons have a positive and negative charge, respectively. Neutrons have no charge.

KEY POINTS

Protons and neutrons are tightly packed in the nucleus of the atom. Electrons orbit the atom in various shells or energy levels.

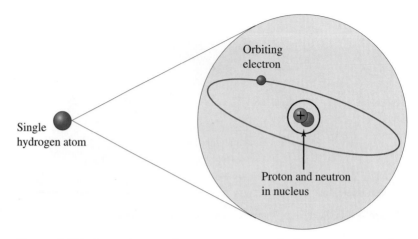

Figure 1-21. Atoms have a planetary structure similar to our own solar system.

NET ATOMIC CHARGE

Unless an atom has been altered by the absorption of external energy, there will always be the same number of protons in the nucleus as there are electrons in the surrounding orbits. Now, since the charges of the electrons and protons are equal in magnitude, but opposite in polarity, the overall net charge of an atom is zero. This is a stable condition unless external energy is introduced.

MODEL OF THE ATOM

Let us now study the behavior of the planetary electrons and understand how atoms of different elements differ from each other. We shall build our discussion around a model that was proposed by Neils Bohr, a Danish physicist, in 1913.

SHELLS

First, it can be shown that the orbits of the electrons as they go around the nucleus occur at certain definite altitudes. Thus, an electron may orbit at one level or another, but no electrons will orbit in between these two levels. These levels, called shells, are pictured in Figure 1-22.

Each electron within a given shell has a definite energy level. The more energy an electron acquires, the higher (further from the nucleus) the orbit will be. Figure 1-23 illustrates the concept of energy levels.

ENERGY EMISSION AND ABSORPTION

In order for an electron to move from one shell to a higher shell, two conditions must be met:

- there must be room in the next higher shell, and
- the electron must absorb a definite amount of energy.

The energy needed to cause an electron to move to a higher orbit can come from several external sources. These sources include heat energy, light energy, nuclear radiation, magnetic fields, electrostatic fields, or impact with another particle.

KEY POINTS

An electron will move to a higher orbit when it absorbs energy or a lower orbit when it loses energy.

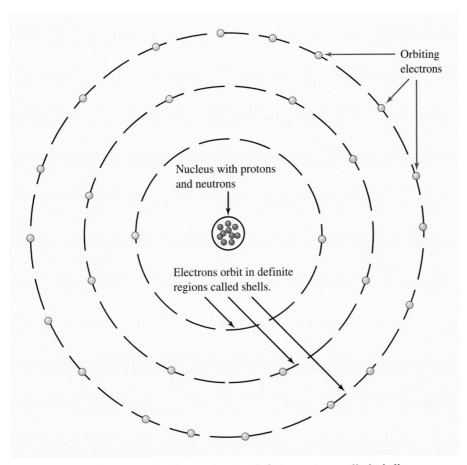

Figure 1-22. Electrons orbit the nucleus in definite regions called *shells*.

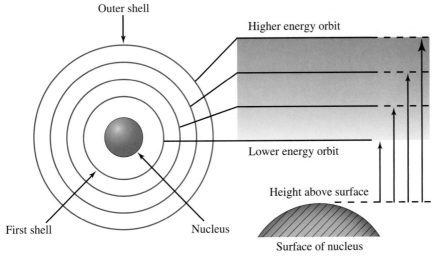

Figure 1-23. Higher orbits have electrons with higher energy levels.

Similarly, in order for an electron to fall to a lower orbit, both of the following conditions must be satisfied:

- there must be room in the next lower orbit, and
- the electron must give up a definite amount of energy.

The energy that is given up when an electron falls to a lower level often takes the form of light energy. The wavelength (color) of the light depends on the specific element being studied. This characteristic is useful for identifying unknown elements. It also forms the basis for light amplification by stimulated emission of radiation (LASER) operation. Figure 1-24 illustrates the relationship between energy emission/absorption and electron orbit changing.

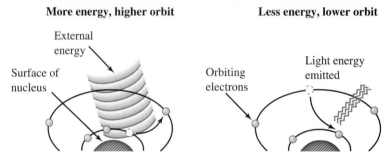

Figure 1-24. An electron absorbs energy to move to a higher orbit. Energy is released when an electron falls to a lower orbit.

Valence Electrons

Now let us narrow our discussion of the atom to the outermost orbit of electrons. These outer-orbit electrons are called **valence electrons.** Since the valence electrons are the farthest away from the nucleus, they have the least attraction. If a valence electron absorbs sufficient energy, it will attempt to move to a higher orbit. However, in the process of doing so, it can escape the attraction of the atom's nucleus and become a **free electron.** The amount of energy required to free an electron from an atom is determined by the type of element being considered. Figure 1-25 demonstrates the production of free electrons.

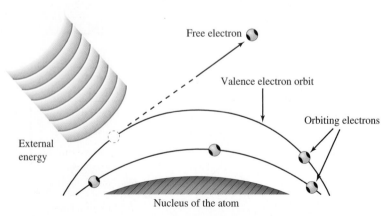

Figure 1-25. Valence electrons can absorb enough energy to escape the hold of the nucleus and become free electrons.

Ions

Earlier it was noted that the overall net charge of a stable atom is zero. This is because the number of protons in the nucleus is the same as the number of orbiting electrons. Also, the protons and electrons have equal but opposite polarity charges.

When a valence electron absorbs energy and becomes a free electron as illustrated in Figure 1-25, the charge balance of the atom is upset. If one valence electron leaves the orbit of a neutral atom, then the atom will be left with more protons (+) than electrons (–). Thus, the overall net charge of the atom is now positive. We call this positively charged atom a **positive ion.**

Once a valence electron (–) has escaped the orbit of an atom and becomes free, it no longer has the counterbalancing effect of a corresponding proton (+) in the nucleus. The free electron may lose energy and fall into orbit around a neutral atom. When this occurs, the atom will then have more orbiting electrons (–) than corresponding protons (+) in the nucleus. The overall net charge on the atom, then, will be negative. We call this negatively charged atom a **negative ion.**

Positive ions, negative ions, and free electrons are all important particles in the study of electronics. Figure 1-26 shows the formation of positive and negative ions. The act of forming ions is called **ionization.**

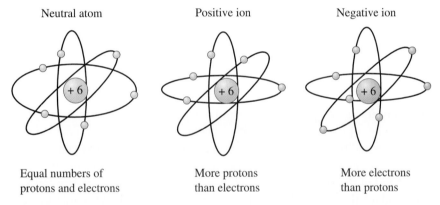

Neutral atom	Positive ion	Negative ion
Equal numbers of protons and electrons	More protons than electrons	More electrons than protons

Figure 1-26. When an electron moves from one neutral atom to another, the original atom becomes a positive ion and the receiving atom becomes a negative ion.

CURRENT FLOW

Figure 1-27 shows a cross-sectional view of a length of copper wire. Copper wire is used extensively in electronic circuits to interconnect the various components. The inset in Figure 1-27 shows a simplified representation of a copper atom. It has a total of 29 protons and 29 orbiting electrons. The 29 electrons are distributed into four major shells having 2, 8, 18, and 1 electrons each. The outermost orbit has a single valence electron, which is very loosely held by the atom. With only small amounts of energy, many of these valence electrons escape their parent atoms and become free electrons. At room temperature, there is enough thermal energy present to ionize many of the copper atoms. Thus, it follows that at room temperature there is an abundance of free electrons within the body of the copper wire.

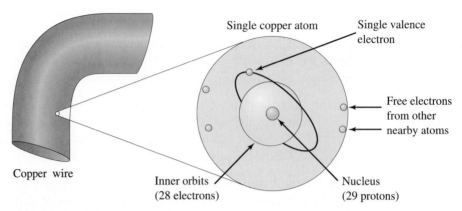

Figure 1-27. Cross-sectional view of a copper wire showing the valence electrons and the free electrons.

Try to visualize being inside the copper wire. Imagine what you would see. First, you should notice that there is a lot of open space. The atoms and free electrons in the copper wire are tiny compared to the vast space between copper atoms. If, for example, you visualize the nucleus of the copper atom as being the size of a tennis ball, then the valence electron's orbit will have about a 6-mile diameter. The electron itself will be a very tiny object. As you mentally travel through the copper wire, you will see many valence electrons skipping out of orbit and zooming off as free electrons. Sometimes one of the free electrons approaches a positively charged copper atom (i.e., one whose valence electron has already left), and falls into its valence orbit for a while. At any given time, however, there are many free electrons sailing randomly through the vast space between the copper atoms.

Figure 1-28 illustrates the results when we put a negative charge (many excess electrons) on one end of the copper wire and a positive charge (positive ions) on the other end. The charges are generated by the chemical action in the battery. The positive charge on one end of the wire attracts all of the electrons in the copper wire. Similarly, the negative charge on the opposite end of the wire repels every electron in the length of wire. The copper atoms are massive and relatively immobile so they remain in position. Further, the charges on the

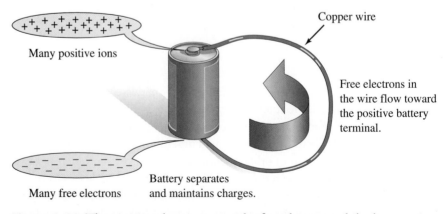

Figure 1-28. The positive charge attracts the free electrons while the negative charge repels them. This causes electron current flow.

ends of the wire are not sufficiently high to pull the lower-level electrons out of orbit. But whenever one of the valence electrons escapes the orbit of a copper atom, it is immediately affected by the charges on the ends of the wire. Since the electron has little mass and is very mobile (i.e., free), it races toward the positive terminal of the battery.

As electrons in the copper wire leave the vicinity of the negative battery terminal, they leave positive ions behind. These positive ions attract other electrons from the abundance of free electrons on the negative end of the battery. Similarly, when a free electron reaches the positive end of the copper wire, it is attracted by the positive ions in the battery and leaves the wire. The continuous directed movement of electrons is called **electron current flow.** This process will continue as long as there is an excess of free electrons on one end of the wire and an abundance of positive ions on the other end. As you will study in a later chapter, this is exactly what a battery does. It maintains a supply of free electrons at the negative terminal of the battery and an abundance of positive ions at the positive terminal.

Figure 1-29 shows a mechanical analogy to further clarify the concept of current flow in a conductor. Here, we see a tube representing a length of copper wire and rubber balls representing electrons. A "battery" continually separates the "electrons" after they recombine with the "positive ions." This maintains the supply of free "electrons" while ensuring that there are plenty of "positive ions" to catch the electrons as they exit the "wire." After "electrons" leave the battery, they travel through the "wire" in a directed motion and exit the opposite end of the "wire." They are collected here by the "positive ions." The ability to visualize models or equivalent mechanisms is a valuable asset toward understanding abstract electronics concepts. As you study new ideas, always try to create a mental picture of the activity being described.

> **KEY POINTS**
>
> When a charge difference is applied and maintained across a length of copper, the free electrons move in a directed manner, forming what we call electron current flow.

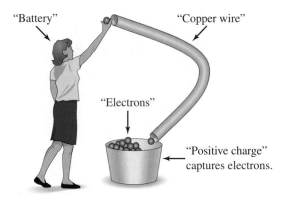

Figure 1-29. A mechanical analogy to illustrate electron current flow.

Exercise Problems 1.2

1. Name the three states of matter and give an example that illustrates each state.
2. Which is the larger building block, an atom or an electron?
3. Which is the larger building block, an element or a molecule?
4. An element is composed of building blocks called molecules. (True or False)

5. Compare the mass of a proton to that of an electron.
6. What is a positive ion and how is it created?
7. What would happen to the average number of free electrons in a length of copper wire if the temperature of the wire were increased?
8. A copper wire has many free electrons at room temperature. (True or False)
9. If a certain neutral atom has eleven orbiting electrons, how many protons will be found in the nucleus?

1.3 Units of Measure and Technical Notation

Units of Measure

There are many quantities that must be measured or expressed in the field of electronics. Each quantity has a unit of measurement (e.g., feet, inches, pounds), an abbreviation, and a symbol. As a technician, you will need to know the unit, abbreviation, and symbol for each quantity used in your field of interest.

Table 1-1 lists many of the basic electrical quantities along with their unit, abbreviation, and symbol. Table 1-1 also provides a representative expression for each quantity that illustrates the use of the abbreviation and symbol. Some of the abbreviations shown in the example column have subscripts. These are used to distinguish multiple instances of the same quantity. For example, V_1, V_2, and V_0 might represent three different voltages.

QUANTITY	ABBREVIATION	UNIT	SYMBOL	EXAMPLE
Capacitance	C	farad	F	$C_3 = 0.002$ F
Charge	Q	coulomb	C	$Q_1 = 2.6$ C
Conductance	G	siemens	S	$G_5 = 0.01$ S
Current	I	ampere	A	$I_2 = 3.9$ A
Frequency	f	hertz	Hz	$f = 60$ Hz
Impedance	Z	ohm	Ω	$Z_3 = 2700\ \Omega$
Inductance	L	henry	H	$L_1 = 0.005$ H
Power	P	watt	W	$P_2 = 150$ W
Reactance	X	ohm	Ω	$X_L = 257\ \Omega$
Resistance	R	ohm	Ω	$R_S = 3900\ \Omega$
Time	t	second	s	$t_0 = 0.003$ s
Voltage	V	volt	V	$V_4 = 12$ V

Table 1-1. Basic Electrical Quantities with Their Abbreviations, Units, Symbols, and an Example of Usage

EXAMPLE SOLUTION

Express the following using the proper abbreviation and symbol:
a. Voltage of 25.2 volts
b. Frequency of 375 hertz
c. Power of 5 watts

EXAMPLE **SOLUTION**

a. $V = 25.2$ V
b. $f = 375$ Hz
c. $P = 5$ W

Practice Problems

Express the following using the proper abbreviation and symbol:

1. Impedance of 3,000 ohms
2. Time of 0.09 seconds
3. Current of 12.5 amperes

Answers to Practice Problems

1. $Z = 3,000\ \Omega$ 2. $t = 0.09$ s   3. $I = 12.5$ A

Technical Notation

Electronics requires the use of extremely small and extremely large numbers—often within the same problem. One police radar frequency, for example, is 10,525,000,000 hertz. Some of the voltages found in a radar receiver, however, might be in the range of 0.0000034 volt. The solution of problems using numbers with this wide range can be very cumbersome—even with a calculator. The following sections present two forms of technical notation that will greatly simplify your calculations. The two forms, scientific notation and engineering notation, are considered separately for clarity, but it is important to realize that they are fundamentally similar. They are both based on **powers of ten.**

POWERS OF TEN

The terminology *power of ten* refers to the value that results when the number ten is raised to a power. Table 1-2 shows several powers of ten, their numerical values, and their equivalent forms.

Any number can be expressed as a series of digits multiplied times a power of ten. The following procedure will allow you to express any number using powers of ten:

1. Move the decimal point to any desired position while counting the number of positional changes.

POWER OF TEN	VALUE	EQUIVALENT
10^{-3}	0.001	$\dfrac{1}{10 \times 10 \times 10}$
10^{-2}	0.01	$\dfrac{1}{10 \times 10}$
10^{-1}	0.1	$\dfrac{1}{10}$
10^{0}	1	1
10^{1}	10	10
10^{2}	100	10×10
10^{3}	1,000	$10 \times 10 \times 10$

Table 1-2. Some Examples of Powers of Ten

2. If the decimal point was moved to the left, the exponent will be positive. If the decimal point was moved to the right, the exponent will be negative.

3. The magnitude of the resulting number of the exponent will be the same as the number of positional changes made by the decimal point.

This is a *very* important concept. Let's work some example problems to illustrate the conversion process.

EXAMPLE SOLUTION

Convert the following numbers to equivalent numbers expressed in power-of-ten form.
a. 123.5
b. 0.0042
c. 0.0226
d. 78,200,000
e. −2.77

EXAMPLE **SOLUTION**

There are an infinite number of correct answers to the preceding problems. Let's look at several alternatives.
a.

ORIGINAL NUMBER	DIRECTION	COUNT	RESULT
123.5	left	2	1.235×10^{2}
123.5	right	2	12350×10^{-2}
123.5	left	5	0.001235×10^{5}
123.5	none	0	123.5×10^{0}
123.5	right	3	123500×10^{-3}

b.

ORIGINAL NUMBER	DIRECTION	COUNT	RESULT
0.0042	right	4	42×10^{-4}
0.0042	right	6	4200×10^{-6}
0.0042	left	2	0.000042×10^{2}

c. Possible correct answers include: 22.6×10^{-3}, 0.000226×10^{2}, and 0.0226×10^{0}

d. Possible correct answers include: 78.2×10^{6}, 0.782×10^{8}, and $78{,}200{,}000{,}000 \times 10^{-3}$

e. Possible correct answers include: -0.277×10^{1}, -2770×10^{-3}, and $-2{,}770{,}000{,}000 \times 10^{-9}$.

Note that the sign of the original number is simply assigned to the converted number.

Sometimes the original number is expressed as a power of ten, and it is desirable to express it as a different power of ten. The process here is identical to that just described except that your count starts with the current power of ten. The procedure follows:

1. Subtract the current exponent of ten from the new exponent of ten.

2. If the result of step 1 is positive, then move the decimal point to the left.

3. If the result of step 1 is negative, then move the decimal point to the right.

4. In either case (steps 2 or 3), move the decimal point the number of positions indicated by the magnitude of the result in step 1.

EXAMPLE SOLUTION

Make the indicated conversions:

a. 144×10^{2} converts to   _____ $\times 10^{4}$

b. -0.09×10^{-1} converts to   _____ $\times 10^{-5}$

c. $327{,}000 \times 10^{3}$ converts to   _____ $\times 10^{6}$

d. 0.002×10^{4} converts to $200{,}000 \times 10^{-}$

EXAMPLE **SOLUTION**

a. 144×10^{2} converts to 1.44×10^{4}

b. -0.09×10^{-1} converts to -900×10^{-5}

c. $327{,}000 \times 10^{3}$ converts to 327×10^{6}

d. 0.002×10^{4} converts to $200{,}000 \times 10^{-4}$

Practice Problems

Make the indicated conversions:

1. 0.005×10^{6} converts to   _____ $\times 10^{0}$

2. 100.6×10^{-4} converts to   _____ $\times 10^{-6}$

3. 100.6×10^{-4} converts to   _____ $\times 10^6$

4. -247×10^3 converts to $-2.47 \times 10^{-}$

Answers to Practice Problems

1. $5,000 \times 10^0$ **2.** $10,060 \times 10^{-6}$

3. $0.00000001006 \times 10^6$ **4.** -2.47×10^5

SCIENTIFIC NOTATION

KEY POINTS

Scientific notation is an application using powers of ten that requires that the base of the converted number be between 1 and 10.

The previous section gave us the basic procedure for manipulating numbers expressed as powers of ten. Scientific notation uses these same procedures. The only difference is that scientific notation requires the final form to be expressed as a number between one and ten times a power of ten. The procedure is as follows:

1. Move the decimal point until it is between the two leftmost nonzero digits.

2. Increase or decrease the exponent by one for each positional change the decimal point was moved left or right, respectively.

EXAMPLE SOLUTION

Express the following numbers in scientific notation:
a. 214.6
b. 0.0087
c. -12.68×10^3
d. 134.6×10^{-2}
e. 0.25

EXAMPLE **SOLUTION**

We move the decimal to form a number between one and ten (either positive or negative as required) with a corresponding power of ten to indicate the direction and amount of decimal movement.
a. 2.146×10^2
b. 8.7×10^{-3}
c. -1.268×10^4
d. 1.346×10^0 or simply 1.346
e. 2.5×10^{-1}

Practice Problems

Express the following numbers in scientific notation:
1. 55.75
2. $-125,000$
3. 0.00055
4. 265.3×10^2
5. -0.002×10^9

Answers to Practice Problems

1. 5.575×10^1 **2.** -1.25×10^5

3. 5.5×10^{-4} **4.** 2.653×10^4

5. -2×10^6

ENGINEERING NOTATION

Technicians should understand scientific notation, but engineering notation is another form more commonly used in electronics for expressing large or small numbers. In the case of scientific notation, we have a fixed decimal point position (between the two leftmost nonzero digits) and a variable exponent. For engineering notation, we can move the decimal point to any convenient position *as long as the resulting exponent is either zero or a multiple of 3* (e.g., –6, –3, 0, 3, 6, 9). The procedure can be stated as follows:

1. Write the number using powers of ten.
2. Move the decimal point left while increasing the exponent or right while decreasing the exponent.
3. The final exponent must be zero or a number that is evenly divisible by three.

This is an important way of expressing numbers in electronics. As will be discussed in Section 1.4, powers of ten that are evenly divisible by 3 have prefixes that can replace them to further simplify the expression of quantities. Any given number can be expressed several different ways and still qualify as a valid form of engineering notation. The best choice for an exponent is determined by the problem being solved and the technician's preference. For now, let us apply the practice of selecting a base number between 1 and 999.

> **KEY POINTS**
>
> Engineering notation is also an application using powers of ten, but the exponent of the converted number must be zero or a number that is evenly divisible by three.

> **KEY POINTS**
>
> The base of numbers expressed in engineering notation is often between 1 and 999.

> **EXAMPLE** SOLUTION

Convert the following numbers to engineering notation. For purposes of this exercise, choose the form that produces base numbers between 1 and 999.
a. 129×10^5
b. 0.0056×10^{-1}
c. -33.9×10^7

> EXAMPLE **SOLUTION**

a. We move the decimal one place to the left, which causes our exponent to count up to six. The result expressed in engineering notation form is 12.9×10^6. This can also be expressed as 12.9 meg, as discussed in Section 1.4.
b. The preferred base number in this case is 560. The result is 560×10^{-6}. This can also be written as 560 micro, as described in Section 1.4.
c. In this case we shall move the decimal point one place to the right, which will decrease our exponent to six. The result in engineering notation is -339×10^6. This is equivalent to –339 meg, as described in Section 1.4.

RULES FOR CALCULATION

We are now ready to solve problems that utilize numbers expressed in a powers-of-ten form. We shall consider the basic arithmetic operations: addition, subtraction, multiplication, and division. Let us first consider the rule for adding and subtracting numbers that are expressed in a power-of-ten form:

1. Express the two numbers in any convenient power-of-ten form that results in *identical* exponents.

2. Add or subtract the base numbers.

3. Assign an exponent that is the same as the exponent of the two numbers.

4. Readjust the decimal point and exponent to any preferred form.

EXAMPLE SOLUTION

Perform the following calculations:
a. $12.9 \times 10^5 + 6.8 \times 10^5$
b. $550.3 \times 10^2 - 2800 \times 10^1$
c. $-140 \times 10^6 + 0.0007 \times 10^{10}$
d. $25 \times 10^1 + 5 \times 10^{-1}$

EXAMPLE **SOLUTION**

a. Since the exponents are already identical, we simply add the two base numbers and attach the common power of ten as follows:

$$12.9 \times 10^5$$
$$\underline{+\ 6.8 \times 10^5}$$
$$19.7 \times 10^5$$

b. The exponents must be made equal before the subtraction can be performed. We can change either or both numbers as long as the two exponents are made equal. Let us change $2,800 \times 10^1$ to have an exponent of 2. We must move the decimal point one place to the left to get 280.0×10^2. Now we can do the subtraction portion of the problem.

$$550.3 \times 10^2$$
$$\underline{-\ 280.0 \times 10^2}$$
$$270.3 \times 10^2$$

c. Let's convert 0.0007×10^{10} to 7.0×10^6 and then perform the arithmetic as follows:

$$-140.0 \times 10^6$$
$$\underline{+\ 7.0 \times 10^6}$$
$$-133.0 \times 10^6$$

d. Since the exponents are different, we will have to adjust at *least* one of the numbers to obtain common exponents. Let us change them both to zero exponents. The number 25×10^1 converts to 250×10^0, and 5×10^{-1} converts to 0.5×10^0. The addition step follows:

$$250.0 \times 10^0$$
$$\underline{+\ 0.5 \times 10^0}$$
$$250.5 \times 10^0, \text{ or simply } 250.5$$

Practice Problems

Perform the indicated calculations.

1. $233.2 \times 10^{-4} + 9.6 \times 10^{-3}$
2. $0.0056 \times 10^{2} - 12.2 \times 10^{-3}$
3. $25 \times 10^{9} + 0.000075 \times 10^{14}$
4. $0.2 \times 10^{2} - 14.0$
5. $0.5 \times 10^{4} + 10 \times 10^{5}$

Answers to Practice Problems

1. 32.92×10^{-3} 2. 547.8×10^{-3} 3. 32.5×10^{9}

4. 6 5. 10.05×10^{5}

Multiplication and division of numbers expressed in a power-of-ten form are more direct than are addition and subtraction. The procedure follows:

1. Multiply or divide the two base numbers as usual.
2. When multiplying, the power of ten in the result is the sum of the powers of ten for the two original numbers.
3. When dividing, the power of ten in the result is obtained by subtracting the exponent of the divisor from the exponent of the dividend.

> **KEY POINTS**
>
> Multiplication and division with powers of ten consists of multiplying or dividing the base numbers as usual. The exponent of the result is found by adding or subtracting (for multiplication and division, respectively) the exponents of the two operands.

EXAMPLE SOLUTION

Perform the indicated calculations.

a. $75 \times 10^{8} \div 25 \times 10^{3}$
b. $400 \times 10^{-5} \div 5 \times 10^{2}$
c. $0.006 \times 10^{9} \div 1,200 \times 10^{4}$
d. $125 \times 10^{2} \times 0.2 \times 10^{-1}$
e. $0.22 \times 10^{-1} \times 119.6 \times 10^{5}$

EXAMPLE SOLUTION

a. First, we divide the base numbers as usual: $75/25 = 3$. Next, we subtract the exponent in the divisor from the exponent in the dividend to obtain the exponent in the result: $8 - 3 = 5$. Our final result is 3×10^{5}.

b. $\dfrac{400 \times 10^{-5}}{5 \times 10^{2}} = 80 \times 10^{(-5-2)} = 80 \times 10^{-7}$

c. $\dfrac{0.006 \times 10^{9}}{1,200 \times 10^{4}} = 0.000005 \times 10^{5} = 0.5$

d. $(125 \times 10^{2}) \times (0.2 \times 10^{-1}) = 25 \times 10^{1} = 250$

e. $0.22 \times 10^{-1} \times 119.6 \times 10^{5} = 26.312 \times 10^{4}$

Practice Problems

Perform the indicated calculations.

1. $100 \times 10^6 \div 20 \times 10^2$

2. $0.5 \times 10^2 \times 40 \times 10^{-9}$

3. $2.25 \times 10^{-3} \div 1,250 \times 10^9$

4. $1,405 \times 2.6 \times 10^6$

Try to apply the rules for multiplication and division to these next two problems.

5. $5 \times 10^a \times 8 \times 10^b$

6. $60 \times 10^a \div 12 \times 10^b$

Answers to Practice Problems

1. 5×10^4 **2.** 2×10^{-6} **3.** 1.8×10^{-15}

4. 3.653×10^9 **5.** $40 \times 10^{a+b}$ **6.** $5 \times 10^{a-b}$

CALCULATOR SEQUENCES

This section provides a brief summary with respect to calculator operations using powers of ten.

EXAMPLE SOLUTION

Perform the following calculations on an engineering calculator:

a. $125.8 \times 10^3 \times 0.6$

b. $62.9 \times 10^{-6} \div 29 \times 10^{-5}$

c. $0.7 \times 10^4 + 81.2 \times 10^3$

d. $219 \times 10^{-2} - 0.5$

EXAMPLE **SOLUTION**

Figure 1–30 illustrates a calculator sequence as performed on a standard engineering calculator. Figure 1–31 shows a similar sequence for engineering calculators such as those made by Hewlett Packard that use Reverse Polish Notation (RPN).

Figure 1-30. Calculator sequence for powers-of-ten calculations using a standard calculator.

a. [1] [2] [5] [.] [8] [EEX] [3] [ENTER] [.] [6] [×]

b. [6] [2] [.] [9] [EEX] [+/−] [6] [ENTER] [2] [9] [EEX] [+/−] [5] [÷]

c. [.] [7] [EEX] [4] [ENTER] [8] [1] [.] [2] [EEX] [3] [+]

d. [2] [1] [9] [EEX] [+/−] [2] [ENTER] [.] [5] [−]

Figure 1-31. Calculator sequence for powers-of-ten calculations using an RPN calculator.

Exercise Problems 1.3

1. Express each of the following electrical quantities using the proper abbreviation and symbol:
    a. Frequency of 250 hertz
    b. Inductance of 0.0002 henry
    c. Voltage of 13.6 volts
2. Express the following numbers in scientific notation:
    a. 443
    b. 0.028
    c. 12,778,220
    d. 0.0009×10^8
    e. 547×10^3
3. Convert the following to engineering notation (use the nearest acceptable exponent):
    a. 266×10^{-4}
    b. 77.5×10^8
    c. 0.007×10^{-5}
    d. 52×10^4
4. Perform the following calculations. Express all answers in engineering notation.
    a. $44.5 \times 10^3 + 0.5 \times 10^4$
    b. $0.008 \times 10^6 - 7,000,000 \times 10^{-4}$
    c. $244 \times 10^7 \div 0.4 \times 10^8$
    d. $0.99 \times 10^4 \times 338 \times 10^2$

1.4 Multiple and Submultiple Units

The quantities measured and discussed in electronics have extreme ranges of values. Just as powers of ten help simplify numerical calculations involving large or small numbers, prefixes simplify written and spoken expressions of electrical quantities. Table 1-3 shows the powers of ten that are most commonly used in electronics along with the equivalent prefixes, symbols, and magnitudes. You will note that most of the powers of ten are those used for engineering notation (i.e., multiples of three).

KEY POINTS

All of the powers of ten that are used with engineering notation have corresponding prefixes and symbols.

POWER OF TEN	PREFIX	SYMBOL	MAGNITUDE
10^{-15}	femto	f	one-quadrillionth
10^{-12}	pico	p	one-trillionth
10^{-9}	nano	n	one-billionth
10^{-6}	micro	μ	one-millionth
10^{-3}	milli	m	one-thousandth
10^{-2}	centi	c	one-hundredth
10^{0}	Whole units need no prefix or symbol.		
10^{3}	kilo	k	thousand
10^{6}	mega	M	million
10^{9}	giga	G	billion
10^{12}	tera	T	trillion

Table 1-3. The Powers of Ten Used in Electronics Have Equivalent Prefixes and Symbols

When you first begin working with powers of ten and prefixes, they often seem to complicate, rather than simplify, the operation. However, if you practice, you will not only master the techniques, but you will soon realize their labor-saving effects. A technician must be proficient with prefixes and powers of ten, or she will be working in a world that speaks a foreign language.

We can use a prefix to express a quantity by applying the following procedure:

1. Write the quantity in engineering notation.
2. Replace the power of ten with the equivalent prefix.

Let's master this important process by example and practice.

EXAMPLE SOLUTION

Write the following quantities using the nearest standard prefix as listed in Table 1-3.

a. 120×10^{3} volts
b. 47×10^{-12} farads

c. 200×10^{-5} henries
d. 2.9×10^{-2} meters

e. 1.02×10^{10} hertz

EXAMPLE **SOLUTION**

a. 120 kilovolts
b. 47 picofarads

c. 2000 microhenries
d. 2.9 centimeters

e. 10.2 gigahertz

EXAMPLE SOLUTION

Write the following quantities using the nearest standard symbol as listed in Table 1-3.

a. 125×10^{-3} V

b. 5×10^{3} Hz

c. 0.001×10^{-11} F

d. 10.2×10^{-9} s

e. 3300×10^{5} Ω

EXAMPLE **SOLUTION**

a. 125 mV

b. 5 kHz

c. 0.01 pF

d. 10.2 ns

e. 330 MΩ

Practice Problems

Write the following quantities using the nearest standard symbol listed in Table 1-3.

1. 56×10^{3} Ω
2. 22×10^{-4} A
3. 500×10^{4} W
4. 12×10^{-6} s
5. 1.75×10^{8} Hz

Answers to Practice Problems

1. 56 kΩ
2. 2.2 mA
3. 5,000 kW
4. 12 μs
5. 0.175 GHz

We can also convert from one prefix to another through the following procedure:

1. Replace the given prefix with the equivalent power of ten.
2. Relocate the decimal (left or right) while you adjust the power of ten.
3. When the power of ten matches the value of the desired prefix, replace it with the prefix.

EXAMPLE SOLUTION

Express 0.1 milliamperes as an equivalent number of microamperes.

EXAMPLE **SOLUTION**

First, we replace the given prefix with the equivalent power of ten as follows:

$$0.1 \text{ mA} = 0.1 \times 10^{-3} \text{ A}$$

Next, we move the decimal until the power of ten matches the micro prefix (i.e., 10^{-6}). In this case, we will have to move the decimal point three places to the right:

$$0.1 \times 10^{-3} \text{ A} = 100 \times 10^{-6} \text{ A}$$

Finally, we substitute the equivalent prefix:

$$100 \times 10^{-6} \text{ A} = 100 \text{ μA}$$

Exercise Problems 1.4

1. Write the standard prefix for each of the following powers of ten:
   a. 10^{-3} **b.** 10^{6} **c.** 10^{9}
   d. 10^{-12} **e.** 10^{-2}

2. Write the following quantities using the nearest standard prefix as listed in Table 1-3.
   a. 22.5×10^{-6} F   **b.** 250×10^{5} Hz
   c. 27×10^{-5} H   **d.** 122×10^{-4} s

3. Write the following quantities using the nearest standard symbol as listed in Table 1-3.
   a. 220×10^{-5} H   **b.** 2500×10^{-4} W
   c. 0.008×10^{-7} F   **d.** 5.99×10^{4} Hz

4. Express the following quantities in the forms indicated.
   a. Express 12,500 hertz as kilohertz.
   b. Express 0.0005×10^{-4} farads as microfarads.
   c. Express 0.0025 milliseconds as microseconds.
   d. Express 390 milliamperes as amperes.

Chapter Summary

• Electronics is a field with an incredibly wide range of applications. Electronic devices are used as important tools in most other career fields. Due to the wide range and endless numbers of electronics applications in today's world, there is also a vast range of career opportunities for electronics technicians. Although the details of the various jobs performed by electronics technicians may vary, they all require a thorough understanding of basic electronics components, fundamental mathematics, and electronics principles.

• It is important for an electronics technician to be able to identify an electrical or electronic component by its physical appearance. This task is complicated by two characteristics of these components. First, there is a wide range of physical shapes and sizes for a given type of component. Second, some radically different components look fundamentally similar. The more experience a technician has with a wide variety of electronic components, the easier identification becomes.

- This chapter discussed the composition of matter. Matter is composed of a number of progressively smaller entities or building blocks. A sample quantity of any given matter can be roughly classified as a mixture, a compound, or an element. A mixture is composed of two or more compounds or elements. A compound is composed of molecules. Molecules are formed when two or more atoms of different elements are chemically combined.

- We divided the molecule into its component atoms and then studied the atom itself. The atom consists of a nucleus made of densely packed particles and orbited by smaller particles. The primary particles in the nucleus are protons and neutrons. Protons have a positive charge. Neutrons have no charge. The orbiting particles are called electrons. They have a negative charge. Although the positive charge of the proton is equal in magnitude to the negative charge of an electron, the proton has 1,836 times more mass than the tiny electron.

- A model of the atom was studied that showed how the electron orbits occurred at definite "altitudes" called energy levels. Several closely related orbits are often collectively referred to as a shell. The energy level of an electron determines its specific orbit. If an electron moves to a lower orbit, it will release energy. Similarly, if an electron absorbs sufficient energy, it will move to a higher orbit.

- Electrons in the outer orbit are called valence electrons. If a valence electron absorbs enough energy, it can escape the orbit of the atom and become a free electron. With no external force, the movement of free electrons in a substance such as copper wire is essentially random. However, if an external field that causes one end of the wire to be positive and the other end negative is applied, then the free electrons no longer travel in random directions. The external charges cause the free electrons to drift toward the positive end of the wire. This directed flow of electrons is called electron current flow.

- Current flow and charge are only two of many electrical quantities that must be measured and expressed in the field of electronics. This chapter presented several other fundamental quantities, their units of measurement, an abbreviation for each quantity, and a symbol for each unit of measurement.

- Many of the quantities measured and expressed in electronics require the use of extremely small and extremely large numbers. To simplify the expression and manipulation of these numbers, we use some form of technical notation. Three forms were discussed in this chapter: powers of ten, scientific notation, and engineering notation. Scientific notation and engineering notation are just special instances of the more general powers-of-ten notation. Scientific notation requires that all numbers be expressed as a number between one and ten times an appropriate power of ten. Engineering notation is simply an application of powers of ten. Here, the power of ten must be zero or an exponent that is evenly divisible by three. The base number is often between 1 and 999.

- To add or subtract numbers that are expressed in a power-of-ten format, the decimal point on either or both numbers must be adjusted until the exponents of the power of ten are identical. Addition or subtraction of the base numbers is then performed as usual. The power of ten for the result will be identical to the common exponent used for the two operands. Multiplication and division of numbers expressed in a power-of-ten form are more straightforward. Perform the calculation on the base numbers as usual. For multiplication, the power of ten in the result will be the sum of the multiplicand and multiplier powers of ten. For division, the power of ten for the result will be the difference of the divisor and dividend exponents (dividend exponent minus divisor exponent). Powers-of-ten calculations may be done quickly with an electronic calculator.

• The use of prefixes as substitutes for certain specific powers of ten greatly simplifies the written and spoken expression of electronic quantities. Those powers of ten that are even multiples of three are the ones most often substituted with a prefix.

Review Questions

Section 1.1: Electronics in Today's World

1. Name three general applications for computers.
2. Name three ways that electronics is used in the automotive industry.
3. Name three ways that electronics is used in the medical field.
4. Name at least three classes of jobs that require training in electronics technology.
5. Identify at least one major difference between the job environment of a bench technician and that of a field service technician.
6. What is a major purpose of resistors?
7. How can you determine the value of a resistor?
8. What is the name of a component that is used to increase or decrease alternating voltages?
9. What is the primary purpose of a switch in an electrical circuit?
10. A common lightbulb is an example of the general class of components called _____.

Section 1.2: Atomic Structure

11. Matter can be found in three physical states. Name the three states and give an example of matter found naturally in each of the states.
12. Assume you have the smallest quantity of a mixture that it is possible to have and still be representative of the original mixture. Describe the nature of the quantity.
13. What is the smallest quantity of a compound that still exhibits the characteristics of the compound?
14. What is the smallest part of an element that still exhibits the characteristics of the original element?
15. Is it possible to examine smaller and smaller quantities of salt water until you finally are examining a single atom of salt water? Explain your reasoning.
16. If you could examine an atom, where would you expect to find the electrons?

17. Of the three most fundamental subatomic particles, which one is the smallest and has the least mass?

18. If you knew the total mass of the electrons and protons in an atom, would that always be close to the total mass of the entire atom? Explain your reasoning.

19. If you knew the total mass of the protons and neutrons in an atom, would that always be close to the total mass of the entire atom? Explain your reasoning.

20. If two electrons were close to each other, they would: a) attract each other; b) repel each other; c) be unaffected by each other.

21. If two protons were close to each other, they would: a) attract each other; b) repel each other; c) be unaffected by each other.

22. If two neutrons were close to each other, they would: a) attract each other; b) repel each other; c) be unaffected by each other.

23. If a proton and an electron were close to each other, they would: a) attract each other; b) repel each other; c) be unaffected by each other.

24. A molecule consists of two or more _____ that are chemically bonded.

25. A certain type of subatomic particle orbits the nucleus of an atom. What determines the altitude of a particular particle?

26. If an electron and a neutron were close to each other, they would: a) attract each other; b) repel each other; c) be unaffected by each other.

27. What must occur in order for an electron to move to a lower orbit?

28. If an electron gains enough energy, it can escape the atom and become a _____ electron.

29. What name is given to describe the electrons in the outermost shell of an atom?

30. If an atom has more protons than corresponding electrons, we called the atom a _____ _____.

31. If you could closely examine the atomic structure of matter, how could you identify a negative ion?

32. Describe the meaning and nature of current flow in a copper wire.

Section 1.3: Units of Measure and Technical Notation

33. Complete the following table:

QUANTITY	UNIT	ABBREVIATION	SYMBOL
Frequency			
	volts		
		R	
			W
Current			

34. Fill in the blanks to make the following equations correct:

    a. $22.5 = \underline{\hspace{2cm}} \times 10^{-1}$ b. $9{,}805 = \underline{\hspace{2cm}} \times 10^{3}$

    c. $0.0014 = \underline{\hspace{2cm}} \times 10^{-5}$ d. $0.062 = \underline{\hspace{2cm}} \times 10^{2}$

    e. $5.7 \times 10^{4} = 570 \times 10^{-}$ f. $0.009 \times 10^{-1} = \underline{\hspace{2cm}} \times 10^{-6}$

35. Convert the following numbers to scientific notation form:

    a. 127   b. 0.0034   c. 12,567,000,000   d. 0.0000564

    e. 100×10^{2} f. 0.00227×10^{3} g. 42.6×10^{-6} h. 0.05×10^{-2}

36. Convert the following numbers to engineering notation form.

    a. 334×10^{5} b. $12{,}500{,}000{,}000 \times 10^{-7}$ c. 0.0025×10^{4}

    d. 0.00000098×10^{8} e. 34×10^{4} f. 887.2×10^{-4}

37. Perform the following calculations without the use of a calculator. Show your work.

    a. $32.5 \times 10^{3} + 14.5 \times 10^{2}$ b. $0.005 \times 10^{-9} - 270 \times 10^{-14}$

    c. $55.6 \times 10^{-4} + 10 \times 10^{-4}$ d. $150 \times 10^{12} - 60 \times 10^{12}$

    e. $5 \times 10^{1} + 5 \times 10^{-1}$ f. $0.002 - 50 \times 10^{-5}$

38. Perform the following calculations without the use of a calculator. Show your work.

    a. $(5 \times 10^{5}) \times (20 \times 10^{2})$ b. $(-25 \times 10^{-3}) \times (4 \times 10^{8})$

    c. $(150 \times 10^{6}) \div (50 \times 10^{4})$ d. $(40 \times 10^{-2}) \div (8 \times 10^{3})$

    e. $(-10 \times 10^{10}) \div (10 \times 10^{-10})$ f. $(10 \times 10^{10}) \times (10 \times 10^{-10})$

39. Perform the following calculations with the aid of an engineering calculator. Express your answer in scientific notation form.

    a. $53.25 \times (157 \times 10^{-2})$ b. $-1.98 \times (0.006 \times 10^{2})$

    c. $(5.7 \times 10^{-3}) \times (0.029 \times 10^{6})$ d. $100 \times 10^{-4} + 27 \times 10^{-2}$

    e. $0.005 \times 10^{-4} - 40 \times 10^{-6}$ f. $(27 \times 10^{5}) \div (0.05 \times 10^{2})$

    g. $0.006 \times 10^{5} - 0.00007 \times 10^{7}$ h. $23.25 + (12{,}500 \times 10^{3})$

40. Perform the following calculations with the aid of an engineering calculator. Express your answer in engineering notation form.

    a. $22.75 \times (0.7 \times 10^{-2})$ b. $0.06 \times (22.96 \times 10^{2})$

    c. $(3.2 \times 10^{-7}) \times (125.9 \times 10^{3})$ d. $0.3 \times 10^{5} + 2.8 \times 10^{6}$

    e. $0.123 \times 10^{-3} - 72 \times 10^{-4}$ f. $(91 \times 10^{-2}) \div (0.67 \times 10^{3})$

    g. $1{,}266 \times 10^{7} - 0.545 \times 10^{6}$ h. $83.13 + (19.82 \times 10^{2})$

Section 1.4: Multiple and Submultiple Units

41. Complete the following table:

POWER OF TEN	PREFIX	SYMBOL	MAGNITUDE
10^{-9}			
	kilo		
		m	
			one-trillionth
10^{-6}			

42. Write the following quantities using the nearest standard prefix:

    a. $345 \times 10^4 \ \Omega$
    b. $29.6 \times 10^{-1} \ \text{Hz}$
    c. $0.005 \times 10^{-7} \ \text{F}$
    d. $100 \times 10^{-5} \ \text{H}$
    e. $1830 \times 10^5 \ \Omega$
    f. $0.33 \times 10^{-4} \ \text{A}$

43. Write the following quantities using the nearest standard symbol:

    a. $78 \times 10^4 \ \Omega$
    b. $0.77 \times 10^{-5} \ \text{F}$
    c. $55.2 \times 10^5 \ \text{Hz}$
    d. $10.7 \times 10^{-4} \ \text{W}$
    e. $25 \times 10^{-5} \ \text{V}$
    f. $286.7 \times 10^{-4} \ \text{s}$

44. Make the following conversions:

    a. $225{,}500 \ \Omega = $ _____ $\text{k}\Omega$
    b. $25.6 \ \text{mA} = $ _____ A
    c. $0.0000068 \ \text{F} = $ _____ μF
    d. $850 \ \text{pA} = $ _____ μA

45. Classify each of the following as tiny, average, or enormous quantities:

    Height of a human: 5,500 millifeet   (Tiny, Average, or Enormous)
    Price of a candy bar: 0.05 kilocents   (Tiny, Average, or Enormous)
    Weight of a sandwich: 0.25 kilopounds   (Tiny, Average, or Enormous)
    Gas mileage on a car: 125 μmiles per gallon   (Tiny, Average, or Enormous)
    Speed of a plane: 200 femtomiles per kilosecond (Tiny, Average, or Enormous)

TECHNICIAN CHALLENGE

Your company has just hired a new technician and has assigned you as his trainer. Your first assignment for the new technician involves a number of calculations using large and small numbers.

The new technician took an excessive amount of time performing the calculations—even though he used a calculator. Upon closer investigation, you learn that the technician does not use engineering notation or any powers of ten. When you question him, he indicates that he believes he can work faster without using powers of ten.

Write a brief explanation that will convince the new technician of the value of powers of ten. Show why they are important, how they can save time, and how they help reduce errors in calculation. Be sure to provide definite examples.

charge
conductor
current
electromotive force
insulator
ionization voltage
pole

potentiometer
power
precision resistor
resistance
resistor tolerance
rheostat
short circuit

superconductivity
throw
voltage
voltage breakdown

objectives

After completing this chapter, you should be able to:

1. Explain the nature and basic laws of charge.

2. Explain how voltage and resistance affect current flow.

3. Describe the characteristics of conductors and insulators.

4. Contrast a variable resistor used as a rheostat with one used as a potentiometer.

5. Name several types of technologies used to manufacture resistors.

6. Determine the value of resistors by interpreting the resistor color code.

7. Identify and properly connect several types of switches.

8. State and solve problems with three forms of Ohm's Law.

9. State and solve problems with the power formulas.

10. Describe the ratings and applications for fuses, circuit breakers, and lamps.

11. State several safety techniques for electronics technicians.

12. Describe the characteristics of voltage and current sources.

13. Draw schematic symbols for the following:

    circuit breaker   neon lamp
    current source   potentiometer
    fixed resistor   rheostat
    fuses   switches
    incandescent lamp   voltage source
    LED

Electric Quantities and Components

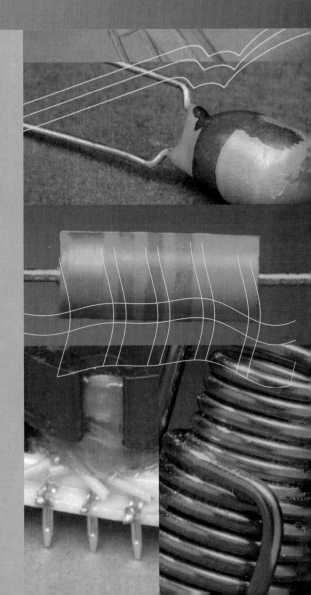

Charge, voltage, current, resistance, and power are some of the most fundamental and important electrical quantities you will ever study in electronics. This chapter defines these quantities and introduces calculations using them. This chapter also examines several electrical components in detail and presents the standard resistor color code.

2.1 Charge

An electrical **charge** is created when a material has more (or less) electrons than protons. Recall that electrons are negatively charged particles, whereas protons are positively charged particles. If we cause any substance (for example, a piece of paper) to have an excess of electrons, then we say the substance has a negative charge. Similarly, if a substance has a deficiency of electrons (i.e., more protons than electrons), then we say the material is positively charged. This idea is illustrated in Figure 2-1.

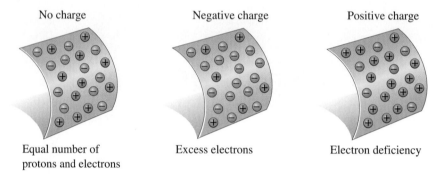

Figure 2-1. A material with excess electrons is negatively charged. A material with a deficiency of electrons is positively charged.

Attraction and Repulsion

You may recall an important and fundamental electrical law from your early science studies or from the brief mention in Chapter 1:

> Like charges repel each other. Unlike charges attract each other.

Figure 2-2 reinforces this important concept. The balls shown in Figure 2-2 are very lightweight spheres called pith balls. They are shown suspended by threads so they are free to swing. Figure 2-2 clearly shows the balls repel each other when they have similar charges, but attract each other when they have opposite charges. Also shown is the case where both balls are neutral (i.e., have no charge). In this case there is no deflection due to attraction or repulsion.

Figure 2-2. Unlike charges attract each other, while like charges repel.

Unit of Charge

A technician needs to be able to express charge in definite units in order to compare relative magnitudes of charge and to perform calculations involving charge. The unit of electric charge is the coulomb. One coulomb is equal to the charge caused by an accumulation of 6.25×10^{18} electrons. This is illustrated in Figure 2-3 and expressed mathematically by Equation 2-1.

● **KEY POINTS**

Charge is measured in coulombs.

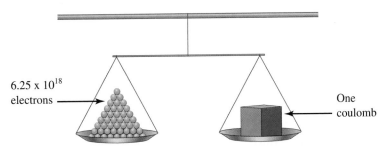

6.25 x 10^{18} electrons

One coulomb

Figure 2-3. One coulomb is equal to the amount of charge produced by 6.25×10^{18} electrons.

$$1\ C = 6.25 \times 10^{18}\ \text{electrons} \qquad (2\text{-}1)$$

Even though a coulomb is defined in terms of electrons (a negative charge), it specifies magnitude without regard to polarity. So, a deficiency of 6.25×10^{18} electrons would also be a charge of 1 C, but it would be a positive charge.

EXAMPLE SOLUTION

Calculate the number of electrons required to produce a charge of 5 C.

EXAMPLE **SOLUTION**

No. of electrons $= 5 \times 6.25 \times 10^{18} = 31.25 \times 10^{18}$

EXAMPLE SOLUTION

If a certain material has a deficiency of 76.25×10^{18} electrons, what is its charge?

EXAMPLE **SOLUTION**

$$
\begin{aligned}
\text{Charge} &= \frac{\text{No. of electrons}}{6.25 \times 10^{18}} \\
&= \frac{76.25 \times 10^{18}}{6.25 \times 10^{18}} \\
&= 12.2\ C
\end{aligned}
$$

1. Compute the number of electrons required to produce a charge of 2.7 C.
2. If a certain material has an excess of 30.625×10^{18} electrons, what is its charge?

Answers to Practice Problems

1. 16.875×10^{18} 2. $4.9\,C$

Coulomb's Law

The unit of charge (coulomb) was named in honor of the French scientist Charles Coulomb. His experiments, which were aimed at quantifying the forces of attraction and repulsion of charges, led to the formulation of Coulomb's Law. It can be stated as follows:

> The force of attraction or repulsion between two charged bodies is directly proportional to the product of the two charges and inversely proportional to the square of the distance between the bodies. The forces lie on a straight line connecting the two charges. The force is attractive if the charges are of opposite polarity and repulsive if the charges have the same polarity.

This law can also be stated in formula form as follows:

$$F = k\frac{Q_1 Q_2}{d^2} \tag{2-2}$$

where Q_1 and Q_2 are the charges in coulombs, d is the distance between the charges in meters, k is a constant value of 9×10^9, and F is the resulting force in newtons (1 N = 0.2248 lb).

Electrostatic Fields

KEY POINTS

Three types of fields are:

• Electrostatic
• Magnetic
• Electromagnetic

There are three types of fields that play major roles in electronics: electrostatic, magnetic, and electromagnetic. To appreciate the concept of a field, let us consider the earth's gravitational field. It is invisible, but it clearly exerts a force on nearby objects. Additionally, the amount of force diminishes with greater distances.

KEY POINTS

There is an electrostatic field associated with every charged particle.

The term **electrostatic field** is used to describe the region near a charged body where another charged body is affected (i.e., is attracted or repelled). The strength of the field at any point is given by Equation 2-2. It is helpful to visualize the electrostatic field as an array of lines (similar to the lines of force used to represent the magnetic field of a bar magnet). Figure 2-4 shows a sketch of the electrostatic field surrounding some charged bodies. A drawing of this type conveys two distinct types of information:

- The direction of each line at any given point indicates the direction in which a positively charged particle would move if it were released at the specified point.
- The density (closeness) of the lines indicates the relative strength of the electrostatic field.

It is important for a technician to understand this method of representing electrostatic fields, since it is frequently used in textbooks and other technical literature to explain the operation of devices that utilize electrostatic fields. Devices such as capacitors, oscilloscopes, radar displays, televisions, and many computer displays utilize electrostatic fields in their operation.

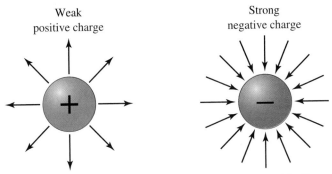

Figure 2-4. An electrostatic field can be represented by lines of force surrounding charged bodies.

Exercise Problems 2.1

1. If two positively charged bodies are placed close to each other, will they attract or repel one another?
2. If a certain object has an excess of 50×10^{17} electrons, express its charge in coulombs.
3. How many electrons are required to produce a charge of 6.8 C?
4. How many electrons must be lost by a body to develop a positive charge of 500 μC?
5. Name three types of fields similar in concept to an electrostatic field.
6. Figure 2-5 shows an electrostatic field between two charged plates. If a free electron is released within the field, which plate will attract it?

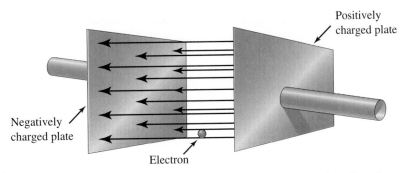

Figure 2-5. Which way will a free electron move if released within the electrostatic field?

2.2 Voltage

When two bodies with unequal charges are separated by some distance, there is a potential for doing work. If, for example, a charged particle were released within the electrostatic field associated with the two bodies, the particle would be moved by the electrostatic field. When there is a *difference* in charge potential between two points, we refer to it as a *difference of potential*. Thus, a difference of potential has the ability to do work—it can move charged particles.

Work

Work, by definition, occurs when a force applied to an object causes the object to move in the direction of the force. If, for example, we applied one pound of force to an object as we moved it over a linear distance of one foot, we could say that we did one foot-pound of work. In electronics, force is often measured in newtons (1 N = 0.2248 lb) and distance is measured in meters. If we applied one newton of force to an object as we moved it over a linear distance of one meter, we could say that we did one newton-meter of work. More probably, we would say we did one *joule* of work, since one joule is equal to one newton-meter. The relationship between joules and foot-pounds is illustrated in Figure 2-6.

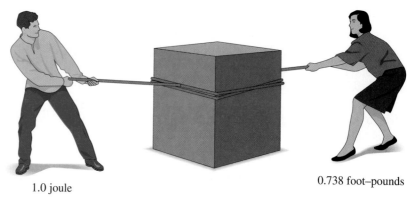

1.0 joule

0.738 foot–pounds

Figure 2-6. Work can be measured in joules or foot-pounds.

Unit of Measurement for Voltage

Now let us look at an important unit of measurement for electronics. If we expend one joule of work (energy) in the process of moving one coulomb (6.25×10^{18} electrons) of charge from one point to another, then the two points have a difference of potential of one *volt* (V). We will use the terms voltage and volt throughout this book to express the difference in potential between two points.

Electromotive Force versus Potential Difference

When a charge difference exists between two bodies or points, we refer to this charge gradient as a difference in potential or **voltage.** It is measured in volts. If a path is provided between the two charges that will permit the flow of charged particles (positive ions, negative ions, or free electrons), the two charges will neutralize each other. Electrons, for example, may leave the negative charge and travel through the path to the positive charge. Every time an electron makes this trip, the difference in potential is reduced. Eventually, the difference of potential will be zero and no further directed movement of charges will occur.

Now, for most practical electronic applications, we need a sustained movement of charges. That is, we want the potential difference to remain constant as the charges are transferred. This requires energy to be expended. This external energy may be supplied by a battery (chemical energy), a solar cell (solar energy), or perhaps by a generator (mechanical energy). In any case, if the potential difference is maintained as charges are transferred, we label the potential difference as an **electromotive force,** or simply emf. Both emf and potential difference are measured in volts.

Voltage Sources

Voltage sources are used to provide a constant source of potential difference (emf) regardless of the current that flows through the device. The battery in your car is an example of a voltage source. It supplies approximately 12 V to the circuits in your car. You will study several types of voltage sources.

IDEAL VOLTAGE SOURCE

A theoretically perfect voltage source would be capable of supplying a constant voltage for an infinite period of time. Additionally, the voltage would not change when more or less current was drawn from the voltage source. There are no known perfect voltage sources, but several components and circuits are capable of maintaining a reasonably constant voltage, for a given period of time, with reasonable changes in current. The subjective term *reasonable* must be defined with respect to a specific application. For example, the voltage delivered by your car battery probably varies from as low as 8 to as high as 14 V. This variation would be absurd for many applications, but it can be viewed as a constant voltage for purposes of the noncritical devices in your car.

Figure 2-7 shows the schematic symbol for a constant voltage source. Technically, Figure 2-7(a) represents a single cell (like a D cell in your flashlight), and Figure 2-7(b) represents a group of interconnected cells called a battery. In practice, however, either symbol can be used to represent a voltage source *without regard to the physical nature of the voltage source.* The cell and battery symbols consist of long and short lines. The shorter line identifies the negative terminal of the voltage source. Except where specifically noted, all voltage sources used in this text can be considered ideal.

PRACTICAL VOLTAGE SOURCES

The output voltage of a practical voltage source varies as the current in the circuit varies. It can be considered to be an ideal voltage source that has an in-line resistance or opposition to current flow. This resistance is called the internal resistance of the voltage source. The available voltage from a practical voltage source decreases as current in the circuit increases. The higher the value of internal resistance, the greater the decrease in voltage for a given current increase.

Exercise Problems 2.2

1. When a force is applied to an object and causes it to move in the direction of the force, we say that we have done _____.
2. If 350 pounds of force is being applied to a certain object, express this force in terms of newtons.
3. List three different units of measure for work.
4. List two units of measure for force.
5. What is the unit of measure for potential difference?
6. What is the unit of measure for electromotive force?
7. Explain the difference between emf and potential difference.
8. Can you offer a brief explanation of why the lights in your car get dim whenever you try to start the car?
9. Draw a schematic symbol for a voltage source.
10. Explain how ideal and practical voltage sources differ.

2.3 Current

KEY POINTS

Current is the directed movement of charged particles—free electrons or positive and negative ions.

In Chapter 1, we took a short preview of electron current flow. You will recall that a conductor, such as copper wire, has many loosely bound valence band electrons. Many of the electrons escape orbit even at room temperature. If we connect a conductor between two points that have a difference in potential, then charges will move from one point through the conductor to the other point. In the case of a metallic conductor, the moving charges are electrons, and we call the resulting directed movement electron current flow.

In some materials other than metallic conductors, positive and/or negative ions may travel from one charged point to another. This movement of charges is also called current flow. You will study devices in electronics that utilize several types of current flow. In all cases, however, current flow is the directed movement of charges. It is the specific type of charge (e.g., electron, ion, and so on) that varies from one form of current to another.

Conventional Current Flow

For many applications, in particular when a circuit is modeled mathematically, it is not essential to be concerned about the physical nature of the moving charges that form current flow. Rather, it is convenient to speak in terms of current without specifying the exact nature of the current.

KEY POINTS

Conventional current flow moves from positive to negative.

Scientists, physicists, and engineers have agreed that the general term *current flow* will refer to the movement of positive charges. Current identified in this way is called *conventional current flow*. It follows that conventional current would be considered as flowing from positive to negative since that is the direction in which a positive charge would move. Conventional current flow is represented in Figure 2-8.

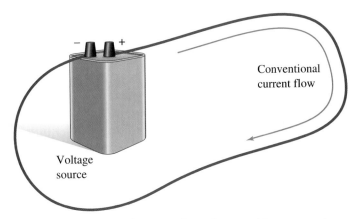

Figure 2-8. Conventional current flows from positive to negative since it represents the direction a positive charge would move.

Electron Current Flow

Many technicians refer to current flow as the movement of electrons. Current identified in this way is called *electron current flow*. It follows that electron current flow would be considered as flowing from negative to positive since that is the direction in which an electron would move. Electron current flow is represented in Figure 2-9.

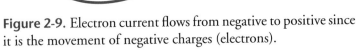

Figure 2-9. Electron current flows from negative to positive since it is the movement of negative charges (electrons).

The Great Debate

Hopefully, the two preceding sections have shown you the nonsense in arguing over which way current flows. Nevertheless, it is a common source for a pointless argument. You are encouraged to understand both conventions, and use the one that is most appropriate for the work you are doing and the environment you are in.

To avoid cumbersome discussions in the remainder of this text, we shall adopt a single convention for current flow. Unless it is specifically noted, you may assume that all references to current flow (including arrows drawn on diagrams) are referring to electron current flow.

Unit of Measurement for Current

You will recall that charge is a measure of the number of electrons (excess or lack) in a particular body. **Current** is a movement of charges. Its measure will be the number of electrons per second. More specifically, the unit of measure for current flow is the ampere (A), and it is the amount of current (I) that flows when one coulomb (Q) flows past a point in one second (t). This is expressed mathematically by Equation 2-3.

$$I = \frac{Q}{T} \qquad (2\text{-}3)$$

where I is in amperes, Q is in coulombs, and t is in seconds. This is an important formula that relates current, time, and charge movement. It provides the basis for explaining the operation of several components (e.g., capacitors) and circuits.

Current Sources

Just as a voltage source is designed to maintain a constant voltage, a current source is designed to maintain a constant current. The schematic representation for a current source is shown in Figure 2-10. An ideal current source will maintain a specified value of current *regardless of the amount of resistance or opposition in its path*. Practical current sources, which are usually electronic circuits, can maintain a constant current over a limited range of resistance values. Unless otherwise noted, all current sources used in this book are considered to be ideal.

Exercise Problems 2.3

1. If someone told you that current flows from positive to negative, would you say they are right? Explain your answer.
2. If someone told you that current flows from negative to positive, would you say they are right? Explain your answer.
3. The unit of charge measures number of electrons. What measures number of electrons per second?
4. How much current is flowing in a wire if 200 mC pass a given point every 350 μs?
5. If the appliances in your home caused 42 A to flow for one hour, how much charge would be transferred?

Figure 2-10.
Schematic symbol for a current source.

2.4 Resistance

As current flows through a material, it encounters opposition. In the case of metallic conductors, for example, the free electrons have frequent collisions as they find their way between the atoms in the wire. These collisions and the attraction of the various atoms present an opposition to current flow. This opposition to current flow is called **resistance.** In order to produce a current with a given magnitude through a material, an electromotive force that has sufficient energy to overcome the resistance of the material must be applied across the material.

Classification of Materials

All materials used in electronics can be roughly categorized into four classes based on their effective resistance: insulators, conductors, semiconductors, and superconductors.

INSULATORS

Insulators are materials that have very few free electrons. The valence band electrons are tightly bonded to the nuclei. When an electrostatic field is applied across an insulator material, the orbits of the individual atoms are somewhat distorted as the electrons strain to break free of the atom. Under normal conditions, very few electrons break free of their parent atoms. Thus, there are very few free electrons available to form a current flow. Insulators, therefore, have a very high resistance.

If a sufficiently high voltage is connected across an insulator, the electrons will be ripped out of orbit. As the newly freed electrons rush toward the positive battery potential, they may collide with atoms and free more electrons. The net effect is that current flow will go from near zero to a very high value almost instantaneously. The insulator may be damaged by the high current flow. This phenomenon is called **voltage breakdown** and is illustrated in Figure 2-11.

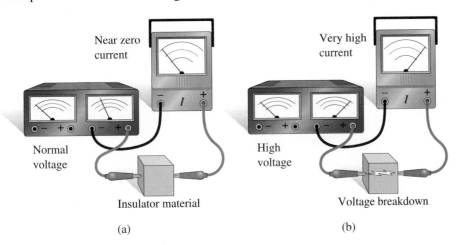

Near zero current

Very high current

Normal voltage

High voltage

Insulator material

Voltage breakdown

(a)

(b)

Figure 2-11. An insulator normally has high resistance but will break down if enough voltage is applied to it.

For most purposes, insulators are selected to have voltage breakdown values that are substantially greater than any voltage normally applied to the insulator. This ensures that the insulator will continue to have a very high resistance under all expected conditions. Glass, ceramic, paper, air, and oil are examples of common insulating materials.

CONDUCTORS

Conductors have many free electrons that can participate in current flow. It takes a very small external voltage to produce a substantial current flow in a conductor. A conductor, therefore, is said to have a very low resistance. Copper, aluminum, and gold are three common conductive materials.

SEMICONDUCTORS

Semiconductors, as their name implies, have a resistance between that of a good conductor and a good insulator. Semiconductors are used in the fabrication of several electronic components, including transistors and integrated circuits. If a voltage is applied to a semiconductor material, then current will flow. The value of current will be higher than would be expected of an insulating material but substantially less than the current that would flow through a conductor. Silicon and germanium are two common semiconductor materials.

SUPERCONDUCTORS

Certain materials exhibit a property or state called **superconductivity.** When a material enters a superconductivity state, its resistance drops to zero; not about zero, not near zero, but truly zero. For reasons that are not yet completely understood, as these materials are cooled, they reach a certain point where their resistance drops quickly to zero. Initially, this phenomenon could only be demonstrated near absolute zero (–273.15°C). Recently, however, advances in ceramic superconductors have produced materials that exhibit superconductivity at much higher temperatures (above –196°C). Liquid nitrogen is relatively inexpensive and can be used to cool these high-temperature (relative to absolute zero) ceramic superconductors. Substantial research effort is under way to develop room-temperature superconductors. Mercury, lead, and some ceramic materials are examples of materials that can exhibit superconductivity. Superconductors can be used to make very strong electromagnets. Superconductor wire can carry enormous amounts of current with essentially no loss.

Unit of Measurement for Resistance

Resistance is measured in ohms (Ω). An ohm is defined as the amount of resistance required to limit the current flow through a material to one ampere when one volt is applied across the material. Practical values of resistance used in electronics range from milliohms to hundreds of megohms. This range is not meant to be inclusive; there are applications that require smaller and larger resistances.

Exercise Problems 2.4

1. Write a definition for electrical resistance.
2. What class of material has no electrical resistance?
3. What class of material has a very high resistance?
4. Is it possible to cause current to flow through an insulator? Explain your answer.
5. What is the unit of measure for electrical resistance? What is its symbol?

2.5 Power

Power is a measure of the rate at which energy is used. Mechanical power is the rate of doing mechanical work. It is measured in horsepower. Electrical power is the rate of using electrical energy. Electrical power is often evidenced as heat. All electrical devices such as lightbulbs, computers, microwave transmitters, and so on dissipate power.

Unit of Measurement for Electrical Power

Electrical power is measured in watts (W) and is the rate of using electrical energy. It can be expressed mathematically by Equation 2-4:

$$\text{Power} = \frac{\text{energy}}{\text{time}} \qquad (2\text{-}4)$$

where power is measured in watts (W), energy is measured in joules (J), and time is measured in seconds (s).

EXAMPLE SOLUTION

What is the power if 250 J of energy are used over a period of 3 s?

EXAMPLE **SOLUTION**

We apply Equation 2-4 as follows:

$$\text{Power} = \frac{\text{energy}}{\text{time}}$$

$$= \frac{250 \text{ J}}{3 \text{ s}} = 83.33 \text{ W}$$

Practice Problems

1. How much energy is required to produce 10 W of power in a 100-ms time period?
2. What is the power if 100 J of energy are expended over a 100-ms interval?
3. During what time interval must 25 J of energy be used if 100 W are to be generated?

Answers to Practice Problems

1. 1.0 J   2. 1.0 kW   3. 250 ms

Comparison of Electrical and Mechanical Power

Mechanical power is measured in horsepower and expresses how many foot-pounds of work are done in one second. More specifically, 1 hp = 550 ft-lb per second.

Electrical power expresses how many joules of work are done in one second. Since both electrical and mechanical powers are rates of doing work, we can equate them with Equation 2-5.

$$1 \text{ hp} = 746 \text{ W} \qquad (2\text{-}5)$$

Figure 2-12 illustrates the relationship between mechanical and electrical power.

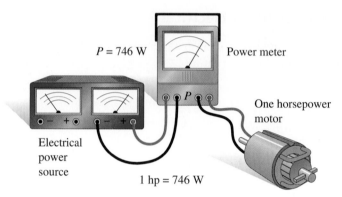

Figure 2-12. Electrical and mechanical power both measure the rate of doing work.

EXAMPLE SOLUTION

If a certain electrical motor delivers 2.5 hp, what is the required electrical power?

EXAMPLE **SOLUTION**

We can apply Equation 2-5 as follows:

electrical power (W) = mechanical power (hp) × 746
= 2.5 hp × 746 = 1,865 W = 1.865 kW

This calculation assumes a perfect motor. Real motors consume additional electrical energy in the form of heat.

Practice Problems

1. Express 5.8 kW as an equivalent number of horsepower.
2. Express 300 hp as an equivalent number of watts.
3. How much electrical power is required to deliver 5 hp?

Answers to Practice Problems

1. 7.77 hp   2. 224 kW   3. 3.73 kW

When working with electronic circuits, it is generally more convenient to compute power in terms of more common circuit quantities (voltage, current, and resistance). Section 2.6 presents these important alternatives.

Exercise Problems 2.5

1. If 120 J of energy are used over a period of 500 ms, how much power is expended?
2. How many joules of energy are required to produce 400 mW of power in a 275-μs time period?
3. Express 35 horsepower as an equivalent number of watts.
4. Express 2 kW as an equivalent number of horsepower.
5. If 75 J of energy are expended during a 100-ms period, how much power is expended?
6. How many joules of energy are required to produce 100 W of power in a 500-ms time period?
7. Express 300 hp as an equivalent number of watts.
8. Express 1 MW as an equivalent number of horsepower.
9. If 2,000 J are used over a period of 2.5 s, how much power is expended?

2.6 Ohm's Law

Possibly the most important relationship you will ever study in electronics is the relationship between voltage, current, and resistance in a circuit. This relationship was formally expressed by Georg Simon Ohm, a German teacher, in 1827. It is now called Ohm's Law and can be stated as follows:

> The current that flows in a circuit is directly proportional to the voltage across the circuit and inversely proportional to the resistance in the circuit.

This relationship is shown mathematically in Equation 2-6.

$$I = \frac{V}{R}, \text{ or}$$
$$I = \frac{E}{R} \qquad (2\text{-}6)$$

The selection of V or E to represent voltage is largely a personal choice. In most cases, the "correct" choice is defined by the person who signs your paycheck! For the purpose of this text, we shall generally use V as the symbol for voltage.

Current Is Directly Proportional to Voltage

Let us first examine the part of the law that says current is directly proportional to voltage. Figure 2-13(a) shows an electric circuit composed of a voltage source and a resistance. It has a certain amount of current, which could be computed with Equation 2-6.

Figure 2-13(b) shows the same basic circuit, but the applied voltage is higher. The higher voltage causes more current to flow in the circuit. This should seem reasonable to you since voltage is the force that causes current to flow and resistance is what limits current flow. If there is an increased force trying to cause current flow, but the opposition to current flow (resistance) remains constant, then it should seem logical that the current flow will increase. Figure 2-13(c) illustrates the effect of reducing the voltage to a value below that in Figure 2-13(a).

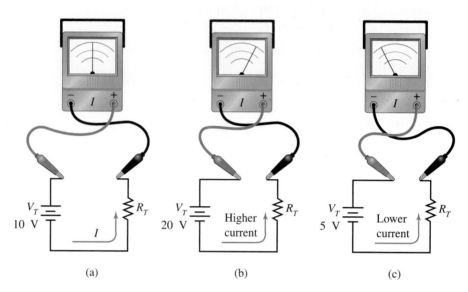

(a)   (b)   (c)

Figure 2-13. Current in a circuit is directly proportional to applied voltage.

The relationship between current and voltage is called *linearly proportional.* This important relationship is shown graphically in Figure 2-14. It is common for technical information to be expressed graphically; a technician must be able to interpret and understand graphical information. The horizontal scale of the graph shown in Figure 2-14 represents the voltage in the circuit. The vertical scale shows the resulting current flow. From any given voltage value, you can move vertically (as indicated by the dotted lines in Figure 2-14) until you intersect the resistance line, and then move horizontally from that point until you intersect the current axis. The point of intersection on the current axis indicates the amount of current that flows with the given applied voltage and the value of resistance represented by the resistance line.

Current Is Inversely Proportional to Resistance

Figure 2-15 examines the second part of Ohm's Law, which states that current flow is inversely proportional to resistance. Figure 2-15(a) shows that a voltage source connected across a certain resistance will cause a given amount of current to flow. We could compute the value of current with Equation 2-6. Figure 2-15(b) shows that current becomes less when the resistance in the circuit is increased. This should seem reasonable since resistance

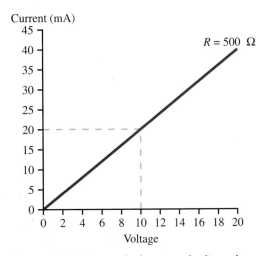

Figure 2-14. A graph showing the linearly proportional relationship between applied voltage and the resulting current flow in a fixed resistance circuit.

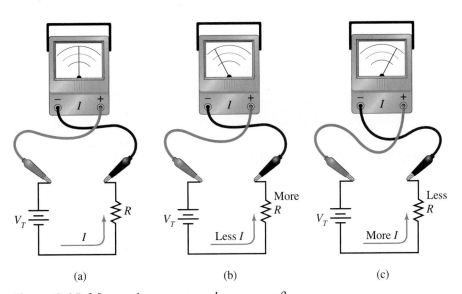

Figure 2-15. More resistance causes less current flow.

is a measure of the opposition to current flow. Finally, Figure 2-15(c) illustrates that current in the circuit will increase if the opposition (resistance) is reduced.

The inverse relationship between current and resistance is shown graphically in Figure 2-16. As you move to the right along the horizontal scale (more resistance), you can see that the corresponding values of current flow on the vertical scale are less.

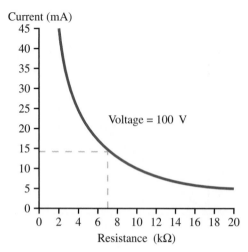

Figure 2-16. Current is inversely proportional to resistance.

Alternate Forms of Ohm's Law

We can manipulate the basic Ohm's Law equation (Equation 2-6), as we can with any equation, to obtain the alternate forms shown as Equations 2-7 and 2-8.

$$R = \frac{V}{I} \tag{2-7}$$

$$V = IR \tag{2-8}$$

COMPUTE I WHEN V AND R ARE KNOWN

If a technician knows the values of voltage and resistance in a circuit, then he can calculate the value of current by applying Equation 2-6. The units of measure for voltage and resistance are volts and ohms, respectively. The unit of measure for the resulting current is amperes.

EXAMPLE SOLUTION

How much current flows through a 25-Ω resistor with 10 V across it?

EXAMPLE **SOLUTION**

We apply Ohm's Law (Equation 2-6) as follows:

$$I = \frac{V}{R}$$

$$= \frac{10}{25} = 0.40 \text{ A}$$

If an electric circuit has 120 V applied and has a total resistance of 20 Ω, how much current is flowing in the circuit?

We apply Ohm's Law (Equation 2-6) as follows:

$$I = \frac{V}{R}$$

$$= \frac{120}{20} = 6.0 \text{ A}$$

COMPUTE R WHEN V AND I ARE KNOWN

If the current and voltage in an electric circuit are known, then the resistance can be computed by applying Equation 2-7. Again, the units of measure for voltage, current, and resistance are volts, amperes, and ohms, respectively.

If a certain resistor allows 250 mA to flow when 35 V are across it, what is its resistance?

Ohm's Law (Equation 2-7) provides our answer as follows:

$$R = \frac{V}{I}$$

$$= \frac{35}{250 \times 10^{-3}}$$

$$= 140 \text{ }\Omega$$

How much resistance is required in an electric circuit to restrict the current flow to 8.5 A when 240 V are applied?

We apply Equation 2-7.

$$R = \frac{V}{I}$$

$$= \frac{240 \text{ V}}{8.5 \text{ A}} = 28.24 \text{ }\Omega$$

COMPUTE V WHEN I AND R ARE KNOWN

If a technician knows the current and resistance values in an electric circuit, then she can calculate the value of applied voltage with Ohm's Law (Equation 2-8).

EXAMPLE SOLUTION

How much voltage must be connected across a 1.2-kΩ resistor to cause 575 μA of current to flow?

EXAMPLE **SOLUTION**

We apply Equation 2-8 as follows:

$$V = IR$$
$$= 575 \times 10^{-6} \times 1.2 \times 10^{3}$$
$$= 690 \text{ mV}$$

EXAMPLE SOLUTION

How much voltage is required to cause 20 A of current to flow through a 5-Ω resistor?

EXAMPLE **SOLUTION**

$$V = IR$$
$$= 20 \text{ A} \times 5 \text{ Ω} = 100 \text{ V}$$

Practice Problems

Apply Ohm's Law to solve the following problems.

1. How much current flows through a 680-kΩ resistor that has 12 V across it?
2. If a 10-kΩ resistor has 100 μA flowing through it, how much voltage is across it?
3. What value of resistance is required in a circuit to limit the current to 2.7 A when 220 V are applied?

Answers to Practice Problems

1. 17.6 μA   2. 1.0 V   3. 81.5 Ω

Power Calculations Based on Ohm's Law

When working with electronic circuits, it is generally more convenient to compute power in terms of more common circuit quantities (voltage, current, and resistance). There are three basic formulas that we will use extensively for computing the electrical power in a circuit. First, the power dissipated in a component or an entire circuit is given by the product of voltage and current. This is expressed formally as Equation 2-9.

$$P = VI \qquad \qquad (2\text{-}9)$$

We can use this basic power equation and the Ohm's Law relationships previously cited to develop two more important power relationships.

$$P = IV$$
$$= \frac{V}{R} \times V \text{ (substitute } \frac{V}{R} \text{ for } I)$$
$$= \frac{V^2}{R}$$

and,

$$P = IV$$
$$= I \times IR \text{ (substitute } IR \text{ for } V)$$
$$= I^2 R$$

These two important power relationships are given formally as Equations 2-10 and 2-11.

$$P = \frac{V^2}{R} \qquad (2\text{-}10)$$

$$P = I^2 R \qquad (2\text{-}11)$$

Equations 2-9, 2-10, and 2-11 are the basic power formulas that we will use throughout the remainder of this text. We will refer to these collectively as the power formulas.

In many cases, these calculations are used to determine the amount of electrical power that is converted to heat. For example, when electron current flows through a conductor, the free electrons lose energy as they collide with atoms in the material. Much of the energy loss is converted to heat, so the material heats up. Figure 2-17 shows the application of these equations to an electrical circuit.

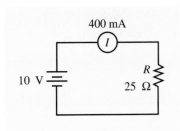

$P = IV$	$P = I^2R$
$P = 400 \text{ mA} \times 10 \text{ V}$	$P = (400 \text{ mA})^2 \times 25 \text{ } \Omega$
$P = 4 \text{ W}$	$P = 4 \text{ W}$

$$P = \frac{V^2}{R}$$
$$P = \frac{(10 \text{ V})^2}{25 \text{ } \Omega}$$
$$P = 4 \text{ W}$$

Figure 2-17. The amount of electrical power in a circuit is directly related to the voltage, current, and resistance in the circuit.

Your electric toaster or electric oven are two excellent examples of electrical energy being consumed as power. When current passes through the heating elements, it encounters opposition or resistance. Much of the electrical energy supplied to the heating element is converted to heat; some is converted to light energy as the element glows red hot.

EXAMPLE SOLUTION

How much power is expended in an electrical circuit when 50 V is connected across a 25-Ω resistance?

EXAMPLE SOLUTION

We can apply Equation 2-10 directly as follows:

$$P = \frac{V^2}{R}$$

$$= \frac{50^2 \text{ V}}{25 \ \Omega}$$

$$= 100 \text{ W}$$

EXAMPLE SOLUTION

How much current must flow through a 100-kΩ resistance in order to produce 5 W of power?

EXAMPLE SOLUTION

We apply Equation 2-11 as follows:

$$P = I^2 R, \text{ or}$$

$$I = \sqrt{\frac{P}{R}}$$

$$= \sqrt{\frac{5 \text{ W}}{100 \times 10^3 \Omega}} = 7.07 \text{ mA}$$

Practice Problems

1. How much power is dissipated when 300 mA flows through a 50-Ω resistance?
2. How much voltage is required to produce 25 W in a 150-Ω resistance?
3. If a certain electrical device has 120 V applied and has 3.5 A of current flow, how much power is dissipated?

Answers to Practice Problems

1. 4.5 W   2. 61.2 V   3. 420 W

Exercise Problems 2.6

1. How much current would flow through a 10-kΩ resistor if a 12-V source were connected across it?
2. Calculate the amount of current through a 100-Ω resistor that has 1.5 V across it.

3. What value of resistance is required to limit the current in a circuit to 25 mA with 100 V applied?

4. It takes 1.2 kΩ of resistance to limit the current flow to 10 mA when a 25-V source is applied. (True or False)

5. How much voltage must be connected across a 27-kΩ resistor to cause 5 mA of current to flow?

6. Which circuit has the most resistance: a) a circuit with 12 V applied and 100 mA of current flow, or b) a circuit with 100 V applied and a current of 5 mA?

7. If 6.3 V are impressed across a 20-Ω resistance, how much power is expended?

8. How much current flows through a 51-Ω resistance if it is dissipating 200 W?

9. How much voltage is needed to cause 350 mW to be dissipated in a 2.7-kΩ resistance?

10. How much power is dissipated by an automobile radio that operates on 12 V and draws 1.2 A?

11. Which resistance would produce the most heat: 300 Ω with 25 V across it or 100 Ω with 2.5 A of current through it?

2.7 Resistors

A resistor is one of the most fundamental components used in electronic circuits. A resistor is constructed to have a specific amount of resistance to current flow. Resistors range in value from less than 1 Ω to well over 20 MΩ. They also vary in size from microscopic through devices that are too large to carry. In general, we can classify resistors into two categories: fixed and variable. Fixed resistors have a single value of resistance. Variable resistors are made to be adjustable so that they can provide different values of resistance. We will examine both classes of resistors in the following paragraphs.

Fixed Resistors

There are many types of fixed resistors, but they all have at least three characteristics that are important to a technician:

- value
- power rating
- tolerance

All of these characteristics are determined at the time the resistor is manufactured. They cannot be altered by the user. The value of the resistor indicates the nominal or ideal amount of resistance exhibited by the resistor. Like all resistance, it is measured in ohms. Figure 2-18 shows several types of fixed resistors.

RESISTOR POWER RATING

When current flows through a resistor, it encounters opposition, which creates heat. If a resistor dissipates too much heat, it will be damaged. The damage may range from a slight change in the resistance value to complete physical disintegration of the resistor body.

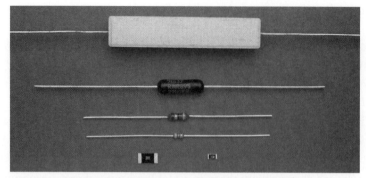

Figure 2-18. There are many types of fixed resistors.

KEY POINTS

The power rating of a resistor specifies the maximum power that can be safely dissipated by the resistor without damage.

Every resistor has a power rating. This indicates the amount of power that can be dissipated for an indefinite amount of time without degrading the performance of the resistor.

The power rating of a resistor is largely, but not solely, determined by the physical size of the resistor. The greater the surface area of the resistor, the more power it can dissipate. Thus, as a general rule, the physical size of a resistor is an indication of its power rating. Resistor power ratings range from less than 0.1 W to many hundreds of watts. The most common power ratings are $^1/_8$, $^1/_4$, $^1/_2$, 1, and 2 W. Figure 2-19 shows resistors with several different power ratings.

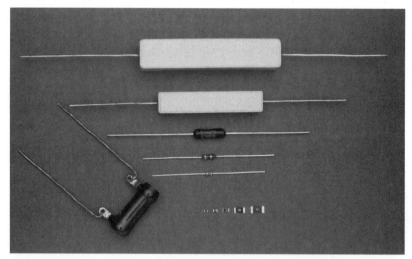

Figure 2-19. The power rating of a resistor is largely affected by its physical size.

Resistor Tolerance

KEY POINTS

Resistors cannot be manufactured to exact values, so they are given a tolerance rating.

Manufacturers cannot make resistors with exactly the right value. There is a certain degree of variation between resistors that are ideally the same value. The manufacturer guarantees that the actual value of the resistor will be within a certain percentage of its nominal or marked value. The allowable variation—expressed as a percent—is called the **resistor tolerance.**

The most common resistor tolerances are 1%, 2%, 5%, 10%, and 20%. As a general rule, tighter tolerance resistors are more expensive. Resistors with tolerances lower than 2% are often called **precision resistors.** Resistors with a 20% tolerance are rarely used in modern designs. The tolerance may be either positive or negative. That is, the actual resistor value may be smaller (negative tolerance) or larger (positive tolerance) than the marked value. The maximum deviation between actual and marked values for a particular resistor is computed by multiplying the tolerance percentage by the marked value of the resistor. This computation is given by Equation 2-12.

$$\text{maximum deviation} = \text{tolerance} \times \text{marked value} \quad \text{(2-12)}$$

We can determine the highest resistance value that a particular resistor can have and still be within tolerance by adding the maximum deviation to the marked value. The minimum acceptable resistance value is computed by subtracting the maximum deviation from the marked value. Thus, the range of resistance values is given by Equation 2-13.

$$\text{resistance range} = \text{marked value} \pm \text{maximum deviation} \quad \text{(2-13)}$$

EXAMPLE SOLUTION

A certain resistor is marked as 1,000 Ω with a 10% tolerance. Compute the maximum deviation and the range of possible resistance values.

EXAMPLE **SOLUTION**

First we apply Equation 2-12 to determine the maximum deviation.

$$\text{maximum deviation} = \text{tolerance} \times \text{marked value}$$
$$= 0.1 \times 1,000 \ \Omega = 100 \ \Omega$$

We now apply Equation 2-13 to calculate the range of resistor values that are within the tolerance specification.

$$\text{resistance range} = \text{marked value} \pm \text{maximum deviation}$$
$$\text{lowest value} = 1,000 - 100 = 900 \ \Omega$$
$$\text{highest value} = 1,000 + 100 = 1,100 \ \Omega$$

Practice Problems

1. A certain resistor is marked as 330 Ω with a 20% tolerance. Compute the maximum deviation and the range of possible resistance values.
2. Determine the highest and lowest resistor value that a 22-kΩ resistor can have, if it has a 5% tolerance.
3. If a resistor that was marked as 39 kΩ, 10% actually measured 37,352 Ω, would it be within tolerance?

Answers to Practice Problems

1. maximum deviation: 66 Ω;
range: 264 Ω to 396 Ω

2. 20.9 kΩ and 23.1 kΩ

3. Yes.

RESISTOR TECHNOLOGY

We shall examine four major classes of fixed resistor technology:

- carbon-composition resistors
- film resistors
- wirewound resistors
- surface-mount technology (SMT)

CARBON-COMPOSITION RESISTORS

> **KEY POINTS**
>
> Carbon-composition resistors use a mixture of carbon and an insulating filler as the resistive element; this mixture is pressed into a cylindrical form.

Carbon-composition resistors are one of the oldest types of resistors used today. They are far from obsolete, however. Figure 2-20 shows a cutaway view of a carbon-composition resistor. They are manufactured by making a slurry of finely ground carbon (fairly low-resistance material), a powdered filler (high-resistance material), and a liquid binder. The slurry is pressed into cylindrical forms and the leads are attached. Finally, the resistor is coated with a hard nonconductive coating and then color banded to indicate its value. Resistors of different values are made by altering the ratio of carbon to filler material. The higher the percentage of carbon, the lower the resistance of the resistor. The power rating is made higher by increasing the physical size of the resistor.

Figure 2-20. A cutaway view of a carbon-composition resistor.

FILM RESISTORS

> **KEY POINTS**
>
> Film resistors are made by depositing a thin layer of resistive material on an insulating rod.

Film resistors come in several varieties. A film resistor is made by depositing a thin layer of resistive material onto an insulating tube or rod called the substrate. The general range of resistance is established by the resistivity of the material used. Leads are attached to end caps which contact the ends of the resistive film.

Once the resistive layer has been deposited, it is trimmed with a high-speed industrial laser. As the laser etches away portions of the deposited film, the resistance of the device increases. The laser normally trims the resistor to the correct value by cutting a spiral pattern along the length of the resistor body. This method of trimming is called **spiraling.** The laser, in combination with automatic test fixtures can produce resistors that are very close to the desired value.

Several types of materials are used as the film material in film resistors: carbon (carbon film), nickel-chromium (metal film), metal and glass mixture (metal glaze), metal and oxide (metal oxide). Each technology provides advantages over the others, including such things as power ratings, ability to withstand momentary surges, reaction to environmental extremes, resistance to physical abuse, and changes in resistance with aging. Figure 2-21 shows a cutaway view of a typical film resistor.

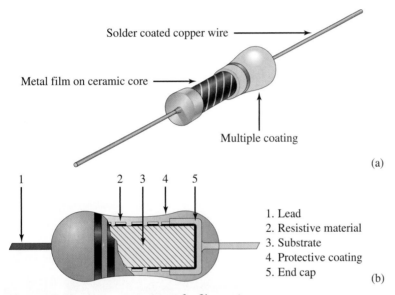

Figure 2-21. A cutaway view of a film resistor.

WIREWOUND RESISTORS

Wirewound resistors are made by winding resistive wire around an insulating rod. The ends of the wire are connected to leads, and the body of the resistor is coated with a hard insulative jacket. The value of a particular wirewound resistor is determined by the type of wire, diameter of the wire, and the length of the wire used to wind the resistor. Wirewound resistors are generally used for their high power ratings, but they are sometimes selected because of their precise values (precision resistors). Figure 2-22 shows a cutaway view of a wirewound resistor. Sometimes a wide colored band is painted on the resistor body to identify it as a wirewound type.

SURFACE-MOUNT RESISTORS

Because there is a continuing demand for smaller electronic products, there is an associated demand for smaller electronic components. One common method for reducing the size of an electronic product is to utilize SMT. Surface-mount resistors have no leads. Their contacts solder directly to pads on the printed circuit board. Surface-mount resistors are manufactured by spreading a resistive layer onto a ceramic substrate. The resistive

KEY POINTS

Wirewound resistors are made by wrapping resistive wire around an insulating rod.

KEY POINTS

Wirewound resistors have higher power ratings for a given physical size than the other resistor types.

KEY POINTS

Surface-mount resistors have no external leads and are used when small size is an important consideration.

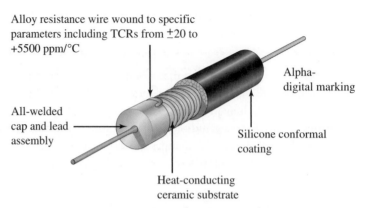

Figure 2-22. A cutaway view of a wirewound resistor.

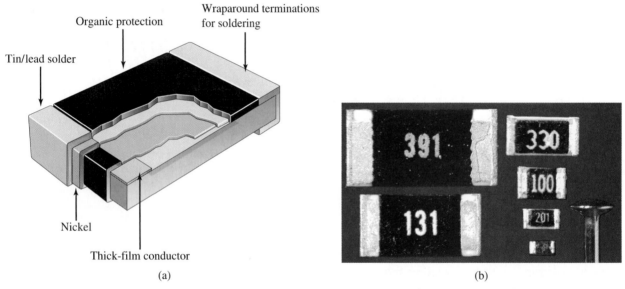

Figure 2-23. (a) A cutaway view of a surface-mount resistor. (b) Typical surface mount resistors. The head of a straight pin is shown for perspective.

coating is then covered with a layer of glass for protection. Figure 2-23 shows a cutaway view of a surface-mount resistor.

RESISTOR COLOR CODE (THREE- AND FOUR-BAND)

Since many resistors are physically small, it is impractical to print their value with numbers big enough to read. Instead, manufacturers often mark resistors with three to five colored bands to indicate their value. The colors are assigned according to a code standardized by the Electronic Industries Association (EIA). It is essential for a technician to memorize this code and know how to use it.

Table 2-1 shows the relationship between a particular digit and its corresponding color in the standard color code. Figure 2-24 shows how to interpret the colored bands on a three- or four-band resistor.

BAND TYPE	COLOR	DIGIT VALUE	MULTIPLIER VALUE	TOLERANCE VALUE
Digit/Multiplier Bands	Black	0	10^0	—
	Brown	1	10^1	—
	Red	2	10^2	±2%
	Orange	3	10^3	—
	Yellow	4	10^4	—
	Green	5	10^5	—
	Blue	6	10^6	—
	Violet	7	10^7	—
	Gray	8	—	—
	White	9	—	—
Multiplier/Tolerance Bands	Silver	—	10^{-2}	±10%
	Gold	—	10^{-1}	±5%
	No band	—	—	±20%

Table 2-1. Resistor Color Code for Three- and Four-Band Resistors

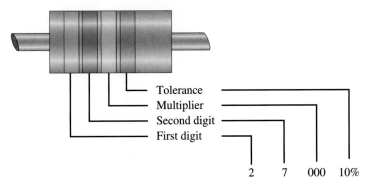

Figure 2-24. Interpretation of the resistor color-code bands.

Three- or four-band resistors can be interpreted by applying the following procedure:

1. The first two bands represent the first two digits of the resistance value.
2. Multiply the digits obtained in step one by the value of the third (multiplier) band.
3. If there is a fourth band, it indicates the tolerance, otherwise tolerance is ±20%.

EXAMPLE SOLUTION

Determine the value of each of the resistors pictured in Figure 2-25.

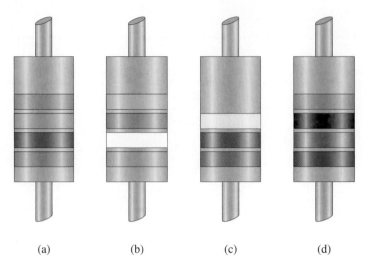

(a)   (b)   (c)   (d)

Figure 2-25. Determine the value of these resistors.

EXAMPLE **SOLUTION**

a. The first band is the one nearest to an end of the resistor. In this case, the resistor value is decoded as shown in Table 2-2.

BAND	COLOR	VALUE
First	Red	2
Second	Violet	7
Third (multiplier)	Orange	000
Fourth (tolerance)	Silver	10%

Table 2-2. Decoding the Value of the Resistor in Figure 2-25(a)

The final decoded value is 27,000 Ω, ±10%.

b. The first two bands (orange and white) give us the first two digits of 39. The next band (red) is the multiplier and tells us to add two zeros. This gives us a base value of 3,900 Ω. Finally, the fourth band is gold, which identifies the tolerance as 5%. Thus, we have a 3.9 kΩ, ±5% resistor.

c. The first two bands (green and blue) give us the digits 56. The multiplier band is yellow so we add four zeros to produce our base resistance value of 560,000 Ω. There is no tolerance band, which means that the resistor has a 20% tolerance. The final decoded value is 560 kΩ, ±20%.

d. The first two bands (brown and gray) give us the digits 18. The multiplier band is black so we add zero 0s (i.e., add nothing). The silver multiplier band indicates a 10% tolerance rating. The final converted value is 18 Ω, ±10%.

Practice Problems

1. Determine the values of the resistors pictured in Figure 2-26.

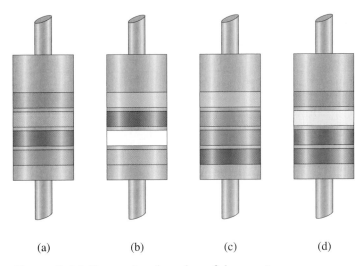

(a)   (b)   (c)   (d)

Figure 2-26. Determine the value of these resistors.

Answers to Practice Problems

1. a. 27 kΩ, ±10%
    b. 390 Ω, ±5%
    c. 1.5 kΩ, ±5%
    d. 680 kΩ, ±10%

RESISTOR COLOR CODE (FIVE-BAND)

Figure 2-27 shows a resistor with five coding bands. The bands are interpreted as follows:

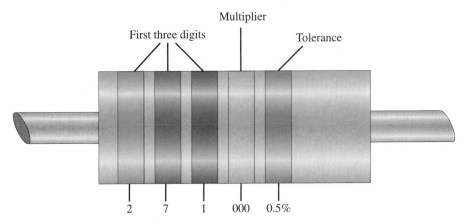

Multiplier

First three digits   Tolerance

2   7   1   000   0.5%

Figure 2-27. A resistor that uses a five-band color code

1. The first three bands determine the first three digits of the resistance value.
2. The fourth band is the multiplier.
3. The fifth band indicates tolerance.

Resistors that use this coding scheme are often called precision resistors. The color values shown in Table 2-1 can be used to decode the first four bands. The tolerance band is decoded as shown in Table 2-3.

TOLERANCE BAND COLOR	TOLERANCE VALUE
Brown	1%
Red	2%
Gold	5%
Silver	10%
Green	0.5%
Blue	0.25%
Violet	0.1%
Gray	0.05%

Table 2-3. Interpretation of Tolerance Band Values for a Five-Band Resistor Code

It should be noted that the color interpretation listed in Table 2-3 reflects the EIA standard number RS-196-A. Some resistors use different five-band color code schemes. One such alternative code is shown in Figure 2-28. Here the first four bands are interpreted according to the four-band color code standard (Table 2-1). The fifth band is used to indicate the reliability (actually failure rate) of the resistor. Reliability is expressed as a percentage of failures in 1,000 hours of operation. Thus, a value of 0.01% means that if one million resistors were operated for 1,000 hours, then no more than 100 resistors would go out of tolerance. The fifth-band colors and their interpretation are shown in Table 2-4.

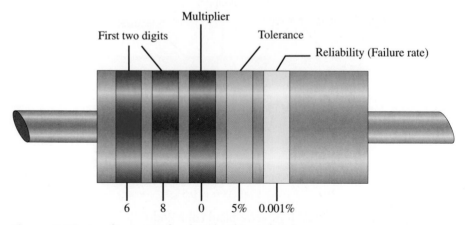

Figure 2-28. An alternative five-band color code scheme.

RELIABILITY BAND COLOR	FAILURES PER 1,000 HOURS
Brown	1.0%
Red	0.1%
Orange	0.01%
Yellow	0.001%

Table 2-4. Interpretation of the Reliability Band on a Five-Band Resistor Code

SURFACE-MOUNT RESISTOR MARKINGS

Many surface-mount resistors have no visible markings to indicate their resistance value. The technician must use test equipment (ohmmeter) to measure the resistance. Some manufacturers indicate the resistance value of surface-mount resistors with a numerical code where the first digit or digits in the code represents the digits in the resistance value, and the last digit in the code indicates the number of trailing zeros. For example, a surface-mount resistor that is coded as 103 would be a 10,000-ohm resistor.

SCHEMATIC SYMBOL

Electronics technicians use schematic diagrams to represent the electrical connections in a circuit. Every component has a symbol that is used to represent it on a schematic diagram. The schematic shows how the various components are interconnected. Figure 2-29 shows the schematic symbol for a fixed resistor.

-\/\/\-

Figure 2-29.
Schematic symbol for a fixed resistor.

Variable Resistors

It sometimes is desirable to change the value of a resistor once it has been installed in a circuit. For these applications, we can use a variable resistor. The resistance of a variable resistor can be adjusted by turning a knob or rotating a screw. There are two major classes of variable resistors: rheostats and potentiometers. These two classes often are actually the same physical device; it is the electrical connection that distinguishes the two types of variable resistors.

KEY POINTS

Variable resistors can be adjusted to provide a desired resistance value.

RHEOSTATS

A **rheostat** is a two-terminal device whose resistance can be changed. Figure 2-30 illustrates one method for constructing a variable resistance. The effective resistance of the rheostat is that portion of the total resistance that appears between the two connections. The movable contact—often called a wiper arm—slides to allow greater and lesser portions of the total resistance to be inserted between the external contacts. Figure 2-31 shows the schematic symbol for a variable resistor configured as a rheostat. The use of an arrow to indicate a variable quantity is a common practice in electronics.

KEY POINTS

Rheostats are two-terminal variable resistors.

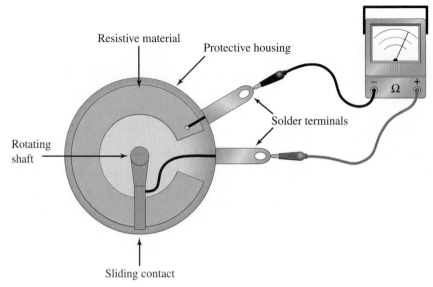

Figure 2-30. Cutaway view of a variable resistor configured as a rheostat.

Figure 2-31.
Schematic symbol for
a rheostat.

POTENTIOMETERS

A **potentiometer** (called a pot for short) is a three-terminal device that is constructed much like a rheostat. Figure 2-32 shows one method for constructing a potentiometer. Two of the three contacts connect to the ends of the total resistance. The third contact is connected to the wiper arm. The total end-to-end resistance of the device shown in Figure 2-32 can be labeled R_{AC}. The resistance from the wiper arm to the ends can be labeled as R_{AB} and R_{BC}. It is important to see that as the wiper arm is adjusted, R_{AC} remains constant and is equal to the total resistance of the device. As the wiper arm is moved closer to point A, the resistance R_{AB} decreases and will be near zero if the wiper arm moves all the way to point A. As R_{AB} decreases, resistance R_{BC} increases. If R_{AB} becomes zero, then R_{BC} will be equal to the total resistance of the rheostat. This can be expressed mathematically by Equation 2-14.

$$R_{AC} = R_{AB} + R_{BC} \qquad (2\text{-}14)$$

This relationship will be valid regardless of the position of the wiper arm. Figure 2-33 shows the schematic symbol for a potentiometer.

EXAMPLE SOLUTION

The end-to-end resistance (R_{AC}) of a certain potentiometer is 10 kΩ. In its present position, the resistance from the wiper arm to one end (R_{AB}) is 6.5 kΩ. What is the resistance between the wiper arm and the other end (R_{BC})?

EXAMPLE **SOLUTION**

We apply Equation 2-14 as follows:

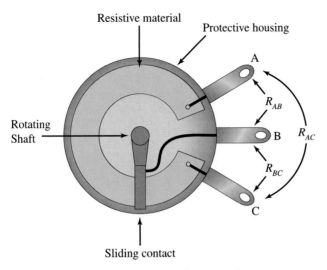

Figure 2-32. A cutaway view of a potentiometer.

Figure 2-33.
Schematic symbol for
a potentiometer.

$R_{AC} = R_{AB} + R_{BC}$, or

$R_{BC} = R_{AC} - R_{AB}$

$\quad = (10 \times 10^3) - (6.5 \times 10^3) = 3.5 \text{ k}\Omega$

Figure 2-34 shows a number of potentiometer types. Some have internal gear mechanisms. This allows multiple turns of the adjusting screw to have small effects on the wiper arm position. This allows the resistor to be set to more precise values.

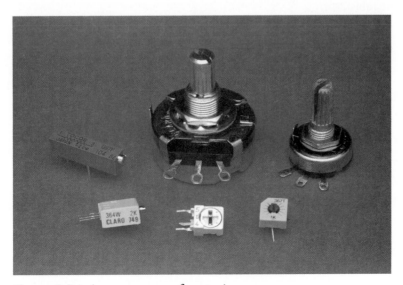

Figure 2-34. An assortment of potentiometer types.

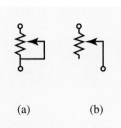

(a)   (b)

Figure 2-35.
A three-terminal
potentiometer can be
connected as a two-
terminal rheostat.

CONNECTING A POTENTIOMETER AS A RHEOSTAT

A three-terminal potentiometer can be connected to operate as a two-terminal rheostat. Two such methods are shown schematically in Figure 2-35. In Figure 2-35(a), the wiper arm is connected to one of the end connections, thus forming a two-terminal rheostat. In Figure 2-35(b), the wiper arm and one end connection are used as the two connections of a rheostat.

Exercise Problems 2.7

1. Name three ratings that must be evaluated when choosing a resistor for a particular application.
2. Resistors can be divided into two general classes called _____ and _____.
3. What is the highest value a 2.2-kΩ, ±5% resistor can measure and still be within its rated tolerance?
4. What is the maximum deviation from the marked value that a 180-kΩ, ±1% resistor can have and still be within its rated tolerance?
5. If a 220-Ω, ±5% resistor measures 205 Ω, is it within tolerance?
6. What is the general name given to resistors that are made by depositing a thin layer of resistive material on an insulating rod?
7. What type of resistor has no wire leads?
8. For a given physical size, what type of resistor has the highest power rating?
9. Interpret the information provided by the color codes on the resistors described in Table 2-5.

FIRST BAND	SECOND BAND	THIRD BAND	FOURTH BAND	FIFTH BAND	INDICATED VALUE
Red	Violet	Brown	Gold	—	
Violet	Green	Orange	—	—	
Orange	Orange	Silver	Gold	—	
Brown	Red	Orange	Brown	Silver	
Gray	Red	Yellow	Gold	Orange	

Table 2-5. Determine the Value of These Resistors

10. The end-to-end resistance of a rheostat changes when the rheostat is adjusted. (True or False)

11. The end-to-end resistance of a potentiometer changes when the potentiometer is adjusted. (True or False)

12. A rheostat is a _____-terminal device, but a potentiometer is a _____-terminal device.

2.8 Switches

A switch is another basic, but important, component used in electronic circuits. Its purpose is to break (open) or make (close) circuit connections. You are familiar with the switches in your home that are used to control the lighting. When you operate the switch to turn the lights on, you are closing the circuit and connecting the lightbulb to the power line. When the switch is off, the lightbulb is isolated from the power line.

● **KEY POINTS**

Switches are used to open or close the path for current flow in an electrical or electronic circuit.

Basic Switch Operation

Figure 2-36 shows a basic switch circuit. The type of switch shown in the picture is called a knife switch. Although it is not often used in electronic circuits, this type of switch is easy to understand, and its operation is representative of other, less obvious, switch types. When the switch is in the upper position, as in Figure 2-36(a), the circuit is open and the lamp remains dark. When the switch closes the circuit, as in Figure 2-36(b), then electrons can travel from the battery, through the lamp, through the switch, and return to the positive side of the battery. This simple switch represents a general class of switches called single-pole single-throw (SPST). The schematic symbol for a SPST switch is shown in Figure 2-37 along with the symbols for several other types of switches.

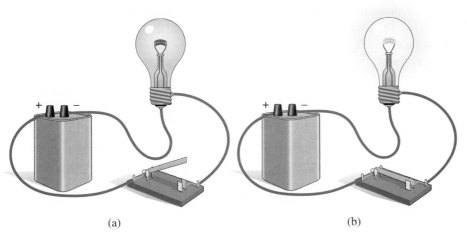

(a)   (b)

Figure 2-36. A basic switch circuit.

The term **pole** refers to the movable portion of the switch. For example, a double-pole switch has two movable arms; it acts as two *electrically* separate switches that are *mechanically* linked so they operate simultaneously. The pole of a switch is often identified with an arrowhead on the switch symbol. The term **throw** identifies the number of circuits that are

● **KEY POINTS**

In a multiple-pole switch, one mechanical link controls more than one electrical switch.

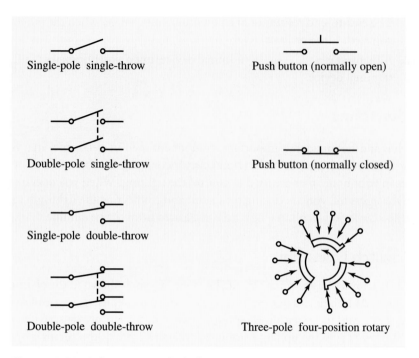

Figure 2-37. Schematic symbols for several types of switches.

opened or closed by each pole when the switch is operated. For example, a double-throw switch opens or closes two circuits on each pole.

Figure 2-38 illustrates how to connect a single-pole double-throw (SPDT) switch into a lamp circuit. In Figure 2-38(a), the red lamp is lit while the blue lamp is off. In the switch position shown in Figure 2-38(b), the circuit for the blue lamp is closed and the red lamp is extinguished.

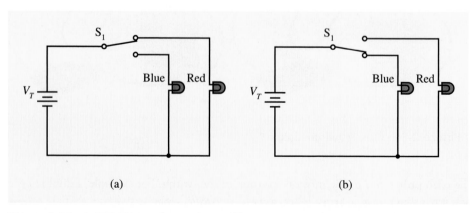

Figure 2-38. A SPDT switch can alternately connect power to two different circuits.

Switching Mechanism

Any of the basic switch types (e.g., SPST, SPDT, DPST, DPDT, and so on) can be made with different mechanical means for operating the switch. Some of the more common means for switch operation are listed:

- toggle
- rocker
- push button
- rotary
- slide

There is a tremendous range of switch forms based on these basic classes of mechanisms. Figure 2-39 shows a sampling of switch types.

Figure 2-39. There many types of switching mechanisms available from switch manufacturers.

Momentary-Contact Switches

Some switches open and/or close a circuit when you operate the switch, but return to the original state as soon as the switch button is released. This type of switch is called a momentary-contact switch. Doorbell and car horn switches are two examples of momentary-contact switches.

Technicians must know which contacts are closed and which ones are open when the switch is in its normal state. There are two ways this is indicated on schematic diagrams. First, the poles for all switch contacts are drawn in the normal (not operated) position. Second, the letters NC and NO are sometimes written near the contacts to indicate normally closed and normally open contacts, respectively.

Ganged Switches

When switches have multiple poles, the individual poles are mechanically linked, or ganged, together so that they operate simultaneously. This mechanical link is shown on

KEY POINTS

Switches may be constructed with momentary contacts, which change states when activated but return to their normal state when the switch is released.

KEY POINTS

The normal state of the switch is shown on the schematic symbol.

schematic diagrams as a dotted line. The multiple-pole switches shown in Figure 2-37 have dotted lines to show the mechanical link between sections of the switch.

Switch Specifications

Switch manufacturers provide extensive catalogs that detail the physical and electrical characteristics of their switches. A technician or engineer needs this information to select a particular switch for a specific application. Some of the factors to be considered in switch selection are cited:

- contact form (i.e., SPDT, DPDT, and so on)
- switching mechanism (e.g., push button, toggle, rocker, and so on)
- voltage rating
- current rating
- environmental performance (e.g., moisture immunity, ruggedness, and so on)

Exercise Problems 2.8

1. How many electrical connections are there on a SPST switch?
2. How many electrical terminals are there on a DPDT switch?
3. How many electrical connections are there on a two-pole, six-position rotary switch?
4. Describe the symptoms if an electrician accidentally installed a momentary-contact switch as a light switch in your home.
5. Figure 2-40 shows an electrical circuit with three switches and three lamps. Each switch has its positions labeled as "A" and "B." Analyze the circuit and determine which lamps are on for each of the conditions listed in Table 2-6.

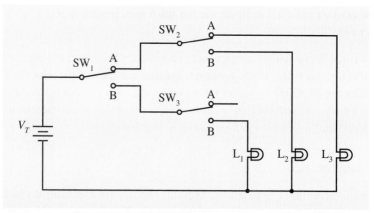

Figure 2-40. A switch circuit for Exercise Problem 5.

POSITION OF SWITCH			STATE OF LAMP		
SW_1	SW_2	SW_3	L_1	L_2	L_3
A	A	A			
A	A	B			
A	B	A			
A	B	B			
B	A	A			
B	A	B			
B	B	A			
B	B	B			

Table 2-6. Relationship Between Switch Position and Lamp State for the Circuit Shown in Figure 2-40

2.9 Fuses and Circuit Breakers

Most electrical and electronic circuits are susceptible to damage by excessive current flow. The increased current normally flows as a result of a defective component or an accidental **short circuit.** When a normal path for current flow is bypassed by a much lower resistance path, we call the low-resistance path a short circuit. Short circuits are generally characterized by increased current flow. Fuses and circuit breakers provide protection against damage from excessive current flow.

Fuses

Fuses are connected such that the current flowing through the protected circuit also flows through the fuse. Figure 2-41 shows the operation of a fuse. There is a resistive link inside the fuse body that heats up when current flows through it. If the current is sufficiently high, the resistive link burns open, which stops all current flow in the circuit. When a fuse has burned open, we say it is blown.

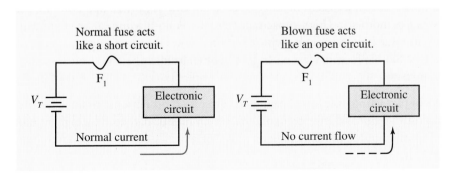

Figure 2-41. A fuse protects circuits or devices from excessive current flow.

Fuses have three electrical ratings, which are particularly important to a technician:

- current rating
- voltage rating
- response time

CURRENT RATING

The current rating of a fuse indicates the maximum *sustained* current that can flow through the fuse without causing the element to burn open. The actual current required to blow the fuse may be less than the rating if the fuse is operated in a high-temperature environment. A fuse normally has a substantially higher current rating than the value of normal operating current in the circuit to be protected.

VOLTAGE RATING

The voltage rating on a fuse specifies the minimum amount of voltage required to arc across the fuse *after it has blown*. The voltage rating of the fuse should exceed the highest voltage to be expected in the protected circuit. If the circuit voltage exceeds the fuse rating, then the voltage may arc (like a miniature lightning bolt) across the open fuse link. This effectively reconnects the circuit and defeats the purpose of the fuse.

RESPONSE TIME

When the current through a fuse exceeds its current rating, the fuse link will burn open. However, it takes a certain amount of time for the resistive element to heat up to the point of disintegration. The time required for the link to burn open is called response time. Different types of fuses have different response times. The response time of any given fuse decreases (opens faster) as the percentage of overcurrent increases. Typical response times vary from several milliseconds to several seconds.

SLO-BLO, NORMAL-BLO, AND FAST-BLO FUSES

Manufacturers classify fuses into three general categories based on the approximate response times. Slo-blo fuses are designed to withstand currents that greatly exceed their current rating as long as the overcurrent condition is only momentary. A short-duration current burst, called a transient or surge, is characteristic of normal operation for some devices such as motors and high-capacitance circuits. A slo-blo fuse can open quickly if the overcurrent value is high enough. Typical response times for slo-blo fuses vary from one-0.1 s to 10 s. Slo-blo fuses in glass packages can often be identified by a coiled spring inside the fuse housing.

Fast-blo, or fast-acting, fuses are designed to have very short response times. They are often used to protect semiconductor components, which may be damaged by high current transients. Typical response times include the range of submillisecond through hundreds of milliseconds.

Normal-blo fuses have response times in between those of either slo-blo or fast-acting fuses. Remember, though, the response time of all three types of fuses varies dramatically with the percentage of overload. Figure 2-42 shows several types of fuses.

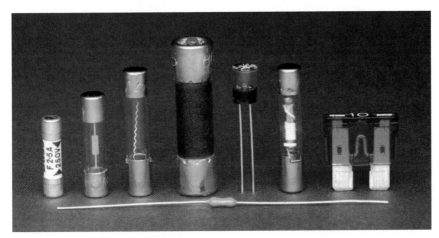

Figure 2-42. Some representative types of fuses.

Circuit Breakers

Fuses have a distinct disadvantage of having to be physically replaced once they have blown. Circuit breakers also provide overcurrent protection for circuits and devices, but they do not have to be replaced after they have opened the circuit. When a circuit breaker opens in response to excessive current flow, we say that it has been tripped. Once a circuit breaker has been tripped, it must be reset before the current path can be restored to the circuit. Most circuit breakers require the user to manually reset the breaker, but some are designed to reset automatically after a time delay.

Figure 2-43 shows one method of circuit breaker construction. The circuit current flows through and heats a bimetallic strip. The two dissimilar metals used in the strip have different coefficients of expansion, which causes the strip to bend when it is heated. Since one of the two breaker contacts is mounted on the bending strip, the circuit is opened if the strip bends sufficiently. Once the circuit has opened, the current flow stops and the bimetallic strip cools. If the strip is allowed to return to its original position as it cools, we say the breaker is automatically reset. This is undesirable for many applications. Manually reset breakers have a spring-loaded lever that prevents the contacts from reconnecting when the strip cools. When a reset button is pressed, the spring-loaded arm is momentarily pushed out of the way, so the contacts can come together.

Although circuit breakers have the advantage of reusability, they have a disadvantage for certain applications since they generally have longer response times than fuses. Figure 2-44 shows several circuit breakers.

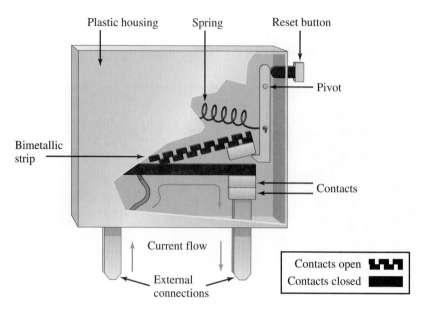

Figure 2-43. A cutaway view of a circuit breaker.

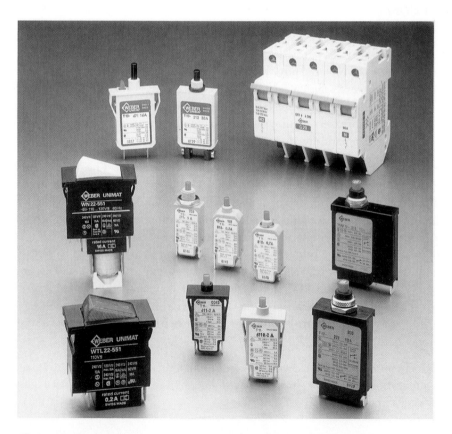

Figure 2-44. Some representative examples of circuit breakers. (Courtesy of Weber US)

Exercise Problems 2.9

1. Explain the meaning of the current rating of a fuse.
2. Why do fuses have voltage ratings?
3. What would be the primary reason for choosing a slo-blo fuse?
4. What type of protection device would you choose to protect a sensitive semiconductor device?

2.10 Optical Indicators

There are thousands of types of indicators, which vary in purpose, size, shape, and technology. Optical indicators can be loosely classified into two groups. One group is designed to serve primarily as a light source or single-point indicator. The second group provides alphanumeric information and/or graphic information. We shall limit our discussion to the light-source group, and we will further divide this group into incandescent, neon, and solid state.

Incandescent Light Sources

Figure 2-45 shows the construction and schematic symbol for an incandescent lamp. The overall construction consists of a tungsten filament supported within an evacuated glass bulb. This has not changed substantially since the time of Thomas Edison. When current flows through the tungsten filament, it becomes hot and glows bright white.

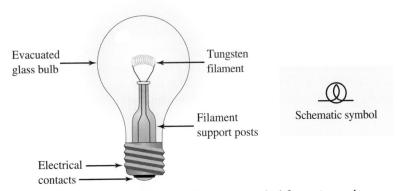

Figure 2-45. Construction and schematic symbol for an incandescent lamp.

KEY POINTS

The incandescent lamp is useful as an indicator (point source) or as a source of illumination (diffused source).

KEY POINTS

The lifetime of an incandescent lamp is often shortened by high-voltage transients, inrush currents, and mechanical shock.

Incandescent lamps are used as point indicators (e.g., a power-on indicator) and as light sources. Most of the lightbulbs used in your home are incandescent lamps. The range of voltage requirements extends from less than 1 V to more than 120 V. Current requirements range from a few milliamperes to many amperes.

Incandescent lamps are very susceptible to mechanical vibrations or shock. It should be noted that the rated lifetimes are specified for a shock-free environment. Actual lifetimes may be much shorter. Incandescent lamps can also be damaged by transient voltages or currents. Transient voltages frequently occur on commercial power lines. Additionally, every time the lamp is turned on it is subjected to a current transient in the form of an inrush current. This occurs because the resistance of a cold tungsten filament is much less than its hot resistance. This allows the initial current to be much higher than the normal operating current. You may have noticed that the lightbulbs in your home burn out most frequently when they are first turned on. Figure 2-46 shows several representative incandescent lamps.

Figure 2-46. Some typical incandescent lamps.

Neon Indicators

Figure 2-47 shows the construction of a simple neon indicator and a circuit diagram showing its connection. The indicator consists of a neon-filled glass bulb with two sealed electrodes passing through the glass. If sufficient voltage (75 V and up) is connected between the two electrodes, then the neon gas is ionized. You will recall that ions can participate in current flow. The voltage required for ionization is called the **ionization voltage,** or firing voltage, and varies between different types of bulbs.

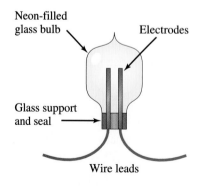
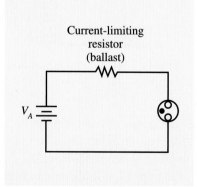

Figure 2-47. The construction and application of a neon indicator.

Prior to firing, the resistance of the neon gas is quite high. Once the gas has ionized, the resistance between the two electrodes drops drastically. This can allow damaging currents (that will actually melt the electrodes) to flow. To prevent electrode damage, all neon lamps must be operated with a resistance (sometimes called a ballast) connected in line with the lamp. This is shown in Figure 2-47. Normal operating currents vary from hundreds of microamperes to tens of milliamperes.

It takes a higher voltage to ionize the neon gas than it does to maintain the ionized state. Therefore, once the lamp has fired, the applied voltage must be reduced to a lower voltage to extinguish the lamp. The ionization voltage varies with the amount of external energy passing through the glass; older neons could not be ionized if they were cold and dark. Modern lamps have a small amount of radioactive material to provide constant radiation and reduce this effect. Figure 2-48 shows two types of neon lamps.

Solid-State Light Sources

The solid-state light sources to be considered here are called light-emitting diodes (LEDs). Figure 2-49 shows a cutaway view of an LED and the method of connection. Two types of semiconductor materials are formed into a small chip that is about 0.01 inch on a side. The junction of the two materials emits light when subjected to the proper voltage conditions. Tiny (0.001-inch diameter) gold wire and conductive epoxy are used to connect the semiconductor to the leads. An epoxy housing is formed around the leads and semiconductor to serve as the body of the LED. The housing is transparent or translucent so that light can pass through it. LEDs are available in red, green, yellow, and blue. Some manufacturers even make LED housings with more than one color LED in the same package.

KEY POINTS

Neon bulbs require current-limiting resistors to protect them from damage by excess current.

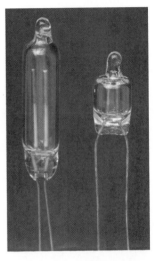

Figure 2-48.
Some typical neon indicators.

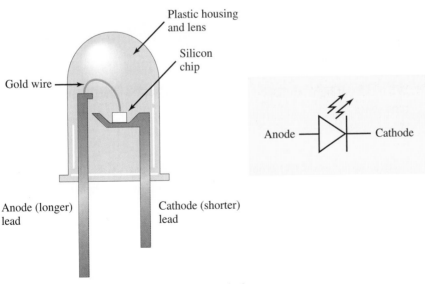

Figure 2-49. LED construction and symbol.

KEY POINTS

LEDs generally operate with less than 2 V across the semiconductor material.

LEDs typically require from 1 to 30 mA at a potential of slightly less than 2 V. The LED is a polarized component. That is, it must be connected in a particular direction. More specifically, the negative potential must be connected to a terminal called the cathode, and the positive potential must be connected to the anode terminal. The cathode lead is generally shorter than the anode lead and/or the body of the LED has one flat side near the cathode lead so a technician can identify the leads. LEDs are easily damaged by excessive current flow so they must be operated with an in-line resistor (or some equivalent method) to limit the maximum current flow. Figure 2-50 shows several representative LEDs.

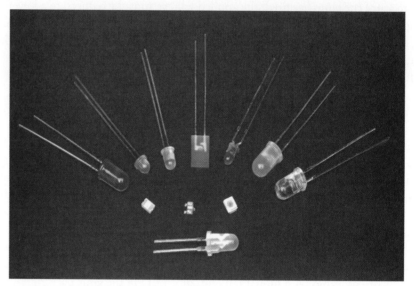

Figure 2-50. Several example LEDs.

Exercise Problems 2.10

1. What class of light source do the standard screw-in lightbulbs used in a home represent?
2. Which class or classes of light sources require an in-line resistance to limit current flow?
3. Which type of indicator must be connected in a particular direction?
4. If the operating voltage on an incandescent lamp were higher than its rated value, what would happen to its life expectancy?

2.11 Safety Practices for Technicians

Electrical and electronic circuits sometimes have lethal voltages and currents. Therefore, it is essential for a technician to have a solid knowledge of appropriate safety practices. While it is certainly important for a technician to respect dangerous voltages and currents, it is also important not to fear them. Fear can obstruct your thinking, interfere with the performance of your job, and actually create a more dangerous situation. Fear often stems from lack of knowledge, so learn all you can about potentially dangerous circuits.

Effects of Voltage or Current

The human body offers a certain resistance to current flow. Ohm's Law can easily confirm that the higher the voltage applied to the body, the higher the current flow. However, it is surprising how small the voltages and currents can be and still affect the body. Voltages of only a few volts and currents of only a few milliamperes can cause serious injury under the right conditions. As the value of current through the body increases, the effects get progressively more severe and include such things as mild tingling sensations, involuntary muscle contractions, difficulty in breathing, heartbeat irregularities, burned skin, and death. One of the most dangerous voltage sources that all technicians work around is the 120-V power line. It is very powerful and can inflict the full spectrum of effects listed previously.

Safety Guidelines

Following are some specific safety guidelines that represent good technician practices:

1. Never work alone on circuits with lethal voltages and currents.
2. Always know the location of circuit breakers, and be sure other people in the area also know where the breakers are located.
3. If practical, remove the power to circuits before working on them.
4. Never trust another person's statement that power has been removed from a circuit; verify it yourself.
5. Work with one hand as much as possible. Keep the other hand in your pocket or otherwise restrained so that it cannot inadvertently contact the circuit.
6. Wear shoes with insulative soles or stand on a rubber mat.
7. Only use tools with insulated handles.
8. Remove all neckties and jewelry before working on equipment.
9. NEVER allow horseplay or practical jokes in the vicinity of a live circuit.

This last rule deserves extra attention since its effects are not always obvious. Suppose a technician is working around a high-voltage circuit. He is not scared, but he is rightfully alert with very focused attention. A would-be prankster sneaks up behind the technician and slams a book closed. Of course, it scares the technician, but what about serious injury? In his eagerness to get away from the circuit, the technician may inadvertently contact a high-voltage point. In the process of jumping back from the circuit, the technician may trip and fall into other equipment, causing injuries. And finally, the technician may be unable to find the humor in such mindless acts and will report the prankster to his supervisor. This behavior would be grounds for termination in some companies.

Chapter Summary

- When a material has an unequal number of electrons and protons, we say it has a charge. An excess of electrons in a body produces a negative charge; a deficiency of electrons (excess protons) in a body produces a positive charge. Charge is measured in coulombs (C) where 1 C is the amount of charge provided by 6.25×10^{18} electrons. A charged body is surrounded by an electrostatic field, which is a region that can exert a force on another charged body. Unlike charges attract while like charges repel.

- When two unequal charges are separated by some distance, there is a potential for doing work. This potential difference, called voltage and measured in volts (V), is capable of causing charged particles to move. The movement of charged particles is called current flow. If the potential difference between two charges is sustained even though charged particles are moving from one point to the other, then we call the potential difference an electromotive force or emf.

- An ideal voltage source provides a constant voltage over an infinite time period regardless of the value of current flowing in the circuit. Practical voltage sources have internal resistance. The internal resistance causes the available voltage to decrease as current flow increases.

- Current is generally the movement of charged particles. When 1 C of charge moves past a given point in a 1-s interval, we say there is 1 ampere (A) of current flow. Many technicians think of current flow as the movement of electrons. Current viewed in this way is called electron current flow and moves through a circuit from negative to positive.

- Ideal current sources are capable of maintaining a constant current flow regardless of the amount of resistance in a circuit. Practical current sources provide reasonably constant currents over a range of resistances.

- Resistance is the opposition to current flow and is measured in ohms (Ω). One ohm of resistance will limit the current to 1 A if 1 V is applied to the circuit. Different materials have different amounts of resistance. Insulators have high resistance, conductors have low resistance, semiconductors have moderate resistance, and superconductors have no resistance at all.

- Power is a measure of the rate at which we use energy and often takes the form of heat in electrical components. Power is measured in watts (W). Electrical power measured in watts can be compared to an equivalent mechanical energy expressed in horsepower by using the relationship 1 hp = 746 W. Electrical power is usually calculated with the power formulas which use current, voltage, and resistance as factors in the equation.

- Resistors are components that are designed to provide a certain amount of resistance. Fixed resistors are manufactured with a specific amount of resistance. Variable resistors can be adjusted for different amounts of resistance. Rheostats (two-terminal devices) and potentiometers (three-terminal devices) are two forms of variable resistors. All resistors, fixed and variable, have power ratings, which indicate the amount of electrical power (heat) they can dissipate without damage. Fixed resistors often use a color code to indicate their value. Resistors may have three, four, or five bands. The bands provide information such as resistance value, tolerance, and reliability.

- Switches are used to open or close an electrical circuit. The pole of a switch is the movable element and is often drawn as an arrow on schematic diagrams. The throw of a switch is a term that identifies how many circuits are switched by a given pole. The various contact forms (SPST, DPDT, SPDT, and so on) can be made with different types of actuator mechanisms: toggle, rocker, push button, or rotary. Momentary-contact switches change states while they are being activated, but they automatically return to the normal state when they are released. Schematic symbols always show the normal or relaxed state.

- Fuses and circuit breakers are devices that protect circuits from excessive current flow. Fuses are usually faster than circuit breakers, but they must be replaced with a new fuse once they have blown. Circuit breakers can be reset (manually or automatically) after they have tripped.

- Incandescent, neon, and LED indicators provide sources of light. Incandescent sources emit light from a heated tungsten filament. The light may be used as a point source or as a diffused source. Neon bulbs require higher voltages to operate. They emit light when the neon gas becomes ionized. LEDs are solid-state devices that require low voltages and currents and emit red, yellow, green, or blue light. Both neon bulbs and LEDs must have an external current-limiting resistor, and both are used as point sources of light.

- Ohm's Law ($I = V/R$) describes a very important relationship. It states that the current in a circuit is directly proportional to the voltage across the circuit and inversely proportional to the amount of resistance in the circuit.

- Ohm's Law and the power formulas are so important, they should be committed to memory. However, some technicians use the following memory aid when they are first learning these relationships.

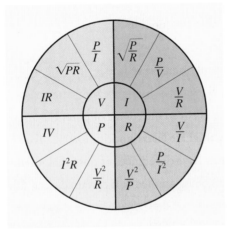

The Ohm's Law and power equation memory wheel

Review Questions

Section 2.1: Charge

1. When a body has more electrons than protons it has a (*positive, negative*) charge.

2. What polarity of charge is on a proton?

3. An electron is a (*positively, negatively*) charged particle.

4. If two electrons were near each other would they attract, repel, or have no effect on each other?

5. How many electrons must be removed from a neutral body to create 1 C of charge?

6. Calculate the number of electrons required to produce a charge of 230 μC.

7. If a certain material has a deficiency of 2.5×10^{19} electrons, what is the polarity and magnitude of its charge?

8. Calculate the force exerted by two bodies separated by 0.25 m, if each body has a negative charge of 3.75 C.

9. Will the bodies described in Question 8 attract or repel one another?

10. How much force is exerted by two charged bodies with charges of +3 C and –2.7 C, if they are separated by a distance of 1.75 m?

11. If a body with a +2-mC charge is separated from another charged body by a distance of 0.15 m, how much charge must be on the second body in order to produce an attraction of 375 mN?

12. The arrows representing an electrostatic field indicate the direction in which a (*positive, negative*) charge would move.

13. What does the closeness of line spacing indicate on a sketch of an electrostatic field?

Section 2.2: Voltage

14. Difference of potential is measured in _____.

15. Voltage is measured in _____.

16. Electromotive force (emf) is measured in _____.

17. Work can be measured in foot-pounds, newton-meters, or _____.

18. Explain the difference between difference of potential and electromotive force.

19. All differences of potential are electromotive forces. (True or False)

20. All electromotive forces have differences of potential. (True or False)

21. Five pounds of force is being applied to a certain object. Express the force in terms of newtons.

22. An ideal voltage source maintains a _____ voltage across its terminals.

23. The output voltage of a practical voltage source varies with current flow because of its _____ resistance.

Section 2.3: Current

24. Current can be defined as the movement of _____.

25. If current is considered to be a flow of electrons, it is called _____ current and flows from (*positive, negative*) to (*positive, negative*).

26. Conventional current is considered to be moving in the same direction as (*positive, negative*) charge.

27. If 7.5 C pass a certain point in a conductor every second, how much current is flowing in the wire?

28. How much current is flowing in a conductor if 400 μC pass a given point in 500 ms?

29. If a current of 400 mA flows past a certain point for 3 s, how much charge will be transferred?

30. How much charge is transferred if 2.7 A flows in a wire for 2 minutes?

31. An ideal current source maintains a _____ current through its terminals.

32. A practical current source is relatively unaffected by changes of in-line resistance. (True or False)

Section 2.4: Resistance

33. Write a brief definition of resistance as it applies to electrical circuits.

34. Insulators have (*high, low, moderate, zero*) resistance.

35. Superconductors have (*high, low, moderate, zero*) resistance.

36. Semiconductors have (*high, low, moderate, zero*) resistance.

37. Conductors have (*high, low, moderate, zero*) resistance.

38. Describe the meaning of the term *breakdown voltage* with reference to insulators.

39. Resistance is measured in _____.

Section 2.5: Power

40. Mechanical power is the rate of doing _____.

41. Mechanical power is measured in _____.

42. Electrical power is the rate of using electrical _____.

43. Electrical power is measured in _____.

44. How much power is dissipated if 175 J of energy are used over a period of 2.7 s?

45. During what time interval must 5 J of energy be expended if 500 mW are to be produced?

46. Express 350 W as an equivalent number of horsepower.

47. Express 75 hp as an equivalent number of watts.

48. How much power is dissipated in an electrical circuit when 75 V is connected across a 5-Ω resistance?

49. How much current must flow through a 2.7-kΩ resistor in order to produce 375 mW of power?

50. How much voltage is required to produce 200 mW in a 270-Ω resistor?

Section 2.6: Ohm's Law

51. Current is (*directly, inversely*) proportional to the resistance in a circuit.

52. Current is (*directly, inversely*) proportional to the voltage in a circuit.

53. Write the three forms of Ohm's Law.

54. How much current flows through a 6.8-kΩ resistor that has 500 mV across it?

55. How much current flows through a 2-MΩ resistor that has 100 V across it?

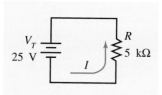

Figure 2-51.
How much current is flowing in this circuit?

56. If a 390-kΩ resistor has 200 mA flowing through it, how much voltage is across it?

57. If a 180-Ω resistor has 10.5 mA flowing through it, how much voltage is across it?

58. How much resistance is required to limit the current to 150 µA when 6.2 V are applied?

59. How much resistance does a circuit have if 150 V causes 200 mA of current to flow?

60. Determine the current (I) that is flowing in Figure 2-51.

61. What is the value of the resistor shown in Figure 2-52?

Section 2.7: Resistors

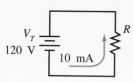

Figure 2-52.
What is the value of the resistor?

62. The two general classes of resistors are _____ and _____.

63. Since resistors cannot be manufactured with exactly the right value, they are given a _____ rating.

64. What characteristic or rating of a resistor is largely determined by its physical size?

65. If a 39-kΩ resistor has a ±5% tolerance, what is the maximum deviation between the actual and marked values of the resistor that is considered to be within tolerance?

66. What is the highest resistance a 27-kΩ, ±20% resistor can measure and still be within tolerance?

67. What is the lowest a 3.3-MΩ, ±2% resistor can measure and still be within tolerance?

68. What type of resistor has no leads and is soldered directly to pads on a printed circuit board?

69. Name three factors used by the manufacturer to control the value of a wirewound resistor.

70. What type of resistor is often associated with high power ratings?

71. In your own words, explain how to interpret the color code of a three-band resistor.

72. Explain how to interpret the fourth band of a four-band resistor.

73. If the fifth band of a certain resistor is silver, what is its tolerance?

74. If the fourth band of a four-band resistor is silver, what is its tolerance?

75. What is indicated by the tolerance band of a resistor if the band is red?

76. Interpret the information provided by the color codes on the resistors listed in Table 2-7.

77. Figure 2-53 shows a variable resistor connected as a _____.

78. A potentiometer is a _____-terminal device.

79. A rheostat is a _____-terminal device.

80. Does the resistance between the end terminals of a rheostat vary when the control is adjusted?

81. Does the resistance between the end terminals of a potentiometer vary when the control is adjusted?

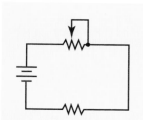

Figure 2-53.
How is the variable resistor connected?

FIRST BAND	SECOND BAND	THIRD BAND	FOURTH BAND	FIFTH BAND	INDICATED VALUE
Yellow	Violet	Brown	Gold	—	
Green	Blue	Orange	—	—	
Orange	Orange	Silver	Silver	—	
Brown	Gray	Orange	Red	Silver	
Blue	Red	Green	Silver	Brown	

Table 2-7. Determine the Value of These Resistors

Section 2.8: Switches

82. How many connections would be on a SPST switch?

83. How many connections would be on a three-pole, four-position rotary switch?

84. Explain the meaning of the labels NO and NC, which are sometimes used with reference to switches.

85. If two switch symbols on a schematic diagram are linked with a dotted line, what does this mean?

Section 2.9: Fuses and Circuit Breakers

86. Explain the purpose of the voltage rating on a fuse.

87. Explain the purpose of a current rating on a fuse.

88. When might a slo-blo fuse be a better choice than a fast-acting fuse?

89. Name one major advantage of a circuit breaker over a fuse.

Section 2.10: Optical Indicators

90. Name three factors that can cause an incandescent lamp to burn out before its rated lifetime.

91. What name is given to describe the amount of voltage required to illuminate a neon bulb?

92. LEDs require an in-line current-limiting resistance for proper operation. (True or False)

Section 2.11: Safety Practices for Technicians

93. Carelessness around electrical/electronic circuits can result in death. (True or False)

94. It takes voltages in excess of 500 V to be considered dangerous. (True or False)

The lights in your shop are presently controlled by two switches. The lights can be turned on or off by either switch. This operation is similar to the light switches at the top and bottom of stairs in houses. Your supervisor has located a schematic that he believes shows how the switches are connected. The schematic is shown in Figure 2-54.

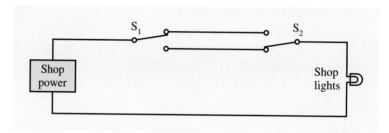

Figure 2-54. A schematic showing how two SPDT switches can control a light from either of two locations.

Your supervisor has given you two assignments. First, he wants you to verify that the schematic shown in Figure 2-54 really does provide the intended operation. Second, a friend has told him that the lights can be controlled from a third location if a DPDT switch is added between the two SPDT switches already installed. It is your job to determine how to connect the third switch. Figure 2-55 provides a schematic view of the problem.

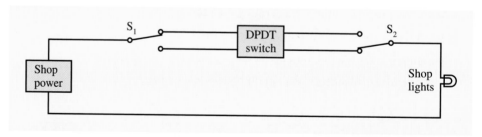

Figure 2-55. How can a DPDT switch be connected in order to provide lamp control from any of three locations?

Equation List

(2-1) $1\ C = 6.25 \times 10^{18}$ electrons

(2-2) $F = k\dfrac{Q_1 Q_2}{d^2}$

(2-3) $I = \dfrac{Q}{t}$

(2-4) $\text{Power} = \dfrac{\text{energy}}{\text{time}}$

(2-5) $1\ \text{hp} = 746\ \text{W}$

(2-6) $I = \dfrac{V}{R}$, or

$I = \dfrac{E}{R}$

(2-7) $R = \dfrac{V}{I}$

(2-8) $V = IR$

(2-9) $P = VI$

(2-10) $P = \dfrac{V^2}{R}$

(2-11) $P = I^2 R$

(2-12) $\text{maximum deviation} = \text{tolerance} \times \text{marked value}$

(2-13) $\text{resistance range} = \text{marked value} \pm \text{maximum deviation}$

(2-14) $R_{AC} = R_{AB} + R_{BC}$

ammeter   ohmmeter
complex circuit   parallel circuit
DMM   series circuit
DVM   series-parallel circuit
interpolate   VOM

objectives

After completing this chapter, you should be able to:

1. State the requirements for current flow in a circuit.

2. Determine the direction of electron current flow in a resistive dc circuit.

3. Demonstrate the correct method for measuring current, voltage, and resistance.

4. Classify a resistive dc circuit into one of the following circuit categories:

   - series
   - parallel
   - series-parallel
   - complex

5. Trace electron current flow through any of the following classes of circuit configurations:

   - series
   - parallel
   - series-parallel

Electric Circuits

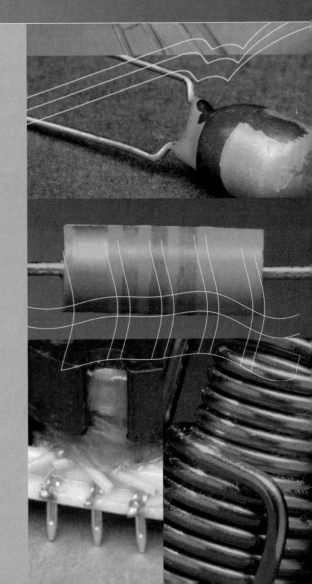

Technicians must be able to analyze electrical and electronic circuits to determine the expected values of circuit quantities such as voltage, current, and resistance. A technician must also be able to measure the actual circuit quantities with electronic test equipment. Finally, a technician must be able to apply the principles of basic electronics theory to account for any differences between the expected and the actual circuit values. This chapter presents all of the fundamental configurations of electrical circuits, shows how to identify each form, and provides a technique for tracing current flow through a given circuit. Procedures for measuring actual current, voltage, and resistance values are also given.

3.1 Requirements for Current Flow

Before a technician can trace current or measure circuit values, it is important to be able to identify a circuit capable of having current flow. The term current flow, as used in this chapter, refers to *sustained* current flow. Momentary or transient currents are considered in a later chapter. There are two basic requirements for sustained current flow in a circuit: electromotive force and a complete path for current.

Electromotive Force

You will recall from Chapter 2, that electromotive force is a difference in potential that does not decay as charges are transferred. In most cases, we are interested in the electrons that leave the negative side of the electromotive force (power source) and flow through the external circuit. As the electrons return from their trip through the circuit, they enter the positive side of the power source. Internal to the power source, the electrons are moved from the positive side of the source to the negative side. This requires an expenditure of energy. This energy may be provided chemically, as in the case of a battery, or perhaps from the energy supplied by the power company.

It is the electromotive force that maintains the potential difference and provides the electrical pressure that causes the electrons to flow through the circuit. Ohm's Law describes the relationship of applied voltage and current flow as $I = V/R$. If there were no electromotive force in a circuit, then the V in the Ohm's Law equation would be zero. This would produce a resulting current flow of zero regardless of the value of resistance in the circuit. Figure 3-1 illustrates the requirement for electromotive force. This figure also provides a hydraulic analogy.

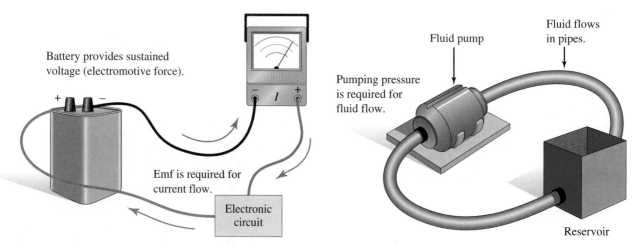

Figure 3-1. A circuit must have an electromotive force (emf) to have current flow. Emf in an electrical circuit is similar to pressure in a hydraulic circuit.

Complete Path for Current

The second requirement for current flow is a complete path or closed loop between the two terminals of the electromotive force. For most purposes, you can consider an open circuit to have an infinite resistance. Thus, if there is any open in a circuit, the current flow will be essentially zero since no amount of electromotive force can cause current to flow through an infinite opposition.

EXAMPLE SOLUTION

Which of the circuits shown in Figure 3-2 have complete paths for current flow?

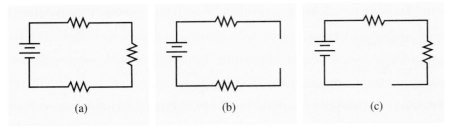

Figure 3-2. Which circuits have complete current paths?

EXAMPLE **SOLUTION**

If we begin at the negative battery terminal and proceed toward the positive terminal in Figure 3-2(a), we find that we can complete the entire journey. Therefore, Figure 3-2(a) has a complete path for current.

As we progress around the circuits shown in Figures 3-2(b) and (c), we encounter an open circuit. No current can flow through an open circuit; there is no complete path for current.

Practice Problems

1. Which of the circuits in Figure 3-3 have complete paths for current flow?

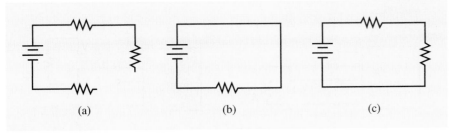

Figure 3-3. Which circuits have complete paths for current flow?

Answers to Practice Problems

1. b and c

Direction of Current Flow

As discussed in Chapter 2, electron current moves from negative to positive since electrons are negatively charged particles. We are now ready to expand this concept to include relative differences in potential.

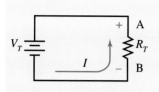

Figure 3-4.
A simple electrical circuit.

⊙ **KEY POINTS**

Electrons leave the nega-
tive side of the power
source, flow through the
external circuit, and
return to the positive terminal of the power
source.

Figure 3-4 shows an electrical circuit consisting of a voltage source and a resistor. We know that current leaves the negative side of the voltage source and flows through the resistor (from point B to point A) and returns to the positive side of the voltage source. In this figure, we can see that point A is positive and point B is negative since they are connected to the positive and negative terminals, respectively, of the voltage source. We can say that point A is positive with respect to point B. We could just as well say that point B is nega-tive with respect to point A. Either of these points can be positive or negative with respect to some other point, but as long as they have the indicated polarity, then current will flow from point B to point A.

Figure 3-5(a) shows an electromotive source with four different voltage terminals. All of the voltages are measured with reference to the common line. If we connect a resistor between the +5-V output and common, as shown in Figure 3-5(b), then we expect current to flow from point B to point A. Similarly, if we connect a resistor between the −5-V output and common, as shown in Figure 3-5(c), we expect current to flow from point A to point B since point A is now the most negative. Finally, Figure 3-5(d) illus-trates a less obvious possibility. If we connect a resistor between the +10- and +5-V outputs, as shown in Figure 3-5(d), then current will flow from point B to point A. In this case, point A is *more positive* than point B. The current through the resistor will be the same as if it had a simple 5-V source across it. This is a very important concept. Be sure you understand it.

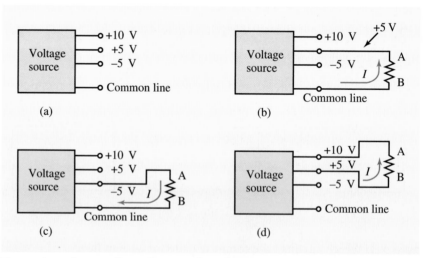

Figure 3-5. Current flows from a *relatively* negative potential toward a point that is *relatively* positive.

Figure 3-6 further illustrates this idea. Here, a scale of electrical potentials is compared to the rungs on a ladder. For example, +45 V is negative *with respect to* +50 V and positive *with respect to* +30 V, just as 8 feet above ground is low *with respect to* a point that is 12 feet above ground, but high *with respect to* a point that is 2 feet above ground.

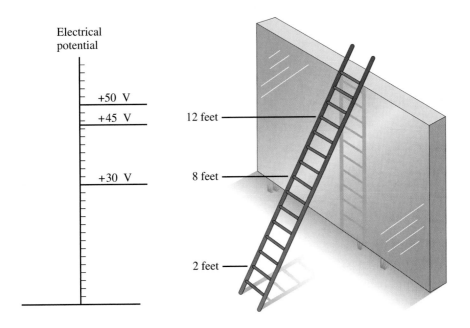

Figure 3-6. An analogy to explain *relative polarities*.

EXAMPLE SOLUTION

Determine the direction of current flow through each of the circuits in Figure 3-7.

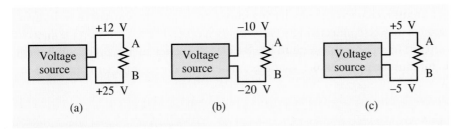

Figure 3-7. Determine the direction of current flow in each resistor.

EXAMPLE **SOLUTION**

Electron current always flows toward a more positive (less negative) potential. In Figure 3-7(a), point B (+25 V) is more positive than point A (+12 V). Current will flow from point A toward point B. Point B, in Figure 3- 7(b), is more negative than point A, so current will leave point B and flow toward point A. Finally, since point A (+5 V) is more positive than point B (–5 V) in Figure 3-7(c), we know that electron current must be flowing toward point A.

Exercise Problems 3.1

1. Explain why an electromotive force is one of the requirements for current flow.
2. Explain why current cannot flow through an open circuit.

3. Which of the circuits shown in Figure 3-8 will have current flow?

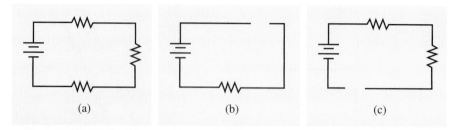

(a)   (b)   (c)

Figure 3-8. Which circuits have current flow?

3.2 Types of Circuit Configurations

No matter how complex electrical and electronic circuits may be, they can still be classified into one of four categories: series, parallel, series-parallel, and complex. Each of these circuit configurations requires different circuit analysis methods to calculate the normal values of voltage, current, and resistance in the circuit. However, before any calculations can be accomplished, the technician must be able to reliably identify a particular circuit. The following sections present methods for positive identification of each of the four types of circuits.

Series Configurations

A **series circuit** is characterized by having only a single path for current flow. Individual components may be connected in series with each other, or the entire circuit may be a series circuit. In order for a circuit to be classified as a series circuit, every component in the circuit must be in series such that there is only a single path for current in the entire circuit.

SERIES COMPONENTS

In order for two components to be in series, there must be one and only one current path that includes both components. Current must enter the first component, flow through both components, and then exit the last component without encountering any branches in the circuit.

EXAMPLE SOLUTION

Which of the components shown in Figure 3-9 are in series?

EXAMPLE **SOLUTION**

If we trace through R_1 in Figure 3-9(a), we encounter a branch at the junction of R_1, R_2, and R_3. Therefore, these components are not in series. There are no branches in Figures 3-9(b) or (c), which means that the components in these circuits are in series. Finally, we can see that the circuit in Figure 3-9(d) has a branch where R_1, R_2, and R_3 join. Thus, the components in Figure 3-9(d) are not in series.

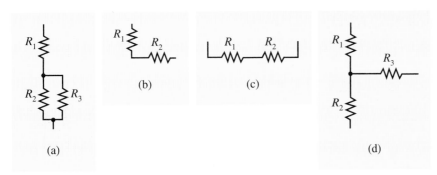

Figure 3-9. Which of these components are connected in series with each other?

Practice Problems

1. Which of the components shown in Figure 3-10 are connected in series?

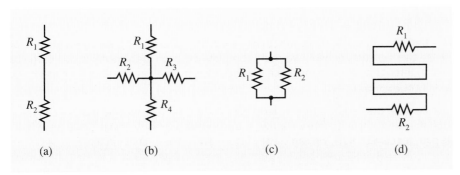

Figure 3-10. Which of these components are in series with each other?

Answers to Practice Problems

1. a. R_1 and R_2
    b. None
    c. None
    d. R_1 and R_2

SERIES CIRCUITS

If every electron that leaves the negative terminal of the power source has only one path for current and flows through every component before returning to the positive side of the power source, then the circuit is a series circuit. Every component in a series circuit is in series with every other component in the sense that they are part of the common, single current path.

KEY POINTS

If every component in a series circuit shares a single common current path, then the entire circuit can be classified as a series circuit.

EXAMPLE SOLUTION

Which of the circuits shown in Figure 3-11 can be classified as series circuits?

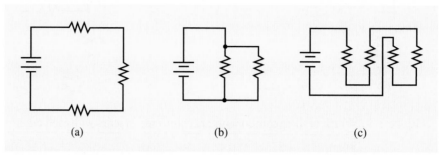

(a)   (b)   (c)

Figure 3-11. Which of these circuits can be classified as series circuits?

EXAMPLE **SOLUTION**

If we begin at the negative side of the power source in Figure 3-11(a) and move through the circuit toward the positive side of the battery, we find that there are no alternate routes. That is, there is only one path for current, and it includes every component. Therefore, the circuit in Figure 3-11(a) is a series circuit.

If we trace the current path for the circuit shown in Figure 3-11(b), we find we have a branch where the two resistors connect. We could complete our path to the positive side of the battery by passing through either resistor. Since the circuit has more than one path for current, it does not qualify as a series circuit.

If we begin at the negative battery terminal in Figure 3-11(c), we find that we can trace through the entire circuit before we arrive at the positive supply terminal. Since there is only a single path for current flow, we can classify Figure 3-11(c) as a series circuit.

Practice Problems

1. Which of the circuits shown in Figure 3-12 can be classified as series circuits?

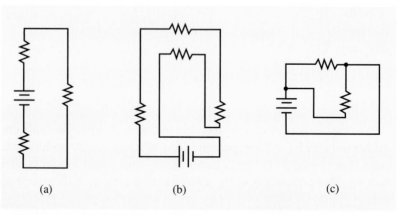

(a)   (b)   (c)

Figure 3-12. Which of these circuits can be classified as series circuits?

Answers to Practice Problems

1. a and b

Parallel Configurations

A **parallel circuit** is characterized by having the same voltage across every component in the circuit. Sets of components may be connected in parallel, or the entire circuit may be connected as a parallel circuit.

PARALLEL COMPONENTS

Two components are in parallel with each other if both ends of both components connect *directly* together. Since the components are connected directly across each other, they will inherently have the same voltage across them.

EXAMPLE SOLUTION

Which of the sets of components shown in Figure 3-13 are connected in parallel?

EXAMPLE SOLUTION

The ends of the resistors in Figure 3-13(a) connect directly together; therefore, they are in parallel. The resistors shown in Figure 3-13(b) are also connected in parallel, since both ends of both components connect directly together. We can see by inspection of Figure 3-13(c), that the resistors do *not* have both ends connected together, therefore these resistors are not in parallel.

Practice Problems

1. Which of the sets of components shown in Figure 3-14 are connected in parallel?

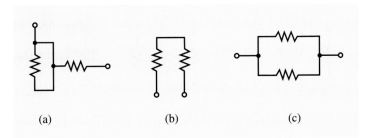

(a)   (b)   (c)

Figure 3-14. Which components are connected in parallel?

Answers to Practice Problems

1. c

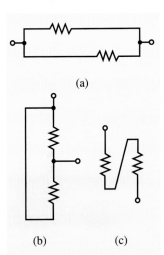

(a)

(b)   (c)

Figure 3-13. Which components are connected in parallel?

PARALLEL CIRCUITS

If every component in a circuit is connected directly across every other component in the circuit, then the entire circuit may be classed as a parallel circuit. Every component in a parallel circuit will have the same voltage across its terminals as every other component in the circuit.

EXAMPLE SOLUTION

Which of the circuits shown in Figure 3-15 can be classified as parallel circuits?

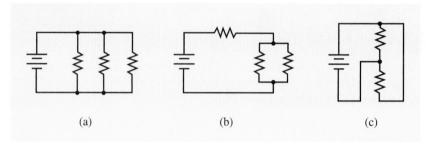

(a)   (b)   (c)

Figure 3-15. Which of these circuits are parallel circuits?

EXAMPLE **SOLUTION**

Since every component in Figure 3-15(a) is connected directly across every other component, we can classify the circuit as a parallel circuit. The circuit shown in Figure 3-15(b) is not a parallel circuit, because not all of the components are connected directly across each other. Finally, we can classify the circuit shown in Figure 3-15(c) as a parallel circuit, since all three components are connected directly across each other.

Practice Problems

1. Which of the circuits shown in Figure 3-16 can be classified as parallel circuits?

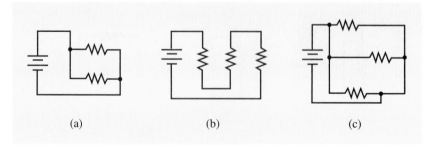

(a)   (b)   (c)

Figure 3-16. Which of these circuits are parallel circuits?

Answers to Practice Problems

1. a and c

Series-Parallel Circuits

Series-parallel circuits are actually combinations of pure series and pure parallel circuits. Some of the components are in series with each other, some are in parallel, and some are not immediately evident. We will need a procedure for positively identifying a series-parallel circuit. If the circuit cannot be classified as a pure series or a pure parallel circuit, then

1. Replace all truly series components with a single resistor symbol.
2. Replace all truly parallel components with a single resistor symbol.
3. Repeat the first two steps on the newly formed drawing until one of the following occurs:
    a. the circuit is simplified to a single resistor symbol and a power source, or
    b. the circuit cannot be simplified to a single resistor and power source by combining sets of series or sets of parallel components.

If you are able to reduce the circuit to a single resistor symbol and a single power source, then the original circuit is a series-parallel circuit.

EXAMPLE SOLUTION

Can the circuit shown in Figure 3-17 be classified as a series-parallel circuit?

EXAMPLE **SOLUTION**

First, we must rule out the possibility of the circuit being either a pure series or pure parallel circuit. We can eliminate the series circuit option since electrons have more than one possible path. After they leave the negative side of the source, they encounter a branch at the bottom of R_2 and R_3. Some of the electrons will go through R_2, while the rest will go through R_3. All of the electrons recombine at the upper end of R_2 and R_3 and return to the positive side of the power source via R_1. We can also rule out a pure parallel configuration, since not all of the components are connected directly across each other. We are now ready to apply the previously described procedure to determine if it is a series-parallel circuit.

Figure 3-18(a) shows the original circuit. Resistors R_2 and R_3 are in parallel with each other so we shall replace them with a single resistor symbol. Let's label the equivalent resistor symbol as $R_{2,3}$ so we will know how it was formed. In Figure 3-18(b), resistors R_2 and R_3 have been replaced by a single resistor symbol ($R_{2,3}$).

We now apply the simplification procedure to our newly formed circuit. It is redrawn in Figure 3-18(c) where resistors R_1 and $R_{2,3}$ can be seen to be in series. According to the procedure, we will replace these two resistors with a single resistor symbol. We shall label the replacement resistor as $R_{1,2,3}$. Our new, and final, circuit is shown in Figure 3-18(d). Since it has been simplified to a single resistor symbol and a single power source, we can positively assert that the original circuit was a series-parallel circuit.

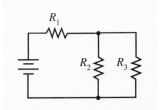

Figure 3-17.
Is this a series-parallel circuit?

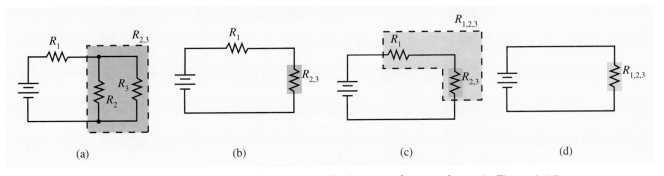

Figure 3-18. Steps of the simplification procedure to identify the type of circuit shown in Figure 3-17.

EXAMPLE SOLUTION

Can the circuit shown in Figure 3-19 be classified as a series-parallel circuit?

EXAMPLE **SOLUTION**

The circuit is definitely not a pure series circuit since there is more than one path for current flow. Further, it cannot be classed as a pure parallel circuit because not all of the components connect directly across each other. So we apply the simplification procedure.

Figure 3-20(a) shows the original circuit with two sets of parallel resistors identified. That is, resistors R_1 and R_2 are in parallel with each other, so we will replace them with a single resistor symbol labeled $R_{1,2}$. Similarly, we replace parallel resistors R_4 and R_5 with a single symbol labeled $R_{4,5}$. This step is shown in Figure 3-20(b).

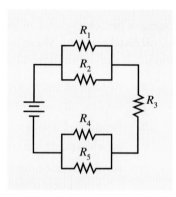

Figure 3-19.
Is this a series-parallel circuit?

Figure 3-20(b) can be classified as a series circuit since there is only one path for current flow. We replace all three series resistors with a single equivalent resistor symbol labeled $R_{1,2,3,4,5}$. This last step, shown in Figure 3-20(c), results in a single-resistor circuit, so the original circuit (Figure 3-19) can definitely be classified as a series-parallel circuit.

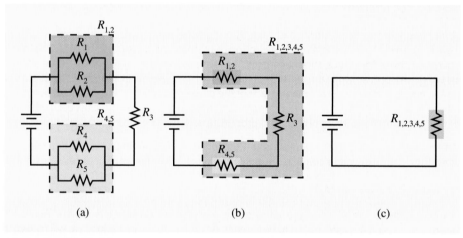

(a)   (b)   (c)

Figure 3-20. Simplification steps for the circuit shown in Figure 3-19.

EXAMPLE SOLUTION

Is the circuit shown in Figure 3-21 a series-parallel circuit?

EXAMPLE **SOLUTION**

We can quickly see that not all of the components are in series, nor are they all in parallel. This rules out the pure series and pure parallel configurations. Careful examination of the circuit will show that resistors R_5 and R_6 are in series with each other and can be replaced with a single resistor symbol. This step is shown in Figure 3-22(a). In Figure 3-22(b), no other resistors are directly in series with each other. Similarly, no resistors are directly in parallel

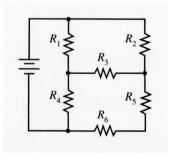

Figure 3-21.
Is this a series-parallel circuit?

with each other. Therefore, since we cannot reduce the circuit to a single resistor symbol, we can positively conclude that the original circuit (Figure 3-21) is *not* a series-parallel circuit.

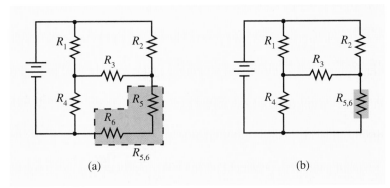

(a)   (b)

Figure 3-22. Simplification of the circuit shown in Figure 3-21.

Practice Problems

1. Is the circuit shown in Figure 3-23 a series-parallel circuit?

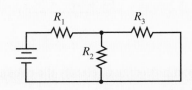

Figure 3-23. Is this a series-parallel circuit?

2. Is the circuit shown in Figure 3-24 a series-parallel circuit?
3. Is the circuit shown in Figure 3-25 a series-parallel circuit?

Answers to Practice Problems

1. Yes   **2.** Yes   **3.** Yes

Complex Circuits

The procedure for identifying a **complex circuit** is merely a continuation of the procedure for classifying a series-parallel circuit. The final step of the series-parallel identification requires us to evaluate whether or not the circuit can be reduced to a single resistor symbol and a single power source. If so, then the original circuit can be classed as a series-parallel circuit. If not, then the circuit can be positively classified as a complex circuit. The circuit shown in Figure 3-21 and simplified in Figure 3-22 is a complex circuit since it could not be reduced to a single resistor symbol and a single power source by replacing sets of series and sets of parallel components.

Figure 3-24.
Is this a series-parallel circuit?

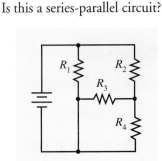

Figure 3-25.
Is this is series-parallel circuit?

○ **KEY POINTS**

If the circuit cannot be reduced to a single resistor and a single power source, then it is a complex circuit.

2. Refer to Figure 3-29. Identify all components that are definitely in parallel with each other.

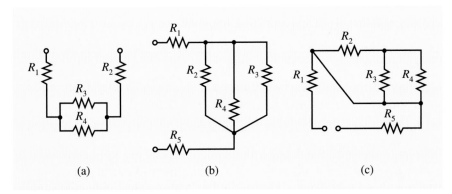

Figure 3-29. Which components are in parallel with each other?

3. Classify the circuit shown in Figure 3-30.
4. Classify the circuit shown in Figure 3-31.
5. Classify the circuit shown in Figure 3-32.
6. Classify the circuit shown in Figure 3-33.

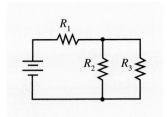

Figure 3-30.
What type of circuit is this?

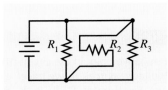

Figure 3-31.
What type of circuit is this?

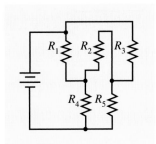

Figure 3-32.
What type of circuit is this?

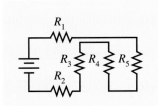

Figure 3-33.
What type of circuit is this?

3.3 Tracing Current Flow in a Circuit

The ability to trace current through an unfamiliar electronic circuit is an important skill for an electronics technician. The rules are very simple, but to effectively implement the rules requires substantial practice. The rules are as follows:

1. Start at the most negative point in the circuit and move toward the most positive point.
2. Trace through the circuit using any path with the following conditions:
   a. never violate the basic behavior of a component, and
   b. never return to a point in the circuit where you have already traced without first passing through a power source.

Your goal will be to return to the positive side of the power source. If you can make the complete trip, then you know there is current flow in the circuit, and you know the path where it flows. The first condition under item two cautions against violating the basic behavior of a component. This often refers to components that are polarized (LEDs, for example). You would be violating the basic behavior if you traced through a polarized component in the reverse direction.

EXAMPLE SOLUTION

Trace all current paths through the circuit shown in Figure 3-34.

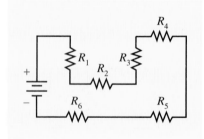

Figure 3-34.
Trace the current in this circuit.

EXAMPLE SOLUTION

We start at the most negative point in the circuit (the negative side of the voltage source). As we trace through the circuit, we pass through R_6, R_5, R_4, R_3, R_2, and R_1. We then complete our journey by returning to the positive side of the voltage source. Since there was only a single path for current, we know this must be a series circuit. The traced current path is shown in Figure 3-35.

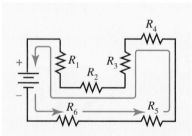

Figure 3-35. Current leaves the negative side of the voltage source, flows through the circuit, and returns to the positive side of the source.

EXAMPLE SOLUTION

Trace all current paths through the circuit shown in Figure 3-36.

EXAMPLE SOLUTION

We start at the negative side of the voltage source. As we move upward, we come to a decision point. According to the rules, we can go in either direction as long as we don't violate one of the stated conditions. Let us first go to the left through R_1, R_2, and R_3. After passing through R_3, we are faced with another decision. Do we continue on through R_6 or do we return to the positive side of the voltage source? Since our goal is to reach the positive terminal of the voltage source, we elect this option. It is interesting to note, however, that if we had continued through R_6, R_5, and R_4, we would have returned to a point where we had already traced without having gone through a power source. This would have violated one of our tracing rules, so that is not an acceptable path.

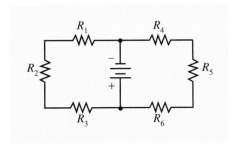

Figure 3-36.
Trace the current paths in this circuit.

Now let us return to the junction of R_1 and R_4 and trace to the right through R_4, R_5, and R_6. After tracing through R_6, we can return to the positive side of the battery. Note that this last segment does not represent a violation of our tracing rules since we have made it to the positive voltage terminal. Figure 3-37 shows our traced paths.

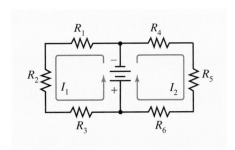

Figure 3-37.
Tracing of the current paths for the circuit shown in Figure 3-36.

EXAMPLE SOLUTION

Trace all current paths for the circuit shown in Figure 3-38.

EXAMPLE SOLUTION

We can start at the negative side of the voltage source and move toward the lower ends of R_4 and R_5. Here the current will split. One portion will flow upward through R_4, while the remaining portion will flow through R_5. At the upper ends of R_4 and R_5, the two branch currents will recombine and flow to the left toward R_1, R_2, and R_3. When the current reaches the junction of R_1 and R_3, it will split. Part of it will flow through R_1. The rest of the current will flow through R_3 and R_2.

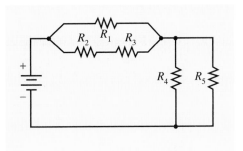

Figure 3-38.
Trace the current paths for this circuit.

After exiting resistors R_1 and R_2, the current recombines and returns to the positive side of the voltage source. These current paths are shown in Figure 3-39.

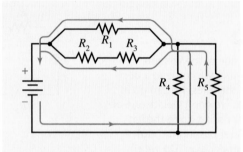

Figure 3-39. Current paths for the circuit shown in Figure 3-38.

Practice Problems

1. Trace the current paths for the circuit shown in Figure 3-40.

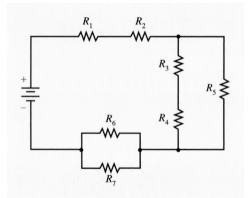

Figure 3-40. Trace the current through this circuit.

2. Trace the current paths for the circuit shown in Figure 3-41.

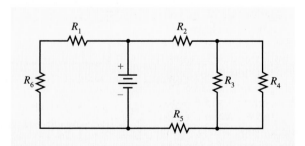

Figure 3-41. Trace the current through this circuit.

3. Trace the current paths for the circuit shown in Figure 3-42.

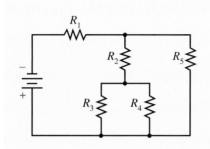

Figure 3-42. Trace the current through this circuit.

Answers to Practice Problems

1.

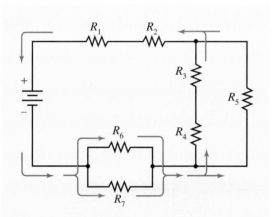

2.

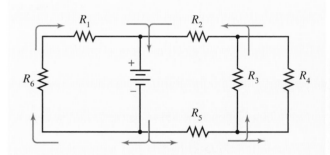

3.

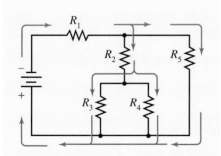

Exercise Problems 3.3

A light-emitting diode (LED) will only allow current to flow in one direction. If it is connected such that the electron current tries to flow against the arrow on the schematic symbol, it offers little opposition to current flow. If it is connected in the reverse direction, it prevents current flow through the LED. Use this information and your knowledge of circuits to work the following exercise problems.

1. Trace the current paths for the circuit shown in Figure 3-43.

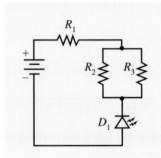

Figure 3-43. Trace the current through this circuit.

2. Trace the current paths for the circuit shown in Figure 3-44.

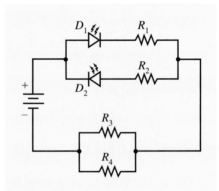

Figure 3-44. Trace the current through this circuit.

3. Trace the current paths for the circuit shown in Figure 3-45.

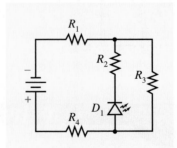

Figure 3-45. Trace the current through this circuit.

3.4 Circuit Measurements

Since measurement of circuit values is often a fundamental part of a technician's job, it is important to know how to measure circuit values correctly. Additionally, if the test equipment is connected to the circuit incorrectly, then the equipment may be damaged.

Meter Types

There are many different kinds of instruments that can be used to measure the basic circuit quantities of voltage, current, and resistance. They can generally be classified into two groups: analog and digital.

ANALOG METERS

Figure 3-46 shows a representative analog meter that is capable of measuring voltage, current, and resistance. It is often called a volt-ohm-milliammeter, or simply **VOM**. The most apparent characteristic of an analog meter is the use of a meter scale and pointer. The circuit quantity being measured causes a certain deflection of the pointer across the meter scale. The value of the measured quantity can be read on the scale directly beneath the pointer.

Figure 3-46. An analog meter that can measure voltage, current, and resistance. (Courtesy of Triplett Corporation)

KEY POINTS

Analog meters have one or more scales on the face of the meter, and a pointer that is deflected by the circuit quantity being measured.

KEY POINTS

Analog meters capable of measuring voltage, current, and resistance are called volt-ohm-milliammeters (VOMs).

A technician must be able to interpret the value associated with a particular pointer position. In general, there are two factors that must be considered:

- position of the pointer on the meter scale, and
- which scale to read.

The meter in Figure 3-46 has several sets of scales printed on its face. Only one scale is used for a particular measurement. There are generally two factors that determine the correct scale to use:

- mode of the meter (e.g., voltage, current, resistance, and ac or dc), and
- position of range switch.

It usually is easy to tell which set of scales (e.g., dc scales) to read if you carefully note the scale labels. For most general-purpose VOMs, some of the scales are marked for use with dc measurements, some for ac measurements, and still others for resistance measurements. The upper scale of the VOM shown in Figure 3-46 is used when measuring resistance. The next two lower scales are used for ac voltage and dc current and voltage measurements. Choose the set of scales that is appropriate for the setting of the meter controls and the type of test you are making.

The value of the marks on a selected scale is determined by the range setting on the VOM. For example, the dc scale immediately below the resistance scale on the VOM in Figure 3-46 has 10 as the highest value. This same scale is used for the 1-, 10-, 100-, and 1000-V ranges. The technician must mentally move the decimal point to make the full-scale value read as 1, 10, 100, or 1000 V.

Once a technician has identified the correct scale, the pointer position can then be interpreted. If the pointer falls directly above one of the scale marks, then interpretation is fairly straightforward. When the pointer falls between two of the scale markings, the technician must **interpolate** the meter scale to obtain the correct reading. Interpolation requires the technician to mentally insert additional marks between the printed marks on the scale. This provides the technician with increased resolution (fineness) of pointer position. Analog meters are discussed in greater detail in Chapter 9.

Practice Problems

1. Interpret the value indicated by the meter face shown in Figure 3-47. The meter is in the dc voltage mode and is on the 10-V range.

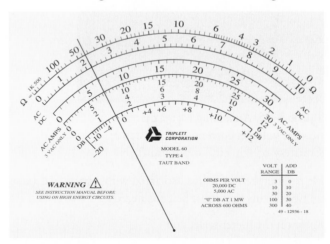

Figure 3-47. Interpret the value indicated by this meter. (Courtesy of Triplett Corporation)

2. Interpret the value being indicated by the meter face shown in Figure 3-48. The meter is in the dc current mode and is on the 100-mA range.

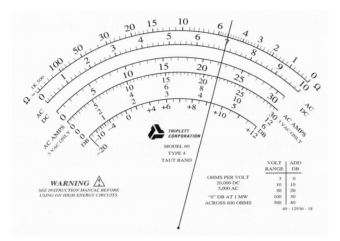

Figure 3-48. What value is indicated by this meter? (Courtesy of Triplett Corporation)

Answers to Practice Problems

1. 2.2 V   **2.** 70 mA

DIGITAL METERS

Digital meters are more common in industry than analog meters for general-purpose measuring equipment. Figure 3-49 shows a representative digital meter that can measure voltage, current, and resistance. It is often called a digital multimeter, or simply, **DMM**. The abbreviation **DVM** stands for digital voltmeter, but common usage does not always distinguish between DVM and DMM.

KEY POINTS

Digital meters can be digital voltmeters (DVMs) or digital multi-meters (DMMs).

Figure 3-49. A digital multimeter (DMM) that can measure voltage, current, and resistance. (Courtesy of Tektronix, Inc.)

A digital meter is generally easier to operate and read than an analog type. Once the technician has selected the quantity to be measured (voltage, current, or resistance), then the meter will generally adjust itself for the correct range. The measured value is displayed directly on the display; no interpolation is required. Digital meters are discussed in greater detail in Chapter 9.

Voltage Measurements

Voltage measurements are probably the easiest type of circuit measurement to make. The following procedure can be used to measure voltage between two points in a circuit:

1. Select the correct voltage mode (ac or dc) on the meter.
2. Select a range that is higher than the expected value of circuit voltage.
3. Connect the leads of the meter between the two points being measured.

Figure 3-50 shows the proper way to measure voltage in a circuit. An encircled V is used to represent a voltmeter on a schematic diagram. In most cases, the black (–) lead of the meter should be connected to the most negative of the two points being measured. The red (+) lead of the meter connects to the most positive potential. If an analog meter is connected in the reverse direction, it may be damaged since the pointer will try to deflect off of the left side of the scale. A digital meter will normally operate correctly with either polarity; however, the polarity indications on the meter may confuse the technician unless the meter is used in a consistent manner (e.g., black lead to the most negative potential).

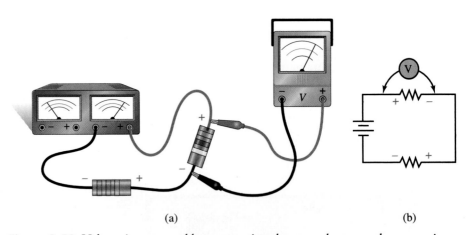

(a)   (b)

Figure 3-50. Voltage is measured by connecting the meter between the two points being monitored.

Current Measurements

Current measurements are more difficult than voltage measurements, because the circuit must be broken in order to insert the meter. Recall that current in a wire is actually a flow of electrons. In order for the current meter to determine the number of electrons moving in the circuit, the flow of electrons must be diverted from the wire through the meter. Figure 3-51 shows how to perform the following steps to measure current in a circuit:

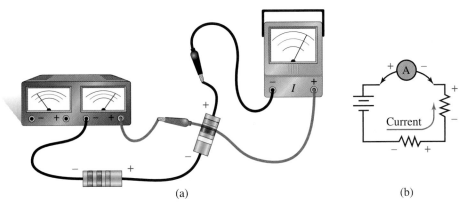

Figure 3-51. The circuit must be opened to insert the current meter.

1. Remove power to the circuit.
2. Open the circuit at the point where current is to be measured.
3. Connect the current meter across the newly formed open (negative meter lead toward negative source terminal and positive meter lead toward positive source terminal).
4. Set the meter for current measurements.
5. Select a current range higher than the expected reading.
6. Reapply power to the circuit.

As shown in Figure 3-51, a current meter (**ammeter**) is represented on a schematic diagram by an encircled A. Figure 3-51 shows another very important consideration. When the meter is inserted across the opened circuit, it is essential to connect the meter such that electrons enter the negative side of the meter, flow through the meter, and return to the circuit via the positive lead of the meter. It is often helpful to run your finger along the wire in the direction electrons will flow. Your path should enter the negative side of the current meter, go through the meter, and find its way back to the circuit.

Figure 3-52 shows two common mistakes that you must avoid. In Figure 3-52(a), the technician is trying to measure the current through a resistor, but has connected the

KEY POINTS

Care must be used to connect a current meter such that electrons enter the negative side of the meter, flow through the meter, and then return to the circuit via the positive meter lead.

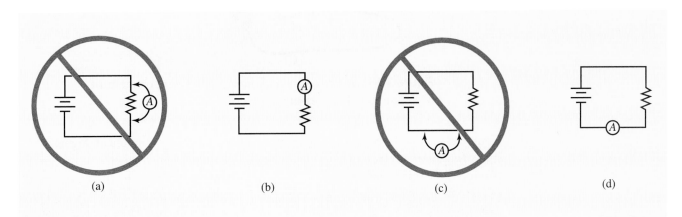

Figure 3-52. Two common errors when measuring current in a circuit.

current meter across the resistor instead of in-line with it. Figure 3-52(b) shows the correct way to measure resistor current. Figure 3-52(c) shows a current meter that has been incorrectly connected to measure current through a wire. Again, the technician has forgotten to break the circuit. Figure 3-52(d) shows the correct way to measure the current.

Resistance Measurements

The following procedure can be followed to measure the resistance of a component in a circuit:

1. Remove power from the circuit.
2. Disconnect at least one end of the component to be measured.
3. Connect the **ohmmeter** across the component whose resistance is to be measured.

The range control on analog ohmmeters is called a multiplier. The positional values are $R \times 1$, $R \times 10$, $R \times 100$, $R \times 1k$, and so on. You should select a range that produces a pointer deflection on the expanded (right-hand) half of the meter scales. Pointer deflections on the compressed (left-hand) side are difficult to interpret since the numbers are so close together; the scale is nonlinear. The actual resistance value is determined by multiplying the value indicated on the scale by the value of the multiplier dial. If, for example, the pointer indicates 5.6 and the multiplier is set for $R \times 1k$, then the actual resistance being measured is $5.6 \times 1k$ or $5.6 \text{ k}\Omega$.

The range switch on digital ohmmeters simply sets the highest resistance that can be measured in a particular position. The resistance is displayed directly and requires no mental arithmetic. You should choose the lowest resistance range that still allows the measurement to be taken.

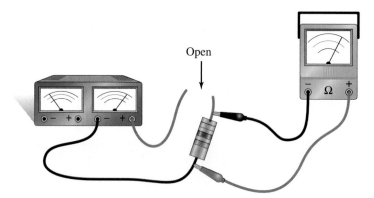

Open

Figure 3-53. *At least* one end of the measured component must be disconnected to measure resistance.

Figure 3-53 shows the correct way to measure the resistance of a component. Figure 3-54 shows two common errors that are made when measuring the resistance of a component. In Figure 3-54(a), the technician has forgotten to turn off the power to the circuit. As a minimum, this will produce wrong readings on the ohmmeter. There is also a possibility of damage to the meter. Figure 3-54(b) shows another common error. Here the technician is trying to measure the resistance of a component without removing it from the circuit first.

While this causes no damage to the meter, it may not measure the correct resistance value. In a later chapter, you will learn that in-circuit resistance checks are a recommended technique, but the technician must compensate for the effects of other components in the circuit. Chapter 9 discusses the operation of an ohmmeter in greater detail.

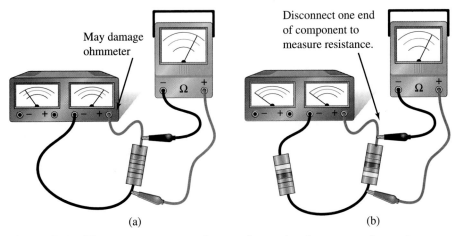

May damage ohmmeter

Disconnect one end of component to measure resistance.

(a)   (b)

Figure 3-54. Two common errors that can be made when measuring resistance.

Exercise Problems 3.4

1. What is the name of the class of meters that has a scale and pointer to indicate the measured value?
2. What is the abbreviation for volt-ohm-milliammeter?
3. Refer to Figure 3-55. What value is indicated if the meter is set to read dc voltage and is on the 30-V range?

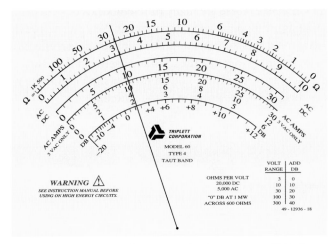

Figure 3-55. A typical meter scale. (Courtesy of Triplett Corporation)

4.   Refer to Figure 3-55. What value is indicated if the meter is set to read resistance and is on the $R \times 100k$ scale?

5.   Refer to Figure 3-55. What value is indicated if the meter is set to read dc current and is on the 100-mA scale?

6.   What type of circuit measurement must be taken with the power turned off?

7.   What type of circuit measurement requires the circuit to be opened to insert the meter in line with the component being measured?

8.   What type of circuit measurement requires at least one end of the measured component to be disconnected from the circuit?

3.5 Troubleshooting Techniques

Armed with nothing more than an ohmmeter and a solid understanding of electronics, a technician can determine the condition of many kinds of components while troubleshooting an electronic circuit. Let's briefly examine some devices that can be checked with an ohmmeter.

Fuses

The resistance of a good fuse is very low (typically less than 5 Ω). If a fuse is defective, it becomes an open circuit and has infinite resistance. An ohmmeter can quickly classify a suspected fuse as good or bad.

Lamps

The filament of an incandescent bulb has a relatively low resistance if it is good. Normal resistance values will generally be less than 50 Ω. If the filament in the lamp becomes open then its resistance will measure near infinity. Again, it is easy to identify a defective lamp with an ohmmeter.

Switches

As you studied in Chapter 2, there are many different kinds of switches. In all cases, however, they are designed to open and close a circuit. Since an ohmmeter can easily distinguish between open and shorted conditions, it provides an effective way to test the condition of a switch. Simply measure the resistance between the contacts on the switch as you operate the switching mechanism. The ohmmeter should indicate either infinite or near zero ohms depending on the position of the switch.

Exercise Problems 3.5

1.   If a fuse measures 0.75 Ω, it is probably _____.
2.   If the filament of an incandescent lamp measures infinity, it is probably _____.
3.   If a SPST switch measures 0 Ω in both positions, it is probably _____.

Chapter Summary

- In order for sustained current to flow in a circuit, there must be an electromotive force, and there must be a complete (closed) path for current flow. Electron current always flows from a given point toward a point that is more positive (less negative).

- Measurement of a circuit's voltage, current, and resistance values can be done with either an analog or a digital meter. Analog meters use a pointer and scale to indicate measured values. Digital meters display the measured quantity on a digital (numerical) display. An analog meter that measures voltage, current, and resistance is called a volt-ohm-milliammeter (VOM). A digital meter capable of measuring the three basic circuit quantities is called a digital multimeter (DMM).

- Voltage measurements are made by connecting the voltmeter directly across the component whose voltage is to be measured. Current measurements require the circuit to be broken, so that the current meter can be inserted in series with the component to be monitored. Resistance measurements are made with the circuit power removed. Additionally, at least one end of the measured component must be disconnected from the circuit.

- Electrical and electronic circuits can be categorized into four classes: series, parallel, series-parallel, and complex. Series components have one and only one current path. If all of the components in a circuit share a common current path, then the entire circuit can be called a series circuit.

- Parallel components are connected directly across each other. They have identical voltages, since the terminals from the parallel components are connected together. If all of the components in a given circuit are connected in parallel, then the overall circuit is a parallel circuit.

- Series-parallel circuits are composed of combinations of series and parallel components. Positive identification of a series-parallel circuit can be made by replacing sets of series resistors and sets of parallel resistors with a single resistor symbol to form a simpler equivalent circuit. Sets of series or parallel resistors in the simplified drawing are then replaced with a single resistor symbol to form a more simplified schematic. This process continues until one of the following situations exists:

    - the circuit has been reduced to a single resistor symbol and a single power source, or
    - the circuit cannot be reduced to a single resistor and power source.

- In the first case, you have positively identified a series-parallel circuit. In the second case, the original circuit can be classified as a complex circuit.

- Current can be traced through a circuit by starting at the most negative point in the circuit and progressing through any available path toward a more positive potential. As you trace the current, you must not violate basic component behavior (e.g., going backward through a polarized component or passing through an open switch), and you must not return to a point in the circuit that has already been traced without first passing through a power source.

Review Questions

Section 3.1: Requirements for Current Flow

1. Explain why an electromotive force must be present in order to have sustained current flow in a circuit.

2. Besides electromotive force, what else is required for current flow in a circuit?

3. Which way will current flow in Figure 3-56?

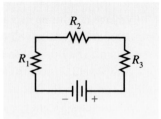

Figure 3-56. Which way does current flow?

4. Which way will current flow in Figure 3-57?

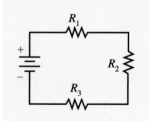

Figure 3-57. Which way does current flow in this circuit?

5. Will there be any current in Figure 3-58? Explain.

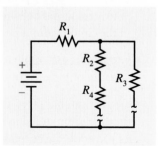

Figure 3-58. Does current flow in this circuit?

6. Does current flow in the circuit shown in Figure 3-59?

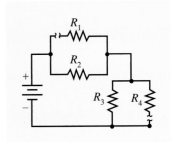

Figure 3-59. Is there any current flow in this circuit?

7. In which direction does current flow in the circuit shown in Figure 3-60?

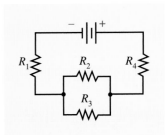

Figure 3-60. Which direction will current flow?

Section 3.2: Types of Circuit Configurations

8. Classify the components shown in Figure 3-61 as series or parallel.

Figure 3-61. Classify these components as series or parallel.

9. Classify the circuit shown in Figure 3-62 as series, parallel, series-parallel, or complex.

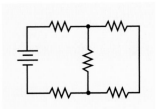

Figure 3-62. Classify this circuit.

10. Classify the circuit shown in Figure 3-63 as series, parallel, series-parallel, or complex.

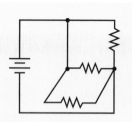

Figure 3-63. Classify this circuit.

11. Classify the circuit shown in Figure 3-64 as series, parallel, series-parallel, or complex.

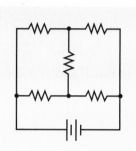

Figure 3-64. Classify this circuit.

12. Classify the circuit shown in Figure 3-65 as series, parallel, series-parallel, or complex.

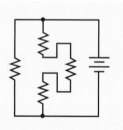

Figure 3-65. Classify this circuit.

13. Classify the circuit shown in Figure 3-66 as series, parallel, series-parallel, or complex.

Figure 3-66. Classify this circuit.

Section 3.3: Tracing Current Flow in a Circuit

14. Trace all current paths through the circuit shown in Figure 3-67.

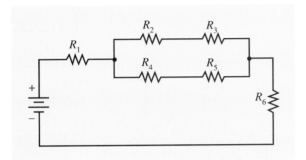

Figure 3-67. Trace all current paths in this circuit.

15. Trace all current paths through the circuit shown in Figure 3-68.

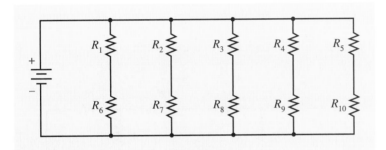

Figure 3-68. Trace all current paths in this circuit.

16. Trace all current paths through the circuit shown in Figure 3-69.

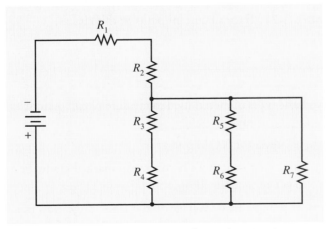

Figure 3-69. Trace all current paths in this circuit.

17. Trace all current paths through the circuit shown in Figure 3-70.

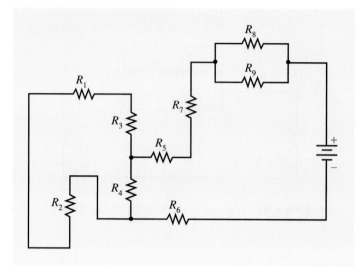

Figure 3-70. Trace all current paths in this circuit.

18. Trace all current paths through the circuit shown in Figure 3-71.

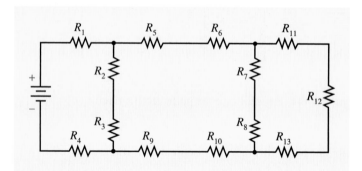

Figure 3-71. Trace all current paths in this circuit.

Section 3.4: Circuit Measurements

19. There are two basic types of meters used to measure circuit quantities. They are called _____ and _____.

20. What does the abbreviation VOM stand for?

21. If the pointer on a meter scale stops between two marked graduations, what must a technician do to get an accurate measurement?

22. If the meter is set to measure dc voltage on the 10-V range, how much voltage is indicated by the meter shown in Figure 3-72?

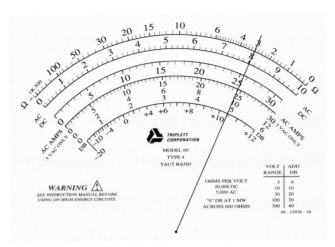

Figure 3-72. A typical meter reading. (Courtesy of Triplett Corporation)

23. If the meter is set to measure resistance on the $R \times 10k$ range, how much resistance is indicated by the meter shown in Figure 3-72?

24. If the meter is set to measure dc current on the 100-mA range, how much current is indicated by the meter shown in Figure 3-72?

25. How much resistance is indicated by the meter shown in Figure 3-73, if the meter is set to the $R \times 100$ position?

26. How much current is indicated by the meter shown in Figure 3-73, if the meter is set to the 10-mA dc range?

27. How much voltage is indicated by the meter in Figure 3-73, if the meter is set to the 30-V dc range?

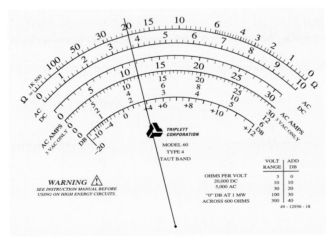

Figure 3-73. A typical meter reading. (Courtesy of Triplett Corporation)

28. Which of the circuits shown in Figure 3-74 shows the proper way to read the voltage (V_{R_4}) across R_4?

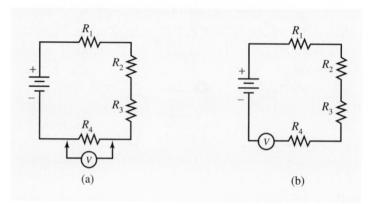

Figure 3-74. Which is the correct way to measure V_{R_4}?

29. Which of the circuits shown in Figure 3-75 shows the proper way to measure the resistance of R_2?

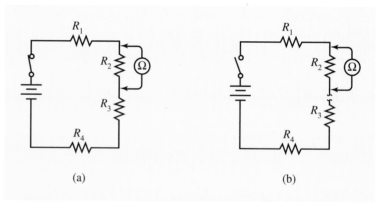

Figure 3-75. Which is the correct way to measure the resistance of R_2?

30. Which of the circuits shown in Figure 3-76 shows the correct way to measure the current (I_3) through R_3?

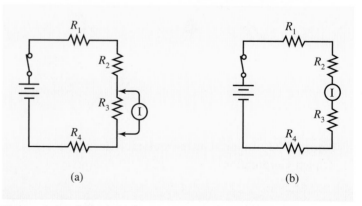

Figure 3-76. Which is the correct way to measure I_3?

Section 3.5: Troubleshooting Techniques

31. If a fuse has infinite resistance it is probably defective. (True or False)

32. If a DPDT switch measures infinite resistance between all terminals in all switch positions, it is probably defective. (True or False)

33. If the filament of an incandescent lamp has a very low resistance, it is probably defective. (True or False)

TECHNICIAN CHALLENGE

Your company has been asked to contribute a demonstration setup for use at a local science fair. Your supervisor has given you the task of designing and building the demonstrator. Your supervisor wants the demonstrator to perform as follows:

- There will be two independent circuits.
- Circuit 1 will demonstrate series and parallel circuits.
- Circuit 2 will demonstrate series-parallel and complex circuits.
- Both circuits must use a switch to change modes. When the switch is operated in circuit 1, the circuit changes from series to parallel. The switch in circuit 2 changes the circuit from series-parallel to complex.
- Circuit 1 is to consist of one voltage source, two resistors, and a switch.
- Circuit 2 will consist of one voltage source, one switch, and as many as six resistors.
- You can use any of the following switch types, which are available in your company's inventory:
  - SPST
  - SPDT
  - DPDT

Draw the schematic diagrams for your implementation of circuit 1 and circuit 2.

complex voltage source   series-aiding voltage sources
fully specified circuit   series-opposing voltage sources
ground   unloaded voltage divider
ground plane   voltage divider
loaded voltage divider   voltage drop
partially specified circuit

objectives

After completing this chapter, you should be able to:

1. Identify a series circuit.

2. State the relationship of the currents through each component in a series circuit.

3. State the relationship between the voltage source and the voltage drops in a series circuit.

4. State and apply Kirchhoff's Voltage Law.

5. Analyze a series circuit to compute the value of current, voltage, resistance, and power for every component.

6. Explain the concept of circuit ground.

7. Express voltages that are referenced to ground or measured between two nonground points.

8. Compute the effective voltage when several voltage sources are connected in series.

9. Analyze an unloaded voltage divider circuit.

10. Design an unloaded voltage divider to provide specific voltages.

11. Troubleshoot a series circuit.

Analyzing Series Circuits

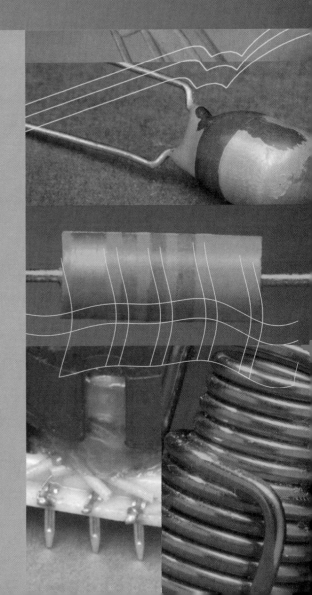

Chapter 3 presented a method that could be used to positively iden-
tify a series circuit. You will recall that if every electron that leaves
the negative terminal of the power source has only one path for
current and flows through every component before returning to the
positive side of the power source, then the circuit is a series circuit.
Every component in a series circuit is in series with every other
component. They are part of the common, single current path. We
are now ready to explore series circuits in greater detail to learn
their basic characteristics, to learn how to analyze them numeri-
cally, and to design circuits that use them.

4.1 Current Through Each Component

Since the current in a series circuit has only a single path, it follows that the current will be the same through all components. If the current is 2.5 A at one point in a series circuit, then it will be 2.5 A at every other point in the circuit. This is a fundamental principle of electronics:

> The current in a series circuit is the same value at every point in the circuit.

Figure 4-1 illustrates this principle on a schematic diagram. All of the current meters will read the same value. Figure 4-2 illustrates the same principle on a pictorial view of a series circuit. It may help to visualize the circuit as a hollow pipe. As the electrons flow through the pipe, there is only one path so the flow must be the same everywhere

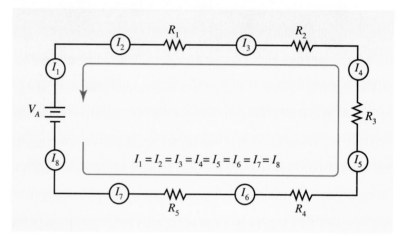

$$I_1 = I_2 = I_3 = I_4 = I_5 = I_6 = I_7 = I_8$$

Figure 4-1. The current is the same in all parts of a series circuit.

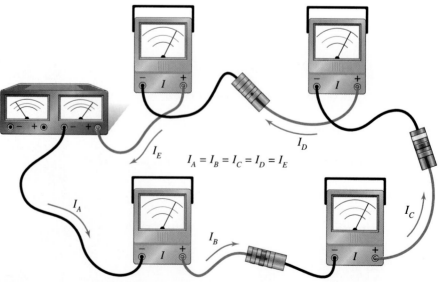

$$I_A = I_B = I_C = I_D = I_E$$

Figure 4-2. A pictorial drawing that shows the current in a series circuit.

EXAMPLE SOLUTION

Determine the current through R_4 in Figure 4-3.

EXAMPLE SOLUTION

First we identify the circuit as a series circuit, since there is only one possible path for current flow. Next we note that the current through R_1 is given as 150 mA. Since the current is the same in all parts of a series circuit, we know the current through R_4 must also be 150 mA.

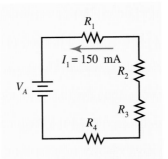

Figure 4-3.
What is the value of current through R_4?

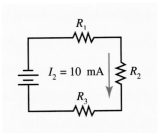

Figure 4-4.
How much current flows through R_3?

Exercise Problems 4.1

1. How much current flows through R_2 in Figure 4-3?
2. What is the value of current through R_3 in Figure 4-4?

4.2 Voltage Across Each Resistor

In a later section, we will discuss methods for calculating the voltage across each resistor in a series circuit. First, however, it is important to develop an intuitive appreciation for how the voltage is distributed in a series circuit.

Voltage Is Proportional to Resistance

Ohm's Law tells us that the value of voltage across a resistance is proportional to the resistance and to the value of current flowing through the resistance ($V = IR$). In a series circuit, the current is the same through all components. It follows then that the voltage across a given resistor must be proportional to the value of the resistor. If a particular resistor is larger than another one in a series circuit, then the larger one will have more voltage across it.

EXAMPLE SOLUTION

Refer to Figure 4-5. Which resistor has the most voltage across it?

EXAMPLE SOLUTION

The circuit has a single path and is therefore a series circuit. In a series circuit, the largest resistor has the most voltage across it. In this case, resistor R_2 has the most voltage across it.

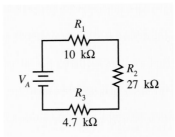

Figure 4-5. Which resistor has the most voltage across it?

Practice Problems

1. Which resistor in Figure 4-6 has the most voltage across it?

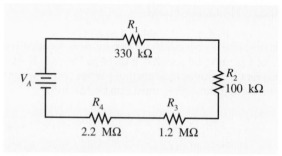

Figure 4-6. Which resistor has the most voltage?

2. Which resistor in Figure 4-7 has the most voltage across it?

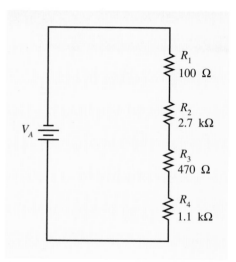

Figure 4-7. Which resistor has the most voltage?

3. Which resistor in Figure 4-8 has the least voltage across it?

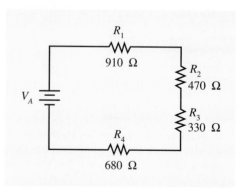

Figure 4-8. Which resistor has the least voltage across it?

Answers to Practice Problems

1. R_4 2. R_2 3. R_3

Resistor Voltage Drops

The voltage across a resistor is often called a **voltage drop.** If, for example, a certain resistor has 10 V across it, we can say it has a voltage drop of 10 V. As you will soon see, the voltage drops in a series circuit have a definite relationship to the applied voltage.

Exercise Problems 4.2

1. Which of the resistors shown in Figure 4-9 has the highest voltage drop?
2. In a series circuit, the voltage drop across a resistance is _____ proportional to the value of the resistor.
3. In a series circuit containing several resistors with different values, how can a technician quickly determine which one has the largest voltage drop? The least voltage drop?
4. If all of the resistors in a series circuit have identical resistances, what can be said about the *relative* voltage drops across the various resistors?

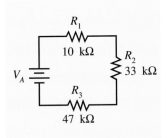

Figure 4-9. Identify the highest voltage drop.

4.3 Kirchhoff's Voltage Law

Kirchhoff's Voltage and Current Laws represent fundamental principles of electronics. It is essential for technicians to understand Kirchhoff's Laws in order to be proficient at analyzing and understanding unfamiliar circuits. Kirchhoff's Laws can also be powerful troubleshooting aids. Kirchhoff's Current Law is presented in Chapter 5. This section introduces Kirchhoff's Voltage Law.

Polarity of Voltage Drops

When current flows through a resistor, there is a corresponding voltage drop across that resistor. Figure 4-10 shows the polarity of voltage drops in a series circuit. As shown in Figure 4-10, the resistor voltage is negative on the end where electron current enters the resistor. The end where electron current exits the resistor is the more positive of the two ends.

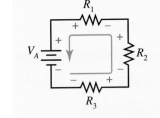

Figure 4-10. The polarity of voltage drops in a series circuit.

EXAMPLE SOLUTION

Label the polarities of the voltage drops on the resistors shown in Figure 4-11.

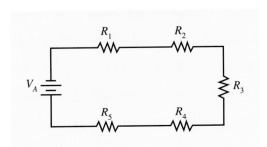

Figure 4-11. Label the resistor polarities.

EXAMPLE **SOLUTION**

Figure 4-12 shows the polarity for each of the resistors originally shown in Figure 4-11. First, we determine the direction of current flow. It leaves the negative side of the power source, flows through the circuit, and returns to the positive side of the power source. Next, we label each resistor such that the negative side of its voltage drop is on the end of the resistor where the current enters the resistor. The end where the current exits is labeled as the positive side of the voltage drop.

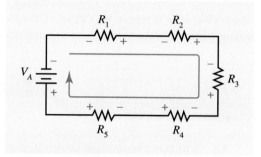

Figure 4-12. The polarity of each resistor originally shown in Figure 4-11.

Practice Problems

1.   Label the polarity for the voltage drop across each resistor shown in Figure 4-13.

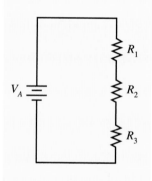

Figure 4-13. Label the polarity of all voltage drops.

Answers to Practice Problems

1.

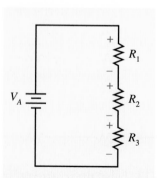

Resistor Voltage Drops and Source Voltage Relationships
SUM OF THE VOLTAGE DROPS EQUALS SOURCE VOLTAGE

The voltage drops have been labeled on the series circuit shown in Figure 4-14. If we add the voltage drops across the resistors, we should get a sum that is equal to the value of applied voltage. This is Kirchhoff's Voltage Law:

> The sum of all the voltage drops in a closed loop is equal to the value of applied voltage.

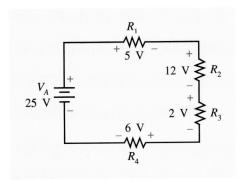

Figure 4-14. The sum of the voltage drops must always equal the value of applied voltage.

We can express Kirchhoff's Voltage Law as an equation:

$$V_A = V_1 + V_2 + V_3 + \ldots + V_N \qquad (4\text{-}1)$$

where V_N represents any number of voltage drops. In the case of the circuit shown in Figure 4-14, we can write the Kirchhoff's Voltage equation as follows:

$$V_A = V_1 + V_2 + V_3 + V_4$$
$$25\text{ V} = 5\text{ V} + 12\text{ V} + 2\text{ V} + 6\text{ V}$$
$$25\text{ V} = 25\text{ V}$$

This calculation is essentially an identity, but its usefulness is not immediately obvious. In the case of the circuit shown in Figure 4-14, all voltage values were given on the schematic. In most cases, one or more of the values will be unknown. It is then that the value of Kirchhoff's Voltage Law becomes apparent.

EXAMPLE SOLUTION

What is the value of applied voltage for the circuit shown in Figure 4-15?

EXAMPLE **SOLUTION**

We can apply Kirchhoff's Voltage Law directly as follows:

$$V_A = V_1 + V_2 + V_3$$
$$= 3 \text{ V} + 10 \text{ V} + 2 \text{ V}$$
$$= 15 \text{ V}$$

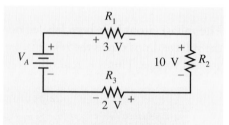

Figure 4-15. What is the applied voltage?

EXAMPLE SOLUTION

Calculate the value of the voltage drop across R_2 in Figure 4-16.

EXAMPLE **SOLUTION**

We can transpose the basic form of Kirchhoff's Voltage Law to solve for V_2 as follows:

$$V_A = V_1 + V_2 + V_3, \text{ or}$$
$$V_2 = V_A - V_1 - V_3$$
$$= 30 \text{ V} - 4 \text{ V} - 18 \text{ V}$$
$$= 8 \text{ V}$$

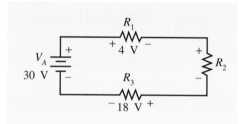

Figure 4-16. Determine the voltage across R_2.

SUM OF VOLTAGE DROPS AND SOURCES

Figure 4-17 shows another way to apply Kirchhoff's Voltage Law. Here all of the voltage drops *and* the voltage source can be added together to produce a sum of zero. This is another important way to express Kirchhoff's Voltage Law:

> The algebraic sum of all the voltage drops and all the voltage sources in any closed loop equals zero.

This is one of the most useful forms of Kirchhoff's Voltage Law. It will be useful to you when analyzing unfamiliar circuits, troubleshooting defective circuits, and even when designing new circuits.

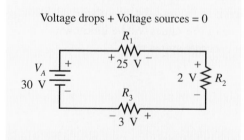

Figure 4-17. The sum of the voltage drops and sources is zero.

WRITING CLOSED-LOOP EQUATIONS

Since correct application of Kirchhoff's Voltage Law requires computation of an algebraic sum, it is essential for you to write the correct polarities for each voltage drop and each voltage source. The following sequential procedure can be used to produce a correctly labeled closed-loop equation:

1. Label the polarity of all voltage sources and all voltage drops.
2. Start at any point in the circuit and write the voltages (including polarity) as you progress around the loop in either direction. Stop when you have completed the loop.

The polarity of the voltage sources will either be given, or you can tell the polarity from the voltage source symbol (recall that the shortest line on the symbol for a voltage source is the negative terminal). The polarity of the voltage drops is determined by the direction of current flow. Recall that the negative end of a voltage drop will be the point where current enters the component.

Step two is the actual writing of the equation. The recommended method is stated as follows: *Move around the loop in either direction, and write the voltages with the polarity that is on the exit end of the component.* Some people prefer to use the polarity of the voltage drop that is on the end where you enter the component. Although this second method will produce the same mathematical answer for the type of problem currently being discussed, it is inconsistent with other techniques discussed at a later point in the text.

EXAMPLE SOLUTION

Use Kirchhoff's Voltage Law to determine whether the voltage drops shown on the circuit in Figure 4-18 are correct.

EXAMPLE SOLUTION

We begin (arbitrarily) at the negative side of the voltage source and write the Kirchhoff's Voltage Law equation for the closed circuit loop as we move clockwise (arbitrarily):

$$V_A - V_1 - V_2 - V_3 = 0$$
$$25 \text{ V} - 5 \text{ V} - 2 \text{ V} - 18 \text{ V} = 0$$
$$25 \text{ V} - 25 \text{ V} = 0$$

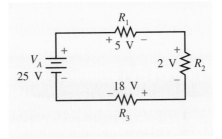

Figure 4-18. Are the voltage drops correct?

Since this latter statement is true, we can conclude that the labeled voltages are most likely correct. If the equation was invalid (e.g., 5 = 9), then we would know for sure that an error had been made.

EXAMPLE SOLUTION

Apply Kirchhoff's Voltage Law to determine whether the voltages shown in Figure 4-19 are labeled correctly.

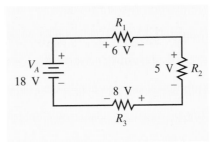

Figure 4-19. Are the voltage drops correctly labeled?

EXAMPLE **SOLUTION**

Let us arbitrarily begin at the negative side of the voltage source and move in a clockwise direction. Remember, you can start at any point and move in either direction. The closed-loop equation is written as

$$V_A - V_1 - V_2 - V_3 = 0$$
$$18\ V - 6\ V - 5\ V - 8\ V = 0$$
$$18\ V - 19\ V = 0\ [error]$$

Since this last equation is clearly erroneous, we can safely say that the voltages labeled on Figure 4-19 are incorrect. We don't know which one is wrong, but there is *definitely* an error.

EXAMPLE SOLUTION

Are the voltages in Figure 4-20 labeled correctly?

EXAMPLE SOLUTION

The Kirchhoff's Voltage Law equation for the circuit shown in Figure 4-20 (starting at the junction of R_1 and R_2 and moving counterclockwise) is

$$V_1 - V_A + V_3 + V_2 = 0$$
$$35\ V - 100\ V + 20\ V + 50\ V = 0$$
$$5\ V = 0\ [error]$$

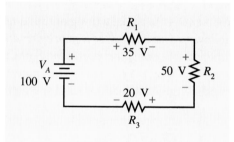

Figure 4-20. Are the voltages labeled correctly?

Practice Problems

1. Refer to Figure 4-21. Are the voltages labeled correctly?

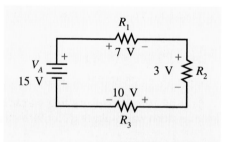

Figure 4-21. Are the voltages correct?

2. Apply Kirchhoff's Voltage Law to the circuit shown in Figure 4-22 to determine whether the voltages are labeled correctly.

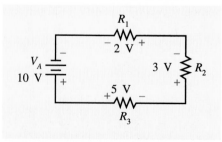

Figure 4-22. Are the voltages correct?

3. Are the voltages in Figure 4-23 labeled correctly?

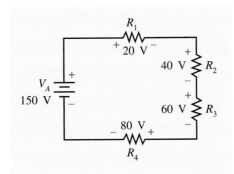

Figure 4-23. Are the voltages correct?

Answers to Practice Problems

1. No   **2.** Yes   **3.** No

Exercise Problems 4.3

1. What is the value of applied voltage in Figure 4-24?

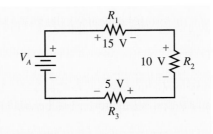

Figure 4-24. What is the value of V_A?

2. Are the voltages labeled correctly in Figure 4-25?

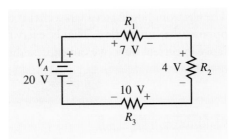

Figure 4-25. Are the voltages labeled correctly?

3. Apply Kirchhoff's Voltage Law to calculate the voltage drop across R_3 in Figure 4-26.

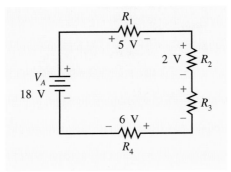

Figure 4-26. What is the value of V_3?

4. What is the voltage drop across R_1 in Figure 4-27?

5. Kirchhoff's Voltage Law states that the _____ of the voltage drops and voltage _____ in a _____ loop is equal to _____.

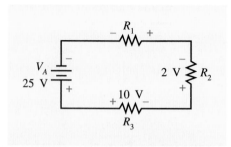

Figure 4-27. How much voltage is dropped across R_1?

4.4 Computing Voltage, Current, Resistance, and Power

We are now ready to apply the power equations, the basic theory of series circuits, Kirchhoff's Voltage Law, and Ohm's Law to the solution of series circuit problems. Sometimes problems require a technician to simply substitute numbers into an equation and compute the result. In other cases, the technician must manipulate the available data in a creative way to obtain solutions to the circuit. In all cases, though, Ohm's Law, Kirchhoff's Law, and the power equations will be valid.

Total Resistance

KEY POINTS

The total resistance in a series circuit can be computed by summing the values of individual resistors.

The electrons flowing in a series circuit must pass through every component. Each resistor opposes the current flow. That is, every resistor in a series circuit increases the total resistance to current flow. We can express this idea mathematically as

$$R_T = R_1 + R_2 + R_3 + \dots + R_N \qquad (4\text{-}2)$$

EXAMPLE SOLUTION

Determine the total resistance of the circuit shown in Figure 4-28.

EXAMPLE **SOLUTION**

We apply Equation 4-2.

$$R_T = R_1 + R_2 + R_3 + R_4$$
$$= 1.2 \text{ k}\Omega + 2.2 \text{ k}\Omega + 5.6 \text{ k}\Omega + 3.9 \text{ k}\Omega$$
$$= 12.9 \text{ k}\Omega$$

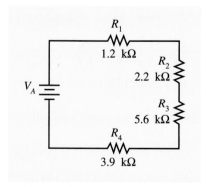

Figure 4-28. Calculate the value of total resistance.

EXAMPLE SOLUTION

Calculate the value of total resistance for the circuit shown in Figure 4-29.

EXAMPLE SOLUTION

$$R_T = R_1 + R_2 + R_3$$
$$= 1.8 \text{ k}\Omega + 680 \text{ }\Omega + 910 \text{ }\Omega$$
$$= 3.39 \text{ k}\Omega$$

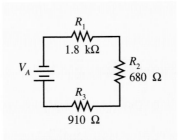

Figure 4-29. How much total resistance is in this circuit?

Practice Problems

1. Calculate the total resistance in Figure 4-30.

2. What is the total opposition to current flow for the circuit in Figure 4-31?

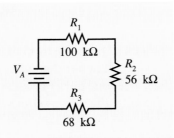

Figure 4-30. What is the total resistance in this circuit?

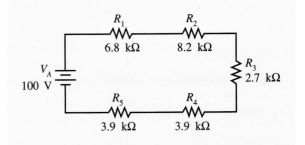

Figure 4-31. Calculate the total opposition to current flow.

3. How much total resistance is in the circuit shown in Figure 4-32?

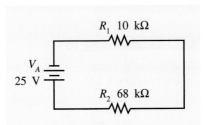

Figure 4-32. What is the value of total resistance?

Answers to Practice Problems

1. 224 kΩ   2. 25.5 kΩ   3. 78 kΩ

Current

You will recall that current in a series circuit is the same at all points in the circuit. Thus, for example, the current through R_1 (I_1) will be the same as the value of total current (I_T). We can express this as an equation (Equation 4-3).

KEY POINTS

The current is the same in all parts of a series circuit. For this reason, current is an important quantity to calculate when analyzing a series network.

$$I_T = I_1 = I_2 = I_3 = \ldots = I_N \qquad (4\text{-}3)$$

This equation tells us that if we know the value of current at any point in a series circuit, then we immediately know the value of current at every other point in the circuit.

EXAMPLE SOLUTION

Calculate the value of current through R_1 and the value of total current in Figure 4-33.

EXAMPLE **SOLUTION**

The value of current through R_3 in Figure 4-33 is given as 200 mA. Since the circuit is a series circuit, we immediately know that the current must be 200 mA at all other points. In equation form, we have

$$I_T = I_1 = I_2 = I_3 = 200 \text{ mA}$$

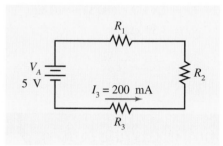

Figure 4-33. How much current flows through R_1?

Total Power

Electrical power consumption often takes the form of heat dissipation. So it is with the power dissipated in a resistor. All power that is dissipated in a circuit must be supplied by the power source. The total power consumed is simply a summation of the individual powers dissipated by the various circuit components. We can state this in equation form as

$$P_T = P_1 + P_2 + P_3 + \ldots + P_N \qquad (4\text{-}4)$$

Total circuit power can also be computed with any one of the basic power formulas presented in Chapter 2 provided total circuit values are used in the calculations.

EXAMPLE SOLUTION

What is the total power dissipated by the circuit shown in Figure 4-34?

EXAMPLE **SOLUTION**

Method 1: One way to compute the total power in a circuit is to find the sum of the powers dissipated by the individual resistors.

$$P_T = P_1 + P_2 + P_3$$
$$= 2.5 \text{ W} + 4 \text{ W} + 1.5 \text{ W}$$
$$= 8 \text{ W}$$

Method 2: Another way to compute total power in a circuit is to simply apply one of the basic power equations with total circuit values used as factors.

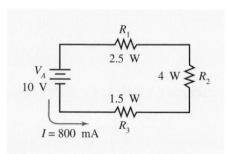

Figure 4-34. Compute total power in this circuit.

$$P_T = I_T V_T$$
$$= 800 \times 10^{-3} \text{ A} \times 10 \text{ V}$$
$$= 8 \text{ W}$$

The subscripts A and T are equivalent when referring to all (applied) circuit values.

Practice Problems

1. How much power is dissipated by the circuit shown in Figure 4-35?

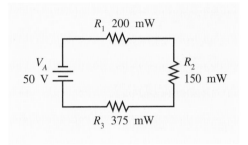

Figure 4-35. Calculate the power in this circuit.

2. What is the total power consumption for the circuit shown in Figure 4-36?

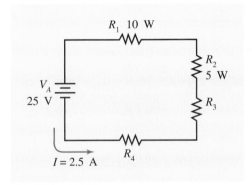

Answers to Practice Problems

1. 725 mW   2. 62.5 W

Figure 4-36. How much power is dissipated by this circuit?

Fully Specified Circuits

If the value of every component in a circuit is known, then the circuit is called a **fully specified circuit.** In these cases, the circuit parameters (e.g., voltage, current, and power) can be computed by substituting the appropriate numbers into a relevant equation. Ohm's Law, Kirchhoff's Voltage Law, and the power equations can be used to compute values for individual components (e.g., the voltage drop across R_6 or the power dissipated by R_2). The same equations can be used to compute total circuit values (e.g., total current or total power).

The first step in computing a circuit value is to select an appropriate equation. Selection of an equation is accomplished as follows:

1. Choose an equation where the value to be calculated is the only unknown factor.
2. Use known values associated with a given component to calculate other values for that same component.
3. Use total circuit values to calculate other total circuit values.

Failure to consistently apply these rules is the most common source of errors in circuit analysis. Let's work some examples.

EXAMPLE SOLUTION

Calculate the value of voltage across R_2 in Figure 4-37.

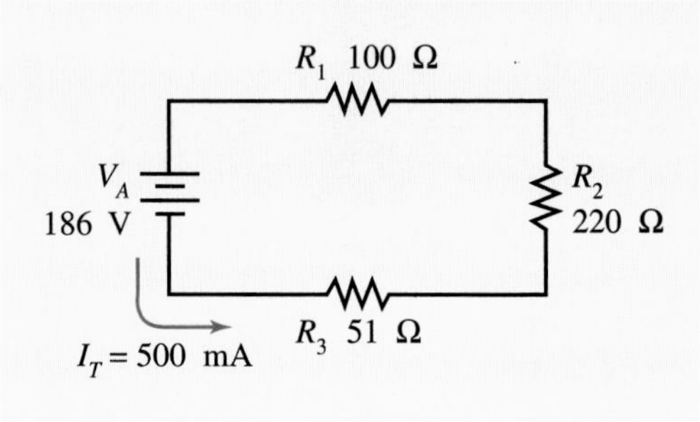

EXAMPLE **SOLUTION**

Since we are required to compute a voltage value for R_2, we need to select an equation where V_2 is the only unknown factor. We know the value of current through R_2, since the total current is given as 500 mA, and since we know the current is the same through all components in a series circuit (Equation 4-3). The value of R_2 is also given on the schematic diagram. We now select the Ohm's Law equation $V = IR$ because current and resistance are known. To avoid using the wrong quantity type, it is important to add subscripts to all quantities. Since we are calculating values for R_2, we can only use known quantities associated with R_2.

$$V_2 = I_2 R_2$$
$$= 500 \times 10^{-3} \text{ A} \times 220 \text{ }\Omega$$
$$= 110 \text{ V}$$

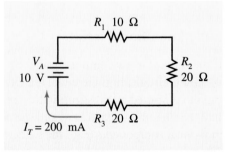

Figure 4-37. Calculate V_2 in this circuit.

EXAMPLE SOLUTION

How much total power is dissipated by the circuit in Figure 4-38?

EXAMPLE **SOLUTION**

Since we are computing total power, we will utilize other total values in the selected equation.

$$P_T = I_T V_T$$
$$= 200 \times 10^{-3} \text{ A} \times 10 \text{ V}$$
$$= 2 \text{ W}$$

Figure 4-38. Compute the total power in this circuit.

EXAMPLE SOLUTION

What is the value of current through R_3 in Figure 4-39?

EXAMPLE SOLUTION

Since this is a series circuit, we know that the current is the same in all parts of the circuit. If we can find the value of current at any point, then we will also know the value of I_3. Let us compute the value of I_1, since we know two other factors (voltage and resistance) associated with R_1.

$$I_1 = \frac{V_1}{R_1}$$

$$= \frac{20 \text{ V}}{10 \text{ k}\Omega}$$

$$= 2 \text{ mA}$$

Since I_3 is the same value as I_1 (Equation 4-3), we know that $I_3 = 2$ mA.

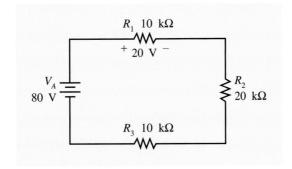

Figure 4-39. What is the value of I_3?

Practice Problems

1. How much current flows through R_2 in Figure 4-40?
2. What is the total power in Figure 4-40?

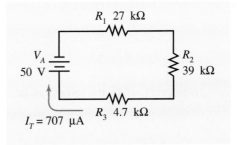

Figure 4-40. Calculate I_2 in this circuit.

3. What is the voltage drop across R_3 in Figure 4-41?

Answers to Practice Problems

1. 707 µA 2. 35.35 mW

3. 1.6 V

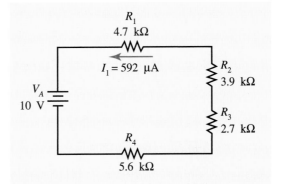

Figure 4-41. Calculate V_3 in this circuit.

Partially Specified Circuits

KEY POINTS

Partially specified circuits have one or more unknown component values. The solution of this type of circuit requires creative application of basic circuit theory in conjunction with Ohm's Law, Kirchhoff's Voltage Law, and the power formulas.

There are many times when a technician does not know the value of every circuit component. A circuit with one or more unspecified component values is called a **partially specified circuit.** In general, this type of circuit is more challenging to analyze, since direct application of an equation is not always possible. To solve such a problem, a technician must creatively apply a series of equations. Each equation provides additional information. Finally, enough information will be known to allow computation of the specific quantity in question. It takes substantial practice to be able to identify the optimum sequence of calculations. When you are uncertain of the next step, and until you gain the experience needed to streamline your calculations, you are advised to calculate any value that can be calculated. Each calculated value provides you with additional data that can be used in subsequent calculations.

EXAMPLE SOLUTION

Determine the voltage drop across R_2 in Figure 4-42.

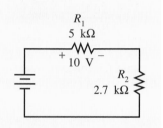

Figure 4-42. What is the voltage drop across R_2?

EXAMPLE **SOLUTION**

We cannot compute V_2 directly because we have no equation with enough known values to solve. Since the value of R_2 is given, we could compute V_2 if we knew the value of I_2 (i.e., $V_2 = I_2R_2$). When analyzing series circuits, it is usually a good idea to compute current, since this value is common to every component.

$$I_1 = \frac{V_1}{R_1} = \frac{10 \text{ V}}{5 \text{ k}\Omega} = 2 \text{ mA}$$

This current will be the same at all points in the circuit (e.g., I_T, I_2, I_3, and so on).

We can now calculate V_2 by applying Ohm's Law.

$$V_2 = I_2R_2 = 2 \text{ mA} \times 2.7 \text{ k}\Omega = 5.4 \text{ V}$$

Figure 4-43. Compute all missing circuit values.

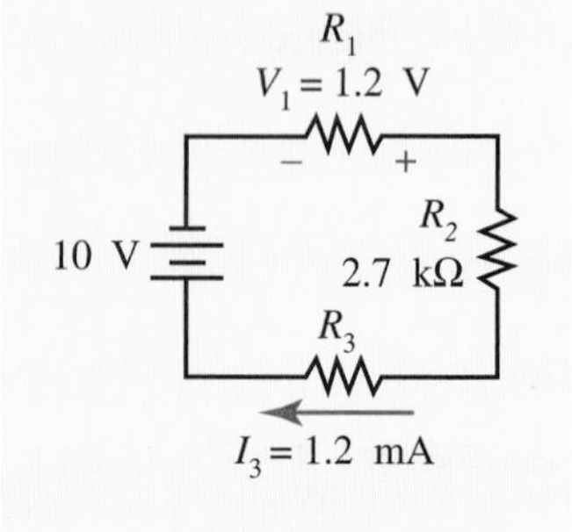

EXAMPLE SOLUTION

Calculate all missing voltages, currents, powers, and resistances for the circuit shown in Figure 4-43. Table 4-1 provides a partially filled solution matrix.

COMPONENT	RESISTANCE	VOLTAGE	CURRENT	POWER
R_1		1.2 V		
R_2	2.7 kΩ			
R_3			1.2 mA	
Total		10 V		

Table 4-1. Partial Solution Matrix for Figure 4-43

EXAMPLE **SOLUTION**

It is important to realize two facts: First, there are many ways (sequences) to solve the problem. Second, there is not necessarily a single best sequence. With these thoughts in mind, let's solve this problem by computing any circuit value for which we have sufficient data.

In a series circuit, it always a good strategy to find the value of current, since it is common to all components. The value of I_3 is given as 1.2 mA, so we immediately know the value of all currents (Equation 4-3).

$$I_T = I_1 = I_2 = I_3 = 1.2 \text{ mA}$$

Next, we might choose to calculate the value of R_1 with Ohm's Law.

$$R_1 = \frac{V_1}{I_1}$$

$$= \frac{1.2 \text{ V}}{1.2 \text{ mA}}$$

$$= 1.0 \text{ k}\Omega$$

We can apply one of the power formulas to compute P_1.

$$P_1 = I_1 V_1$$

$$= 1.2 \text{ mA} \times 1.2 \text{ V}$$

$$= 1.44 \text{ mW}$$

Since we know the resistance of R_2 and the current through it, we can calculate V_2 and P_2 as follows:

$$V_2 = I_2 R_2$$

$$= 1.2 \text{ mA} \times 2.7 \text{ k}\Omega$$

$$= 3.24 \text{ V}$$

and,

$$P_2 = I_2 V_2$$

$$= 1.2 \text{ mA} \times 3.24 \text{ V}$$

$$= 3.89 \text{ mW}$$

KEY POINTS

A good strategy to use with series circuits is to determine the value of current as early as possible.

We now know all but one of the voltages in the closed loop. We can apply Kirchhoff's Voltage Law to compute V_3.

$$0 = V_A - V_3 - V_2 - V_1$$
$$0 = 10\text{ V} - V_3 - 3.24\text{ V} - 1.2\text{ V}$$
$$V_3 = 10\text{ V} - 1.2\text{ V} - 3.24\text{ V}$$
$$V_3 = 5.56\text{ V}$$

We can now find the value of R_3 by applying Ohm's Law.

$$R_3 = \frac{V_3}{I_3}$$
$$= \frac{5.56\text{ V}}{1.2\text{ mA}}$$
$$= 4.63\text{ k}\Omega$$

The power dissipated by R_3 can be calculated as

$$P_3 = I_3^2 R_3$$
$$= (1.2\text{ mA})^2 \times 4.63\text{ k}\Omega$$
$$= 6.67\text{ mW}$$

We can compute total power dissipation as

$$P_T = I_T V_T$$
$$= 1.2\text{ mA} \times 10\text{ V}$$
$$= 12\text{ mW}$$

Finally, we can find the total resistance (Equation 4-2) in the circuit as follows:

$$R_T = R_1 + R_2 + R_3$$
$$= 1.0\text{ k}\Omega + 2.7\text{ k}\Omega + 4.63\text{ k}\Omega$$
$$= 8.33\text{ k}\Omega$$

The results of our calculations are summarized in Table 4-2.

COMPONENT	RESISTANCE	VOLTAGE	CURRENT	POWER
R_1	1.0 kΩ	1.2V	1.2 mA	1.44 mW
R_2	2.7 kΩ	3.24 V	1.2 mA	3.89 mW
R_3	4.63 kΩ	5.56 V	1.2 mA	6.67 mW
Total	8.33 kΩ	10 V	1.2 mA	12 mW

Table 4-2. Completed Solution Matrix for the Circuit in Figure 4-43

Practice Problems

1. Complete a solution matrix similar to Table 4-2 for the circuit shown in Figure 4-44.

2. Complete a solution matrix for the circuit shown in Figure 4-45.

3. Complete a solution matrix for the circuit shown in Figure 4-46.

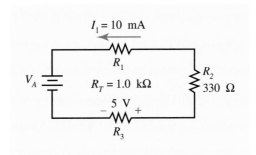

Figure 4-44. Complete a solution matrix for this circuit.

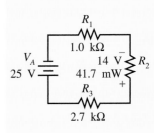

Figure 4-45. Complete a solution matrix for this circuit.

Answers to Practice Problems

1. Completed solution matrix for the circuit in Figure 4-44:

COMPONENT	RESISTANCE	VOLTAGE	CURRENT	POWER
R_1	170 Ω	1.7 V	10 mA	17 mW
R_2	330 Ω	3.3 V	10 mA	33 mW
R_3	500 Ω	5 V	10 mA	50 mW
Total	1.0 kΩ	10 V	10 mA	100 mW

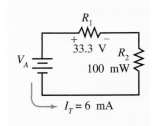

Figure 4-46. Complete a solution matrix for this circuit.

2. Completed solution matrix for the circuit in Figure 4-45:

COMPONENT	RESISTANCE	VOLTAGE	CURRENT	POWER
R_1	1.0 kΩ	2.98 V	2.98 mA	8.88 mW
R_2	4.7 kΩ	14 V	2.98 mA	41.7 mW
R_3	2.7 kΩ	8.04 V	2.98 mA	23.90 mW
Total	8.4 kΩ	25 V	2.98 mA	74.48 mW

Note that some values are rounded.

3. Completed solution matrix for the circuit in Figure 4-46:

COMPONENT	RESISTANCE	VOLTAGE	CURRENT	POWER
R_1	5.55 kΩ	33.3 V	6 mA	199.8 mW
R_2	2.78 kΩ	16.67 V	6 mA	100 mW
Total	8.33 kΩ	49.97 V	6 mA	299.8 mW

Note that some values are rounded.

Exercise Problems 4.4

1. The total resistance in a series circuit is equal to the _____ of the individual resistances.

2. If the resistances in a three-resistor series circuit are 10 kΩ, 22 kΩ, and 18 kΩ, what is the total resistance in the circuit?

3. If the total resistance of a three-resistor circuit is 100 kΩ, R_1 is 47 kΩ, and R_2 is 22 kΩ, what is the value of R_3?

4. If the current through R_2 in a series circuit is 350 µA, what is the value of current through R_1?

5. Refer to Figure 4-47. Calculate the total power dissipated by the circuit.

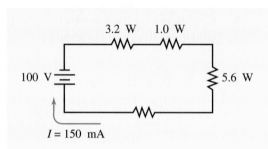

Figure 4-47. What is the total power in this circuit?

6. Complete a solution matrix for the circuit shown in Figure 4-48.

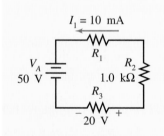

Figure 4-48. Compute all unknown circuit values.

4.5 Ground and Other Reference Points

Anytime a voltage is measured, the voltmeter is actually indicating the voltage that is present at one point in the circuit with reference to another point. We briefly discussed this idea in Chapter 3 with regard to relative polarities.

The Concept of Ground

In most electronic circuits, there is a point to which all circuit voltages are measured. This common point is called the circuit **ground.** Most often, but certainly not always, ground is considered to be one side of the circuit's power source. In all cases, ground is considered to be 0 V. This is sensible, since ground is the reference point for all measurements, and since any voltage is zero with reference to itself (i.e., there is no difference in potential). Figure 4-49(a) shows a series circuit where the negative side of the voltage source is considered to be the circuit ground. Note the symbol ($\perp$) used to represent ground. Figure 4-49(b) shows exactly the same circuit as Figure 4-49(a), but two ground symbols are used. *Both ground symbols represent the same electrical point in the circuit.* The interconnecting wire is assumed but not shown. The omission of the interconnecting ground wires on schematic diagrams is the normal practice since it removes unnecessary clutter from otherwise complex diagrams.

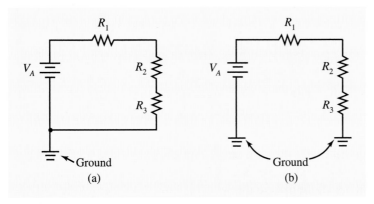

Figure 4-49. Ground is a reference point from which other voltages are measured.

When a circuit is physically constructed, ground can take a number of different forms. If the circuit has a metal chassis or frame associated with it, then the frame is usually connected to circuit ground. This is especially true for high-frequency devices or high-speed digital circuits. Most circuits are constructed on printed circuit boards (PCBs). The conductors on a PCB are traces of copper bonded to an insulating material. The copper traces connect the various components together just as wire might do in a laboratory circuit. Ground on a PCB is often a very wide trace that wanders throughout the board. If the PCB has more than one layer of traces, then one or more of the layers is frequently dedicated as a **ground plane.** This means that the entire PCB layer (sheet of copper) is connected to ground.

Voltages with Reference to Ground

Figure 4-50 shows a series circuit with one point labeled as the ground connection. The remaining points in the circuit are labeled A through E. When we measure the voltages at points A through E, we connect one lead of the voltmeter to ground and the other lead to the point being measured. For example, suppose we wanted to measure the voltage at point C in Figure 4-50 with reference to ground. We would connect one side of the voltmeter to ground and the other to point C. You can see from inspection of Figure 4-50, that the meter is essentially connected across R_3, so it will indicate 6 V. We label the voltage at point C as V_C. So in the present example, $V_C = 6$ V.

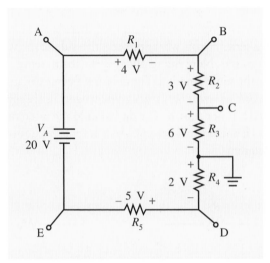

Figure 4-50. Voltages can be measured with respect to ground.

The resulting measurement is less obvious if we were to measure the voltage at point E. There is a simple procedure, however, that will allow you to determine the correct theoretical voltage at any point in a circuit.

1. Start at the reference point and write the voltage drops as you move toward the measured point. Use the polarity of the voltage drop that is on the end where you exit a component.

2. When you reach the measured point, the algebraic sum of the accumulated voltage drops will be the measured value.

The polarities of the voltage drops are determined in exactly the same manner as if we were writing the Kirchhoff's Voltage Law equation for the loop. The only differences are that we do not complete the loop, and we do not set the voltages equal to zero. (If you prefer to use the polarity where you *enter* a component, then you should start at the measured point and move to the reference point.)

EXAMPLE SOLUTION

Determine the voltage at point E in Figure 4-50.

EXAMPLE SOLUTION

We begin at the reference point (ground) and move toward the measured point (point E). As we pass through R_4, we will write its voltage drop as –2 V. Continuing through R_5, we find a drop of –5 V. Since we have reached our destination, the measured value will be equal to the algebraic sum of our accumulated voltage drops. In this particular case, the measured voltage will be

$$V_E = -2\ V + -5\ V = -7\ V$$

It is important to realize that you may move in either direction as you progress from the reference point toward the measured point. Although the individual voltage drops may be different, the algebraic sums of the two directions will be identical. If, in the present example, we had moved in the opposite direction, we would have accumulated the following sequence of voltages: +6 V, +3 V, +4 V, and –20 V (the voltage source). The algebraic sum of these voltages is still –7 V.

EXAMPLE SOLUTION

Determine the voltage that would be measured between ground and point A in Figure 4-50.

EXAMPLE SOLUTION

As we start at ground and move toward point A, we generate the following sequence of voltages: +6 V, +3 V, and +4 V. This yields an algebraic total of V_A = +13 V.

Practice Problems

1. Refer to Figure 4-51, and determine the following voltages: V_A, V_B, V_C, and V_D.

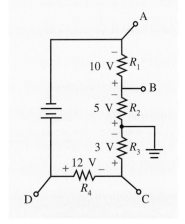

Figure 4-51. Determine the voltages at points A, B, C, and D with reference to ground.

Answers to Practice Problems

1. a. V_A = –15 V

   b. V_B = –5 V

   c. V_C = +3 V

   d. V_D = +15 V

Nonground References

In some cases, a technician must measure a voltage that is referenced to some point other than ground. Suppose, for example, we want to determine the voltage that would be measured at point B with reference to point D in Figure 4-52. We apply the exact same procedure that we used for ground references. In this case, we begin at the reference point (point D) and move toward point B. As we go, we will generate the following sequence of voltages: +10 V and +4 V. The algebraic sum of these voltages is +14 V.

We must be careful to specify the reference point when writing the voltage at a certain point in a circuit. Figure 4-53 shows the accepted method for indicating voltage references. In the case of a ground reference, the second subscript is omitted and ground is assumed to be the reference point. Exceptions to this labeling method are usually obvious by inspection. In our preceding example, we would label the voltage at point B that was measured with respect to point D as V_{BD} = +14 V.

KEY POINTS

Voltages that are measured with reference to some point other than ground are generally identified with a double subscript (e.g., V_{AD}). The first subscripted letter indicates the measured point in the circuit; the second subscript indicates the reference point.

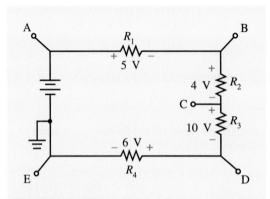

Figure 4-52. Nonground references are also used by technicians.

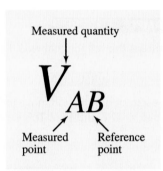

Figure 4-53. Method for labeling nonground references.

EXAMPLE SOLUTION

Refer to Figure 4-54. What is the voltage at point E with reference to point C?

EXAMPLE SOLUTION

We begin at point C and move toward point E as we generate the following sequence of voltages: +1 V, +4 V, and +10 V. The algebraic sum is $V_{EC} = +15$ V.

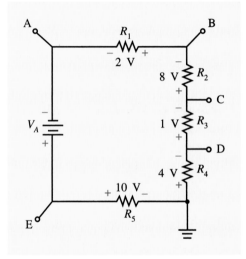

EXAMPLE SOLUTION

Refer to Figure 4-54. What is the value of V_{AB}?

EXAMPLE SOLUTION

When moving from point B to point A, we obtain a voltage drop of –2 V. Thus, $V_{AB} = -2$ V.

Figure 4-54. Determine the value of V_{EC} and V_{AB}

Practice Problems

1. What is the value of V_{DA} in Figure 4-54?
2. What is the value of V_{CE} in Figure 4-54?
3. Find the value of V_{AE} in Figure 4-54.

Answers to Practice Problems

1. +11 V   2. –15 V   3. –25 V

Other Types of Ground References

There are several alternate symbols that are used to represent ground (i.e., 0 V) reference points on schematic diagrams. Some of these are illustrated in Figure 4-55. Although the "correct" use of a particular symbol has been documented by standards committees, the actual use of the symbols by manufacturers seems to be somewhat arbitrary. The symbol shown in Figure 4-55(a) is generally used to represent earth ground. That is, a point in the circuit that is connected to the safety ground of the 120 Vac power line and ultimately connects to a metal stake driven into the soil. This is a particularly important reference point for several reasons. If properly implemented, it provides some degree of increased safety with respect to shock from the 120 Vac power line. In particular, if the metal equipment chassis and other metal parts that are accessible to the equipment user are connected to ground, then no shock hazard will exist if an internal wire comes loose or inadvertently touches the chassis.

The symbol shown in Figure 4-55(b) is used to represent circuit ground by some manufacturers. In some cases, a given circuit may be designed to have multiple grounds that are essentially isolated from each other. In these cases, the ground symbol shown in Figure 4-55(c) can be labeled to identify the various ground connections. Figure 4-55(c) shows the symbol labeled with the letter "A."

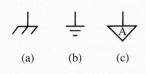

Figure 4-55.
Several alternative ground symbols are used in the industry.

Exercise Problems 4.5

1. Refer to Figure 4-56. Determine each of the following voltages:
   a. V_A b. V_B c. V_C d. V_D e. V_E f. V_F
2. Refer to Figure 4-56. Determine each of the following voltages:
   a. V_{AB} b. V_{EA} c. V_{DA} d. V_{FB} e. V_{BF} f. V_{DF}

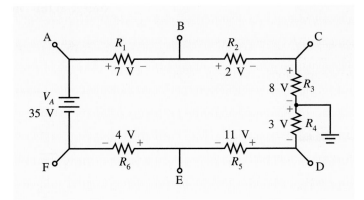

Figure 4-56. Determine the voltages in this circuit.

4.6 Multiple Voltage Sources

Many electronic circuits require more than one voltage source for proper operation. A typical application, for example, might require +5 V, +15 V, and −15 V power sources.

Let us now consider the effects of multiple voltage sources that are connected as part of a series circuit.

Series-Aiding Voltage Sources

Figure 4-57 shows two voltage sources that are connected in the same polarity. They are said to be **series-aiding,** since both voltage sources cause current flow in the same direction. The effective voltage in the circuit is the sum of the series-aiding voltages.

$$V_T = V_{A_1} + V_{A_2} + \ldots + V_{A_N} \qquad (4\text{-}5)$$

It is important to note that the double subscripts (e.g., V_{A_1}) are used to distinguish the various voltage sources and do *not* indicate a nonground reference point.

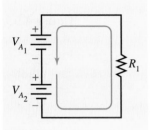

Figure 4-57.
Series-aiding
voltage sources.

KEY POINTS

In a circuit with series-aiding voltage sources, all of the voltage sources have the same polarity. The net voltage in the circuit is simply the sum of the individual series-aiding voltage sources.

EXAMPLE SOLUTION

What is the effective applied voltage for the circuit shown in Figure 4-58?

EXAMPLE **SOLUTION**

First, we note that both voltage sources are connected in the same polarity; they are series-aiding. The total voltage is the sum of the two individual sources (Equation 4-5):

$$V_T = V_{A_1} + V_{A_2}$$

$$= +5 \text{ V} + 15 \text{ V} = +20 \text{ V}$$

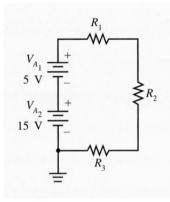

Figure 4-58. What is the total voltage in this circuit?

EXAMPLE SOLUTION

How much total voltage is applied to the circuit in Figure 4-59?

EXAMPLE **SOLUTION**

We apply Equation 4-5 as follows:

$$V_T = V_{A_1} + V_{A_2} + V_{A_3}$$

$$= 2 \text{ V} + 10 \text{ V} + 5 \text{ V} = 17 \text{ V}$$

The equivalent voltage source will be negative *with respect to ground* by virtue of the ground connection on the positive side of the voltage sources.

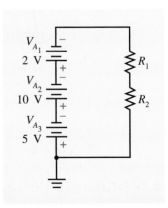

Figure 4-59. What is the total voltage in this circuit?

Series-Opposing Voltage Sources

Figure 4-60 shows two voltage sources that are connected in the opposite polarity. That is, they try to cause current to flow in opposite directions. This configuration of voltage sources is called **series-opposing.** The effective voltage of two series-opposing voltages sources is the difference between the two individual sources (Equation 4-6).

$$V_T = V_{A_1} - V_{A_2} \qquad (4\text{-}6)$$

The polarity of the net voltage will be the same as the larger of the two opposing sources. The direction of current flow is determined by the relative magnitudes of the series voltage sources. Current flows in the direction determined by the polarity of the net effective voltage.

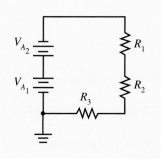

Figure 4-60. Series-opposing voltage sources.

EXAMPLE SOLUTION

Calculate the effective voltage for the circuit in Figure 4-61 and determine the direction of current flow.

EXAMPLE SOLUTION

The effective voltage is computed with Equation 4-6 as follows:

$$V_T = V_{A_1} - V_{A_2}$$
$$= 25\ V - 10\ V = 15\ V$$

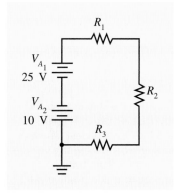

Figure 4-61. Determine the net voltage and direction of current in this circuit.

○ **KEY POINTS**

Series-opposing voltage sources have opposite polarities. The net voltage for two series-opposing sources is the difference between the two opposing sources with the polarity being the same as the larger of the two.

The polarity of the net voltage is the same as the polarity of V_{A_1}. Therefore, the current will flow in the direction determined by V_{A_1}. In the present case, the current will flow in a counterclockwise direction. Figure 4-62 shows an equivalent circuit.

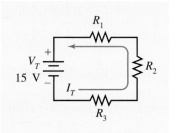

Figure 4-62. An equivalent diagram for the circuit shown in Figure 4-61.

Complex Voltage Sources

It is also important for a technician to be able to analyze circuits that involve combinations of series-aiding and series-opposing voltage sources. A circuit that has both series-aiding and series-opposing voltages sources is said to have a **complex voltage source.** Figure 4-63 shows a circuit with a complex voltage source.

○ **KEY POINTS**

A complex voltage source is composed of both series-aiding and series-opposing voltage sources. The effective voltage is the algebraic sum of the individual voltages.

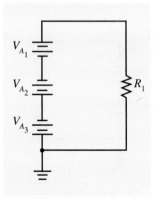

Figure 4-63.
A complex voltage source has both series-aiding and series-opposing voltage sources.

The effective applied voltage for a complex voltage source configuration is the algebraic sum of the individual voltage sources (Equation 4-7):

$$V_T = V_{A_1} \pm V_{A_2} \pm \ldots \pm V_{A_N} \qquad (4\text{-}7)$$

The sign (plus or minus) for a given polarity of voltage source is essentially arbitrary as long as we apply our choice consistently throughout the problem.

EXAMPLE SOLUTION

Determine the effective applied voltage for the circuit shown in Figure 4-64, and indicate the direction of current flow.

EXAMPLE **SOLUTION**

We shall determine the value of net voltage by applying Equation 4-7. No ground reference is shown in Figure 4-64, so it is not clear where to begin combining the voltages. We will arbitrarily choose to assign positive values to voltage sources that are the same polarity as V_{A_1} and negative values to the opposite polarity (e.g., V_{A_2}). Let us now apply Equation 4-7 to the circuit shown in Figure 4-64 to determine the net applied voltage:

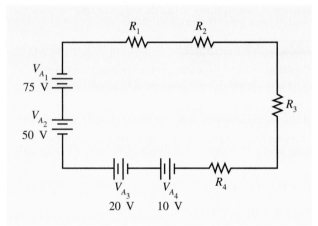

Figure 4-64. What is the net voltage in this circuit?

$$V_T = V_{A_1} - V_{A_2} - V_{A_3} - V_{A_4}$$
$$= +75 \text{ V} - 50 \text{ V} - 20 \text{ V} - 10 \text{ V} = -5 \text{ V}$$

Since the result is negative, we know the net voltage has a polarity that is the same as V_{A_2} (based on our original assignment of polarity). We can then conclude that current must be flowing in a counterclockwise direction. Our results are shown in Figure 4-65.

It should be noted that the voltage sources may be distributed around the series circuit. The sources do not have to appear immediately adjacent to each other on the schematic diagram. The problem solution, however, is identical to our example.

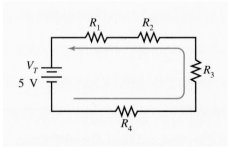

Figure 4-65. An equivalent diagram for the circuit shown in Figure 4-64.

1. Determine the net voltage and direction of current flow for the circuit shown in Figure 4-66.

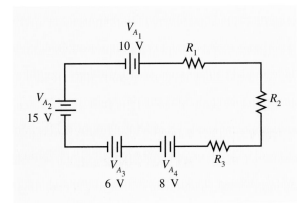

Figure 4-66. What is the effective voltage in this circuit?

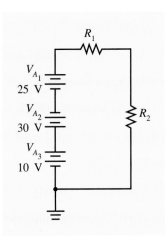

2. Find the effective applied voltage and the direction of current flow in the circuit shown in Figure 4-67.

3. Compute the net applied voltage and determine the direction of current flow in the circuit shown in Figure 4-68.

Figure 4-67. Find the net applied voltage.

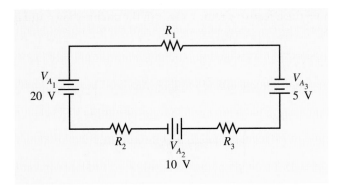

Figure 4-68. Compute the net applied voltage in this circuit.

Answers to Practice Problems

1. 11 V. Electron current flows counterclockwise.

2. 5 V. Electron current flows counterclockwise.

3. 15 V. Electron current flows counterclockwise.

1. If three voltage sources having voltages of 5 V, 25 V, and 50 V are connected in a series-aiding configuration, what is the effective circuit voltage?

2. What is the net applied voltage in the circuit shown in Figure 4-69?

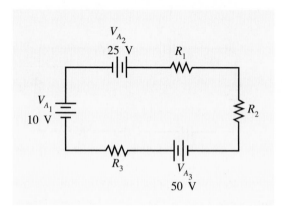

Figure 4-69. Find the net applied voltage in this circuit.

3. Refer to Figure 4-69. Which way does current flow in the circuit?

4. If a 12-V battery and a 24-V battery were connected in a series-opposing configuration, what would be the value of the net circuit voltage? Describe the polarity of the effective voltage.

5. Determine the effective circuit voltage in the circuit shown in Figure 4-70.

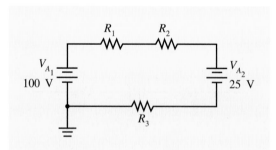

Figure 4-70. What is the effective circuit voltage?

6. In which direction does current flow in the circuit shown in Figure 4-70?

7. Classify the voltage sources in Figure 4-70 (series-aiding, series-opposing, or complex).

8. Classify the voltage sources in Figure 4-71 (series-aiding, series-opposing, or complex).

9. Calculate the effective voltage in the circuit shown in Figure 4-71.

10. In which direction does current flow in Figure 4-71?

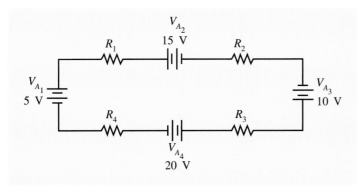

Figure 4-71. A multiple-voltage-source circuit.

4.7 Applied Technology: Voltage Dividers

Several resistors can be connected in series to form a resistive **voltage divider.** The purpose of a voltage divider is to divide (or reduce) a source voltage into one or more lower voltages for use by another circuit or device. For example, suppose you have a transistor radio that requires 9 V for proper operation. You can construct a voltage divider circuit that will allow the 9-V radio to operate from a higher voltage (e.g., a 12-V car battery).

Voltage Divider Analysis

Figure 4-72 shows the schematic diagram of a resistive voltage divider that divides a 12-V source into three lower voltages (+2 V, +5 V, and +9 V). This circuit is more completely described as an **unloaded voltage divider.** An unloaded voltage divider is strictly a series circuit; all current flow is within the series circuit. In Chapter 6, we will learn to analyze **loaded voltage dividers.** These are similar to the unloaded voltage dividers discussed in this section, but they are capable of supplying current to other circuits or devices.

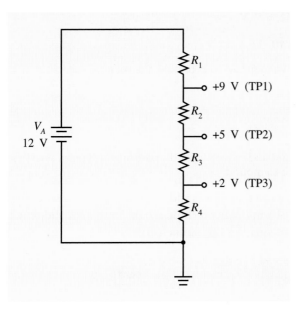

Figure 4-72.
An unloaded voltage divider with three reduced voltage taps.

KEY POINTS

Since an unloaded volt-
age divider is really just a
series circuit, all of the
analytical techniques
associated with series circuits can also be used to
analyze unloaded voltage
dividers.

One of the first things you should understand about unloaded voltage divider circuits
is that they are identical to the series circuits previously studied. Unloaded voltage
dividers are not a new circuit configuration. No new techniques are required to analyze
them.

EXAMPLE SOLUTION

Analyze the unloaded voltage divider circuit
shown in Figure 4-73 to determine the
following:

- voltage that is available at each tap
  (TP1 and TP2)
- total current flow

EXAMPLE **SOLUTION**

First, we might choose to compute total
resistance with Equation 4-2:

$$R_T = R_1 + R_2 + R_3$$
$$= 700 \ \Omega + 300 \ \Omega + 1{,}500 \ \Omega$$
$$= 2.5 \ k\Omega$$

Next, we might want to compute total
current with Ohm's Law:

$$I_T = \frac{V_T}{R_T}$$
$$= \frac{25 \ V}{2.5 \ k\Omega} = 10 \ mA$$

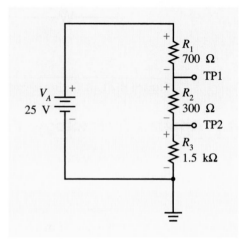

Figure 4-73. Analyze this voltage divider
circuit.

We can now compute the voltage drops across each of the series resistors by applying Ohm's Law:

$$V_{R_1} = I_{R_1}R_1 = 10 \ mA \times 700 \ \Omega = 7 \ V$$
$$V_{R_2} = I_{R_2}R_2 = 10 \ mA \times 300 \ \Omega = 3 \ V$$
$$V_{R_3} = I_{R_3}R_3 = 10 \ mA \times 1.5 \times 10^3 \ \Omega = 15 \ V$$

Finally, we can apply Kirchhoff's Voltage Law to determine the voltage available at TP1 and
TP2 as follows:

$$V_{TP1} = +15 \ V + 3 \ V = 18 \ V$$
$$V_{TP2} = +15 \ V$$

In both cases, we started at ground and moved toward a particular test point since both voltages (V_{TP1} and V_{TP2}) are referenced to ground.

Although it is important to understand that Ohm's and Kirchhoff's Laws can be applied to
solve unloaded voltage divider problems, many technicians prefer to use a shortcut that is
generally called the voltage divider equation (Equation 4-8):

$$V_X = \left(\frac{R_X}{R_T}\right)V_T \qquad\qquad (4\text{-}8)$$

where R_X is the resistance (one or more resistors) across which the voltage drop (V_X) is to be computed, R_T is the total series resistance, and V_T is the total applied voltage.

EXAMPLE SOLUTION

Apply the voltage divider equation to compute the voltages available at TP1 and TP2 in Figure 4-73.

EXAMPLE **SOLUTION**

First, we compute total resistance (Equation 4-2):

$$R_T = R_1 + R_2 + R_3$$
$$= 700\ \Omega + 300\ \Omega + 1{,}500\ \Omega$$
$$= 2.5\ k\Omega$$

By inspection of the circuit in Figure 4-73, we can see that the voltage felt at TP1 is a combination of the voltages across R_2 and R_3. Thus, the total resistance of R_2 and R_3 will be used as R_X, and V_{TP1} will be used as V_X in Equation 4-8:

$$V_{TP1} = \left(\frac{R_2 + R_3}{R_T}\right) V_A$$

$$= \left(\frac{300\ \Omega + 1{,}500\ \Omega}{2{,}500\ \Omega}\right) 25\ V = 18\ V$$

The voltage available at TP2 is simply the voltage drop across R_3:

$$V_{TP2} = \left(\frac{R_3}{R_T}\right) V_A$$

$$= \left(\frac{1{,}500\ \Omega}{2{,}500\ \Omega}\right) 25\ V = 15\ V$$

Practice Problems

1. Use the voltage divider equation to compute the voltage at TP1 in Figure 4-74.

2. Compute the voltage at TP2 in Figure 4-74 by applying the voltage divider equation.

3. Apply the voltage divider equation to determine the voltage at TP3 in Figure 4-74.

4. Compute the voltage across R_3 in Figure 4-74 by applying the voltage divider equation.

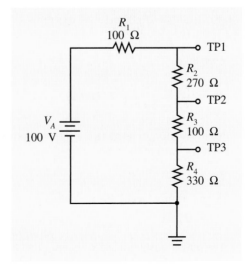

Figure 4-74. An unloaded voltage divider.

Answers to Practice Problems

1. 87.5 V   **2.** 53.75 V   **3.** 41.25 V   **4.** 12.5 V

Voltage Divider Design

You can design an unloaded voltage divider by applying Ohm's Law, Kirchhoff's Voltage Law, power equations, and basic series circuit theory. The application normally will dictate the following circuit values:

- applied voltage
- required voltage divider taps (divided voltages)
- allowable supply current

This leaves the values of individual resistors for computation by the technician.

EXAMPLE SOLUTION

Design an unloaded voltage divider that will provide voltages of +10 V and +5 V when powered from a +20-V source. The maximum allowable supply current is 50 mA.

EXAMPLE SOLUTION

Figure 4-75 shows a schematic representation of the problem. We shall use the stated maximum value of 50 mA for this example, but you may choose to use a lower value of current if desired. To compute the values of resistors R_1, R_2, and R_3, we shall apply Ohm's Law as follows:

$$R_1 = \frac{V_1}{I_1} = \frac{20\ V - 10\ V}{50\ mA} = 200\ \Omega$$

Similarly,

$$R_2 = \frac{V_2}{I_2} = \frac{10\ V - 5\ V}{50\ mA} = 100\ \Omega$$

And, finally,

$$R_3 = \frac{V_3}{I_3} = \frac{5\ V}{50\ mA} = 100\ \Omega$$

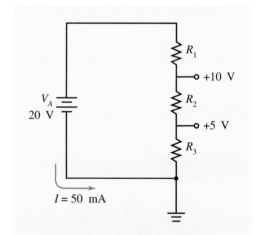

Figure 4-75. A voltage divider design problem.

Next, we must determine the required power rating for each of the resistors. We can use any convenient form of the power equations.

$$P_1 = I_1 V_1$$
$$= 50\ mA \times (20\ V - 10\ V)$$
$$= 0.5\ W$$

The power dissipation in R_2 is computed as follows:

$$P_2 = I_2 V_2$$
$$= 50 \text{ mA} \times 5 \text{ V} = 0.25 \text{ W}$$

Since R_3 has the same current and is the same resistance value as R_2, we know it will dissipate the same amount of power. The results of our design are shown in Figure 4-76. In a design for a real application, we would normally use resistors with power ratings about twice the calculated value.

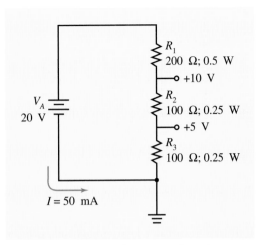

Figure 4-76. Final schematic of a voltage divider design.

Exercise Problems 4.7

1. Refer to the circuit shown in Figure 4-77. Calculate the voltages that are present at testpoints TP1, TP2, and TP3.

2. Determine the total current flow in the circuit shown in Figure 4-77.

3. How much power is dissipated by R_2 in the voltage divider circuit shown in Figure 4-77?

4. Design an unloaded voltage divider that will provide +7 V and +12 V when powered from a +30-V source. Design for a supply current of 10 mA.

5. Design an unloaded voltage divider that will provide –20 V and –40 V when powered from a –100-V source. Design for a supply current of 25 mA.

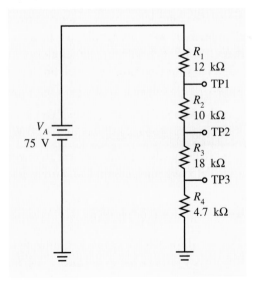

Figure 4-77. Calculate the voltage divider outputs.

6. Design an unloaded voltage divider that will provide +15 V and –15 V when powered from a 30-V power source. Design for a supply current of 100 mA. (Hint: Do not connect ground to the 30-V source.)

4.8 Troubleshooting Series Circuits

One of the most important and fundamental skills of an electronics technician is the ability to diagnose and repair a defective electronic circuit. Whether you work as a field service technician, an engineering technician, a manufacturing technician, or one of endless other technician positions, troubleshooting will very likely be a central part of your job. This is an area where you will really want to excel.

Basic Concepts

The ability to locate defects in a series circuit is not only important so that you can trouble-shoot series circuits, but the techniques also provide the foundation for troubleshooting more complex circuit configurations. It is important that you practice only logical trouble-shooting procedures because these same procedures will help you diagnose more advanced circuits and systems in the future. When you are first learning to troubleshoot, the circuits are comparatively simple (2 to 10 resistors). Even if you make wild guesses or test every component, you will fix the problem fairly fast. But beware! If you let yourself practice poor habits such as guessing or making excessive measurements now, then you will probably be unable to effectively diagnose more complex systems. In a practical electronic system, there may be thousands of possible defects. In cases like this, it is unlikely that you can guess where the actual defect is located, and it is physically impractical to measure all of the components.

If you develop a logical and systematic troubleshooting technique early in your training, then not only will you be able to quickly repair simple circuits, but you will also have the skills needed to diagnose complex electronics systems. You will learn more specific tech-niques in your laboratory manual, but for now, you should have at least three definite goals as you troubleshoot:

- Make a measurement only if you know what a "good" reading would be.
- Make as few measurements as possible.
- Rely mostly on voltage measurements. Use ohmmeter checks as the final proof of a defective component. Use current measurements only in very special cases.

Know What Is Normal

KEY POINTS

Make a measurement *only* if you know what a "good" reading would be.

There is absolutely no sense in measuring the voltage at some point in a circuit if you do not already know what constitutes a "normal" value. Each measurement you take should get you closer to the problem, but if you cannot classify a measurement as "good" or "bad," then it provides no useful information.

Minimize the Number of Measurements

KEY POINTS

Make as few measure-ments as possible.

You should make as few measurements as possible. When you are in the laboratory, most of your components are easily accessible and readily measured. In a real situation, however, it often takes substantial effort just to locate a component that you want to measure. In many cases, portions of the system have to be disassembled to gain access to certain components. So, you should practice making few, but smart measurements.

USE THE BEST TOOL

A voltmeter is your most powerful diagnostic tool for the type of circuits presented in this chapter. You can measure any component without disturbing the circuit (except for possible voltmeter loading, which is discussed in Chapter 6), and it is not necessary to turn off power. When troubleshooting an actual electronic system, opening a circuit for a current measurement or lifting a component for a resistance test requires desoldering components on a delicate (and expensive) printed circuit board. It should be avoided whenever possible. Once you have identified what you think is the defective component, then you can desolder it, verify your assertion by measuring the resistance of the suspected component, and then replace it with a new component.

Intuitive Troubleshooting

The first step toward becoming a skilled troubleshooter of series circuits is to develop a solid intuitive (i.e., nonmathematical) understanding of circuit behavior. For this, you must learn to make mental approximations, and be able to classify circuit quantities as high, low, or normal. Let's begin by developing a list of basic rules that will form the basis for all of our troubleshooting efforts. These rules should not only be committed to memory, they must be understood intuitively. You are encouraged to first memorize the rules, and then go on to understand why they are valid and be able to visualize them in your mind as you troubleshoot a circuit. We shall consider the effects of open components, shorted components, and components that have changed values. In general, opens are more common than shorts. When a short occurs, however, it can cause extensive damage to other components in the circuit due to the resulting increase in current flow.

EFFECTS OF AN OPEN COMPONENT

Figure 4-78 shows a series resistive circuit that has an open in one of the resistors. If there is an open at any point in a series circuit, then the current will be zero at all points in the circuit. With no current flow, there can be no voltage drop across the various resistors except for the one that is open. The open resistor will have the total applied voltage across it.

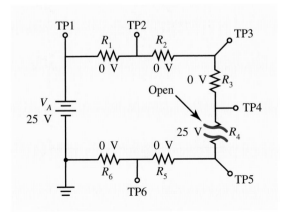

Figure 4-78. An open in a series circuit will have the applied voltage across it.

When troubleshooting circuits with a voltmeter, it is common practice to connect one side of the voltmeter to ground. The various testpoints are then probed with the remaining voltmeter lead. The following rule tells us how to interpret the measured values:

> If the open circuit appears between a monitored point and ground, then the meter will indicate the full supply voltage, otherwise the meter will read zero.

We can further refine this rule by considering the direction of increase or decrease in voltage at the measured point as a result of the open circuit:

> When a circuit develops an open, all voltages in the circuit will measure the same as the supply voltage terminal on the same side of the open.

EXAMPLE SOLUTION

Refer to the circuit shown in Figure 4-78. What voltage will be measured at each of the indicated testpoints (TP1 to TP6) with respect to ground?

EXAMPLE **SOLUTION**

Since testpoints TP1 through TP4 are on the same side of the open as the positive supply terminal, they will all read +25 V. Testpoints TP5 and TP6 are on the same side of the open as the negative terminal of the voltage source. Therefore, they will all read 0 V, because the negative side of the voltage source is connected directly to ground (0 V with respect to ground).

EXAMPLE SOLUTION

Refer to Figure 4-79. Determine the voltage that would be measured at each testpoint with respect to ground.

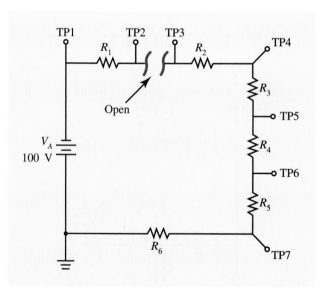

Figure 4-79. Determine the voltage at every testpoint.

Since testpoints TP1 and TP2 are on the same side of the open as the negative supply terminal, they will both measure –100 V. All other testpoints (TP3 to TP7) are on the side of the open that is connected to the positive supply terminal. They will all measure 0 V.

EXAMPLE SOLUTION

Refer to Figure 4-79. How much voltage would be measured between each of the following sets of testpoints?

   a. TP1 and TP2   **b.** TP2 and TP3   **c.** TP3 and TP4

   d. TP4 and TP5   **e.** TP5 and TP6   **f.** TP6 and TP7

EXAMPLE **SOLUTION**

If there is an open in a series circuit, then there will be no current flow. With no current flow, there will be no voltage drops around the circuit except for across the open. The open will measure the full supply voltage. In the case of Figure 4-79, we would measure the following:

   a. TP1 and TP2: 0 V   **b.** TP2 and TP3: 100 V   **c.** TP3 and TP4: 0 V

   d. TP4 and TP5: 0 V   **e.** TP5 and TP6: 0 V   **f.** TP6 and TP7: 0 V

Practice Problems

Refer to the circuit shown in Figure 4-80.

   1. How much voltage would be measured between testpoints TP1 and TP2?

   2. How much voltage would be measured between testpoints TP2 and TP3?

   3. How much voltage would be measured between testpoints TP3 and TP4?

   4. How much voltage would be measured between testpoints TP4 and TP5?

   5. How much voltage would be measured between testpoints TP5 and TP6?

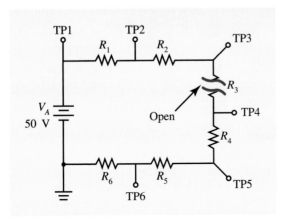

Figure 4-80. Determine the voltage at all testpoints.

Answers to Practice Problems

1. 0 V **2.** 0 V **3.** 50 V **4.** 0 V **5.** 0 V

EFFECTS OF A SHORTED COMPONENT

Figure 4-81 shows a series-resistive circuit with a shorted resistor. When a component shorts in a series circuit, the current will be higher through all components. Since the current is higher, the voltage drops across all but the shorted component will also be higher. The increased voltages are shown in Figure 4-81 as a voltage label followed by an arrow symbol (↑) pointing up. The voltage drop across the shorted component will be zero.

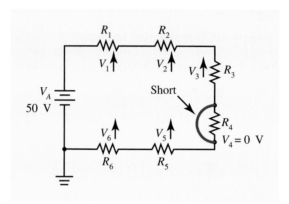

Figure 4-81. The voltage across a short circuit is zero. All other component voltage drops are higher than normal.

Considering that we have one lead of our voltmeter grounded in accordance with standard practice, we can state the following rule about shorts in series circuits:

> When a circuit develops a short, all voltages in the circuit (with respect to ground) will measure nearer to the value of the applied voltage terminal on the opposite side of the short.

EXAMPLE SOLUTION

Refer to Figure 4-82. Describe the relative changes in voltage at each of the indicated testpoints as a result of the short circuit.

EXAMPLE **SOLUTION**

Testpoints TP2 and TP3 will measure nearer to the negative side of the applied voltage. Since the negative side of the applied voltage is connected directly to ground (0 V), then we can say that the voltage at testpoints TP2 and TP3 will decrease (i.e., get closer to zero). The voltage at TP1 will not be affected by the short circuit, since TP1 connects directly to $+V_A$. By contrast, testpoints TP4 and TP5 are on the opposite side of the short and will measure nearer to the value of the positive supply terminal. We can say that the voltage at TP4 and TP5 will increase (i.e., become more positive).

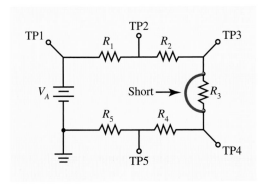

Figure 4-82. What relative voltage changes are caused by the short circuit?

EXAMPLE SOLUTION

Refer to Figure 4-83. Describe the relative changes in voltage at each of the indicated testpoints as a result of the short circuit.

EXAMPLE SOLUTION

Testpoint TP2 will measure nearer to the positive supply terminal. Since the positive supply terminal is grounded, we can say that TP2 become less negative (i.e., closer to zero) as a result of the short circuit. The voltage at TP1 will be unaffected by the short circuit, because TP1 is connected directly to $-V_A$. Testpoints TP3 to TP5 will measure closer to the value of the negative supply terminal. We can say that TP3 to TP5 become more negative as a result of the short circuit.

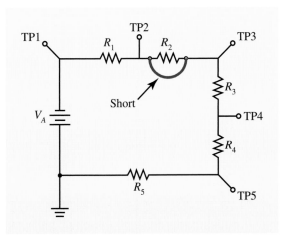

Figure 4-83. What relative voltage changes are caused by the short circuit?

EXAMPLE SOLUTION

Refer to Figure 4-83. Determine the relative direction of change caused by the short circuit for each of the voltage measurements that are taken between two testpoints.

   a. TP1 and TP2   **b.** TP2 and TP3   **c.** TP3 and TP4   **d.** TP4 and TP5

EXAMPLE SOLUTION

The short circuit will cause an increased current through all circuit components except R_2. This means that resistors R_1, R_3, R_4, and R_5 will have an increased voltage drop. R_2 will have no voltage drop since it is shorted. These observations allow us to determine the following testpoint measurements:

   a. TP1 and TP2: increased voltage   **b.** TP2 and TP3: 0 V

   c. TP3 and TP4: increased voltage   **d.** TP4 and TP5: increased voltage

Practice Problems

Refer to Figure 4-84 for questions 1 to 6.

1. Describe the relative changes in voltage at each of the indicated testpoints as a result of the short circuit.

2. Determine the relative direction of change caused by the short circuit for the voltage measurement taken between testpoints TP1 and TP2.

3. Determine the relative direction of change caused by the short circuit for the voltage measurement taken between testpoints TP2 and TP3.

4. Determine the relative direction of change caused by the short circuit for the voltage measurement taken between testpoints TP3 and TP4.

5. Determine the relative direction of change caused by the short circuit for the voltage measurement taken between testpoints TP4 and TP5.

6. Determine the relative direction of change caused by the short circuit for the voltage measurement taken between testpoints TP5 and TP6.

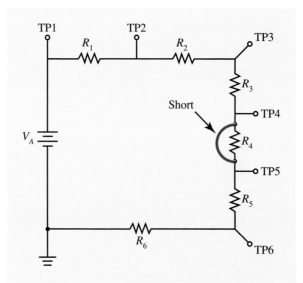

Figure 4-84. What relative voltage changes are caused by the short circuit?

Answers to Practice Problems

1. TP1 constant. TP2 decreases. TP3 decreases. TP4 decreases. TP5 increases. TP6 increases.

2. Increase

3. Increase

4. Increase

5. Decrease (0 V)

6. Increase

WRONG VALUE

In the preceding paragraphs, we considered the effects of opened and shorted components in a series circuit. By definition, an open component has a resistance that approaches infinity while a shorted component has a resistance that approaches zero. We shall now consider the effects when a component changes values (perhaps due to aging or overheating).

The rules are simple. If a component has increased in resistance, then it causes effects similar to those an open component would exhibit, but the symptoms are not as extreme. The voltage across the increased-value component will increase. All points in the circuit will measure nearer to the value of the applied voltage on the same side of the defect.

A component that has decreased in value will cause symptoms similar to a shorted component, although the voltage changes will not be as extreme. Thus, the basic rules for locating an incorrectly valued component are identical to the rules for opened and shorted components. These rules for series circuits are summarized in Table 4–3.

TYPE OF DEFECT	VOLTAGE		
	WITH RESPECT TO GROUND	ACROSS DEFECT	ACROSS OTHER COMPONENTS
Open	Nearer or equal to supply terminal on same side of defect	Full supply	Zero
Increased resistance		Increased	Decreased
Shorted	Nearer or equal to supply terminal on opposite side of defect	Zero	Increased
Decreased resistance		Decreased	Increased

Table 4-3. Summary of Troubleshooting Rules for Series Circuits

If your test equipment is connected directly across a voltage source when measuring a particular testpoint, then that voltage will normally be unaffected by either shorts or opens unless the short circuit damages the voltage source. Every time you take a measurement as you troubleshoot a series circuit, you should mentally cross-reference the measured voltage to an entry in Table 4-3. With only a very few measurements, you will have isolated the defect.

Troubleshooting Practice

Since troubleshooting is such an important skill for electronics technicians, this book is designed to provide you with many opportunities for troubleshooting practice. The

KEY POINTS

When a resistor increases or decreases in value, the symptoms are similar to those caused by an opened or shorted circuit, respectively, but not as extreme.

KEY POINTS

Opened or increased-value components cause the voltages at all points in the circuit to measure nearer to the value of the supply terminal that is on the same side of the open.

KEY POINTS

Shorted or decreased-value components cause the voltages at all points in the circuit to measure nearer to the value of the supply terminal on the opposite side of the short.

method used is called *paper troubleshooting* or *PShooting*. Be sure you understand how the procedure works, so you can maximize your learning.

Each PShooter circuit has an associated table that lists the various testpoints in the circuit along with the normal value of voltage and resistance. The table also provides a bracketed [ ] number for each measurable entry. The bracketed number identifies a specific entry in the PShooter lookup table provided in Appendix A, which gives you the actual measured value of the circuit quantity. The following list will help you interpret the measured values:

- All testpoint (TPxx) measurements are made with reference to ground.
- All nontestpoint values are measured directly across the component.
- Resistance values for all testpoints (TPxx) are measured with the applied voltage disconnected. All other components remain connected.
- Resistance tests for all nontestpoint measurements are made with the specified component removed from the circuit.

EXAMPLE SOLUTION

Troubleshoot the circuit shown in Figure 4-85. Table 4-4 provides the measurement data.

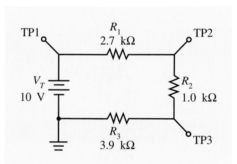

Figure 4-85. Troubleshoot this circuit.

EXAMPLE SOLUTION

The laboratory manual provides additional details regarding a systematic troubleshooting procedure, but it is usually good practice to start with a measurement near the center of the circuit. Suppose we decide to measure the voltage at TP2. Table 4-4 lists [146] as the reference number to the actual value. Appendix A tells us that this corresponds to a voltage of 10 V. Since the normal voltage for TP2 is listed as 6.45 V in Table 4-4, we know that the voltage at TP2 is higher than normal. By applying our rules for troubleshooting series circuits, we can infer that one of the following conditions exists:

- There is an open circuit between TP2 and the negative side of the voltage source.
- There is a short circuit between TP2 and the positive side of the voltage source.

Measuring the voltage across R_2 will provide us with valuable information about the location and nature of the defect. Table 4-4 lists a normal voltage of 1.32 V for R_2. However, the reference number [22] in Appendix A lists the actual voltage for R_2 as 0 V. By applying our series circuit troubleshooting rules and some logical thinking, we can make the following conclusions:

- The nature of the defect is an open.
- R_2 is not the open component (unless the circuit has multiple defects).

TESTPOINT	VOLTAGE		RESISTANCE	
	NORMAL	ACTUAL*	NORMAL	ACTUAL*
TP1	10 V	[35]	7.6 kΩ	[80]
TP2	6.45 V	[146]	4.9 kΩ	[99]
TP3	5.13 V	[208]	3.9 kΩ	[138]
R_1	3.55 V	[67]	2.7 kΩ	[91]
R_2	1.32 V	[22]	1.0 kΩ	[43]
R_3	5.13 V	[104]	3.9 kΩ	[5]
V_T	10 V	[49]	—	—

* Numbers in brackets refer to value numbers in Appendix A.

Table 4-4. PShooting Data for the Circuit in Figure 4–85

Since we already know that the open lies between TP2 and the negative side of the voltage source, and since R_2 is not the defective component, we know that the defect must be R_3 open or an open interconnecting wire. Let's measure the resistance of R_3 to confirm our suspicions.

Table 4-4 lists the normal value for R_3 as 3.9 kilohms, but the reference number [5] in Appendix A lists the actual value as infinity (i.e., open). So, R_3 is the defective component.

Exercise Problems 4.8

1. Refer to the circuit shown in Figure 4-86. If TP2 measures 0 V, which of the following are possible defects (circle all that apply):

   a. R_1 shorted   **b.** R_1 opened   **c.** V_T is zero

   d. R_3 opened   **e.** R_2 decreased in resistance

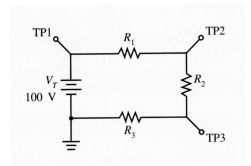

Figure 4-86. A defective series circuit.

2. Refer to the circuit shown in Figure 4-86. If TP2 measures 100 V, which of the following are possible defects (circle all that apply):

   a. R_1 shorted   **b.** R_1 opened   **c.** V_A is zero

   d. R_3 opened   **e.** R_2 decreased in resistance

3. Refer to the circuit shown in Figure 4-86. If the voltage at TP3 is more positive than normal, which of the following are possible causes (circle all that apply):

   a. R_2 opened   **b.** R_1 shorted

   c. R_3 increased in value   **d.** R_2 increased in value

4. Troubleshoot the circuit shown in Figure 4-87 using the PShooter data provided in Table 4-5.

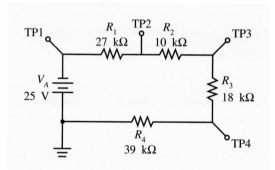

Figure 4-87. Troubleshoot this circuit.

TESTPOINT	VOLTAGE		RESISTANCE	
	NORMAL	ACTUAL*	NORMAL	ACTUAL*
TP1	−25 V	[95]	94 kΩ	[205]
TP2	−17.8 V	[110]	67 kΩ	[107]
TP3	−15.1 V	[120]	57 kΩ	[8]
TP4	−10.4 V	[76]	39 kΩ	[149]
R_1	7.18 V	[220]	27 kΩ	[45]
R_2	2.66 V	[83]	10 kΩ	[15]
R_3	4.79 V	[141]	18 kΩ	[67]
R_4	10.4 V	[102]	39 kΩ	[86]
V_A	25 V	[127]	—	—

* Numbers in brackets refer to value numbers in Appendix A.

Table 4-5. PShooting Data for the Circuit in Figure 4-87

5. Troubleshoot the circuit shown in Figure 4-87 using the PShooter data provided in Table 4-6.

TESTPOINT	VOLTAGE		RESISTANCE	
	NORMAL	ACTUAL*	NORMAL	ACTUAL*
TP1	−25 V	[38]	94 kΩ	[113]
TP2	−17.8 V	[41]	67 kΩ	[131]
TP3	−15.1 V	[88]	57 kΩ	[93]
TP4	−10.4 V	[151]	39 kΩ	[73]
R_1	7.18 V	[51]	27 kΩ	[58]
R_2	2.66 V	[27]	10 kΩ	[24]
R_3	4.79 V	[144]	18 kΩ	[135]
R_4	10.4 V	[188]	39 kΩ	[89]
V_A	25 V	[70]	—	—

* Numbers in brackets refer to value numbers in Appendix A.

Table 4-6. PShooting Data for the Circuit in Figure 4-87

Chapter Summary

• A series circuit is characterized by having one and only one path for current flow. The current is the same through every component in every part of the circuit. Each resistor in a series circuit drops a portion of the applied voltage. The portion of voltage dropped is in direct proportion to the resistance of the resistor as compared to the total resistance in the circuit. Total resistance in a series circuit is simply the sum of the individual resistances. Kirchhoff's Voltage Law states that the algebraic sum of the voltages (drops and sources) in any closed loop is equal to zero. This means that the sum of the voltage drops across the circuit resistance must always be equal to the value of applied voltage.

• Each resistor in a series circuit dissipates a certain amount of power, which is supplied by the applied voltage source. Total power consumption from the voltage source is equal to the sum of the individual resistor power dissipations.

• Fully specified series circuits can generally be analyzed by direct substitution of component values into an applicable equation. Analysis of partially specified circuits requires creative application of Ohm's Law, Kirchhoff's Voltage Law, power equations, and basic series circuit theory. In order to compute a circuit quantity, you must select an equation with only one unknown value. It is essential that properly subscripted values be used in all calculations to avoid accidental mixing of quantity types.

• Ground is a point in the circuit from which voltage measurements are referenced. Ground generally takes the physical form of a metal chassis, a wide PCB trace, or an entire plane in a multilayer PCB.

- A series circuit may have more than one voltage source. If the multiple voltage sources all have the same polarity, the sources are said to be series-aiding. The effective voltage of series-aiding sources is the sum of the individual sources. When two voltage sources are connected with opposite polarities, they are said to be series-opposing. The net voltage of two series-opposing voltage sources is the difference between the individual source voltages. The polarity of the resulting voltage is the same as the larger of the two individual sources. If a series circuit has both series-aiding and series-opposing voltage sources, then the circuit has a complex voltage source. The effective voltage of a complex voltage source is equal to the algebraic sum of the individual source voltages.

- An unloaded voltage divider is an application of a series circuit. Voltage dividers are used to provide one or more lower voltages from a single higher source voltage. Unloaded voltage dividers cannot supply current to other circuits or devices. Because they are series circuits, unloaded voltage dividers can be analyzed and designed using the same equations and techniques used with other series circuits.

- Logical, systematic troubleshooting of series circuits is an important technician skill since it provides the foundation for troubleshooting of more complex circuits. The possible defects in a series circuit can be loosely categorized into two classes: opened (or increased resistance) and shorted (decreased resistance). When a resistor increases in value in a series circuit, all voltages in the circuit will measure nearer to the value of the supply terminal on the same side of the open. When a resistor decreases in value, all voltages in the circuit will measure nearer to the value of the supply terminal on the opposite side of the defect.

Review Questions

Section 4.1: Current Through Each Component

1. Current is the same in all parts of a series circuit. (True or False)
2. If a certain series circuit has 1.2 A leaving the voltage source, what percentage of this current will flow through the first resistor in the series chain? The last resistor in the series chain?

Section 4.2: Voltage Across Each Resistor

3. The voltage across a resistor is (*directly, indirectly*) proportional to its resistance value.
4. If one resistor in a series circuit is increased, what happens to the amount of voltage dropped across it?
5. If one resistor in a series circuit is increased, what happens to the amount of voltage dropped across the remaining resistors?
6. The largest resistor in a series circuit will always drop the (*least, most*) voltage.
7. The (*smallest, largest*) resistor in a series circuit will drop the least voltage.
8. Equal-valued resistors in a series circuit will have equal voltage drops. Explain why this is true.

Section 4.3: Kirchhoff's Voltage Law

9. Electron current exits the (*positive, negative*) end of a resistor.

10. Electron current returns to the (*positive, negative*) side of a voltage source.

11. The algebraic sum of the voltage drops and voltage sources in a closed loop is always equal to _____.

12. What is the relative magnitude of the source voltage in a series circuit as compared to the sum of the resistor voltage drops?

13. Write the closed-loop equation to describe the circuit shown in Figure 4-88.

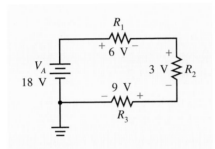

Figure 4-88. Circuit for Problem 13.

14. Use Kirchhoff's Voltage Law to determine whether the voltages shown on the circuit in Figure 4-89 are correct.

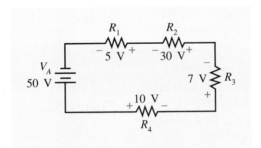

Figure 4-89. Circuit for Problem 14.

15. Label the polarities of the voltage drops in Figure 4-90.

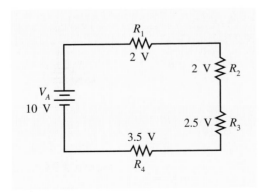

Figure 4-90. Circuit for Problem 15.

16. Use Kirchhoff's Voltage Law to determine whether the voltages shown on the circuit in Figure 4-91 are correct.

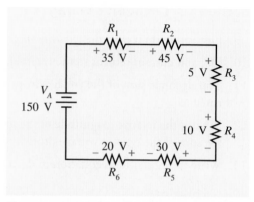

Figure 4-91. Circuit for Problem 16.

Section 4.4: Computing Voltage, Current, Resistance, and Power

17. A certain series circuit has the following resistors: 10 kΩ, 4.7 kΩ, 5.6 kΩ, and 8.2 kΩ. Calculate the value of total circuit resistance.

18. If a series circuit is composed of three 27-kΩ resistors, what is the total circuit resistance?

19. How much current flows in the circuit shown in Figure 4-92?

20. A certain series circuit has three 10-kΩ resistors. Each resistor has 325 mA of current flowing through it. What is the value of total current through the voltage source?

21. A circuit has a 5-kΩ resistor and a 10-kΩ resistor in series. If the 5-kΩ resistor has 100 μA of current flow, how much current will flow through the 10-kΩ resistor?

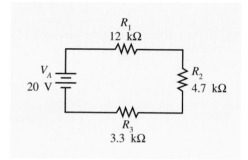

Figure 4-92. Circuit for Problem 19.

22. The resistors in a particular series circuit dissipate 1.2 W, 3.5 W, and 2.0 W each. How much power must be supplied by the voltage source?

23. Compute the total power in the circuit shown in Figure 4-93.

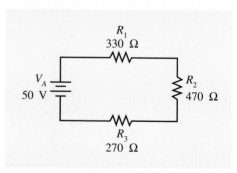

Figure 4-93. Circuit for Problem 23.

24. Refer to Figure 4-94 and complete the solution matrix shown in Table 4-7.

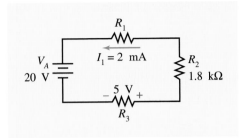

Figure 4-94. Circuit for Problem 24.

COMPONENT	RESISTANCE	VOLTAGE	CURRENT	POWER
R_1			2 mA	
R_2	1.8 kΩ			
R_3		5 V		
Total		20 V		

Table 4-7. Solution Matrix for the Circuit in Figure 4-94

25. Refer to Figure 4-95 and complete the solution matrix shown in Table 4-8.

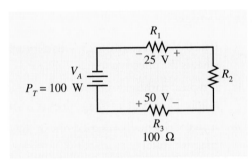

Figure 4-95. Circuit for Problem 25.

COMPONENT	RESISTANCE	VOLTAGE	CURRENT	POWER
R_1		25 V		
R_2				
R_3	100 Ω	50 V		
Total				100 W

Table 4-8. Solution Matrix for the Circuit in Figure 4-95

Section 4.5: Ground and Other Reference Points

26. Ground is the reference point from which other voltages are measured. (True or False)

27. If $V_A = +2$ V and $V_B = -4$ V, what is the value of V_{AB}?

28. If $V_{ED} = 10$ V and $V_E = 50$ V, what is the value of V_D?

29. If $V_{BC} = -6$ V and $V_C = +20$ V, what is the value of V_B?

Refer to Figure 4-96 for Problems 30 through 35.

30. What is the value of V_B?

31. What is the value of V_F?

32. Determine the voltage at point B with reference to point C.

33. Determine the value of V_{BE}.

34. What is the value of V_{FA}?

35. Determine the value of V_{EB}.

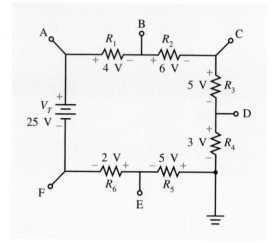

Figure 4-96. Circuit for Problems 30 through 35.

Section 4.6: Multiple Voltage Sources

36. If a 10-V battery and a 25-V battery are connected in a series-opposing configuration, what is the effective source voltage?

37. A typical flashlight uses two 1.5-V cells (D cells) connected in a series-aided configuration. How much voltage does this arrangement provide for the lamp?

38. If you accidentally reversed one of the cells in a two-cell flashlight, what would be the effective voltage?

39. A certain electronic game requires four D cells (1.5 V each) to be connected as series-aiding voltage sources. If one of the cells is accidentally reversed, what will be the effective voltage available to the game circuitry?

40. Suppose you have three 28-V batteries and you need to obtain the highest possible voltage. How would you connect the batteries?

41. Determine the effective voltage for the circuit shown in Figure 4-97.

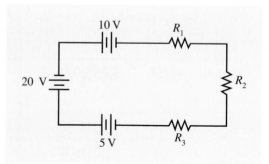

Figure 4-97. What is the effective voltage in this circuit?

42. Determine the effective voltage for the circuit shown in Figure 4-98.

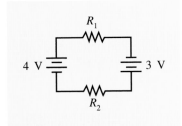

Figure 4-98. What is the effective voltage in this circuit?

Section 4.7: Applied Technology: Voltage Dividers

43. Calculate the current in the circuit shown in Figure 4-99.

44. What is the voltage at TP2 in Figure 4-99?

45. How much voltage is available at TP1 in Figure 4-99?

46. Design a voltage divider similar to the circuit shown in Figure 4-99, but with the following design requirements:

- Supply voltage will be 18 V.
- Supply current will be 10 mA.
- TP1 must supply +12 V.
- TP2 must supply +8 V.

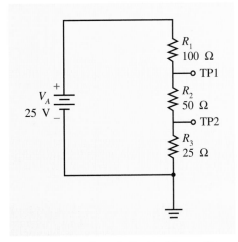

Figure 4-99. Circuit for Problems 43 through 45.

Section 4.8: Troubleshooting Series Circuits

47. Explain why it makes no sense to measure the voltage at a point while troubleshooting, unless you already know what a normal voltage would be.

48. Explain why a technician should strive to make the fewest measurements possible while troubleshooting a circuit.

49. We know that the value of current in a series circuit is an important quantity. Why, then, does an experienced technician normally avoid measuring current when troubleshooting electronic systems?

Refer to Figure 4-100 for Problems 50 through 55.

50. What happens to the value of V_B if R_4 increases?

51. What effect does a short across R_3 have on the voltage V_{AB}?

52. Does an open in R_2 have any effect on the voltage V_A? Explain your response.

53. What happens to the value of V_{CD} if resistor R_1 increases in value?

54. What happens to the value of V_{CD} if resistor R_4 increases in value?

55. If resistor R_3 decreases in resistance, what happens to the voltage V_{DA}?

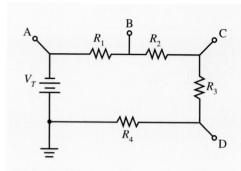

Figure 4-100. Circuit for Problems 50 through 55.

56. Troubleshoot the circuit shown in Figure 4-101 by using the PShooter data provided in Table 4-9.

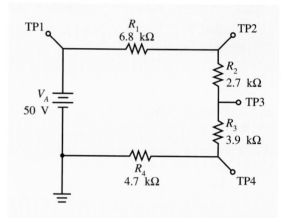

Figure 4-101. Troubleshoot this circuit.

	VOLTAGE		RESISTANCE	
TESTPOINT	NORMAL	ACTUAL*	NORMAL	ACTUAL*
TP1	50 V	[2]	18.1 kΩ	[31]
TP2	31.2 V	[156]	11.3 kΩ	[1]
TP3	23.8 V	[122]	8.6 kΩ	[33]
TP4	13 V	[47]	4.7 kΩ	[12]
R_1	18.8 V	[29]	6.8 kΩ	[39]
R_2	7.5 V	[18]	2.7 kΩ	[84]
R_3	10.8 V	[124]	3.9 kΩ	[55]
R_4	13 V	[60]	4.7 kΩ	[100]
V_A	50 V	[63]	—	—

* Numbers in brackets refer to value numbers in Appendix A.

Table 4-9. PShooting Data for the Circuit in Figure 4-101

57. Troubleshoot the circuit shown in Figure 4-101 by using the PShooter data provided in Table 4-10.

TESTPOINT	VOLTAGE		RESISTANCE	
	NORMAL	ACTUAL*	NORMAL	ACTUAL*
TP1	50 V	[160]	18.1 kΩ	[181]
TP2	31.2 V	[133]	11.3 kΩ	[196]
TP3	23.8 V	[22]	8.6 kΩ	[211]
TP4	13 V	[151]	4.7 kΩ	[53]
R_1	18.8 V	[72]	6.8 kΩ	[116]
R_2	7.5 V	[19]	2.7 kΩ	[36]
R_3	10.8 V	[176]	3.9 kΩ	[153]
R_4	13 V	[200]	4.7 kΩ	[65]
V_A	50 V	[111]	—	—

* Numbers in brackets refer to value numbers in Appendix A.

Table 4-10. PShooting Data for the Circuit in Figure 4-101

58. Troubleshoot the circuit shown in Figure 4-101 by using the PShooter data provided in Table 4-11.

TESTPOINT	VOLTAGE		RESISTANCE	
	NORMAL	ACTUAL*	NORMAL	ACTUAL*
TP1	50 V	[158]	18.1 kΩ	[162]
TP2	31.2 V	[90]	11.3 kΩ	[48]
TP3	23.8 V	[52]	8.6 kΩ	[77]
TP4	13 V	[147]	4.7 kΩ	[97]
R_1	18.8 V	[137]	6.8 kΩ	[105]
R_2	7.5 V	[109]	2.7 kΩ	[214]
R_3	10.8 V	[106]	3.9 kΩ	[166]
R_4	13 V	[165]	4.7 kΩ	[26]
V_A	50 V	[219]	—	—

* Numbers in brackets refer to value numbers in Appendix A.

Table 4-11. PShooting Data for the Circuit in Figure 4-101

As a computer repair technician, you are troubleshooting a graphics display monitor. You have determined that the brightness control circuit is defective. Unfortunately, the required components cannot be purchased because the manufacturer has gone out of business. The portion of the circuit that is defective is essentially an unloaded voltage divider. It uses a potentiometer as a movable tap to vary the available voltage. You know that the current through the divider circuit was originally designed to be 20 mA. You also know that the divider is supplied by a 25-V source. In order to control the monitor properly, the potentiometer must be able to adjust the tapped voltage between the limits of +5 V and +10 V.

You have decided to design your own voltage divider circuit to use in the defective monitor. Figure 4-102 shows the schematic that you plan to use. Determine the resistance required for the potentiometer and for the two fixed resistors. Be sure to specify the minimum power rating for all three resistances.

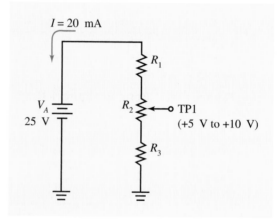

Figure 4-102. A voltage divider design problem using a potentiometer.

Equation List

(4-1) $V_A = V_1 + V_2 + V_3 + \ldots + V_N$

(4-2) $R_T = R_1 + R_2 + R_3 + \ldots + R_N$

(4-3) $I_T = I_1 = I_2 = I_3 = \ldots = I_N$

(4-4) $P_T = P_1 + P_2 + P_3 + \ldots + P_N$

(4-5) $V_T = V_{A_1} + V_{A_2} + \ldots + V_{A_N}$

(4-6) $V_T = V_{A_1} - V_{A_2}$

(4-7) $V_T = \pm V_{A_1} \pm V_{A_2} \pm \ldots \pm V_{A_N}$

(4-8) $V_X = \left(\dfrac{R_X}{R_T} \right) V_T$

objectives

After completing this chapter, you should be able to:

1. Identify a parallel circuit.

2. State the relationship of voltage across each component in a parallel circuit.

3. State the relationship between total current and the branch currents in a parallel circuit.

4. State and apply Kirchhoff's Current Law.

5. Analyze a parallel circuit to compute the value of current, voltage, resistance, and power for every component.

6. State the effect of connecting multiple voltage sources in parallel.

7. Analyze a current divider circuit.

8. Troubleshoot a parallel circuit.

Analyzing Parallel Circuits

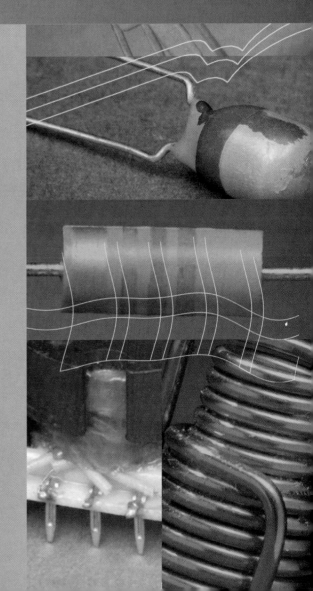

Chapter 3 introduced the basic definition of parallel circuits and provided a means for positive identification of parallel components. You will recall that two components are in parallel if both ends of both components connect directly together. If all components in a circuit are connected in parallel (also called **shunt**), then the entire circuit can be classified as a parallel circuit. Many electrical and electronic applications are essentially parallel circuits. Much of the wiring in your home or your car is configured as a parallel circuit. We are now ready to explore parallel circuits in greater detail to learn their basic characteristics, how to analyze them numerically, and how to design circuits that use them.

5.1 Voltage Drops Across Resistors

Figure 5-1 shows a basic parallel circuit with voltmeters connected across each component. Since the meters are connected directly together with low-resistance wire, they will all measure the same value of voltage. This is an important characteristic of a parallel circuit:

The voltage is the same across every component in a parallel circuit.

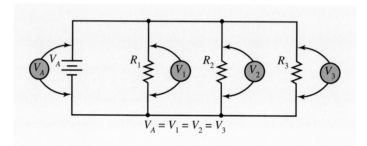

$$V_A = V_1 = V_2 = V_3$$

Figure 5-1. The voltage is the same across all parallel components.

This can be expressed mathematically with Equation 5-1:

$$V_A = V_1 = V_2 = \ldots = V_N \qquad (5\text{-}1)$$

EXAMPLE SOLUTION

Determine the voltage drop across R_1 in Figure 5-2.

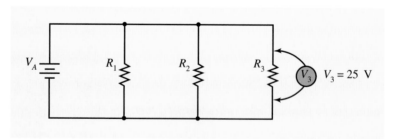

Figure 5-2. How much voltage is across R_1?

EXAMPLE **SOLUTION**

Since the voltage is the same across all components in a parallel circuit, R_1 must have the same voltage as R_3, which is listed as 25 V. We could express this mathematically as

$V_1 = V_3$

$V_1 = 25$ V

Exercise Problems 5.1

1. Determine the value of applied voltage for the circuit shown in Figure 5-3.

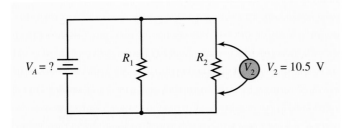

Figure 5-3. What is the value of applied voltage?

2. How much voltage is across resistor R_3 in Figure 5-4?

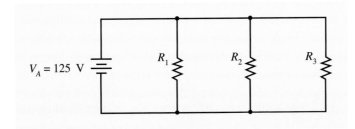

Figure 5-4. Determine the value of V_3.

3. Express the value of voltage across R_2 in Figure 5-5.

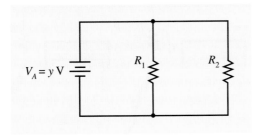

Figure 5-5. Express the value of V_2.

5.2 Current Through Branches

Figure 5-6 shows a parallel circuit with three branches across the voltage source. The current through each of the branches is indicated with arrows. As shown in Figure 5-6, total current leaves the negative terminal of the voltage source and flows toward the three resistors. At the lower end of R_1, part of the current splits off and flows upward through R_1. The rest of the current continues to the right toward R_2 and R_3. At the bottom of R_2, the current splits again with part of it flowing through R_2 and the rest through R_3. These two currents recombine at the upper end of R_2 as they flow toward R_1. Finally, the current

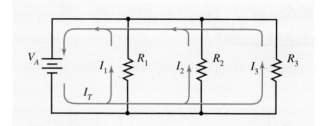

Figure 5-6. Total current divides, flows through the branches, and then recombines.

through R_1 merges with the combined currents of R_2 and R_3 at the upper end of R_1 to become the total current that flows into the positive side of the voltage source.

Branch Current Is Inversely Proportional to Resistance

The current divides among the various resistors according to their resistances. Branches with higher resistance have less current flow; those with lower resistance have proportionally more current flow. This should seem logical to you, since it is consistent with what you know about Ohm's Law. This fundamental characteristic of a parallel circuit is stated as follows and is illustrated in Figure 5-7:

> Branch current in a parallel circuit is inversely proportional to branch resistance.

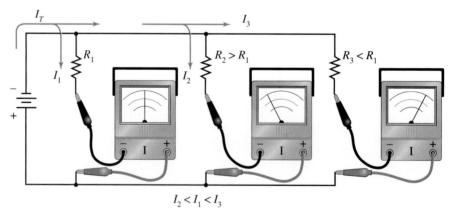

Figure 5-7. Branch current is inversely proportional to branch resistance.

Branch Currents Are Independent

The current through a particular branch is determined by the value of applied voltage and the value of the branch resistance. It is totally unaffected by the resistance of the other parallel branches. If the resistance in a branch changes (e.g., a resistor burns open), then the current in that branch will change, but the current in the remaining parallel branches will remain constant. Since the total current in the circuit is composed of the various branch currents, we would expect the total current to change if one of the branch currents changed. In the case of an open resistor, for example, the total current will be less by the amount previously drawn by the resistor that has become open.

EXAMPLE SOLUTION

Figure 5-8 shows a parallel circuit consisting of a voltage source and four parallel resistors. All of the resistors are the same value, so they each draw the same amount of current (2 A). Explain what happens to the value of total current if resistor R_3 becomes open.

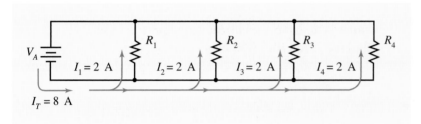

Figure 5-8. What is the effect on total current if R_3 becomes open?

EXAMPLE **SOLUTION**

If one of the resistors in a parallel circuit opens, then total current will decrease by the amount of current previously drawn by the now open resistor. The other branch currents will be unaffected. This is illustrated in Figure 5-9.

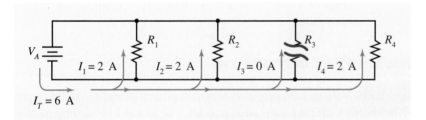

Figure 5-9. Total current decreases when a branch opens.

Exercise Problems 5.2

1. The lights in the various rooms of your home are basically connected in parallel. Explain why the light in one room is unaffected by turning off a light in another room.

2. Label the currents in Figure 5-10 to indicate the relative magnitudes. Label the highest value of current as "HIGHEST," the lowest current value as "LOWEST," and the remaining current as "BETWEEN."

3. Refer to Figure 5-11. Describe the effects on the three current meter readings as the variable resistor is adjusted. Assume the variable resistor is never allowed to be zero ohms.

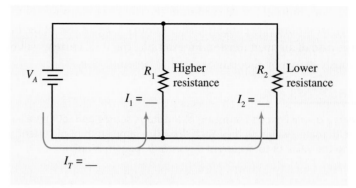

Figure 5-10. Label the current magnitudes as LOWEST, HIGHEST, and BETWEEN.

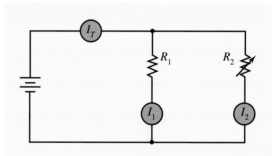

Figure 5-11. What is the effect on I_T, I_1, and I_2 as R_2 is adjusted?

5.3 Kirchhoff's Current Law

Just as Kirchhoff's Voltage Law provided us with a powerful tool for analyzing series circuits, Kirchhoff's Current Law is an important tool for use in analyzing parallel circuits. In a preceding section, we saw how the total current divided into the various branches, flowed through the branches, and then recombined into total current. Kirchhoff's Current Law essentially describes this splitting and recombination process more precisely. The following is one way to express Kirchhoff's Current Law:

> The total current entering any point in a circuit must be equal to the total current leaving at that point.

Figure 5-12 illustrates this principle through some examples. In all cases, the sum of the current entering a particular point must be equal to the sum of the current leaving that point.

Branch Currents and Source Current Relationship

Figure 5-12(c) not only illustrates Kirchhoff's Current Law, but it describes another fundamental characteristic of a parallel circuit:

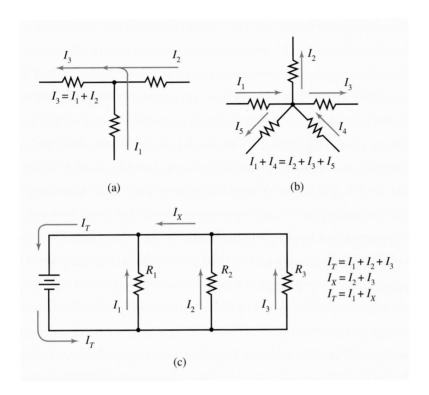

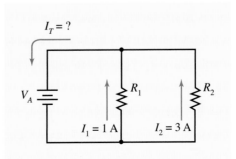

Figure 5-12. Examples of Kirchhoff's Current Law.

> The total current in a parallel circuit is equal to the sum of the branch currents.

We can express this important relationship mathematically as follows:

$$I_T = I_1 + I_2 + \ldots + I_N \qquad (5\text{-}2)$$

EXAMPLE SOLUTION

Compute the total current for Figure 5-13.

EXAMPLE SOLUTION

We apply Equation 5-2 as follows:

$$I_T = I_1 + I_2$$
$$= 1\ A + 3\ A = 4\ A$$

Figure 5-13. Find the total current in this circuit.

EXAMPLE SOLUTION

Determine the value of current that flows through R_2 in Figure 5-14.

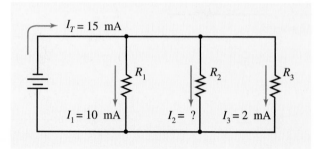

Figure 5-14. What is the value of I_2?

EXAMPLE **SOLUTION**

We apply Equation 5-2 as follows:

$$I_T = I_1 + I_2 + I_3, \text{ or}$$
$$I_2 = I_T - I_1 - I_3$$
$$= 15 \text{ mA} - 10 \text{ mA} - 2 \text{ mA}$$
$$= 3 \text{ mA}$$

Practice Problems

1. Find the value of total current in Figure 5-15.
2. What is the total current in Figure 5-16?
3. How much current flows through R_1 in Figure 5-17?
4. What is the value of I_3 in Figure 5-18?

Answers to Practice Problems

1. 21 μA 2. 65 μA 3. 80 μA 4. 40 mA

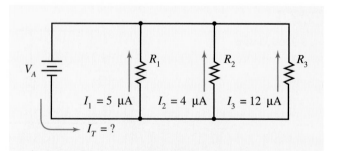

Figure 5-15. Calculate total current.

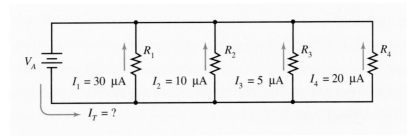

Figure 5-16. Find total current.

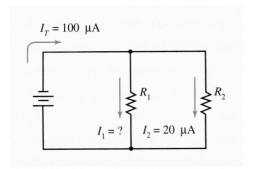

Figure 5-17. What is the value of I_1?

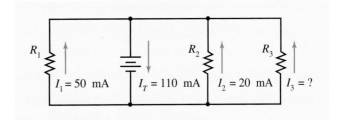

Figure 5-18. Find I_3.

Polarity of Current Flow

When we express voltages as positive or negative, the terms are meaningless without having a reference point. A point is only positive or negative with reference to some other point. It is also useful to speak in terms of current polarity, but in this case, we have no obvious reference point. Current polarity simply refers to the direction of current flow. If we arbitrarily assigned the label of "positive" to current flowing in one direction, then current flowing in the opposite direction would be labeled "negative." It is customary to label current that is entering (i.e., flowing toward) a given point as positive. Similarly, current that leaves (i.e., flows away from) that same point is considered to be negative. Having established these conventions, we can express Kirchhoff's Current Law in another way:

> The algebraic sum of the current entering and leaving a point is zero.

We can express this mathematically with Equation 5-3:

$$I_{IN_1} + I_{IN_2} + \ldots + I_{IN_N} + I_{OUT_1} + I_{OUT_2} + \ldots + I_{OUT_N} = 0 \quad (5\text{-}3)$$

where subscripts IN label current entering a point and OUT subscripts label current leaving a point. Remember that all OUT currents are assumed to be negative.

KEY POINTS

It is customary to consider currents that enter a point as positive and currents that leave a point as negative.

Writing Node Equations

For purposes of our present discussion, a circuit **node** is simply a point where two or more components connect together. We will want to express the current that enters and leaves every node in a circuit.

EXAMPLE SOLUTION

Use Kirchhoff's Current Law to write the node equation for the labeled nodes in Figure 5-19.

EXAMPLE SOLUTION

Let's begin with node A. First we assign positive values to currents entering the node (I_1 and I_X) and negative values to currents flowing away from the node (I_T). We then apply Kirchhoff's Current Law as follows:

$$\text{Node A: } I_1 + I_X - I_T = 0$$

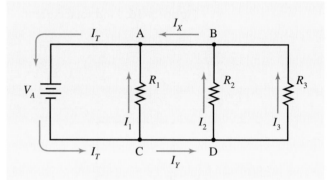

Figure 5-19. Write the node equations for this circuit.

A similar process yields the following equations for nodes B through D:

$$\text{Node B: } I_2 + I_3 - I_X = 0$$
$$\text{Node C: } I_T - I_1 - I_Y = 0$$
$$\text{Node D: } I_Y - I_2 - I_3 = 0$$

Practice Problems

1. Write the Kirchhoff's Current Law equation for each labeled node in Figure 5-20.
2. Write the Kirchhoff's Current Law equation for the labeled nodes in Figure 5-21.

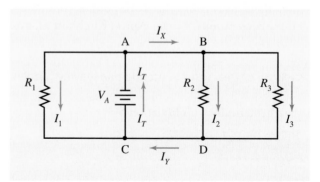

Figure 5-20. Write the node equations for this circuit.

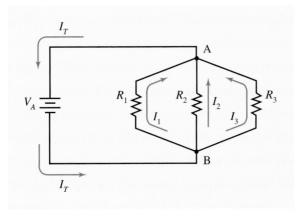

Figure 5-21. Write the node equations for this circuit.

Answers to Practice Problems

1. Node A: $I_T - I_1 - I_X = 0$
      Node B: $I_X - I_2 - I_3 = 0$
      Node C: $I_Y + I_1 - I_T = 0$
      Node D: $I_2 + I_3 - I_Y = 0$

2. Node A: $I_1 + I_2 + I_3 - I_T = 0$
      Node B: $I_T - I_1 - I_2 - I_3 = 0$

The circuit shown in Figure 5-21 is a parallel circuit that is electrically similar to the circuits shown in Figures 5-19 and 5-20 (except polarity). In Figure 5-21, however, all of the points that are connected *directly* together are considered to be a single node. That is, a node is considered to be an *electrical* point where two or more components connect together. If several components are electrically connected together at some point, then that electrical point is considered to be a single common node. This definition of a node is nearly always used in more advanced analysis. For purposes of learning and applying Kirchhoff's Current Law and to help us understand parallel circuit principles, however, we shall consider a node to be a point where two or more components connect together, but we will continue to permit a single electrical node to be viewed as two or more independent nodes as we did in Figures 5-19 and 5-20.

Exercise Problems 5.3

1. Draw and label all of the current paths for the circuit shown in Figure 5-22.
2. If a certain parallel circuit has branch currents of 120 mA, 40 mA, and 25 mA, what is the value of total current?
3. If the total current in a three-branch parallel circuit is 250 µA, branch 1 current is 100 µA, and branch 3 current is 10 µA, what is the value of current through branch 2?

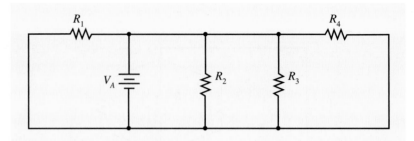

Figure 5-22. Show all current paths.

4. Figure 5-23 shows the representation of a circuit node with six currents entering and leaving the node. The direction and value of currents I_1 through I_5 are labeled on the schematic. Determine the value and direction of I_6.

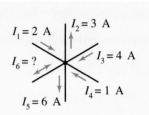

5. Write the node equations for the labeled nodes in Figure 5-24.

Figure 5-23. What is the value of I_6 and which way does it flow?

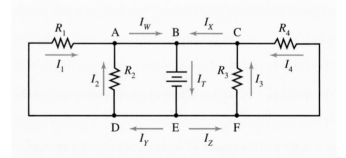

Figure 5-24. Write the node equations.

5.4 Computing Voltage, Current, Resistance, and Power

We are now ready to apply Ohm's Law, Kirchhoff's Current Law, and our understanding of parallel circuits to calculate the various circuit values.

Total Resistance

We know that current in the various branches of a parallel circuit is unaffected by the resistance in the other branches. We know that the voltage across all components in a parallel circuit is the same as the applied voltage. As we add additional branches to a parallel circuit, total current must increase, but the applied voltage is constant. Based on our understanding of basic electronics, if the current in a circuit increases while the voltage remains the same, then we know that resistance must decrease. So it is with parallel circuits. As we add more branches, the total circuit resistance decreases. This effect is exactly opposite the effect of adding resistors in series.

The preceding observations allow us to make an important conclusion about parallel circuits:

> The total resistance in a parallel circuit is always less than the smallest branch resistance.

EXAMPLE SOLUTION

What can be said about the total circuit resistance in Figure 5-25 when the switch S_1 is closed?

EXAMPLE SOLUTION

With the switch open, total resistance is determined by resistors R_1 and R_2. The total current in the circuit is given as $I_T = I_1 + I_2$. When the switch is closed, the applied voltage stays the same, but the total

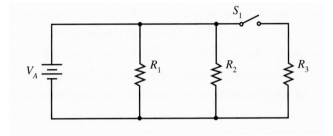

Figure 5-25. What happens to total resistance when the switch is closed?

current flow increases: $I_T = I_1 + I_2 + I_3$. Since the current is higher, but the applied voltage is still the same, we know the total resistance must be less.

If you have difficulty understanding why total resistance decreases when additional parallel branches are added, then try it another way. Recall that resistance is, by definition, opposition to current flow. When parallel branches are added, there are alternate paths for current flow. Since the electrons now have more ways of flowing around the circuit, there will be less opposition (i.e., less resistance).

It might also help to consider the hydraulic analogy shown in Figure 5-26. As long as the pressure can be maintained (equivalent of applied voltage in an electrical circuit), the existing branch currents will remain constant when new pipes (branches) are added. The *total* flow, however, will increase with each additional pipe. The "resistance," or opposition, to water flow of the system decreases as more pipes are added.

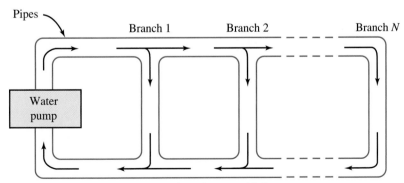

Total water flow increases as more branches are added.

Figure 5-26. A hydraulic analogy to show decreased resistance with more parallel paths.

OHM'S LAW

We can use Ohm's Law to calculate the value of total resistance if we know total current and the value of applied voltage.

EXAMPLE SOLUTION

Calculate the value of total resistance for the circuit shown in Figure 5-27.

EXAMPLE **SOLUTION**

We apply Ohm's Law as follows:

$$R_T = \frac{V_A}{I_T}$$

$$= \frac{50 \text{ V}}{200 \text{ mA}}$$

$$= 250 \ \Omega$$

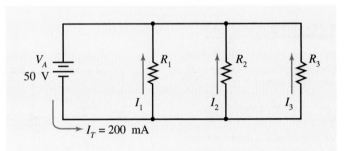

Figure 5-27. Calculate the total resistance in this circuit.

EXAMPLE SOLUTION

What is the value of total resistance for the circuit shown in Figure 5-28?

EXAMPLE **SOLUTION**

In order to apply Ohm's Law to find total resistance, we must know total voltage and total current. Total voltage (V_A) is given as 10 V. We can determine total current by applying Kirchhoff's Current Law as follows:

$$I_T = I_1 + I_2$$

$$= 10 \text{ mA} + 40 \text{ mA}$$

$$= 50 \text{ mA}$$

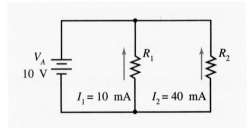

Figure 5-28. Calculate total resistance in this circuit.

We can now find total resistance by applying Ohm's Law:

$$R_T = \frac{V_A}{I_T}$$

$$= \frac{10 \text{ V}}{50 \text{ mA}}$$

$$= 200 \ \Omega$$

RECIPROCAL FORMULA

It is often necessary for a technician to calculate total resistance in a circuit when the total voltage is unknown. Consider the case shown in Figure 5-29.

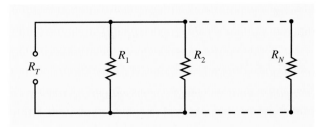

Figure 5-29. Find R_T for this circuit.

We can calculate total resistance in a case like this by applying Equation 5-4:

$$R_T = \frac{1}{\dfrac{1}{R_1} + \dfrac{1}{R_2} + \ldots + \dfrac{1}{R_N}}, \text{ or}$$

$$\frac{1}{R_T} = \frac{1}{R_1} + \frac{1}{R_2} + \ldots + \frac{1}{R_N}$$

(5-4)

This equation is called the reciprocal formula and can be used with any number of parallel branches. The derivation is based on Ohm's Law.

EXAMPLE SOLUTION

What is the total resistance of a 10-kΩ and a 20-kΩ resistor that are connected in parallel?

EXAMPLE **SOLUTION**

We apply Equation 5-4 as follows:

$$R_T = \frac{1}{\dfrac{1}{R_1} + \dfrac{1}{R_2}}$$

$$= \frac{1}{\dfrac{1}{10\text{ k}\Omega} + \dfrac{1}{20\text{ k}\Omega}}$$

$$= \frac{1}{(0.1 \times 10^{-3}) + (0.05 \times 10^{-3})}$$

$$= \frac{1}{0.15 \times 10^{-3}} \approx 6.67\text{ k}\Omega$$

Be sure to notice that this value is smaller than either of the individual resistors, which is one of the basic characteristics of parallel circuits.

EXAMPLE SOLUTION

Show the calculator key sequence to determine the value of a 22-kΩ resistor in parallel with a 47-kΩ and an 18-kΩ resistor. Apply the reciprocal formula.

EXAMPLE SOLUTION

Figure 5-30 shows one possible sequence of keystrokes when this problem is solved using a standard engineering calculator. The keystroke sequence for solving the same problem on a Reverse Polish Notation (RPN) engineering calculator is shown in Figure 5-31.

Figure 5-30. Determination of total resistance using a standard engineering calculator.

Figure 5-31. Determination of total resistance using an RPN engineering calculator.

Practice Problems

1. Calculate the total resistance if a 200-Ω resistor is connected in parallel with 470-Ω and 330-Ω resistors to form a three-branch circuit.
2. What is the total resistance of four 100-Ω resistors in parallel?
3. If a 2.2-kΩ resistor is connected in parallel with a 3.9-kΩ resistor and then that combination is connected across a 2.7-kΩ resistor, what is the total resistance of the three resistors?

Answers to Practice Problems

1. 98.44 Ω   **2.** 25 Ω   **3.** 924.8 Ω

CONDUCTANCE METHOD

In Chapter 1, we learned the name of an electrical quantity called **conductance** (G). Recall that the unit of measurement for conductance is siemens. Conductance is essentially the opposite of resistance and describes the ease with which current flows. It is defined mathematically by Equation 5-5:

$$\text{Conductance} = \frac{1}{\text{Resistance}} \qquad (5\text{-}5)$$

EXAMPLE SOLUTION

What is the conductance of a 4.7-kΩ resistor?

[EXAMPLE] **SOLUTION**

We apply Equation 5-5 as follows:

$$G = \frac{1}{R}$$

$$= \frac{1}{4.7 \text{ k}\Omega}$$

$$= 212.8 \text{ } \mu\text{S}$$

Since conductance describes the ease with which current flows, it follows that total conductance in a parallel circuit will increase as we add parallel branches. We can state this relationship mathematically as

$$G_T = G_1 + G_2 + \ldots + G_N \qquad (5\text{-}6)$$

This equation allows us to compute total circuit resistance by using the conductance values of the various branch resistors.

EXAMPLE [SOLUTION]

What is the resistance of a 4.7-kΩ resistor and a 6.8-kΩ resistor in parallel?

[EXAMPLE] **SOLUTION**

First, we calculate the individual conductances with Equation 5-5:

$$G_1 = \frac{1}{R_1} = \frac{1}{4.7 \text{ k}\Omega} = 212.8 \text{ } \mu\text{S}$$

$$G_2 = \frac{1}{R_2} = \frac{1}{6.8 \text{ k}\Omega} = 147.1 \text{ } \mu\text{S}$$

Next, we find total circuit conductance by applying Equation 5-6:

$$G_T = G_1 + G_2$$

$$= 212.8 \text{ } \mu\text{S} + 147.1 \text{ } \mu\text{S}$$

$$= 359.9 \text{ } \mu\text{S}$$

Finally, we apply Equation 5-5 to compute total resistance:

$$G_T = \frac{1}{R_T}, \text{ or}$$

$$R_T = \frac{1}{G_T}$$

$$= \frac{1}{359.9 \text{ } \mu\text{S}}$$

$$= 2.78 \text{ k}\Omega$$

Although this method may seem cumbersome at first, it is important for at least two reasons. First, there are types of circuit analysis problems where you will know the conductance values, but not the resistance values. Second, if you carefully examine Equation 5-4, you can see that it is essentially the same as the calculations in the previous example.

PRODUCT-OVER-THE-SUM FORMULA

Equation 5-4 will work for any number of parallel resistors. Many times, however, a technician will have to determine the resistance of two parallel resistors. In this special case, Equation 5-7 can be used:

$$R_T = \frac{R_1 R_2}{R_1 + R_2} \qquad (5\text{-}7)$$

This is sometimes called a shortcut, but it is misnamed because it actually takes more calculator keystrokes to apply Equation 5-7 than it does to use the reciprocal formula (Equation 5-4).

EQUAL-VALUED RESISTORS

When it is necessary to determine the combined resistance of several parallel resistors that are all the same value, many technicians use Equation 5-8 as a genuine shortcut:

$$R_T = \frac{R_N}{N} \qquad (5\text{-}8)$$

where R_N is the common resistor value and N is the total number of similar resistors.

EXAMPLE SOLUTION

Determine the total resistance if ten 150-kΩ resistors are connected in parallel.

EXAMPLE **SOLUTION**

We apply Equation 5-8 as follows:

$$R_T = \frac{R_N}{N}$$
$$= \frac{150 \text{ k}\Omega}{10}$$
$$= 15 \text{ k}\Omega$$

Practice Problems

1. What is the value of eleven 33-kΩ resistors connected in parallel?
2. What is the total resistance if five 20-kΩ resistors are connected in parallel?
3. How many 100-kΩ resistors must be connected in parallel to produce a total resistance of 25 kΩ?

Answers to Practice Problems

1. 3 kΩ   **2.** 4 kΩ   **3.** Four

ASSUMED VOLTAGE

Now let us examine one final alternative for finding total resistance in a parallel circuit. If the voltage is unknown (or even if it is known), we can temporarily *assume* a value of applied voltage that is easily divisible by each of the branch resistances. Consider, for example, the circuit shown in Figure 5-32. If we temporarily assume an applied voltage of 100 V, then we can apply Ohm's Law to each branch to find the branch currents and total current. In the present case, we would have the following results:

$$I_1 = \frac{V_1}{R_1} = \frac{100 \text{ V}}{20 \text{ }\Omega} = 5 \text{ A}$$

$$I_2 = \frac{V_2}{R_2} = \frac{100 \text{ V}}{25 \text{ }\Omega} = 4 \text{ A}$$

$$I_3 = \frac{V_3}{R_3} = \frac{100 \text{ V}}{50 \text{ }\Omega} = 2 \text{ A}$$

$$I_T = I_1 + I_2 + I_3 = 5 \text{ A} + 4 \text{ A} + 2 \text{ A} = 11 \text{ A}$$

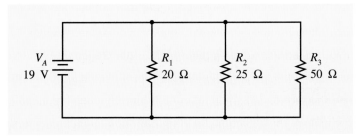

Figure 5-32. We can temporarily assume a voltage that is evenly divisible by all resistors to help determine total resistance.

It is important to realize that these currents do not really flow in the original circuit, since we have *assumed* 100 V as the applied voltage. Nevertheless, the "error" will cancel itself out in the following step. Next, we compute total resistance by applying Ohm's Law:

$$R_T = \frac{V_A}{I_T}$$

$$= \frac{100 \text{ V}}{11 \text{ A}}$$

$$= 9.09 \text{ }\Omega$$

When you are using an electronic calculator, there is very little advantage of one method over another. But, if you are out on a job and must do a pencil-and-paper calculation, the preceding method can greatly simplify the calculations that must be done. The first four calculations can probably be done in your head. This leaves only one calculation that must be done longhand.

Total Current

We have already seen one method (Equation 5-2) that can be used to determine total current in a parallel circuit. However, this method requires us to calculate the individual branch currents. Another alternative is to apply Ohm's Law to the total voltage and total resistance values in the circuit as indicated by Equation 5-9:

$$I_T = \frac{V_A}{R_T} \tag{5-9}$$

EXAMPLE SOLUTION

What is the total current for the circuit shown in Figure 5-33?

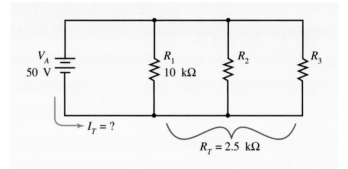

Figure 5-33. Compute the total current in this circuit.

EXAMPLE **SOLUTION**

We can apply Equation 5-9 as follows:

$$I_T = \frac{V_A}{R_T}$$

$$= \frac{50 \text{ V}}{2.5 \text{ k}\Omega}$$

$$= 20 \text{ mA}$$

Practice Problems

1. Calculate the total current for the circuit shown in Figure 5-34.

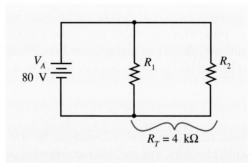

Figure 5-34. What is the total current in this circuit?

2. What is the total current for the circuit shown in Figure 5-35?

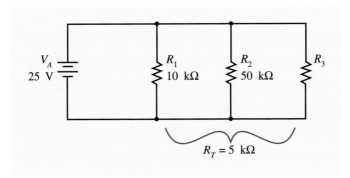

Figure 5-35. Calculate the total current for this circuit.

Answers to Practice Problems

1. 20 mA   **2.** 5 mA

Total Power

There are two simple ways that can be used to calculate the total power dissipated in a parallel circuit. We can apply the basic power formulas to the total circuit values (voltage, current, and resistance) or we can simply add all of the individual resistor power dissipations.

POWER FORMULAS TO COMPUTE TOTAL POWER

If we know any two of the total circuit values (voltage, current, or resistance), then we can apply the power formulas presented in Chapter 2 (Equations 2-9, 2-10, and 2-11) to determine the total circuit power.

EXAMPLE SOLUTION

What is the total power dissipation for the circuit in Figure 5-36?

EXAMPLE SOLUTION

Since total voltage (V_A) and total current (I_T) are known, we can apply one of the power formulas to determine total circuit power as follows:

$$P_T = I_T V_A$$
$$= 100 \text{ mA} \times 25 \text{ V}$$
$$= 2.5 \text{ W}$$

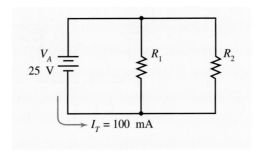

Figure 5-36. Find the total power dissipation in this circuit.

EXAMPLE SOLUTION

What is the total power dissipation for the circuit in Figure 5-37?

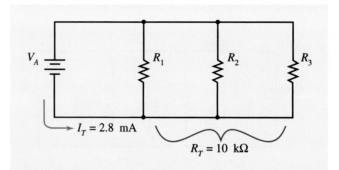

Figure 5-37. How much power is dissipated by this circuit?

EXAMPLE SOLUTION

We can apply one of the power formulas as follows:

$$P_T = I_T^2 R_T$$
$$= (2.8 \text{ mA})^2 \times 10 \text{ k}\Omega$$
$$= 78.4 \text{ mW}$$

SUMMATION OF BRANCH POWERS

You will recall that power dissipation in a resistive circuit is essentially a heat loss. Therefore, regardless of whether the resistors are connected in series or parallel, the total power dissipation (i.e., heat loss) can be found by adding all of the individual resistor power dissipations as indicated by Equation 5-10:

$$P_T = P_1 + P_2 + \dots + P_N \qquad \text{(5-10)}$$

EXAMPLE SOLUTION

Calculate the total power dissipation for the circuit shown in Figure 5-38.

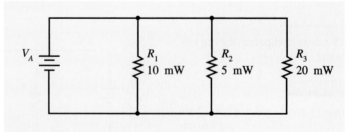

Figure 5-38. Find the total power dissipation in this circuit.

EXAMPLE SOLUTION

We can apply Equation 5-10 as follows:

$$P_T = P_1 + P_2 + P_3$$
$$= 10 \text{ mW} + 5 \text{ mW} + 20 \text{ mW}$$
$$= 35 \text{ mW}$$

Branch Currents

A prior section indicated that Ohm's Law could be applied to the individual branches of a parallel circuit to determine branch current. More specifically, we can use any form of Ohm's Law or any form of the power formulas for computing branch current. As with series circuit analysis, however, it is essential that proper subscripts be used to avoid using improper values. If, for example, you are computing I_3, then be sure to use values such as V_3, P_3, and R_3. It is easy to make mistakes unless you use correct subscripts.

EXAMPLE SOLUTION

Calculate the branch currents for the circuit shown in Figure 5-39.

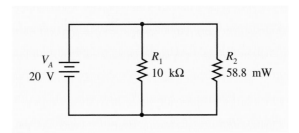

Figure 5-39. Determine the branch currents in this circuit.

EXAMPLE SOLUTION

As with series circuits, if we know any two values for a given component, then we can calculate all other values for that component. In the case of R_1, we know voltage ($V_1 = V_A = 20$ V) and we know resistance (10 kΩ). We can use Ohm's Law to compute the branch current for R_1 as follows:

$$I_1 = \frac{V_1}{R_1}$$

$$= \frac{20 \text{ V}}{10 \text{ k}\Omega}$$

$$= 2 \text{ mA}$$

In the case of R_2, we know voltage and we know power dissipation, so we can apply one of the power formulas as follows:

$$P_2 = V_2 I_2, \text{ or}$$

$$I_2 = \frac{P_2}{V_2}$$

$$= \frac{58.8 \text{ mW}}{20 \text{ V}}$$

$$= 2.94 \text{ mA}$$

Fully Specified Circuits

Sometimes a technician is required to analyze a parallel circuit where the values of all components are known. We call this a fully specified circuit. Computation of circuit values for a fully specified circuit generally consists of direct substitution of known values into one of the basic circuit equations (Ohm's Law, Kirchhoff's Current Law, power formulas, and so on).

KEY POINTS

For fully specified circuits, any given circuit quantity can be found by direct substitution of known quantities into an appropriate equation.

EXAMPLE SOLUTION

Calculate the voltage, current, and power for every branch of the circuit shown in Figure 5-40. Complete the solution matrix provided in Table 5-1.

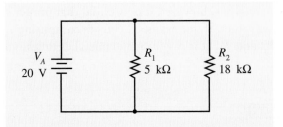

Figure 5-40. Compute all circuit values.

COMPONENT	RESISTANCE	VOLTAGE	CURRENT	POWER
R_1	5 kΩ			
R_2	18 kΩ			
Total		20 V		

Table 5-1. Solution Matrix for Figure 5-40

EXAMPLE SOLUTION

There are numerous sequences that can be used to complete the solution matrix. In general, we want to choose an item to compute and then select an equation where the selected quantity is the only unknown quantity. A good place to start might be to compute the values for the resistor voltage drops. Since the components are in parallel, we know the resistor voltages will be the same as the applied voltage (Equation 5-1). That is, $V_A = V_1 = V_2 = 20$ V.

Next, we might want to calculate the branch currents by applying Ohm's Law as follows:

$$I_1 = \frac{V_1}{R_1} = \frac{20 \text{ V}}{5 \text{ k}\Omega} = 4 \text{ mA}$$

$$I_2 = \frac{V_2}{R_2} = \frac{20 \text{ V}}{18 \text{ k}\Omega} = 1.11 \text{ mA}$$

We can now apply Equation 5-2 to determine total current.

$$I_T = I_1 + I_2$$
$$= 4 \text{ mA} + 1.11 \text{ mA}$$
$$= 5.11 \text{ mA}$$

Total resistance can be found in several ways. Let us apply Ohm's Law, since we know the values of total voltage and total current in the circuit.

$$R_T = \frac{V_A}{I_T}$$

$$= \frac{20 \text{ V}}{5.11 \text{ mA}}$$

$$= 3.91 \text{ k}\Omega$$

We can now compute the individual branch power dissipations by applying our choice of power formulas. The following is one possible alternative:

$$P_1 = \frac{V_1^2}{R_1} = \frac{(20 \text{ V})^2}{5 \text{ k}\Omega} = 80 \text{ mW}$$

$$P_2 = \frac{V_2^2}{R_2} = \frac{(20 \text{ V})^2}{18 \text{ k}\Omega} = 22.22 \text{ mW}$$

Finally, we can compute total power with any one of our power formulas:

$$P_T = V_A I_T$$

$$= 20 \text{ V} \times 5.11 \text{ mA}$$

$$= 102.2 \text{ mW}$$

The completed solution matrix is shown in Table 5-2.

COMPONENT	RESISTANCE	VOLTAGE	CURRENT	POWER
R_1	5 kΩ	20 V	4 mA	80 mW
R_2	18 kΩ	20 V	1.11 mA	22.22 mW
Total	3.91 kΩ	20 V	5.11 mA	102.2 mW

Table 5-2. Completed Solution Matrix for Figure 5-40

Practice Problems

1. Complete the solution matrix shown in Table 5-3 for the circuit shown in Figure 5-41.

COMPONENT	RESISTANCE	VOLTAGE	CURRENT	POWER
R_1	4.7 kΩ			
R_2	6.8 kΩ			
R_3	3.9 kΩ			
Total		120 V		

Table 5-3. Solution Matrix for Figure 5-41

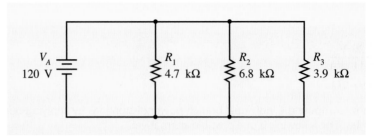

Figure 5-41. Analyze this circuit.

2. Complete a solution matrix similar to Table 5-3 for the circuit shown in Figure 5-42.

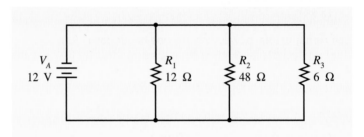

Figure 5-42. Analyze this circuit.

Answers to Practice Problems

1.

COMPONENT	RESISTANCE	VOLTAGE	CURRENT	POWER
R_1	4.7 kΩ	120 V	25.53 mA	3.06 W
R_2	6.8 kΩ	120 V	17.65 mA	2.12 W
R_3	3.9 kΩ	120 V	30.77 mA	3.69 W
Total	1.62 kΩ	120 V	73.95 mA	8.87 W

2.

COMPONENT	RESISTANCE	VOLTAGE	CURRENT	POWER
R_1	12 Ω	12 V	1 A	12 W
R_2	48 Ω	12 V	250 mA	3 W
R_3	6 Ω	12 V	2 A	24 W
Total	3.69 Ω	12 V	3.25 A	39 W

Partially Specified Circuits

Many times a technician must analyze a circuit where not all of the component values are known. This type of circuit is called a partially specified circuit, and its solution requires creative application of Ohm's Law, Kirchhoff's Law, power formulas, and basic circuit theory. Each calculation provides additional information. There are always multiple ways a problem can be solved, and there is not always a single best method. The more you practice solving this type of problem, the better you will be at identifying a direct solution sequence. When you are uncertain of the next step and until you gain the experience needed to streamline your calculations, you are advised to calculate any value that can be calculated. Each calculated value provides you with additional data that can be used in subsequent calculations.

> **KEY POINTS**
>
> For partially specified circuits, if you know any two quantities (voltage, current, resistance, or power) associated with a given component, then you can calculate all other values for that component.

EXAMPLE SOLUTION

Complete the solution matrix shown in Table 5-4 for the circuit shown in Figure 5-43.

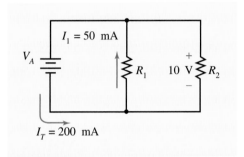

Figure 5-43. Analyze this circuit.

COMPONENT	RESISTANCE	VOLTAGE	CURRENT	POWER
R_1			50 mA	
R_2		10 V		
Total			200 mA	

Table 5-4. Solution Matrix for Figure 5-43

EXAMPLE SOLUTION

One possible first step is to recognize that the voltage is the same across components in a parallel circuit. Since V_2 is given as 10 V, we immediately know that V_A and V_1 will also be 10 V.

We know two values for R_1 (I_1 and V_1), so we can calculate the other values. Let's use Ohm's Law to compute the value of R_1.

$$R_1 = \frac{V_1}{I_1} = \frac{10 \text{ V}}{50 \text{ mA}} = 200 \ \Omega$$

We can calculate total resistance in a similar manner.

$$R_T = \frac{V_A}{I_T} = \frac{10 \text{ V}}{200 \text{ mA}} = 50 \ \Omega$$

Next, we might choose to calculate the current through R_2. We can apply Equation 5-2.

$$I_T = I_1 + I_2$$
$$I_2 = I_T - I_1$$
$$= 200 \text{ mA} - 50 \text{ mA} = 150 \text{ mA}$$

We can now calculate the value of R_2 by applying Ohm's Law.

$$R_2 = \frac{V_2}{I_2} = \frac{10 \text{ V}}{150 \text{ mA}} = 66.67 \text{ }\Omega$$

We are now in a position to compute all of the power dissipation in the circuit. We can apply our choice of power formulas. One possible sequence follows:

$$P_1 = V_1 I_1 = 10 \text{ V} \times 50 \text{ mA} = 500 \text{ mW}$$
$$P_2 = V_2 I_2 = 10 \text{ V} \times 150 \text{ mA} = 1.5 \text{ W}$$
$$P_T = P_1 + P_2 = 500 \text{ mW} + 1.5 \text{ W} = 2\text{W}$$

The completed solution matrix for this problem is shown in Table 5-5.

COMPONENT	RESISTANCE	VOLTAGE	CURRENT	POWER
R_1	200 Ω	10 V	50 mA	500 mW
R_2	66.67 Ω	10 V	150 mA	1.5 W
Total	50 Ω	10 V	200 mA	2.0 W

Table 5-5. Completed Solution Matrix for Figure 5-43

EXAMPLE SOLUTION

Complete the solution matrix shown in Table 5-6 for the circuit shown in Figure 5-44.

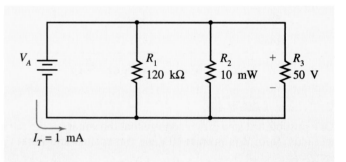

Figure 5-44. Analyze this circuit.

COMPONENT	RESISTANCE	VOLTAGE	CURRENT	POWER
R_1	120 kΩ			
R_2				10 mW
R_3		50 V		
Total			1 mA	

Table 5-6. Solution Matrix for Figure 5-44

EXAMPLE SOLUTION

Since we know the voltage for one component (R_3) in the parallel circuit, we immediately know the voltage (50 V) for all others. We now know two values for R_1, R_2, and total. We can calculate all other values for these rows in Table 5-6. One possible sequence follows:

$$I_1 = \frac{V_1}{R_1} = \frac{50 \text{ V}}{120 \text{ k}\Omega} = 416.7 \text{ μA}$$

$$P_1 = V_1 I_1 = 50 \text{ V} \times 416.7 \text{ μA} = 20.84 \text{ mW}$$

Next, we calculate R_2 values.

$$R_2 = \frac{V_2^2}{P_2} = \frac{(50 \text{ V})^2}{10 \text{ mW}} = 250 \text{ k}\Omega$$

$$I_2 = \frac{V_2}{R_2} = \frac{50 \text{ V}}{250 \text{ k}\Omega} = 200 \text{ μA}$$

Now, we calculate the missing total values.

$$R_T = \frac{V_A}{I_T} = \frac{50 \text{ V}}{1 \text{ mA}} = 50 \text{ k}\Omega$$

$$P_T = V_A I_T = 50 \text{ V} \times 1 \text{ mA} = 50 \text{ mW}$$

Since we only have one known value (V_3) for R_3, we cannot compute the missing values directly. Let's choose to apply Equation 5-2 to compute the current for R_3.

$$I_T = I_1 + I_2 + I_3, \text{ or}$$

$$I_3 = I_T - I_1 - I_2$$
$$= 1 \text{ mA} - 416.7 \text{ μA} - 200 \text{ μA}$$
$$= 383.3 \text{ μA}$$

We can calculate R_3 and P_3 by direct equation substitution.

$$R_3 = \frac{V_3}{I_3} = \frac{50 \text{ V}}{383.3 \text{ μA}} = 130.4 \text{ k}\Omega$$

$$P_3 = V_3 I_3 = 50 \text{ V} \times 383.3 \text{ μA} = 19.17 \text{ mW}$$

The completed solution matrix is shown in Table 5-7.

COMPONENT	RESISTANCE	VOLTAGE	CURRENT	POWER
R_1	120 kΩ	50 V	416.7 μA	20.84 mW
R_2	250 kΩ	50 V	200 μA	10 mW
R_3	130.4 kΩ	50 V	383.3 μA	19.17 mW
Total	50 kΩ	50 V	1.0 mA	50 mW

Table 5-7. Completed Solution Matrix for Figure 5-44

Practice Problems

1. Complete a solution matrix similar to Table 5-7 for the circuit shown in Figure 5-45.

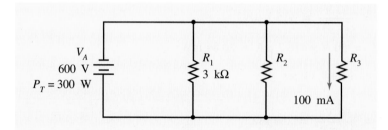

Figure 5-45. Complete a solution matrix for this circuit.

2. Complete a solution matrix similar to Table 5-7 for the circuit shown in Figure 5-46.

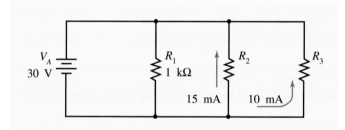

Figure 5-46. Complete a solution matrix for this circuit.

Answers to Practice Problems

1.

COMPONENT	RESISTANCE	VOLTAGE	CURRENT	POWER
R_1	3 kΩ	600 V	200 mA	120 W
R_2	3 kΩ	600 V	200 mA	120 W
R_3	6 kΩ	600 V	100 mA	60 W
Total	1.2 kΩ	600 V	500 mA	300 W

2.

COMPONENT	RESISTANCE	VOLTAGE	CURRENT	POWER
R_1	1 kΩ	30 V	30 mA	900 mW
R_2	2 kΩ	30 V	15 mA	450 mW
R_3	3 kΩ	30 V	10 mA	300 mW
Total	545.5 Ω	30 V	55 mA	1.65 W

Designing a Parallel Circuit

Figure 5-47 shows one of the most common situations that a technician may encounter that requires designing a parallel circuit. The technician may need a particular value of resistance that is not readily available. If two resistors are connected in parallel, the desired "effective" resistance may be obtained. There are many possible combinations of resistances that will produce the desired total resistance, but it is sometimes difficult to determine values that are standard. The following design procedure will generally produce acceptable results:

1. Select one standard resistor (R_1) that is about 20% higher than the targeted total resistance (R_T).
2. Apply Equation 5-11 to determine the value of the second resistor (R_2).
3. Determine the percentage of error by applying Equation 5-12.

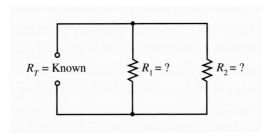

Figure 5-47. Two parallel resistors can be used to obtain a desired value of total resistance.

Equation 5-11 is obtained by transposing Equation 5-7 (product-over-the-sum formula):

$$R_2 = \frac{R_T R_1}{R_1 - R_T} \qquad (5\text{-}11)$$

Equation 5-12 is used to determine the amount of error between the desired total resistance and the actual resulting resistance:

$$\% \text{ error} = \frac{R_{ACTUAL} - R_{DESIRED}}{R_{DESIRED}} \times 100 \qquad (5\text{-}12)$$

EXAMPLE SOLUTION

Refer to Figure 5-48. Determine standard values for resistors R_1 and R_2 that will produce a total resistance of 4 kΩ ±5%.

EXAMPLE **SOLUTION**

We apply the procedure listed previously. First, we compute the value of R_1 as follows:

$$R_1 = 1.2 R_T$$
$$= 1.2 \times 4 \text{ k}\Omega$$
$$= 4.8 \text{ k}\Omega$$

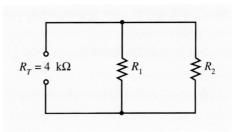

Figure 5-48. What standard values are needed for R_1 and R_2?

The nearest standard value (refer to Appendix B) is 4.7 kΩ.

Next, we compute the required value for R_2 by applying Equation 5-11:

$$R_2 = \frac{R_T R_1}{R_1 - R_T}$$

$$= \frac{4 \text{ k}\Omega \times 4.7 \text{ k}\Omega}{4.7 \text{ k}\Omega - 4 \text{ k}\Omega}$$

$$= 26.86 \text{ k}\Omega$$

The nearest standard value is 27 kΩ. The total resistance of our resulting combination can be computed with Equation 5-7 as:

$$R_T = \frac{R_1 R_2}{R_1 + R_2}$$

$$= \frac{4.7 \text{ k}\Omega \times 27 \text{ k}\Omega}{4.7 \text{ k}\Omega + 27 \text{ k}\Omega}$$

$$= 4.003 \text{ k}\Omega$$

Finally, we verify that our computations have produced a total resistance within the required range by applying Equation 5-12:

$$\% \text{ error} = \frac{R_{\text{ACTUAL}} - R_{\text{DESIRED}}}{R_{\text{DESIRED}}} \times 100$$

$$= \frac{4.003 \text{ k}\Omega - 4 \text{ k}\Omega}{4 \text{ k}\Omega} \times 100$$

$$= 0.075\%$$

Practice Problems

1. Calculate a set of two parallel resistors that will produce a total resistance of 108 kΩ ±5%.

2. Determine the values of two parallel resistors that will produce a total resistance of 143 Ω ±5%.

3. Calculate a set of two parallel resistors that will combine to produce a total resistance of 17.5 kΩ ±5%.

Answers to Practice Problems

1. One possible answer: R_1 = 130 kΩ
   R_2 = 620 kΩ

2. One possible answer: R_1 = 180 Ω
   R_2 = 680 Ω

3. One possible answer: R_1 = 22 kΩ
   R_2 = 82 kΩ

Exercise Problems 5.4

1. What is the total resistance if the following resistors are connected in parallel: 1.2 kΩ, 3.9 kΩ, 3.3 kΩ, and 4.7 kΩ?

2. The total current of a three-resistor parallel circuit is 1.667 mA. One of the resistors is 22 kΩ and the applied voltage is 15 V. What is the value of total resistance?

3. What is the conductance of a 33-kΩ resistor?

4. What is the total conductance of a parallel circuit composed of the following resistors: 27 kΩ, 36 kΩ, and 51 kΩ?

5. Use the product-over-the-sum formula to calculate the combined resistance of a 120-kΩ and a 180-kΩ resistor that are connected in parallel.

6. If 50 resistors are connected in parallel, and each resistor has a value of 1.0 MΩ, what is the total resistance of the combination?

7. Calculate the current through R_3 for the circuit shown in Figure 5-49.

8. Determine the total power dissipation for the circuit shown in Figure 5-50.

9. Complete the solution matrix provided in Table 5-8 for the circuit shown in Figure 5-51.

10. Complete the solution matrix provided in Table 5-9 for the circuit shown in Figure 5-52.

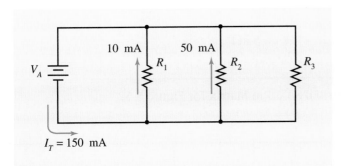

Figure 5-49. What is the value of I_3?

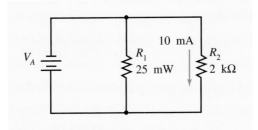

Figure 5-50. Calculate the total power dissipation for this circuit.

COMPONENT	RESISTANCE	VOLTAGE	CURRENT	POWER
R_1	68 kΩ			
R_2	82 kΩ			
Total		25 V		

Table 5-8. Partial Solution Matrix for Figure 5-51

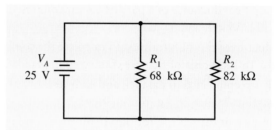

Figure 5-51. Complete a solution matrix for this circuit.

COMPONENT	RESISTANCE	VOLTAGE	CURRENT	POWER
R_1	22 kΩ			
R_2			35 mA	
R_3		100 V		
R_4	5 kΩ			
Total			75 mA	

Table 5-9. Partial Solution Matrix for Figure 5-52

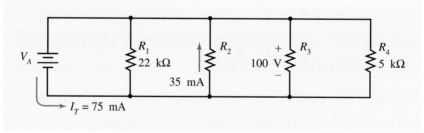

Figure 5-52. Complete a solution matrix for this circuit.

KEY POINTS

If a particular application requires more current than a given voltage source can supply, then multiple voltage sources may be connected in parallel.

5.5 Multiple Voltage Sources

Sometimes more power is needed for a circuit than can be supplied by a single voltage source. Multiple voltage sources can be connected in parallel to provide additional power. Figure 5-53 illustrates this technique. Since the voltage is the same for all compo-

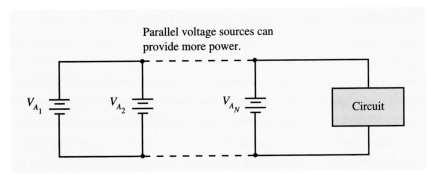

Figure 5-53. Voltage sources can be connected in parallel.

nents in a parallel circuit, it is essential that each of the parallel-connected sources have the same nominal voltage. If unequal voltage sources are connected in parallel, current will flow in the reverse direction through the lower voltage source. This may or may not damage the voltage sources, but in any case, it does not accomplish the desired goal. If the voltage sources are batteries, then reverse current may cause a violent explosion if it is excessive.

The total voltage for multiple sources connected as shown in Figure 5-53, is the same as any one of the sources. The current *capacity*, however, is increased. Be careful not to confuse current capacity of a voltage source with total current from the source. Total current is the actual amount of current that flows. It can be computed with Ohm's Law. Current capacity of the source, by contrast, is the maximum amount of current that can be supplied by a voltage source. If this value is exceeded, then the supply voltage will decrease, and the voltage source may be damaged.

Theoretically, the total current capacity of parallel voltage sources is the sum of the individual sources. This is expressed by Equation 5-13.

$$I_{\text{CAPACITY(TOTAL)}} = I_{\text{CAPACITY(1)}} + I_{\text{CAPACITY(2)}} + \ldots + I_{\text{CAPACITY}(N)} \quad (5\text{-}13)$$

In practice, however, the total current capacity is generally somewhat less than this calculated value, since the voltage sources will not have exactly the same terminal voltage under varying current requirements.

> **KEY POINTS**
>
> Each of the parallel sources must have the same terminal voltage.

> **KEY POINTS**
>
> The total current capacity of the parallel combination is approximately equal to the sum of the individual sources.

Exercise Problems 5.5

1. A certain application requires 12 V and has a current requirement of 50 A. How many 12-V cells are required to supply this circuit if each cell has a 12 V @ 20 A rating?

2. If a certain laboratory dc power supply can provide from 0 to 15 V and currents as high as 500 mA, explain how two of the power supplies can be used to supply 10 V at 800 mA.

5.6 Applied Technology: Current Dividers

Just as series circuits could be viewed as voltage divider circuits, so parallel circuits can be viewed as **current divider** circuits. It is important to keep in mind that the term current divider is just another way to view a parallel circuit. It is not a new circuit. It requires no new methods or techniques. All of the analytical tools that you have learned for solving parallel circuit problems are applicable for current divider circuits.

Figure 5-54 illustrates the meaning of the term current divider. The total power supply current is 3 A. This total current divides between R_1 and R_2. The ratio of R_1 and R_2 determines how the current will divide. As you might expect, the smaller resistor will have the greater portion of the current. We can express this mathematically with Equation 5-14 (current divider equation):

$$I_X = \left(\frac{R_T}{R_X}\right)I_T \qquad\qquad (5\text{-}14)$$

where I_X is the current through a particular parallel resistor (R_X), R_T is the total resistance of the parallel network, and I_T is the total current flowing to the parallel network.

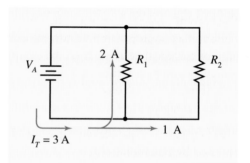

Figure 5-54. A basic current divider.

EXAMPLE SOLUTION

Three resistors having values of 20 Ω, 40 Ω, and 60 Ω are connected in parallel. The total line current to the parallel network is 10 A. How much current flows through the 40-Ω resistor?

EXAMPLE **SOLUTION**

Before we can apply Equation 5-14, we will need to compute total resistance. We can use Equation 5-4 as follows:

$$R_T = \frac{1}{\dfrac{1}{R_1} + \dfrac{1}{R_2} + \dfrac{1}{R_3}}$$

$$= \frac{1}{\dfrac{1}{20\ \Omega} + \dfrac{1}{40\ \Omega} + \dfrac{1}{60\ \Omega}}$$

$$= 10.91\ \Omega$$

We can now use Equation 5-14 to compute the current through the 40-Ω resistor.

$$I_X = \left(\frac{R_T}{R_X}\right)I_T$$

$$= \left(\frac{10.91 \ \Omega}{40 \ \Omega}\right) \times 10 \ A = 2.73 \ A$$

Many technicians prefer to use the conductance version of the current divider equation, since it is more similar to the voltage divider equation used with series circuits. It can also be used for any number of parallel resistors. This version of the current divider equation is provided by Equation 5-15:

$$I_X = \left(\frac{G_X}{G_T}\right)I_T \qquad \text{(5-15)}$$

where I_X and G_X are the current and conductance in the branch of interest, and G_T is the total circuit conductance.

Exercise Problems 5.6

1. Use the current divider equation to find the current through R_1 and R_2 in Figure 5-55.

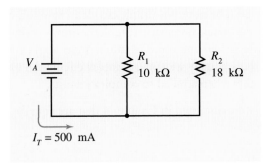

Figure 5-55. Find the current through R_1 and R_2.

2. Use the current divider equation to determine the current through R_1 and R_2 in Figure 5-56.

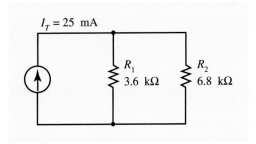

Figure 5-56. Calculate I_1 and I_2.

3. Use the conductance version of the current divider equation to calculate the currents through R_1, R_2, and R_3 in Figure 5-57.

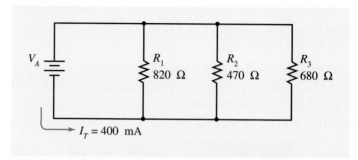

Figure 5-57. Determine I_1, I_2, and I_3.

5.7 Troubleshooting Parallel Circuits

Parallel circuits are one of the most basic circuit configurations, but they are far from being easy to troubleshoot in a professional manner. It is important to practice only logical, systematic troubleshooting methods while you are learning to troubleshoot. Because the circuits in this text and the associated lab manual have relatively few components, it is very tempting to utilize unprofessional troubleshooting methods. Unprofessional methods (such as guessing) may even be faster for simple circuits. However, if you discipline yourself to practice only professional, systematic troubleshooting—even on simple circuits—then you will be rewarded when you begin to troubleshoot more complex systems. A logical procedure will work equally well on a complex circuit. An unprofessional method, by contrast, fails miserably when applied to complex circuits.

Three important rules were discussed in Chapter 4 that are equally applicable to troubleshooting of parallel circuits:

- Know what is normal.
- Minimize the number of measurements.
- Use the best tool.

The first two items were adequately discussed in Chapter 4 and will not be repeated here. The third item deserves special consideration with reference to parallel circuits. First, a voltmeter is of only minimal value when troubleshooting a purely parallel circuit, since the voltage across all components (even open or shorted ones) is the same. Second, a current meter is an excellent diagnostic tool in theory, but it is impractical for most real-world troubleshooting situations. The primary limitation of the current meter is that the circuit must be opened to measure the current. This increases the likelihood of damage to an expensive printed circuit board (PCB). Finally, an ohmmeter is of minimal value. Unless the circuit is broken, all components will measure the same value (i.e., they are all connected in parallel). The following paragraphs examine two alternative methods that can help you locate a defect in a parallel circuit.

Troubleshooting with a Current Probe

Many manufacturers of test equipment make digital multimeters (DMMs) that have **current probe** options. Current probes that work with most standard DMMs can also be purchased. A representative current probe is shown in Figure 5-58. A current probe senses the magnetic field produced by a current-carrying conductor and produces a corresponding voltage. The output voltage is displayed by the DMM and is a direct function of current flow. Some current probes have jaws that open like a pair of pliers. The jaws must be clamped around the conductor where the current is to be measured. This type of probe is excellent for troubleshooting circuits that have discrete interconnecting wires. Unfortunately, a clamp-on probe is of little value for troubleshooting PCBs, since the probe cannot be clamped around a PCB trace. Additionally, the clamp-type probe generally responds to alternating currents only, so it is of no value when troubleshooting dc circuits.

KEY POINTS

A current probe is an attachment for a DMM that allows current to be measured in a conductor without having to break the circuit.

Another type of current probe—such as the one pictured in Figure 5-58—has the advantage of not having to be clamped around the measured conductor. It can sense the magnetic field associated with the current flow just by being near the conductor. Its primary disadvantage is that the output is a *relative* indication of current. That is, the technician cannot measure the exact amount of current, but can usually tell when one branch is more or less than another branch. This type of current probe also responds to dc current.

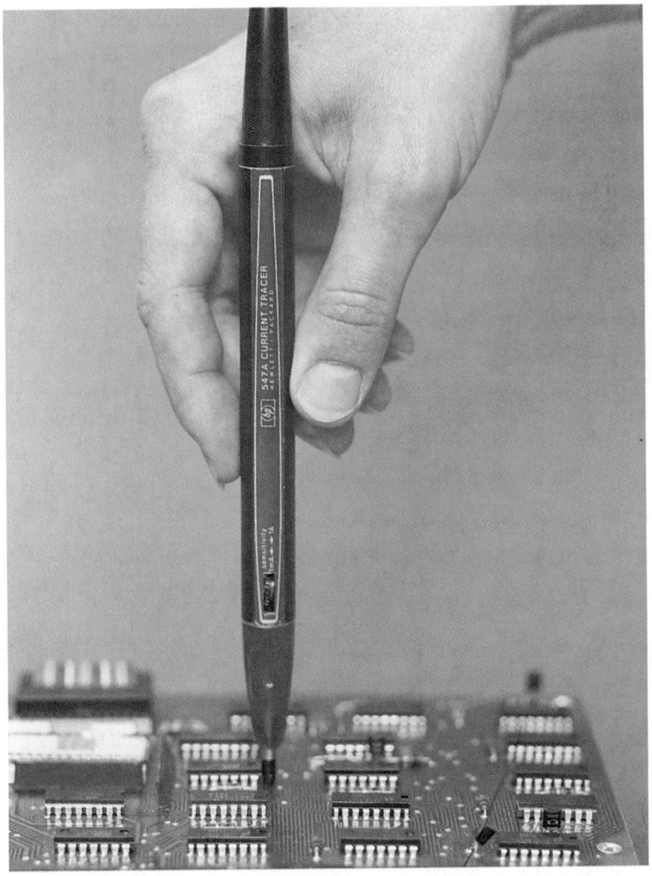

Figure 5-58. A representative current probe. (Courtesy of Hewlett-Packard Company)

Both types of current probes have the advantage of not having to break the circuit to get an indication of the current through a conductor. This provides us with a tool for locating defects in a parallel circuit.

EXAMPLE SOLUTION

Show how a current probe could be used to locate an open component in a parallel circuit.

EXAMPLE **SOLUTION**

Figure 5-59 illustrates one possible technique. The current probe is used to monitor the currents in the various parallel branches. The good branches have a normal value of current flow. The open branch has no current and is easily detected. If the physical implementation of the circuit will allow, then the technician can use the split-half method to reduce the number of measurements needed to locate the defect.

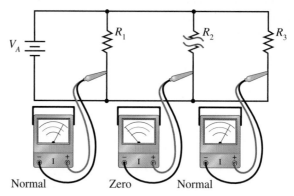

Figure 5-59. A current probe is a valuable tool for troubleshooting parallel circuits without having to break the circuit.

EXAMPLE SOLUTION

Describe how a current probe could be used to locate a resistor whose resistance is substantially lower than normal.

EXAMPLE **SOLUTION**

The procedure illustrated in Figure 5-59 also applies to locating resistances that are lower than normal. In this case, however, the defective branch would be identified by its excessive current flow.

Troubleshooting with a Temperature Probe

A **temperature probe** is another attachment that you can buy for a DMM. It generally has a small, metal surface that is used to touch an object. The temperature of the contacted object is displayed on the DMM readout. Figure 5-60 shows a representative temperature probe.

Since power dissipation in an electronic component produces heat, we can measure the relative temperatures of the components in the branches of a parallel circuit and locate a defect. The exact temperature of a component is usually not an important troubleshooting

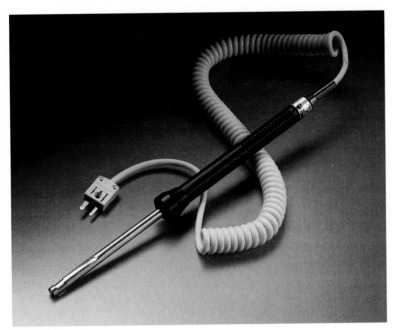

Figure 5-60. A representative temperature probe. (Courtesy of Hewlett-Packard Company)

consideration. But its temperature relative to normal and relative to the other branches is a valuable diagnostic tool for technicians.

If a component in a parallel circuit becomes open, it will dissipate no power. Consequently, the temperature probe will indicate a temperature much lower than normal. Similarly, if the resistance of a particular branch becomes lower than normal, it will dissipate more power. The temperature probe can sense the resulting increase in temperature.

Neither the current probe nor the temperature probe is the ideal tool for every application, but they are very helpful for many types of circuits that technicians encounter.

Exercise Problems 5.7

1. What disadvantage does a standard current meter have when troubleshooting parallel circuits?
2. Explain how the use of a current probe overcomes the disadvantage described in Problem 1.
3. When troubleshooting a parallel circuit with a current probe, explain how a technician might detect an open circuit in one of the parallel branches.
4. If a technician were using a temperature probe to help troubleshoot a parallel circuit, explain how a low-resistance (lower than normal) branch could be detected.

Chapter Summary

- This chapter discussed the operation, analysis, and troubleshooting of parallel circuits. Parallel circuits are characterized by having all circuit components connected directly across one another. This shunt connection causes the voltage to be the same across each of the parallel branches. The current through a given branch is inversely proportional to resistance (Ohm's Law). The current from the various branches combines to form the total current that flows through the power source. The relationship of branch currents is defined more precisely by Kirchhoff's Current Law, which says that "the current entering a point in a circuit must be equal to the current leaving that point." Kirchhoff's Current Law may be alternatively expressed as "the algebraic sum of the current entering and leaving a point is zero."

- Circuit computations for component and total voltages, currents, powers, and resistances rely heavily on Kirchhoff's Current Law, Ohm's Law, and basic circuit theory. It was shown that total resistance in a parallel circuit is always less than any single branch resistance. This is the equivalent of saying that the total conductance of a parallel circuit is greater than any individual branch.

- Total power in a parallel circuit is found by applying the various power formulas presented in Chapter 2, or by simply summing the individual branch powers. Power dissipation is essentially a heat loss and is additive regardless of the type of circuit configuration (e.g., series, parallel, and so on).

- Calculation of circuit values for a partially specified circuit is generally more challenging than similar calculations for a fully specified circuit. In the case of a fully specified circuit, all component values are known. Calculation of circuit values consists of selecting an appropriate formula and substituting component values. Partially specified circuits, by contrast, require creative application of circuit theory. If at least two quantities (voltage, current, resistance, conductance, or power) are known for a given component, then all remaining values can be determined by applying a basic formula. The "correct" formula is one where the quantity to be computed is the only unknown quantity.

- Technicians sometimes need to build a two-resistor parallel circuit to obtain a specific (perhaps nonstandard) resistance. This chapter presented a formal method for making the resistor selections that eliminates guesswork and trial-and-error alternatives.

- Voltage sources can be connected in parallel to increase the total current capacity available to a circuit. The resulting current capacity of parallel voltage sources is theoretically equal to the sum of the individual current capacities of the voltage sources. In practice, however, the current capacity of the parallel combination is somewhat less than the sum of the individual current capacities.

- A parallel circuit can be viewed as a current divider. This concept is particularly useful when the circuit is powered by a current source. In any case, the branch currents divide according to their relative resistances. Lower-value resistors have greater currents.

- Parallel circuits can be difficult to troubleshoot with conventional equipment, since voltage is the same across all components (even defective ones). Troubleshooting is further complicated by the fact that current and resistance checks—while excellent measurements in theory—require breaking of the circuit. A technician should always strive to avoid desoldering of components to open a circuit. DMMs equipped with current or temperature probes can be valuable tools for troubleshooting parallel circuits.

Review Questions

Section 5.1: Voltage Drops Across Resistors

1. The voltage in a parallel circuit is (*identical, different*) across every branch.

2. A certain parallel circuit is composed of three resistors and a voltage source. Each of the resistors has 10 V across it. What is the value of the voltage source?

3. If the various branches of a parallel circuit have unequal resistances, the voltage will be different across each branch. (True or False)

4. Determine the value of applied voltage for the circuit shown in Figure 5-61.

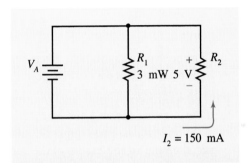

Figure 5-61. What is the value of applied voltage?

Section 5.2: Current Through Branches

5. In a parallel circuit, a branch with the lowest resistance will have the highest current. (True or False)

6. If the resistance of a particular branch in a parallel circuit increases, what happens to the current through that branch?

7. If the resistance of branch 1 in a three-branch parallel circuit increases, what happens to the value of current through branch 2? Repeat for branch 3.

8. Refer to Question 7. What is the effect on total current in the circuit?

9. Calculate the value of total current in Figure 5-62.

10. What happens to the value of current through R_2 in Figure 5-62 if resistor R_3 develops an open circuit?

11. What happens to the value of total current in Figure 5-62 if resistor R_3 increases in resistance?

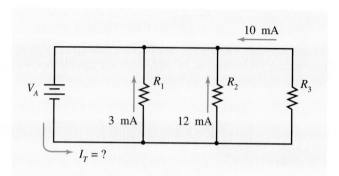

Figure 5-62. Compute total current.

Section 5.3: Kirchhoff's Current Law

12. Determine the direction and value for current I_4 in Figure 5-63.

13. If 300 mA is flowing toward a given point in a circuit, how much current must be flowing away from that same point?

14. Current that flows toward a given node is generally assigned a positive polarity. (True or False)

15. If the currents entering and leaving a given node are assigned opposite polarities, then the algebraic sum of the currents will be _____.

16. Write the node equation for nodes A and B in Figure 5-64.

17. Write the node equations for nodes C and D in Figure 5-64.

18. If the total current in a two-branch parallel circuit is 350 mA and branch 1 current is 100 mA, what is the value of current through branch 2?

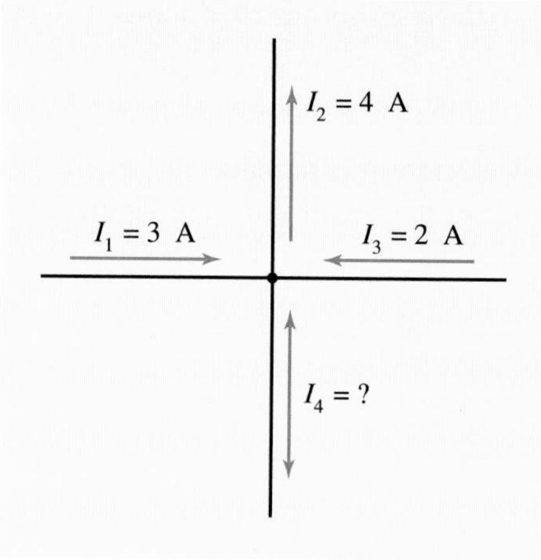

Figure 5-63. Find I_4.

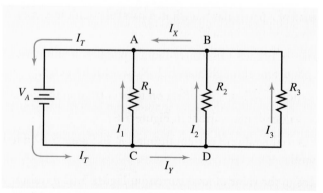

Figure 5-64. Write node equations.

19. Determine the current through R_3 in Figure 5-65.

20. What is the value of I_X in Figure 5-65?

21. Calculate I_Y in Figure 5-65.

22. Determine the value of I_Z in Figure 5-65.

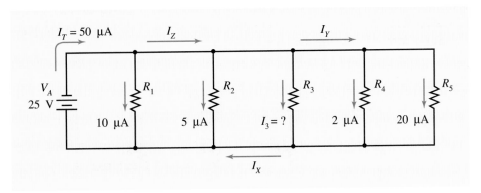

Figure 5-65. Find I_3.

Section 5.4: Computing Voltage, Current, Resistance, and Power

23. The total resistance in a parallel circuit is always less than the smallest branch resistance. (True or False)

24. Total resistance in a parallel circuit can be found by adding the values of individual branch resistances. (True or False)

25. Refer to Figure 5-66. What combination of switch positions results in the lowest value of total resistance?

26. Refer to Figure 5-66. What is the value of R_T if SW_1 is in position B and SW_2 is in position B?

27. What combination of switch positions produces the least total current flow for the circuit shown in Figure 5-66?

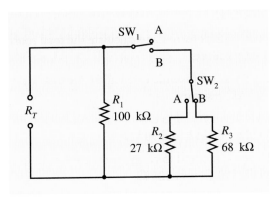

Figure 5-66. A switchable circuit.

28. What is the total resistance of the following parallel resistors: 680 Ω, 470 Ω, 910 Ω, and 1.2 kΩ?

29. If resistor R_1 in Figure 5-67 is 10 Ω, describe the relative change in total resistance when switch SW_1 is moved to the opposite position: major change, minor change, or can't say without knowing the value of R_2.

30. If resistor R_2 in Figure 5-67 is 22 MΩ, describe the relative change in total resistance when switch SW_1 is moved to the opposite position: major change, minor change, or can't say without knowing the value of R_1.

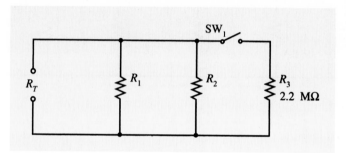

Figure 5-67. A switchable circuit.

31. Use the reciprocal formula to find the total resistance of the following parallel resistors: 10 kΩ, 22 kΩ, and 18 kΩ.

32. Use the reciprocal formula to find the total resistance of the following parallel resistors: 150 kΩ, 120 kΩ, 68 kΩ, and 220 kΩ.

33. What is the conductance of a 3.3-kΩ resistor?

34. What is the conductance of a 270-kΩ resistor?

35. What is the total conductance of a circuit consisting of a 4.7-kΩ resistor and a 12-kΩ resistor in parallel?

36. Use the product-over-the-sum formula to calculate the value of the following parallel resistors: 27 kΩ and 39 kΩ.

37. Use the product-over-the-sum formula to calculate the value of the following parallel resistors: 100 Ω and 91 Ω.

38. Use the equal-valued resistor formula to calculate the total resistance produced by paralleling ten 27-kΩ resistors.

39. Use the equal-valued resistor formula to calculate the total resistance produced by paralleling four 100-kΩ resistors.

40. Use the assumed-voltage method to calculate the total resistance of the following parallel resistors: 180 kΩ, 360 kΩ, and 90 kΩ.

41. Use the assumed-voltage method to calculate the total resistance of the following parallel resistors: 220 Ω, 390 Ω, and 820 Ω.

42. Calculate the total current for the circuit shown in Figure 5-68.

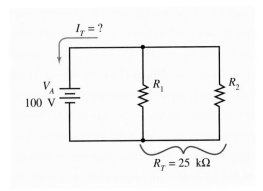

Figure 5-68. Example circuit.

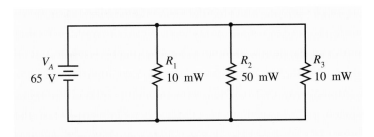

Figure 5-69. Example circuit.

43. How much total power is dissipated by the circuit in Figure 5-68?
44. Calculate the total power dissipation for the circuit shown in Figure 5-69.
45. Calculate the current through R_2 in Figure 5-69.
46. What is the total resistance of the circuit shown in Figure 5-69?
47. Calculate the conductance of resistor R_3 in Figure 5-69.
48. Complete Table 5-10 for the circuit shown in Figure 5-70.
49. Complete a solution matrix (similar to Table 5-10) for the circuit shown in Figure 5-71.

COMPONENT	RESISTANCE	VOLTAGE	CURRENT	CONDUCTANCE	POWER
R_1	10 kΩ				
R_2	18 kΩ				
R_3	18 kΩ				
Total		25 V			

Table 5-10. Solution Matrix for Figure 5-70

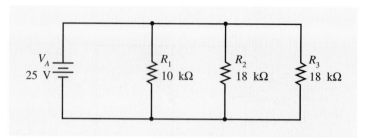

Figure 5-70. Complete a solution matrix for this circuit.

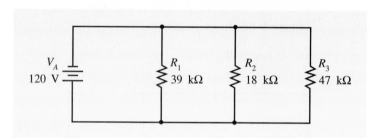

Figure 5-71. Complete a solution matrix for this circuit.

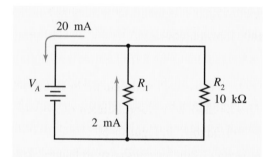

Figure 5-72. Complete a solution matrix for this circuit.

COMPONENT	RESISTANCE	VOLTAGE	CURRENT	CONDUCTANCE	POWER
R_1			2 mA		
R_2	10 kΩ				
Total			20 mA		

Table 5-11. Solution Matrix for Figure 5-72

50. Complete Table 5-11 for the circuit shown in Figure 5-72.

51. Apply the technique presented in this chapter to determine the values for two resistors that can produce a total resistance of 139 kΩ when connected in parallel.

52. Choose two standard resistor values (Appendix B) that will produce a total resistance of 129.7 Ω ±5% when connected in parallel.

Section 5.5: Multiple Voltage Sources

53. Explain why a technician might connect two voltage sources in parallel.

54. Describe a precaution that must be observed when paralleling voltage sources.

55. If each of two voltage sources has a rating of 25 V @ 3 A, what is the theoretical rating of the combined source if they are connected in parallel?

56. A certain type of voltage source is rated at 1.5 V and can supply 200 mA. Draw a schematic diagram that shows how to obtain a 1.5-V source that can supply at least 1.1 A.

57. The practical total current capacity of parallel voltage sources is usually (*more, less*) than the theoretical rating.

Section 5.6: Applied Technology: Current Dividers

58. Apply the current divider equation (Equation 5-14) to compute the current through R_2 in Figure 5-73.

59. Calculate the current through R_1 in Figure 5-73 by using the current divider equation (Equation 5-14).

60. Use the conductance version of the current divider equation (Equation 5-15) to compute the current through R_2 in Figure 5-74.

61. Find I_3 in Figure 5-74 by applying Equation 5-15.

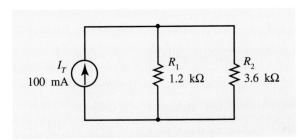

Figure 5-73. Apply the current divider equation to this circuit.

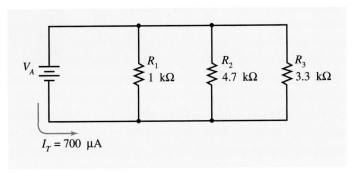

Figure 5-74. Example circuit.

Section 5.7: Troubleshooting Parallel Circuits

62. Describe at least one advantage and one disadvantage of a DMM with a current probe as compared to a standard DMM for measuring current.

63. If a branch resistance in a parallel circuit increases, then a temperature probe could be used to detect (*an increased, a decreased*) temperature in the component.

64. Troubleshoot the circuit shown in Figure 5-75. Table 5-12 provides the PShooter measurement data. The PShooter values listed in Table 5-12 are included in Appendix A.

65. Troubleshoot the circuit shown in Figure 5-75. Table 5-13 provides the PShooter measurement data. The PShooter values listed in Table 5-13 are included in Appendix A.

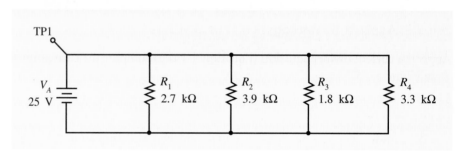

Figure 5-75. Troubleshoot this circuit.

TESTPOINT	VOLTAGE		CURRENT		RELATIVE TEMPERATURE
	NORMAL	ACTUAL	NORMAL	ACTUAL	
V_A	25 V	[78]	37.1 mA	[139]	[79]
R_1	25 V	[103]	9.26 mA	[81]	[87]
R_2	25 V	[154]	6.41 mA	[50]	[98]
R_3	25 V	[163]	13.9 mA	[142]	[161]
R_4	25 V	[217]	7.58 mA	[157]	[215]

Table 5-12. PShooter Data for Figure 5-75

TESTPOINT	VOLTAGE		CURRENT		RELATIVE TEMPERATURE
	NORMAL	ACTUAL	NORMAL	ACTUAL	
V_A	25 V	[92]	37.1 mA	[183]	[202]
R_1	25 V	[46]	9.26 mA	[198]	[207]
R_2	25 V	[174]	6.41 mA	[167]	[170]
R_3	25 V	[193]	13.9 mA	[182]	[74]
R_4	25 V	[212]	7.58 mA	[3]	[145]

Table 5-13. PShooter Data for Figure 5-75

You are the lead technician in the instrumentation lab of a major electronics corporation. Your supervisor has given you the task of designing a test fixture to be used on the production line by the assembly workers. The fixture acts like a switch-selectable resistor. It has two switches; one is labeled "coarse" and the other is labeled "fine." There are two terminals for connection of test leads. The required resistance of the test fixture for each switch position is shown in Table 5-14. The resistance of the fixture must be within 1.0% of the ideal value in every case.

SW_1 POSITION	SW_2 POSITION	RESISTANCE
A	A	10.5 Ω
A	B	20.5 Ω
B	A	10.5 kΩ
B	B	20.5 kΩ

Table 5-14. Required Resistance Values for an Instrument Test Fixture

A proposed schematic diagram is shown in Figure 5-76. Calculate the resistor values that will be required. An alternate design may be used, but be certain to use only standard resistor values. Verify your calculations in a laboratory if possible.

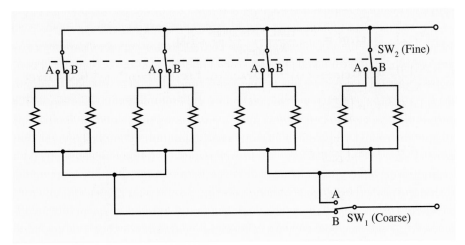

Figure 5-76. Proposed schematic diagram for an instrument test fixture.

Equation List

(5-1) $\quad V_A = V_1 = V_2 = \ldots = V_N$

(5-2) $\quad I_T = I_1 + I_2 + \ldots + I_N$

(5-3) $\quad I_{\text{IN}_1} + I_{\text{IN}_2} + \ldots + I_{\text{IN}_N} + I_{\text{OUT}_1} + I_{\text{OUT}_2} + \ldots + I_{\text{OUT}_N} = 0$

(5-4)
$$R_T = \frac{1}{\dfrac{1}{R_1} + \dfrac{1}{R_2} + \ldots + \dfrac{1}{R_N}}, \text{ or}$$
$$\frac{1}{R_T} = \frac{1}{R_1} + \frac{1}{R_2} + \ldots + \frac{1}{R_N}$$

(5-5) $\quad \text{Conductance} = \dfrac{1}{\text{Resistance}}$

(5-6) $\quad G_T = G_1 + G_2 + \ldots + G_N$

(5-7) $\quad R_T = \dfrac{R_1 R_2}{R_1 + R_2}$

(5-8) $\quad R_T = \dfrac{R_N}{N}$

(5-9) $\quad I_T = \dfrac{V_A}{R_T}$

(5-10) $\quad P_T = P_1 + P_2 + \ldots + P_N$

(5-11) $\quad R_2 = \dfrac{R_T R_1}{R_1 - R_T}$

(5-12) $\quad \% \text{ error} = \dfrac{R_{\text{ACTUAL}} - R_{\text{DESIRED}}}{R_{\text{DESIRED}}} \times 100$

(5-13) $\quad I_{\text{CAPACITY(TOTAL)}} = I_{\text{CAPACITY(1)}} + I_{\text{CAPACITY(2)}} + \ldots + I_{\text{CAPACITY}(N)}$

(5-14) $\quad I_X = \left(\dfrac{R_T}{R_X}\right) I_T$

(5-15) $\quad I_X = \left(\dfrac{G_X}{G_T}\right) I_T$

balanced bridge   loading
bleeder current   thermistor
load current   voltage regulation
load resistor   voltmeter loading
loaded voltage divider

After completing this chapter, you should be able to:

1. Positively identify a series-parallel circuit.

2. Determine a simplified equivalent circuit for a given series-parallel circuit.

3. Determine the voltage, current, and power dissipation of every component in a series-parallel circuit.

4. Classify a given bridge circuit as balanced or unbalanced.

5. Compute the voltage, current, and power dissipation for every component in a balanced bridge circuit.

6. Describe the relative effects on voltage and current throughout a series-parallel circuit when a component is increased or decreased in value.

7. Describe the effects and causes of voltmeter loading.

8. Apply a logical troubleshooting procedure to locate a defect in a series-parallel circuit.

Analyzing Series-Parallel Circuits

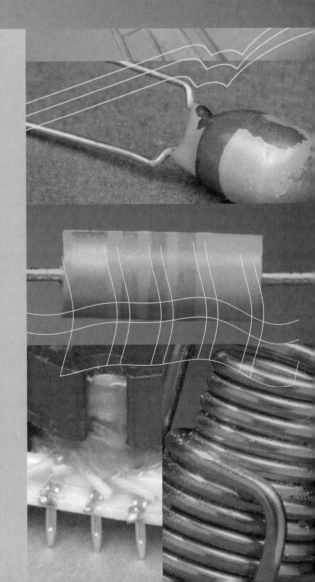

Series-parallel circuits are simply a combination of the pure series and pure parallel circuits that you studied in Chapters 4 and 5. A series-parallel circuit is analyzed by applying series circuit techniques to the series portions of the circuit and parallel circuit techniques to the parallel parts of the circuit. Since most practical electronic circuits are series-parallel circuits, it is important for a technician to be able to analyze and troubleshoot series-parallel circuits.

6.1 Positive Identification Method

Before we can begin analyzing series-parallel circuits, we must have a way to determine if a particular circuit is indeed series-parallel. In Chapter 3, we briefly discussed the four classes of circuits: series, parallel, series-parallel, and complex.

In Chapter 4, you learned how to identify and analyze series circuits. You will recall that series circuits have all components in series with each other. For components to be in series, they must have one, and only one, end connected together, with no other components or wires connected to the common point.

In Chapter 5, you learned to identify and analyze parallel circuits. You will recall that a parallel circuit has both ends of all components connected directly together.

Once you have determined that a circuit is definitely not series and not parallel, then you need only distinguish between series-parallel and complex to positively identify the circuit. To distinguish between series-parallel and complex circuits, apply the following procedure. On a scratch paper, redraw the schematic to be classified, but replace all sets of series resistors with a single resistor. Similarly, replace all truly parallel resistors with a single resistor. Repeat this procedure for the newly drawn figure. Continue until one of the following situations occurs:

- The circuit is reduced to a single resistor.
- The circuit cannot be reduced to a single resistor.

In the first case, you have identified a series-parallel circuit. In the second case, you have identified a complex circuit.

EXAMPLE SOLUTION

Classify the circuit shown in Figure 6-1 as series-parallel or complex.

EXAMPLE **SOLUTION**

Figure 6-2 shows the scratch-paper steps used to classify the circuit. The first sketch in Figure 6-2 shows the original circuit (Figure 6-1). The second step replaces parallel resistors R_2 and R_3 with a single resistor. Finally, the third step replaces the two series resistors (formed in the second step) with a single resistor. Since the circuit has been reduced to a single resistor, this circuit can be classified as a series-parallel circuit.

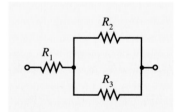

Figure 6-1. Is this a series-parallel or a complex circuit?

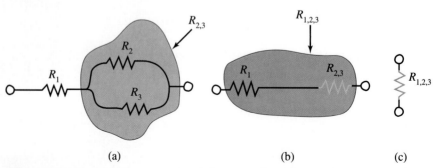

(a) (b) (c)

Figure 6-2. The scratch-paper simplification of the circuit shown in Figure 6-1.

EXAMPLE SOLUTION

Classify the circuit shown in Figure 6-3 as series-parallel or complex.

EXAMPLE SOLUTION

The scratch-paper steps are shown in Figure 6-4. The first sketch in Figure 6-4 is the original circuit. The second sketch shows the parallel resistor pairs R_1/R_2 and R_4/R_5 replaced with single resistors. This produces a pure series circuit, which is replaced in the third sketch by a single resistor. Since the circuit has been reduced to a single resistor by replacing sets of series and parallel components, we can classify it as a series-parallel circuit.

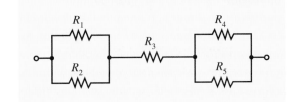

Figure 6-3. Is this a series-parallel or a complex circuit?

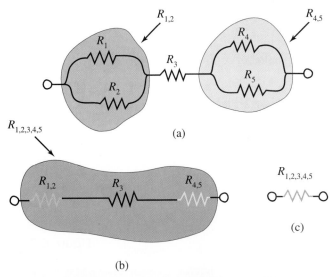

(a)

(b)

(c)

Figure 6-4. The scratch-paper simplification of the circuit shown in Figure 6-3.

EXAMPLE SOLUTION

Classify the circuit shown in Figure 6-5 as series-parallel or complex.

EXAMPLE SOLUTION

First, we look for any components that are truly in series. There are none. Next, we try to locate all parallel components. There are none. Since this circuit cannot be reduced to a single resistor by replacing series and parallel components with a single resistor, then this circuit must be classified as complex. We will learn to analyze this type of circuit in Chapter 7. The important skill to learn at this time is how to identify a series-parallel circuit and how to tell it from a series, a parallel, or a complex circuit.

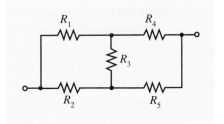

Figure 6-5. Classify this circuit as series-parallel or complex.

1. Classify the circuit shown in Figure 6-6 as series-parallel or complex.

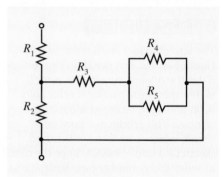

Figure 6-6. Classify this circuit.

2. Classify the circuit shown in Figure 6-7 as series-parallel or complex.

Answers to Practice Problems

1. Series-parallel   2. Series-parallel

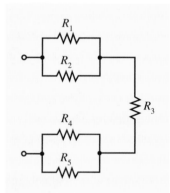

Figure 6-7. Classify this circuit.

1. Use the scratch-pad simplification method to positively identify the circuit shown in Figure 6-8 as either series-parallel or complex. Be sure to show each intermediate sketch in your scratch-pad simplification procedure.

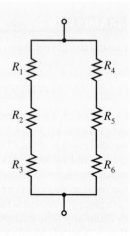

Figure 6-8. Simplify and classify this circuit.

2. Simplify and classify the circuit shown in Figure 6-9.

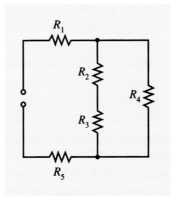

Figure 6-9. Simplify and classify this circuit.

3. Simplify and classify the circuit shown in Figure 6-10.

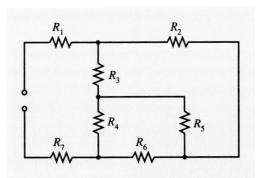

Figure 6-10. Simplify and classify this circuit.

6.2 Simplification with Equivalent Circuits

The next step toward understanding series-parallel circuits is to determine the total resistance and total current in the circuit. Any series-parallel circuit, as shown in Figure 6-11, can be reduced to a single resistor having a value that is equal to the total resistance of the series-parallel circuit. Once we know the total resistance and the applied voltage, it is a simple matter of applying Ohm's Law to determine the total current flow.

KEY POINTS

Series or parallel resistors are combined in a given equivalent circuit and shown as a single resistor.

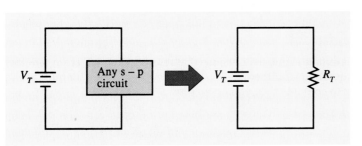

Figure 6-11. A simplified series-parallel circuit.

With one exception, the procedure used to determine the value of total resistance in a series-parallel circuit is identical to the process we used in Section 6.1 to distinguish between series-parallel and complex circuits. The only added thing we need to do is include the value of each single resistor used to replace two or more series or parallel resistors. That is, every time we replace two or more series resistors with a single resistor in the next scratch-pad drawing, we simply compute its value. Similarly, if we replace some parallel resistors with a single resistor in the next scratch-pad drawing, then we also compute its value.

When computing the value for each equivalent resistor, we apply the procedures that were presented in Chapters 4 and 5. If you are replacing a series circuit with an equivalent resistance, then apply the rules for series circuits presented in Chapter 4. If it is a parallel portion of the circuit that is being simplified, then apply the rules for parallel circuits that were presented in Chapter 5. Table 6-1 shows the two equations that must be applied to the series and parallel portions of the series-parallel circuit, respectively.

TYPE OF CIRCUIT SECTION BEING SIMPLIFIED	EQUATION FOR TOTAL RESISTANCE
Series	$R_T = R_1 + R_2 + \ldots + R_N$
Parallel	$R_T = \dfrac{1}{\dfrac{1}{R_1} + \dfrac{1}{R_2} + \ldots + \dfrac{1}{R_N}}$

Table 6-1. Series Resistance Equation and Parallel Resistance Equation

EXAMPLE SOLUTION

Determine the total resistance and total current for the circuit shown in Figure 6-12.

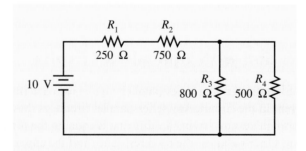

Figure 6-12. Determine the total resistance and total current in this circuit.

EXAMPLE **SOLUTION**

When we first examine Figure 6-12, we see that resistors R_1 and R_2 are in series with each other. We will circle these and label them as $R_{1,2}$. Since they are in series, we must use the series resistance equation to compute their equivalent resistance.

$$R_{1,2} = R_1 + R_2$$
$$= 250\ \Omega + 750\ \Omega$$
$$= 1\ \text{k}\Omega$$

The scratch-pad procedure is shown in Figure 6-13.

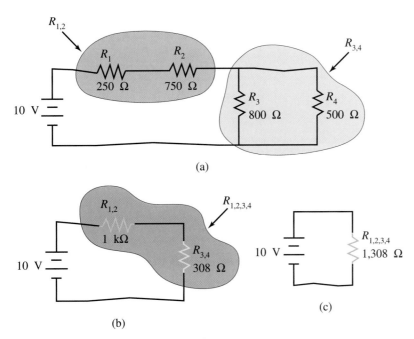

Figure 6-13. Scratch-paper procedure.

Similarly, we can see by inspection of Figure 6-12 that resistors R_3 and R_4 are in parallel with each other. Let's circle these and label them as $R_{3,4}$. The equivalent resistance value can be found with the parallel resistance equation.

$$
\begin{aligned}
R_{3,4} &= \cfrac{1}{\cfrac{1}{R_3} + \cfrac{1}{R_4}} \\
&= \cfrac{1}{\cfrac{1}{800\ \Omega} + \cfrac{1}{500\ \Omega}} \\
&\approx 308\ \Omega
\end{aligned}
$$

The second sketch in Figure 6-13 shows the results of the simplification process to this point.

Now we examine our partially simplified scratch-pad drawing and see that resistors $R_{1,2}$ and $R_{3,4}$ are in series with each other. We circle these and label them as $R_{1,2,3,4}$. Since this is the final step in the simplification process, we know that $R_{1,2,3,4}$ is equal to the total resistance (R_T) of the original circuit. Its value is found by applying the series resistance equation.

$$
\begin{aligned}
R_{1,2,3,4} = R_T &= R_{1,2} + R_{3,4} \\
&= 1{,}000\ \Omega + 308\ \Omega \\
&= 1{,}308\ \Omega
\end{aligned}
$$

These last results are shown in the final sketch in Figure 6-13.

KEY POINTS

The process of combining series and parallel combinations of resistors and forming progressively simpler equivalent circuits continues until we have a single resistor representing the entire circuit.

KEY POINTS

Once you have determined the total resistance, apply Ohm's Law to compute total current in the circuit.

The total current for the original circuit in Figure 6-12 is identical to the total current in the final sketch of Figure 6-13. The current in Figure 6-13 (final sketch) is easily computed with Ohm's Law.

$$I_T = \frac{V_T}{R_T}$$

$$= \frac{10\ V}{1{,}308\ \Omega}$$

$$= 7.65\ mA$$

EXAMPLE SOLUTION

Verify that the circuit shown in Figure 6-14 is a series-parallel circuit. If it is, then find the total resistance and total current in the circuit.

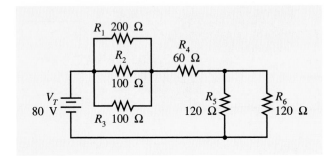

Figure 6-14. Compute total resistance and total current in this circuit.

EXAMPLE **SOLUTION**

First, we note that resistors R_1 through R_3 are connected in parallel. We use the parallel resistance equation to compute the value of equivalent resistance.

$$R_{1,2,3} = \frac{1}{\dfrac{1}{R_1} + \dfrac{1}{R_2} + \dfrac{1}{R_3}}$$

$$= \frac{1}{\dfrac{1}{200\ \Omega} + \dfrac{1}{100\ \Omega} + \dfrac{1}{100\ \Omega}}$$

$$= 40\ \Omega$$

We can also see that resistors R_5 and R_6 are in parallel with each other. We can apply the parallel resistance equation to find the equivalent resistance. Since resistors R_5 and R_6 have the same value, we can apply the short cut discussed in Chapter 5. That is,

$$R_{5,6} = \frac{R_{value}}{N}$$

$$= \frac{120\ \Omega}{2}$$

$$= 60\ \Omega$$

The results of our simplification to this point are shown in Figure 6-15(b).

Now we examine our partially simplified circuit (Figure 6-15(b)) and identify any series or parallel components. We can see that all three resistors are in series, so we will use the series resistance equation and compute the equivalent resistance.

$$R_T = R_{1,2,3,4,5,6} = R_{1,2,3} + R_4 + R_{5,6}$$
$$= 40 \ \Omega + 60 \ \Omega + 60 \ \Omega$$
$$= 160 \ \Omega$$

The results of our final simplification are shown in Figure 6-15(c).

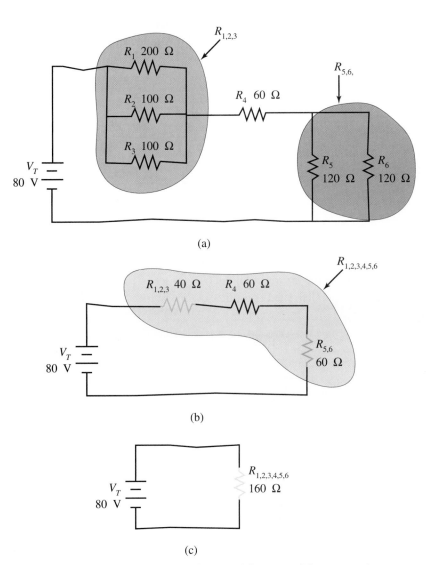

Figure 6-15. Scratch-paper steps for simplification of the circuit shown in Figure 6-14.

We can now apply Ohm's Law to the simplified circuit to determine the value of total current.

$$I_T = \frac{V_T}{R_T}$$

$$= \frac{80 \text{ V}}{160 \text{ } \Omega}$$

$$= 500 \text{ mA}$$

Exercise Problems 6.2

1. Classify the circuit shown in Figure 6-16 as series-parallel or complex. If the circuit is series-parallel, then compute total resistance and total current.

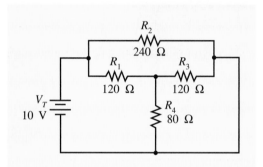

Figure 6-16. Classify this circuit.

2. Classify the circuit shown in Figure 6-17 as series-parallel or complex. If the circuit is series-parallel, then compute total resistance and total current.

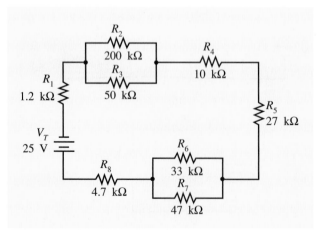

Figure 6-17. Classify this circuit.

3. Classify the circuit shown in Figure 6-18 as series-parallel or complex. If the circuit is series-parallel, then compute total resistance and total current.

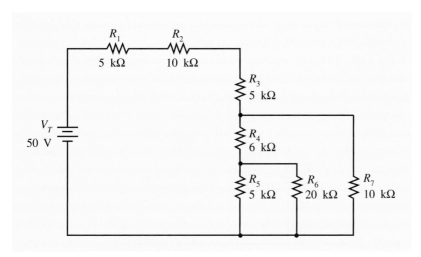

Figure 6-18. Classify this circuit.

4. Classify the circuit shown in Figure 6-19 as series-parallel or complex. If the circuit is series-parallel, then compute total resistance and total current.

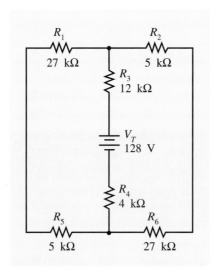

Figure 6-19. Classify this circuit.

6.3 Application of Kirchhoff's Voltage and Current Laws

Chapter 4 presented Kirchhoff's Voltage Law and used it in the solution of series circuit problems. In a similar way, Chapter 5 showed how to use Kirchhoff's Current Law to aid in the solution of parallel circuit problems. Both of these techniques can be applied to help us analyze series-parallel circuits.

Kirchhoff's Voltage Law

Refer to Figure 6-20 as we apply Kirchhoff's Voltage Law. In Figure 6-20, the voltage drops for each resistor are labeled. There are three possible complete loops in the circuit. They are

KEY POINTS

To use Kirchhoff's Voltage Law, draw as many complete loops on the circuit diagram as needed to include each component in at least one loop.

labeled as L_1, L_2, and L_3. Actually, loop L_3 is unnecessary since all of its voltage drops and sources are contained in one of the other loops. In general, you only need as many loops as it takes to include every voltage drop and every voltage source in at least one loop.

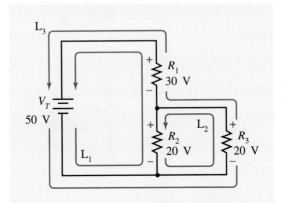

Figure 6-20. Apply Kirchhoff's Voltage Law to this circuit.

Let's write the Kirchhoff's Voltage Law equation for each of the loops in Figure 6-20.

$$L_1: V_2 + V_1 - V_T = 0$$
$$+ 20 \text{ V} + 30 \text{ V} - 50 \text{ V} = 0$$
$$L_2: V_3 - V_2 = 0$$
$$20 \text{ V} - 20 \text{ V} = 0$$
$$L_3: V_3 + V_1 - V_T = 0$$
$$+ 20 \text{ V} + 30 \text{ V} - 50 \text{ V} = 0$$

Since all of these equations do indeed equal zero, then we can be assured that the labeled voltages are correct. If a particular equation is shown to be incorrect (i.e., not equal to zero) then we know that one or more voltages is incorrect.

EXAMPLE SOLUTION

Write the loop equations for each of the labeled loops in Figure 6-21. Determine whether or not the voltages are labeled correctly.

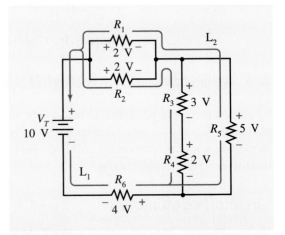

Figure 6-21. Use Kirchhoff's Voltage Law to determine if the voltages are labeled correctly.

When we apply Kirchhoff's Voltage Laws to the circuit in Figure 6-21, we get the following results.

$$L_1: +4 \text{ V} + 2 \text{ V} + 3 \text{ V} + 2 \text{ V} - 10 \text{ V} = 0$$
$$11 \text{ V} - 10 \text{ V} = 0$$
$$1 \text{ V} = 0 \text{ [Error]}$$
$$L_2: +4 \text{ V} + 5 \text{ V} + 2 \text{ V} - 10 \text{ V} = 0$$
$$11 \text{ V} - 10 \text{ V} = 0$$
$$1 \text{ V} = 0 \text{ [Error]}$$

Both of these equations produce incorrect results (i.e., 1 V ≠ 0). This means that one or more of the labeled voltages are wrong. In section 6.4, we will learn to compute the voltages and currents in all parts of a series-parallel circuit. After you have completed the computations, you can apply Kirchhoff's Laws to verify the correctness of your calculations.

EXAMPLE SOLUTION

Apply Kirchhoff's Voltage Law to the circuit in Figure 6-22 to determine if the labeled voltages are correct.

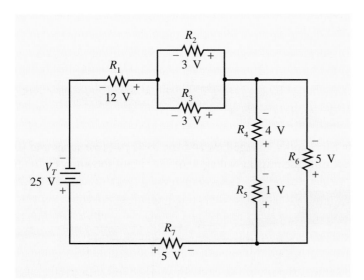

Figure 6-22. Are the voltages and polarities correctly labeled?

Figure 6-23 shows one possible set of loops that includes every voltage source and every voltage drop. The equations for loops L_1 and L_2 are as follows:

$$L_1: -V_7 - V_5 - V_4 - V_3 - V_1 + V_T = 0$$
$$-5 \text{ V} - 1 \text{ V} - 4 \text{ V} - 3 \text{ V} - 12 \text{ V} + 25 \text{ V} = 0$$
$$-25 \text{ V} + 25 \text{ V} = 0$$
$$L_2: -V_7 - V_6 - V_2 - V_1 + V_T = 0$$
$$-5 \text{ V} - 5 \text{ V} - 3 \text{ V} - 12 \text{ V} + 25 \text{ V} = 0$$
$$-25 \text{ V} + 25 \text{ V} = 0$$

Since the equations produce valid equalities, we can be fairly confident that the labeled voltages and polarities are correct.

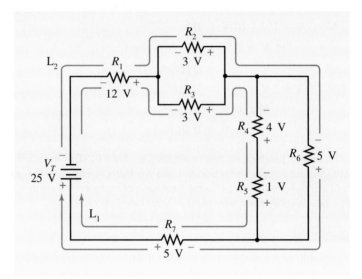

Figure 6-23. The indicated loops include every voltage source and every voltage drop.

Practice Problems

KEY POINTS

If we know all of the voltages but one in any given loop, then we can write the Kirchhoff's Voltage Law equation for that loop and solve for the unknown voltage value.

1. Apply Kirchhoff's Voltage Law and the basic rules for series and parallel circuits to determine the missing voltages for the circuit shown in Figure 6-24.

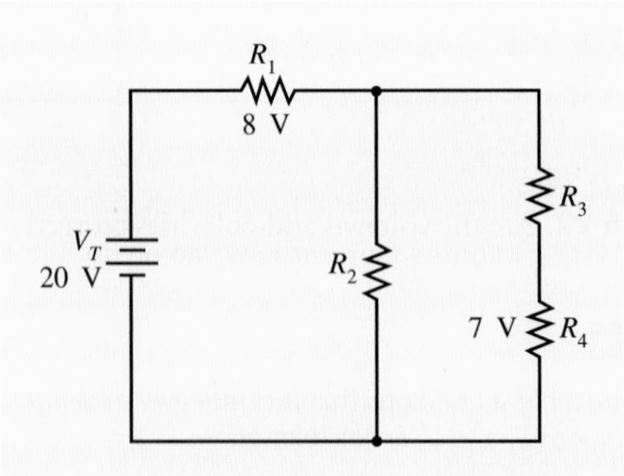

Figure 6-24. Use Kirchhoff's Voltage Law to determine the missing voltage drops.

2. Apply Kirchhoff's Voltage Law to determine the missing voltages in the circuit shown in Figure 6-25.

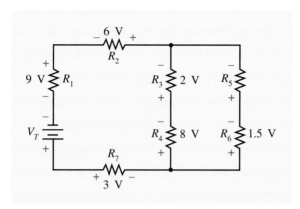

Figure 6-25. Determine the missing voltages with Kirchhoff's Voltage Law.

Answers to Practice Problems

1. $V_{R_2} = 12$ V; $V_{R_3} = 5$ V **2.** $V_{R_5} = 8.5$ V; $V_T = 28$ V

Kirchhoff's Current Law

Kirchhoff's Current Law can also be applied to help in the solution of series-parallel circuit problems. Consider the circuit shown in Figure 6-26 where some currents are unknown. We can apply Kirchhoff's Current Law to determine the missing values.

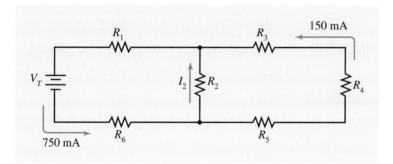

Figure 6-26. Kirchhoff's Current Law can be used to find all unknown currents.

First, we note that resistors R_3, R_4, and R_5 are in series with each other. There is only one path for current flow; therefore, $I_3 = I_4 = I_5 = 150$ mA.

Second, we examine the node at the lower end of R_2. We now know all of the currents entering and leaving this node except one (I_2). Kirchhoff's Current Law can be used to determine the current (I_2) through R_2 as follows.

$$750 \text{ mA} - 150 \text{ mA} - I_2 = 0$$

$$I_2 = 600 \text{ mA}$$

We now know all of the currents in the circuit with the single exception of I_1. We can determine I_1 in either of two ways. First, since R_1, V_T, and R_6 are in series with each other, their currents must all be the same value (750 mA). Alternatively, we can write the Kirchhoff's Current Law equation for the node on the upper end of R_2 since we know all but one of the currents entering and leaving this node. The Kirchhoff's Current Law equation follows.

$$I_2 + I_3 - I_1 = 0$$
$$600 \text{ mA} + 150 \text{ mA} - I_1 = 0$$
$$I_1 = 750 \text{ mA}$$

EXAMPLE SOLUTION

Determine the values of current through every component in the circuit shown in Figure 6-27.

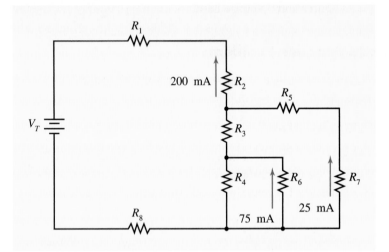

Figure 6-27. Determine the current through every component.

EXAMPLE SOLUTION

We can see that R_2, R_1, V_T, and R_8 are all in series with each other. Therefore, they all have the same value of current. That is,

$$I_1 = I_2 = I_8 = I_T = 200 \text{ mA}$$

Now let's write the Kirchhoff's Current Law equation for the node at the right end of R_8, since we know the value of all but one current (I_4) entering and leaving that point.

$$I_8 - I_4 - I_6 - I_7 = 0$$
$$200 \text{ mA} - I_4 - 75 \text{ mA} - 25 \text{ mA} = 0$$
$$I_4 = 100 \text{ mA}$$

We can now use Kirchhoff's Current Law to compute the current through R_3. That is,

$$I_4 + I_6 - I_3 = 0$$
$$100 \text{ mA} + 75 \text{ mA} - I_3 = 0$$
$$I_3 = 175 \text{ mA}$$

We could also have found the current through R_3 by writing the Kirchhoff's Current Law equation for the node at the upper end of R_3. In this case, we would have

$$I_3 + I_5 - I_2 = 0$$

$$I_3 + 25 \text{ mA} - 200 \text{ mA} = 0$$

$$I_3 = 175 \text{ mA}$$

In general, if you know all but one current entering or leaving a given node in a circuit, then you can find the remaining current by applying Kirchhoff's Current Law to that node.

Exercise Problems 6.3

1. Determine the value of voltage across R_3 in Figure 6-28.

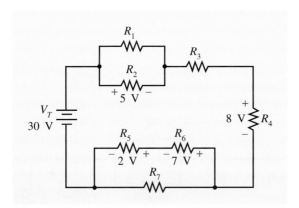

Figure 6-28. Compute the voltage across R_3.

2. How much current flows through R_6 in Figure 6-29?

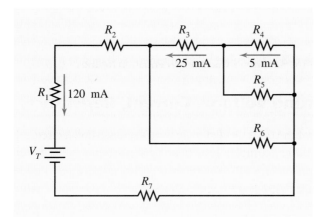

Figure 6-29. Calculate the current through R_6.

3. Is the value of current through R_1 and R_5 the same in Figure 6-30? Explain why or why not.

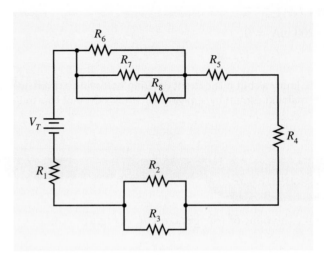

Figure 6-30. Do R_1 and R_5 have equal currents?

4. Determine the voltage across every component in Figure 6-31.

5. Determine the current through every component in Figure 6-31.

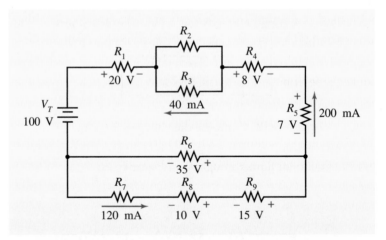

Figure 6-31. Calculate all component voltages.

6.4 Computing Voltage, Current, Resistance, and Power

You now have all of the tools needed to analyze a series-parallel circuit. In this section, you will apply those tools to completely analyze a given series-parallel circuit.

Fully Specified Circuits

The basic sequence of activities required to completely analyze a series-parallel circuit with all component values given follows:

1. Draw a series of progressively simpler circuits by replacing series and parallel components with an equivalent resistance.

2. Compute the value of total resistance.

3. Apply Ohm's Law and Kirchhoff's Voltage and Current Laws to determine all voltages and currents as you begin with the fully simplified circuit and work back through the intermediate sketches to end with the original schematic.

KEY POINTS

All circuit values such as voltage, current, and power can be found in a series-parallel circuit by forming simpler circuits and then applying Ohm's and Kirchhoff's Laws.

EXAMPLE SOLUTION

Analyze the circuit shown in Figure 6-32 to determine all voltage, current, resistance, and power values.

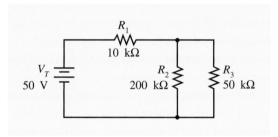

Figure 6-32. Analyze this circuit.

EXAMPLE **SOLUTION**

Our first task will be to combine series and parallel components (as discussed in Sections 6.1 and 6.2) until we have a single resistor representing total resistance. We can see that resistors R_2 and R_3 are in parallel. We will loop these components and label the loop as $R_{2,3}$. The value of the parallel combination can be found with the parallel resistance equation.

$$R_{2,3} = \frac{1}{\frac{1}{R_2} + \frac{1}{R_3}}$$

$$R_{2,3} = \frac{1}{\frac{1}{200 \text{ k}\Omega} + \frac{1}{50 \text{ k}\Omega}}$$

$$R_{2,3} = 40 \text{ k}\Omega$$

Our progress so far is shown in Figure 6-33(a) and (b).

Next, we see that resistors R_1 and $R_{2,3}$ in Figure 6-33(b) are in series. We loop these and label the equivalent resistance as $R_{1,2,3}$. We apply the series resistance equation to determine the value of effective resistance.

$$R_{1,2,3} = R_T = R_1 + R_{2,3}$$

$$R_T = 10 \text{ k}\Omega + 40 \text{ k}\Omega$$

$$R_T = 50 \text{ k}\Omega$$

Our final simplified circuit is shown in Figure 6-33(c).

We are now ready to work our way back through the intermediate circuits until we reach the original circuit. At each intermediate step, we will calculate as many circuit values as we can.

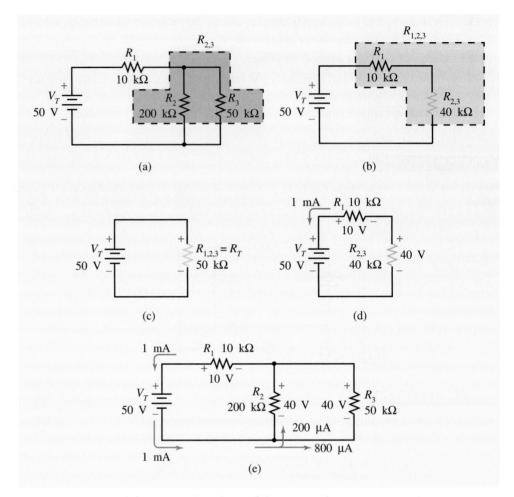

Figure 6-33. Simplification and analysis of the circuit shown in Figure 6-32.

Let's begin with Figure 6-33(c), our most simplified circuit. The circuit in Figure 6-33(c) provides us with total voltage and total resistance. We can apply Ohm's Law to find total current.

$$I_T = \frac{V_T}{R_T}$$

$$I_T = \frac{50 \text{ V}}{50 \text{ k}\Omega}$$

$$I_T = 1 \text{ mA}$$

We can also apply the power equation to find total power.

$$P_T = I_T V_T$$

$$P_T = 1 \text{ mA} \times 50 \text{ V} = 50 \text{ mW}$$

There are no more circuit values that can be found on this diagram, so we move to the previous sketch (Figure 6-33(b)). The first step when we begin work on a new schematic is to transfer all of our values from the previous schematic. Figure 6-33(b) is redrawn in Figure 6-33(d), and we have added the 1-mA total current value computed on the prior schematic. Note that resistor $R_{1,2,3}$ had 1 mA through it in Figure 6-33(c). Therefore, the series resistors R_1 and $R_{2,3}$ in Figure 6-33(d) must have 1 mA through each of them. If we know the current through a resistor, then we can apply Ohm's Law to calculate the voltage drop. In the case of resistor R_1, we have

$$V_1 = I_1 R_1$$
$$V_1 = 1 \text{ mA} \times 10 \text{ k}\Omega$$
$$V_1 = 10 \text{ V}$$

The voltage drop across $R_{2,3}$ can be found in the same way, or we can apply Kirchhoff's Voltage Law, since we know all voltages in the loop but one. Let's choose the latter approach.

$$V_{2,3} + V_1 - V_T = 0$$
$$V_{2,3} + 10 \text{ V} - 50 \text{ V} = 0$$
$$V_{2,3} = 40 \text{ V}$$

The results of the last two calculations are shown in Figure 6-33(d). We can also compute the power dissipated in resistor R_1 as follows.

$$P_1 = \frac{V_1^2}{R_1}$$
$$P_1 = \frac{(10 \text{ V})^2}{10 \text{ k}\Omega}$$
$$P_1 = 10 \text{ mW}$$

Now, let's transfer our newly computed information to the next schematic which, in our present example, is the original diagram. In Figure 6-33(e) we have labeled V_1 and I_1. Additionally, we have split resistor $R_{2,3}$ into its component parts (R_2 and R_3). Since resistor $R_{2,3}$ had 40 V across it, then parallel components R_2 and R_3 must each have 40 V across them. This is also labeled in Figure 6-33(e).

Since we now know the voltage across resistors R_2 and R_3, we can apply Ohm's Law to calculate the current through each of them. In the case of R_2, we have

$$I_2 = \frac{V_2}{R_2}$$
$$I_2 = \frac{40 \text{ V}}{200 \text{ k}\Omega}$$
$$I_2 = 200 \text{ }\mu\text{A}$$

Similarly, we can calculate the current through R_3 as

$$I_3 = \frac{V_3}{R_3}$$
$$I_3 = \frac{40 \text{ V}}{50 \text{ k}\Omega}$$
$$I_3 = 800 \text{ }\mu\text{A}$$

To complete our analysis, we can compute the power dissipation in resistors R_2 and R_3. In the case of R_2, we have

$$P_2 = \frac{V_2^2}{R_2}$$

$$P_2 = \frac{(40\ \text{V})^2}{200\ \text{k}\Omega}$$

$$P_2 = 8\ \text{mW}$$

And, in the case of resistor R_3, we calculate power dissipation as

$$P_3 = I_3 V_3$$

$$P_3 = 800\ \mu\text{A} \times 40\ \text{V}$$

$$P_3 = 32\ \text{mW}$$

This completes the analysis of the circuit originally shown in Figure 6-32. The final results (except for power dissipation) are labeled on Figure 6-33(e).

You can solve any series-parallel circuit analysis problem by applying a process similar to that just completed. Although the details of the analysis will vary from one circuit to the next, the basic steps remain the same.

EXAMPLE SOLUTION

Compute the current, voltage drop, and power dissipation for every component in the circuit shown in Figure 6-34.

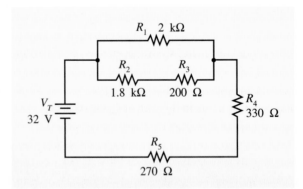

Figure 6-34. Compute all voltages, currents, and power dissipations for this circuit.

EXAMPLE **SOLUTION**

One possible first step is to recognize that resistors R_2 and R_3 are in series. The combined resistance is computed with the series resistance equation.

$$R_{2,3} = R_2 + R_3$$

$$= 1.8\ \text{k}\Omega + 200\ \Omega$$

$$= 2\ \text{k}\Omega$$

The results of the simplification up to this point are shown in Figure 6-35(b).

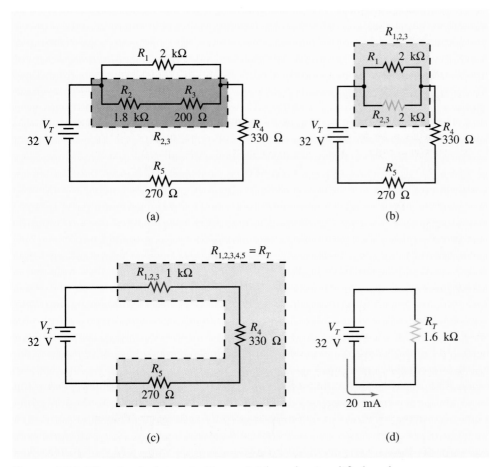

Figure 6-35. The circuit shown in Figure 6-34 can be simplified as shown.

We can see in Figure 6-35(b) that resistors R_1 and $R_{2,3}$ are in parallel. We use the parallel resistance equation to compute their combined resistance.

$$R_{1,2,3} = \frac{2 \text{ k}\Omega}{2} = 1 \text{ k}\Omega$$

The results of this portion of the simplification are shown in Figure 6-35(c).

Next, we see that resistors $R_{1,2,3}$, R_4, and R_5 are all in series. We apply the series resistance equation to determine their combined resistance, which in this case will also be the total resistance of the circuit.

$$R_T = R_{1,2,3,4,5} = R_{1,2,3} + R_4 + R_5$$
$$= 1 \text{ k}\Omega + 330 \ \Omega + 270 \ \Omega = 1.6 \text{ k}\Omega$$

We can now find the value of total current in the circuit by applying Ohm's Law as follows:

$$I_T = \frac{V_T}{R_T}$$
$$= \frac{32 \text{ V}}{1.6 \text{ k}\Omega} = 20 \text{ mA}$$

We can also compute the total power dissipation for the circuit by applying any one of the various forms of the power equation.

$$P_T = I_T V_T$$
$$= 20 \text{ mA} \times 32 \text{ V} = 640 \text{ mW}$$

The fully simplified schematic is shown in Figure 6-35(d).

We are now ready to begin working backward through the various intermediate schematics, computing values as we do. We can transfer the current value computed in Figure 6-35(d) to the preceding circuit of Figure 6-35(c). Since all three resistors in Figure 6-35(c) are in series, they will all have the same current (I_T) flowing through them. We can compute their individual voltage drops with Ohm's Law as follows:

$$V_{1,2,3} = I_{1,2,3} R_{1,2,3}$$
$$= 20 \text{ mA} \times 1 \text{ k}\Omega = 20 \text{ V}$$
$$V_4 = I_4 R_4$$
$$= 20 \text{ mA} \times 330 \ \Omega = 6.6 \text{ V}$$
$$V_5 = I_5 R_5$$
$$= 20 \text{ mA} \times 270 \ \Omega = 5.4 \text{ V}$$

Our results so far are shown in Figure 6-36(a).

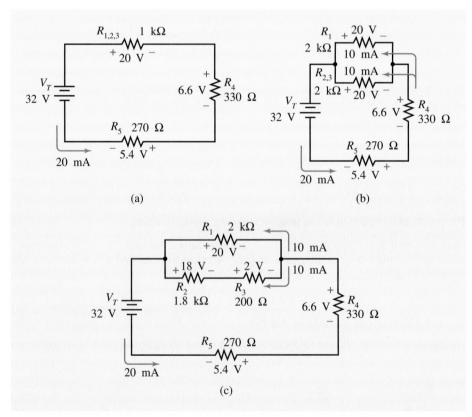

Figure 6-36. Computational steps to analyze the circuit shown in Figure 6-34.

While we are still working with Figure 6-36(a), let's compute the power dissipation in resistors R_4 and R_5.

$$P_4 = I_4^2 R_4$$
$$= (20 \text{ mA})^2 \times 330 \ \Omega = 132 \text{ mW}$$

$$P_5 = \frac{V_5^2}{R_5}$$
$$= \frac{(5.4 \text{ V})^2}{270 \ \Omega} = 108 \text{ mW}$$

Since there are no more circuit values that we can compute with Figure 6-36(a), let us move to the next intermediate schematic shown in Figure 6-35(b). Here, resistor $R_{1,2,3}$ has been broken into two resistors (R_1 and $R_{2,3}$). When we transfer the previously calculated circuit values to Figure 6-35(b), we find that resistors R_1 and $R_{2,3}$ each have 20 V across them, since they are in parallel. Knowing this enables us to compute the individual currents for these two resistances.

$$I_1 = \frac{V_1}{R_1}$$
$$= \frac{20 \text{ V}}{2 \text{ k}\Omega} = 10 \text{ mA}$$

$$I_{2,3} = \frac{V_{2,3}}{R_{2,3}}$$
$$= \frac{20 \text{ V}}{2 \text{ k}\Omega} = 10 \text{ mA}$$

We can also compute the power dissipation in resistor R_1.

$$P_1 = \frac{V_1^2}{R_1}$$
$$= \frac{(20 \text{ V})^2}{2 \text{ k}\Omega} = 200 \text{ mW}$$

The results of our computations to this point are shown in Figure 6-36(b).

We are now ready to transfer our computed values to the original circuit shown in Figure 6-35(a). Here, we are breaking resistor $R_{2,3}$ into its component parts (R_2 and R_3). Since these resistors are in series, they will have the same value of current ($I_{2,3}$). We can now compute their individual voltage drops.

$$V_2 = I_2 R_2$$
$$= 10 \text{ mA} \times 1.8 \text{ k}\Omega = 18 \text{ V}$$
$$V_3 = I_3 R_3$$
$$= 10 \text{ mA} \times 200 \ \Omega = 2 \text{ V}$$

We can also compute the power dissipation for resistors R_2 and R_3.

$$P_2 = \frac{V_2^2}{R_2}$$
$$= \frac{(18 \text{ V})^2}{1.8 \text{ k}\Omega} = 180 \text{ mW}$$

$$P_3 = \frac{V_3^2}{R_3}$$

$$= \frac{(2\ \mathrm{V})^2}{200\ \Omega} = 20\ \mathrm{mW}$$

This completes the analysis of our circuit. The final results (except power dissipations) are shown in Figure 6-36(c). It is always a good idea to verify your calculations with Kirchhoff's Voltage and Current Laws. Do the voltage drops and sources in every closed loop add up to zero? Do all of the currents entering and leaving every node in the circuit equal zero? If they do, then you have probably solved the problem correctly. If, on the other hand, either of Kirchhoff's Laws indicates a problem, then you can be certain that an error was made.

Partially Specified Circuits

The analysis presented earlier is applicable to any series-parallel circuit that has all of the component values given. In some cases, however, only partial information is available. Although the exact method of solution will vary dramatically from one problem to the next, the basic procedure follows:

1. Simplify the circuit when possible by replacing sets of series and parallel components with equivalent resistances.

2. Compute as many circuit values as possible in the simplified circuit.

3. Work backward through the intermediate drawings, computing as many circuit values as possible on each schematic.

These steps are very similar to the procedure described earlier, but because not all component values are provided, you will be unable to fully simplify the circuit initially. Most of your effort will normally be spent on the second step. Success in this area depends on your creative application of Ohm's Law, Kirchhoff's Voltage and Current Laws, and your basic understanding of series circuits and parallel circuits.

Suppose, for example, our goal is to determine the value of R_2 in Figure 6-37. First, note that we have no direct information about R_2. We don't know its voltage, current, or power so we cannot immediately calculate its resistance. Further, we cannot simplify the circuit by replacing series or parallel components, since we would be unable to compute the equivalent resistance of R_2 and R_3. So, we must begin with the second step; compute as many circuit values as possible.

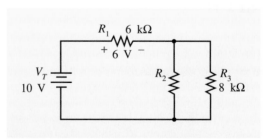

Figure 6-37. Determine the value of R_2.

In general, try to locate components that have two known circuit values (e.g., voltage and current, current and resistance, power and voltage, and so on). If you know any two values, then you can easily compute all others with Ohm's Law or one of the power equations.

In the present case, we have only one component (R_1) that has two known circuit values. We can compute the current through R_1 with Ohm's Law.

$$I_1 = \frac{V_1}{R_1}$$

$$= \frac{6 \text{ V}}{6 \text{ k}\Omega} = 1 \text{ mA}$$

We can also find the power dissipation in R_1.

$$P_1 = \frac{V_1^2}{R_1}$$

$$= \frac{(6 \text{ V})^2}{6 \text{ k}\Omega} = 6 \text{ mW}$$

Since our goal is to find the value of R_2, it may not be necessary to compute the power dissipation in R_1. However, until you gain enough experience to be certain that a particular circuit value will not be needed, it is advisable to compute all parameters (i.e., current, voltage, resistance, and power).

Now let's apply Kirchhoff's Voltage Law to the circuit. We will choose a loop where the voltage across R_2 is the only unknown value. Figure 6-38 shows the chosen loop. The voltage equation and solution for this loop are

$$V_2 + V_1 - V_T = 0$$
$$V_2 + 6 \text{ V} - 10 \text{ V} = 0$$
$$V_2 = 4 \text{ V}$$

Now that we have the voltage across R_2, we need only find one more parameter for R_2 in order to be able to compute its value. Let's work on finding the current through R_2. If we knew the current through R_3, then we could apply Kirchhoff's Current Law to the lower

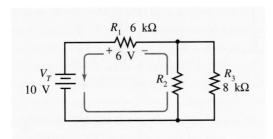

Figure 6-38. Application of Kirchhoff's Voltage Law to the circuit shown in Figure 6-37.

node and determine the current through R_2. We previously computed the voltage across R_2 as 4 V. Since R_2 is paralleled by R_3, then they must both have 4 V across them. The current through R_3 can be found with Ohm's Law.

$$I_3 = \frac{V_3}{R_3}$$

$$= \frac{4 \text{ V}}{8 \text{ k}\Omega} = 500 \text{ } \mu A$$

We now know all but one current (I_{R_2}) for the lower node. Applying Kirchhoff's Current Law to the lower node gives us the following.

$$I_T - I_2 - I_3 = 0$$
$$1 \text{ mA} - I_2 - 500 \mu A = 0$$
$$I_2 = 500 \text{ } \mu A$$

Now that we know two circuit values associated with R_2, we can determine its resistance.

$$R_2 = \frac{V_2}{I_2}$$

$$= \frac{4 \text{ V}}{500 \text{ } \mu A} = 8 \text{ k}\Omega$$

There are many alternate ways that could have been used to find the value of R_2. As you become more experienced, you will more quickly identify the most effective method for a given problem. But as you are gaining the experience, it is not possible for you to calculate too much. That is, if you are uncertain about the next step, or you are unable to visualize all of the steps needed for a solution, then calculate all of the values that you can whether or not they appear to be needed. This availability of information will often make the solution more apparent.

Now let's try to determine the value of applied voltage (V_T) in Figure 6-39. No immediate simplifications (i.e., series and parallel equivalents) are possible, so we will try to get more information by calculating some circuit values.

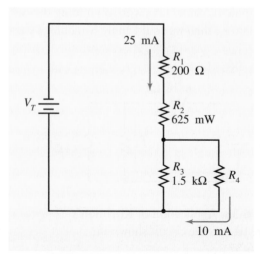

Figure 6-39. What is the value of applied voltage?

We know the resistance and current for R_1, so we can calculate the voltage drop with Ohm's Law.

$$V_1 = I_1 R_1$$
$$= 25 \text{ mA} \times 200 \ \Omega = 5 \text{ V}$$

Since R_1 and R_2 are in series, they must have the same current (25 mA). We can use this current value and the given power dissipation to compute the voltage drop across R_2.

$$P_2 = I_2 V_2, \text{ or}$$
$$V_2 = \frac{P_2}{I_2}$$
$$= \frac{625 \text{ mW}}{25 \text{ mA}} = 25 \text{ V}$$

If we knew the voltage across R_3, then we would know all voltages in the loop containing V_T with the exception of V_T. Kirchhoff's Voltage Law could be used to find V_T. So let's work on finding V_3.

We can apply Kirchhoff's Current Law to the lower node, since we know all currents except I_3.

$$-I_T + I_3 + I_4 = 0$$
$$-25 \text{ mA} + I_3 + 10 \text{ mA} = 0$$
$$I_3 = 15 \text{ mA}$$

We can now calculate the voltage drop across R_3 with Ohm's Law.

$$V_3 = I_3 R_3$$
$$= 15 \text{ mA} \times 1.5 \text{ k}\Omega = 22.5 \text{ V}$$

We can now apply Kirchhoff's Voltage Law to the loop shown in Figure 6-40.

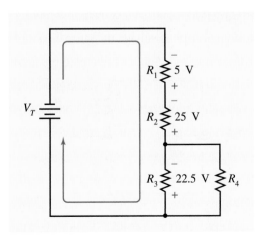

Figure 6-40. Application of Kirchhoff's Voltage Law to the analysis of the circuit shown in Figure 6-39.

$$+ V_1 + V_2 + V_3 - V_T = 0$$
$$+ 5\ \text{V} + 25\ \text{V} + 22.5\ \text{V} - V_T = 0$$
$$V_T = 52.5\ \text{V}$$

This completes our problem, which was to find the value of applied voltage. Be reminded that there are many alternate ways to solve this problem, but the overall procedure is similar. Again, it is recommended that you verify your final results by applying Kirchhoff's Voltage and Current Laws to the circuit.

Practice Problems

1. Calculate the value of current through resistor R_6 in Figure 6-41.

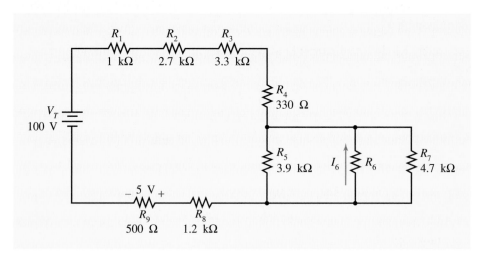

Figure 6-41. Calculate the current (I_6) through resistor R_6.

2. Calculate the value of applied voltage in Figure 6-42.

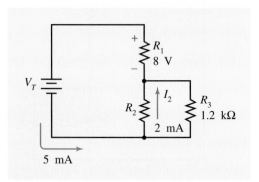

Figure 6-42. Find the value of V_T.

Answers to Practice Problems

1. $I_6 = 5.449$ mA   **2.** $V_T = 11.6$ V

1.   Refer to Figure 6-43. Compute total resistance and total current.

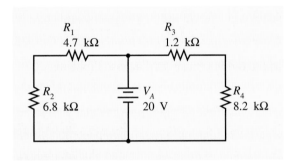

Figure 6-43. Compute total resistance and total current.

2.   Refer to Figure 6-44. Determine the values of total resistance (R_T) and total current (I_T).

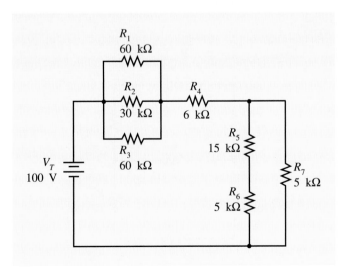

Figure 6-44. Circuit for Exercise Problem 2.

3. Refer to Figure 6-45 and complete Table 6-2.

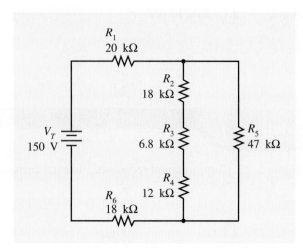

Figure 6-45. Circuit for Exercise Problem 3.

	R_1	R_2	R_3	R_4	R_5	R_6	TOTAL
Voltage							150 V
Current							
Power							

Table 6-2. Complete Table for the Circuit in Figure 6-45

4. Refer to Figure 6-46. Calculate the values of V_3, I_5, and P_T.

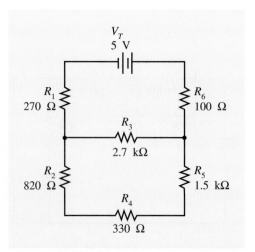

Figure 6-46. Circuit for Exercise Problem 4.

5. Refer to Figure 6-47. Determine values for R_T, I_T, V_3, and I_4.

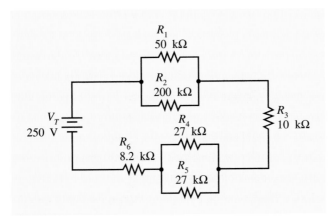

Figure 6-47. Circuit for Exercise Problem 5.

6. Refer to Figure 6-48. Compute the value of resistor R_3.

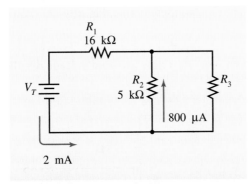

Figure 6-48. Circuit for Exercise Problem 6.

6.5 Applied Technology: Series-Parallel Circuits

Nearly all practical electronic circuits are composed of series-parallel circuits. Sometimes, the series-parallel nature of the circuit is obvious, as in the cases of loaded voltage dividers and bridge circuits, which are discussed in this section. In other cases, the series-parallel nature of the circuit is concealed by other electronic devices. That is, portions of the circuit may be internal to a more complex component such as a transistor or an integrated circuit. But, when these circuits are analyzed, they are often viewed as series-parallel circuits. So it is important for you to be able to confidently analyze series-parallel circuits.

Loaded Voltage Dividers

In Chapter 4, we discussed voltage dividers. Figure 6-49 shows a simple voltage divider that provides voltages of +5 and +15 V from a +20-V source. A circuit of this type is more correctly called an unloaded voltage divider, since no other components are connected to the +5- and +15-V connections.

● **KEY POINTS**

Voltage dividers are circuits that divide a given source voltage into one or more lower voltages.

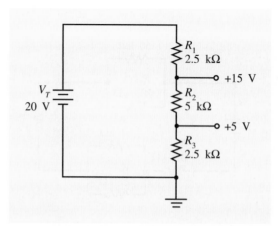

Figure 6-49. A basic unloaded voltage divider.

In order for the divided voltages (+5 V and +15 V in the case of Figure 6-49) to be useful, we must connect other circuitry to these points. Figure 6-50 shows resistors R_4 and R_5 connected to the +5- and +15-V points, respectively. These resistors are called **load resistors** and the current they draw is called **load current.** The voltage divider circuit consisting of V_T and R_1 through R_3 is now a **loaded voltage divider.**

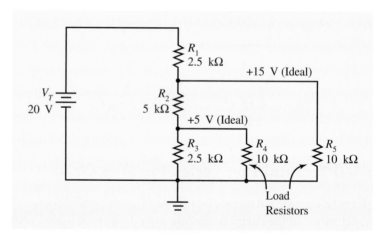

Figure 6-50. A loaded voltage divider supplies current to other components.

As you can see by examining the circuit in Figure 6-50, a loaded voltage divider is really just a basic series-parallel circuit. We have already discussed methods for analyzing this type of circuit, but let us make some observations by contrasting the circuits shown in Figures 6-49 and 6-50.

Consider the value of current flow through R_1 before and after the load resistors are connected. Prior to connecting the load resistors, the only current flowing through R_1 is the current through R_2 and R_3. When one or more load resistors are connected, R_1 will still have the current that flows through R_2 and R_3, but it will also have the current that

KEY POINTS

When a load resistor is connected to a voltage divider, the total current in the circuit increases and the divided voltages decrease.

flows through the load resistors. Figure 6-51 shows the current paths for the loaded voltage divider. Now, since R_1 has more current with the loads connected, it must drop more voltage (Ohm's Law). If R_1 drops more voltage, then there will be less voltage available for the remainder of the circuit (Kirchhoff's Voltage Law). Therefore, we would expect the +5- and +15-V sources to become less. Let's confirm these observations numerically by analyzing and comparing the circuits shown in Figure 6-49 and 6-50.

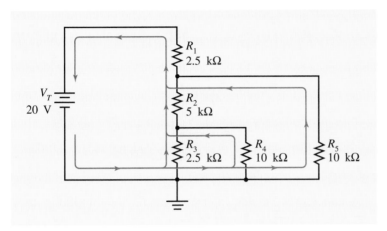

Figure 6-51. Current paths for a loaded voltage divider.

Figure 6-49 is a simple series circuit. We can find total resistance by adding the individual resistances.

$$R_T = R_1 + R_2 + R_3$$
$$= 2.5 \text{ k}\Omega + 5 \text{ k}\Omega + 2.5 \text{ k}\Omega$$
$$= 10 \text{ k}\Omega$$

Total current (and the current through R_1) can be calculated with Ohm's Law.

$$I_T = \frac{V_T}{R_T}$$
$$= \frac{20 \text{ V}}{10 \text{ k}\Omega} = 2 \text{ mA}$$

And, of course, the voltage across R_1 can also be found with Ohm's Law as

$$V_1 = I_1 R_1$$
$$= 2 \text{ mA} \times 2.5 \text{ k}\Omega = 5 \text{ V}$$

The voltages across R_2 and R_3 can be found with Ohm's Law or with the voltage divider equation presented in Chapter 4. That is,

$$V_2 = \frac{R_2}{R_T} V_T$$
$$= \frac{5 \text{ k}\Omega}{10 \text{ k}\Omega} \times 20 \text{ V}$$
$$= 10 \text{ V}$$

$$V_3 = \frac{R_3}{R_T} V_T$$

$$= \frac{2.5 \text{ k}\Omega}{10 \text{ k}\Omega} \times 20 \text{ V}$$

$$= 5 \text{ V}$$

Now, let's analyze Figure 6-50 and determine the following values: I_1, V_1, V_2, V_3, and the two divided voltages. Figure 6-52 shows the scratch-pad sketches used to simplify the circuit. The calculations are as follows. First, the parallel combination of R_3 and R_4 is computed.

$$R_{3,4} = \frac{1}{\dfrac{1}{R_3} + \dfrac{1}{R_4}}$$

$$= \frac{1}{\dfrac{1}{2.5 \text{ k}\Omega} + \dfrac{1}{10 \text{ k}\Omega}}$$

$$= 2 \text{ k}\Omega$$

The series combination of $R_{3,4}$ and R_2 can be combined, as shown in Figures 6-52(a) and (b), by applying the series resistance equation.

$$R_{2,3,4} = R_2 + R_{3,4}$$

$$= 5 \text{ k}\Omega + 2 \text{ k}\Omega$$

$$= 7 \text{ k}\Omega$$

Combining $R_{2,3,4}$ with the parallel resistor R_5 is shown in Figures 6-52(b) and (c) and computed as follows.

$$R_{2,3,4,5} = \frac{1}{\dfrac{1}{R_{2,3,4}} + \dfrac{1}{R_5}}$$

$$= \frac{1}{\dfrac{1}{7 \text{ k}\Omega} + \dfrac{1}{10 \text{ k}\Omega}}$$

$$\approx 4.12 \text{ k}\Omega$$

And finally, total resistance is found by combining the series resistance of R_1 and $R_{2,3,4,5}$ as shown in Figure 6-52(c) and (d).

$$R_T = R_1 + R_{2,3,4,5}$$

$$= 2.5 \text{ k}\Omega + 4.12 \text{ k}\Omega$$

$$= 6.62 \text{ k}\Omega$$

Ohm's Law can be used to compute total current as

$$I_T = \frac{V_T}{R_T}$$

$$= \frac{20 \text{ V}}{6.62 \text{ k}\Omega}$$

$$\approx 3.02 \text{ mA}$$

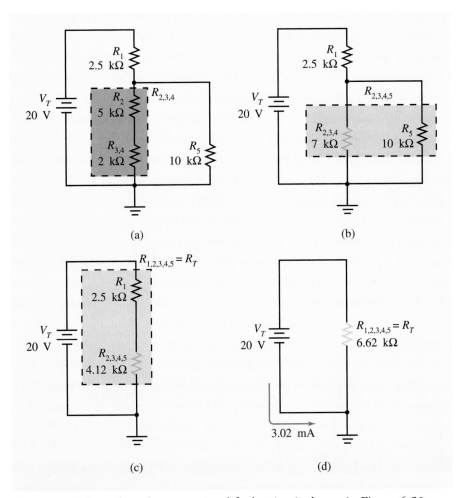

(a)

(b)

(c)

(d)

Figure 6-52. Scratch-pad steps to simplify the circuit shown in Figure 6-50.

The preceding steps and results are shown in Figure 6-52. We are now ready to work our way back through the simplified circuits in the reverse direction to compute all of the individual circuit values. Let us begin by moving to the circuit in Figure 6-52(c). Here, we can calculate V_1 and $V_{2,3,4,5}$. Let's use the voltage divider equation (or Ohm's Law if you prefer) to calculate V_1.

$$V_1 = \frac{R_1}{R_T}V_T$$

$$= \frac{2.5 \text{ k}\Omega}{6.62 \text{ k}\Omega} \times 20 \text{ V}$$

$$\approx 7.55 \text{ V}$$

We can compute $V_{2,3,4,5}$ by applying Kirchhoff's Voltage Law.

$$+V_{2,3,4,5} + V_1 - V_T = 0$$

$$V_{2,3,4,5} + 7.55\text{V} - 20 \text{ V} = 0$$

$$V_{2,3,4,5} = 12.45 \text{ V}$$

We can also note by comparing Figures 6-52(b) and (c), that this is the same voltage that is across R_5 and is one of the divided voltages (ideally +15 V).

Now, let's transfer our information to Figure 6-52(b) and compute $I_{2,3,4}$.

$$I_{2,3,4} = \frac{V_{2,3,4}}{R_{2,3,4}}$$

$$= \frac{12.45 \text{ V}}{7 \text{ k}\Omega}$$

$$\approx 1.78 \text{ mA}$$

If we now transfer this current to the circuit in Figure 6-52(a), we can calculate V_2 and $V_{3,4}$. We will use Ohm's Law to find V_2.

$$V_2 = I_2 R_2$$

$$= 1.78 \text{ mA} \times 5 \text{ k}\Omega$$

$$= 8.9 \text{ V}$$

Similarly, we can calculate $V_{3,4}$.

$$V_{3,4} = I_{3,4} R_{3,4}$$

$$= 1.78 \text{ mA} \times 2 \text{ k}\Omega$$

$$= 3.56 \text{ V}$$

We can note that this latter calculation also gives us the value for one of our divided voltages (ideally +5 V).

Now, let's contrast the results of these calculations with those associated with Figure 6-49. Both sets of results are shown in Table 6-3.

	UNLOADED DIVIDER (FIGURE 6-49)	LOADED DIVIDER (FIGURE 6-50)
I_1	2 mA	3.02 mA
V_1	5 V	7.55 V
V_2	10 V	8.9 V
V_3	5 V	3.56 V
+5-V source	+5 V	+3.56 V
+15-V source	+15 V	+12.45 V

Table 6-3. Comparison of Loaded and Unloaded Voltage Dividers

KEY POINTS

A decrease in voltage when a load resistor draws current from a voltage divider is called *loading*.

First, observe that our initial beliefs that the current through R_1 and the voltage across it would increase were confirmed by our calculations. And, as we expected, all other voltages in the circuit decreased. The divider outputs dropped to +3.56 V and +12.45 V for the +5- and +15-V outputs, respectively. This decrease in load voltage when a load resistor draws current is called **loading.**

So, in general, the divided output voltages of a voltage divider circuit will decrease when the voltage divider is loaded. The greater the load (i.e., more current flow), the lower the divided voltages become.

Voltage Regulation

Ideally, the voltage provided by a voltage divider would remain constant with changes in load current. In practice, however, the divided voltage varies with load current as shown in our previous example. The ability of a circuit to maintain a constant load voltage under conditions of changing load current is called **voltage regulation.** How well a circuit regulates load voltage can be expressed as a percentage and is computed with Equation 6-1.

$$\% \text{ regulation} = \frac{V_{\text{no load}} - V_{\text{full load}}}{V_{\text{full load}}} \times 100 \quad (6\text{-}1)$$

Let's determine the voltage regulation for the voltage divider shown in Figure 6-50. The computed values for $V_{\text{no load}}$ and $V_{\text{full load}}$ are shown in Table 6-3 and are +5 V and +3.56 V, respectively, for the +5-V output. The +15-V output had values of +15 V and +12.45 V for $V_{\text{no load}}$ and $V_{\text{full load}}$, respectively. Computing the voltage regulation for the +5-V output gives us

$$\% \text{ regulation} = \frac{V_{\text{no load}} - V_{\text{full load}}}{V_{\text{full load}}} \times 100$$

$$= \frac{5 \text{ V} - 3.56 \text{ V}}{3.56 \text{ V}} \times 100$$

$$= 40.45\%$$

A similar calculation gives us the voltage regulation for the +15-V output.

$$\% \text{ regulation} = \frac{V_{\text{no load}} - V_{\text{full load}}}{V_{\text{full load}}} \times 100$$

$$= \frac{15 \text{ V} - 12.45 \text{ V}}{12.45 \text{ V}} \times 100$$

$$= 20.48\%$$

Neither of these values qualifies as "good" voltage regulation. Let's try to improve the circuit.

Figure 6-51 shows the current paths for a loaded voltage divider. The current through R_3 is called **bleeder current** (I_B) and flows with or without a load connected. The greater the bleeder current relative to the load current, the better the voltage regulation. That is, if we increase the amount of bleeder current in a particular voltage divider design, we will improve the voltage regulation.

Figure 6-53 shows another voltage divider that is designed to provide +5- and +15-V outputs from a single 20-V source. It differs from the circuit in Figure 6-50 in that it was designed with a substantially higher bleeder current. If we analyze this circuit in the same way we did for the circuit in Figure 6-50, we will get the results shown in Table 6-4. Compare these results with those in Table 6-3.

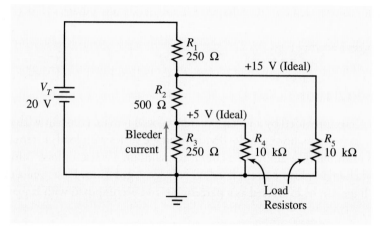

Figure 6-53. Higher bleeder current provides improved voltage regulation.

	UNLOADED DIVIDER	LOADED DIVIDER
I_1	20 mA	21.22 mA
V_1	5 V	5.31 V
V_2	10 V	9.87 V
V_3	5 V	4.82 V
+5-V source	+5 V	+4.82 V
+15-V source	+15 V	+14.69 V

Table 6-4. Comparison of Loaded and Unloaded Voltage Dividers with Increased Bleeder Current

The change in current through R_1 between no load and full load is now a much lower percentage. Consequently, the voltage across R_1 does not change as much. Similarly, all of the remaining voltages still decrease when a full load current is drawn, but the decrease is substantially less.

The practical limit on how much we can increase bleeder current to improve voltage regulation is determined by the power dissipation in resistors R_1 through R_3 in Figure 6-53. The greater the value of bleeder current, the higher the power dissipation in these resistors.

Voltmeter Loading

In Chapter 9, you will learn that a voltmeter is ideally an open circuit (i.e., infinite internal resistance). You will also learn that a practical voltmeter has a much lower internal resistance, depending on the type of meter and, in some cases, on the range selected. The internal resistance of the meter may range from about 1,000 Ω to several megohms.

Because the voltmeter has less than infinite resistance, it can act as a load when connected into a circuit. As with the voltage divider circuits previously studied, the connection of a load (in this case a voltmeter) causes the voltage to decrease. In the case of a voltmeter, this decrease in voltage when the meter is connected is called **voltmeter loading.** Figure 6-54 illustrates how voltmeter loading can cause erroneous voltage indications in a circuit.

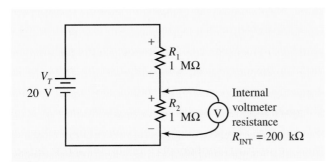

Figure 6-54. A voltmeter can load the circuit and produce erroneous indications.

Before the voltmeter is connected in Figure 6-54, the two equal series resistors divide the applied voltage equally with 10 V across each 1-MΩ resistor. But when the voltmeter is connected to measure the voltage drop across R_2, its internal resistance (200 kΩ in this case) appears in parallel with R_2. If we analyze this "new" circuit as a three-resistor, series-parallel circuit, we can determine that the voltage across R_2 and the meter will now be 2.86 V.

Thus, before the meter is connected, the voltage across R_2 is 10 V. After the meter is connected, the voltage across R_2 is only 2.86 V. If you were troubleshooting the circuit, you might be led to believe there is a malfunction at this point when, in fact, the voltmeter is simply loading the circuit.

You should be particularly aware of this problem when measuring high-resistance circuits. In general, if the resistance of the voltmeter is less than ten times the value of the resistance you are measuring across, then you can expect voltmeter loading to be a problem. You can resolve the problem in either of two ways:

- Select a voltmeter with a higher internal resistance.
- Compute the "correct" voltage by viewing the voltmeter as a resistance and analyzing the resulting series-parallel circuit.

Bridge Circuits

Figure 6-55 shows the schematic diagram of a simple bridge circuit. If you momentarily ignore the current meter that is connected across the center of the bridge, you can see that the basic bridge is a series-parallel circuit. That is, resistors R_1 and R_2 are in series with each other as are resistors R_3 and R_4. These two sets of series resistors ($R_{1,2}$ and $R_{3,4}$) are in parallel with each other.

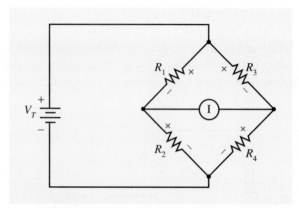

Figure 6-55. A basic bridge circuit.

When the current meter (or other component) is connected across the center of the bridge, the circuit becomes a complex circuit (discussed in Chapter 7). For purposes of the immediate discussion, however, we will examine a specific circuit condition called a **balanced bridge.** This condition occurs whenever the voltage across R_2 is the same as the voltage across R_4. That is,

$$V_2 = V_4$$

Of course, the voltage across either of these resistors (without the current meter connected) can be computed with the basic voltage divider equation. Thus, we could express V_2 and V_4 as

$$V_2 = \frac{R_2}{R_1 + R_2} V_T$$

$$V_4 = \frac{R_4}{R_3 + R_4} V_T$$

Under balanced conditions, these two expressions must be equal. Let's set them equal to each other and derive an important relationship:

$$V_2 = V_4$$

$$\frac{R_2}{R_1 + R_2} V_T = \frac{R_4}{R_3 + R_4} V_T, \text{ or}$$

$$\frac{R_1}{R_2} = \frac{R_3}{R_4} \qquad (6\text{-}2)$$

KEY POINTS

If the bridge is balanced, the ratio of the resistances in one leg of the bridge is equal to the ratio of resistances in the second leg.

The result tells us that the bridge will be balanced if the ratio of the resistors on one leg of the bridge (e.g., R_1 and R_2) is equal to the ratio of the resistors in the second leg (e.g., R_3 and R_4).

KEY POINTS

There is no voltage developed across the center points of a balanced bridge.

When the bridge is balanced, there is no difference in potential across the center of the bridge. So if we connect a current meter across the center of the bridge, there will be no current flow. In other words, according to Ohm's Law, if there is no difference in potential (i.e., voltage), then there can be no current flow.

Now suppose we upset the balance of the bridge by making R_4 slightly larger. Since R_4 has more resistance, it will drop more voltage, which will cause the right-hand side of the current meter to become more positive than the left-hand side. We will now have current flow through the current meter from left to right. This unbalanced condition and the resulting current flows are shown in Figure 6-56. If voltage V_4 were less than V_2, then the left side of the bridge would be more positive, and current would flow from right to left through the current meter.

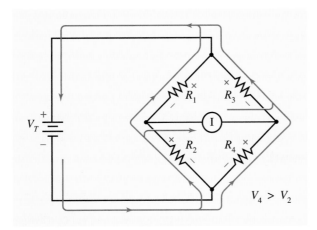

Figure 6-56. If the bridge is unbalanced, current will flow through the current meter.

So, in an unbalanced condition, there will be current through the current meter that is connected across the center of the bridge. The meter current may be in either direction depending on the nature of the imbalance (i.e., $V_2 > V_4$ or $V_4 > V_2$). The magnitude of the meter current is determined by the degree of imbalance.

EXAMPLE SOLUTION

Determine whether the bridge circuit in Figure 6-57 is balanced or unbalanced.

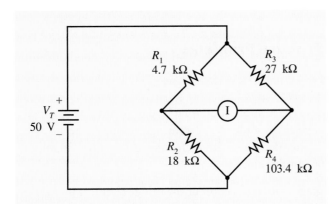

Figure 6-57. Is the bridge balanced?

EXAMPLE SOLUTION

In order for a bridge to be balanced, the following equality (Equation 6-2) must be true:

$$\frac{R_1}{R_2} = \frac{R_3}{R_4}$$

In the present case, we have the following relationship:

$$\frac{R_1}{R_2} \stackrel{?}{=} \frac{R_3}{R_4}$$

$$\frac{4.7 \text{ k}\Omega}{18 \text{ k}\Omega} \stackrel{?}{=} \frac{27 \text{ k}\Omega}{103.4 \text{ k}\Omega}$$

$$0.261 \stackrel{?}{=} 0.261$$

The equality is true, therefore the bridge is balanced.

Practice Problems

1. Determine the value of resistor R_X in Figure 6-58 required to balance the bridge.

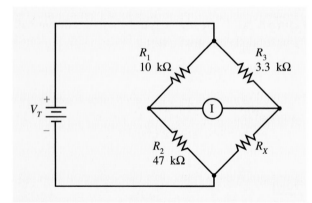

Figure 6-58. What value of R_X is needed to balance the bridge?

Answers to Practice Problems

1. 15.51 kΩ

KEY POINTS

Bridge circuits are widely used in instrumentation circuits that monitor manufacturing processes.

Figure 6-59 shows a typical application for a bridge circuit. Here, a thermistor is used to monitor the temperature of some manufacturing process. A **thermistor** is essentially a resistor whose resistance decreases with increasing temperature. In this case, the thermistor serves as one leg of a bridge circuit.

Recall that the voltage across the center of the bridge circuit is proportional to the degree of imbalance in the bridge. In the case of the circuit shown in Figure 6-59, the voltage across the bridge will be proportional to temperature.

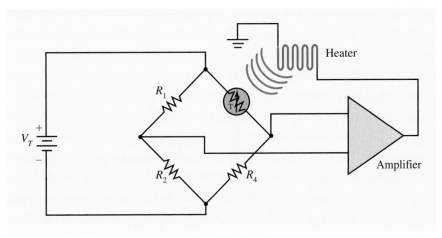

Figure 6-59. A bridge circuit can be used as part of a temperature control system.

The voltage developed across the bridge circuit is routed to an amplifier. You will study amplifiers later in your electronics career, but for now you must know that the amplifier simply boosts (i.e., amplifies) the power of the bridge output. This increased signal is then used to control the current through an electrical heating element. The heating element controls the temperature of the process being monitored by the thermistor.

If the temperature of the process tries to increase, the thermistor will sense the increased temperature, and its resistance will decrease. This changes the degree of imbalance in the bridge, which changes the magnitude (and possibly polarity) of the voltage sent to the amplifier. The output of the amplifier will decrease the current through the heating element, thus reducing the temperature of the process. An opposite response occurs if the temperature of the process tries to decrease. Thus, the process temperature is not only being monitored (as a thermometer might do), but it is being controlled or regulated at a constant temperature. If one of the fixed resistors in Figure 6-59 were made variable, then it could be used to establish the desired operating temperature of the system.

Exercise Problems 6.5

1. Calculate the current through resistor R_1 in the loaded voltage divider circuit shown in Figure 6-60.
2. With the load connected in Figure 6-60, calculate the amount of bleeder current in the circuit.
3. Refer to Figure 6-61. Calculate the value of load voltage.
4. How much total power must be supplied by the source in Figure 6-61?

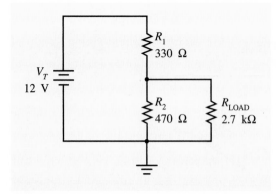

Figure 6-60. Circuit for Exercise Problems 1 and 2.

Figure 6-61. Circuit for Exercise Problems 3 and 4.

5. Refer to Figure 6-62. Compute the value of voltage available for each of the load resistors.

6. Suppose you are troubleshooting the circuit shown in Figure 6-63. Will voltmeter loading probably cause bad measurements in this circuit if the resistance of the voltmeter is 20 MΩ? Explain your answer.

7. Compute the voltage across R_3 in Figure 6-64 before and after the voltmeter is connected. Is voltmeter loading a problem in this circuit?

8. Determine the value of R_B in Figure 6-65 that is required to balance the bridge. What is the voltage across R_E when the bridge is balanced? How much current flows through R_E when the bridge is balanced?

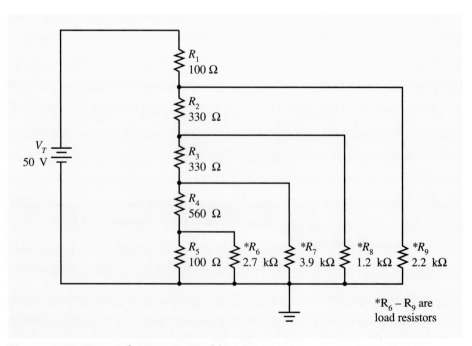

Figure 6-62. Circuit for Exercise Problem 5.

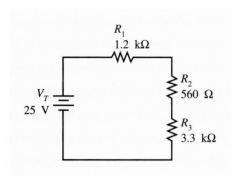

Figure 6-63. Circuit for Exercise Problem 6.

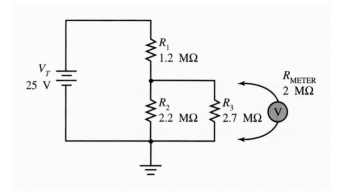

Figure 6-64. Circuit for Exercise Problem 7.

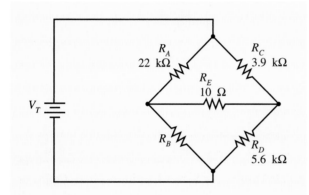

Figure 6-65. Circuit for Exercise Problem 8.

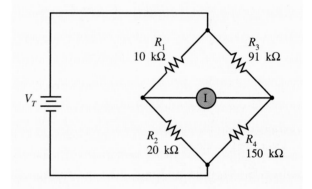

Figure 6-66. Circuit for Exercise Problem 9.

9. Is the bridge circuit shown in Figure 6-66 balanced? If there is a current through the current meter, indicate the direction of current flow.

10. Determine one possible set of components that can be used for resistors R_1 and R_2 in Figure 6-67 to cause the bridge to be balanced. All resistor values must be greater than 3.3 kΩ and less than 270 kΩ. At least one of the selected resistors (R_1 or R_2) must be a standard 5% value.

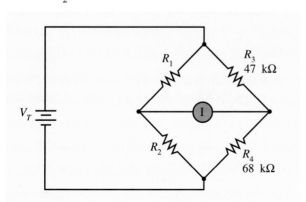

Figure 6-67. Circuit for Exercise Problem 10.

6.6 Troubleshooting Series-Parallel Circuits

We are now in a position to apply our knowledge of series-parallel circuits to the trouble-shooting of defective circuits. Troubleshooting is one of the most important job skills a technician can possess.

The first step toward becoming a skilled troubleshooter of series-parallel circuits is to develop a solid intuitive (i.e., nonmathematical) understanding of circuit behavior. For this you must learn to make mental approximations and be able to classify circuit quantities as high, low, or normal.

Let's begin by briefly reviewing the troubleshooting chart presented in Chapter 4 with reference to series circuits. It is repeated in Table 6-5.

TYPE OF DEFECT	VOLTAGE		
	WITH RESPECT TO GROUND	ACROSS DEFECT	ACROSS OTHER COMPONENTS
Open	Nearer or equal to supply terminal on same side of defect	Full supply	Zero
Increased resistance		Increased	Decreased
Shorted	Nearer or equal to supply terminal on opposite side of defect	Zero	Increased
Decreased resistance		Decreased	Increased

Table 6-5. Troubleshooting Chart Indicating Relative Voltage Values in a Series Circuit with an Open or Shorted Component

As you troubleshoot a circuit, every time you take a measurement, you should mentally cross-reference the measurement to an entry in Table 6-5. With only a very few measurements, you will have isolated the defect.

In the case of series-parallel circuits, the methods presented in Chapter 4 and the use of Table 6-5 is still applicable. The only significant difference is that you must mentally view parallel combinations of resistors as a single resistor. Once a defect has been isolated to a bank of parallel resistors, then apply the troubleshooting methods discussed in Chapter 5 to identify the specific defect.

EXAMPLE SOLUTION

Identify the defective component in Figure 6-68 with as few measurements as you can. Table 6-6 gives the measurement data for the circuit. Use the foldout sheet in Appendix A to determine any particular value.

EXAMPLE **SOLUTION**

One possible first step would be to measure the voltage at approximately midpoint in the circuit. Let's select TP3. Table 6-6 indicates 10 V as the normal voltage on TP3, but lists 20 V as the actual value. According to Table 6-5, TP3 is on the same side of an open as the positive supply terminal, or it is on the opposite side of a short circuit (e.g., a short may be between TP3 and the positive supply terminal).

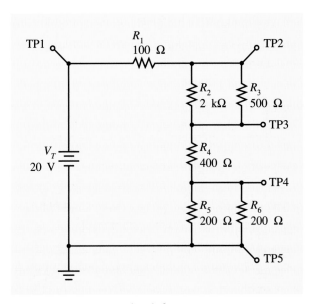

Figure 6-68. Locate the defect.

TESTPOINT	VOLTAGE		RESISTANCE	
	NORMAL	ACTUAL*	NORMAL	ACTUAL*
TP1	20 V	[230]	1 kΩ	[80]
TP2	18 V	[225]	900 Ω	[58]
TP3	10 V	[221]	500 Ω	[80]
TP4	2 V	[22]	100 Ω	[56]
TP5	0 V	[41]	0 Ω	[22]
R_1	2 V	[57]	100 Ω	[224]
R_2	8 V	[27]	2 kΩ	[226]
R_3	8 V	[137]	500 Ω	[229]
R_4	8 V	[223]	400 Ω	[99]
R_5	2 V	[151]	200 Ω	[227]
R_6	2 V	[188]	200 Ω	[218]
V_T	20 V	[228]	—	—

* Numbers in brackets refer to value numbers in Appendix A.

Table 6-6. Circuit Values for the Circuit in Figure 6-68

For our next measurement, we will select a point midway through the area of potential defects. Let us select TP4. Table 6-6 lists 2 V as normal and 0 V as the actual voltage at TP4. Table 6-5 can be used to conclude that we must now be measuring on the opposite side of the defect from TP3. Since the defect is now known to be below TP3 and above TP4, resistor R_4 must be open. We should measure its resistance to confirm our assertion. Table 6-6 lists the normal value of R_4 as 400 Ω and the actual value as infinite. So, R_4 is the defective component and we located it in two measurements (plus one for confirmation).

EXAMPLE SOLUTION

Troubleshoot the circuit shown in Figure 6-69. The measurement data for the circuit is given in Table 6-7.

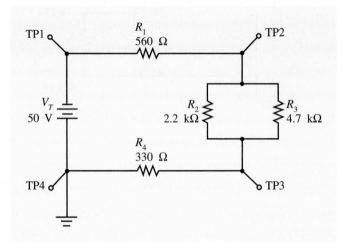

Figure 6-69. Troubleshoot this circuit.

	VOLTAGE		RESISTANCE	
TESTPOINT	NORMAL	ACTUAL*	NORMAL	ACTUAL*
TP1	−50 V	[235]	2.39 kΩ	[238]
TP2	−38.3 V	[240]	1.83 kΩ	[237]
TP3	−6.9 V	[231]	330 Ω	[234]
TP4	0 V	[72]	0 Ω	[176]
R_1	11.7 V	[232]	560 Ω	[233]
R_2	31.4 V	[169]	2.2 kΩ	[137]
R_3	31.4 V	[41]	4.7 kΩ	[100]
R_4	6.9 V	[236]	330 Ω	[239]
V_T	50 V	[111]	—	—

* Numbers in brackets refer to value numbers in Appendix A.

Table 6-7. Circuit Values for the Circuit in Figure 6-69

[EXAMPLE SOLUTION]

For our first check, it is good to select a point approximately midway through the circuit. Let's choose TP2. Table 6-7 indicates a normal reading of –38.3 V, but the actual voltage is listed as –18.54 V. By referring to Table 6-5, we can determine that TP2 must either be between an open (or increased resistance value) and the negative supply terminal, or there must be a short (or reduced resistance value) between TP2 and the positive supply terminal.

Let's next measure the voltage on TP3. Table 6-7 lists the normal value as –6.9 V, but the actual value is listed as –18.54 V. By referring to Table 6-5, we know that TP3 is either on the same side of an open as the negative supply terminal, or on the opposite side in the case of a short. Comparing this with the previously obtained information about TP2 leads us to the conclusion that TP2 and TP3 must be on opposite sides of a short. Therefore, either R_2 or R_3 must be shorted.

If we arbitrarily measure the resistance of either R_2 or R_3 (out of the circuit) we will know the exact location of the defect. Let us measure the resistance of R_3. Table 6-7 indicates both normal and actual values of 4.7 kΩ. Therefore, it must be R_2 that is defective.

Practice Problems

1. The circuit shown in Figure 6-70 has a malfunction. Use the circuit value data supplied in Table 6-8 to locate the defective component in as few steps as you can.

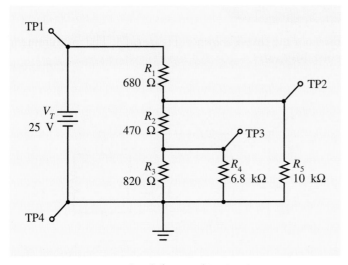

Figure 6-70. Locate the defect in this circuit.

TESTPOINT	VOLTAGE		RESISTANCE	
	NORMAL	ACTUAL*	NORMAL	ACTUAL*
TP1	25 V	[70]	1.75 kΩ	[58]
TP2	15.29 V	[22]	1.07 kΩ	[247]
TP3	9.3 V	[132]	684 Ω	[250]
TP4	0 V	[57]	0 Ω	[72]
R_1	9.71 V	[217]	680 Ω	[138]
R_2	5.99 V	[27]	470 Ω	[112]
R_3	9.3 V	[137]	820 Ω	[246]
R_4	9.3 V	[151]	6.8 kΩ	[39]
R_5	15.29 V	[144]	10 kΩ	[15]
V_T	25 V	[51]	—	—

* Numbers in brackets refer to value numbers in Appendix A.

Table 6-8. Circuit Values for the Circuit in Figure 6-70

Answers to Practice Problems

1. R_1 is open.

Exercise Problems 6.6

1. Troubleshoot the circuit shown in Figure 6-71. The measurement values are given in Table 6-9.

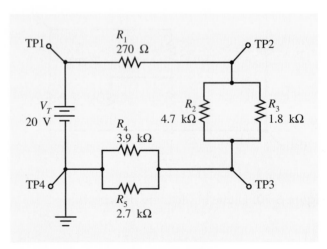

Figure 6-71. Troubleshoot this circuit.

TESTPOINT	VOLTAGE		RESISTANCE	
	NORMAL	ACTUAL*	NORMAL	ACTUAL*
TP1	20 V	[230]	3.17 kΩ	[243]
TP2	18.3 V	[221]	2.9 kΩ	[245]
TP3	10 V	[249]	1.6 kΩ	[260]
TP4	0 V	[144]	0 Ω	[57]
R_1	1.7 V	[105]	270 Ω	[169]
R_2	8.2 V	[241]	4.7 kΩ	[12]
R_3	8.2 V	[248]	1.8 kΩ	[251]
R_4	10.1 V	[242]	3.9 kΩ	[153]
R_5	10.1 V	[244]	2.7 kΩ	[91]
V_T	20 V	[233]	—	—

* Numbers in brackets refer to value numbers in Appendix A.

Table 6-9. Circuit Data for Exercise Problem 1

2. Troubleshoot the circuit shown in Figure 6-72. Circuit values are provided in Table 6-10.

3. The circuit shown in Figure 6-72 has a malfunction. Use the circuit values provided in Table 6-11 to isolate the defective component.

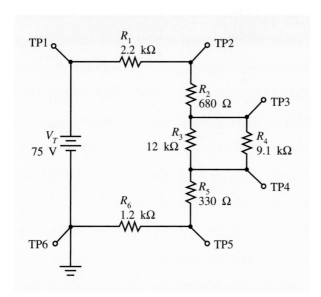

Figure 6-72. Troubleshoot this circuit.

TESTPOINT	VOLTAGE		RESISTANCE	
	NORMAL	ACTUAL*	NORMAL	ACTUAL*
TP1	75 V	[136]	9.59 kΩ	[254]
TP2	57.8 V	[255]	7.39 kΩ	[256]
TP3	52.5 V	[257]	6.71 kΩ	[258]
TP4	11.97 V	[252]	1.53 kΩ	[253]
TP5	9.4 V	[259]	1.2 kΩ	[270]
TP6	0 V	[132]	0 Ω	[22]
R_1	17.2 V	[261]	2.2 kΩ	[272]
R_2	5.3 V	[269]	680 Ω	[278]
R_3	40.5 V	[105]	12 kΩ	[263]
R_4	40.5 V	[151]	9.1 kΩ	[88]
R_5	2.58 V	[280]	330 Ω	[234]
R_6	9.4 V	[271]	1.2 kΩ	[264]
V_T	75 V	[262]	—	—

* Numbers in brackets refer to value numbers in Appendix A.

Table 6-10. Circuit Data for Exercise Problem 2

TEST POINT	VOLTAGE		RESISTANCE	
	NORMAL	ACTUAL*	NORMAL	ACTUAL*
TP1	75 V	[70]	9.59 kΩ	[276]
TP2	57.8 V	[273]	7.39 kΩ	[266]
TP3	52.5 V	[268]	6.71 kΩ	[275]
TP4	11.97 V	[277]	1.53 kΩ	[267]
TP5	9.4 V	[2.65]	1.2 kΩ	[279]
TP6	0 V	[88]	0 Ω	[151]
R_1	17.2 V	[274]	2.2 kΩ	[299]
R_2	5.3 V	[297]	680 Ω	[282]
R_3	40.5 V	[284]	12 kΩ	[288]
R_4	40.5 V	[292]	9.1 kΩ	[283]
R_5	2.58 V	[295]	330 Ω	[290]
R_6	9.4 V	[281]	1.2 kΩ	[293]
V_T	75 V	[78]	—	—

* Numbers in brackets refer to value numbers in Appendix A.

Table 6-11. Circuit Data for Exercise Problem 3

4. Troubleshoot the circuit shown in Figure 6-73. Circuit values are provided in Table 6-12.

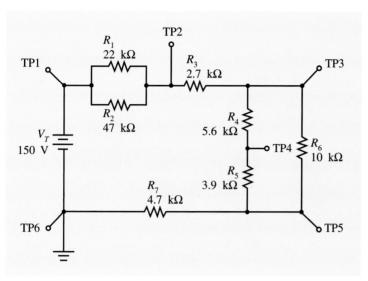

Figure 6-73. Troubleshoot this circuit.

TESTPOINT	VOLTAGE		RESISTANCE	
	NORMAL	ACTUAL*	NORMAL	ACTUAL*
TP1	150 V	[309]	27.3 kΩ	[138]
TP2	67.5 V	[300]	12.3 kΩ	[36]
TP3	52.7 V	[291]	9.6 kΩ	[119]
TP4	36.9 V	[320]	7.8 kΩ	[181]
TP5	25.9 V	[294]	4.7 kΩ	[204]
TP6	0 V	[105]	0 Ω	[72]
R_1	82.5 V	[132]	22 kΩ	[315]
R_2	82.5 V	[169]	47 kΩ	[55]
R_3	14.9 V	[188]	2.7 kΩ	[91]
R_4	15.8 V	[200]	5.6 kΩ	[298]
R_5	11 V	[22]	3.9 kΩ	[312]
R_6	26.8 V	[41]	10 kΩ	[15]
R_7	25.9 V	[286]	4.7 kΩ	[113]
V_T	150 V	[304]	—	—

* Numbers in brackets refer to value numbers in Appendix A.

Table 6-12. Circuit Data for Exercise Problem 4

Chapter Summary

- The first step in analyzing a series-parallel circuit is to be certain the circuit being analyzed really is a series-parallel circuit. This positive identification process consists of replacing sets of parallel resistors and sets of series resistors with single resistor symbols to create a more simplified schematic. This same substitution procedure is then applied to the simplified schematic. The process of forming progressively simpler circuits continues until one of two situations exists:

  - the circuit has been reduced to a one-resistor circuit, or
  - the circuit cannot be reduced to a one-resistor circuit.

- In the first case, a series-parallel circuit has been identified, and the analytical techniques discussed in this chapter can be applied. In the second case, a complex circuit has been identified, and the circuit must be analyzed using the techniques presented in Chapter 7.

- Any series-parallel circuit can be reduced to a single-resistor circuit using the preceding technique and computing the value of equivalent resistance at each intermediate step. That is, whenever a set of series resistors is replaced with a single resistor, then compute the value of that equivalent resistor by applying the series resistor equation presented in Chapter 4. Similarly, when sets of parallel resistors are replaced, their equivalent resistance is computed using the parallel resistor equation discussed in Chapter 5. When the circuit is finally reduced to a single-resistor circuit, the value of that equivalent resistor will be equal to the total resistance in the original circuit.

- To thoroughly analyze a series-parallel circuit when all component values are known requires two major processes:

  - simplify the circuit to a one-resistor circuit by sketching equivalent circuits as described.
  - work backward through the intermediate sketches while applying Ohm's and Kirchhoff's Laws.

- If at least two circuit values (i.e., voltage, current, resistance, and power) are known for a single component, then all remaining values can be calculated with Ohm's Law and the relevant power equations. If all but one voltage (both sources and drops) in any closed loop are known, then Kirchhoff's Voltage Law can be applied to calculate the unknown voltage. Similarly, if all but one current value is known for any single node in the circuit, then Kirchhoff's Current Law can be used to compute the value of the unknown current.

- When working backward through the intermediate sketches, calculate as many circuit values as possible for each sketch. The computed values are then transferred to the next intermediate drawing where additional calculations can be made. The process continues until all circuit values on the original schematic are known.

- If an equivalent resistor in a particular intermediate sketch represents a set of series resistors, then all of these resistors will have the same current as the equivalent resistor. In a similar manner, if an equivalent resistor represents a set of parallel resistors, then all of these resistors will have the same voltage as the equivalent resistor.

- If some component values are unknown for a given series-parallel circuit, then the analysis of the circuit relies on your creative application of Ohm's and Kirchhoff's

Laws, the power equations, and your understanding of the basic rules of series and parallel circuits. In general, the basic process is the same for completely or partially specified circuits, with the exception that it may not be possible to compute the value of total resistance in the partially specified circuit until late in the analysis.

• Loaded voltage dividers and balanced bridge circuits are two practical applications of series-parallel circuits. When a voltage divider (like those studied in Chapter 4) provides power to an external load resistor, we say the voltage divider is a loaded voltage divider. When a voltage divider is loaded, its divided voltages decrease from their unloaded values. The amount of voltage change between loaded and unloaded can be expressed as a percentage and is called percent of voltage regulation. Voltage regulation can be improved in a loaded voltage divider by increasing the amount of bleeder current. Bleeder current is the current that flows through the unloaded voltage divider. Voltage divider circuits are series-parallel circuits and require no special techniques for analysis.

• Bridge circuits are widely used in electronic instrumentation systems. The bridge circuit can be in either of two conditions: balanced or unbalanced. In order for a bridge circuit to be balanced, the ratio of resistances in the two parallel legs of the bridge must be equal. There is no voltage developed across the midpoints of a bridge circuit when it is balanced. A balanced bridge can be analyzed by removing any components connected across the midpoints of the bridge and then applying series-parallel analysis techniques. An unbalanced bridge is a complex circuit and must be analyzed using the techniques presented in Chapter 7.

• A defective series-parallel circuit can be quickly diagnosed by applying the trouble-shooting procedures detailed in Chapters 4 and 5. In the case of series-parallel circuits, all parallel branches are mentally eliminated. The circuit is then diagnosed as a simple series circuit. If the malfunction is isolated to a set of parallel components, then the techniques described in Chapter 5 for parallel circuit troubleshooting can be used to identify the exact defect.

• Troubleshooting is a critical skill for electronics technicians. Much effort and practice should be devoted to becoming an excellent troubleshooter. Most of the trouble-shooting task in series-parallel circuits involves voltage measurements. Current and resistance measurements are appropriate only after the malfunction has been local-ized. In order to develop skills that will continue to be effective when troubleshooting complex systems, always avoid guessing.

Review Questions

Section 6.1: Positive Identification Method

1. Name the four classes of circuit configurations.

2. State the rule for ensuring that two or more components are definitely in parallel.

3. State the rule for ensuring that two or more components are definitely in series.

4. If a circuit is not series and not parallel, then describe the procedure for positively identifying its type.

Section 6.2: Simplification with Equivalent Circuits

5. Any series-parallel circuit can be reduced to a _____-resistor equivalent circuit having a resistance equal to the _____ resistance of the original circuit.

6. What class of equations are utilized when computing circuit values in parallel portions of a series-parallel circuit?

7. What class of equations are utilized when computing circuit values in series portions of a series-parallel circuit?

8. If a portion of a series-parallel circuit consists of three series resistors with values of 10 kΩ, 27 kΩ, and 18 kΩ, compute the value of equivalent resistance.

9. Three parallel resistors having values of 560 Ω, 330 Ω, and 1.2 kΩ form one portion of a series-parallel circuit. Compute the value of equivalent resistance.

10. Simplify the circuit shown in Figure 6-74 and compute the value of total resistance.

11. Refer to Figure 6-74. How much total current flows in the circuit?

12. In Figure 6-74, how much total power must be supplied by the battery?

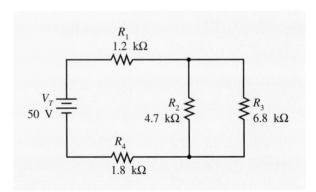

Figure 6-74. Circuit for Problems 10 through 12.

Section 6.3: Application of Kirchhoff's Voltage and Current Laws

13. Write Kirchhoff's Voltage Law in your own words.

14. Express Kirchhoff's Current Law in your own words.

15. If Kirchhoff's Voltage Law is to be used to determine a component voltage drop in a particular circuit loop, how many other voltage drops and sources in the loop can be unknown?

16. If currents of 1.2 mA and 6.8 mA are entering a particular node in a series-parallel circuit, how much current will be leaving that same node?

17. A certain complete loop in a series-parallel circuit contains three resistors and a voltage source of 8 V. If one resistor drops 2 V, how much total voltage is dropped by the remaining resistors?

18. There are three possible loops in Figure 6-75. Write the Kirchhoff's Voltage Law equation for each of the three loops.

19. Apply Kirchhoff's Voltage Law to determine the voltage drop across R_3 in Figure 6-75.

20. Refer to Figure 6-75. Does the sum of the voltage drops across R_3 and R_4 equal the value of the applied voltage? Explain why or why not.

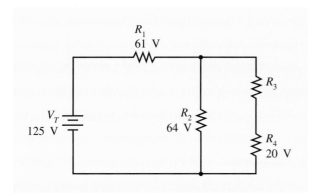

Figure 6-75. Circuit for Problems 18 through 20.

21. Use Kirchhoff's Voltage Law to determine the voltage drop across R_4 in Figure 6-76.

22. Refer to Figure 6-76. Apply Kirchhoff's Voltage Law to determine the voltage drop across R_5.

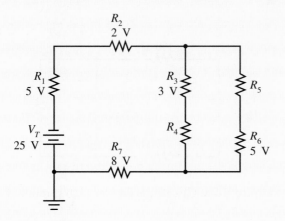

Figure 6-76. Circuit for Problems 21 through 22.

23. Use Kirchhoff's Current Law to calculate the current through resistor R_6 in Figure 6-77.

24. What is the value of current through resistor R_5 in Figure 6-77?

25. How much current flows through resistor R_2 in Figure 6-77?

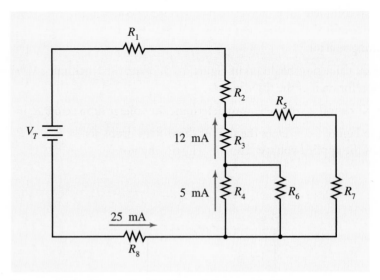

Figure 6-77. Circuit for Problems 23 through 25.

Section 6.4: Computing Voltage, Current, Resistance, and Power

26. Determine the value of total resistance for the circuit in Figure 6-78.

27. What is the value of total current in the circuit shown in Figure 6-78?

28. Compute the amount of power dissipation for R_2 in Figure 6-78.

29. Refer to Figure 6-78. What is the voltage drop across R_3?

30. Which resistor in Figure 6-78 dissipates the most power?

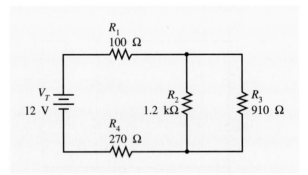

Figure 6-78. Circuit for Problems 26 through 30.

31. Refer to Figure 6-79. How much current flows through resistor R_3?

32. Which resistor dissipates the most power in Figure 6-79?

33. Calculate the voltage drop across R_2 in Figure 6-79.

34. Determine the voltage drop across resistor R_1 in Figure 6-79.

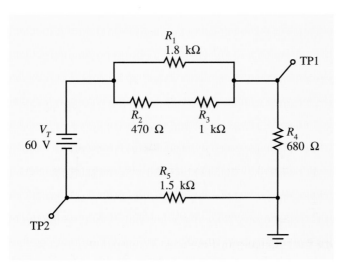

Figure 6-79. Circuit for Problems 31 through 36.

35. What is the voltage (with reference to ground) at TP1 in Figure 6-79?
36. What is the voltage (with reference to ground) at TP2 in Figure 6-79?
37. Compute the current for R_1 in Figure 6-80.
38. Determine the value of R_2 in Figure 6-80.
39. Which of the resistors in Figure 6-80 dissipates the most power?

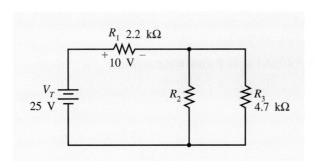

Figure 6-80. Circuit for Problems 37 through 39.

40. Refer to Figure 6-81. Calculate the value of resistor R_7.
41. How much current flows through resistor R_4 in Figure 6-81?
42. What is the voltage at the top of R_7 in Figure 6-81 when measured with reference to ground?
43. What is the value of applied voltage for the circuit in Figure 6-81?

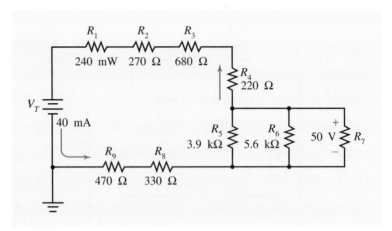

Figure 6-81. Circuit for Problems 40 through 43.

Section 6.5: Applied Technology: Series-Parallel Circuits

44. When the current supplied by a voltage divider increases, what happens to the value of divided voltage?

45. Refer to the loaded voltage divider circuit shown in Figure 6-82. What is the value of voltage across R_4?

46. What is the value of voltage across R_5 in Figure 6-82?

47. Compute the percent of voltage regulation for each of the loads in Figure 6-82 if both loads are removed or connected at the same time.

48. What is the value of bleeder current for the voltage divider circuit shown in Figure 6-82 when both loads are removed?

49. If resistors R_1, R_2, and R_3 in Figure 6-82 were all reduced to one-half of their present value, what would happen to the percent of voltage regulation?

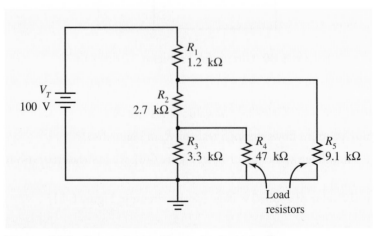

Figure 6-82. Circuit for Problems 45 through 49.

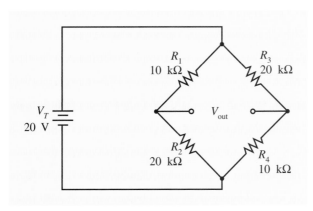

Figure 6-83. Circuit for Problems 50 and 51.

50. Classify the bridge circuit shown in Figure 6-83 as balanced or unbalanced.

51. What is the value of voltage (V_{out}) across the midpoints of the bridge circuit in Figure 6-83?

52. Refer to Figure 6-84. What value of R_3 is required to balance the bridge circuit?

53. If the bridge circuit in Figure 6-84 is balanced, what is the voltage (V_{out}) across the midpoints of the bridge?

54. If resistor R_3 in Figure 6-84 were made smaller than the value required to balance the bridge circuit, which side (left or right) of the bridge midpoints would be the most positive?

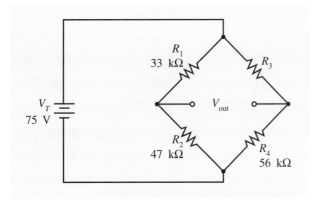

Figure 6-84. Circuit for Problems 52 through 54.

Section 6.6: Troubleshooting Series-Parallel Circuits

55. When troubleshooting series-parallel resistor circuits, which type of test equipment should be used most frequently? (voltmeter, ohmmeter, or ammeter)

56. When troubleshooting a certain circuit, you discover that the voltage across a particular resistor is higher than normal. If this resistor is defective, describe the type of defect.

57. If a circuit has a shorted component and you are making a voltage measurement from ground to a point between the defect and the negative battery terminal, describe the relative value of voltage that will be measured.

58. Table 6-13 gives the circuit values for the circuit shown in Figure 6-85. Troubleshoot the circuit and identify the defective component.

TESTPOINT	VOLTAGE		RESISTANCE	
	NORMAL	ACTUAL*	NORMAL	ACTUAL*
TP1	25 V	[127]	2.24 kΩ	[119]
TP2	23.9 V	[22]	2.14 kΩ	[302]
TP3	15.5 V	[176]	1.39 kΩ	[285]
TP4	4.3 V	[41]	389 Ω	[301]
TP5	0 V	[27]	0 Ω	[61]
R_1	1.12 V	[154]	100 Ω	[80]
R_2	8.37 V	[105]	2 kΩ	[226]
R_3	8.37 V	[72]	1.2 kΩ	[279]
R_4	11.2 V	[132]	1 kΩ	[43]
R_5	4.35 V	[188]	680 Ω	[278]
R_6	4.35 V	[169]	910 Ω	[303]
V_T	25 V	[51]	—	—

* Numbers in brackets refer to value numbers in Appendix A.

Table 6-13. Circuit Data for Problem 58

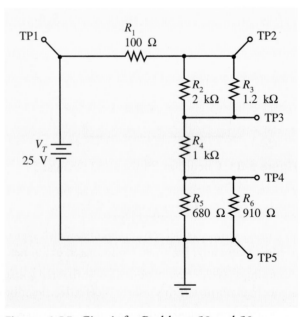

Figure 6-85. Circuit for Problems 58 and 59.

59. Table 6-14 lists the circuit values for a certain defect in the circuit shown in Figure 6-85. Locate the defect in as few measurements as possible.

TESTPOINT	VOLTAGE		RESISTANCE	
	NORMAL	ACTUAL*	NORMAL	ACTUAL*
TP1	25 V	[103]	2.24 kΩ	[308]
TP2	23.9 V	[287]	2.14 kΩ	[285]
TP3	15.5 V	[319]	1.39 kΩ	[289]
TP4	4.3 V	[307]	389 Ω	[311]
TP5	0 V	[88]	0 Ω	[200]
R_1	1.12 V	[305]	100 Ω	[56]
R_2	8.37 V	[22]	2 kΩ	[27]
R_3	8.37 V	[88]	1.2 kΩ	[264]
R_4	11.2 V	[310]	1 kΩ	[314]
R_5	4.35 V	[306]	680 Ω	[282]
R_6	4.35 V	[296]	910 Ω	[313]
V_T	25 V	[180]	—	—

* Numbers in brackets refer to value numbers in Appendix A.

Table 6-14. Circuit Data for Problem 59

60. Locate the defective component in the circuit shown in Figure 6-86. The circuit values are provided in Table 6-15.

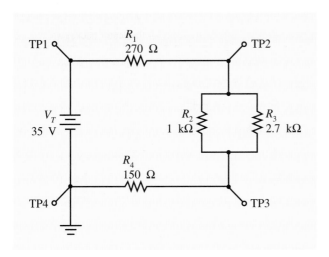

Figure 6-86. Circuit for Problem 60.

TESTPOINT	VOLTAGE		RESISTANCE	
	NORMAL	ACTUAL*	NORMAL	ACTUAL*
TP1	35 V	[333]	1.15 kΩ	[325]
TP2	26.8 V	[321]	880 Ω	[335]
TP3	4.56 V	[329]	150 Ω	[328]
TP4	0 V	[72]	0 Ω	[132]
R_1	8.22 V	[176]	270 Ω	[188]
R_2	22.2 V	[326]	1 kΩ	[330]
R_3	22.2 V	[316]	2.7 kΩ	[84]
R_4	4.57 V	[324]	150 Ω	[322]
V_T	35 V	[331]	—	—

* Numbers in brackets refer to value numbers in Appendix A.

Table 6-15. Circuit Data for Problem 60

TECHNICIAN CHALLENGE

You are the senior technician for a company that manufactures solid-state indicating devices. Their latest product contains three light-emitting diodes (red, green, and yellow). The plant manager has asked you to develop a test fixture that can be used to quickly test the device as part of a quality control check. The electrical requirements for the indicating device are shown in Figure 6-87.

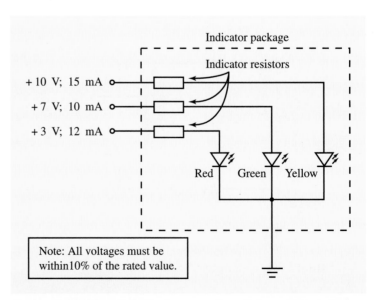

Figure 6-87. Electrical requirements for a light-emitting diode tester.

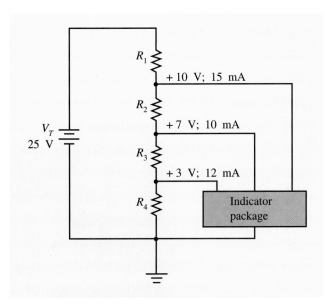

Figure 6-88. A voltage divider can be used as the heart of the light-emitting diode tester.

After evaluating the task, you decide to utilize a loaded voltage divider circuit as the test fixture. The basic diagram is shown in Figure 6-88. It makes use of the 25-V supply voltage that is available in the manufacturing area. Maximum total power that can be consumed by the entire test fixture and indicator package is 4 W.

Apply your knowledge of series-parallel circuits to compute the required values for the voltage divider resistances (R_1 through R_4). If possible, you should try to use standard resistor values. Be sure to calculate the required wattage rating for each resistor, and be certain that the entire circuit consumes less than the stated 4-W limit.

Equation List

(6-1) $\% \text{ regulation} = \dfrac{V_{\text{no load}} - V_{\text{full load}}}{V_{\text{full load}}} \times 100$

(6-2) $\dfrac{R_1}{R_2} = \dfrac{R_3}{R_4}$

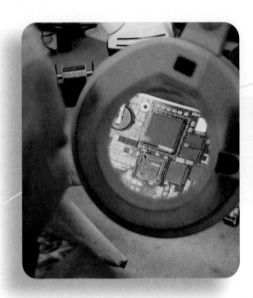

Complex Circuits and Network Analysis

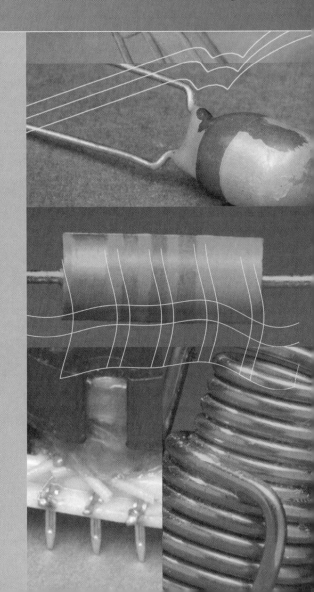

This chapter presents several methods that can be used to simplify and analyze various circuit configurations. The techniques presented in this chapter can be applied to any circuit configuration including complex circuits. Our initial objective will be to master the tools that will enable us to analyze complex circuits. It is important to recognize, however, that many of these same techniques can be applied to the solution of other circuit configurations (e.g., series-parallel).

7.1 Positive Identification Method

KEY POINTS

A complex circuit can be positively identified through a series of replacements:

- Each set of parallel or series resistances is replaced by an equivalent resistance.
- Each set of series voltage sources is replaced by an equivalent voltage source.
- Sets of parallel voltage sources can be replaced by an equivalent source *if they all have the same voltage.*

The replacement process continues until no further replacement of series or parallel components can be done.

Chapter 3 introduced a procedure for positive identification of series-parallel circuits. This same procedure is used to positively identify a complex circuit. Identification can be accomplished by performing the following procedure: On a scratch paper, redraw the schematic to be classified, but replace all sets of series resistors with a single resistor. Similarly, replace all truly parallel resistors with a single resistor. Repeat this procedure for the newly drawn figure. Continue until one of the following situations occurs:

- The circuit is reduced to a single resistor.
- The circuit cannot be reduced to a single resistor.

In the first case, you have identified a series-parallel circuit as discussed in Chapter 6. In the second case, you have identified a complex circuit.

We can also extend the complexity of a circuit to include multiple power sources. In Chapters 4 and 5, you learned how to combine series and parallel voltage sources. You will recall that series voltage sources are added (algebraically) to determine the equivalent voltage. The current capacity of several parallel voltage sources is found by adding the current capacities of the individual voltage sources. If a circuit has multiple voltage sources, then simplify the sources by replacing series or parallel combinations with an equivalent voltage source according to the following rules:

- Replace series voltage sources with a single voltage source having the voltage and polarity determined by the algebraic sum of the individual voltage sources.
- Replace parallel voltage sources that have the same voltage rating with a single voltage source having the same voltage value.

If the procedures described in this section are used to simplify a circuit, then a complex circuit can be positively identified when the resulting simplified circuit has multiple resistances or multiple voltage sources. We can state this more formally as follows:

> If a given resistive circuit with one or more voltage sources cannot be reduced to a single equivalent resistance and a single power source by simple replacement of series or parallel components with a single equivalent component, then the original circuit can be classified as complex.

EXAMPLE SOLUTION

Can the circuit shown in Figure 7-1 be positively identified as a complex circuit?

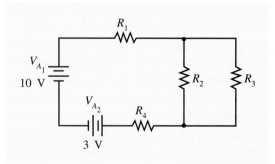

Figure 7-1. Is this a complex circuit?

EXAMPLE SOLUTION

We simplify the circuit by replacing all sets of series or parallel components with a single equivalent component. Figure 7-2 shows the simplification steps.

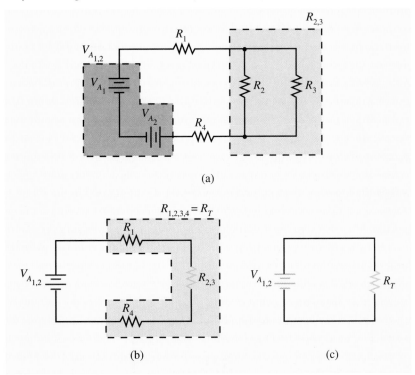

(a)

(b)   (c)

Figure 7-2. Simplification steps for the circuit shown in Figure 7-1.

The original circuit is repeated in Figure 7-2(a). We see that the two voltage sources are in series, so we replace them in Figure 7-2(b) with a single equivalent voltage source having a value equal to the algebraic sum of the two original sources. The polarity of the equivalent source will be the same as the larger of the two series sources. Resistors R_2 and R_3 are in parallel and can be replaced by a single equivalent resistance ($R_{2,3}$) as shown in Figure 7-2(b).

The three resistors in Figure 7-2(b) are in series and can be replaced by a single equivalent resistor. Since this last replacement produces the one resistor and one voltage source equivalent circuit shown in Figure 7-2(c), we know that the original circuit was definitely not a complex circuit.

EXAMPLE SOLUTION

Can the circuit shown in Figure 7-3 be classified as a complex circuit?

EXAMPLE SOLUTION

When we try to apply our simplification procedure, we find that there are no series or parallel resistances and no series or parallel voltage sources. Since the circuit has more than one voltage source and resistance, we know the circuit is a complex circuit.

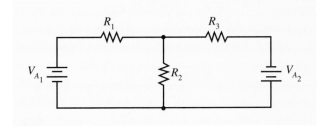

Figure 7-3. Is this a complex circuit?

EXAMPLE SOLUTION

Refer to the circuit shown in Figure 7-4. Is this a complex circuit?

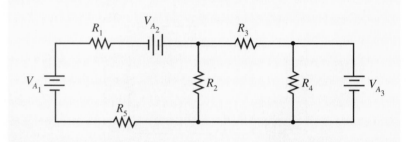

Figure 7-4. Is this a complex circuit?

EXAMPLE **SOLUTION**

First, we note that voltage sources V_{A_1} and V_{A_2} are in series (there is only one possible path for current through both sources). They can be replaced by an equivalent voltage source ($V_{A_{1,2}}$). Similarly, resistors R_1 and R_5 are in series and can be replaced by a single equivalent resistor. Both of these replacements are shown in Figure 7-5.

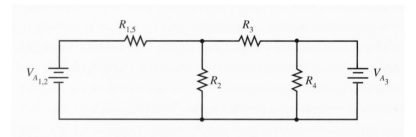

Figure 7-5. Simplification of the circuit shown in Figure 7-4.

The simplified circuit shown in Figure 7-5 has no series or parallel resistors and no series or parallel voltage sources, and so it cannot be simplified further by simple replacement. The original circuit can be classified as complex, since our replacement procedure resulted in a circuit with multiple resistances and multiple voltage sources.

Practice Problems

1. Can the circuit shown in Figure 7-6 be classified as complex?

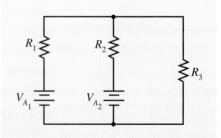

Figure 7-6. Is this a complex circuit?

2. Is the circuit shown in Figure 7-7 a complex circuit?

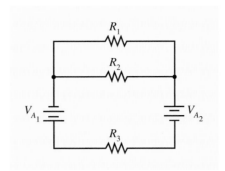

Figure 7-7. Is this a complex circuit?

3. Can the circuit shown in Figure 7-8 be identified as a complex circuit?

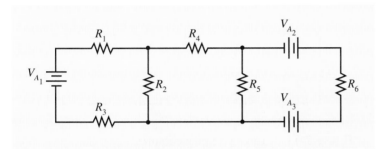

Figure 7-8. Is this a complex circuit?

Answers to Practice Problems

1. Yes   **2.** No   **3.** Yes

Exercise Problems 7.1

1. Can the circuit shown in Figure 7-9 be classified as a complex circuit?

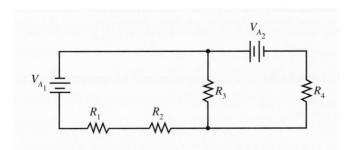

Figure 7-9. Is this a complex circuit?

2. Is the circuit shown in Figure 7-10 a complex circuit?

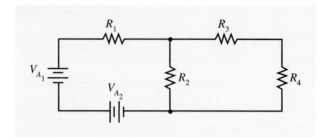

Figure 7-10. Is this a complex circuit?

3. Can the circuit shown in Figure 7-11 be identified as a complex circuit?

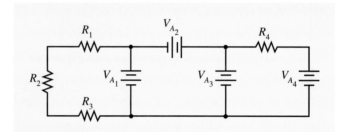

Figure 7-11. Is this a complex circuit?

7.2 Mesh Analysis

Mesh analysis is an analytical tool that will enable us to analyze complex circuits. We shall examine it in two parts. First, we will learn to label the circuit and to write the algebraic expressions needed to describe the circuit. Second, we will solve the circuit equations to determine the actual circuit quantities (i.e., current, voltage, and so on).

Writing Mesh Equations

Mesh analysis is essentially an extended application of Kirchhoff's Voltage Law. In Chapter 6, we used Kirchhoff's Voltage Law to verify the correctness of our circuit calculations. We drew a series of loops that included every component at least once, and then wrote the equation for the closed loop. Mesh analysis is very similar, but the loops are called meshes, and the loop equations are written using an algebraic form for the voltage drops.

EXAMPLE SOLUTION

Draw the mesh currents for the circuit shown in Figure 7-12, and write the mesh equations.

EXAMPLE SOLUTION

There is usually more than one possible set of mesh currents. Select a set that has the fewest possible loops or mesh currents. Also be sure that each loop has one or more resistances that are

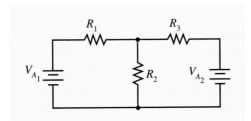

Figure 7-12. Draw the mesh currents and write the mesh equations for this circuit.

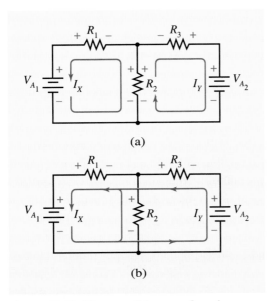

(a)

(b)

Figure 7-13. Two possible sets of mesh currents.

part of another mesh. The actual direction of the mesh currents is arbitrary. The actual direction of current flow will become evident at a later point in the analysis. Figure 7-13 shows two possible sets of mesh currents for our present circuit. Let's use the set shown in Figure 7-13(a) to write the loop equations.

We begin at any point in a particular mesh and progress around the loop. As we go, we write the polarity of the voltage drops or sources *as we exit* the component. Since no absolute voltages are labeled on the schematic, we will simply substitute IR for each voltage drop. Thus, if I_X flows through R_1, it will produce a voltage drop of $I_X R_1$ according to Ohm's Law. When we have completed the loop, we set the equation equal to zero (i.e., Kirchhoff's Voltage Law). In the case of Figure 7-13(a), we have the following equations:

Mesh 1: $I_X R_2 + I_Y R_2 + I_X R_1 - V_{A_1} = 0$

Mesh 2: $I_Y R_2 + I_X R_2 + I_Y R_3 - V_{A_2} = 0$

Note that R_2 has two voltage drops to be accounted for in each loop; one is produced by I_X and the other by I_Y.

EXAMPLE SOLUTION

Write a set of mesh equations for the circuit shown in Figure 7-14.

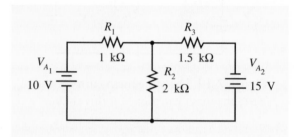

Figure 7-14. Write a set of mesh equations for this circuit.

EXAMPLE **SOLUTION**

Figure 7-15 shows one possible set of mesh currents. Be reminded that the mesh current is not a real current, and so you can draw it as flowing in any direction you choose. Based on the mesh currents shown in Figure 7-15, we can write the following mesh equations:

$$\text{Mesh 1: } I_X R_2 - I_Y R_2 + I_X R_1 - 10 = 0$$
$$2{,}000 I_X - 2{,}000 I_Y + 1{,}000 I_X - 10 = 0$$
$$3{,}000 I_X - 2{,}000 I_Y - 10 = 0$$
$$\text{Mesh 2: } -15 + I_Y R_3 + I_Y R_2 - I_X R_2 = 0$$
$$-15 + 1{,}500 I_Y + 2{,}000 I_Y - 2{,}000 I_X = 0$$
$$-2{,}000 I_X + 3{,}500 I_Y - 15 = 0$$

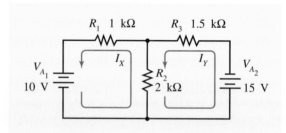

Figure 7-15. Write a set of mesh equations for this circuit.

Solving Mesh Equations

The solution of the mesh equations requires us to solve multiple equations with multiple unknowns. The preceding example yielded two equations (Mesh 1 and Mesh 2) with two unknowns (I_X and I_Y) in each equation. These are called simultaneous equations. There are several methods for solving this type of equation. For the purposes of this chapter, we shall use determinants. Although determinants may seem strange at first, they are actually much easier to use than equivalent algebraic manipulation. If, however, you prefer another method, then you are encouraged to use it as long as the end result is the same.

Solve the following simultaneous mesh equations (from the preceding example).

$$3,000I_X - 2,000I_Y - 10 = 0$$
$$-2,000I_X + 3,500I_Y - 15 = 0$$

EXAMPLE **SOLUTION**

We use determinants as follows.

Mesh 1:

$$I_X = \frac{\begin{vmatrix} 10 & -2,000 \\ 15 & 3,500 \end{vmatrix}}{\begin{vmatrix} 3,000 & -2,000 \\ -2,000 & 3,500 \end{vmatrix}} = \frac{35,000 + 30,000}{10,500,000 - 4,000,000} = 10 \text{ mA}$$

$$I_Y = \frac{\begin{vmatrix} 3,000 & 10 \\ -2,000 & 15 \end{vmatrix}}{\begin{vmatrix} 3,000 & -2,000 \\ -2,000 & 3,500 \end{vmatrix}} = \frac{45,000 + 20,000}{10,500,000 - 4,000,000} = 10 \text{ mA}$$

Since both currents are positive, we know that the direction of our assumed mesh currents was correct. If a computed current is negative, then its magnitude is correct, but it flows opposite the assumed mesh current direction.

If more than one mesh current flows through a single component (e.g., R_2 in Figure 7-15), then the actual current is found by combining the two mesh currents algebraically. In the present example, the two currents I_X and I_Y are equal, but they flow in the opposite direction. Therefore, the actual current through R_2 is zero.

KEY POINTS

If a particular mesh current is determined to be negative, then it actually flows in the opposite direction than the initially assumed mesh current. The voltage drops associated with a negative mesh current must also be reversed.

KEY POINTS

If two or more mesh currents flow through a single component, then the *real* current is equal to the algebraic sum of the component mesh currents. The resulting direction is determined by the larger of the currents (or combined currents).

KEY POINTS

Once the mesh currents have been found, Ohm's Law can be used to determine all of the component voltage drops. The power formulas can be used to compute all power dissipations.

Practice Problems

1. Write a set of mesh equations for the circuit shown in Figure 7-16.

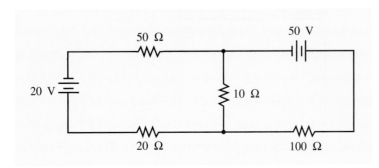

Figure 7-16. Write a set of mesh equations for this circuit.

2. Solve the following mesh equations to determine the currents I_X and I_Y.

$$80I_X - 10I_Y - 20 = 0, \text{ and}$$
$$-10I_X + 110I_Y + 50 = 0$$

Answers to Practice Problems

1. Note: Assume I_X and I_Y flow clockwise.

   $$80I_X - 10I_Y - 20 = 0, \text{ and}$$
   $$10I_X - 110I_Y + 50 = 0$$

2. $I_X = 195.4$ mA
   $I_Y = -436.8$ mA

Applied Mesh Analysis

Now let us use mesh analysis to solve a practical electronics problem. Suppose we want to analyze the bridge circuit shown in Figure 7-17 to determine the amount of current flowing through R_5. By applying Equation 6-2, we can determine that the bridge is unbalanced, so it qualifies as a complex circuit and cannot be solved with the methods presented in previous chapters.

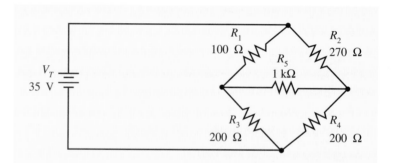

Figure 7-17. Mesh analysis can be used to analyze a bridge circuit.

First, we label a set of mesh currents. Figure 7-18 shows one possible set of mesh currents. Note that we must create three mesh currents in order to include all components at least once. Also note that each resistor has been marked to show the polarity of the voltage drop as a result of each mesh current. The real polarities cannot be determined until after we find the values of I_X, I_Y, and I_Z.

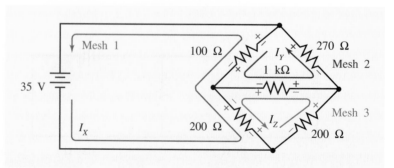

Figure 7-18. One possible way to label mesh currents in a bridge circuit.

We can now write our mesh equations. For Mesh 1, we have

$$200I_X - 200I_Z + 100I_X - 100I_Y - 35 = 0$$
$$300I_X - 100I_Y - 200I_Z - 35 = 0$$

The Mesh 2 equations are

$$270I_Y + 100I_Y - 100I_X + 1{,}000I_Y - 1{,}000I_Z = 0$$
$$-100I_X + 1{,}370I_Y - 1{,}000I_Z = 0$$

And, finally, we write the equation for Mesh 3.

$$200I_Z + 1{,}000I_Z - 1{,}000I_Y + 200I_Z - 200I_X = 0$$
$$-200I_X - 1{,}000I_Y + 1{,}400I_Z = 0$$

These three equations can now be solved with determinants to compute the values of I_X, I_Y, and I_Z. In this case, however, our goal is to calculate the current through R_5 (in Figure 7-17). We need only calculate mesh currents I_Y and I_Z, since these are the only two that flow through R_5. These calculations follow. First, we find the determinant of the system:

$$\Delta = \begin{vmatrix} 300 & -100 & -200 \\ -100 & 1{,}370 & -1{,}000 \\ -200 & -1{,}000 & 1{,}400 \end{vmatrix} = 535.4 \times 10^6 - (368.8 \times 10^6) = 166.6 \times 10^6$$

Next, we find I_Y:

$$I_Y = \frac{\begin{vmatrix} 300 & 35 & -200 \\ -100 & 0 & -1{,}000 \\ -200 & 0 & 1{,}400 \end{vmatrix}}{\Delta} = \frac{7 \times 10^6 - (-4.9 \times 10^6)}{166.6 \times 10^6} = 71.43 \text{ mA}$$

Now we compute I_Z:

$$I_Z = \frac{\begin{vmatrix} 300 & -100 & 35 \\ -100 & 1{,}370 & 0 \\ -200 & -1{,}000 & 0 \end{vmatrix}}{\Delta} = \frac{3.5 \times 10^6 - (-9.59 \times 10^6)}{166.6 \times 10^6} = 78.57 \text{ mA}$$

Since both currents are positive, we know the original directions of mesh currents shown in Figure 7-18 are correct. The current through R_5 will be the difference between mesh currents I_Y and I_Z, since they flow in opposite directions. We compute I_5 as follows:

$$I_5 = I_Y - I_Z$$
$$= 71.43 \text{ mA} - 78.57 \text{ mA}$$
$$= -7.14 \text{ mA}$$

The minus sign has no significance in this case. Our result is negative because I_Z happened to be larger than I_Y. In any case, the *magnitude* of current through R_5 is 7.14 mA. Its direction will be the same as the larger of the two mesh currents (I_Y and I_Z). So, in our case, R_5 will have 7.14 mA that will flow from right to left (same as I_Z). Figure 7-19 shows the bridge circuit with our computed values.

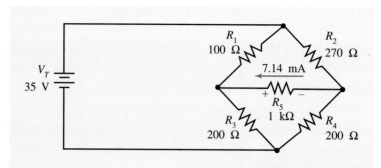

Figure 7-19. The current through R_5 has been determined.

In this example, we only computed the values that were needed to determine the current through R_5. In many cases, we would want to compute the remaining currents and voltages also, and then apply Kirchhoff's Laws to verify the correctness of our calculations.

Practice Problems

1. Calculate the voltage across R_3 in Figure 7-20 using mesh analysis.

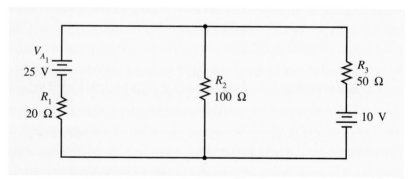

Figure 7-20. What is the voltage across R_3?

2. Calculate the current through R_5 in Figure 7-21 using mesh analysis.

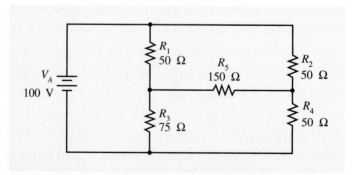

Figure 7-21. Find the current through R_5?

Answers to Practice Problems

1. 8.125 V   **2.** 48.4 mA

Exercise Problems 7.2

1. What is the least number of mesh currents required to analyze the circuit shown in Figure 7-22?

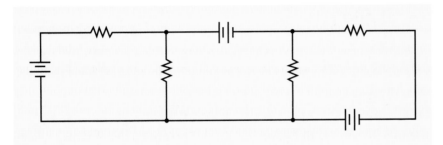

Figure 7-22. How many mesh currents are required?

2. Write the mesh equations for the circuit shown in Figure 7-23. Use the mesh currents shown on the schematic.

3. Compute the voltage drops across each resistor for the circuit shown in Figure 7-23.

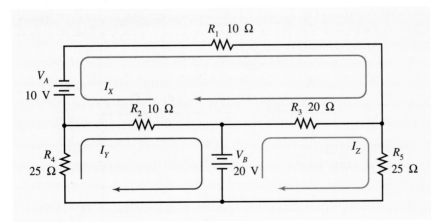

Figure 7-23. A circuit for analysis.

4. What is the power dissipation for R_3 in Figure 7-24?

5.   How much current flows through R_1 in Figure 7-24?

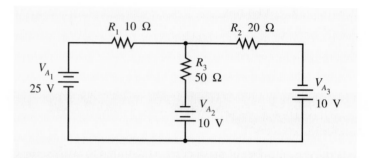

Figure 7-24. A circuit for analysis.

7.3 Nodal Analysis

Nodal analysis is another tool that allows us to analyze complex circuits. Nodal analysis is particularly convenient when analyzing a circuit composed of multiple voltage sources that are referenced to a common point (e.g., ground). Nodal analysis also simplifies the solution of circuits with one or more current sources. Just as mesh analysis is based on Kirchhoff's Voltage Law, nodal analysis is based on Kirchhoff's Current Law. In this context, a **node** will be considered to be any electrical point in a circuit where two or more currents combine.

Writing and Solving Nodal Equations

Figure 7-25 shows a complex circuit with two voltage sources. The circuit also has a ground reference. In order to apply nodal analysis, all node voltages must be referenced to a common point. If no ground point were indicated, then we could choose an arbitrary node as the reference node. If the voltage V_X were known, then all other voltages and currents could be computed with Ohm's Law. To compute V_X, we write a Kirchhoff's Current Law equation that sums the currents associated with the V_X node. Figure 7-26 shows the labeled currents (I_1, I_2, and I_3). All currents are assumed to flow *toward* the node. Of course, we know that at least one current must ultimately flow away from the node, but for now we assume otherwise.

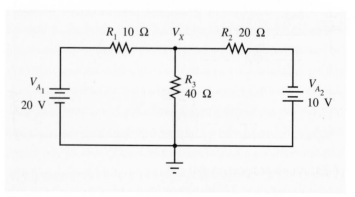

Figure 7-25. A complex circuit for analysis.

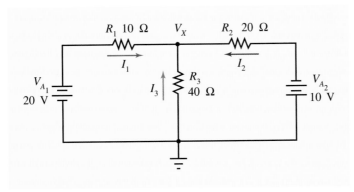

Figure 7-26. Nodal currents are assumed to flow toward the associated node.

The Kirchhoff's Current Law equation for the V_X node is written as follows:

$$I_1 + I_2 + I_3 = 0$$

Since $I = V/R$, according to Ohm's Law, we can substitute V/R for each of the currents. The voltage for a particular resistor is simply the difference in potential across it. Since all currents are assumed to flow toward the node, we must also assume that the node voltage (V_X) is more positive than any other voltage. The V/R substitutions are as follows:

$$\frac{V_X - 20}{R_1} + \frac{V_X - (-10)}{R_2} + \frac{V_X}{R_3} = 0, \text{ or}$$

$$\frac{V_X - 20}{10} + \frac{V_X + 10}{20} + \frac{V_X}{40} = 0$$

Since there is only one unknown quantity (V_X) in this equation, we can solve it directly as follows:

$$\frac{V_X - 20}{10} + \frac{V_X + 10}{20} + \frac{V_X}{40} = 0$$

$$\frac{4V_X - 80 + 2V_X + 20 + V_X}{40} = 0$$

$$7V_X = 60$$

$$V_X = 8.57 \text{ V}$$

Now that we have the value for V_X, we can compute any other circuit value. A few examples follow:

$$I_1 = \frac{V_X - V_{A_1}}{R_1} = \frac{8.57 \text{ V} - 20 \text{ V}}{10 \text{ } \Omega} = -1.143 \text{ A}$$

$$I_2 = \frac{V_X - (V_{A_2})}{R_2} = \frac{8.57 \text{ V} + 10 \text{ V}}{20 \text{ } \Omega} = 928.5 \text{ mA}$$

$$I_3 = \frac{V_X}{R_3} = \frac{8.57 \text{ V}}{40 \text{ } \Omega} = 214.3 \text{ mA}$$

KEY POINTS

The nodal equation assumes that all branch currents flow toward the unknown node (i.e., the node is assumed to be the most positive point).

$$V_2 = I_2 R_2 = 928.5 \text{ mA} \times 20 \ \Omega = 18.57 \text{ V}$$

$$P_3 = I_3^2 R_3 = (214.3 \text{ mA})^2 \times 40 \ \Omega = 1.84 \text{ W}$$

In the case of I_1, we have a negative current. This simply tells us that it is flowing in the opposite direction (i.e., *away from* V_X). Its magnitude, however, is correct.

EXAMPLE SOLUTION

Use nodal analysis to analyze the circuit shown in Figure 7-27.

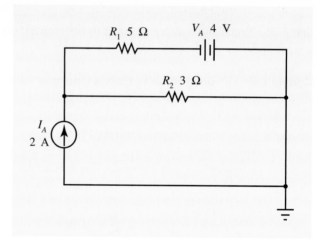

Figure 7-27. Analyze this circuit with nodal analysis.

● **KEY POINTS**

Identify and label all *unknown* nodes. This includes nodes that are neither grounded nor connected directly to a known voltage source.

EXAMPLE **SOLUTION**

First, we label all unknown node voltages. In this case, there is only one, and we have labeled it V_X in Figure 7-28. If we can find the value of V_X, then we will be able to compute all other circuit values. We start by labeling the node currents (I_1, I_2, and I_A in Figure 7-28). We can then apply Kirchhoff's Current Law to set their sum equal to zero:

$$I_1 + I_2 + I_A = 0$$

Next, we substitute the V/R values for the corresponding I values (i.e., $I = V/R$ by Ohm's Law) and solve the resulting equation for V_X:

$$\frac{V_X - V_A}{R_1} + \frac{V_X}{R_2} + I_A = 0$$

$$\frac{V_X - (-4)}{5} + \frac{V_X}{3} + 2 = 0$$

$$\frac{3V_X + 12 + 5V_X + 30}{15} = 0$$

$$8V_X = -42$$

$$V_X = -5.25 \text{ V}$$

● **KEY POINTS**

Once the unknown nodes have been labeled, we write a nodal equation for each unknown node. The nodal equation is simply the Kirchhoff's Current Law equation to describe the node in question.

● **KEY POINTS**

The equation or equations are then solved to determine the node voltages.

Now that we have the value for V_X, we can find any other value of interest by applying the basic circuit equations (e.g., Ohm's Law and power formulas). A few examples follow:

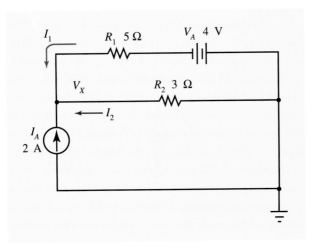

Figure 7-28. Node currents for the circuit shown in Figure 7-27.

$$I_1 = \frac{V_X - V_A}{R_1} = \frac{-5.25\ \text{V} - (-4\ \text{V})}{5\ \Omega} = \frac{-1.25\ \text{V}}{5\ \Omega} = -0.25\ \text{A}$$

$$I_2 = \frac{V_X}{R_2} = \frac{-5.25\ \text{V}}{3\ \Omega} = -1.75\ \text{A}$$

$$V_1 = I_1 R_1 = 0.25\ \text{A} \times 5\ \Omega = 1.25\ \text{V}$$

$$V_2 = I_2 R_2 = 1.75\ \text{A} \times 3\ \Omega = 5.25\ \text{V}$$

$$P_1 = I_1^2 R_1 = (0.25\ \text{A})^2 \times 5\ \Omega = 312.5\ \text{mW}$$

Be sure to note that both I_1, and I_2 are negative. This simply means that they both flow *away* from the V_X node. The calculated magnitude, however, is correct.

Practice Problems

1. Write the nodal equation to describe the V_X node in Figure 7-29.

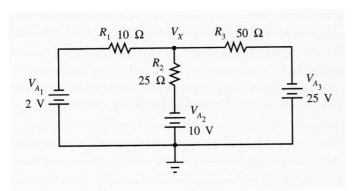

Figure 7-29. Circuit for analysis.

2. Use nodal analysis to calculate the current through R_3 in Figure 7-29.
3. Use nodal analysis in conjunction with basic circuit theory to compute the amount of power dissipated by R_2 in Figure 7-29.
4. Label the polarity of voltage drop across R_1 in Figure 7-29.

Answers to Practice Problems

1.
$$\frac{V_X - V_{A_1}}{R_1} + \frac{V_X - V_{A_2}}{R_2} + \frac{V_X - V_{A_3}}{R_3} = 0$$

$$\frac{V_X + 2\text{ V}}{10\ \Omega} + \frac{V_X + 10\text{ V}}{25\ \Omega} + \frac{V_X + 25\text{ V}}{50\ \Omega} = 0$$

2. 362.5 mA

3. 390.6 mW

4. + on left; – on right

Applied Nodal Analysis

Figure 7-30 shows an unbalanced bridge circuit. Let's use nodal analysis to determine the current through resistor R_5. First, we must assign labels to all ungrounded nodes that do not have a known voltage level. Next, we assign node currents to each of the nodes. These two steps are shown in Figure 7-31.

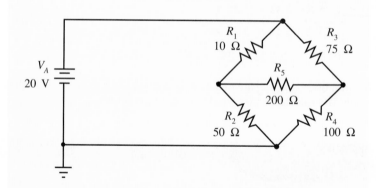

Figure 7-30. An unbalanced bridge circuit for analysis.

We can now write the nodal equation for each labeled node. The equation for the V_X node is determined as follows:

$$I_1 + I_2 + I_5 = 0$$

$$\frac{V_X - V_A}{R_1} + \frac{V_X}{R_2} + \frac{V_X - V_Y}{R_5} = 0$$

$$\frac{V_X - 20\text{ V}}{10\ \Omega} + \frac{V_X}{50\ \Omega} + \frac{V_X - V_Y}{200\ \Omega} = 0$$

$$\frac{20V_X - 400 + 4V_X + V_X - V_Y}{200\ \Omega} = 0$$

$$25V_X - V_Y = 400\text{ V}$$

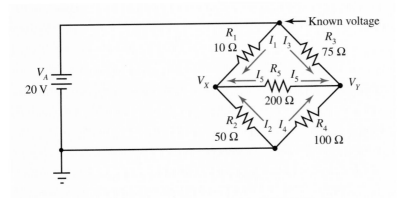

Figure 7-31. All ungrounded nodes without known voltages are assigned labels and node currents.

In a similar manner, we can write the equation for the V_Y node:

$$I_3 + I_4 + I_5 = 0$$

$$\frac{V_Y - V_A}{R_3} + \frac{V_Y}{R_4} + \frac{V_Y - V_X}{R_5} = 0$$

$$\frac{V_Y - 20\text{ V}}{75\text{ }\Omega} + \frac{V_Y}{100\text{ }\Omega} + \frac{V_Y - V_X}{200\text{ }\Omega} = 0$$

$$\frac{8V_Y - 160 + 6V_Y + 3V_Y - 3V_X}{600} = 0$$

$$-3V_X + 17V_Y = 160\text{ V}$$

We now solve the equations to find the node voltages V_X and V_Y. Since we have two equations with two unknown quantities, we will use determinants to solve them. First, we find our system determinant.

$$\Delta = \begin{vmatrix} 25 & -1 \\ -3 & 17 \end{vmatrix} = 425 - 3 = 422$$

Next, we find the value of V_X.

$$V_X = \frac{\begin{vmatrix} 400 & -1 \\ 160 & 17 \end{vmatrix}}{\Delta} = \frac{6{,}800 - (-160)}{422} = 16.49\text{ V}$$

In a similar manner, we compute the value of V_Y.

$$V_Y = \frac{\begin{vmatrix} 25 & 400 \\ -3 & 160 \end{vmatrix}}{\Delta} = \frac{4{,}000 - (-1{,}200)}{422} = 12.32\text{ V}$$

KEY POINTS

If there are two or more equations, then they must be solved using simultaneous methods such as determinants.

With the node voltages calculated, we are now in a position to determine all other circuit values through straightforward application of Ohm's Law and the power formulas. In the present case, we are specifically looking for the value of I_5. We can use Ohm's Law as follows:

$$I_5 = \frac{V_X - V_Y}{R_5}$$

$$I_5 = \frac{16.49\ \text{V} - 12.32\ \text{V}}{200\ \Omega} = 20.85\ \text{mA}$$

Since V_X is more positive than V_Y, we know the current (I_5) must flow from right to left through R_5.

Practice Problems

1. Figure 7-32 shows an equivalent circuit for a certain type of amplifier commonly used in electronic circuits. Apply nodal analysis and basic circuit theory to determine the current through each of the three resistors.

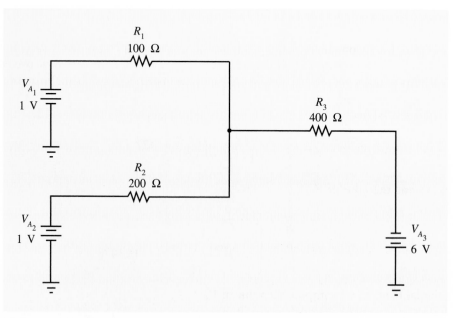

Figure 7-32. Use nodal analysis to compute the resistor currents.

2. Use nodal analysis to find the current through each of the three resistors in Figure 7-33.

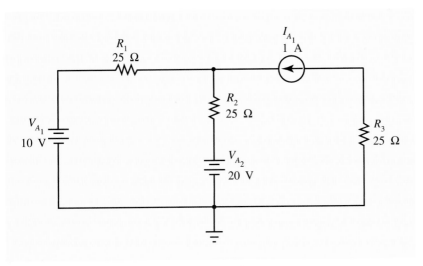

Figure 7-33. Find the current through each resistor.

Answers to Practice Problems

1. $I_1 = 10$ mA; $I_2 = 5$ mA; $I_3 = 15$ mA

2. $I_1 = 300$ mA; $I_2 = 700$ mA; $I_3 = 1$ A

Exercise Problems 7.3

1. How many unknown nodes are in the circuit shown in Figure 7-34?
2. Write the nodal equation(s) for the circuit shown in Figure 7-34.
3. Use nodal analysis to compute the amount of current flowing through R_4 in Figure 7-34.

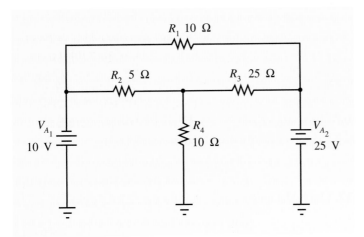

Figure 7-34. Circuit for analysis.

4. Use nodal analysis and basic circuit theory to determine the amount of power dissipated by R_3 in Figure 7-35.

5. Use nodal analysis and basic circuit theory to determine the amount of voltage across R_2 in Figure 7-35.

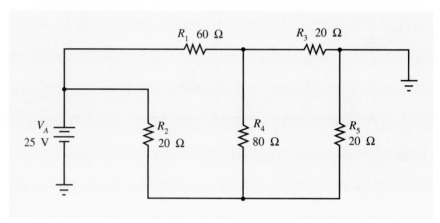

Figure 7-35. A circuit for analysis.

KEY POINTS

The superposition theorem lets us determine the combined effect of multiple voltage and/or current sources by evaluating the effects of each source independently and then algebraically combining the individual effects.

7.4 Superposition Theorem

The superposition theorem permits us to evaluate circuits with multiple sources (current or voltage) by considering the effects of each source independently. Once the effects of the individual sources are known, then the combined effect of all sources can be determined by algebraic summation of the individual effects. We can state the superposition theorem more formally as

> The current through or voltage drop across any component in a linear circuit with multiple current or voltage sources is equal to the algebraic sum of the currents or voltages produced by each source considered independently.

KEY POINTS

The superposition theorem is only applicable to linear circuits and linear quantities.

There are two important points to note about the theorem. First, the theorem only applies to linear circuits. All circuits that have been presented so far in this text are linear circuits. The current through a given component is *directly* proportional to the voltage across the component. At a later point in your studies, you will learn about nonlinear components. In this case, the superposition theorem will not be applicable. The second point to note is that the theorem only applies to voltage and current. Power is specifically omitted, since it is nonlinear (i.e., it varies as the square of current or voltage).

KEY POINTS

The superposition theorem can be used to calculate the net voltage or current in a circuit (linear quantities), but power (nonlinear quantity) must be computed using the standard power formulas.

The procedure to apply the superposition theorem follows:

1. Replace all but one voltage source or one current source with their internal resistances (normally a short for voltage sources and an open for current sources).

2. Compute the various currents and voltages in the circuit based on the effect of the single remaining source.

3. Repeat the first two steps until the currents and voltages have been calculated for each independent source.

4. Algebraically combine the individual currents or voltages associated with a particular component to find the actual current or voltage.

5. Compute any power dissipations that may be required using the net values of current and voltage for a given component.

Consider, for example, the circuit shown in Figure 7-36. Suppose it is our goal to compute the current flow through R_2. Of course, we could find this current with either mesh or nodal analysis, but the superposition theorem gives us an alternative. Since there are two voltage sources, we will replace each one, in turn, with a short circuit, while we compute the effects of the other. Figure 7-37 shows the circuit that results when we short V_{A_2}.

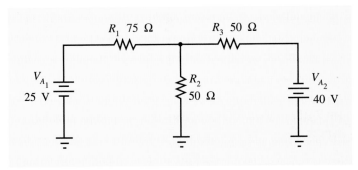

Figure 7-36. Compute the current through R_2.

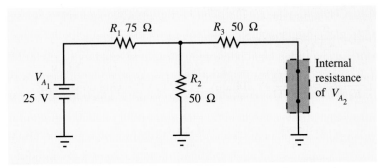

Figure 7-37. Replace all but one source (V_{A_1}) with internal resistances.

It is now a straightforward series-parallel circuit and can be solved with conventional methods. Following is one possible sequence of calculations to determine the current through R_2 as a result of V_{A_1}:

$$R_{2,3} = \frac{R_2 R_3}{R_2 + R_3} = \frac{50\ \Omega \times 50\ \Omega}{50\ \Omega + 50\ \Omega} = 25\ \Omega$$

$$R'_T = R_1 + R_{2,3} = 75\ \Omega + 25\ \Omega = 100\ \Omega$$

$$I'_T = \frac{V_{A_1}}{R'_T} = \frac{25\ \text{V}}{100\ \Omega} = 250\ \text{mA}$$

$$I'_2 = \left(\frac{R_{2,3}}{R_2}\right)I'_T = \left(\frac{25\ \Omega}{50\ \Omega}\right) \times 250\ \text{mA} = 125\ \text{mA (flowing upward)}$$

If we were interested in other currents and voltages besides those associated with R_2, then we would calculate these other values as well. Next, we restore V_{A_2} and replace V_{A_1} with a short circuit. The resulting circuit is shown in Figure 7-38, and the relevant calculations follow:

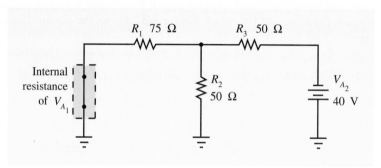

Figure 7-38. Replace all but one source (V_{A_2}) with internal resistances.

$$R_{1,2} = \frac{R_1 R_2}{R_1 + R_2} = \frac{75\ \Omega \times 50\ \Omega}{75\ \Omega + 50\ \Omega} = 30\ \Omega$$

$$R''_T = R_3 + R_{1,2} = 50\ \Omega + 30\ \Omega = 80\ \Omega$$

$$I''_T = \frac{V_{A_2}}{R''_T} = \frac{40\ \text{V}}{80\ \Omega} = 500\ \text{mA}$$

$$I''_2 = \left(\frac{R_{1,2}}{R_2}\right)I''_T = \left(\frac{30\ \Omega}{50\ \Omega}\right) \times 500\ \text{mA} = 300\ \text{mA (flowing downward)}$$

Again, we could calculate other voltages and currents if desired. We are now ready to determine the net current through R_2 by algebraically summing the two currents computed previously. We can assign polarities (i.e., positive or negative) to our currents any way we like provided that currents flowing in opposite directions are assigned opposite polarities. Let us assume current flowing upward through R_2 is positive (e.g., I'_2), which means that current flowing downward through R_2 (e.g., I''_2) must be given a negative polarity. The algebraic summation is

$$I_2(\text{net}) = I'_2 + I''_2$$
$$= 125\ \text{mA} + (-300\ \text{mA})$$
$$= -175\ \text{mA}$$

The negative result, based on our own definitions, means that the net current is flowing from top to bottom through R_2.

1. Use the superposition theorem to compute the voltages across R_1 and R_2 in Figure 7-39.

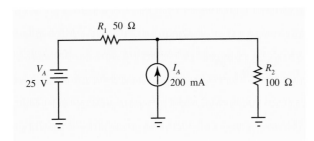

Figure 7-39. What is the voltage across R_1 and R_2?

2. Use the superposition theorem to determine the current through R_2 in Figure 7-40.

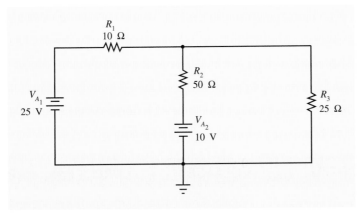

Figure 7-40. Find the current through R_2.

3. Use the superposition theorem to compute the power dissipated by R_2 in Figure 7-41.

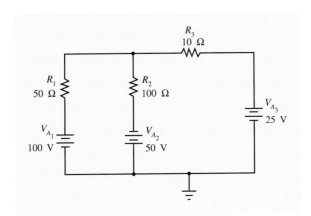

Figure 7-41. Find the power dissipated by R_2.

KEY POINTS

Like voltage polarities are added and opposite polarities are subtracted to determine the actual voltage drop for a given component.

KEY POINTS

After you have analyzed a circuit using the super-position theorem, you should verify the correct-ness of your calculations by applying Kirchhoff's Current and Voltage Laws.

Answers to Practice Problems

1. $V_1 = 15$ V; $V_2 = 10$ V   **2.** $I_2 = 137.5$ mA

3. $P_2 = 65.24$ W

Exercise Problems 7.4

1. Use the superposition theorem to analyze the circuit shown in Figure 7-42. Complete the solution matrix shown in Table 7-1.

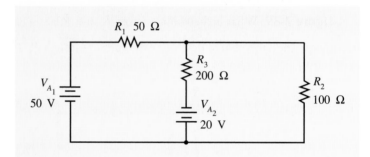

Figure 7-42. Analyze this circuit with the superposition theorem.

COMPONENT	VOLTAGE	CURRENT	POWER
R_1			
R_2			
R_3			

Table 7-1. Solution Matrix for the Circuit in Figure 7-42

2. Use the superposition theorem to calculate the current (magnitude and direction) through R_2 in Figure 7-43.

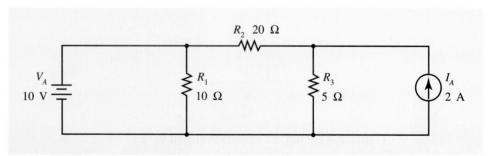

Figure 7-43. How much current flows through R_2?

3. Use the superposition theorem to calculate the voltage drop across R_3 in Figure 7-43.

4. How much power is dissipated by R_2 in Figure 7-43?

5. What is the value of voltage across R_2 in Figure 7-43?

7.5 Thevenin's Theorem

Thevenin's Theorem is a circuit analysis method that allows us to convert any linear circuit, or more often a portion of a circuit, into a simple equivalent circuit. The resulting simplified circuit consists of a constant voltage source and a single series resistor called the Thevenin voltage (V_{TH}) and Thevenin resistance (R_{TH}), respectively. Once the values of the simplified equivalent circuit have been calculated, subsequent analysis of the original circuit becomes much easier. Frequently, Thevenin's Theorem is used to simplify one part of a circuit (consisting of components that will not vary) in order to simplify the analysis of a second part of the circuit (consisting of components whose values change). The point in the circuit that separates these two circuit sections is called the **point of simplification.**

You can Thevenize a circuit by applying the following sequential steps:

1. Short all voltage sources and open all current sources (replace all sources with their internal impedance if it is known). Also open the circuit at the point of simplification.

2. Calculate the value of Thevenin's resistance as seen from the point of simplification.

3. Replace the voltage and current sources with their original values and open the circuit at the point of simplification.

4. Calculate Thevenin's voltage at the point of simplification.

5. Replace the original circuit with the Thevenin equivalent for subsequent analysis of the circuit beyond the point of simplification.

EXAMPLE **SOLUTION**

Refer to the circuit in Figure 7-44. We want to determine the effect on the output of the voltage divider circuit when loads of various values are connected. Apply Thevenin's Theorem to the circuit to simplify the analysis.

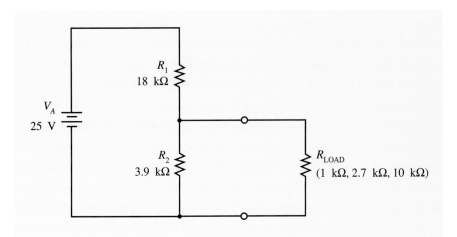

Figure 7-44. A voltage divider circuit with multiple values of load resistance.

First, we will define the point of simplification to be the place where the various load resistors will be connected; components to the left of this point will remain fixed, while those to the right will change. This is shown in Figure 7-45(a). Procedural step number one is shown in Figure 7-45(b). Here the voltage source is shorted, and the circuit is opened at the point of simplification.

We can now calculate the Thevenin resistance (R_{TH}) as listed in procedural step two. By inspection, we can see that the 3.9-kΩ and the 18-kΩ resistors are now in parallel. The Thevenin resistance is found (in this case) by applying the parallel resistor equation:

$$R_{TH} = \frac{R_1 R_2}{R_1 + R_2}$$

$$R_{TH} = \frac{(18 \text{ k}\Omega)(3.9 \text{ k}\Omega)}{(18 \text{ k}\Omega) + (3.9 \text{ k}\Omega)} = 3.2 \text{ k}\Omega$$

Next, we determine the Thevenin voltage by replacing the sources (procedural step three). This is shown in Figure 7-45(c). At this time, we can use any appropriate circuit analysis method to calculate the voltage across R_2, which will also be our Thevenin voltage (V_{TH}). Since we have a

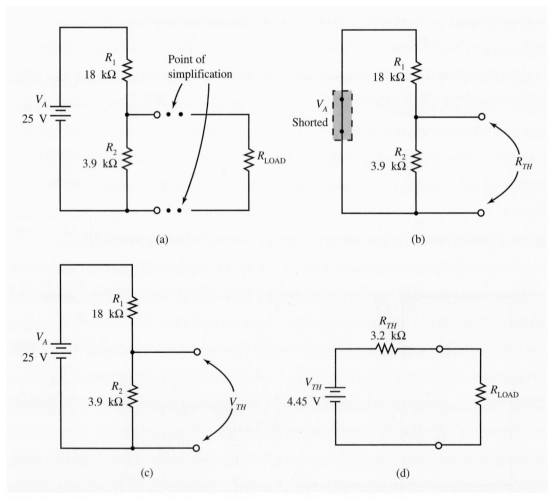

Figure 7-45. Thevenization of the circuit shown in Figure 7-44.

simple two-resistor series circuit, let's apply the voltage divider equation to compute the voltage across R_2:

$$V_{TH} = \left(\frac{R_2}{R_1 + R_2}\right)V_A = \left(\frac{3.9 \text{ k}\Omega}{18 \text{ k}\Omega + 3.9 \text{ k}\Omega}\right) \times 25 \text{ V} = 4.45 \text{ V}$$

Figure 7-45(d) shows the Thevenin equivalent circuit. Calculations for each of the individual load resistances can now be quickly computed by simply applying the voltage divider equation:

$$V_{LOAD(1)} = \left(\frac{R_{LOAD(1)}}{R_{LOAD(1)} + R_{TH}}\right)V_{TH} = \left(\frac{1 \text{ k}\Omega}{1 \text{ k}\Omega + 3.2 \text{ k}\Omega}\right) \times 4.45 \text{ V} = 1.06 \text{ V}$$

$$V_{LOAD(2)} = \left(\frac{R_{LOAD(2)}}{R_{LOAD(2)} + R_{TH}}\right)V_{TH} = \left(\frac{2.7 \text{ k}\Omega}{2.7 \text{ k}\Omega + 3.2 \text{ k}\Omega}\right) \times 4.45 \text{ V} = 2.04 \text{ V}$$

$$V_{LOAD(3)} = \left(\frac{R_{LOAD(3)}}{R_{LOAD(3)} + R_{TH}}\right)V_{TH} = \left(\frac{10 \text{ k}\Omega}{10 \text{ k}\Omega + 3.2 \text{ k}\Omega}\right) \times 4.45 \text{ V} = 3.37 \text{ V}$$

It is important to realize that by reducing the original circuit (less the load resistor) to a single voltage source and a single resistor, *all subsequent calculations are reduced to evaluation of a simple series circuit*. Without a simplification theorem such as Thevenin's Theorem, calculation of circuit values with each new load resistor would require several computations.

EXAMPLE SOLUTION

Figure 7-46 shows a portion of a circuit that is used in digital electronics to convert digital voltages into equivalent analog voltages. Determine the values for a Thevenin equivalent circuit that includes all but R_X, then compute the voltage across R_X.

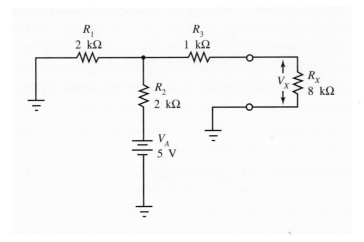

Figure 7-46. Determine the Thevenin equivalent for this circuit.

EXAMPLE SOLUTION

Our first procedural step is shown in Figure 7-47(a). Here we have replaced the voltage source with a short circuit. We have also opened the circuit at the point of simplification. We can now compute the Thevenin resistance (procedural step two). This is the resistance seen when looking *into* the point of simplification. By inspection, we can see that we have a series-parallel circuit; the parallel combination of R_1 and R_2 is in series with R_3. We can find the total resistance as follows:

$$R_{1,2} = \frac{R_N}{N} = \frac{2 \text{ k}\Omega}{2} = 1.0 \text{ k}\Omega$$

$$R_{TH} = R_{1,2} + R_3 = 1.0 \text{ k}\Omega + 1.0 \text{ k}\Omega = 2.0 \text{ k}\Omega$$

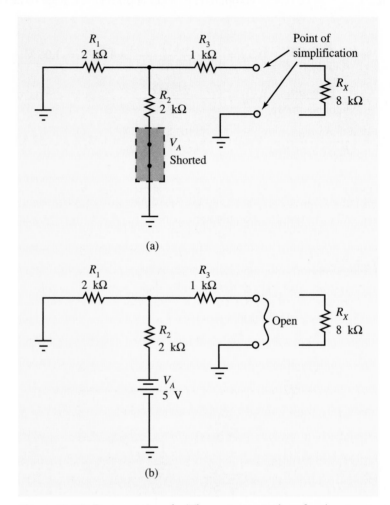

(a)

(b)

Figure 7-47. Determining the Thevenin equivalent for the circuit shown in Figure 7-46.

Next, we restore the voltage source (procedural step three) and compute the Thevenin voltage (procedural step four). Figure 7-47(b) shows how this is accomplished. First, we observe that there is no current flow through R_3 since one end is open circuited. This means there will be no voltage drop across R_3. That is to say, the voltage at one end of R_3 will be the same as the voltage at the other end. So, to find the voltage at the open end of R_3 (our goal), we can simply calculate the voltage at the opposite end; they will be the same. By inspection, we can see that if we were to measure the voltage from the left end of R_3 to ground, we would actually be measuring across R_1. So, if we compute the voltage across R_1, we will have the Thevenin voltage for this circuit. The circuit can be viewed as a simple series circuit consisting of the voltage source and resistors R_1 and R_2. Since R_3 is open, it has no effect on our calculations. We find V_{TH} as follows:

$$V_{TH} = V_1 = \left(\frac{R_1}{R_1 + R_2}\right)V_A$$

$$= \left(\frac{2\ k\Omega}{2\ k\Omega + 2\ k\Omega}\right) \times 5\ V = 2.5\ V$$

Figure 7-48 shows the final Thevenin equivalent circuit reconnected to resistor R_X. It is now a simple series circuit. We can find the voltage across R_X as follows:

$$V_X = \left(\frac{R_X}{R_{TH} + R_X}\right)V_{TH} = \left(\frac{8\ k\Omega}{2\ k\Omega + 8\ k\Omega}\right) \times 2.5\ V = 2.0\ V$$

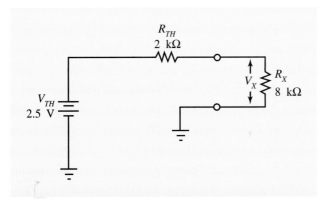

Figure 7-48. Thevenin equivalent for the circuit shown in Figure 7-46.

The real value of Thevenin's theorem becomes apparent when the circuit must be evaluated under changing conditions. Suppose, for example, that a technician needed to determine the voltage across R_X in Figure 7-46 for several different values of R_X. The problem could be solved with standard series-parallel methods, but a complete circuit simplification would be required for each new value of R_X. By contrast, the exact same results can be obtained by analyzing the Thevenin equivalent shown in Figure 7-48. The voltage (V_X) for each new value of R_X now requires only a simple voltage divider calculation; the Thevenized portion of the circuit does not change.

Exercise Problems 7.5

1. Application of Thevenin's Theorem produces an equivalent circuit that consists of a _____ and a _____ .

2. Determine the Thevenin equivalent for the circuit shown in Figure 7-49.

3. Find the Thevenin equivalent for the circuit shown in Figure 7-50.

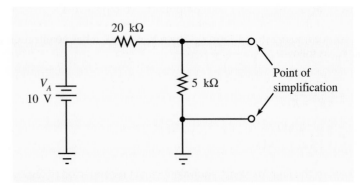

Figure 7-49. Find the Thevenin equivalent for this circuit.

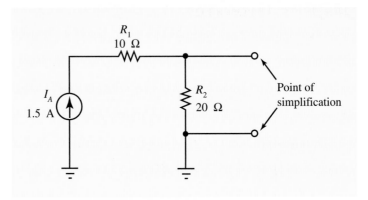

Figure 7-50. Find the Thevenin equivalent for this circuit.

7.6 Norton's Theorem

Norton's Theorem is similar in concept to Thevenin's Theorem, but it produces an equivalent circuit that consists of a current source (I_N) and a parallel resistance (R_N). The value of Norton's Theorem is similar to Thevenin's Theorem. A technician should be able to simplify circuits with both theorems. Although either theorem can be used for any given circuit, one or the other is often more convenient for a specific application.

A Norton equivalent circuit can be found by applying the following procedural steps:

1. Short all voltage sources and open all current sources (replace all sources with their internal impedance if it is known). Also open the circuit at the point of simplification.

2. Calculate the value of Norton's resistance as seen from the point of simplification.

3. Replace the voltage and current sources with their original values and short the circuit at the point of simplification.

4. Calculate Norton's current at the point of simplification.

5. Replace the original circuit component with the Norton's equivalent for subsequent analysis of the circuit beyond the point of simplification.

Determine the Norton equivalent for the circuit shown in Figure 7-51. Include all components but R_X.

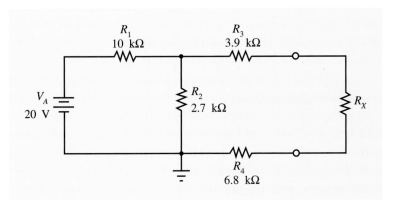

Figure 7-51. Find the Norton equivalent for this circuit.

Procedural steps one and two are the same as for Thevenin's Theorem. We short voltage sources and open current sources, then calculate the Norton resistance (R_N) as seen from the point of simplification. Figure 7-52(a) illustrates this step.

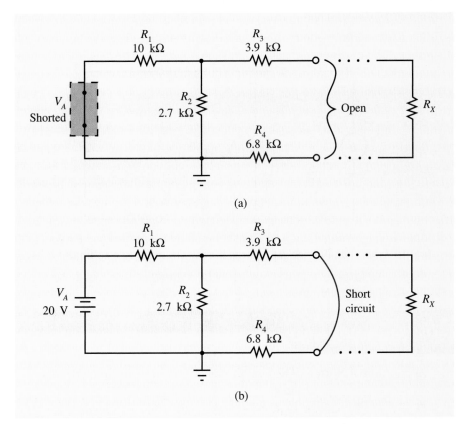

(a)

(b)

Figure 7-52. Obtaining the Norton equivalent for the circuit shown in Figure 7-51.

Once the voltage source has been replaced by a short, we have a simple series-parallel circuit. We can compute total resistance (as seen from the point of simplification) as follows:

$$R_{1,2} = \frac{R_1 R_2}{R_1 + R_2} = \frac{10 \text{ k}\Omega \times 2.7 \text{ k}\Omega}{10 \text{ k}\Omega + 2.7 \text{ k}\Omega} = 2.13 \text{ k}\Omega$$

$$R_N = R_{1,2} + R_3 + R_4 = 2.13 \text{ k}\Omega + 3.9 \text{ k}\Omega + 6.8 \text{ k}\Omega = 12.83 \text{ k}\Omega$$

Procedural steps three and four are illustrated in Figure 7-52(b). We now have a series-parallel circuit, and we need to determine the current through the series combination of R_3 and R_4. One possible sequence of calculations follows:

$$R_{3,4} = R_3 + R_4 = 3.9 \text{ k}\Omega + 6.8 \text{ k}\Omega = 10.7 \text{ k}\Omega$$

$$R_{2,3,4} = \frac{R_2 R_{3,4}}{R_2 + R_{3,4}} = \frac{2.7 \text{ k}\Omega \times 10.7 \text{ k}\Omega}{2.7 \text{ k}\Omega + 10.7 \text{ k}\Omega} = 2.16 \text{ k}\Omega$$

$$R_T = R_1 + R_{2,3,4} = 10 \text{ k}\Omega + 2.16 \text{ k}\Omega = 12.16 \text{ k}\Omega$$

$$I_T = \frac{V_A}{R_T} = \frac{20 \text{ V}}{12.16 \text{ k}\Omega} = 1.65 \text{ mA}$$

$$I_N = I_{3,4} = \left(\frac{R_{2,3,4}}{R_{3,4}}\right) I_T = \left(\frac{2.16 \text{ k}\Omega}{10.7 \text{ k}\Omega}\right) \times 1.65 \text{ mA} = 333.1 \text{ }\mu\text{A}$$

The final Norton equivalent is shown in Figure 7-53. As with a Thevenin equivalent, we could now analyze the effects with several values of R_X. The calculations in each case would be that of a two-resistor current divider circuit.

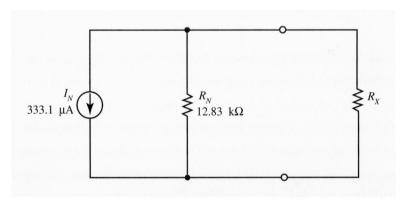

Figure 7-53. The Norton equivalent for the circuit shown in Figure 7-51.

Exercise Problems 7.6

1. Application of Norton's Theorem produces an equivalent circuit that consists of a
_____ and a _____.

2. Determine the Norton equivalent for the circuit shown in Figure 7-54.

3. Find the Norton equivalent for the circuit shown in Figure 7-55.

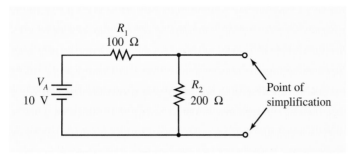

Figure 7-54. Find the Norton equivalent for this circuit.

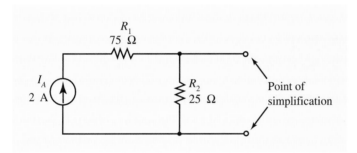

Figure 7-55. Find the Norton equivalent for this circuit.

KEY POINTS

Millman's Theorem is an analysis tool that is particularly well suited to circuits that have several parallel branches and multiple voltage and/or current sources.

7.7 Millman's Theorem

Millman's Theorem is another circuit analysis tool that is helpful to a technician for certain types of problems. Millman's Theorem is particularly well suited to simplifying circuits that have several parallel branches and multiple voltage and/or current sources. Figure 7-56 shows such a circuit. Applying Millman's Theorem results in an equivalent circuit that consists of a single current source (I_M) and a single parallel resistance (R_M).

KEY POINTS

The theorem produces an equivalent circuit that consists of a single current source and a single parallel resistor.

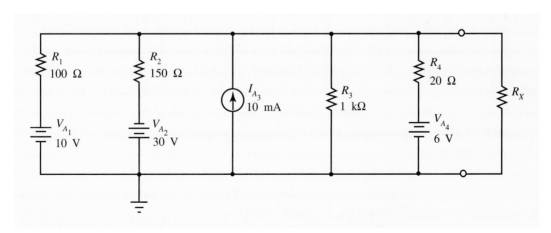

Figure 7-56. A circuit that can be readily simplified with Millman's Theorem.

Millman's Theorem can be applied to a circuit like that in Figure 7-56 by applying the following procedural steps:

1. Convert all branches that have voltage sources and series resistances to equivalent current sources and parallel resistances.
2. Algebraically combine the resulting current sources into a single current source.
3. Combine the various parallel resistances into a single resistance.

EXAMPLE SOLUTION

Apply Millman's Theorem to the circuit shown in Figure 7-56. Include all components except R_X.

EXAMPLE **SOLUTION**

The first procedural step requires us to convert all of the voltage sources and series resistances to equivalent current sources and parallel resistances. This is an easy step. The equivalent parallel resistance in the converted circuit will be the same as the value of series resistance in the original circuit. The value for the equivalent current source in a particular branch is found by applying Ohm's Law. Let's now convert all of the voltage sources to equivalent current sources:

$$I_{A_1} = \frac{V_{A_1}}{R_1} = \frac{10 \text{ V}}{100 \text{ }\Omega} = 100 \text{ mA}$$

$$I_{A_2} = \frac{V_{A_2}}{R_2} = \frac{-30 \text{ V}}{150 \text{ }\Omega} = -200 \text{ mA}$$

$$I_{A_4} = \frac{V_{A_4}}{R_4} = \frac{6 \text{ V}}{20 \text{ }\Omega} = 300 \text{ mA}$$

The circuit at this point in the simplification is shown in Figure 7-57.

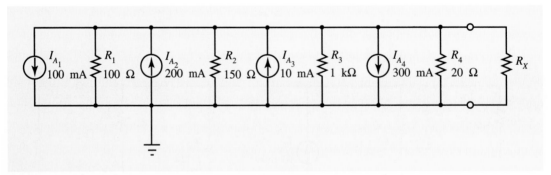

Figure 7-57. Partial simplification of the circuit shown in Figure 7-56 by using Millman's Theorem.

Procedural step two requires us to algebraically combine the various current sources. Recall that current polarity simply refers to direction. So, let us consider a negative current to be one that flows away from ground. A positive current will flow toward ground. Based on this assignment, we know that I_{A_1} and I_{A_4} will be considered to be positive currents, while I_{A_2} and I_{A_3} will be negative currents. We combine them as follows:

$$I_M = I_{A_1} + I_{A_2} + I_{A_3} + I_{A_4}$$
$$= 100 \text{ mA} + (-200 \text{ mA}) + (-10 \text{ mA}) + 300 \text{ mA}$$
$$= 190 \text{ mA}$$

The Millman equivalent resistance is found by combining the parallel branch resistances as follows:

$$R_M = R_1 \| R_2 \| R_3 \| R_4$$
$$= \frac{1}{\dfrac{1}{R_1} + \dfrac{1}{R_2} + \dfrac{1}{R_3} + \dfrac{1}{R_4}}$$
$$= \frac{1}{\dfrac{1}{100 \ \Omega} + \dfrac{1}{150 \ \Omega} + \dfrac{1}{1{,}000 \ \Omega} + \dfrac{1}{20 \ \Omega}}$$
$$= 14.78 \ \Omega$$

The converted circuit is shown in Figure 7-58. Of course, if desired, we could convert this circuit to an equivalent voltage source and series resistance by applying Thevenin's Theorem.

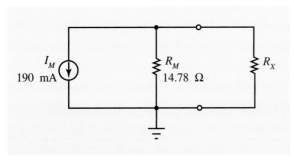

Figure 7-58. The Millman equivalent for the circuit shown in Figure 7-56.

Exercise Problems 7.7

1. Use Millman's Theorem to develop an equivalent circuit for the circuit shown in Figure 7-59. Include all components except R_X.
2. How much current flows through R_X in Figure 7-59?
3. What is the voltage drop across R_X in Figure 7-59?
4. Apply Millman's Theorem to the circuit shown in Figure 7-60. Include all components except R_X.
5. What is the voltage drop across R_X in Figure 7-60?

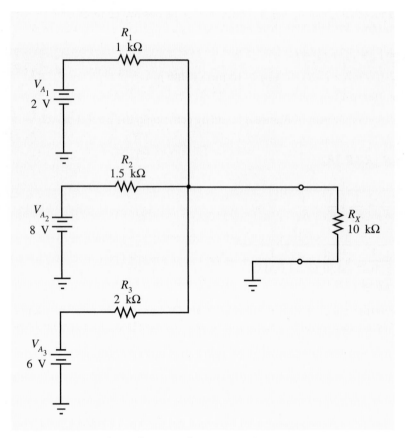

Figure 7-59. Apply Millman's Theorem to this circuit.

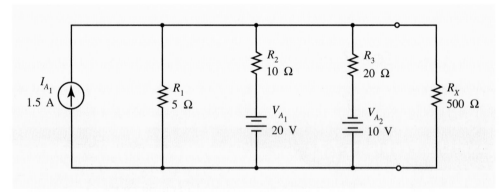

Figure 7-60. Apply Millman's Theorem to this circuit.

7.8 Pi-to-Tee and Tee-to-Pi Conversions

This section introduces a new type of analysis tool. The previous analytical techniques were aimed at obtaining final solutions to problems. The conversion methods described in this section, by contrast, merely change the form of a circuit. By changing the form of a complex circuit, we can sometimes obtain a form that can be completely analyzed without using advanced circuit analysis techniques such as mesh or nodal analysis.

Figure 7-61 shows the two types of three-terminal networks that will be discussed in this section. Figure 7-61(a) shows a connection called a **tee configuration.** It is called a tee because it resembles the letter "T" when drawn on a schematic. Some technicians prefer to draw the same electrical configuration in a slightly different way as shown in Figure 7-61(b). When it is drawn this way, it is often called a **wye configuration.** In general, the terms tee and wye are interchangeable.

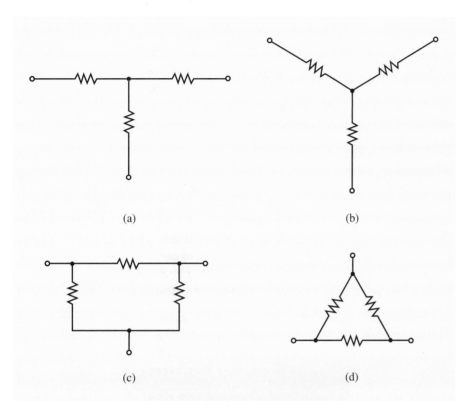

Figure 7-61. Types of three-terminal networks: (a) tee, (b) wye, (c) pi, and (d) delta.

Figure 7-61(c) shows another basic circuit arrangement called a **pi configuration.** Again it is named for its shape, which in this case resembles the Greek letter pi (π). This same electrical configuration is sometimes drawn in a slightly different way as shown in Figure 7-61(d). When drawn this way, it is often called a **delta configuration,** because it resembles the Greek letter delta (Δ). This section presents a method that can be used to convert a tee (or wye) configuration into an electrically equivalent pi (or delta) configuration. Similarly, a method is introduced that can transform a pi (or delta) configuration into an equivalent tee (or wye) configuration.

Pi-to-Tee

Figure 7-62 shows the labeling that will be used in our introductory discussion of pi and tee conversions. In the case of a pi-to-tee conversion, the values for R_{ac}, R_{ab}, and R_{bc} will be known. The corresponding tee network values of R_a, R_b, and R_c must be determined. It is important to realize that once a conversion has been made, the circuit *external* to the tee or pi network will be totally unaffected. That is to say, one network has exactly the same electrical characteristics (voltage between the terminals, current in or out of a terminal, and so

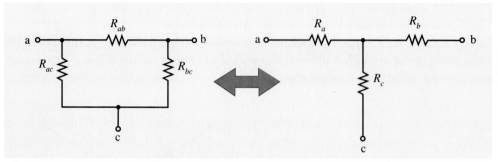

Figure 7-62. The tee and pi networks are labeled to simplify identification.

on) as the other. The labeling method shown in Figure 7-62 makes it easy to remember the relative locations of the various resistors (e.g., R_{bc} is connected between terminals b and c, R_a is connected to terminal a, and so forth).

Any pi network can be converted to an equivalent tee network with the following procedure:

1. Select any terminal and multiply the two resistors in the pi network connected to that terminal.
2. Divide the product obtained in step one by the sum of all three pi-network resistors.
3. Set the value of the resistor in the tee network that connects to the same terminal as that selected in step one to the value computed in step two.
4. Repeat the preceding steps for each of the pi-network terminals.

This procedure can be expressed more formally in equation form. The resistor references used in the following equations correspond to the labeling shown in Figure 7-62.

$$R_a = \frac{R_{ab}R_{ac}}{R_{ab} + R_{ac} + R_{bc}} \qquad (7\text{-}1)$$

$$R_b = \frac{R_{ab}R_{bc}}{R_{ab} + R_{ac} + R_{bc}} \qquad (7\text{-}2)$$

$$R_c = \frac{R_{ac}R_{bc}}{R_{ab} + R_{ac} + R_{bc}} \qquad (7\text{-}3)$$

EXAMPLE SOLUTION

Convert the pi network shown in Figure 7-63 to an equivalent tee network.

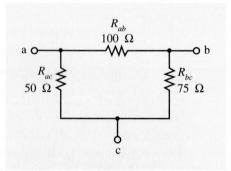

Figure 7-63. Convert this circuit to an equivalent tee network.

We apply Equations 7-1 through 7-3 as follows:

$$R_a = \frac{R_{ab}R_{ac}}{R_{ab} + R_{ac} + R_{bc}}$$

$$= \frac{100\ \Omega \times 50\ \Omega}{100\ \Omega + 50\ \Omega + 75\ \Omega} = 22.22\ \Omega$$

$$R_b = \frac{R_{ab}R_{bc}}{R_{ab} + R_{ac} + R_{bc}}$$

$$= \frac{100\ \Omega \times 75\ \Omega}{100\ \Omega + 50\ \Omega + 75\ \Omega} = 33.33\ \Omega$$

$$R_c = \frac{R_{ac}R_{bc}}{R_{ab} + R_{ac} + R_{bc}}$$

$$= \frac{50\ \Omega \times 75\ \Omega}{100\ \Omega + 50\ \Omega + 75\ \Omega} = 16.67\ \Omega$$

The converted circuit is shown in Figure 7-64.

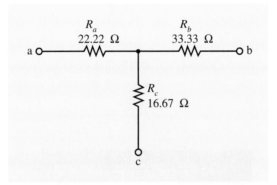

Figure 7-64. The result of converting Figure 7-63 to a tee network.

Practice Problems

1. Convert the pi network shown in Figure 7-65 to an equivalent tee network.

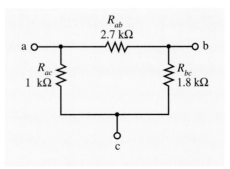

Figure 7-65. Convert this circuit to a tee network.

2. Convert the delta network shown in Figure 7-66 to an equivalent wye network.

Answers to Practice Problems

1. $R_a = 490.9 \ \Omega$
$R_b = 883.6 \ \Omega$
$R_c = 327.3 \ \Omega$

2. $R_a = 107.8 \ \Omega$
$R_b = 26.8 \ \Omega$
$R_c = 44.22 \ \Omega$

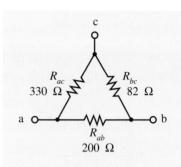

Figure 7-66. Convert this delta network to an equivalent wye network.

Tee-to-Pi

Converting a tee network into an equivalent pi network is also a straightforward process. The procedure can be described as follows:

1. Multiply each of the resistances in the tee network by every other resistance (i.e., $R_a \times R_b$, $R_a \times R_c$, and $R_b \times R_c$).
2. Sum the three products obtained in step one.
3. Select any terminal of the tee and divide that resistance into the result obtained in step two.
4. Repeat step three for each terminal. The calculations associated with a given terminal produce the value of the pi-network resistance *on the opposite side of the network* (i.e., not connected to that terminal).

The preceding procedure can be expressed more concisely as a series of equations:

$$R_{ab} = \frac{R_a R_b + R_a R_c + R_b R_c}{R_c} \qquad (7\text{-}4)$$

$$R_{bc} = \frac{R_a R_b + R_a R_c + R_b R_c}{R_a} \qquad (7\text{-}5)$$

$$R_{ac} = \frac{R_a R_b + R_a R_c + R_b R_c}{R_b} \qquad (7\text{-}6)$$

EXAMPLE SOLUTION

Convert the tee network shown in Figure 7-67 to an equivalent pi network.

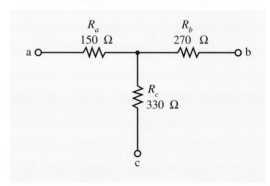

Figure 7-67. Convert this circuit to an equivalent pi network.

We apply Equations 7-4 through 7-6 as follows:

$$R_{ab} = \frac{R_a R_b + R_a R_c + R_b R_c}{R_c}$$

$$= \frac{(150\ \Omega \times 270\ \Omega) + (150\ \Omega \times 330\ \Omega) + (270\ \Omega \times 330\ \Omega)}{330\ \Omega} = 542.7\ \Omega$$

$$R_{bc} = \frac{R_a R_b + R_a R_c + R_b R_c}{R_a}$$

$$= \frac{(150\ \Omega \times 270\ \Omega) + (150\ \Omega \times 330\ \Omega) + (270\ \Omega \times 330\ \Omega)}{150\ \Omega} = 1.19\ k\Omega$$

$$R_{ac} = \frac{R_a R_b + R_a R_c + R_b R_c}{R_b}$$

$$= \frac{(150\ \Omega \times 270\ \Omega) + (150\ \Omega \times 330\ \Omega) + (270\ \Omega \times 330\ \Omega)}{270\ \Omega} = 663.3\ \Omega$$

The resulting pi network is shown in Figure 7-68.

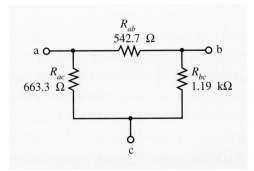

Figure 7-68. The result of converting the circuit shown in Figure 7-67 to an equivalent pi network.

Practice Problems

1. Convert the tee network shown in Figure 7-69 to an equivalent pi network.

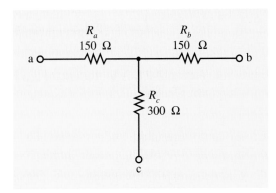

Figure 7-69. Convert this tee circuit to an equivalent pi network.

2. Convert the wye network shown in Figure 7-70 to an equivalent delta network.

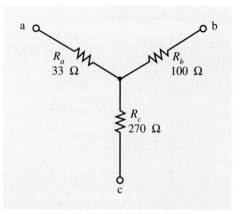

Answers to Practice Problems

1. $R_{ab} = 375 \; \Omega$
   $R_{bc} = 750 \; \Omega$
   $R_{ac} = 750 \; \Omega$

2. $R_{ab} = 145.2 \; \Omega$
   $R_{bc} = 1.188 \; k\Omega$
   $R_{ac} = 392.1 \; \Omega$

Figure 7-70. Convert this wye circuit to an equivalent delta network.

Applied Network Conversions

Now let us take a look at a real application that a technician might encounter where network conversions would be helpful. Suppose you wanted to show one of the technicians in your shop how to analyze the bridge circuit shown in Figure 7-71 to determine total current drawn from the supply. Of course, you have learned several ways to analyze the bridge circuit, but the technician you are trying to assist doesn't know how to work with simultaneous equations or determinants. This eliminates mesh and nodal analysis from consideration.

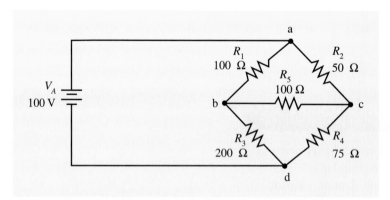

Figure 7-71. A bridge circuit can be simplified with network conversions.

Close examination of the bridge circuit, however, will reveal several interesting subnetworks. Figure 7-72 highlights four "hidden" tee and pi networks. If we apply the tee-to-pi or the pi-to-tee conversion method discussed previously, we can convert the bridge circuit into a series-parallel circuit (i.e., not a complex circuit).

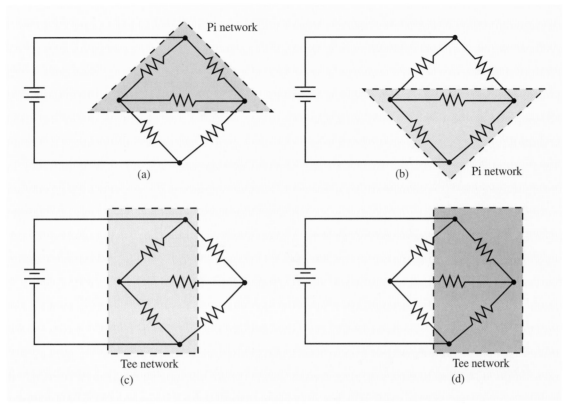

Figure 7-72. Hidden pi and tee networks in a bridge circuit.

[**EXAMPLE** SOLUTION]

Convert the bridge circuit shown in Figure 7-71 to a standard series-parallel circuit and compute total current.

[EXAMPLE **SOLUTION**]

We have four choices (all highlighted in Figure 7-72). Let's choose to convert the upper pi network (Figure 7-72(a)) to an equivalent tee network. Figure 7-73 shows how the conversion will produce a standard series-parallel circuit. We apply Equations 7-1 through 7-3 as follows:

$$R_a = \frac{R_{ab}R_{ac}}{R_{ab} + R_{ac} + R_{bc}}$$

$$= \frac{R_1 R_2}{R_1 + R_2 + R_5}$$

$$= \frac{100\ \Omega \times 50\ \Omega}{100\ \Omega + 50\ \Omega + 100\ \Omega}$$

$$= 20\ \Omega$$

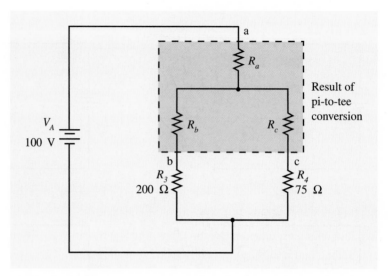

Figure 7-73. Pi-to-tee conversion on a bridge circuit produces a series-parallel circuit.

$$R_b = \frac{R_{ab}R_{bc}}{R_{ab} + R_{ac} + R_{bc}}$$

$$= \frac{R_1 R_5}{R_1 + R_2 + R_5}$$

$$= \frac{100\ \Omega \times 100\ \Omega}{100\ \Omega + 50\ \Omega + 100\ \Omega}$$

$$= 40\ \Omega$$

$$R_c = \frac{R_{ac}R_{bc}}{R_{ab} + R_{ac} + R_{bc}}$$

$$= \frac{R_2 R_5}{R_1 + R_2 + R_5}$$

$$= \frac{50\ \Omega \times 100\ \Omega}{100\ \Omega + 50\ \Omega + 100\ \Omega}$$

$$= 20\ \Omega$$

The converted circuit is shown in Figure 7-74(a). The circuit can now be further simplified by combining series resistances: R_b and R_3, R_c and R_4. The resulting resistances ($R_{b,3}$ and $R_{c,4}$) are in parallel and can be combined. Finally, this last result ($R_{b,c,3,4}$) is in series with R_a and can be combined to find total resistance. These calculations are as follows:

$$R_{b,3} = R_b + R_3 = 40\ \Omega + 200\ \Omega = 240\ \Omega$$

$$R_{c,4} = R_c + R_4 = 20\ \Omega + 75\ \Omega = 95\ \Omega$$

$$R_{b,c,3,4} = \frac{R_{b,3}R_{c,4}}{R_{b,3} + R_{c,4}} = \frac{240\ \Omega \times 95\ \Omega}{240\ \Omega + 95\ \Omega} = 68.1\ \Omega$$

$$R_T = R_a + R_{b,c,3,4} = 20\ \Omega + 68.1\ \Omega = 88.1\ \Omega$$

Finally, we can calculate total current flow by applying Ohm's Law:

$$I_T = \frac{V_A}{R_T}$$

$$= \frac{100 \text{ V}}{88.1 \text{ }\Omega}$$

$$= 1.14 \text{ A}$$

The fully simplified circuit is shown in Figure 7-74(b).

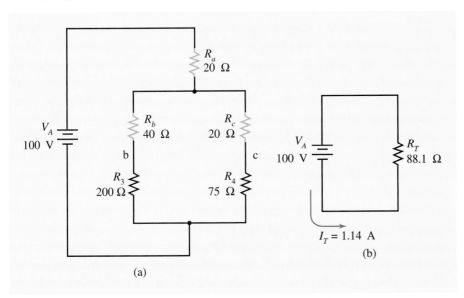

(a)

(b)

Figure 7-74. Simplification of a bridge circuit after pi-to-tee conversion.

Practice Problems

1. Refer to Figure 7-75. Convert the tee network consisting of R_2, R_3, and R_4 to an equivalent pi network. Compute total current flow.

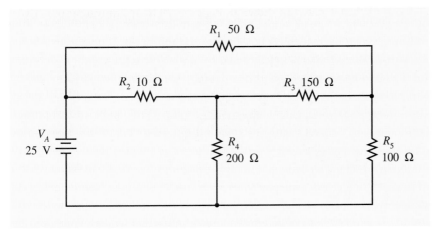

Figure 7-75. Simplify this circuit with tee-to-pi conversion.

2. Refer to Figure 7-76. Convert the pi network consisting of R_1, R_3, and R_4 to an equivalent tee network. Compute total current flow.

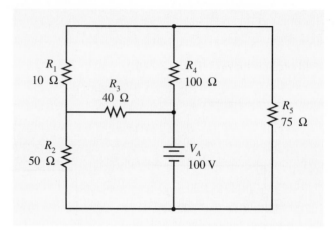

Figure 7-76. Use pi-to-tee conversion to simplify this circuit.

Answers to Practice Problems

1. 296.3 mA   2. 1.7 A

Exercise Problems 7.8

1. Another name for a pi network is a _____ network.
2. Another name for a wye network is a _____ network.
3. Convert the pi network shown in Figure 7-77 to an equivalent tee network.
4. Convert the tee network shown in Figure 7-78 to an equivalent pi network.

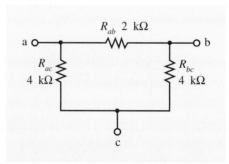

Figure 7-77. Convert to a tee network.

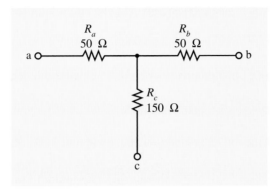

Figure 7-78. Convert to a pi network.

Chapter Summary

- This chapter presented a method for positive identification of complex circuits. Essentially, all sets of series or parallel resistors are replaced by a single equivalent resistor to form a simpler circuit. Repeat this process until no further combination of series and parallel resistors can be done. If the simplified circuit is more complex than a single resistor and a power source, then the original circuit can be classified as a complex circuit.

- Mesh analysis is based on Kirchhoff's Voltage Law and is a tool for analyzing complex circuits. To analyze a circuit with mesh analysis, complete loops or meshes are identified on the circuit, such that each component has at least one mesh current and each mesh has at least one component that is common to another mesh; use as few loops as possible. Next, mark the polarity of the voltage drops across each component that corresponds to the assumed direction of mesh currents. Kirchhoff's Voltage Law equations (mesh equations) are then written for each loop using the *IR* form for the unknown voltage drops. The resulting equations are then solved using any preferred method (e.g., determinants) for solving multiple equations in multiple unknowns (simultaneous equations). Solving the equations produces values for the mesh currents. Negative mesh currents simply flow opposite the assumed direction. If two or more mesh currents flow through a common component, then the real current is the algebraic sum of the associated mesh currents. Ohm's Law is then applied to determine the remaining circuit quantities that are of interest.

- Nodal analysis is another tool that permits analysis of complex circuits. Nodal analysis is based on Kirchhoff's Current Law. To apply nodal analysis, we first label all unknown nodes. Next, we assume that each node is the most positive point in the circuit (i.e., assume that all currents flow toward it) and write the Kirchhoff's Current Law equations (nodal equations) for each point. Solution of the equation(s) will provide a voltage value for the initially unknown nodes. Ohm's Law is then applied to determine the remaining circuit quantities that are of interest.

- The superposition theorem is useful for evaluating the cumulative effect of multiple current and/or voltage sources in a linear circuit. The effects of each source are considered one at a time, while all other sources are replaced with their internal resistances (normally zero ohm for voltage sources and open circuits for current sources). The currents and/or voltages that are produced by the various sources are then algebraically combined in the component(s) of interest to find the net effect of all sources.

- Thevenin's Theorem provides a tool that can convert any linear circuit (or part of a circuit) to an equivalent circuit consisting of a single voltage source and a single series resistance. External to the simplified portion of the circuit, the Thevenin equivalent and the original circuit will appear identical. The Thevenin equivalent is found by replacing all power sources with their internal resistances and calculating the circuit resistance as seen from the point of simplification. This is the Thevenin resistance. Next, the sources are restored, and the voltage at the point of simplification is computed to determine the Thevenin voltage. Thevenin's Theorem is particularly useful when analyzing the effects on a circuit when one component is altered and all other components remain fixed.

- Norton's Theorem is used for the same analytical purposes as Thevenin's Theorem, but the resulting simplified circuit consists of a single current source and a single parallel resistor. The Norton equivalent is found by replacing all power sources with their internal resistances and calculating the circuit resistance as seen from the point of simplification. This is the Norton resistance. Next, the sources are restored, and the short-circuit current is calculated at the point of simplification. This is the Norton current.

- Millman's Theorem is useful for simplifying circuits that have multiple parallel branches with multiple power sources. To utilize Millman's Theorem, all voltage sources and series resistances are converted to equivalent current sources and parallel resistances. The resulting parallel current sources are then combined algebraically, and the resistances are combined with a parallel resistance equation. The simplified circuit consists of a single current source and a single parallel resistance.

- This chapter also introduced two types of three-terminal networks: tee and pi. Tee and pi networks are also called wye and delta networks, respectively. Either type of three-terminal network can be converted to an equivalent network of the opposite form. External to the converted circuit, the electrical characteristics are identical to those of the original circuit. The reason for converting from one form to the other is to simplify subsequent circuit calculations. Often a complex circuit can be converted into a series-parallel circuit through tee-to-pi or pi-to-tee conversion. Either type of conversion consists of direct calculation with appropriate conversion formulas.

Review Questions

Section 7.1: Positive Identification Method

1. Can the circuit in Figure 7-79 be positively classified as a complex circuit?

2. Is the circuit shown in Figure 7-80 a complex circuit?

3. Can the circuit in Figure 7-81 be classified as a complex circuit?

4. Is the circuit shown in Figure 7-82 a complex circuit?

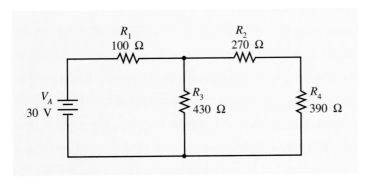

Figure 7-79. Example circuit.

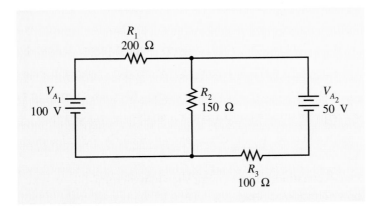

Figure 7-80. Example circuit.

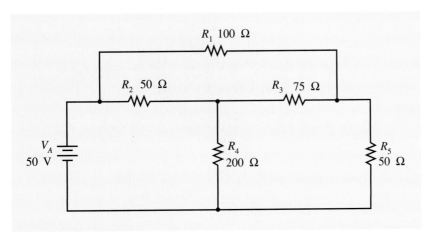

Figure 7-81. Example circuit.

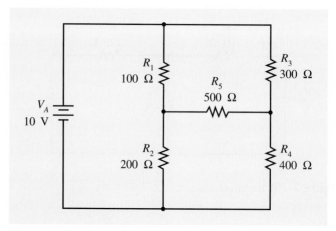

Figure 7-82. Example circuit.

Section 7.2: Mesh Analysis

5. Refer to Figure 7-79. How many mesh currents would be required to analyze this circuit?

6. Refer to Figure 7-80. How many mesh currents would be required to analyze this circuit?

7. How many mesh currents are required to analyze the circuit shown in Figure 7-81?

8. Figure 7-82 requires three mesh currents for analysis. (True or False)

9. Write the mesh equations for the circuit shown in Figure 7-79. Assume that the mesh currents flow counterclockwise.

10. Write the mesh equations for the circuit shown in Figure 7-80. Assume that the mesh currents flow counterclockwise.

11. Write the mesh equations for the circuit shown in Figure 7-81. Assume that the mesh currents flow counterclockwise.

12. Write the mesh equations for the circuit shown in Figure 7-82. Assume that the mesh currents flow counterclockwise.

13. Solve the following mesh equations to determine the values for I_X and I_Y.

$$2I_X - 3I_Y = -12, \text{ and}$$

$$3I_X + I_Y = -7$$

14. Solve the following mesh equations to determine the values for I_X and I_Y.

$$3I_X + 5I_Y + 3 = 0, \text{ and}$$

$$4I_X + 2I_Y - 4 = 0$$

15. Solve the following mesh equations to determine the values for I_X and I_Y.

$$100 + 250I_X - 200I_Y - 50I_Z = 0$$

$$-200I_X + 325I_Y - 75I_Z = 0, \text{ and}$$

$$-50I_X - 75I_Y + 225I_Z = 0$$

16. Use mesh analysis and basic circuit theory to determine the current flow through R_2 in Figure 7-80.

17. How much power is dissipated by resistor R_1 in Figure 7-80?

18. Which of the resistors in Figure 7-80 has the highest current flow?

19. Use mesh analysis and basic circuit theory to determine the current though R_3 in Figure 7-81.

20. What is the voltage drop across R_4 in Figure 7-81?

21. What is the total current for the circuit shown in Figure 7-81?

22. Which way (left to right or right to left) does electron current flow through R_1 in Figure 7-81?

23. Use mesh analysis and basic circuit theory to determine the current through R_5 in Figure 7-82.

24. Which resistor in Figure 7-82 dissipates the most power?

Section 7.3: Nodal Analysis

25. All node currents are assumed to flow (*toward, away from*) the node.

26. Nodal analysis is basically an application of Kirchhoff's _____ Law.

27. Refer to Figure 7-83. How many unknown node voltages are there?

28. Refer to Figure 7-84. How many nodal equations will be required to describe this circuit?

29. Refer to Figure 7-85. How many nodal equations will be required to solve this circuit?

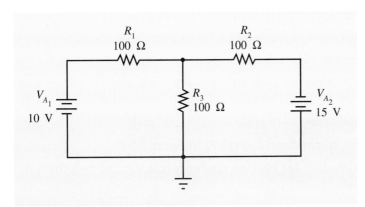

Figure 7-83. Example circuit.

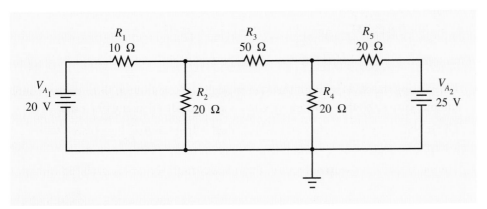

Figure 7-84. Example circuit.

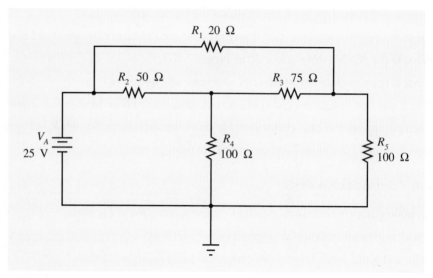

Figure 7-85. Example circuit.

30. Use nodal analysis and basic circuit theory to determine the current through R_2 in Figure 7-83.

31. How much power is dissipated by R_3 in Figure 7-83?

32. Refer to Figure 7-83. Which resistor has the largest voltage drop?

33. Use nodal analysis and basic circuit theory to calculate the voltage drop across R_1 in Figure 7-84.

34. What is the voltage drop across R_3 in Figure 7-84?

35. How much current flows through R_4 in Figure 7-84?

36. Use nodal analysis and basic circuit theory to determine the current through R_3 in Figure 7-85.

37. Calculate the voltage drop across R_1 in Figure 7-85.

38. Which resistor in Figure 7-85 has the greatest voltage drop?

39. Which resistor in Figure 7-85 has the most current?

40. Compute the power dissipation in resistor R_2 in Figure 7-85.

Section 7.4: Superposition Theorem

41. When using the superposition theorem to analyze a circuit with three voltage sources and one current source, how many sources are considered at any one time?

42. The superposition theorem requires that ideal voltage sources be replaced with (*a short, an open*).

43. The superposition theorem requires that ideal current sources be replaced with (*a short, an open*).

44. Use the superposition theorem and basic circuit theory to determine the current through R_2 in Figure 7-86.

45. What is the voltage drop across R_3 in Figure 7-86?

46. How much power is dissipated by R_1 in Figure 7-86?

47. Use the superposition theorem and basic circuit theory to determine the voltage drop across R_4 in Figure 7-87.

48. What is the value of current through V_{A_1} in Figure 7-87?

49. Which resistor in Figure 7-87 has the most current flow?

50. Which resistor in Figure 7-87 has the largest voltage drop?

51. Use the superposition theorem and basic circuit theory to determine the power dissipated by R_3 in Figure 7-88.

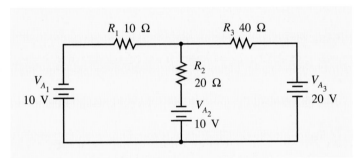

Figure 7-86. Example circuit.

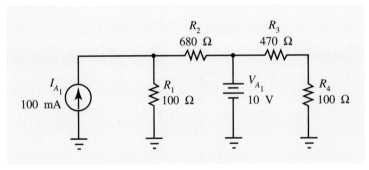

Figure 7-87. Example circuit.

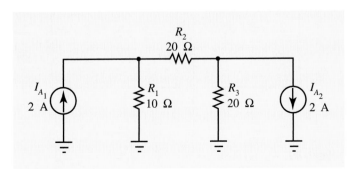

Figure 7-88. Example circuit.

52. How much current flows through R_2 in Figure 7-88?

53. What is the voltage drop across R_1 in Figure 7-88?

54. Which resistor in Figure 7-88 has the greatest current?

55. Which resistor in Figure 7-88 has the highest voltage drop?

Section 7.5: Thevenin's Theorem

56. Thevenin's Theorem can be used to convert a linear circuit into an equivalent circuit consisting of a single _____ source and a single _____ resistance.

57. The circuit is (*opened, shorted*) at the point of simplification to compute the Thevenin voltage.

58. The circuit is (*opened, shorted*) at the point of simplification to compute the Thevenin resistance.

59. What is the value of Thevenin's voltage for the circuit shown in Figure 7-89?

60. What is the value of Thevenin's resistance for the circuit shown in Figure 7-89?

61. Connect a 10-kΩ resistor across the point of simplification terminals in Figure 7-89, and compute the voltage drop across the 10-kΩ resistor.

62. Construct the Thevenin equivalent for the circuit shown in Figure 7-89 by using the values computed previously. Now connect a 10-kΩ resistor across the output terminals, and compute its voltage drop. How do your results compare with the value obtained in the preceding question?

63. What is the value of Thevenin's voltage for the circuit shown in Figure 7-90?

64. What is the value of Thevenin's resistance for the circuit shown in Figure 7-90?

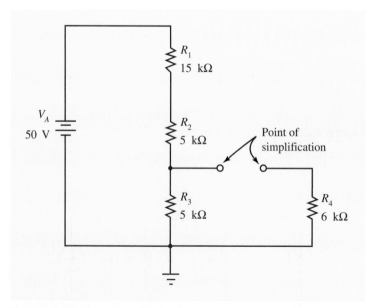

Figure 7-89. An example circuit.

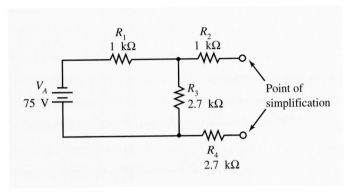

Figure 7-90. An example circuit.

65. If a 4.7-kΩ resistor is connected across the output terminals of the Thevenin equivalent for the circuit shown in Figure 7-90, how much current flows through the 4.7-kΩ resistor?

66. If a 4.7-kΩ resistor is connected across the point of simplification terminals in Figure 7-90, how much current flows through the 4.7-kΩ resistor?

Section 7.6: Norton's Theorem

67. Norton's Theorem can be used to convert a linear circuit into an equivalent circuit consisting of a single _____ source and a single _____ resistance.

68. The circuit is (*opened, shorted*) at the point of simplification to compute the Norton current.

69. The circuit is (*opened, shorted*) at the point of simplification to compute the Norton resistance.

70. What is the value of Norton's current for the circuit shown in Figure 7-89?

71. What is the value of Norton's resistance for the circuit shown in Figure 7-89?

72. Connect an 18-kΩ resistor across the point of simplification terminals in Figure 7-89, and compute the voltage drop across the 18-kΩ resistor.

73. Construct the Norton equivalent for the circuit shown in Figure 7-89 by using the values computed previously. Now connect an 18-kΩ resistor across the output terminals, and compute its voltage drop. How do your results compare with the value obtained in the preceding question?

74. What is the value of Norton's current for the circuit shown in Figure 7-90?

75. What is the value of Norton's resistance for the circuit shown in Figure 7-90?

76. If a 6.8-kΩ resistor is connected across the output terminals of the Norton equivalent for the circuit shown in Figure 7-90, how much current flows through the 6.8-kΩ resistor?

77. If a 6.8-kΩ resistor is connected across the point of simplification terminals in Figure 7-90, how much current flows through the 6.8-kΩ resistor?

Section 7.7: Millman's Theorem

78. When Millman's Theorem is used to simplify a circuit, the equivalent circuit consists of a _____ source and a single _____ resistance.

79. When applying Millman's Theorem, all current sources are converted to equivalent voltage sources. (True or False)

80. Determine the Millman equivalent for the circuit shown in Figure 7-91.

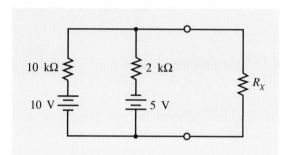

Figure 7-91. An example circuit.

81. Determine the Millman equivalent for the circuit shown in Figure 7-92.

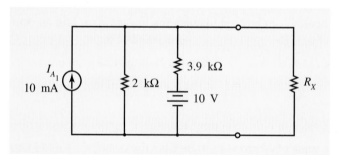

Figure 7-92. An example circuit.

Section 7.8: Pi-to-Tee and Tee-to-Pi Conversions

82. A pi (π) configuration is also called a wye configuration. (True or False)

83. A delta (Δ) configuration is also called a _____ configuration.

84. Convert the circuit shown in Figure 7-93 to an equivalent pi network.

85. Convert the circuit shown in Figure 7-94 to an equivalent tee network.

86. Use pi-to-tee or tee-to-pi conversions along with basic circuit theory to compute the current flow through R_5 in Figure 7-95.

87. Use a tee-to-pi or pi-to-tee conversion to simplify the highlighted portion of the circuit shown in Figure 7-96. Draw a labeled schematic of your simplified circuit.

88. Using the equivalent circuit developed in the prior question, compute total current flow in the circuit.

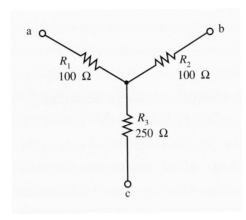

Figure 7-93. Convert this circuit to a pi network.

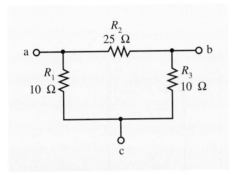

Figure 7-94. Convert this circuit to a tee network.

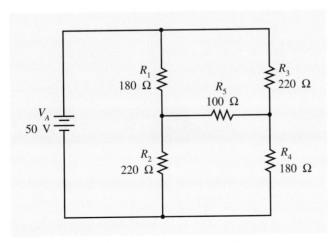

Figure 7-95. Find the current through R_5.

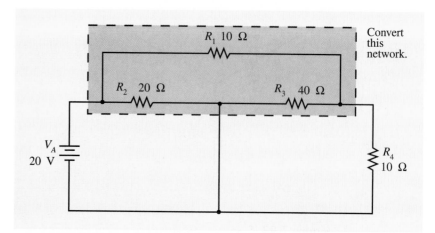

Figure 7-96. An example circuit.

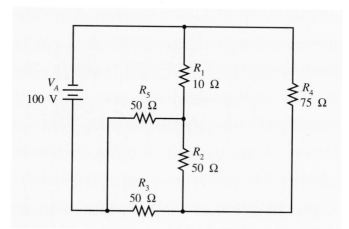

Figure 7-97. An example circuit.

89. Use a tee-to-pi or pi-to-tee conversion and basic circuit theory to determine the amount of total current through the voltage source in Figure 7-97.

90. You could use either a tee-to-pi or a pi-to-tee conversion to help simplify the circuit shown in Figure 7-97. (True or False)

TECHNICIAN CHALLENGE

Following is a brainteaser that has puzzled technicians for many years. This chapter has presented all the techniques that you will need to solve this problem. Can you be one of the technicians that solves this mystery?

The challenge consists of a cube made of resistors as pictured in Figure 7-98. Each edge—twelve total—is formed by one of the resistors. Each of the twelve resistors has exactly the same value; each is 1.0 Ω. As indicated in the figure, you must compute the

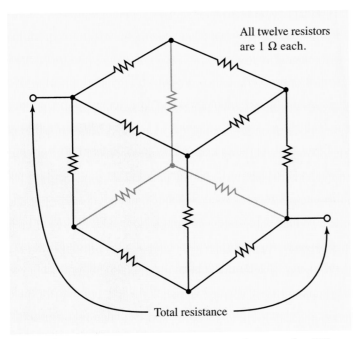

All twelve resistors are 1 Ω each.

Total resistance

Figure 7-98. Twelve resistors (each 1.0 Ω) form a cube. What is the resistance between diagonal corners?

total resistance as measured between any two diagonal (through the cube) corners. It's a tough challenge, but you can do it. (Hint: There are several ways to solve the problem. Multiple tee-to-pi and pi-to-tee conversions can lead to a solution. Regardless of the method chosen, it is essential that you take it one step at a time and carefully draw a partially simplified circuit after each step.)

Equation List

$$(7\text{-}1) \quad R_a = \frac{R_{ab}R_{ac}}{R_{ab} + R_{ac} + R_{bc}}$$

$$(7\text{-}2) \quad R_b = \frac{R_{ab}R_{bc}}{R_{ab} + R_{ac} + R_{bc}}$$

$$(7\text{-}3) \quad R_c = \frac{R_{ac}R_{bc}}{R_{ab} + R_{ac} + R_{bc}}$$

$$(7\text{-}4) \quad R_{ab} = \frac{R_aR_b + R_aR_c + R_bR_c}{R_c}$$

$$(7\text{-}5) \quad R_{bc} = \frac{R_aR_b + R_aR_c + R_bR_c}{R_a}$$

$$(7\text{-}6) \quad R_{ac} = \frac{R_aR_b + R_aR_c + R_bR_c}{R_b}$$

ampere-hours	piezoelectric effect
battery	resistivity
breakdown voltage	shelf life
cryogenics	specific gravity
hydrometer	temperature coefficient
leakage current	thermocouple
memory effect	thermopile

objectives

After completing this chapter, you should be able to:

1. Briefly describe how a cell converts chemical energy into electrical energy.

2. State the differences between primary and secondary cells.

3. Name several types of primary and secondary cell technologies.

4. Explain the ratings on a battery specification sheet.

5. Explain the basic operation of an alternator and a generator.

6. Name at least four technologies that convert other forms of energy directly into electrical energy.

7. Contrast conductors, semiconductors, superconductors, and insulators.

8. Calculate the circular mil area of a wire with a known diameter.

9. Calculate the resistance of a given length of copper wire.

Electrical Power Sources and Electrical Materials

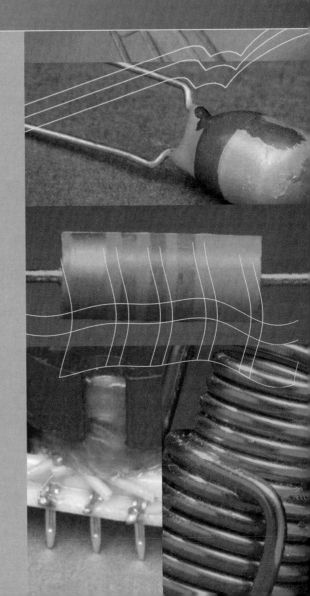

All electrical and electronic circuits require sources of electrical energy and some means of transferring that energy between two points, while preventing its transfer to other areas. We will examine several technologies for producing electricity and contrast the electrical properties of several materials used to transfer or contain electrical energy.

8.1 Cells and Batteries

One of the most common sources of electrical energy is the electrochemical cell. These cells convert chemical energy into electrical energy. A standard 1.5-V cell (D cell) used in a flashlight is an example. Several cells may be connected in series and parallel combinations to obtain higher voltage and current ratings. The combination of cells is called a **battery.** The battery that is used in your car is an example. Many people incorrectly apply the term battery to describe both cells and batteries (e.g., flashlight battery instead of flashlight cell).

There are many ways to classify cells and batteries. One common approach is to classify cells that are not designed to be recharged as primary cells. Those cells that can be recharged (like those used in your car battery) are called secondary cells. Within each of these overall classifications, there are several technologies used to manufacture the cells. Each technology has its own unique advantages and disadvantages relative to the other technologies.

All electrochemical cells, both primary and secondary, have at least three major components in common: anode, cathode, and electrolyte. The anode is the negative terminal of the cell, the cathode is the positive terminal of the cell, and the electrolyte is the material that forms the conducting path between the anode and cathode inside the cell. The electrolyte can be liquid (called a wet cell), a paste (called a dry cell), or a solid. The electrolyte reacts chemically with the anode and cathode to create a difference in potential that is ultimately used to provide power to an external circuit.

There is some disagreement as to the identification of the anode and cathode. Technicians and engineers in electronics typically use the definitions previously given. Chemists, on the other hand, generally label the positive and negative electrodes as the anode and cathode, respectively.

Ratings of Cells and Batteries

Cell/battery manufacturers provide many specifications to describe the performance of a given device. Some of the more important cell/battery specifications include the following:

- terminal voltage
- capacity
- shelf life
- operating/storage temperature ranges
- package style

TERMINAL VOLTAGE

The terminal voltage of a cell is determined by the technology used to construct the cell. It is simply a measure of the voltage produced by the cell and measured across the terminals of the cell. In the case of a multicell battery, the terminal voltage depends on the technology used as well as the connection (i.e., series, parallel, or series-parallel) of the cells in the battery. The voltages of series-connected cells add, and the current capacities of parallel-connected cells add. Terminal voltages for cells are generally less than 3 V, but terminal voltages for batteries can be almost any value.

CAPACITY

Cell/battery capacity describes the amount of energy that can be provided by the device under a given set of conditions. The amount of useful energy varies greatly with the rate of discharge, the cell age, and the temperature. For example, most cells can provide more energy at lower discharge rates and higher (relatively) temperatures.

Cell/battery capacity is specified in **ampere-hours** (Ah). Thus, a 1-Ah cell would have enough stored energy to provide 1 A to an external load for 1 h. A cell rated for 3 Ah might supply 1 A for 3 h or, perhaps, 3 A for 1 h. Although there are definite limits on the maximum current, the capacity rating provides an indication of the cell/battery capacity relative to other devices.

SHELF LIFE

If a fully charged cell or battery remains unused, it will eventually lose its charge because of a loss of chemical energy inside the cell. Manufacturers specify the amount of time a cell or battery can be stored without losing its effectiveness. This time is called the **shelf life** of the device. Shelf life for different devices varies from a few weeks to tens of years.

OPERATING/STORAGE TEMPERATURE RANGES

A cell or battery can only provide its specified voltage and current under certain temperature conditions. The manufacturer specifies these conditions as the operating temperature range. Similarly, if the cell/battery is stored at temperatures outside of its storage temperature range, then its shelf life will be reduced and the cell may even be damaged.

PACKAGE STYLES

Batteries and cells are available in many different package styles, sizes, and terminal configurations. Not all types of cells are available in all of the various package styles, but there is generally more than one package style available for a given cell technology. The 9-V transistor battery and the D cell used in many flashlights are two common package styles.

Primary Cells

Primary cells are characterized by their inability (or at least impracticality) to be recharged. Once the energy in the cell has been expended, the cell is discarded and replaced with a new one. Let us now examine the operation of a representative primary cell in some detail. We will then cite the relative characteristics of several other technologies.

> **● KEY POINTS**
>
> Primary cells are not designed to be recharged.

CARBON-ZINC CELLS

Figure 8-1 shows a cutaway view of a carbon-zinc cell. It consists of a zinc cylinder lined with an absorbent paper. The cylinder is filled with an electrolyte paste consisting of crushed graphite, manganese dioxide, ammonium chloride, and zinc chloride. The ammonium chloride is the active chemical in the electrolyte paste. A carbon rod is inserted into the center of the electrolyte. A metal cap is added to the carbon rod, and the entire cylinder is sealed.

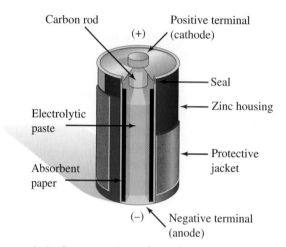

Figure 8-1. Cutaway view of a carbon-zinc cell.

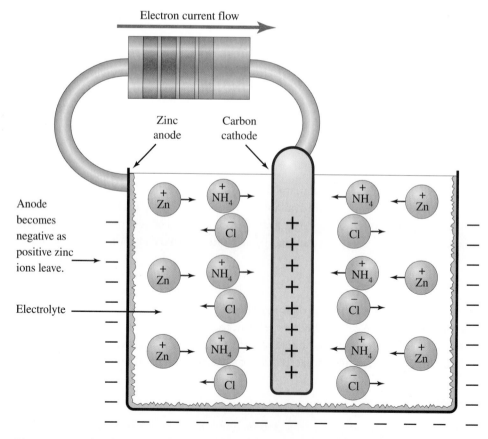

Figure 8-2. The chemical action inside a carbon-zinc cell.

Figure 8-2 illustrates the chemical activity that occurs inside a carbon-zinc cell. The ammonium chloride (electrolyte) breaks down into two types of charged particles called ions. There are ammonium ions (NH^+), which are positively charged, and chlorine ions (Cl^-), which have a negative charge. In the vicinity of the anode, the zinc (Zn) is essentially dissolved or slowly eaten away by the chemical action. This action produces positive zinc ions (Zn^+), which combine with the negative chlorine ions (Cl^-) in the electrolyte to form zinc chloride. The resulting zinc chloride has no electrical charge, but the anode is left with a negative charge as the positive zinc ions are lost to the electrolyte.

The positively charged ammonium ions in the electrolyte are repelled toward the carbon cathode rod by the positive zinc ions before they recombine with the chlorine. In the vicinity of the carbon rod, electrons leave the rod and combine with the positive ammonium ions (NH_4^+) to form ammonia (NH_3) and hydrogen (H^+) gases. The hydrogen gas recombines with the manganese dioxide in the electrolyte to form water. The ammonia combines with some of the free zinc ions to form a more complex ion. As electrons leave the carbon rod and combine with the ammonium ions, the carbon rod is left with a deficiency of electrons. Therefore, it takes on a positive charge.

The aforementioned chemical activity is abbreviated and primarily of interest only to chemists. What is important to our immediate subject, however, is the fact that chemical activity within the cell causes one electrode to take on a negative charge (i.e., an excess of electrons), while the other electrode builds up a positive charge (i.e., a deficiency of electrons). There exists, therefore, a potential difference between the two charged electrodes. The amount of voltage between the two electrodes is determined by the specific chemicals used to construct the cell. In the case of the carbon-zinc cell, the terminal voltage is about 1.5 V, but decreases as the chemical activity slows.

Figure 8-2 illustrates another important point regarding cell operation. Recall that electrons leave the carbon rod (cathode) and combine with the ammonium ions. This action can only occur until there are no more available electrons in the carbon rod. But, as shown in Figure 8-2, if we connect an external circuit path, then the excess electrons that have accumulated on the zinc electrode (anode) can travel through the external circuit to the electron-deficient cathode. This is a continuous process as long as the chemical activity can maintain the difference of potential. When the chemical energy is used up, we commonly say the cell is dead.

Even without an external circuit connected, there is some chemical activity that continues to occur. Thus, it is possible for the chemicals to be expended without ever connecting an external circuit. Of course, it takes much longer to dissipate the stored chemical energy without an external current path. This time is called the shelf life of the cell.

All cells have internal resistance. In Figure 8-2, for example, current in the circuit encounters resistance as it flows from the carbon rod to the zinc container inside the battery. This internal resistance has the effect of decreasing the available cell voltage as current increases. Figure 8-3 shows an equivalent circuit that illustrates the internal resistance of a cell.

Our understanding of series circuits will quickly reveal that part of the 1.5-V source voltage in Figure 8-3 will be dropped across R_I (internal to the cell) and the rest across the external load resistor (R_L). The actual terminal voltage (V_T) of the cell will be less than the ideal voltage by the amount dropped across the internal resistance.

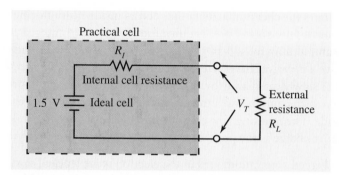

Figure 8-3. All cells have internal resistance, which appears in series with the ideal voltage source.

Clearly, we want the internal resistance of a cell to be as low as possible. A low internal resistance allows greater current flow before the terminal voltage decreases substantially. It also results in lower power dissipation within the cell. One factor that affects the internal resistance of a cell is the surface area of the electrodes. Greater surface area results in lower resistances and greater current capacities.

It is interesting to note that although maximum terminal voltage occurs for the smallest internal resistance, maximum power is developed in the external load when the load resistance is equal to the internal resistance. This is called the maximum power transfer theorem and is discussed in greater detail in Chapter 17.

Alkaline–Manganese Dioxide Cells

Figure 8-4 shows the cutaway view of an alkaline–manganese dioxide cell. This cell uses zinc for its anode as in the zinc-carbon cell. The cathode, however, is manganese dioxide, and the electrolyte is potassium hydroxide. These chemicals produce voltage differences that are comparable to the carbon-zinc cell, but the internal resistance of the alkaline cell is substantially lower. The alkaline cell is capable of providing substantially more power in a similar size package. Additionally, relatively high currents can be drawn for a longer time without overheating the battery, since the internal resistance is lower. The shelf life of an alkaline cell is greater than four years, which is much longer than the shelf life of a carbon-zinc cell. Alkaline-manganese cells are available in many different sizes including AA, AAA, AAAA, C, D, 9 V, and others.

Mercuric-Oxide Cells

The cutaway view of a mercuric-oxide cell is shown in Figure 8-5. The button-cell package, illustrated in Figure 8-5, is probably the most common, but the mercuric-oxide cells (commonly called mercury cells) are also made in AA, 9 V, and other styles. For its physical size, the mercury cell has a high energy output.

The terminal voltage of a mercury cell is characterized as being nearly flat throughout the life of the cell. This characteristic makes it well suited to medical applications (e.g., hearing aids), cameras, and other small devices requiring a constant voltage. A graph contrasting

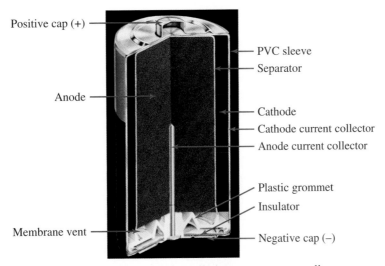

Figure 8-4. Cutaway view of an alkaline-manganese cell. (Courtesy of Duracell, Inc.)

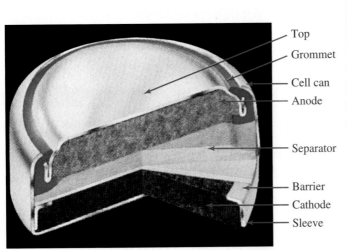

Figure 8-5. Cutaway view of a mercuric-oxide cell. (Courtesy of Duracell, Inc.)

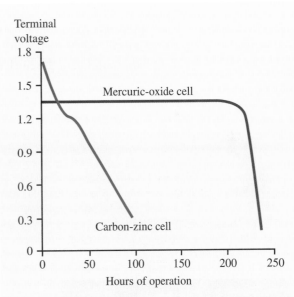

Figure 8-6. The terminal voltage of a mercuric-oxide cell remains fairly constant throughout the life of the cell.

the terminal voltage of a typical carbon-zinc cell with that of a mercuric-oxide cell is shown in Figure 8-6. This graph does not represent the limits of either technology, but rather is meant to contrast the voltage characteristics of the two types of cells.

The mercuric-oxide cell uses a zinc anode and a cathode made of mercuric oxide or mercuric oxide mixed with manganese dioxide. The electrolyte material is potassium hydroxide. Mercuric-oxide cells are being replaced by silver-oxide cells for many applications.

SILVER-OXIDE CELLS

A silver-oxide cell is constructed in the same manner as the cell pictured in Figure 8-5. It uses a zinc anode, a silver-oxide cathode, and potassium hydroxide for the electrolyte. The terminal voltage of a silver-oxide cell is relatively flat throughout its life, as is the mercury cell. The silver-oxide cell, however, provides a higher terminal voltage (1.6 V) and greater current capacity for a given size cell. The silver-oxide cell also has improved performance at low temperatures.

LITHIUM CELLS

There are many different kinds of lithium cells, but Figure 8-7 shows a cutaway view of two representative lithium–manganese dioxide cells. Lithium cells are characterized by their long life. For a given size cell, lithium cells provide more energy than the previously discussed technologies, and they work well at both high and low temperatures. Lithium cells come in a range of package styles including AA, miniature, and cylindrical, and have a shelf life of up to ten years.

Lithium metal is used for the anode in lithium cells. The cathode and electrolyte materials vary between manufacturers and battery types. The lithium–manganese dioxide cell, in particular, uses a conductive organic electrolyte and a manganese dioxide cathode.

Lithium cells are often used to provide the auxiliary power to internal computer memories and computer clocks. The life expectancy in such an application depends on the current consumption of the clock or memory, but it is typically measured in years.

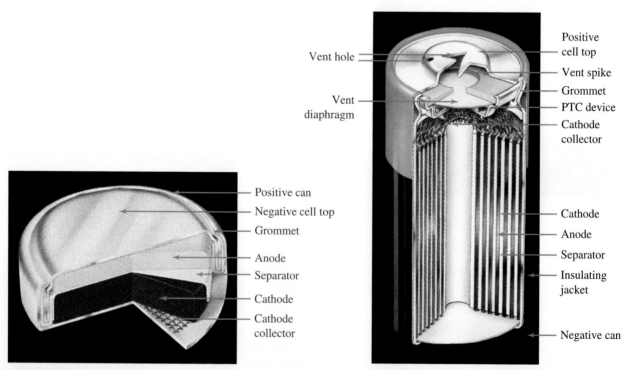

Figure 8-7. The cutaway view of two types of lithium-manganese dioxide cells. (Courtesy of Duracell, Inc.)

ZINC-AIR CELLS

A zinc-air cell provides the greatest energy storage for a given cell size of all the primary cell technologies. It has a flat discharge curve and good low-temperature performance. Figure 8-8 shows the cutaway view of a zinc-air cell.

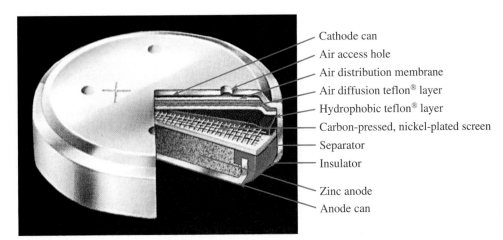

- Cathode can
- Air access hole
- Air distribution membrane
- Air diffusion teflon® layer
- Hydrophobic teflon® layer
- Carbon-pressed, nickel-plated screen
- Separator
- Insulator
- Zinc anode
- Anode can

Figure 8-8. The cutaway view of a zinc-air cell. (Courtesy of Duracell, Inc.)

The zinc-air cell uses powdered zinc as the anode material and potassium hydroxide as the electrolyte. It has no integral cathode, but rather uses oxygen from the surrounding air as the cathode reactant. The air (oxygen) enters through small holes in the cell package and is absorbed by a porous, carbon cathode material. Shelf life is greatly extended by keeping the air holes sealed prior to using the cell. The zinc-air cell is generally made in a button-cell package.

Table 8-1 summarizes and contrasts the characteristics of the primary cells discussed in this section.

Secondary Cells

A cell that is specifically designed to be recharged is called a secondary cell. An electrochemical cell converts chemical energy into electrical energy. But during recharging, electrical energy is converted into chemical energy inside the secondary cell. There are several technologies used to manufacture secondary cells; each has its own relative advantages and disadvantages. We will examine the operation of a lead-acid cell in detail and then briefly contrast the characteristics of other secondary-cell technologies.

KEY POINTS

The energy in a secondary cell can be replenished by recharging.

LEAD-ACID CELLS

The battery used in your car contains several lead-acid cells. It is the oldest and probably the most common rechargeable storage technology in use today. Figure 8-9 shows the cutaway view of a lead-acid battery. It consists of several lead-acid cells. A mixture of sulfuric acid and water is used for the electrolyte. Lead plates are used for both cathode and anode. This seems to defy the underlying rule that says dissimilar materials must react with

KEY POINTS

Lead-acid cells and batteries can provide very high storage capacities, but they are inherently bulky and heavy.

TECHNOLOGY	NO-LOAD VOLTAGE	ANODE MATERIAL	CATHODE MATERIAL	CHARACTERISTICS	APPLICATIONS
Carbon-zinc	1.5 V	Zinc	Carbon	Economical, voltage decays with use, poor at low temperature	Radios, flashlights, small toys, test equipment, smoke detectors
Alkaline–manganese dioxide	1.5 V	Zinc	Manganese dioxide	Excellent shelf life, good at low and high temperatures, low internal resistance	Same as carbon-zinc, but good for higher currents and/or longer life
Mercuric-oxide	1.34 to 1.4 V	Zinc	Mercuric oxide	Flat discharge curve, long shelf life, good at high temperatures	Cameras, watches, hearing aids, calculators
Silver-oxide	1.6 V	Zinc	Silver oxide	Flat discharge curve, high capacity, long shelf life, expensive, good at low and high temperatures	Hearing aids, cameras, watches, calculators
Lithium–manganese dioxide	3.2 V	Lithium	Manganese dioxide	Flat discharge curve, good performance at temperature extremes (−40°F to 140°F), long shelf life, high energy storage	Photography equipment, smoke alarms, calculators, computer memory backup, pagers
Zinc-air	1.4 V	Powdered zinc	Atmospheric oxygen	Flat discharge curve, moderate temperature performance (32°F to 120°F), highest theoretical storage capacity of all primary cells, one-year shelf life (if sealed)	Hearing aids, photography equipment, pagers, medical equipment

Table 8-1. Characteristics of Primary Cells

an electrolyte to produce an electrochemical potential. Actually, the rule is not violated. When the cell is first constructed, there is no voltage produced. The anode and cathode have to be formed as a coating on the zinc plates. They are formed in the process of charging the cell. A cell is charged by forcing current to flow in the reverse direction (i.e., electrons enter the negative electrode). The anode (negative electrode) is formed by a layer of spongy lead, which forms on the surface of the lead plate. Similarly, a layer of lead peroxide forms on the surface of the cathode (positive electrode). Once the anode and cathode have been formed, the cell can then deliver power to an external circuit. As the energy is used up in the cell, the anode and cathode coatings are slowly dissolved. Eventually, the battery will be discharged and will no longer supply power. The electrodes can be reformed by charging the cell. Lead-acid cells can be recharged hundreds or even thousands of times.

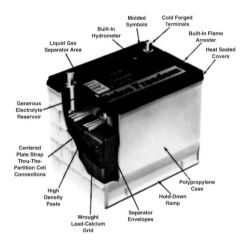

Figure 8-9. A cutaway view of a lead-acid battery.

Figure 8-10 illustrates the chemical activity that occurs in the lead-acid cell throughout its charge/discharge cycle. The discharged condition is shown in Figure 8-10(a). Here, the lead sponge and the lead peroxide coatings on the anode and cathode, respectively, are nearly depleted. The sulfuric acid is very diluted. The degree of concentration of the sulfuric acid in the electrolyte is an indication of the state of charge. The concentration is determined by measuring the **specific gravity** (S.G.) of the electrolyte with a **hydrometer.** A hydrometer measures the relative weight of the electrolyte as compared to water. Discharged cells have specific gravities of less than 1.2 (i.e., 1.2 times heavier than pure water).

The cell is charged (i.e., electrical energy is converted and stored as chemical energy) by connecting it to an external dc source. The polarity must be such that a charging current flows in the reverse direction, as illustrated in Figure 8-10(b). As the charging current flows, the lead peroxide on the cathode and the lead sponge on the anode are restored. In particular, a discharged cell has accumulations of lead sulphate on both electrodes. The charging current causes the sulphate ions to combine with the electrolyte to increase the concentration of the electrolyte, eliminate the sulphate coating, and restore

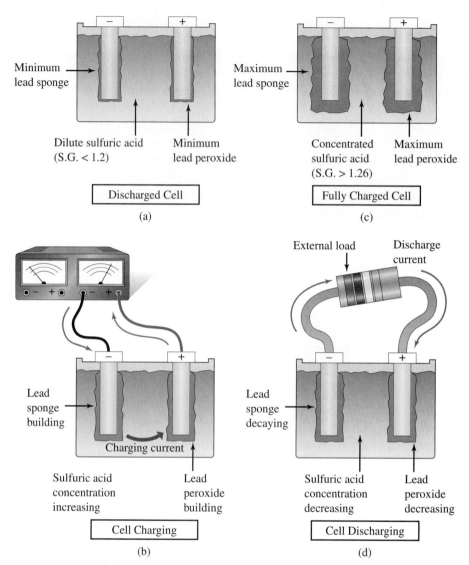

Figure 8-10. The charge/discharge cycle of a lead-acid cell.

the spongy lead on the anode and the lead peroxide on the cathode. The full recharging process may take several hours. The rate of charge (i.e., value of charging current) must be controlled to avoid excessive heat and possible explosion from the hydrogen gas that is formed during the charging process. Figure 8-10(c) shows the fully charged cell. The cathode and anode are fully restored, and the specific gravity of the electrolyte has increased to over 1.26.

Finally, Figure 8-10(d) shows the cell conditions as the cell is discharging. The potential of the cell causes current to flow through the external load resistance. Current continues to flow as long as the cell can maintain a difference in potential. As the cell loses its charge, the lead peroxide coating on the cathode decays, and a lead sulphate coating forms on both electrodes. The lead sulphate accumulation slowly increases the internal resistance of the cell.

Lead-acid cells have extraordinary storage capacities. A typical cell can deliver several amperes for several hours, while maintaining a relatively constant terminal voltage. Many package designs are available for the lead-acid technology, but all are characterized by the relatively heavy weight resulting from the use of lead plates. A typical automotive battery consists of six series-connected lead-acid cells that each produce an unloaded terminal voltage of about 2.1 V. Collectively, the cells provide a battery voltage of about 12.6 V.

NICKEL-CADMIUM CELLS

Nickel-cadmium (or NiCad) is another popular type of rechargeable cell. Figure 8-11 illustrates the basic construction of one common type of NiCad cell. NiCad cell technology has been the most popular choice for portable power technology since the late 1960s.

The positive electrode or cathode is made from nickel hydroxide (and graphite). The negative electrode or anode is made from cadmium (and iron oxide). Potassium hydroxide is used as the electrolyte. The NiCad cell is sealed and can be operated in any position.

The nominal terminal voltage of a NiCad cell is 1.2 V. The cell is characterized by relatively long life (e.g., over 500 charge/discharge cycles), relatively constant voltage over 90 percent of its discharge curve, fast recharge times, and high current capacities. Because of their high current capacity and low internal resistance, they can be damaged if they are shorted.

NiCad cells exhibit a characteristic called the **memory effect.** If the cell is repeatedly charged after only slightly discharging, then it tends to lose much of its capacity. It is as if

KEY POINTS

NiCad cells and batteries still have a relatively high storage capacity, but they are much smaller, lighter, and well-suited to portable equipment.

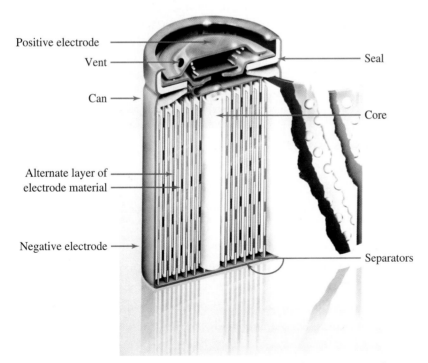

Figure 8-11. Cutaway view of a nickel-cadmium cell. (Courtesy of Eveready Batteries, Inc.)

the cell remembers that it will be charged often. The memory effect can often be corrected by fully discharging and subsequently charging the cell. When several NiCad cells are connected in parallel to form a battery for increased current capacity, they should never be fully discharged. A complete discharge can cause the weakest cell to have reverse currents, which can damage the cell.

There are many applications for NiCad cell technology. Most portable, rechargeable tools, flashlights, camcorders, laptop/palmtop computers and calculators, and many security systems use NiCad cells as the power source. Typical load currents range from a few milli-amperes to well over ten amperes.

NICKEL-METAL-HYDRIDE CELLS

KEY POINTS

A NiMH cell has many of the same characteristics of a NiCad cell, but provides greater storage capacity.

Although the performance of NiCad cells continues to increase (capacities doubled between 1988 and 1990), nickel-metal-hydride (NiMH) cells are a newer technology that promises to surpass NiCads as the choice for portable power. The positive electrode is made from nickel hydroxide, which is essentially the same as the cathode in a NiCad cell. The negative electrode is formed by a metallic alloy, which provides more power for a given cell size. NiMH cells can often provide as much as twice the available energy as comparable sized NiCad cells. The cell voltage and charge/discharge characteristics are similar to the NiCad technology, which encourages upgrading to the newer technology. Since cadmium (used in NiCad cells) has been identified as harmful to the environment, absence of this material is an advantage for NiMH cells.

The primary disadvantage of NiMH cells over comparable NiCad cells is their relatively high cost. Additionally, NiMH cells tend to self-discharge during storage at rates nearly double equivalent NiCad cells. Both of these disadvantages are projected to be overcome in the foreseeable future, which will position NiMH as the cell technology of choice for portable power.

Exercise Problems 8.1

1. What is the shelf life of a cell?
2. Contrast the terms cell and battery.
3. If a car battery is rated at 100 Ah, about how long could it power a stereo that required 2.5 A?
4. A primary cell is designed to be recharged. (True or False)
5. Name at least three primary cell technologies.
6. A lead-acid battery consists of secondary cells. (True or False)
7. High current capacity is a major advantage of a lead-acid cell. (True or False)
8. Mercuric-oxide cells have a relatively flat (i.e., constant voltage) discharge curve. (True or False)
9. What is the nominal terminal voltage of a lead-acid cell?
10. What is the nominal terminal voltage of an alkaline–manganese dioxide cell?

11. Why are NiCad cells and batteries more popular than lead-acid cells for power sources in portable equipment?

12. What is the name of the technology that is expected to replace NiCad cells as the leading power source for portable equipment?

13. NiCad cells can be recharged. (True or False)

14. NiMH cells are secondary cells. (True or False)

15. Carbon-zinc cells are secondary cells. (True or False)

8.2 Other Power Sources

The electrochemical cells discussed in the preceding section produce a voltage by converting chemical energy into electrical energy. Other forms of energy such as mechanical, heat, and light can also be converted into electrical energy. Let us examine some of these power-generating alternatives.

Electromechanical Sources

Mechanical energy can be converted to electrical energy by an electromechanical source. There are two major electromechanical power sources: alternators and generators. Alternators produce alternating current and voltage. Generators produce direct (unidirectional) current and voltage. The basic operation of generators and alternators relies on the fact that a voltage is created in a wire as it moves through a magnetic field. The amount of voltage produced is determined by several factors, including strength of the magnetic field, number of loops on the coil of wire, and speed of the wire, as it passes through the field.

It is possible to produce incredible amounts of electrical power from an electromechanical source. Power companies, for example, utilize alternators that produce millions of watts of electrical power. This electrical power is not created by the alternator; rather it is converted from mechanical energy (the energy required to move the coiled wire through the magnetic field). In the case of the power company alternators, the mechanical energy can be supplied by the force of flowing water or high-pressure steam. The steam is initially heated in fossil fuel-fired boilers or with nuclear energy.

Figure 8-12 shows a picture of a system that uses a steam turbine to cause rotation of the alternator. Water is heated and converted to steam by burning fossil fuels or by a controlled nuclear reaction. In either case, the steam is passed through the blades of a steam turbine. When the steam passes through the turbine blades, it causes them to rotate. The turbine shaft is common to the alternator shaft, which has coils of wire. As the shaft or rotor turns, the coils of wire are moved through the magnetic field of the alternator and electrical energy is produced in the coils. Greater detail on alternator operation is provided in Chapter 11 of this text.

The operation of a dc generator is virtually identical to that of an alternator with one important exception; a generator has an internal mechanical switch called a commutator that converts the alternating voltage produced in the rotating coils into a single-polarity voltage. Details of generator operation are provided in Chapter 10 of this text.

KEY POINTS

Alternators and generators are electromechanical devices that convert mechanical energy into electrical energy.

KEY POINTS

The basic operation of alternators and generators relies on the fact that a voltage is created in a wire as it moves through a magnetic field.

KEY POINTS

The generator has an internal switching device called a commutator that converts the alternating current and voltage to unidirectional current (direct current) and voltage.

Figure 8-12. A steam turbine in a power plant that uses high-pressure steam to provide the mechanical energy needed to rotate a coil of wire through the magnetic field in an alternator. (Courtesy of Florida Power Corporation)

Piezoelectric Sources

Certain crystalline materials such as Rochelle salt and barium titanate produce a voltage across the surface when the crystal is subjected to mechanical stress. The phenomenon is called the **piezoelectric effect.**

Piezoelectric sources are used in some microphones and certain old-style phono cartridges. Sound waves entering the microphone cause a crystal to be mechanically distorted at the frequency of the sound. Each distortion produces a corresponding voltage that is then amplified by subsequent circuitry.

Many butane and propane burners use piezoelectric devices for ignition. The crystal is mechanically stressed by a spring-loaded hammer when the operator presses the igniter button. The voltage produced by the crystal (as much as 20 kV) causes a spark, which ultimately ignites the gas. Although piezoelectric devices can generate very high voltages, they have a very high internal resistance, so the current is limited to a few microamperes.

Solar Cells

A solar or photovoltaic cell is a semiconductor device that converts light energy directly into electrical energy. Figure 8-13 shows that a solar cell is a semiconductor material that is covered with a glass window. When light energy passes through the window and strikes the semiconductor material, valence electrons are dislodged from their parent atoms. Action within the semiconductor material causes the electrons to move toward one side of the solar cell. This causes an accumulation of electrons (negative charge) on one side of the material and a deficiency of electrons (positive charge) on the other side. A metallic ring

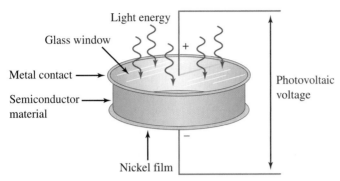

Figure 8-13. A solar cell converts light energy into electrical energy.

on the upper side of the solar cell and a connection to the nickel film on the lower side of the cell provide access to the voltage produced by the cell.

A typical solar cell produces slightly more than 0.5 V in direct sunlight. Several cells can be connected in series to produce higher voltages. The current capability of a solar cell is determined by its physical size. Additionally, multiple cells or branches of series-connected cells can be connected in parallel to increase the available current.

Although solar cells are inherently inefficient with regard to power conversion (less than 25%), they are valuable, since the input power is sunlight and is essentially free. In some applications, solar cells are the only practical form of power. In particular, solar cells provide the electrical power required by many satellites. Substantial research is under way to develop high-output solar cells for earthbound applications, but present cells are somewhat limited by the required physical size. Some calculators, remote battery chargers, toys, and light-sensing equipment (e.g., light meters) utilize solar cells.

Fuel Cells

Fuel cells are electrochemical devices that operate much like the electrochemical cells discussed in a preceding section. The electrodes for the previously discussed cells react chemically with the electrolyte to produce a potential difference. Eventually, the electrodes lose their chemical activity and must be discarded or, in the case of secondary cells, recharged. The electrodes in fuel cells, by contrast, do not react with the electrolyte. Rather, the chemicals used to react with the electrolyte are supplied continuously from an external source. Figure 8-14 illustrates the construction of a basic fuel cell.

As shown in Figure 8-14, a fuel cell consists of two porous metallic electrodes separated by an electrolytic barrier. Ions can pass through the electrolyte, but gas molecules cannot. Oxygen and hydrogen gases (fuel) are supplied to one side of the cathode and anode, respectively. The gases penetrate some distance into the porous metal used in the electrodes. Similarly, the electrolyte penetrates the opposite side of the electrodes.

The oxygen reacts with the electrolyte in the cathode to produce a deficiency of electrons. Similarly, the hydrogen gas reacts with the electrolyte in the anode to produce an excess of electrons. The chemical reactions also produce water as the hydrogen and oxygen react.

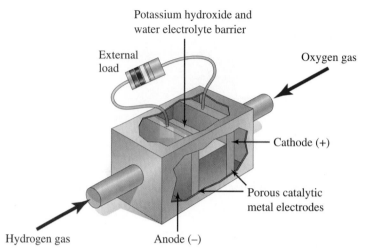

Figure 8-14. A fuel cell provides direct conversion of chemical energy into electrical energy.

The charge created by the chemical reaction produces a difference in potential between the cathode and anode of about 1.2 V.

The fuel cell pictured in Figure 8-14 is called a hydrooxygen cell, since it uses hydrogen and oxygen as the fuel. Some fuel cell designs use other types of fuel. Fuel cells are used extensively in the space program and experimentally to provide power to agricultural vehicles. Fuel cells are extremely efficient and can be used to produce hundreds of kilowatts of power.

Thermoelectric Sources

As the name implies, thermoelectric devices convert heat energy into electrical energy. A **thermocouple** is shown in Figure 8-15. It consists simply of a junction of two dissimilar metals. There are many combinations of materials that are used for thermocouples including the following pairs of dissimilar metals: iron/copper-nickel, copper/copper-nickel, platinum/platinum-rhodium. Although thermocouples do generate a voltage that is proportional to temperature, the value of voltage is very low (a few millivolts). In most cases, thermocouples are not used as power sources, but rather are used to measure temperature—even extreme temperatures. For example, thermocouples can be used to measure temperatures as low as −270°C and as high as 1,700°C.

Figure 8-16 illustrates how thermocouples are used to measure temperature. Two junctions are connected in series. One of the junctions is maintained at a known reference temperature. The temperature of ice (0°C) is often used as the reference temperature. Each thermocouple pictured in Figure 8-16 produces a voltage that is proportional to its temperature. The series connection of the two thermocouples produces a net voltage (i.e., Kirchhoff's Voltage Law) that is proportional to the temperature difference of the two junctions ($T_2 - T_1$). The voltage produced is very low (a few microvolts per degree), but it is sufficient for detection and amplification by electronic circuitry.

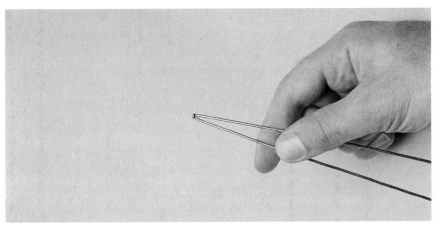

Figure 8-15. A thermocouple converts heat energy into electrical energy.
(© Copyright 1992 Omega Engineering, Inc. All rights reserved. Reproduced with permission of Omega Engineering, Inc, Stamford, CT.

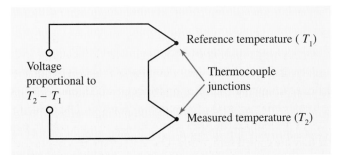

Figure 8-16. Two series-connected thermocouples can be used to accurately-measure temperature.

The voltage produced by a thermocouple, while proportional to temperature, is not always linear. Manufacturers provide calibration tables that can be used to convert the output voltage of a thermocouple to a corresponding temperature. Several thermocouples can be series-connected to provide a more substantial output. When used in this way, the complete assembly is called a **thermopile.**

Exercise Problems 8.2

1. Alternators and generators convert _____ energy into _____ energy.
2. When a coil of wire passes through a magnetic field, a _____ is created in the wire.
3. Alternators are used to supply alternating power. (True or False)
4. Generators are used to supply direct voltage and current (dc). (True or False)
5. What is the purpose of a steam turbine in a power generation plant?

6. Piezoelectric devices convert _____ energy into electrical energy.
7. It is possible to generate several thousand volts with a piezoelectric crystal. (True or False)
8. Solar cells convert _____ energy into _____ energy.
9. What determines the current capacity of a solar cell?
10. Fuel cells are very inefficient devices. (True or False)
11. Fuel cells can only generate powers up to about 500 mW. (True or False)
12. Thermocouples are used to convert _____ energy into _____ energy.
13. Several thermocouples can be connected in series to produce a higher output voltage. The overall connection is called a _____.
14. Thermocouples are often used to provide auxiliary power for space vehicles. (True or False)

8.3 Conductors, Semiconductors, Superconductors, and Insulators

Materials can be classified based on their relative ability to pass electrical current. Materials that pass current easily (i.e., low resistance) are called conductors. Insulators, by contrast, have a very high resistance and essentially prevent the flow of current. Semiconductor materials, as the name implies, have resistance values greater than that of conductors but less than that of insulators. The resistance of a material is affected by temperature. In the case of a semiconductor, for example, the resistance increases as the temperature decreases. The resistance of conductors, by contrast, decreases as the temperature decreases. Certain materials can be made to have virtually zero resistance if they are cooled sufficiently. These materials are classified as superconductors.

Ideally, insulators have infinite resistance, which results in zero current flow regardless of the applied voltage. Practical insulators have less than infinite resistance. The tiny current that flows through an insulator is called **leakage current.** In most cases, leakage current is so small that it cannot be measured on standard current meters.

Resistivity of Materials

The actual resistance of a particular material is determined by several factors:

- physical dimensions
- temperature
- type of material

Consider the resistance of the section of wire shown in Figure 8-17. If the length (l) of the wire is increased, then its resistance will increase because the electrons will encounter greater opposition as they flow through the wire. That is, wire resistance is directly proportional to the length of the wire. If the area (A) of the wire is increased (i.e., larger diameter), then the electrons will have an easier path to travel and the resistance of the wire will be reduced. That is, the resistance of the wire is inversely proportional to the area of the wire.

The type of material used in the wire clearly affects its resistance. The specific resistance or **resistivity** of a material allows materials to be compared based on the inherent opposition

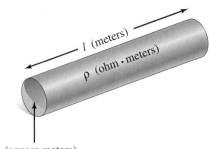

Figure 8-17. The resistance of a wire is determined by its physical dimensions and its resistivity (ρ).

to current flow. Figure 8-18 shows how the resistivity of a material is interpreted. The resistivity of a material is the resistance measured between opposite faces of a cube of the material that is 1 m on each side (1 m³). The symbol for specific resistance or resistivity is the Greek letter rho (ρ). The resistivity of a material varies with temperature and is therefore specified at a particular temperature (20°C or 25°C in most cases). The resistance of the wire shown in Figure 8-17 will be directly proportional to the resistivity of the material used to make the wire. The relationship between physical dimensions, resistivity, and net resistance is stated more formally as Equation 8-1.

$$R = \frac{\rho l}{A} \qquad (8\text{-}1)$$

where R is the overall resistance of the wire, ρ is the resistivity of the material, l is the length of the wire, and A is the cross-sectional area.

Table 8-2 lists several common materials and their resistivity values. Although the SI units are generally preferred, the British units are still used in the industry. As a technician, you

KEY POINTS

Resistivity can vary with temperature.

KEY POINTS

Resistance of a wire is determined by its physical dimensions, temperature, and resistivity. Its resistance increases with the length of the wire and decreases with larger diameters (greater conducting area).

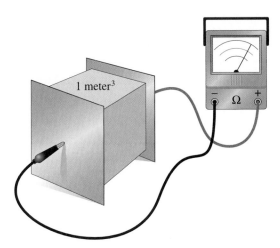

Figure 8-18. Specific resistance (resistivity) of a material is the resistance measured between opposite faces of a one-meter cube of the material.

should be able to work equally well with either system. The units for each of the two systems are given in Table 8-3.

The circular mil unit in the British system of measurement is simply the square of the diameter in mils (i.e., 0.001 inch). This is expressed by Equation 8-2.

$$A_{CMIL} = d^2 \qquad (8\text{-}2)$$

where A_{CMIL} is the circular mil area and d is the diameter of the wire or rod in mils.

MATERIAL	RESISTIVITY AT 20°C	
	SI UNITS	BRITISH UNITS
Aluminum	2.82×10^{-8}	16.96
Annealed copper	1.724×10^{-8}	10.37
Cadmium	7.6×10^{-8}	45.71
Constantan	49×10^{-8}	294.7
Gold	2.44×10^{-8}	14.68
Mica (insulator)	2×10^{10}	12×10^{18}
Nickel	7.8×10^{-8}	46.9
Platinum	10×10^{-8}	60.15
Silicon (semiconductor)	550	3.3×10^{11}
Silver	1.53×10^{-8}	9.2
Tantalum	15.5×10^{-8}	93.2
Tungsten	5.6×10^{-8}	33.68
Zinc	5.8×10^{-8}	34.89

Table 8-2. Resistivity (ρ) Values for Several Common Materials

QUANTITY	SI UNITS	BRITISH UNITS
Resistance	ohms (Ω)	ohms (Ω)
Resistivity	ohm-meter (Ω·m)	ohm-circular mil per foot (Ω·cmil/ft)
Length	meters (m)	feet (ft)
Area	square meters (m²)	circular mils (cmil)

Table 8-3. Units of Measurement for the SI and the British Systems of Measurement

EXAMPLE SOLUTION

What is the circular mil area of a wire that has a diameter of 25 mil (0.025 inch)?

EXAMPLE SOLUTION

We apply Equation 8-2 as follows:

$$A_{CMIL} = d^2$$
$$= (25 \text{ mil})^2 = 625 \text{ cmil}$$

It is important to note that area of wire expressed in circular mils is not the same numeric value as the area expressed in square mils. Figure 8-19 illustrates the relationship between these two units of measure. In Figure 8-19, the shaded area represents 1 cmil. It is superimposed on a 1 mil² area. Clearly, 1 cmil has less cross-sectional area than 1 mil². The use of the circular mil unit of measure simplifies the calculation of wire area ($A = d^2$ instead of $A = \pi r^2$). It can often be estimated with a mental calculation.

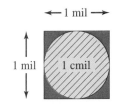

Figure 8-19.
Circular mils and square mils are both units of measure for area, but they have different values.

EXAMPLE SOLUTION

Calculate the resistance of an aluminum rod (at 20°C) that has the following physical dimensions: length = 1.5 m, area = 0.0025 m².

EXAMPLE SOLUTION

The resistivity of aluminum is given in Table 8-2 as 2.82×10^{-8}. We apply Equation 8-1 as follows:

$$R = \frac{\rho l}{A}$$
$$= \frac{2.82 \times 10^{-8} \, \Omega \cdot m \times 1.5 \, m}{0.0025 \, m^2}$$
$$= 0.00001692 \, \Omega = 16.92 \, \mu\Omega$$

EXAMPLE SOLUTION

What is the resistance of a silicon rod (at 20°C) if its length is 0.025 ft and its diameter is 0.005 mils?

EXAMPLE SOLUTION

First we apply Equation 8-2 to determine the circular mil area.

$$A_{CMIL} = d^2$$
$$= (0.005 \text{ mil})^2 = 25 \times 10^{-6} \text{ cmil}$$

Now we can use Equation 8-1 to compute resistance.

$$R = \frac{\rho l}{A}$$
$$= \frac{3.3 \times 10^{11} \, \Omega \cdot cmil/ft \times 0.025 \, ft}{25 \times 10^{-6} \, cmil}$$
$$= 3.3 \times 10^{14} \, \Omega$$

Practice Problems

1. Determine the resistance (at 20°C) of a 3-ft section of silver wire that has a diameter of 1.5 mils.
2. What is the resistance (at 20°C) of a 100-m section of gold wire, if it has an area of 125×10^{-8} m²?
3. Which of the following has the highest resistance at 20°C?
   a. 10 ft of tungsten wire that has a diameter of 3 mils
   b. 0.001 m of nickel wire with an area of 3×10^{-6} m²
   c. 6 in of silver wire with a diameter of 8 mils

Answers to Practice Problems

1. 12.27 Ω   2. 1.952 Ω   3. a

Temperature Coefficients

As stated previously, the resistivity (and therefore resistance) of most materials changes with changes in temperature. In the case of many metals, an increase in temperature causes an increase in the random movement of the free electrons in the wire. Any directed movement of electrons (i.e., current flow) encounters more opposition. That is, the resistance of the metal has increased.

Semiconductors have only a few free electrons at room temperature, so their resistivity is relatively high. As the temperature is increased, the thermal energy causes more electrons to escape the orbits of their parent atoms and become free electrons. With greater numbers of free electrons, the overall resistance of the material decreases.

○ **KEY POINTS**

Materials whose resistance increases with increasing temperature are said to have a positive temperature coefficient.

The **temperature coefficient** of resistance is a factor that identifies the amount and direction of resistance change for a given temperature change. When the direction of increase or decrease in resistance is the same as the direction of increase or decrease in temperature, the temperature coefficient is positive. Many metals have a positive temperature coefficient. When the directions of change for resistance and temperature are opposite (as in the case of many semiconductors), the temperature coefficient is negative.

The temperature coefficient of a material can be defined more formally by Equation 8-3.

$$TC = \frac{R_2 - R_1}{R_1(T_2 - T_1)} \qquad (8\text{-}3)$$

where TC is the temperature coefficient, R_1 is the initial resistance, T_1 is the initial temperature (often 20°C), R_2 is the resistance of the material at the second temperature T_2.

The temperature coefficient of wire and other materials is well known and well documented. It is available in reference handbooks and manufacturers' specification sheets. A more useful relationship can be obtained by transposing Equation 8-3 to determine the resistance (R_2) that results from a given temperature change. This is given by Equation 8-4.

$$R_2 = R_1 + TC \times R_1(T_2 - T_1) \qquad (8\text{-}4)$$

The temperature coefficients for several common materials are provided in Table 8-4.

MATERIAL	TEMPERATURE COEFFICIENT OF RESISTANCE
Aluminum	+0.0039
Brass	+0.002
Carbon	−0.0005
Constantan	+0.00000084
Copper (annealed)	+0.00393
Gold	+0.0034
Manganin	+0.00001
Nickel	+0.006
Platinum	+0.003
Silver	+0.0038

Table 8-4. Temperature Coefficients of Common Materials

EXAMPLE SOLUTION

The resistance of a length of copper wire is 0.5 Ω at 20°C. What is its resistance at 100°C?

EXAMPLE **SOLUTION**

We apply Equation 8-4 as follows:

$$R_2 = R_1 + TC \times R_1 (T_2 - T_1)$$
$$= 0.5 \ \Omega + 0.00393 \times 0.5 \ \Omega \times (100° - 20°)$$
$$= 0.6572 \ \Omega$$

Practice Problems

1. The resistance of a length of gold wire is 0.02 Ω at 20°C. What is its resistance at 75°C?
2. What is the resistance of a length of gold wire at −50°C, if it measures 0.02 Ω at 20°C?
3. If a spool of aluminum wire measures 1.6 Ω at 20°C, what will it measure at 40°C?
4. For a given temperature change and a given initial resistance, which of the materials listed in Table 8-4 would have the smallest absolute change in resistance value?

Answers to Practice Problems

1. 0.02374 Ω   2. 0.01524 Ω   3. 1.725 Ω   4. Constantan

Since the resistance of some materials decreases as we decrease temperature, it follows that at some very low temperature, the resistance of the material will be zero. Just a few years ago, this concept was essentially a theory. Today, however, numerous materials have been shown to have essentially zero resistance at very low temperatures. The study of the behavior of materials as they approach absolute zero (–273.2°C) is called **cryogenics.** Materials that exhibit near-zero resistance at very low temperatures are called superconductors.

Certain ceramic materials have been shown to have virtually zero resistance in the superconducting state. They have literally no resistance (even theoretically). This phenomenon presents a long list of exciting possibilities and opportunities. For example, thousands of amperes of current could be passed through a small wire without heating it up. Power for entire cities might be routed by tiny underground wires.

Figure 8-20 illustrates how the resistivity of some superconducting materials changes with temperature. Throughout much of the usable temperature range, the resistivity varies linearly with temperature. If the temperature falls to the critical temperature (T_c), then the resistivity drops to zero. The material enters the superconducting state.

Although the critical temperature is different for different materials, the critical temperature for all known materials is extremely low. Even "high-temperature" superconductors require cooling by liquid nitrogen (≈–195.8°C). Much research is under way to develop a material that will enter the superconducting state at room temperature. If this goal is achieved, the practical applications for superconductors will be limitless.

Wire Sizes

KEY POINTS

The size (area) of a wire is the dominant factor that determines its ampacity or current-carrying capability.

Technicians must work with many different sizes of wire. Some wires are so thin they can barely be seen with the naked eye. Other wires are so large, they require special tools to work with them. The choice of wire size for a given application is largely determined by the current that must pass through the wire. The current-carrying capacity of a wire is sometimes called its **ampacity.**

If a wire is too small for the current being carried, then it will drop excessive voltage across the resistance in the wire ($V = IR$). Additionally, excessive power dissipation in the wire resis-

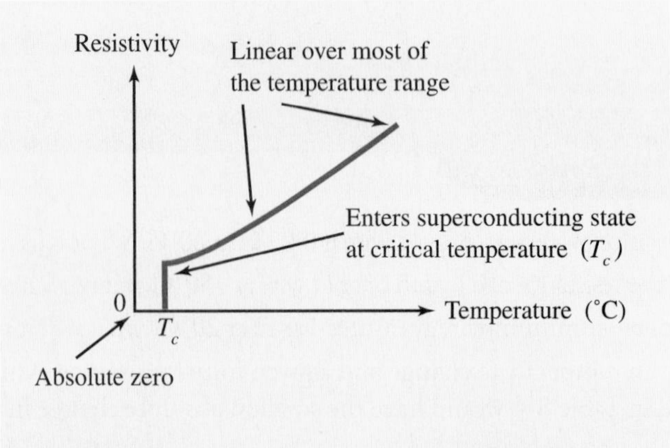

Figure 8-20. Superconductive materials have a drastic drop in resistivity at the critical temperature.

resistance ($P = I^2R$) will cause heating of the wire and may even damage it. This latter event is of particular concern when choosing wire for power distribution in a home or office. If the wire is too small, it may get hot and cause a fire in the walls of the building.

Wire size is specified by gauge numbers. There are at least six different industry standards for assigning a particular wire size to a particular gauge number. They are not the same in all cases. Most technicians in the United States rely on the American Wire Gauge (also called Brown and Sharpe [B&S] gauge) numbers to specify wire size. Table 8-5 shows several of the American Wire Gauge (AWG) numbers along with the approximate diameter and cross-sectional area. Table 8-5 also shows the approximate resistance for a 1,000-ft length of a given wire gauge for annealed copper wire. The complete AWG table includes both smaller and larger wire sizes than listed in Table 8-5.

KEY POINTS

Wire size is specified by its gauge. Higher gauge numbers correspond to physically smaller wire sizes.

AMERICAN WIRE GAUGE NUMBER	DIAMETER (MILS AT 20°C)	CROSS-SECTIONAL AREA (APPROXIMATE CMIL)	OHMS PER 1,000 FEET OF ANNEALED COPPER WIRE AT 20°C
0000	460	211,600	0.04901
000	409.6	167,772	0.06180
00	364.8	133,079	0.07793
0	324.9	105,560	0.09827
1	289.3	83,694	0.1239
2	257.6	66,358	0.1563
3	229.4	52,624	0.1970
4	204.3	41,738	0.2485
5	181.9	33,088	0.3133
6	162.0	26,244	0.3951
7	144.3	20,822	0.4982
8	128.5	16,512	0.6282
9	114.4	13,087	0.7921
10	101.9	10,383	0.9989
11	90.74	8,234	1.26
12	80.81	6,530	1.588
13	71.96	5,178	2.003
14	64.08	4,106	2.525
15	57.07	3,257	3.184

Table 8-5. Wire Sizes for Standard Annealed Copper Wire *(Continues)*

AMERICAN WIRE GAUGE NUMBER	DIAMETER (MILS AT 20°C)	CROSS-SECTIONAL AREA (APPROXIMATE CMIL)	OHMS PER 1,000 FEET OF ANNEALED COPPER WIRE AT 20°C
16	50.82	2,583	4.016
17	45.26	2,048	5.064
18	40.3	1,624	6.385
19	35.89	1,288	8.051
20	31.96	1,021	10.15
21	28.45	809	12.8
22	25.35	643	16.14
23	22.57	509	20.36
24	20.1	404	25.67
25	17.9	320	32.37
26	15.94	254	40.81
27	14.2	202	51.47
28	12.64	160	64.9
29	11.26	127	81.83
30	10.03	101	103.2

Table 8-5. Wire Sizes for Standard Annealed Copper Wire *(Continued)*

Inspection of Table 8-5 shows that larger gauge numbers correspond to smaller wire sizes. You might also note that the cross-sectional area approximately doubles (or halves) for every three gauge numbers.

The maximum current that a given wire size can carry is determined by several factors including the following:

- maximum allowable increase in temperature of the wire
- ambient temperature
- maximum allowable voltage drop ($V = IR$)
- type of insulation on the wire
- how the wire is routed (e.g., free air, bundled, conduit, and so on)

Standards are available for most applications that can be used to determine the correct wire size for a given current. Care should be taken when interpreting manufacturers' wire data. Some manufacturers define maximum current-carrying capacity as the amount of current needed to either melt the wire or the insulation. Neither of these definitions are consistent with the safety-based standards that prevail in the industry. For example, the National Electrical Code cites a maximum current of 20 A (less for certain insulations) for 14-gauge copper wire. By contrast, a leading wire manufacturer lists 27 to 45 A (depending on the type of insulation) for 14-gauge wire.

Types of Wire and Cable

To this point in our discussion, we have been focusing on single-conductor wires with or without a layer of insulation. In practice, there are many types of wire to choose from for a given application. We will discuss some of the major factors that distinguish wire types.

SOLID VERSUS STRANDED

Figure 8-21 illustrates the difference between solid-conductor and stranded wire. A given wire gauge of stranded wire consists of a number of smaller-gauge wires tightly braided together. The overall diameter determines the final gauge size of the stranded wire. Stranded wires are less prone to breaking when they are flexed repeatedly. There are many combinations of strand gauge and number of strands to achieve a given overall gauge size. Some common examples are listed in Table 8-6.

> **KEY POINTS**
>
> Conductors can be solid or stranded, single or multiwire, and either bare or insulated.

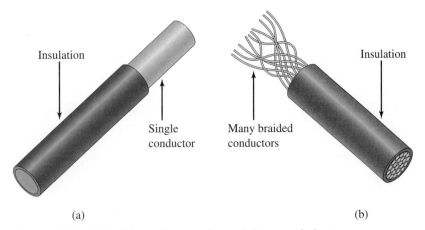

Figure 8-21. (a) Solid-conductor wire and (b) stranded wire.

TYPE OF INSULATION

Most wires used by technicians are covered with some form of insulation. The insulation prevents shorts to other wires or circuit components and reduces the probability of electrical shock. Many different materials are used for wire insulation. They vary in ways that include the following:

- breakdown voltage
- usable temperature range
- resistance to water, acid, alkali, and petroleum products
- resistance to flame

OVERALL GAUGE	NUMBER OF STRANDS	STRAND GAUGE
24	10	34
	19	36
	41	40
18	7	26
	16	30
	41	34
14	7	22
	41	30
	105	34

Table 8-6. Stranded Wire Consists of Several Strands of Smaller-Gauge Wire

- resistance to weathering
- cost
- flexibility

Breakdown voltage is the potential difference that will cause the insulation to break down. When an insulator breaks down, it loses its ability to stop current flow. During breakdown, a high current can flow and the insulator may be permanently damaged. Breakdown voltage depends on the type and thickness of insulation. Table 8-7 contrasts some of the characteristics of several common materials used to make wire insulation.

MATERIAL	BREAKDOWN VOLTAGE (PER MIL THICKNESS)	RESISTANCE TO					FLAME RETARDANCY
		HEAT	WATER	ALKALI	ACID	ABRASION	
Butyl rubber	600	Good	Excellent	Excellent	Excellent	Fair	Poor
Kynar®	260	Excellent	Good	Excellent	Excellent	Good	Excellent
Neoprene	600	Good	Excellent	Good	Good	Excellent	Good
Polypropylene	750	Good	Excellent	Excellent	Excellent	Fair	Poor
Polyurethane	500	Good	Good	Fair	Fair	Excellent	Good
PVC	400	Good	Good	Good	Good	Good	Excellent
Silicon rubber	400	Excellent	Excellent	Good	Good	Fair	Good
Teflon®	500	Excellent	Excellent	Excellent	Excellent	Fair	Excellent

Table 8-7. Comparison of Characteristics for Wire Insulating Materials

It is important for a technician to realize that not all wire insulation materials are as obvious (visually) as the insulation on standard hook-up wire. Some wires (collectively called magnet wire) have a thin, nearly invisible insulation that is applied by dip-coating the conductor in a varnish-like material. Magnet wire is used to wind motors, relays, solenoids, and transformers. The thin insulation permits many turns of wire in a relatively compact space. When soldering this type of wire, care must be used to ensure that the insulation has been removed from the portion being soldered.

SINGLE-CONDUCTOR VERSUS CABLE

Previous discussions have focused on single-conductor wires (either solid or stranded). Figure 8-22 illustrates how two or more individually insulated wires may be bundled in a common outer jacket. The bundled assembly is called a cable.

The number of conductors in a cable varies from as low as two to over one hundred. Most cables use color coding to distinguish the wires. The individually insulated wires may be either solid or stranded. The outer jacket, which bonds the wires into a single cable, is generally an insulator. Multiconductor cables are also available where each of the individual wires is actually a coaxial cable (described next).

COAXIAL CABLE

Figure 8-23 shows the construction of coaxial cable (called coax for short). It consists of an insulated center conductor surrounded by a braided wire shield. Both the center conductor and the braided shield are encased by a tough outer jacket.

The braided shield, which surrounds the center conductor, is connected to ground in a typical application. This provides an electrical barrier, which prevents stray fields from inducing noise voltages into the center conductor. This is important when low-level signals must be routed through an electromagnetically harsh environment. Coaxial cables are also used to prevent high-frequency voltages on the inner conductor from interacting with nearby circuitry or from causing electromagnetic emissions that would cause interference to nearby receivers.

KEY POINTS

Coaxial cable includes an integral shield. Both coaxial cable and foil-wrapped cable reduce coupling between the central conductor(s) and other circuits.

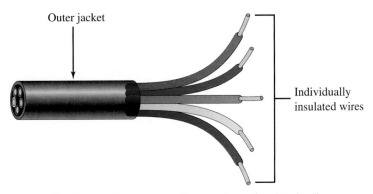

Figure 8-22. A cable consists of a number of individually insulated wires in a common sheath.

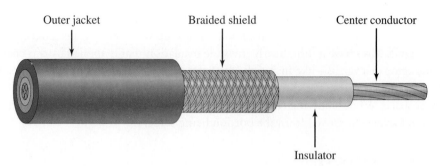

Figure 8-23. Coaxial cable reduces electromagnetic coupling between the center conductor and other nearby circuits.

An alternative to standard coaxial cables that still provides shielding for the center wire(s) consists of a spiral-wrapped metal foil. The foil surrounds the center conductor(s) and is generally connected to ground. An uninsulated wire, called a drain wire, runs along the length of the metal foil and provides a means for connecting to it.

Ribbon Cable

KEY POINTS

Ribbon cable provides a means of connecting to and routing many conductors at the same time through the use of a crimp-on connector.

Figure 8-24 shows a multiconductor cable where the individually insulated wires are side-by-side, forming a flat strap. The bound wires are collectively called a ribbon cable. Ribbon cable is frequently used to interconnect circuit boards and devices in a computer system. Some types have a foil shield that surrounds the entire ribbon cable. The shield and the other conductors are then covered with a protective insulative jacket.

The individual wires in a ribbon cable can be split out and soldered individually if necessary. In most cases, however, crimp-on connectors are applied to the cable ends that make simultaneous connection to all of the wires. The connectors can then be plugged into mating sockets on circuit boards or other devices.

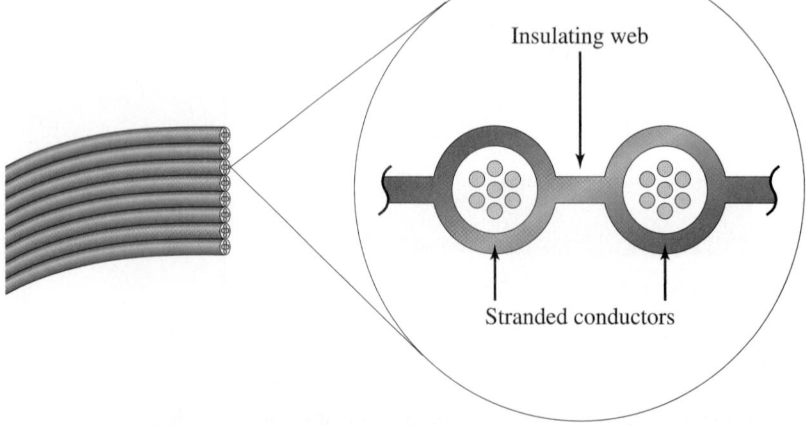

Figure 8-24. Ribbon cable is frequently used in computer systems.

A variation of the basic ribbon cable consists of flat copper conductors (like traces on a printed circuit board) sandwiched between two sheets of plastic laminate. The resulting package provides a flexible, multiconductor connection between two points.

Exercise Problems 8.3

1. A material that has a higher resistivity than a conductor but lower than an insulator is called a _____.

2. What is the name given to materials that exhibit zero resistance at cryogenic temperatures?

3. The resistance of a wire is unaffected by temperature. (True or False)

4. If the resistance of a material decreases as the temperature increases, the material is said to have a (*positive, negative*) temperature coefficient.

5. The resistance of a wire is (*directly, inversely*) proportional to the length of the wire.

6. The resistance of a wire is (*directly, inversely*) proportional to the cross-sectional area of the wire.

7. The resistance of a wire is directly proportional to the resistivity of the material used to make the wire. (True or False)

8. Refer to Table 8-2. A gold and a silver wire are made with identical physical dimensions. Which will have the lower resistance?

9. What is the unit of measure for the cross-sectional area of a wire in the British system?

10. What is the unit of measure for resistivity in the SI system?

11. Compute the circular mil area of a wire that has a diameter of 30 mils.

12. What is the diameter of a wire that has an area of 2,025 cmil.

13. Calculate the resistance of a tungsten wire (at 20°C) that has the following physical dimensions: length = 0.5 m, area = 0.0005 m^2.

14. What is the resistance of a gold wire (at 20°C), if it is 2.5 ft long and has a diameter of 2 mils?

15. Semiconductors often have a (*positive, negative*) temperature coefficient.

16. If an aluminum wire and a nickel wire had similar physical dimensions, which would show the greatest resistance change for a given change in temperature?

17. The resistance of a length of copper wire is 0.12 Ω at 20°C. What is its resistance at 75°C?

18. What is the resistance of a length of aluminum wire at 0°C, if it measures 1 Ω at 20°C?

19. Which of the following wire gauges would have the greatest ampacity: 0, 14, or 30?

20. What is the resistance of 3,000 ft of 18-gauge annealed copper wire at 20°C?

21. What is the cross-sectional area of a 24-gauge wire?

22. Name at least one advantage of stranded wire over solid-conductor wire.

23. If a wire is to be used in a 120-V circuit, then the insulation must have a breakdown voltage rating of 120 V or less. (True or False)

24. What is the approximate breakdown voltage for a 35-mil thickness of PVC insulation?
25. What type of insulation is used on magnet wire?
26. Describe the physical construction of coaxial cable.
27. Describe the physical construction of ribbon cable.
28. What type of cable is used to reduce the effects of intercircuit coupling?

8.4 Troubleshooting Power Sources, Conductors, and Insulators

Locating a defective power source, conductor, or insulator is a straightforward process that is easily described and understood. In practice, however, location of this type of defect can be very challenging to the best technician.

Identifying Defects in Power Sources

All of the power sources presented in this chapter can be viewed (for purposes of trouble-shooting) as their Thevenin equivalent. You will recall from Chapter 7 that the Thevenin equivalent for any circuit consists of a voltage source and a resistance. This equivalency is illustrated in Figure 8-25.

The value of V_{OC} is the open-circuit (i.e., unloaded) voltage of the power source. R_{INT} is the internal resistance of the power source. We will rely on the equivalency illustrated in Figure 8-25 for the following discussion.

Defects in power sources can be grouped into one of the following categories:

- no output
- low output
- excessive output

In each of these cases, the defect may actually lie within the power source, or it may simply be a symptom, the actual cause of which is external to the power source. Consider the example shown in Figure 8-26. A voltmeter is being used to make in-circuit measurements of a power source. The nominal voltage of the source is 3.6 V. The measured value,

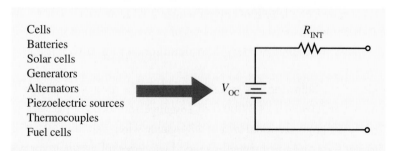

Figure 8-25. All power sources can be viewed as a constant voltage source (V_{OC}) and a series resistance (R_{INT}).

however, is greatly affected by the value of load resistance. The measured voltage can have values ranging from the full 3.6 V (V_{OC}) to 0 V depending on the value of load resistance.

Voltage checks to determine the condition of a power source are valuable, but a technician must be aware of possible loading effects as illustrated in Figure 8-26. Accurate determination of power source condition can be made using the following procedure:

1. Disconnect the power source from the load and measure the open-circuit voltage. Many power sources (e.g., most electrochemical cells, generators, alternators, and solar cells) have an internal resistance (R_{INT}) that is low compared to the resistance of most voltage-measuring devices. In these cases, V_{OC} can be accurately measured. For high-impedance power sources (e.g., piezoelectric sources), incorrect measurements may be made due to voltmeter loading (discussed in Chapter 6). Incorrect open-circuit voltage readings point to a defective power source.

2. Connect a known load other than the one being diagnosed, and measure the full-load voltage. In most cases, the loaded voltage will be lower than the open-circuit voltage due to the voltage divider action illustrated in Figure 8-26. Classification of measured voltages as normal or abnormal can only be made based on a knowledge of the system (i.e., you must know or be able to compute the loaded voltage for a normal power source). Incorrect full-load voltage readings indicate a defective power source. The internal resistance of electrochemical cells, for example, increases as the cell ages. This increased resistance (see Figure 8-26) does not affect the open-circuit voltage, but it causes a reduction in normal full-load voltage.

3. If both of the preceding tests produce normal voltage values, then the defect lies in the external load and not in the power source itself. If the load is shorted or has a low resistance, then the open-circuit voltage may be normal, but the terminal voltage will drop drastically when connected to the circuit. Open or increased-resistance loads can cause the terminal voltage to measure higher than the normal full-load value.

Identifying Defects in Conductors

An open circuit is one of the most probable defects in a conductor. Once you have classified a given conductor as a suspect, then a simple ohmmeter check will show the

KEY POINTS

Defects in power sources can generally be found by using a voltmeter.

KEY POINTS

Symptoms may include low terminal voltage, high terminal voltage, or a terminal voltage that decreases excessively under full-load conditions.

KEY POINTS

A good conductor will measure near zero ohms on a standard ohmmeter; an open conductor will indicate infinite ohms. If a conductor is intermittent, then mechanical flexing of the conductor can cause its resistance to vary from near zero to infinite ohms.

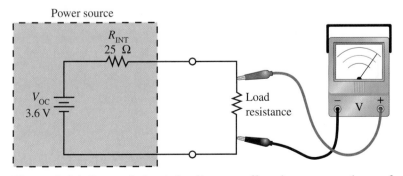

Figure 8-26. External circuit loading can affect the output voltage of power sources.

conductor to be good or bad. A good conductor will read near zero ohms on a standard ohmmeter. We say the conductor has continuity. An open conductor, of course, will indicate infinite ohms and has no continuity. When determining the continuity of a conductor, it is always a good practice to mechanically flex the conductor if possible. Movement sometimes reveals an intermittent conductor. Figure 8-27 summarizes the ohmmeter tests on a conductor.

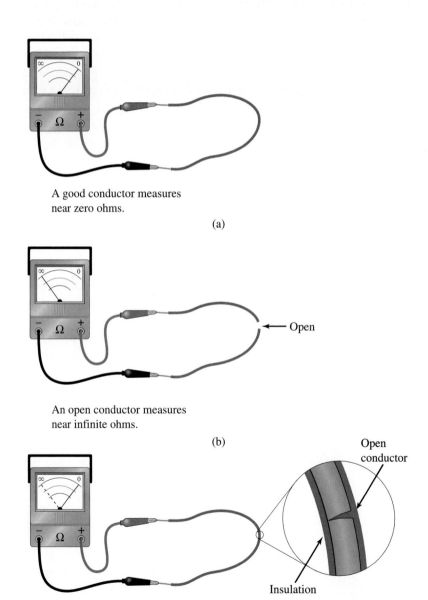

A good conductor measures
near zero ohms.

(a)

An open conductor measures
near infinite ohms.

(b)

An intermittent conductor measures zero and
infinite ohms alternately as the wire is flexed.

(c)

Figure 8-27. An ohmmeter can be used to determine the condition of a conductor.

Identifying Defects in Insulators

Technicians that work in certain specialized areas (e.g., transformer design and manufacturing) have access to test equipment that can actually measure the resistance of an insulator. Most technicians, however, must rely on standard test equipment for troubleshooting. In the case of insulators, an ohmmeter can be used to measure the continuity between two conductive points that are supposed to be separated by the suspected insulator. If the two points have no continuity, then the insulation is probably good. If the two points register continuity, then the insulator has broken down and no longer prevents current flow.

Figure 8-28 shows a common application for insulators. Power transistors get very hot under normal operation. They are often mounted to metal heat sinks to help dissipate the heat safely. Although they must be thermally coupled to the heat sink, the transistors must be electrically isolated for proper operation. A thin plastic or rubber spacer is used to provide the electrical isolation.

Sometimes the insulator will become defective and allow the transistor case to contact the metal heat sink. A technician must be able to detect this unwanted continuity and locate the defective insulator. The actual damage to the insulator may be visible in the form of cracks, or burned pinholes. Even if it is not visible, the ohmmeter will reveal any shorted conditions.

Insulators in high-voltage circuits sometimes lose their ability to withstand high voltages but can prevent current flow at low voltages (like those used inside the ohmmeter). This type of defect is more difficult to identify, since the insulation measures infinite ohms with a standard ohmmeter. In this case, the technician must either have access to a special ohmmeter (often called a megger) that can measure the high resistances associated with insulators, or the defect must be located through observation of the symptoms.

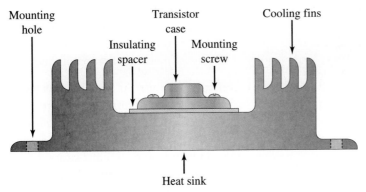

Figure 8-28. A thin insulator is often used to isolate power transistor cases from their metal heat sinks.

Exercise Problems 8.4

1. If the open-circuit voltage of an electrochemical cell is low, then the cell is probably defective. (True or False)

2. If the open-circuit voltage of an electrochemical cell is normal, but the full-load voltage is low, the cell is definitely defective. (True or False)

3. A good conductor measures _____ ohms on an ohmmeter.

4. If a conductor develops an open, it will indicate _____ ohms on an ohmmeter.

5. Can an ohmmeter be used to detect a shorted insulator?

Chapter Summary

- Electrochemical cells convert chemical energy into electrical energy. The available voltage and current from a cell can be increased by connecting several cells together to form a battery. Secondary cells, such as lead-acid, NiMH, and NiCad, are designed to be recharged, which replenishes the chemical energy. Primary cells are not normally rechargeable.

- Alternators and generators are electromechanical devices that convert mechanical energy into electrical energy by moving a conductor through a magnetic field. Alternators produce alternating voltage and current (ac), whereas generators are used as a source of direct voltage and current (dc).

- Piezoelectric sources produce electrical energy in response to mechanical stress. Although the generated voltages can be quite high, the current capability is very low.

- Solar cells are used to convert light energy (typically sunlight) into electrical energy. Although the basic solar cell only generates a fraction of a volt, multiple cells can be series-connected for higher net voltages.

- Fuel cells are a special class of electrochemical cells that have electrodes that do not react with the electrolyte. The reactive chemicals are fed in continuously from an external source.

- A theromocouple converts heat energy into electrical energy. A thermocouple consists of a junction of two dissimilar metals. The output voltages of a thermocouple are extremely low. Multiple thermocouples can be series-connected as a thermopile to produce higher output voltages. Temperature sensing is the major application of thermocouples.

- All materials can be classed as conductors, insulators, semiconductors, or superconductors based on their relative conductivity. Insulators prevent current flow for practical purposes. Conductors readily permit current flow and have a low resistance. Semiconductors have conductivity values that lie between those of conductors and insulators. They are neither good conductors nor good insulators. The resistance of a particular class of materials called superconductors drops to zero as the temperature of the material approaches cryogenic temperatures.

- The actual resistance of a material is determined by its physical dimensions, its resistivity, and the temperature. The resistance of a wire is directly proportional to its length and resistivity. It is inversely proportional to the cross-sectional area of the wire. The cross-sectional area of a wire is often measured in circular mils (cmils) and is computed by squaring the diameter of the wire expressed in mils.

- Materials may have a positive, negative, or zero (over a limited range) temperature coefficient. The resistance of materials with positive temperature coefficients increases as the temperature increases. The resistance of materials with negative temperature coefficients decreases with increasing temperature.

- Wire size is expressed with gauge numbers. Smaller gauge numbers correspond to larger wire sizes. The size (i.e., cross-sectional area) of a wire is the primary factor that determines the maximum allowable current that can be carried by the wire.

- There are many types of wire such as solid, stranded, bare, or insulated. Multiple conductors can be bound together by an overall jacket to form a cable. Each wire in a cable is individually insulated. Coaxial cables have an inner conductor that is surrounded by a braided shield. Ribbon cables have multiple conductors that are side-by-side and physically resemble a wide strap.

- Defective power sources can generally be diagnosed with a voltmeter and a systematic troubleshooting procedure. Shorts or low-resistance paths in the external circuit are easily mistaken for defects in the power source.

- Defects in conductors and insulators can be located with an ohmmeter in most cases. Good conductors measure near zero ohms and good insulators measure infinite ohms with a standard ohmmeter.

Review Questions

Section 8.1: Cells and Batteries

1. Electrochemical cells are used to convert _____ energy into electrical energy.

2. Primary cells are normally designed to be recharged. (True or False)

3. When several cells are connected together to obtain a higher voltage or current capacity, the complete assembly is called a _____.

4. Name the three major components that are found in every electrochemical cell.

5. What is the name given to the negative terminal of a cell?

6. What is the name given to the positive terminal of a cell?

7. Cell capacity is specified in units of _____.

8. A certain cell has a shelf life of five years. Explain what this rating means.

9. An ideal voltage source would have an internal resistance of $0 \ \Omega$. (True or False)

10. The specific gravity of the electrolyte in a lead-acid cell can be used to indicate the state of the cell's charge. (True or False)

11. Specific gravity is measured with a _____.

12. Secondary cells are normally rechargeable. (True or False)

13. Describe the term *memory effect* as it applies to NiCad cells.

14. What is the no-load voltage of a carbon-zinc cell?

15. Name at least one advantage of an alkaline–manganese dioxide cell over a carbon-zinc cell.

16. The terminal voltage of a mercuric-oxide cell gradually decays throughout its useful life. (True or False)

17. Name at least one advantage of a silver-oxide cell as compared to a mercuric-oxide cell.

18. Lithium cells normally have a very long shelf life.

19. What is used as the cathode material in a zinc-air cell?

20. During recharging of an electrochemical cell, _____ energy is converted into _____ energy.

21. Inside an electrochemical cell, electrons flow from the positive terminal to the negative terminal as it supplies energy to an external circuit. (True or False)

22. What is the primary disadvantage of nickel-metal-hydride cells as compared to nickel-cadmium cells?

Section 8.2: Other Power Sources

23. An electromechanical power source converts _____ energy into electrical energy.

24. An alternator produces unidirectional current. (True or False)

25. A commutator is a mechanical device that converts alternating current (or voltage) to direct current (or voltage). (True or False)

26. A piezoelectric source converts _____ energy into electrical energy.

27. A crystal microphone is an example of a _____ voltage source.

28. Piezoelectric sources generally have very high internal resistances. (True or False)

29. Solar cells are made from superconductors. (True or False)

30. A solar cell converts _____ energy into electrical energy.

31. What determines the current capacity of a solar cell?

32. One type of fuel cell uses oxygen and hydrogen as the fuel for the cell. (True or False)

33. Fuel cells are only capable of generating a few milliwatts of power. (True or False)

34. Thermocouples convert _____ energy into electrical energy.

35. The voltage from a typical thermocouple is 1.5 V. (True or False)

36. What is the name of the assembly formed when several thermocouples are connected in series?

37. Thermocouples cannot be used to measure temperatures below freezing (0°C). (True or False)

Section 8.3: Conductors, Semiconductors, Superconductors, and Insulators

38. Conductors have a relatively high resistivity. (True or False)

39. Insulators have a very high resistance. (True or False)

40. The resistance of a wire is directly proportional to its cross-sectional area. (True or False)

41. The current-carrying capacity of a wire is directly proportional to its cross-sectional area. (True or False)

42. The resistance of a 10-ft section of 00-gauge copper wire is greater than a 10-ft section of 36-gauge copper wire. (True or False)

43. If a section of wire is cut in half, the resistance of either section will be less than the total resistance of the original wire. (True or False)

44. Resistivity is also called _____.

45. What is the circular mil area of a wire that has a diameter of 8 mils?

46. What is the circular mil area of a wire that has a 100-mil diameter?

47. Compute the resistance of a gold wire (at 20°C) that has the following physical dimensions: length = 2.75 m, area = 64×10^{-9} m².

48. Calculate the resistance (at 20°C) of an aluminum wire that is 2,500 ft long and has a diameter of 25 mils.

49. A 25-ft copper wire with a diameter of 10 mils has more resistance (at 20°C) than a 1,000-ft section of silver wire with a diameter of 20 mils. (True or False)

50. If the temperature of material that has a positive temperature coefficient is decreased, what happens to its resistance?

51. Many metals have positive temperature coefficients. (True or False)

52. The resistance of a length of copper wire is 2.3 Ω at 20°C. What is its resistance at 85°C?

53. The resistance of a length of aluminum wire is 0.05 Ω at 20°C. Calculate its resistance at –25°C.

54. If a reel of copper wire measures 22.5 Ω at 20°C, what will it measure at 100°C?

55. What is the name given to the study of the behavior of materials as their temperature approaches absolute zero?

56. When a material enters the superconducting state, its resistance becomes extremely high. (True or False)

57. What is the name given to the temperature where a material makes the transition into the superconducting state?

58. The current-carrying capacity of a wire is also called its _____.

59. If a wire gets hot when carrying a certain value of current, a wire with a lower gauge number might be required. (True or False)

60. Which of the following would weigh the most: (a) 1,000 ft of 30-gauge copper wire; (b) 1,000 ft of 000-gauge copper wire?

61. Refer to the previous question. Which would have the lowest resistance?

62. Name an advantage of stranded wire over solid-conductor wire.

63. Name at least three factors that affect the choice of wire insulation for a given application.

64. Explain what is meant by the term *breakdown voltage* with reference to insulation.

65. A cable consists of several bare copper wires bound together by an outer layer of insulation. (True or False)

66. What is the name of the cable type that has a braided shield around a solid (or stranded) center conductor?

67. What is the purpose of a drain wire in a foil-shield cable?

68. What is the name of the wire cable that has several conductors formed into a flat strap?

69. What is the diameter of a wire that has a cross-sectional area of 5,625 cmil?

70. What is the approximate breakdown voltage for a 2-mil thickness of Kynar® insulation?

Section 8.4: Troubleshooting Power Sources, Conductors, and Insulators

71. If the open-circuit voltage of a power source is low, then the power source is probably defective. (True or False)

72. If the full-load voltage of a power source is normal when tested in the laboratory, but is low when connected to the actual circuit being powered, then the power source is probably defective. (True or False)

73. If the open-circuit voltage is normal, but the full-load voltage is low when connected to a test load, the power source is probably defective. (True or False)

74. A good conductor should indicate near _____ ohms on an ohmmeter, whereas an open conductor will indicate _____ ohms.

75. An insulator is being used to isolate two points in a circuit. An ohmmeter shows there is continuity between the two points. The insulator is probably defective. (True or False)

TECHNICIAN CHALLENGE

You are working for an oil exploration company as a field technician and are assigned to a remote location. A severe storm has caused a major power outage that may last for days. The company has portable generators for emergency power, but they were also damaged in the storm. The work site has an emergency two-way radio, but its power pack is dead. There are several dozen 1.5-V alkaline–manganese dioxide cells in the site supplies, but the radio requires 12 V at 1.1 A. The alkaline cells are designed for an extended current drain of 235 mA.

Develop a series-parallel network to make a battery of alkaline–manganese dioxide cells that will provide 12 V and at least 1.1 A so the emergency radio can be powered and you can get help to your fellow employees.

Equation List

(8-1) $\quad R = \dfrac{\rho l}{A}$

(8-2) $\quad A_{\text{CMIL}} = d^2$

(8-3) $\quad TC = \dfrac{R_2 - R_1}{R_1(T_2 - T_1)}$

(8-4) $\quad R_2 = R_1 + TC \times R_1(T_2 - T_1)$

backoff ohmmeter scale
CRT
full-scale current
graticule
half-digit
interpolation
meter accuracy

meter damping
meter resolution
meter sensitivity
meter shunt
ohms-per-volt rating
parallex error
seven-segment display

objectives

After completing this chapter, you should be able to:

1. Contrast the basic operation of digital and analog meters.
2. State the internal resistance of ideal current and voltage meters.
3. Select the correct scale for optimum reading accuracy on a VOM.
4. Explain the operation of the moving-coil and iron-vane meter movements.
5. Describe how to connect a voltmeter, ammeter, and an ohmmeter into a circuit.
6. Calculate values for current meter shunts.
7. Calculate values for voltmeter multipliers.
8. Interpret manufacturers' specifications for voltmeters, ammeters, and ohmmeters.
9. Explain the basic operation of an oscilloscope.
10. Describe how to measure dc circuit values with an oscilloscope.
11. State the purpose of each control/jack on a basic voltmeter, ammeter, ohmmeter, or oscilloscope.

Direct-Current Test Equipment

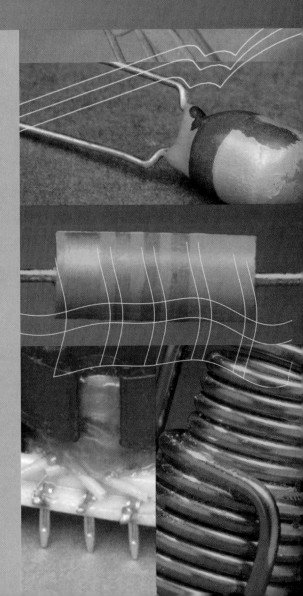

Electronics technicians use voltmeters, ohmmeters, ammeters, and oscilloscopes to troubleshoot and calibrate electronic circuits. This chapter explains how these devices work, and reviews their application to dc circuits.

9.1 Analog Meters

All analog meters (e.g., voltmeters, ammeters, and ohmmeters) have one thing in common—the meter movement. This is the actual mechanism that causes the meter pointer to deflect across the scale. We will look at two types of meter movements: moving-coil and iron-vane.

Moving-Coil

Figure 9-1 shows a pictorial view of a moving-coil meter movement. This movement is also called a d'Arsonval movement. In Chapter 10, we will examine the interaction of the magnetic fields that cause this movement to deflect, but for now let us concentrate on its overall operation.

As shown in Figure 9-1, a coil is mounted to the same shaft as the pointer. A spiral spring holds the meter movement in its deenergized position. In the case of a voltmeter or ammeter, the pointer indicates zero on the scale when the meter is deenergized.

When current is passed through the coil wires, an electromagnetic field is developed around the coil. This field interacts with the magnetic field provided by the permanent magnet. The attraction/repulsion produced by the interaction of the two fields overcomes the force of the spiral spring and causes the pivot shaft to rotate. Rotation of the pivot shaft causes the pointer to be deflected across the meter scale. The amount of deflection is directly proportional to the amount of current flowing in the meter coil.

If current through the coil is interrupted, then the spiral spring forces the pointer back to its deenergized position. If current through the coil is reversed, then the polarity of the electromagnetic field is also reversed. This causes deflection of the meter pointer in the opposite direction (i.e., downscale). In most cases, reverse deflection is undesired and may even cause damage to the meter movement (e.g., it may bend the pointer). The technician must ensure that the meter is connected properly so that current will always go through the meter coil in the right direction.

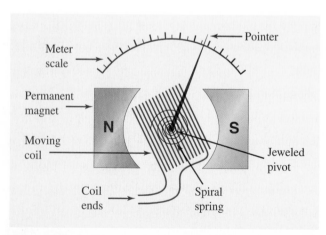

Figure 9-1. A d'Arsonval or moving-coil meter movement.

Iron-Vane

In an iron-vane meter movement, current flows through the meter coil and produces a magnetic field around the coil. The magnetic field attracts a soft-iron vane. Since the vane is physically attached to the pivot shaft, the shaft rotates. The meter pointer then moves upscale, since it is also attached to the pivot shaft.

The force of attraction is proportional to the square of the current. Thus, if the current through the coil is doubled, deflection will increase by a factor of four. This nonlinear relationship causes the meter scale to be nonlinear as well. That is, the scale is compressed near the left side and expanded near the right (or full-scale) side. The iron vane can be shaped in a manner that compensates for the inherent nonlinearity of the meter movement. In this case, the scale would be linear.

The iron-vane meter movement shown in Figure 9-2 uses two iron vanes; one is fixed and one is movable. Both of the vanes are mounted in the region of the magnetic field produced by the coil. Since both vanes are in the same magnetic field, they are affected in the same way. You will recall from your grade-school science classes that similar magnetic fields repel each other. In this case, the fixed and movable iron vanes repel each other. Because the movable vane is mounted to the pivot shaft, it causes the shaft to rotate and deflect the meter pointer.

Iron-vane meter movements are simple in construction and are very rugged. They can be used to measure direct current of either polarity. Upscale deflection occurs regardless of the direction of coil current. Iron-vane movements are generally less accurate than moving-coil movements. Additionally, external magnetic fields have a greater effect on iron-vane movements.

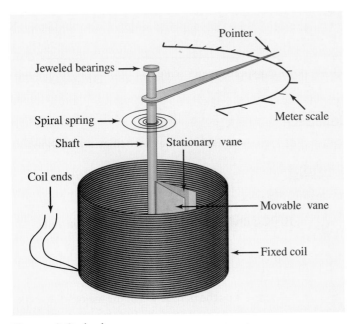

Figure 9-2. An iron-vane meter movement.

Scale Interpretation

Reading a meter scale is just like reading the speedometer scale in your car or the thermometer scale on a mercury thermometer. In each of these cases, the following characteristics must be considered in order to determine the value of the measured quantity:

- full-scale value
- minimum-scale value
- number of major divisions
- number of minor divisions
- location of the pointer or indicating device

DIVISIONAL WEIGHTS

KEY POINTS

The scales on an analog meter are generally divided into major and minor divisions. Each divisional mark is weighted according to the measurement range of the meter and the number of divisional marks.

Figure 9-3 shows a speedometer scale that might be found in a sports car. The minimum value is 0 miles per hour (mph). Its full-scale value is 160 mph. We can see that the speedometer scale is divided into four major divisions. We can determine the weight of each division with Equation 9-1:

$$\text{Division weight} = \frac{\text{Highest division value} - \text{Minimum division value}}{\text{Number of divisions}} \qquad (9\text{-}1)$$

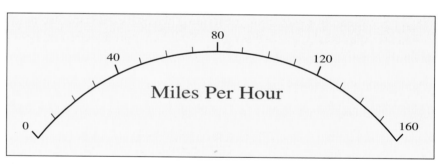

Figure 9-3. A sports car speedometer scale.

EXAMPLE SOLUTION

Compute the weight of each major division for the speedometer scale shown in Figure 9-3.

EXAMPLE **SOLUTION**

We compute the weight as follows:

$$\text{Division weight} = \frac{\text{Highest division value} - \text{Minimum division value}}{\text{Number of divisions}}$$

$$= \frac{160 \text{ mph} - 0 \text{ mph}}{4} = 40 \text{ mph}$$

This value is evident in Figure 9-3 by the numbering of the major scale markings as 0, 40, 80, 120, and 160. Even if the major scale divisions were not marked, we could determine their values by applying Equation 9-1.

Each of the four major regions of the scale shown in Figure 9-3 is further divided into four minor divisions. We can apply Equation 9-1 to determine the value for each of the minor divisions.

EXAMPLE SOLUTION

Determine the weight of each minor division mark on the speedometer scale shown in Figure 9-3.

EXAMPLE **SOLUTION**

We can use the minimum and maximum values for any portion of the scale. For purposes of this example, let us use the segment between 40 and 80 mph. We apply Equation 9-1 as follows:

$$\text{Division weight} = \frac{\text{Highest division value} - \text{Minimum division value}}{\text{Number of divisions}}$$

$$= \frac{80 \text{ mph} - 40 \text{ mph}}{4} = 10 \text{ mph}$$

Thus, for example, two minor marks above 80 mph would be read as 100 mph. Similarly, one minor mark below 160 mph would represent 150 mph.

Many scales have additional levels of scale division. Equation 9-1 can be used as needed to determine the weight of any given divisional grouping.

Practice Problems

Refer to the thermometer scale shown in Figure 9-4 for each of the following questions:

1. What is the weight of each major divisional mark?
2. What is the weight of each minor scale mark?
3. What temperature is indicated by each of the following marks?
   A. _____ B. _____ C. _____ D. _____ E. _____

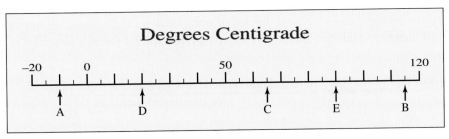

Figure 9-4. A typical thermometer scale.

Answers to Practice Problems

1. $10°$ 2. $5°$ 3. $-10°, 115°, 65°, 20°, 90°$

INTERPOLATION

Many times the meter pointer comes to rest at a point between the scale marks. The technician must still determine the indicated value. Interpreting values that lie between the scale marks is called **interpolation.** The degree of accuracy for scale interpolation depends on the application, the type of meter scale, and the skill of the technician. Consider the speedometer scale shown in Figure 9-5.

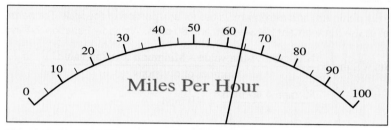

Figure 9-5. A speedometer scale that requires interpolation.

Each major division is labeled, so we can readily see they are weighted at 10 mph. Each of the major segments is further divided into two minor segments. Applying Equation 9-1 reveals that each minor divisional mark is weighted at 5 mph. By examining the scale shown in Figure 9-5, we can see that the pointer is between 60 mph (a major divisional mark) and 65 mph (a minor divisional mark). To determine the actual value indicated by the pointer, we mentally apply the following procedure:

1. Visualize additional scale divisions extending between the scale marks on either side of the pointer. Usually two, four, or five subdivisions are imagined.
2. Determine the weight of each imaginary scale mark by applying Equation 9-1.
3. Determine the indicated value.

Although the procedure sounds complex when written, it quickly becomes second nature to a technician and rarely requires anything but simple mental arithmetic.

In the case of the speedometer shown in Figure 9-5, we might imagine that the portion of the scale between 60 and 65 mph is divided into five smaller subdivisions. We can determine the weight of each new division by applying Equation 9-1 (usually mentally) as follows:

$$\text{Division weight} = \frac{\text{Highest division value} - \text{Minimum division value}}{\text{Number of divisions}}$$

$$= \frac{65 \text{ mph} - 60 \text{ mph}}{5} = 1 \text{ mph}$$

Finally, we determine the value of the imaginary scale mark that is intersected by the pointer. In the case of Figure 9-5, the pointer crosses the imaginary mark corresponding to 64 mph.

In general, the greater the number of scale subdivisions you can imagine, the greater the accuracy of your interpolation effort.

Practice Problems

1. Interpolate the thermometer readings shown in Figure 9-6 for each of the following points on the scale:

A. _____ B. _____ C. _____ D. _____ E. _____

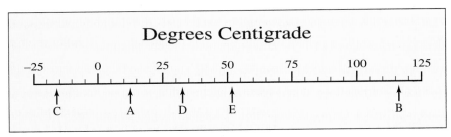

Figure 9-6. Interpolate these thermometer values.

Answers to Practice Problems

1. 12.5°, 116°, −16°, 32.5°, 51°

REDUCING PARALLAX ERRORS

The meter pointer has to be in front of the meter scale in order to allow movement. That is, there has to be a certain physical distance or clearance between the pointer and the printed scale marks. This required clearance leads to a possible source of error called **parallax error.** If a technician views the meter pointer from different angles, different values will be interpreted. Figure 9-7 illustrates how parallax errors occur.

KEY POINTS

Parallax errors can occur because the meter pointer and the printed scale are in two different planes.

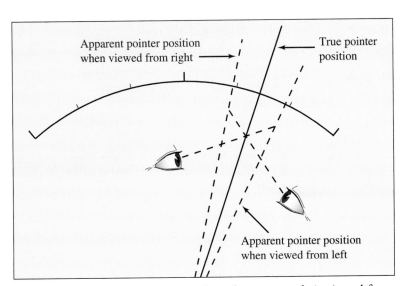

Figure 9-7. Parallax error occurs when the meter scale is viewed from different angles.

To reduce parallax errors, many meter scales have a mirrored band positioned next to the printed scales. If the technician is positioned such that the actual meter pointer and the image of the pointer reflected in the mirrored band are directly in line, then there will be no significant parallax error. It will also help to increase reading accuracy if you close one eye while aligning the pointer with its reflection and while reading the measured value.

Meter Movement Damping

Many meter movements have a **damping** mechanism, which minimizes overshoot and generally stabilizes pointer movement. The damping mechanism can be either mechanical or electrical (electromagnetic). Figure 9-8 shows a mechanical method of meter damping that relies on a miniature air piston. The movable portion of the air piston is fastened to the pivot shaft. Pointer movement is retarded by the action of the piston. The piston works just like a shock absorber on your car. The piston can only change position as fast as air can escape or enter the back side of the piston. By restricting this air flow, we restrict the rate of piston movement and ultimately the rate of pointer movement. A damped meter movement does not respond to sudden voltage or current changes and generally appears to be a more stable and higher quality meter movement. Chapter 10 shows how meter damping can also be accomplished magnetically.

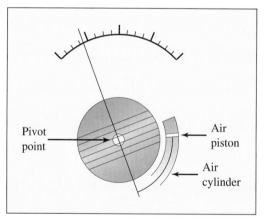

Figure 9-8. An air piston can provide meter movement damping.

Exercise Problems 9.1

1. What is another name for a moving-coil meter movement?
2. The iron vanes in a dual-vane meter movement (*attract, repel*) each other.
3. What type of meter movement is the most rugged?
4. Is the direction of pointer deflection in a moving-coil movement affected by the direction of current through the meter coil?
5. What is the weight of each of the major divisional marks in Figure 9-9?

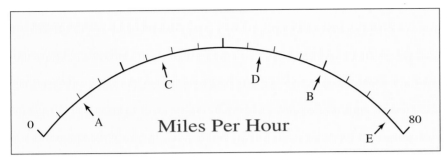

Figure 9-9. A typical speedometer scale.

6. What is the value of each of the minor divisional marks in Figure 9-9?
7. What values are indicated by each of the following points on the speedometer scale shown in Figure 9-9?
   A. _____ B. _____ C. _____ D. _____ E. _____

9.2 Digital Meters

Digital meters are now more popular for most applications than equivalent analog meters, although analog meters are preferred when monitoring fluctuating values. Regardless of the quantity being measured, all digital meters are characterized by the digital display. Figure 9-10 shows a sketch of a typical digital display. The display consists of a number of digit positions. Each position can generally display the numbers 0 to 9. Thus, in combination, the digital display can display any decimal number up to the limit imposed by the number of digits in the display.

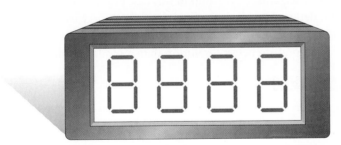

Figure 9-10. A typical digital display.

Many digital displays use **seven-segment displays** for each digit position in the overall digital display. Each segment emits light when it is activated by the controlling circuitry. Figure 9-11 shows how the simultaneous illumination of various segment combinations can produce any decimal digit.

Some digital displays only have the segments that correspond to the "b" and "c" segments in Figure 9-11. These devices can only display the digit one or be blank. This type of device is called a **half-digit**. In addition to the segments that form the display digit, most digital displays have another smaller segment that can be illuminated to represent a decimal point.

> **KEY POINTS**
>
> A digital display consists of a number of digits.

> **KEY POINTS**
>
> In a seven-segment display, each digit is typically formed by illuminating a unique combination of the total seven segments.

> **KEY POINTS**
>
> Sometimes a digit position is only capable of indicating a blank or a one. This type of digit is called a *half-digit*.

> **KEY POINTS**
>
> *Resolution* refers to the smallest input change that can be detected and displayed.

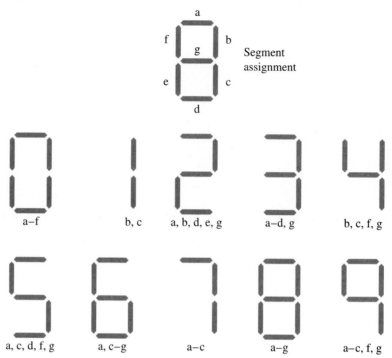

Figure 9-11. Each decimal digit is formed by simultaneous illumination of certain segments.

Resolution of a digital display basically refers to the smallest change in the measured quantity that can be detected and indicated. Consider the case of a two-digit display configured to indicate 0 to 99 seconds. Resolution in this case is one second. Although resolution is, to some extent, determined by the techniques and circuitry used inside the digital measuring device, for most practical purposes, a technician can consider that more digits in a display corresponds to better resolution than a similar display with fewer digits.

EXAMPLE SOLUTION

What is the resolution of a 3 1/2-digit electronic scale display configured to indicate 0 to 1.999 pounds?

EXAMPLE **SOLUTION**

The smallest detectable change is the value represented by a change in the right-most digit. In this case, each change in the right-most digit corresponds to a weight change of 1/1,000th pound.

Practice Problems

1. What is the resolution of a four-digit thermometer configured to display 00.00 to 99.99 degrees?

2. Repeat Question 1 if the thermometer is configured to indicate 000.0 to 999.9 degrees.

3. What is the resolution of a two-digit speedometer display capable of indicating 0 to 99 mph?

Answers to Practice Problems

1. 1/100th degree   **2.** 1/10th degree   **3.** 1 mph

Exercise Problems 9.2

1. If a seven-segment display has segments b, c, e, f, and g illuminated, describe the visible pattern.

2. What is the resolution of a 3 1/2-digit pressure meter that is configured to display pressures ranging from 00.00 to 19.99 pounds per square inch?

3. Sometimes a seven-segment indicator will have a smaller eighth segment that can be illuminated. What is the purpose of this added indicator?

4. What is the resolution of the digital scale display shown in Figure 9-12?

Figure 9-12.
A digital scale display.

9.3 Current Meters

Current meters can be either analog or digital and are used to measure the value of current in a circuit. Usable ranges for current meters include fractional microamps through thousands of amperes. The current meter may be a dedicated device or part of a multipurpose test equipment.

Methods of Connection

The purpose of a current meter is to measure the value of current flowing through the circuit at a particular point. In most cases, the circuit must be opened and the meter inserted across the open. Insertion of the meter completes the circuit so that normal circuit operation is restored. The current, however, now flows through the meter where it can be measured.

SERIES-INSERTION

Figure 9-13 shows the most common way to measure current in a circuit. Here, one end of a component (or wire) is lifted. The meter is then connected between the lifted component and the point in the circuit where the component was originally connected. In the example shown in Figure 9-13(b), the current through R_3 and R_4 is being measured.

When preparing to measure current, a technician should generally turn off power to the circuit before opening the circuit and inserting the meter. Once the meter is in place, power can be reapplied and the current measured.

● KEY POINTS

Both analog and digital current meters are connected in series with the component whose current is to be measured. This generally requires turning off the power to the circuit, opening the circuit, inserting the meter, and reapplying power.

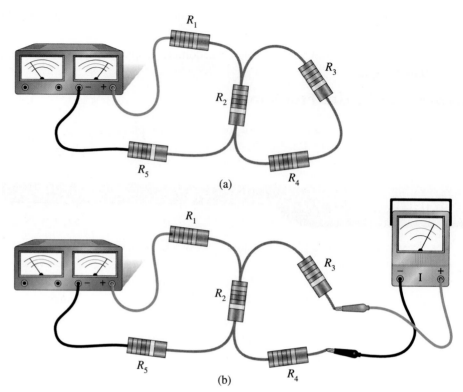

Figure 9-13. Measuring current with a standard current meter requires breaking the circuit.

CURRENT-MEASUREMENT JACK

The value of current in some circuits (some microwave oscillators for example) is a critical circuit quantity and must be monitored frequently by a technician. In cases like this, a jack may be permanently installed in the circuit. Under ordinary conditions, the jack is shorted and current flows through the circuit normally. A technician can use a current meter that has leads connected to a plug that mates to the installed current jack. When the current meter probe is inserted, the short is automatically removed from the jack and current is routed through the current meter. In either case, the circuit functions normally, but the technician can quickly monitor the current by simply plugging a current meter into the permanently installed jack. It is generally not necessary to remove power to the circuit when connecting the meter via a current test jack.

CURRENT PROBES

Current probes provide an alternative way to measure current that does not require the circuit to be opened. A current probe simply clamps around the conductor of the current being measured. In most cases, current probes are used for high values of currents (greater than 1 A).

There are two basic types of current probes. One is called a current transformer and can only be used to measure alternating currents (discussed in Chapter 11). Probes of this type are commonly used by residential and industrial electricians. The output of a current transformer probe is typically 1 mA for every ampere of measured current.

A second type of current probe is called a Hall-effect probe. A constant current is passed through a semiconductor material in the probe tip. This creates a certain voltage drop across the semiconductor. As the current being measured flows near the tip, its associated magnetic field (discussed in Chapter 10) alters the voltage drop across the semiconductor in the probe tip. The amount of voltage change is directly proportional to the strength of the magnetic field which, in turn, is directly proportional to the current being measured. The output from a Hall-effect probe is typically 1 mV for every ampere of measured current. Since the probe essentially responds to the strength of the magnetic field caused by the current being monitored, it can be used to measure either direct current or alternating current (discussed in Chapter 11). Figure 9-14 shows a Hall-effect sensor.

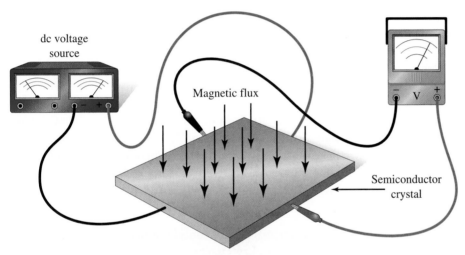

Figure 9-14. A Hall-effect sensor can be used to measure the magnetic field strength and, indirectly, the value of current in a current-carrying conductor.

Range Switch Operation

Most current meters have a range switch. This allows the same physical meter movement to be used for measuring different ranges of currents. The range values normally specify the current required for full-scale deflection. A particular current meter might, for example, have switch-selectable ranges of 1.0 mA, 10 mA, and 100 mA. Each of these numbers indicates the amount of current that is required to make the meter deflect to its full-scale value.

The meter movement itself takes a definite amount of current for full-scale deflection. This value of current is called the **full-scale current** (I_{FS}). This value depends on the physical characteristics of the coil, the permanent magnet (in the case of a d'Arsonval movement), the spiral spring, and other physical factors that cannot be altered once the meter has been manufactured. How then does a range switch allow us to obtain multiple full-scale values from the same movement? Figure 9-15 shows that solution of this problem is essentially an application for parallel circuits.

KEY POINTS

Each current meter has a rating, which indicates the amount of current required to cause full-scale deflection.

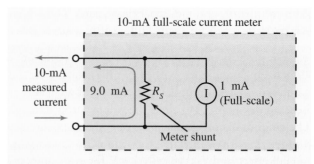

Figure 9-15. A meter shunt can increase the apparent full-scale current value for a current meter.

As shown in Figure 9-15, an internal resistor is connected in parallel (or shunt) with the meter movement. This resistor is called a **meter shunt.** In the case shown, the shunt allows 9.0 mA to flow through it while the meter movement passes the remaining 1.0 mA. Externally, the meter appears to have a full-scale value of 10 mA. In general, the smaller the shunt resistor relative to the meter resistance, the higher the effective full-scale current value of the combination. Figure 9-16 shows how a range switch can be used to select different values for the meter shunt. It is important to note that the range switch must be a shorting-type switch. This means that the movable wiper arm makes contact with the next position before breaking contact with the present one. It essentially shorts the two positions together as the switch is rotated between positions. This ensures that the current meter is never operated without a shunt as the switch is rotated. If contact to the shunts were lost as the switch was rotated, then the meter movement would have to handle 100% of the input current. The meter would likely be damaged.

Values for meter shunts are easily calculated. Just consider it to be a parallel circuit problem and use Ohm's and Kirchhoff's Laws to compute the required shunt value.

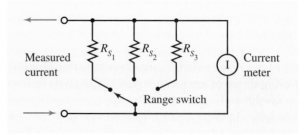

Figure 9-16. A range switch selects different values for the meter shunt.

Practice Problems

1. Calculate the value of shunt needed to extend the range of a 1-mA, 100-Ω meter movement to a full-scale value of 1.0 A.

2. What size shunt is needed to extend the range of a 50-µA meter movement to a full-scale value of 250 mA? The meter resistance is 2,000 Ω.

3. If you want to use a 10-mA, 50-Ω meter movement to measure a maximum current of 5 A, what value should you use for the shunt?

4. Refer to Figure 9-17. Determine the required values for the meter shunts.

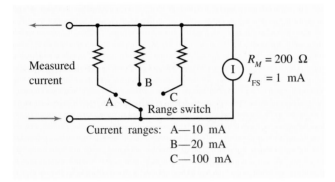

Figure 9-17. What size meter shunts are required?

Answers to Practice Problems

1. 0.1001 Ω   **2.** 0.40008 Ω

3. 0.1002 Ω   **4.** 22.222 Ω, 10.526 Ω, 2.020 Ω

Range Selection

It is important for a technician to choose the correct range when using a multirange current meter. Always start with the meter set on the highest range, or at least on a range that you are certain is higher than the current to be measured. By selecting a higher range, you avoid the risk of damaging the meter when it is inserted into the circuit. Once the meter has been connected, you can reduce the range to obtain an optimum reading. In the case of analog meters, reduce the range until the measured current deflects the pointer into a convenient part (preferably the upper half) of the scale. With a digital current meter, you should use the lowest range possible that does not cause the meter to give an overrange indication. Some digital current meters indicate an overrange condition by flashing the display; others display a unique pattern or message such as "----" or "OFL." Some digital current meters are autoranging. This type of meter automatically selects the proper range for optimum value interpretation.

Interpreting Current Meter Specifications

Technicians are frequently responsible for buying test equipment for companies. It is important to be able to interpret manufacturers' specifications.

ANALOG CURRENT METER SPECIFICATIONS

Full-scale current is probably the most important consideration. This is the value of current required to deflect the pointer to the right-most part of the scale. It is always provided in the manufacturer's specification sheet, but it is also generally obvious from the labeling of the scale.

There is always a trade-off between sensitivity and degree of circuit upset caused by insertion of the current meter. Ideally, a current meter should have an effective resistance of zero ohms, so that it will have no effect on the circuit when it is inserted. In practice, a current meter has some value of internal resistance. This is essentially the parallel combination of the meter and its shunt. To make a meter more sensitive requires many turns of fine wire in the meter coil. This causes the meter movement to have more resistance and appear less ideal.

KEY POINTS

Accuracy is a current meter specification of the largest measurement error that will occur under a given set of conditions. Accuracy is specified as a percentage of the full-scale current for analog meters.

Accuracy indicates the largest meter error that will occur under the specified operating conditions. It is specified as a percentage of full-scale current. So, for example, if a 1-mA current meter has a ±2% accuracy specification, then the indicated current could differ from the actual current by as much as

$$I_{ERROR} = \pm \text{Accuracy} \times I_{FS}$$
$$= \pm 0.02 \times 1 \text{ mA} = \pm 20 \text{ } \mu A$$

Since this same magnitude of error could occur anywhere on the scale, effective error is reduced when the pointer is deflected into the higher regions of the scale.

DIGITAL CURRENT METER SPECIFICATIONS

As with analog movements, the full-scale or maximum current that can be displayed is an important factor in meter selection. It is clearly determined by your intended application.

KEY POINTS

Accuracy of a digital current meter is expressed as a percentage of the displayed reading plus one or more digits.

Accuracy of a digital current meter is normally specified as a two-part value. First, a portion of the accuracy factor is given as a percentage of the reading. This means, for example, that if a current meter had a ±1% accuracy and displayed a value of 8.5 mA, the actual current value could be higher or lower by as much as

$$\text{Error} = \pm \text{Accuracy} \times \text{Reading}$$
$$= \pm 0.01 \times 8.5 \text{ mA} = \pm 85 \text{ } \mu A$$

Most manufacturers add a second factor to the accuracy specification in the form of a number of digits. For example, a particular meter may have an accuracy ±(1% + 2 digits). The digit suffix indicates how many counts the right-most digit may deviate from the correct value.

EXAMPLE SOLUTION

If a 3 1/2-digit current meter has an accuracy specification of ±(0.5% + 1 digit), what is the possible range of measured currents if the display indicates 100.0 mA?

EXAMPLE **SOLUTION**

The first part of the accuracy specification can cause errors as great as

$$\text{Error} = \pm \text{Accuracy} \times \text{Reading}$$
$$= \pm 0.005 \times 100 \text{ mA} = \pm 0.5 \text{ mA}$$

The right-most digit represents tenths of milliamps. The accuracy specification states that the measurement error can vary by ±1 digit or ±0.1 mA. The combined accuracy errors yield the following range of total error:

$$\text{Accuracy} = \pm(0.5 \text{ mA} + 0.1 \text{ mA}) = \pm 0.6 \text{ mA}$$

Resolution of a digital current meter describes how fine a measurement the meter can display. Resolution is closely related to the number of digits in the display. The smallest current change that can be displayed corresponds to the weight of the right-most digit on the display. Some manufacturers provide the resolution specification as a certain number of microamperes on a given scale (e.g., ±1 μA on the 10-mA range).

Exercise Problems 9.3

1. What is the resolution of a 4 1/2-digit pressure meter that has a display configured to indicate 0 to 19.999 pounds per square inch?

2. What is the resolution of a digital fuel gauge whose display indicates 0 to 19.9 gallons?

3. Current meters are generally connected in (*series, shunt*) with the component whose current is to be measured.

4. It is possible to measure current in a circuit without breaking the circuit by using a _____.

5. A Hall-effect current probe can be used to measure direct current. (True or False)

6. Calculate the value of meter shunt required to extend the range of a 50-μA, 2,000-Ω meter movement to a full-scale current of 200 mA.

7. What size shunt is needed to extend the range of a 1-mA, 150-Ω meter movement to a full-scale current of 2 A?

8. If the accuracy of a 1-mA analog meter movement is stated as 1.5%, what is the maximum error due to accuracy that can occur if the pointer is indicating 0.75 mA?

9. If a digital current meter has an accuracy of 0.25% + 2 digits, what is the maximum error when reading 1.255 mA on a 3 1/2-digit meter?

10. What is the internal resistance of an ideal current meter?

9.4 Voltmeters

Analog voltmeters are really analog current meters with a series resistance (multiplier) and a modified scale. Figure 9-18 shows the basic diagram of a voltmeter. Digital voltmeters (DVMs) are similar to digital current meters, but they are calibrated to indicate the value of applied voltage.

Method of Connection

Voltmeters are connected in parallel with the voltage to be measured. Figure 9-19 shows how to connect a voltmeter in a circuit. The meter in Figure 9-19 is connected to measure the voltage across R_3.

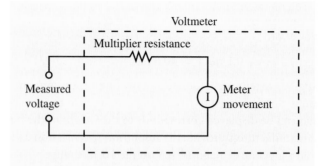

Figure 9-18. A basic voltmeter diagram.

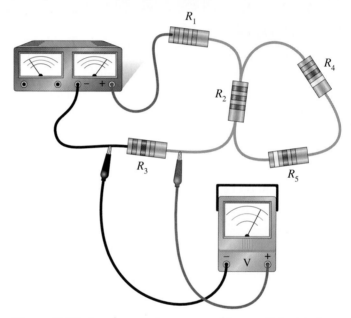

Figure 9-19. A voltmeter is connected in parallel with the voltage being measured.

It is not necessary to remove power from the circuit before connecting a voltmeter, but the technician must ensure that the meter is properly configured before connecting it to the circuit.

Range Switch Operation

A typical voltmeter has more than one voltage range. In the case of an analog meter, the range refers to the voltage required for full-scale deflection. Figure 9-20 shows an analog voltmeter diagram with a range switch.

You will recall that an ammeter range switch used a shorting-type switch. Voltmeters, by contrast, should have a nonshorting-type range switch. If a shorting type were used, then two multiplier resistances would be momentarily paralleled when changing ranges. This

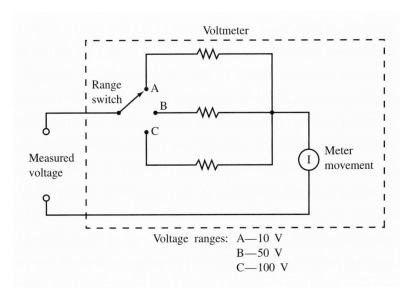

Figure 9-20. A voltmeter with a range switch.

would effectively put the voltmeter into a lower voltage range and might damage the meter.

A digital voltmeter generally specifies the voltage range in a similar, but slightly different way. Most meters use a half digit in the left-most position. This leads to the relationships between voltage range and maximum displayable voltages shown in Table 9-1.

The range switch for a digital voltmeter is shown in Figure 9-21 and is essentially a voltage divider circuit.

A technician should always select a range that is higher than the expected value of voltage to be measured. Once the meter has been connected, a range can be selected that provides a more optimum display. In the case of an analog meter, a range should be selected that causes the pointer to deflect to the upper part of the scale. If a digital voltmeter is being used, then the lowest range that does not produce an overrange condition should be chosen.

The value for the multiplier resistor in an analog voltmeter is easily calculated. View the problem as a simple series circuit, and apply Ohm's and Kirchhoff's Laws.

VOLTAGE RANGE	MAXIMUM DISPLAYABLE VOLTAGE	
	3 1/2-DIGIT METER	4 1/2-DIGIT METER
2 V	1.999 V	1.9999 V
20 V	19.99 V	19.999 V
200 V	199.9 V	199.99 V

Table 9-1. Relationship Between Voltage Range and Maximum Displayable Voltage

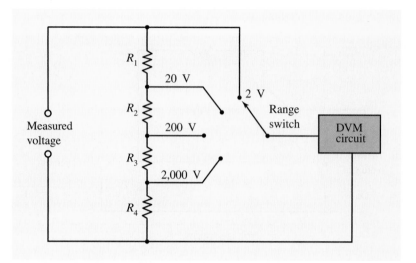

Figure 9-21. The range switch for a digital voltmeter is essentially a voltage divider circuit.

Practice Problems

1. Calculate the required multiplier to convert a 1-mA, 100-Ω meter movement into a 150-V full-scale voltmeter.

2. Refer to Figure 9-22. Calculate the values for multiplier resistors R_1, R_2, and R_3.

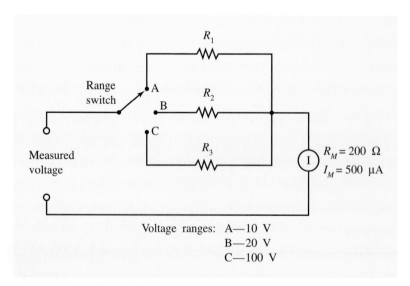

Figure 9-22. Calculate the multiplier resistances.

Answers to Practice Problems

1. 149.9 kΩ 2. $R_1 = 19.8$ kΩ, $R_2 = 39.8$ kΩ, $R_3 = 199.8$ kΩ

Interpreting Voltmeter Specifications

A technician must be able to interpret the manufacturer's specifications in order to select a voltmeter that will provide adequate performance in a particular application. We will consider analog and digital meters separately.

ANALOG VOLTMETER SPECIFICATIONS

The full-scale voltage range is one important consideration when selecting a voltmeter. The meter must have a range that is high enough to accommodate the expected voltages, but it must also have a range that provides adequate accuracy (i.e., permits deflection into the upper regions of the scale).

An ideal voltmeter has infinite internal resistance. That is, it draws no current from the circuit being tested. Of course, a practical voltmeter does draw some current in order to deflect the pointer. In general, the more sensitive the meter movement, the higher the internal resistance of the voltmeter. **Meter sensitivity** is inversely proportional to full-scale current. It is also called the **ohms-per-volt rating** of the meter. The sensitivity rating can be computed with Equation 9-2.

> **KEY POINTS**
>
> To prevent voltmeter loading, an ideal voltmeter would have infinite internal resistance.

$$\text{Sensitivity} = \frac{1}{I_{\text{FS}}} \qquad (9\text{-}2)$$

where I_{FS} is the full-scale meter current. Sensitivity is measured in ohms-per-volt units.

> **KEY POINTS**
>
> Meter sensitivity is measured in ohms per volt and is the reciprocal of full-scale current.

EXAMPLE SOLUTION

Calculate the sensitivity of a voltmeter that uses a 50-μA meter movement.

EXAMPLE **SOLUTION**

We apply Equation 9-2 as follows:

$$\text{Sensitivity} = \frac{1}{I_{\text{FS}}}$$

$$= \frac{1}{50 \ \mu A} = 20{,}000 \ \Omega/\text{V}$$

Practice Problems

1. Calculate the sensitivity of a voltmeter using a 1-mA movement.
2. What is the sensitivity of a voltmeter that has a 100-μA meter movement?

Answers to Practice Problems

1. $1{,}000 \ \Omega/\text{V}$ 2. $10{,}000 \ \Omega/\text{V}$

In Chapter 6 we discussed voltmeter loading. We can use the sensitivity rating of a voltmeter to determine the amount of voltmeter loading that will occur. More specifically, we can determine the total internal resistance of a voltmeter by applying Equation 9-3.

$$\boxed{\text{Internal resistance} = \text{Sensitivity} \times \text{Voltage range}} \quad \textbf{(9-3)}$$

Once we know the internal resistance of the voltmeter, we can view it as a resistor and determine its effects on the circuit under test (i.e., loading effects) as described in previous chapters.

EXAMPLE SOLUTION

What is the total resistance of a voltmeter that has a 20,000 Ω/V sensitivity and is operated on the 50-V range?

EXAMPLE **SOLUTION**

$$\text{Internal resistance} = \text{Sensitivity} \times \text{Voltage range}$$
$$= 20{,}000 \ \Omega/\text{V} \times 50 \ \text{V} = 1.0 \ \text{M}\Omega$$

EXAMPLE SOLUTION

Calculate the effective resistance of a voltmeter that uses a 10-μA meter movement and is set on the 20-V full-scale range.

EXAMPLE **SOLUTION**

First, we determine the sensitivity rating with Equation 9-2.

$$\text{Sensitivity} = \frac{1}{I_{\text{FS}}}$$

$$= \frac{1}{10 \ \mu A} = 100{,}000 \ \Omega/\text{V}$$

Next, we use Equation 9-3 to compute the total internal resistance.

$$\text{Internal resistance} = \text{Sensitivity} \times \text{Voltage range}$$
$$= 100{,}000 \ \Omega/\text{V} \times 20 \ \text{V} = 2.0 \ \text{M}\Omega$$

Analog voltmeters have an accuracy rating that is interpreted the same as the accuracy rating discussed for current meters. Thus, for example, if a certain voltmeter had a ±2% accuracy on the 100-V range, then the actual voltage could vary by as much as ±2 V from the indicated voltage. Further, this ±2-V error could be anywhere on the scale. This means that the effective error is lessened when readings are taken from the upper part of the scale.

Digital Voltmeter Specifications

The accuracy and resolution ratings of a digital voltmeter are interpreted the same way as similar current meter specifications. They are not repeated here.

The internal resistance (generally called impedance) of a digital voltmeter is important for the same reasons discussed for analog meters, but it varies greatly from analog meters. In the case of an analog meter, the internal resistance of the voltmeter was determined by the amount of current required for the meter movement and the selected voltage range. Digital meters, by contrast, are electronic. There is no direct relationship between internal resistance and the selected voltage range. The resistance is given in the manufacturer's specifications. Typical values are in excess of 10 MΩ on all ranges. This high resistance means that most digital voltmeters will cause less voltmeter loading than many analog meters.

1. What value should be used for the multiplier resistance if a 100-µA, 1,200-Ω meter movement is to be used in a voltmeter with a 0- to 50-V scale?

2. Calculate the required multiplier resistance that will make a 20-V full-scale voltmeter out of a 1-mA, 50-Ω meter movement.

3. A 50-µA, 750-Ω meter movement is to be used to make a multirange voltmeter. Calculate the required multipliers to give the voltmeter ranges of 5 V, 10 V, and 50 V. Draw a schematic of the final circuit.

4. Refer to the previous question. Should the range switch be a shorting or non-shorting type?

5. What is the unit of measurement for voltmeter sensitivity?

6. What is the sensitivity of a voltmeter that uses a 200-µA meter movement?

7. A 1-mA movement has a higher sensitivity than a 10-µA movement. (True or False)

8. Compute the internal resistance of a voltmeter that uses a 200-µA movement and is on the 25-V range.

9. What is the internal resistance of a voltmeter that has a 120-µA meter movement and is used on the 150-V range?

10. Digital voltmeters generally have a higher internal resistance than a standard analog voltmeter. (True or False)

9.5 Ohmmeters

Ohmmeters have an internal battery that supplies power to the component or circuit whose resistance is being measured. The meter in the ohmmeter responds to the current that flows through the overall circuit. Figure 9-23 shows a basic ohmmeter circuit. This is clearly a series circuit, which means it can be analyzed with series circuit rules and equations. Some ohmmeters use a form of a parallel circuit and can be analyzed with parallel circuit rules. For purposes of our present discussion, let us narrow our discussion to the circuit shown in Figure 9-23.

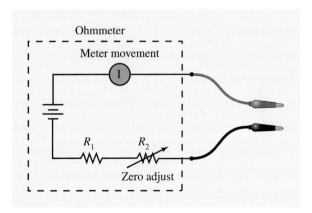

Figure 9-23. A basic ohmmeter circuit.

The variable resistor (R_2) in Figure 9-23 is called the zero adjustment. It is used to calibrate the ohmmeter and to compensate for changes in battery voltage as the battery ages. Proper operation of this control is discussed in the section on ohmeter calibration.

The scale for an ohmmeter like the one shown in Figure 9-23 has two interesting characteristics. First, it is labeled in the reverse direction. That is, zero ohms is on the right side of the scale and infinite ohms is on the extreme left. This is called a **backoff scale**. Second, the meter scale is nonlinear; the divisions are not equally spaced. The numbers on the right portion of the scale are widely spaced, while the numbers on the left part of the scale are compressed or squeezed together. Figure 9-24 illustrates the cause of the nonlinear backoff ohmmeter scale.

In Figure 9-24(a), the meter leads are shorted and the meter is zeroed. This causes full-scale current through the meter (100 μA in this case) and corresponds to zero on the meter scale. The total internal resistance of the meter, for this example, is 15 kΩ. In Figure 9-24(b), an external 15-kΩ resistor is connected to the meter. This reduces meter current to one-half of its full-scale value, which causes half-scale deflection. In Figure 9-24(c), we add another 15 kΩ. If the scale were linear, this would cause a 25% full-scale deflection. As shown in the figure, however, the total internal and external resistance is 45 kΩ, which produces a one-

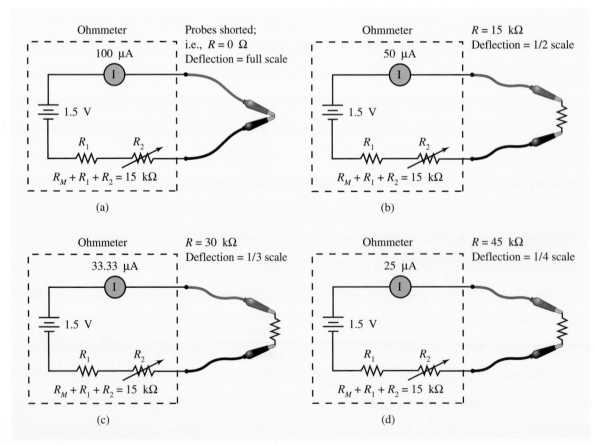

Figure 9-24. Development of a backoff ohmmeter scale.

third scale deflection. Finally, in Figure 9-24(d), we add another 15 kΩ to bring the deflection down to one-fourth scale. Note that the first 15 kΩ caused a 50% current or scale deflection change. The last 15 kΩ increase only produced a change of about 8.33%. We clearly have a nonlinear relationship between resistance and scale position. If we left the meter leads separated (i.e., infinite resistance), there would be no meter current and no deflection. Thus, the left-most point on the scale would represent infinite ohms.

Figure 9-25 shows the scale of a typical analog ohmmeter. The uppermost scale is for resistance measurements. Note the reverse labeling and the nonlinear divisional markings. Also note the infinity symbol (∞) at the extreme left end of the scale.

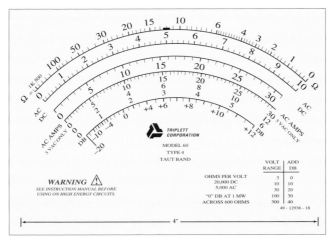

Figure 9-25. A representative ohmmeter scale. (Courtesy of Triplett Corporation)

Methods of Connection

Voltmeters and current meters draw power from the circuit being tested, and so require "live" circuit conditions for proper operation. As discussed previously, ohmmeters have their own internal voltage source. For proper operation, power to the circuit under test must be disconnected. Failure to remove circuit power before connecting an ohmmeter will produce erroneous readings as a minimum and may even damage the ohmmeter.

Another factor to be considered when using an ohmmeter is that the ohmmeter responds to the total current flow caused by the internal voltage source. This leads to errors if a component is measured while it is connected to other components. Figure 9-26 illustrates this problem.

In Figure 9-26, the intent is to measure the resistance of R_1. However, since R_2 and R_3 provide an alternate, or sneak, current path, the total ohmmeter current will be higher than expected. This means that the indicated resistance will be less than expected. In effect, the equivalent resistance of the series-parallel circuit consisting of R_1, R_2, and R_3 is being measured.

A technician must be careful not to touch the ohmmeter probes during a resistance test. The resistance of the human body can cause another sneak path. This is particularly important for high-resistance measurements.

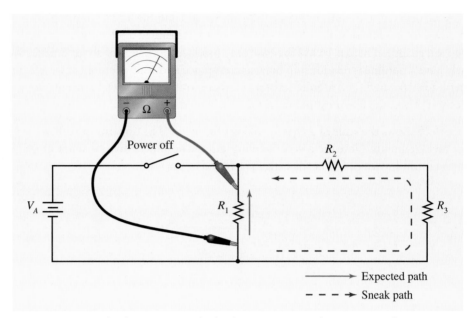

Figure 9-26. Multiple current paths lead to erroneous ohmmeter readings.

REMOVE COMPONENT

The most obvious way to eliminate problems caused by sneak circuit paths is to remove the component being tested. In the case of the example shown in Figure 9-26, we could remove R_1 from the circuit and connect it directly to the ohmmeter leads. This would provide an accurate measurement of the resistance of R_1. The disadvantage of this method is that it generally requires desoldering of a component from a printed circuit board. Since it is so easy to damage an expensive circuit board, desoldering should be kept to a minimum.

DISCONNECT ONE END OF COMPONENT

Figure 9-27 shows an alternative to complete desoldering of a component that still provides an accurate resistance measurement. Here, one end of the component is lifted. In the case of a two-lead component, this method reduces the desoldering exercise by 50%.

In the case of components with three or more leads (e.g., transistors), this method can still help the technician if properly applied. It can be used if either of the following conditions is satisfied:

- all but one component lead is disconnected from the circuit, or
- at least one of the two leads being measured is disconnected from the circuit.

IN-CIRCUIT MEASUREMENTS

Because desoldering of components should be minimized—even if you are very skilled at soldering and desoldering components—many technicians prefer to make in-circuit measurements. As discussed previously, when a component is measured in the circuit, the ohmmeter also responds to any sneak current paths. Now, if the technician can carefully evaluate the circuit theoretically, then in-circuit measurements can be very useful. Consider the case shown in Figure 9-28 as an example.

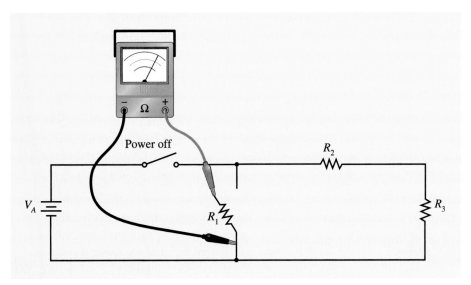

Figure 9-27. Disconnecting one end of a two-lead component effectively isolates it from the rest of the circuit.

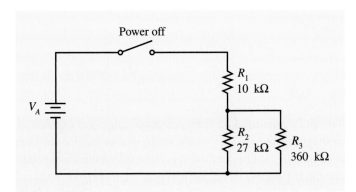

Figure 9-28. In-circuit measurements require careful interpretation.

If the technician wants to measure the resistance of resistor R_1, then there is no need to desolder it, since one end is essentially disconnected due to the open power switch. If the condition of R_2 is to be tested, then it can be done in the circuit also. In this case, R_1 will have no effect on the measurement of R_2, since R_1 is in series with the open switch. Second, resistor R_3 will have a very small effect on the measurement of R_2, because it is in parallel and has more than ten times the resistance of R_2. If the technician is trying to determine the condition of R_3, it can be done without desoldering, but it requires more thought. If R_3 (or R_2) were shorted, this condition could be detected without desoldering; although distinguishing between R_2 and R_3 would require removal of one of them. The "normal" value of an in-circuit resistance measurement of R_3 can be quickly determined by calculating the total resistance of R_2 and R_3 using the parallel resistance equation. If the measured value is correct, then no desoldering is needed. An incorrect reading would indicate a malfunction, but it would not locate the exact defect.

Since most resistance checks made for troubleshooting purposes need not be exact (i.e., we are usually interested in relative values), in-circuit measurements can provide a technician with a powerful tool.

Range Switch Operation

Range switch operation on an ohmmeter is more complex than voltmeter or current meter range switches. In general, the effective value of the internal voltage source is increased on higher resistance ranges, and meter shunts are inserted to obtain lower resistance ranges. All of this switching is generally accomplished by a multiple-section rotary switch.

In the case of analog ohmmeters, the various range positions are normally labeled as $R \times 1$, $R \times 10$, $R \times 100$, $R \times 1k$, and so on. Here, the R represents the value indicated by the pointer. This indicated value must be multiplied by the scale factor to obtain the actual resistance value. For example, if the pointer indicated 2.7, and the meter was on the $R \times 100$ range, then the actual resistance being measured would be 2.7×100 or 2,700 Ω.

EXAMPLE SOLUTION

What resistance is being measured if the pointer of the ohmmeter indicates 39 and the range switch is in the $R \times 10$ position?

EXAMPLE **SOLUTION**

In this case, R is 39, so the actual resistance is 39×10 or 390 Ω.

If a digital ohmmeter is autoranging, then the value of resistance being measured is automatically displayed on the readouts; the decimal point is positioned automatically. In the case where the ohmmeter range is selected manually, the range label normally indicates the highest value that can be measured on that particular range. For example, a 3 1/2-digit voltmeter with a range labeled as "2K" would mean that resistance values as high as 1.999 kΩ could be measured. The decimal is positioned automatically, but the technician must consider the range position when interpreting the displayed value.

EXAMPLE SOLUTION

If the 3 1/2-digit display of a certain digital ohmmeter indicates 0.420 on the 2K range, what would the meter display if the range switch were changed to the 200K position?

EXAMPLE **SOLUTION**

In essence, the display is moved two places to the right. The resulting display would be 0.004.

Practice Problems

1. Describe the display indication on a 4 1/2-digit ohmmeter if it is measuring a 3.9-kΩ resistor on the 20K range.
2. What value is being measured by a 3-1/2 digit ohmmeter that indicates 0.225 on the 2K range?

3. What value is being measured by a 3 1/2-digit ohmmeter that indicates 022.5 on the 200K range?

Answers to Practice Problems

1. 3.900   **2.** 225 Ω   **3.** 22,500 Ω

Range Selection

Although no damage to the ohmmeter will occur by having it on the wrong range, readings can be more accurate on some ranges. In general, a technician should try to select a range that causes a meter deflection in the right half of the scale. In this region, the scale is more expanded and much easier to interpret accurately.

Most digital voltmeters are autoranging, thus optimum range selection is accomplished automatically. In the case of manually selected meters, the lowest range that does not cause an overrange error should be chosen.

Interpreting Ohmmeter Specifications

Accuracy, in the case of an analog ohmmeter, is often specified as a percentage of arc or a percentage of scale length. For example, a certain meter has an accuracy of ±1% of arc. Suppose the scale on this particular meter swings a 100° arc. The specification, then, is saying that the actual pointer position will be within 1° of its theoretically correct position. The accuracy of a backoff ohmmeter cannot be expressed as a percentage of full-scale, since full-scale is zero ohms!

Resolution and accuracy specifications for a digital ohmmeter are interpreted the same as the corresponding current meter and voltmeter specifications. Minimum and maximum resistance ranges are also important considerations when selecting a digital ohmmeter.

Calibrating an Ohmmeter

The calibration of an analog ohmmeter must be checked frequently by a technician. It should always be checked before the first of a series of measurements, and it should be checked if the range is changed.

Calibration requires the technician to check the two extreme points on the ohmmeter scale: zero and infinity. The infinity adjustment takes the form of a small screw near the pivot point of the meter movement. The leads of the ohmmeter are separated (i.e., infinite resistance), and the calibration screw is adjusted until the pointer is aligned with the infinity (∞) symbol. On a multipurpose meter, this position normally coincides with zero on the nonresistance scales of the meter. Once the infinity adjustment has been made, then the leads are shorted together and the meter is **zeroed.** Zeroing the meter means adjusting the zero (sometimes labeled "ohms") adjustment until the pointer indicates zero on the ohms scale. Some analog meters have internal electronic circuitry to make them more ideal. In this case, there may be two accessible controls for the zero and infinity adjustments. These two controls often interact and may require repeated adjustments.

1. Where does zero ohms appear on a backoff ohmmeter scale?
2. What portion of a nonlinear, backoff ohmmeter scale can be read most accurately?
3. Why must power be removed from a circuit before an ohmmeter is connected?
4. Is it possible to verify the condition of a resistor while it is still connected to other components? Defend your answer.
5. Why should a technician try to minimize desoldering of components?
6. What value is being measured if a 4 1/2-digit ohmmeter is on the 200-Ω scale and displays a value of 102.99?
7. What is the largest resistance value that can be displayed on a 6 1/2-digit ohmmeter on the 20K range?

KEY POINTS

A multimeter is a combination measuring device that normally measures current, voltage, and resistance.

KEY POINTS

The analog multimeter is called a VOM.

9.6 Multimeters

A multimeter is one of the technician's favorite tools. It is a single test equipment item that performs a multitude of measurements. Figure 9-29 shows a representative analog multimeter.

This meter is intended to measure volts, ohms, and milliamperes and is frequently called a VOM. Meters of this type normally have a function switch and a range switch, although some meters have the two functions combined into a single switch. The function switch establishes the basic mode of operation. Typical options include dc voltage, ac voltage, dc

Figure 9-29. A representative analog multimeter. (Photo courtesy of Triplett Corporation)

current, and resistance. The range switch sets the meter range for the selected mode. Some meters have multiple jacks. In this case, the technician must move the test leads to special jacks for certain operations. One common use of special jacks is for very high or very low current measurements.

Analog VOMs have several sets of scales. They are always labeled and often color coded to help the technician identify the correct scale to use for a particular measurement.

Figure 9-30 shows a representative digital multimeter (DMM). Digital multimeters nearly always provide current, voltage, and resistance measurements, but many provide a spectrum of other tests as well. Some of the additional measurements include frequency, time delays, temperature, semiconductor tests, and others. Some even include built-in math functions that automatically determine such things as minimum or maximum values, average values, comparison to preset limits, and others.

KEY POINTS

The digital multimeter is called a DMM.

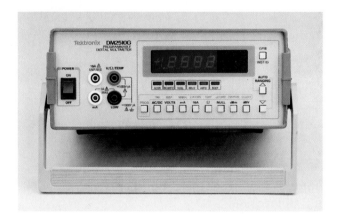

Figure 9-30. A digital multimeter is a versatile tool for electronic technicians. (Courtesy of Tektronix, Inc.)

The operation of a digital multimeter can range from trivial to complex. At the trivial end of the spectrum, a single switch is used to choose the desired quantity to be measured. All other settings (e.g., range selection) are automatic. The complexity of multimeters extends to include devices that have an internal computer that can be programmed to provide customized operations. Although basic operation of these more complex instruments is generally obvious to an experienced technician, full use of all capabilities normally requires studying the user's manual that was written by a technician at the manufacturing company.

Exercise Problems 9.6

1. What is the primary purpose of the mode selection switch on a VOM?
2. If you wanted a multimeter that also measured frequency, would you probably buy a VOM or a DMM?
3. If a certain dc voltage were unstable, describe why a technician might prefer to monitor it with a VOM instead of a DMM.

9.7 Oscilloscopes

An oscilloscope (or simply scope) is probably the most versatile tool used by a technician. Its real power is more obvious when used with ac circuits discussed in subsequent chapters, but it is also widely used for dc measurements, especially during troubleshooting. The following discussion addresses only those issues that are relevant to the measurement of dc circuit values. A scope is most directly suited to the measurement of voltages, but it can indirectly measure current.

Figure 9-31 shows a representative oscilloscope. The number of controls alone is indicative of its versatility. We shall discuss the basic function of several typical controls in the following paragraphs.

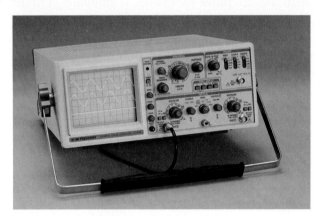

Figure 9-31. A representative oscilloscope. (Courtesy of B+K Precision)

Functional Overview

To understand the operation and application of an oscilloscope, it is helpful to understand the purpose and functional operation of the major sections of the scope. Once we understand how the individual sections operate, we can make more sense of the overall operation.

THE OSCILLOSCOPE SCREEN

Figure 9-32 illustrates one of the most important parts of a scope—the display, or screen as it is called. The scope display is essentially a graph. The horizontal axis or scale generally represents time (e.g., seconds, μs, ns, and so on). The vertical scale normally represents voltage (e.g., V, mV, and so on).

As can be seen in Figure 9-32, both the vertical axis (also called the y axis) and the horizontal axis (also called the x axis) are divided into equally spaced divisions. Each division has a certain value depending on the setting of the scope controls. Newer scopes generally have the value of each division displayed along one edge of the screen. The screen divisions, or **graticule,** are screened or etched onto glass on some scopes. They are visible whether the scope is on or not. Other scopes change the graticule marks as needed. In either case, the screen is actually the physical end of a special electronic device called a cathode-ray tube (**CRT**). The CRT causes an illuminated graph to appear on the screen graticule.

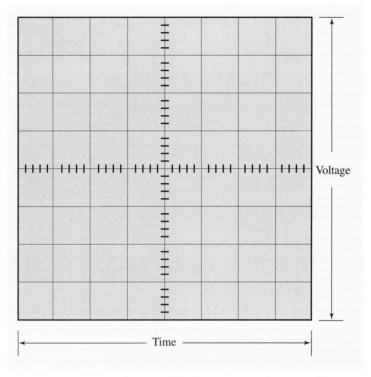

Figure 9-32. An oscilloscope display is essentially a graph.

CATHODE-RAY TUBE FUNDAMENTALS

Figure 9-33 shows the basic construction of a cathode-ray tube. The purpose of the electron-gun assembly is to generate (and maintain) a plentiful supply of free electrons that are ultimately formed into a narrow, highly focused beam. The electrons are freed from their parent atoms at the cathode of the CRT. The cathode is in close proximity to the heater, which gets red hot and imparts thermal energy to the electrons. Once the electrons are free, they are attracted toward the face of the CRT by the presence of very high (thousands of volts) positive voltages near the front of the CRT. The voltage on the control grid is negative and tends to keep the electrons from traveling down the tube. The actual number of electrons that travel down the tube at any given time determines the intensity or brightness of the scope trace and is controlled by the intensity control.

Once the electrons overcome the control grid value, they are accelerated by the potential on the accelerating grid. When the electrons enter the region of the focusing grid, they are squeezed (by electrostatic fields) into a very narrow beam. The beam then continues past the deflection plates and eventually strikes the face of the CRT. The face of the CRT is coated with a phosphorous material that emits light when it is excited by the impacting electrons. The focused beam would cause a tiny spot of light in the center of the screen if it weren't for the deflection plates.

The deflection plates consist of two sets of metal plates. One set, called the vertical deflection plates, is located above and below the focused electron beam. The other set is called the horizontal deflection plates. They are positioned on the left and right sides of the

● KEY POINTS

A CRT has an electron-beam gun that directs high-energy electrons toward the phosphor-coated screen of the tube.

● KEY POINTS

When the electrons strike the screen, the screen emits light.

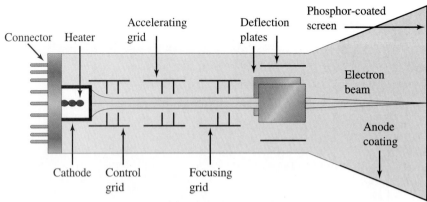

Figure 9-33. A cutaway view of a cathode-ray tube.

beam. Figure 9-34 shows how applying different voltage polarities to the deflection plates can cause the beam to strike any portion of the screen. Recall that a positive potential tends to attract the negative electron beam, whereas a negative potential will repel it. At the time the electrons are passing the deflection plates, they are traveling too fast to deviate from their course enough to contact a positive deflection plate. Their path, however, is bent.

After the electrons strike the face of the CRT and lose their energy, they are collected by a positive coating on the inside of the tube near the screen called the anode or aquadag. They then travel through wires to the power supply and ultimately complete their path back to the cathode.

The CRT construction shown in Figure 9-33 is very fundamental, but it illustrates some important concepts. The display devices used in newer digital oscilloscopes are more complex, but, for our purposes, can be visualized as being similar to Figure 9-33.

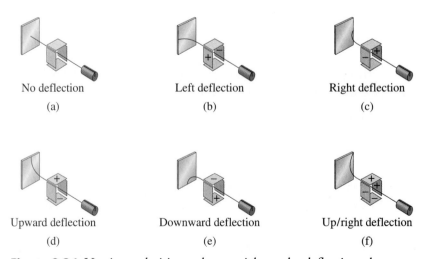

Figure 9-34. Varying polarities and potentials on the deflection plates causes the electron beam in a cathode-ray tube to be deflected.

Now, let's briefly examine the purpose of each control on a general-purpose oscilloscope.

Vertical Channel Controls

Several controls and jacks are dedicated to affecting the behavior of the electron beam in the vertical direction.

INPUT JACK

The voltage to be measured is connected to the vertical input jack. A positive voltage at this point will ultimately cause the beam to be deflected upward on the face of the CRT.

ac/dc SWITCH

This switch must be in the dc position if dc voltages are to be measured. If the switch is accidentally placed in the ac position when trying to measure dc voltages, all measured points will read zero.

VERTICAL POSITION

The vertical position control is a potentiometer that sets the average voltage on the vertical deflection plates. This control positions the beam vertically for optimum viewing.

VERTICAL SENSITIVITY

The vertical sensitivity (also called vertical gain) control determines how much beam deflection will occur for a given voltage at the input jack. It is calibrated in terms of volts per division where a division refers to the large divisions (or squares) on the face of the CRT. If, for example, the vertical sensitivity were set to 2 V per division, then a −5 V potential at the input jack would cause the beam to deflect downward by 2.5 divisions.

In most cases, we observe the amount of deflection and multiply by the sensitivity setting to determine the value of an unknown voltage. This is expressed by Equation 9-4.

$$\text{Voltage} = \text{Number of divisions of deflection} \times \text{sensitivity} \quad (9\text{-}4)$$

EXAMPLE SOLUTION

If the sensitivity control is set to 10 mV per division and the input voltage causes the beam to deflect upward by 3.6 divisions, what is the value of the measured voltage?

EXAMPLE **SOLUTION**

Since the beam deflection was upward, we know the voltage being measured is positive. We calculate its value with Equation 9-4 as follows:

3.6 divisions × 10 mV per division = 36 mV

Practice Problems

1.   Complete Table 9-2.

NUMBER OF DIVISIONS OF DEFLECTION	SENSITIVITY SETTING	VALUE OF MEASURED VOLTAGE
2.8	100 mV per division	
5	1.0 V per division	
8.3	20 mV per division	
	0.5 V per division	6 V
	2 V per division	12.5 V

Table 9-2. Table for Problem 1.

Answers to Practice Problems

1.

NUMBER OF DIVISIONS OF DEFLECTION	SENSITIVITY SETTING	VALUE OF MEASURED VOLTAGE
2.8	100 mV per division	280 mV
5	1.0 V per division	5 V
8.3	20 mV per division	166 mV
12	0.5 V per division	6 V
6.25	2 V per division	12.5 V

Table 9-3. Completed Table

Horizontal Channel Controls

The horizontal controls are of minimal importance with reference to dc measurements. They are, however, critical with regard to ac measurements. We will discuss them in greater detail in a later chapter. For our present purposes, we need to understand that the primary purpose of the horizontal controls is to adjust the deflection (sweep) of the electron beam in the horizontal direction. Some controls determine the rate of deflection; others determine the absolute position. In the case of dc measurements, there is tremendous flexibility in the setting of these controls.

SWEEP SOURCE

The sweep source control is a switch that chooses the origin of the voltage that is to be used to deflect the electron beam horizontally. For dc measurements, it is normally set to the internal position.

SWEEP RATE

When the sweep source is set to internal, the sweep rate control determines how fast the electron beam moves across the screen horizontally. For dc measurements, we normally adjust this control so that the beam moves fast enough to form a horizontal line. That is, the persistency of the screen phosphor coupled with the persistency of the human eye, cause a repetitive, fast-moving point of light to look like a solid line.

HORIZONTAL POSITION

The horizontal position control is used to set the average voltage on the horizontal deflection plates. It is used to locate the painted graph left or right as desired.

Triggering Controls

The triggering controls work together to control the relative timing of the vertical and horizontal deflection systems. For dc measurements, the timing is relatively unimportant, since dc voltages do not vary rapidly with time. We will discuss these controls in greater depth in a later chapter as they apply to the measurement of alternating voltages.

TRIGGER SOURCE

This control determines the origin of the voltage that is used for timing of the horizontal deflection of the electron beam. It is generally set to internal or automatic for dc measurements, but it is not critical.

TRIGGER SLOPE

This has virtually no effect on dc measurements. It will be discussed in a later chapter with regard to the measurement of alternating voltages.

TRIGGER LEVEL

The trigger level control sets the amount of voltage that must appear at the input jack before the horizontal sweep is allowed to start. For purposes of dc measurements, it can be set to any position that allows a trace to be seen.

Display Controls

Display controls are used to set the visible characteristics of the trace for optimum viewing. These controls are not discussed in detail, however, since newer oscilloscopes have few of these controls:

- Intensity/brightness
- Focus

> **KEY POINTS**
>
> The triggering controls determine the relative timing of the vertical and horizontal deflection.

- Astigmatism
- Scale illumination

Measuring dc Circuit Values

The dc voltage measurements can be performed with an oscilloscope by applying the following procedure:

1. Connect the input jack to ground (sets the zero-volt reference).
2. Adjust the scope controls to obtain steady trace (horizontal line).
3. Use the vertical position to set the trace to the bottom graticule line (for positive voltages), the top graticule line (for negative voltages), or the center graticule line (for mixed polarity or unknown measurements).
4. Connect the voltage to be measured to the input jack.
5. Adjust the vertical sensitivity until the trace deflects as much as possible without disappearing from the screen.
6. Count the number of divisions of deflection from the beam's initial or reference position.
7. Multiply the deflection by the sensitivity setting to obtain the value of measured voltage.

The scope can be used to measure current indirectly. For this, a technician will break the circuit and insert a very small resistance (e.g., 1 Ω). The scope is then used to measure the voltage across this current-sensing resistor. If the sensing resistor is 1 Ω and the scope is set for 10 mV per division, then each vertical division corresponds to 10 mA of measured current ($I = V/R$).

Exercise Problems 9.7

1. The voltage to be measured is connected to the _____ of an oscilloscope.
2. The sweep controls affect the (*vertical, horizontal*) deflection of the beam.
3. The triggering controls affect the relative _____ of the vertical and horizontal deflection.
4. When preparing to measure positive dc voltages, why would the technician position the trace at the bottom line of the graticule?

Chapter Summary

- Analog meter movements are generally one of two types: moving-coil or iron-vane. A moving-coil movement—also called a d'Arsonval movement—relies on the interaction of the magnetic field of a permanent magnet and the electromagnetic field of the movable coil. The interaction of fields causes the coil to rotate and move the pointer across the scale of the meter. This motion is opposed and restricted by the action of a spiral spring. A moving-coil meter is polarity sensitive and must be connected such that upscale deflection occurs.

- There are two types of iron-vane movements. A single-vane movement relies on the attraction of a soft-iron vane by the electromagnetic field of the meter coil. A dual-vane movement has two soft-iron vanes positioned within a common meter coil. The two vanes assume a similar magnetic polarity and then repel each other. One vane is fixed; the other is attached to the scale pointer and is movable. Iron-vane movements are generally very rugged and not polarized.

- A meter scale is divided into a number of divisions and subdivisions that permit more accurate interpretation of pointer position. Each scale division has a definite weight that is determined by the number of divisions and the position of the range switch. If the pointer comes to rest between scale marks, then the technician must interpolate the reading. A mirrored scale helps reduce errors caused by parallax.

- Digital meters consist of a number of decimal digits that display the measured quantity. More digits generally correspond to higher resolution. Full digits are capable of displaying any digit from zero to nine. Half digits are often used in the left-most position and can only display blank (off) or the digit one.

- Current meters are generally placed in series with the current to be measured, although current probes permit measurement without circuit interruption. A meter shunt is used to increase the effective full-scale current value of a particular meter. Multiple-range current meters may have multiple shunts. Full-scale current and accuracy are two important current meter specifications. Digital meter specifications also include resolution. Ideal current meters have zero internal resistance.

- Voltmeters are connected in parallel with the voltage to be measured. Ideal voltmeters have infinite internal resistance. The actual internal resistance of an analog meter consists of the resistance of the meter movement and a series multiplier resistance. The value of total internal resistance can be found by multiplying the sensitivity (ohms-per-volt rating) by the setting of the range switch. Digital voltmeters normally have constant internal resistance in excess of 10 MΩ.

- Power must be disconnected from the circuit under test before connecting an ohmmeter to the circuit. An ohmmeter has its own internal voltage source. Resistance measurements can be made without removing any components if the technician can adequately determine what a normal measurement would be. Otherwise, the component (or at least all but one lead) is disconnected from the circuit to eliminate sneak current paths that would otherwise cause the measured value to be less than the actual value. Analog ohmmeter scales are normally nonlinear and numbered from right to left (backoff scale). More accurate measurements can be made when a range is selected that causes deflection into the right half of the scale.

- An ohmmeter, a voltmeter, and a current meter are often combined into a single measurement device called a multimeter. Digital multimeters frequently provide additional measurement capabilities.

• The oscilloscope relies on the controlled vertical and horizontal deflection of an electron beam to paint a graph on the face of a CRT. In the case of dc voltage measurements, the amount of vertical deflection is directly proportional to the magnitude of the measured voltage. Additionally, the direction of deflection is up or down for positive and negative input voltages respectively.

Review Questions

Section 9.1: Analog Meters

1. Which type of meter movement is more rugged: moving-coil/iron-vane?

2. Which meter movement is sensitive to polarity: moving-coil/iron-vane?

3. What is the purpose of the spiral spring in a d'Arsonval meter movement?

4. Why is the iron-vane in an iron-vane meter movement shaped irregularly?

5. In a two-vane meter movement, the vanes (*attract, repel*) each other.

6. Refer to the speedometer scale shown in Figure 9-35. How many miles per hour does each major division represent?

7. How many miles per hour are represented by each minor division on the speedometer shown in Figure 9-35?

8. What speed is indicated by point A in Figure 9-35?

9. What speed is indicated by point B in Figure 9-35?

10. Point C in Figure 9-35 represents 25 mph. (True or False)

11. Point D in Figure 9-35 represents 80 mph. (True or False)

12. Briefly explain what causes parallax errors and how they can be reduced.

13. If a meter movement seems to be erratic and jumps quickly when a voltage is connected or disconnected to the meter, then the meter probably has very little damping. (True or False)

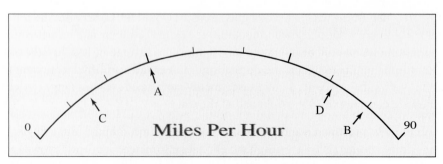

Figure 9-35. A speedometer scale.

Section 9.2: Digital Meters

14. Excluding the decimal point, how many segments are normally used to display the digits 0 to 9 in a digital display?

15. If a half digit is used in a digital display, where will it be positioned?

16. If a 3 1/2-digit wind velocity meter is configured to indicate from 0 to 199.9 mph, what is the resolution of the display?

17. What is the resolution of a 4 1/2-digit timer configured to display 0 to 19.999 seconds?

18. If a 3 1/2-digit display used as the speedometer in a car has a 1/10th mph resolution, what is the highest speed it could indicate?

19. Name an application where an analog meter would be preferred over a digital meter.

20. Which segments of a seven-segment display would have to be illuminated to display the letter *F*?

21. Is it possible to display the letter *L* on a seven-segment display?

22. Is it possible to display the letter *T* on a seven-segment display?

Section 9.3: Current Meters

23. A current meter is normally inserted in (*series, parallel*) with the component whose current is being measured.

24. An ideal current meter has (*zero, infinite*) internal resistance.

25. If a 100-μA meter movement is to be used in a 0- to 10-mA meter, a shunt must be connected in series with the meter movement. (True or False)

26. Which of the following would have the lowest internal resistance: a 50-μA, 2,000-Ω meter movement or the same movement with a shunt for 1-mA full-scale operation?

27. Is it necessary to remove power to the circuit when inserting a current meter into a permanently installed current-monitoring jack?

28. Is it necessary to remove power to the circuit when measuring current with a current probe?

29. When inserting a current meter into the circuit, a technician should set the range switch for a value that is higher than expected. (True or False)

30. What type of current probe can sense direct currents?

31. What name is given to the value of current required to cause the meter to deflect fully?

32. Range switches on a current meter should be a (*shorting, nonshorting*) type.

33. Calculate the value of meter shunt required to extend the range of a 120-μA, 1,000-Ω meter movement to a full-scale current of 5 mA.

34. What value shunt is required to convert a 50-μA, 1,500-Ω movement into a meter with a full-scale current of 500 mA?

35. If you want to use a 100-μA, 500-Ω meter movement to measure a maximum current of 10 A, what value should you use for the shunt?

36. If a 10-mA current meter has an accuracy of ±2% and is measuring an actual current of 4 mA, what are the highest and lowest values of the indicated current that will still be within the manufacturer's specifications?

37. Your job is to test the calibration of a 50-μA meter movement. You connect it to a precision current source adjusted to produce 25 μA. The meter actually indicates 25.1 μA. If the meter has a ±3% accuracy rating, is the calibration acceptable?

38. A certain digital current meter has an accuracy specification of ±0.5% of the reading and can measure 19.99 mA maximum. If the indicated current is 5 mA, what is the largest error that could be caused by the meter accuracy rating?

39. If a 3 1/2-digit current meter has an accuracy specification of ±(0.7% + 2 digits), what is the possible range of measured currents if the display indicates 075.2 mA?

Section 9.4: Voltmeters

40. An ideal voltmeter has (*zero, infinite*) internal resistance.

41. Voltmeters are connected in (*series, parallel*) with the voltage being measured.

42. The range switch on a voltmeter must be a (*shorting, nonshorting*) type.

43. A multiplier resistance is connected in (*series, parallel*) with a meter movement to form a voltmeter.

44. What size multiplier resistor is needed to make a voltmeter with a 75-V full-scale range? The meter movement has a full-scale current of 100 μA and a resistance of 800 Ω.

45. What size multiplier is required to convert a 125-μA, 150-Ω meter movement into a 100-V full-scale voltmeter?

46. Calculate the values for R_1, R_2, and R_3 in Figure 9-36.

47. What is the unit of measurement for voltmeter sensitivity?

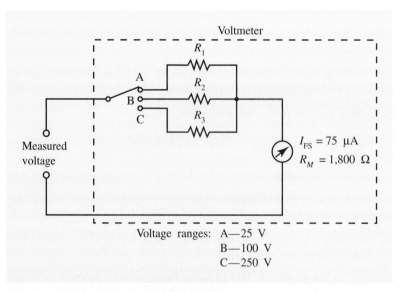

Figure 9-36. Find the multiplier values.

48. What is the sensitivity of a voltmeter that uses a 120-μA meter movement?

49. What is the sensitivity of a voltmeter that has a 75-μA meter movement?

50. A 10-mA meter movement has a lower sensitivity than a 100-μA movement. (True or False)

51. Most digital voltmeters have internal resistance that exceeds 10 MΩ. (True or False)

52. Standard analog meters generally have a higher internal resistance than most digital meters. (True or False)

53. What is the internal resistance of an analog meter that uses a 50-μA meter movement and is operated on the 100-V range?

Section 9.5: Ohmmeters

54. Power must be removed from the circuit before an ohmmeter can be used. (True or False)

55. Zero ohms causes (*full, no*) deflection in a backoff ohmmeter.

56. Most analog ohmmeters have linear scales. (True or False)

57. What does the "∞" symbol mean when it appears on an ohmmeter scale?

58. Describe what is meant by the expression *sneak current path* when discussing in-circuit resistance tests.

59. What resistance is being measured if the pointer of an ohmmeter indicates 43 and the range switch is in the $R \times 100$ position?

60. What resistance is being measured if the pointer of an ohmmeter indicates 2.6 and the range switch is in the $R \times 10K$ position?

61. If the 3 1/2-digit display of a certain digital ohmmeter indicates 1.257 on the 2K range, what resistance is being measured?

62. Describe the display on a 3 1/2-digit ohmmeter if it is measuring a 27-kΩ resistor on the 200K range.

63. Explain how to interpret an ohmmeter accuracy specification of ±1.5% of arc.

Section 9.6: Multimeters

64. What is meant by the term *VOM*?

65. What is meant by the term *DMM*?

Section 9.7: Oscilloscopes

66. What is meant by the expression *CRT*?

67. What is the purpose of the control grid in a CRT?

68. What is the purpose of the deflection plates in a CRT?

69. What element's voltage in a CRT is adjusted with the intensity control?

70. When a technician adjusts the vertical position control, the average voltage of the _____ _____ _____ is being changed.

71. If the vertical sensitivity is set to 50 mV per division and a certain dc voltage causes the trace to move downward by 2.5 divisions, what is the value of voltage being measured?

72. If the vertical sensitivity is set to 5 V per division, describe the effect on the scope trace when the scope is connected to a positive 18-V source.

TECHNICIAN CHALLENGE

Your supervisor has asked you to design and build a combination voltmeter/ammeter that will be used as a built-in tester for a new product manufactured by your company. It must use a single meter (with multiple scales), a two-position mode switch, and a three-position range switch. It must satisfy the requirements listed in Table 9-4.

RANGE SWITCH POSITION	FULL-SCALE VALUE	
	MODE = VOLTAGE	MODE = CURRENT
A	5 V	1 mA
B	25 V	5 mA
C	50 V	25 mA

Table 9-4. Requirements for Voltmeter/Ammeter Design

Based on availability of parts and the required performance, you have decided to use the schematic shown in Figure 9-37. Calculate the required values for resistors R_1 through R_6.

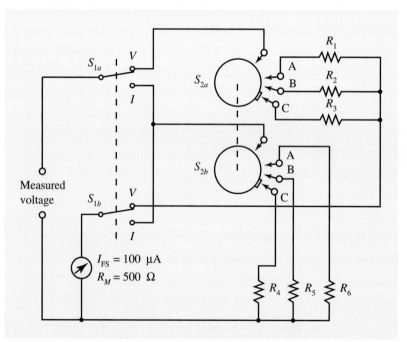

Figure 9-37. A combination voltmeter/ammeter design.

Equation List

(9-1) Division weight $= \dfrac{\text{Highest division value} - \text{Minimum division value}}{\text{Number of divisions}}$

(9-2) Sensitivity $= \dfrac{1}{I_{FS}}$

(9-3) Internal resistance $=$ Sensitivity $\times$ Voltage range

(9-4) Voltage $=$ Number of divisions of deflection $\times$ sensitivity

objectives

After completing this chapter, you should be able to:

1. Describe the nature of magnetic and electromagnetic fields.

2. Describe the interaction of like or opposite magnetic poles.

3. Apply the left-hand rule to a current-carrying conductor and a solenoid to determine the direction of the magnetic fields.

4. Apply the left-hand rule to demonstrate basic generator or motor action.

5. Solve problems using magnetic units.

6. State the "Ohm's Law" of magnetic circuits.

7. Explain the operation of several devices or systems that rely on magnetism.

8. Interpret a schematic diagram that uses interconnected relays.

9. Explain how a magnetic shield works.

Magnetism and Electromagnetism

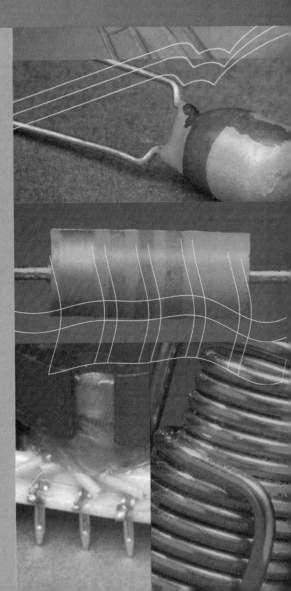

Magnetic and electromagnetic devices are used in most electronic equipment. These devices include speakers, magnetic recorders, electrically operated latches, security sensors, motors, and many more. This chapter explores the theoretical and practical aspects of magnetism and electromagnetism.

10.1 Magnetic Fields

We have all experienced some of the effects of magnetic fields. We know that a magnet will attract pieces of metal when they are brought near the magnet. We know that two magnets affect each other *even before they touch each other.* These effects are due to the presence of an invisible, but measurable region around a magnet called a magnetic field.

Characteristics of Magnetic Fields

Although the magnetic field is not a physical object, it is helpful to visualize the field as being made up of many individual lines. Figure 10-1 shows a pictorial representation of the magnetic lines around a bar magnet.

MAGNETIC FLUX

The lines around the magnet in Figure 10-1 are called **lines of flux** or magnetic flux lines. The lines are invisible, but the effects of the magnetic flux are easily demonstrated. Each flux line is elastic but continuous. Additionally, flux lines never intersect or touch each other.

FIELD DISTORTION

The external field of a magnet can be distorted by the fields from another magnet or by the effects of a nearby object. Some types of materials pass flux lines much easier than air. These materials are classified as **ferromagnetic.** Figure 10-2 shows the field distortion that results when a ferromagnetic object is placed in a magnet's field.

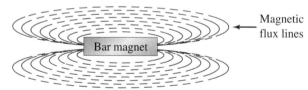

Figure 10-1. The magnetic field around a bar magnet.

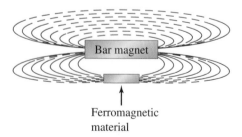

Figure 10-2. A magnetic field can be distorted by the introduction of a ferromagnetic material.

Materials that have essentially no effect on a magnetic field are called **paramagnetic**. Still other types of material oppose (although only slightly) the passing of magnetic lines of force. These materials are classified as **diamagnetic.**

MAGNETIC POLES

As shown in Figure 10-3, the flux lines are concentrated near the ends of the magnet. The regions where the external flux lines are most concentrated are called the **poles** of the magnet. Not all magnetic materials have poles; in some cases (e.g., a toroid transformer) the flux lines are completely contained within the magnetic material.

Magnetic Polarities

Since the two poles of a magnet exhibit different characteristics, it is necessary to distinguish between them. This allows consistent and repeatable designs and analyses.

LABELING OF MAGNETIC POLES

Figure 10-4(a) shows the labeling of the pole polarities. Figure 10-4(b) illustrates the origin of the labeling. If a magnet is suspended (as shown in Figure 10-4b), so that it may rotate with minimum friction, it will align itself with the earth's magnetic field. The end of the magnet that points toward the north pole of the earth is called the north-seeking pole or simply north pole. The other end of the magnet is called the magnet's south pole. North and south poles always appear in pairs. No one has ever shown that a single, isolated magnetic pole can exist.

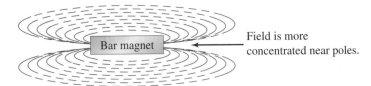

Figure 10-3. The flux lines are concentrated at the poles of the magnet.

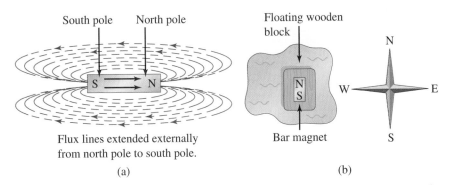

Figure 10-4. A magnet's poles are distinguished by referring to them as north and south poles.

Figure 10-4 also shows another characteristic of a magnetic field. The flux lines are considered to extend from the north pole to the south pole external to the magnet. Internal to the magnet, the flux extends from the south pole to the north pole.

EXAMPLE SOLUTION

Label the magnetic poles of the bar magnet shown in Figure 10-5.

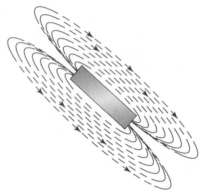

Figure 10-5. Label the poles on the bar magnet.

EXAMPLE **SOLUTION**

Since external lines of flux go from north pole to south pole, the polarity of the magnet must be like that shown in Figure 10-6.

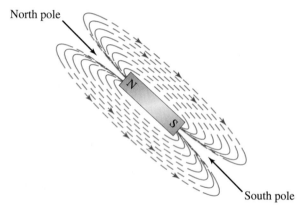

Figure 10-6. Correct labeling of the magnetic poles.

Practice Problems

1. Label the poles on the horseshoe magnet shown in Figure 10-7.

Figure 10-7. Label the poles on this magnet.

Answers to Practice Problems

ATTRACTION AND REPULSION BY POLES

When the fields from two magnets try to occupy the same general region in space, there is an interaction of the fields. Figure 10-8 shows the various possibilities and the results.

In Figure 10-8(a), two north poles are brought near each other. Recall from an earlier discussion that magnetic lines of force cannot touch or intersect. The result, as shown in

Like poles repel.

(a)

Like poles repel.

(b)

Unlike poles attract.

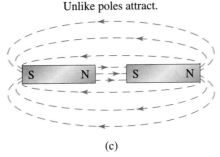

(c)

Figure 10-8. Two magnets in the same general space will interact.

Figure 10-8(a), is that the fields of both magnets are distorted and the two magnets physically repel each other. If the magnets have strong fields, then the force of repulsion can be too strong for a person to push them into contact with each other. Figure 10-8(b) shows a similar result when two south poles are brought together.

When unlike poles are brought near each other, the two magnets attract one another. As shown in Figure 10-8(c), the flux flows from the north pole of one magnetic to the south pole of the other. The two magnets essentially act as a single magnet. The principles illustrated in Figure 10-8 can be summarized as follows:

> 1. Like poles repel.
>
> 2. Unlike poles attract.

In either case, the force of attraction or repulsion is inversely proportional to the distance between the poles and directly proportional to the strength of the magnetic fields.

Magnetic Domains

Some materials can be magnetized; others cannot. The properties of a material that allow it to be magnetized result from the spins of the outer orbit electrons. Materials that can be magnetized have regions called **magnetic domains.** Each magnetic domain acts like a miniature bar magnet. The length of a domain is only a fraction of a millimeter. Figure 10-9(a) shows that the domains in an unmagnetized material are pointed in random directions. Their fields do not aid each other.

Figure 10-9(b) shows the same material after it has been magnetized. Here, the domains have all rotated to point in the same direction. Their fields are now additive and the overall material takes on the same magnetic polarity as the collective polarity of the individual domains.

It takes energy to cause a domain to rotate or change orientation. This energy loss is often evidenced as heat in electromagnetic devices such as transformers and motors.

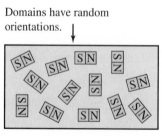

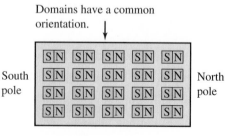

Domains have random orientations.

No net magnetic field in unmagnetized material.

(a)

Domains have a common orientation.

South pole   North pole

Net magnetic field has same polarity as domains.

(b)

Figure 10-9. Magnetic materials have magnetic domains.

Magnetic versus Electromagnetic Fields

Our discussion thus far has been limited to the magnetic fields associated with **permanent magnets.** A permanent magnet retains its magnetism after the magnetizing field has been removed. Many electromechanical devices rely on the use of **temporary magnets.** These magnets exhibit the characteristics of a magnet as long as they are in a magnetizing field. However, if the magnetizing field is removed, then the domains in the material return to their original, random arrangement. We say that the material is no longer magnetized. Temporary magnets generally rely on the flow of electrons through a wire to create the magnetizing field.

DIRECTION OF ELECTROMAGNETIC FIELD

Figure 10-10 shows that when current flows through a wire, a magnetic field that surrounds the wire is produced. As indicated in Figure 10-10, the presence of the field can be detected with a magnetic compass. The compass always points in the direction of the magnetic field. As it is positioned in various places near the wire, it becomes evident that the field is circular and is perpendicular to the current flow.

Figure 10-11 shows that the direction of the magnetic field is determined by the direction of current flow. Additionally, the intensity of the magnetic field is determined by the magnitude of the current.

For the purposes of this chapter, we shall consider an electromagnetic field to be a magnetic field that is produced by the flow of electron current. The terms magnetic field and electromagnetic field are often used interchangeably when discussing electromagnetic devices.

Many technicians use a memory aid called the *left-hand rule,* which can be used to determine the direction of the magnetic field around a current-carrying conductor. Figure 10-12 illustrates the left-hand rule for conductors.

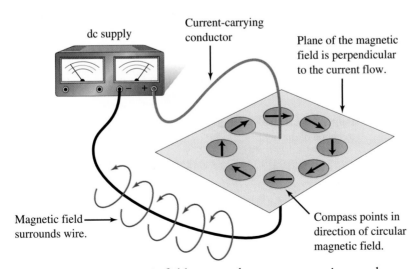

Figure 10-10. A magnetic field surrounds a current-carrying conductor.

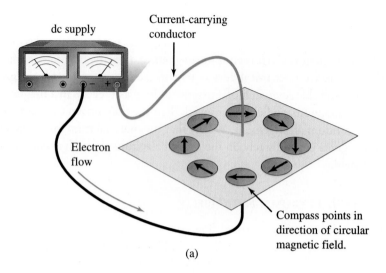

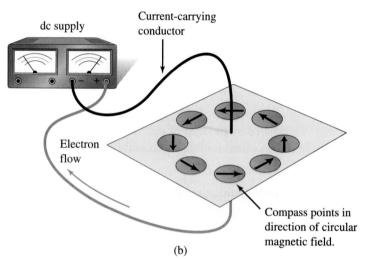

Figure 10-11. The direction of the magnetic field is determined by the direction of the electron current.

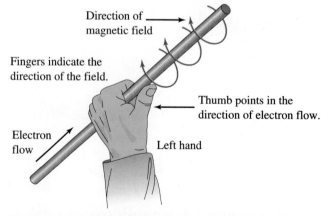

Figure 10-12. The left-hand rule for conductors.

The left-hand rule for conductors works as follows:

1. Grasp the wire with the left hand such that the thumb is pointing in the direction of electron current.
2. As the fingers curl around the wire, they will point in the direction of the magnetic field.

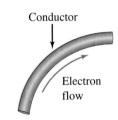

Figure 10-13.
What is the direction of the magnetic field?

EXAMPLE SOLUTION

Determine the direction of the magnetic field for the conductor shown in Figure 10-13.

EXAMPLE **SOLUTION**

Mentally grasp the wire with the left thumb pointing in the direction of electron current flow. Your fingers will indicate the direction of the magnetic field. In this case, the field goes over the top and down behind the wire as illustrated in Figure 10-14.

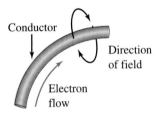

Figure 10-14.
Magnetic field polarity is determined by the direction of current flow.

DOT AND CROSS SYMBOLS

When discussing magnetic circuits, technicians often have to indicate the direction of current flow on diagrams. Where practical, the use of arrows is employed. However, many electromagnetic diagrams need to illustrate a conductor whose current is flowing directly toward or directly away from the reader. In these cases, a standard arrow cannot be used to label the current direction; rather, dot and cross symbols are used as shown in Figure 10-15.

As shown in Figure 10-15(a), a dot is used to indicate that electron current is flowing directly toward the reader (i.e., out of the page). This is easy to remember if you imagine that the dot represents the end of an arrow as it comes toward you. Figure 10-15(b) shows how a cross is used to indicate that electron current is flowing away from the reader (i.e., into the page). You can remember this by considering the cross to be the feathers of an arrow as it flies away from you.

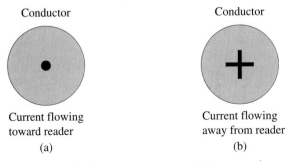

Figure 10-15. Dot and cross symbols are used to indicate the direction of current flow.

Practice Problems

1. Determine the direction of the magnetic field for each conductor shown in Figure 10-16.

2. Determine and mark the direction of current flow for each conductor shown in Figure 10-17.

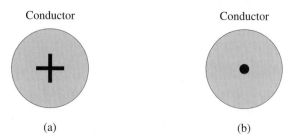

Figure 10-16. What is the direction of the magnetic field?

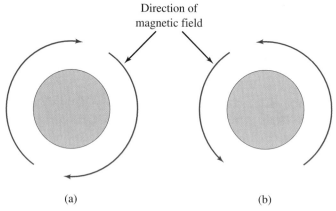

Figure 10-17. Which way does current flow in these conductors?

Answers to Practice Problems

1. a. ccw b. cw   2. a. out (·) b. in (+)

INTERACTION OF ELECTROMAGNETIC FIELDS

Electromagnetic fields exhibit the same properties as magnetic fields produced by a permanent magnet. For example, the lines of flux appear to be elastic, they are continuous, and they never intersect one another. Further, if two electromagnetic fields are in the same general space, they will influence each other.

Figure 10-18 shows the results when current flows in two adjacent conductors. The left-hand rule can be used to determine the direction of the magnetic field. As indicated in Figure 10-18(a), lines of flux that are going in the same direction (between the conductors) tend to repel each other. Magnetic flux lines going in opposite directions between the two conductors, as shown in Figure 10-18(b), will attract each other. The net effect is that the conductors in Figure 10-18(a) will be repelled by each other, while those in Figure

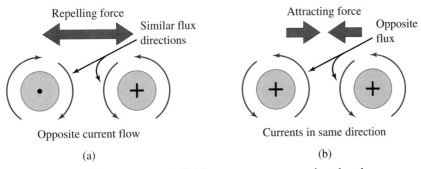

Figure 10-18. Electromagnetic fields can attract or repel each other.

10-18(b) will be attracted to each other. If the fields are sufficiently strong and the wires are free to move, then you can actually observe a physical movement of the wires.

EXAMPLE SOLUTION

Determine the direction of force (attraction or repulsion) that will be present between the two conductors shown in Figure 10-19.

EXAMPLE SOLUTION

The left-hand rule can be used to determine the directions of the two magnetic fields. The fields are shown in Figure 10-20. Since the flux direction of the two fields between the two conductors is opposite, we can conclude that the two conductors will attract each other and will tend to move together.

If a conductor is wound into a spiral coil as shown in Figure 10-21, then the fields from adjacent loops interact as discussed previously. The net effect is to distort the individual fields to form a composite field as shown in Figure 10-21. A coil wound in this manner is called a **solenoid,** and exhibits magnetic poles at the ends of the coil.

Figure 10-19.
Will these conductors
attract or repel each other?

Figure 10-20.
The conductors
will attract each other.

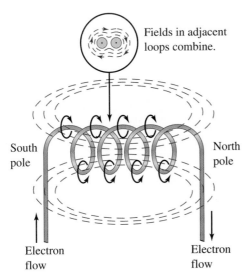

Figure 10-21. Individual loop fields
combine to form a composite coil field.

🔴 **KEY POINTS**

The field strength of a current-carrying conductor can be increased by winding it into a spiral coil called a solenoid.

🔴 **KEY POINTS**

The magnetic fields from the individual loops of a solenoid combine to form a composite field that is stronger and more concentrated than the original distributed field of the straight wire.

🔴 **KEY POINTS**

The solenoid has magnetic poles; the magnetic polarity can be determined with the left-hand rule for solenoids.

The resulting composite field of a coil is stronger and more useful than the distributed fields associated with a straight-line conductor. It can be made even stronger by wrapping the wire around a magnetic material.

Technicians use another version of the left-hand rule to determine the magnetic polarity of a solenoid. Figure 10-22 illustrates the rule. It can be stated as:

> Grasp the solenoid with the left hand such that the fingers curl around the coil in the same direction as the electron current is flowing. The thumb will point to the north pole of the solenoid.

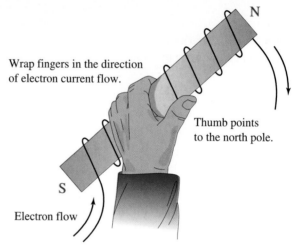

Figure 10-22. The left-hand rule for solenoids.

EXAMPLE SOLUTION

Determine the magnetic polarity of the solenoid shown in Figure 10-23.

EXAMPLE **SOLUTION**

Figure 10-24 shows the application of the left-hand rule for solenoids.

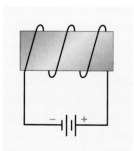

Figure 10-23. Determine the magnetic polarity of this solenoid.

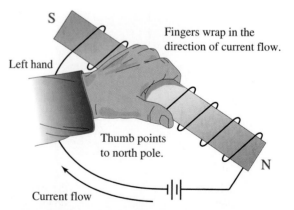

Figure 10-24. The left-hand rule can be used to determine magnetic polarity of a solenoid.

Practice Problems

1. Determine the magnetic polarity for each solenoid shown in Figure 10-25.

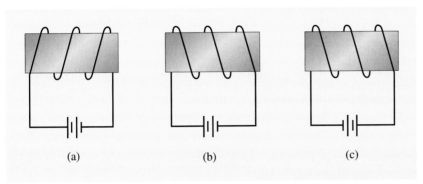

(a)   (b)   (c)

Figure 10-25. Determine the magnetic polarity for these solenoids.

2. Determine the direction of current flow for each solenoid shown in Figure 10-26.

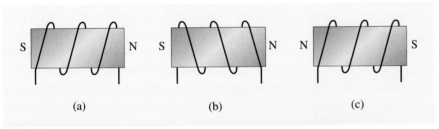

(a)   (b)   (c)

Figure 10-26. Which way does current flow through the solenoids?

Answers to Practice Problems

1. a. N–S b. S–N c. N–S;

2. a. left-to-right through solenoid
   b. right-to-left through solenoid
   c. right-to-left through solenoid

ELECTROMAGNETIC FIELD STRENGTH

The strength of the magnetic field produced by a solenoid whose length is at least ten times its diameter is determined by the following factors:

- the amount of current through the coil
- the number of turns on the coil
- the physical length of the coil
- the magnetic characteristics of the material used as the core

More specifically, the strength of the solenoid field is directly proportional to the number of turns, the amount of current, and the "magnetic conductance" (called permeance and discussed in a later section). It is inversely proportional to the length of the coil.

EXAMPLE SOLUTION

Name two ways to increase the strength of a solenoid's magnetic field.

EXAMPLE **SOLUTION**

Increasing the amount of current, adding more turns, or increasing the magnetic conductance of the core would increase the strength of the field. Alternatively, the same number of turns could be wound over a shorter length.

Figure 10-27.
Determine the direction of the magnetic field.

Exercise Problems 10.1

1. An object must physically touch a magnet to experience the effects of magnetism. (True or False)
2. A magnetic field is composed of imaginary lines called _____ __ _____.
3. A material that passes flux lines easily is classified as _____.
4. A paramagnetic material can be used to distort a magnetic field. (True or False)
5. Magnetic poles always appear in pairs. They are labeled _____ and _____ poles.
6. Unlike magnetic poles _____.
7. Like magnetic poles _____.
8. Magnetic materials have tiny regions that act like miniature bar magnets. These regions are called _____ _____.
9. The magnetic field around the conductor shown in Figure 10-27 would be (*cw, ccw*).
10. Draw the direction of the current flow in the conductor shown in Figure 10-28.
11. If the magnetic fields around the two conductors shown in Figure 10-29 interact, the two conductors will (*attract, repel*).
12. Determine the magnetic polarity of the solenoid shown in Figure 10-30.
13. The strength of the magnetic field of a solenoid is determined by the direction of current flow. (True or False)
14. The number of turns on a solenoid directly affects the strength of the magnetic field. (True or False)
15. The value of current through a solenoid affects the strength of the magnetic field. (True or False)

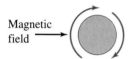

Figure 10-28.
Which way does current flow?

Figure 10-29.
Will these conductors attract or repel?

10.2 Units of Measurement

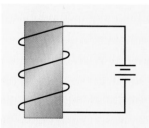

Figure 10-30.
Determine the polarity of this solenoid.

Just as there is more than one system used to express linear measurements (e.g., feet, meters, cubits), there are multiple systems used to express magnetic quantities. For purposes of this chapter, we will limit our discussions to the use of the more popular Système International (SI), also called the meter-kilogram-second (MKS) system. For reference purposes, a conversion table is provided in Appendix C that can be used to convert to the older centimeter-gram-second (CGS) system.

Lines of Flux

The magnetic field is considered to be composed of a number of invisible lines called flux lines. Even though flux lines cannot be seen as discrete entities, it is useful to characterize the magnetic field in terms of its flux. We are interested in both total flux and flux density.

UNIT OF FLUX

The unit of magnetic flux is the *weber*. One weber corresponds to 1×10^8 flux lines. Because a weber is such a large unit, we generally express flux in terms of milliwebers (100,000 lines) or microwebers (100 lines). The abbreviation for weber is Wb. The symbol for flux is the Greek letter phi (Φ). The concept of magnetic flux provides us with a convenient tool for thinking about magnetic fields, but it is also possible to express the amount of flux contained in one weber without referencing individual flux lines. This alternative is presented in a later section.

FLUX DENSITY

Another important characteristic of a magnetic field is the **flux density**. This measurement is not concerned with total flux, but rather how many flux lines appear in a certain area (i.e., how closely packed are the lines?). The unit of measurement for flux density is the *tesla* (T). One tesla corresponds to a flux density of one weber per square meter. The symbol for flux density is B. Figure 10-31 further clarifies the meaning of flux density.

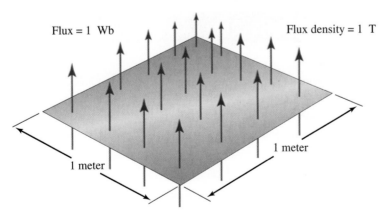

Figure 10-31. If a one-square-meter area has one weber of flux passing through it, then we define the flux density as one tesla.

We can express this mathematically with Equation 10-1.

$$B = \frac{\Phi}{A} \qquad (10\text{-}1)$$

where Φ is flux measured in webers and A is area measured in square meters.

EXAMPLE SOLUTION

If a 1.5 m² area contains 5.75 Wb of flux, what is the flux density?

EXAMPLE **SOLUTION**

We apply Equation 10-1 as follows:

$$B = \frac{\Phi}{A}$$
$$= \frac{5.75 \text{ Wb}}{1.5 \text{ m}^2}$$
$$= 3.83 \text{ T}$$

Magnetomotive Force

Magnetomotive force (mmf) refers to the energy source that actually creates the flux. It is the equivalent of electromotive force (emf) in electrical circuits. The symbol for magnetomotive force is $\mathscr{F}$.

You will recall that the total flux generated by a current-carrying conductor is proportional to the value of current (I). You will also recall that we can increase the magnetic field strength by winding the wire into a coil or solenoid. Field strength, in this case, is increased as we increased the number of turns (N) in the coil. Since field strength is affected by both current and number of turns, it is reasonable that the units for magnetomotive force are ampere-turns (A · t). This is the simple product of current × number of turns. It is expressed formally as Equation 10-2.

$$\mathscr{F} = NI \tag{10-2}$$

EXAMPLE SOLUTION

What is the magnetomotive force produced when 3 A of current flow through a 20-t coil?

EXAMPLE **SOLUTION**

We compute the magnetomotive force as follows:

$$\mathscr{F} = NI$$
$$= 20 \text{ t} \times 3 \text{ A} = 60 \text{ A} \cdot \text{t}$$

Practice Problems

1. What mmf is produced when 3.5 A flow through 25 turns of wire?
2. If 500 mA flow through a 250-t coil, what is the resulting mmf?
3. How many turns must be wound on a coil if 2.5 A is to produce an mmf of 2,500 A · t?
4. How much current must pass through a 100-t coil to produce an mmf of 25 A · t?

Answers to Practice Problems

1. $87.5 \, \text{A} \cdot \text{t}$ 2. $125 \, \text{A} \cdot \text{t}$

3. $1{,}000 \, \text{t}$ 4. $250 \, \text{mA}$

Magnetizing Force

Magnetizing force, also called magnetic field intensity, is closely related to magnetomotive force, but it includes a physical dimension. It is convenient to consider the closed loop formed by flux lines as a magnetic circuit. If the length of that circuit is increased, then we know intuitively that the field intensity will be reduced. This relationship is expressed mathematically by Equation 10-3.

$$H = \frac{\mathscr{F}}{l} \qquad\qquad \textbf{(10-3)}$$

where H is the symbol used to represent magnetizing force, $\mathscr{F}$ is the mmf, and l is the length of the total magnetic path measured in meters. The unit of measure for magnetizing force is ampere-turns per meter ($\text{A} \cdot \text{t/m}$).

The magnetizing force measurement of a magnetic circuit is similar to electrostatic field strength (electric field intensity) in an electrical circuit. In Chapter 2, we saw that the strength of an electrostatic field is inversely proportional to the distance between the charges. In a magnetic circuit, the magnetic field strength is inversely proportional to the distance the flux has to travel to complete the magnetic loop. Figure 10-32 illustrates the relationship between magnetic path length and magnetizing force. Figures 10-32(a) and (b) both have the same magnetomotive force, since they have the same current and the same number of turns on the coils. Figure 10-32(a) has a higher field intensity, however, because the magnetic circuit is shorter. In Figure 10-32(b), the field intensity is less, as a result of the increased magnetic path length.

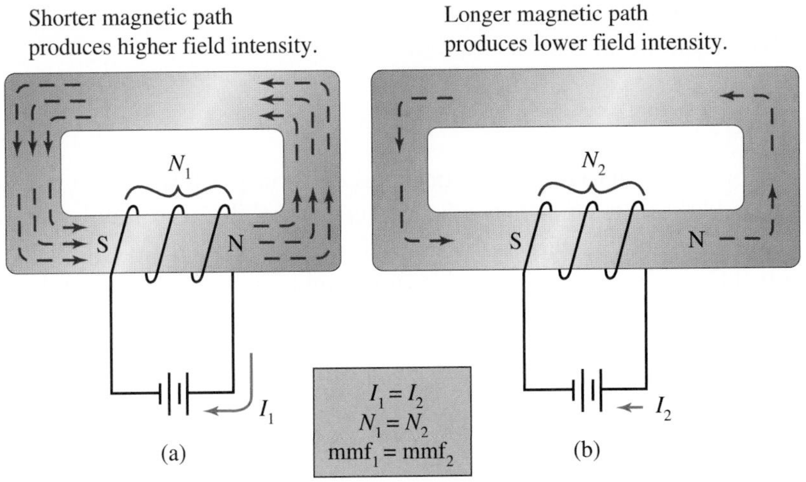

> ## KEY POINTS
>
> The term magnetizing force (H), or magnetic field intensity, is a measure of the mmf per meter. Its units are ampere-turns per meter ($\text{A} \cdot \text{t/m}$). Field intensity decreases as the length of the magnetic path is increased.

Figure 10-32. Magnetizing force, or magnetic field intensity, is inversely proportional to the length of the magnetic path.

EXAMPLE SOLUTION

What is the magnetizing force in a magnetic circuit that is 45 cm long and has a mmf of 150 A · t?

EXAMPLE **SOLUTION**

We apply Equation 10-3 as follows:

$$H = \frac{\mathcal{F}}{l}$$

$$= \frac{150 \text{ A} \cdot \text{t}}{45 \times 10^{-2} \text{ m}}$$

$$= 333.3 \text{ A} \cdot \text{t/m}$$

Practice Problems

1. Calculate the magnetizing force produced by a 50 A · t magnetomotive force with a magnetic path length of 0.5 m.
2. Calculate the magnetizing force in a magnetic circuit that is 250 cm long and has a mmf of 25 A · t.
3. If a field intensity of 50 A · t/m is produced by a mmf of 5 A · t, what is the length of the magnetic circuit?

Answers to Practice Problems

1.   100 A · t/m   **2.**   10 A · t/m   **3.**   0.1 m or 10 cm

Permeance

Permeance is the ease with which a magnetic field can be established in a particular material. Its symbol is $\wp$, and its units of measure are webers per ampere · turn (Wb/A · t). It is comparable to conductance in an electrical circuit.

PERMEABILITY

The ability of a material to pass magnetic flux is called its **permeability.** It is easier to pass a magnetic field through a material with a higher permeability. Permeability in a magnetic circuit is similar to conductivity in an electrical circuit. The standard symbol for permeability is μ. This is unfortunate, since this same symbol is used to represent the prefix micro. Care must be used to avoid confusing the use of this symbol.

We can compute the permeability of a material with Equation 10-4.

$$\mu = \frac{B}{H} \tag{10-4}$$

where B is the flux density in tesla and H is the magnetizing force in A · t/m. The units of measure for permeability computed in this manner is webers per ampere · turn · meter (Wb/A · t · m). It is important to note that the permeability of a ferromagnetic material is *not* a constant value.

It varies with the amount of magnetizing force. This factor will be explored further in a later section.

EXAMPLE SOLUTION

What is the permeability of a material that produces a flux density of 2 T when a magnetizing force of 200 A · t/m is applied?

EXAMPLE **SOLUTION**

We apply Equation 10-4 as follows:

$$\mu = \frac{B}{H}$$

$$= \frac{2 \text{ T}}{200 \text{ A} \cdot \text{t/m}}$$

$$= 0.01 \text{ Wb/A} \cdot \text{t} \cdot \text{m}$$

RELATIVE PERMEABILITY

When permeability is computed with Equation 10-4, the result is called *absolute permeability*. It is more common to express the permeability of a material in terms of *relative permeability*. More specifically, the permeability of the material is compared to the permeability of a vacuum (or air for practical purposes). The permeability of a vacuum (μ_v) is a physical constant and has the value of approximately

$$\mu_V = 1.257 \times 10^{-6} \text{ Wb/A} \cdot \text{t} \cdot \text{m}$$

The relative permeability (μ_r) of a material, then, describes how easily the material passes magnetic flux as compared to a vacuum. We can calculate the relative permeability of a material with Equation 10-5.

$$\mu_r = \frac{\mu}{\mu_v} \qquad\qquad (10\text{-}5)$$

The units of measure cancel in this division problem. That is, relative permeability of a material is a simple ratio and has no units of measure. Values of typical materials range from slightly less than one for diamagnetic materials to values as high as 100,000 for some ferromagnetic alloys.

EXAMPLE SOLUTION

What is the relative permeability of a material that has a flux density of 5 T with a magnetizing force of 5,000 A · t/m?

EXAMPLE **SOLUTION**

First, we apply Equation 10-4 to determine the absolute permeability of the material.

$$\mu = \frac{B}{H}$$

$$= \frac{5 \text{ T}}{5,000 \text{ A} \cdot \text{t/m}}$$

$$= 1 \times 10^{-3} \text{ Wb/A} \cdot \text{t} \cdot \text{m}$$

> **KEY POINTS**
>
> It is common to express permeability as a value relative to the absolute permeability of a vacuum (μ_v). When expressed in this way, the permeability is called relative permeability (μ_r), and it indicates how much easier it is to establish a field in the material as compared to air.

> **KEY POINTS**
>
> There are no units of measure for relative permeability.

Next, we apply Equation 10-5 as follows:

$$\mu_r = \frac{\mu}{\mu_v}$$

$$= \frac{1 \times 10^{-3}\ \text{Wb/A} \cdot \text{t} \cdot \text{m}}{1.257 \times 10^{-6}\ \text{Wb/A} \cdot \text{t} \cdot \text{m}}$$

$$= 795.5$$

Practice Problems

1. A material used as the core in a solenoid has a flux density of 10 T with a magnetizing force of 200 A · t/m. What is its relative permeability?
2. If a certain material has a relative permeability of 2,000, what magnetizing force is required to produce a flux density of 0.5 T?
3. What flux density is produced by a coil with a 1,000 A · t/m magnetizing force, if its core has a relative permeability of 650?

Answers to Practice Problems

1. 39,777   2. 198.9 A · t/m   3. 0.817 T

Reluctance

Reluctance is to magnetic circuits like resistance is to electrical circuits. That is, reluctance opposes the passage of magnetic flux. Thus, a material with a higher reluctance will have less flux for a given magnetomotive force. The symbol for reluctance is $\Re$. Its value is affected by three factors: the permeability of the material, the cross-sectional area, and the length of the magnetic path. This is expressed by Equation 10-6.

$$\Re = \frac{l}{\mu A} \qquad (10\text{-}6)$$

where l is the length of the magnetic path in meters, μ is the permeability of the material, and A is the cross-sectional area of the magnetic path in square meters. After cancellation of units in Equation 10-6, we are left with A· t per weber as the unit of measurement for reluctance.

EXAMPLE SOLUTION

What is the reluctance of a 25-cm bar that measures 5 cm × 5 cm across the end, if the permeability of the material is 2.5×10^{-3} Wb/A · t · m?

EXAMPLE **SOLUTION**

We convert the centimeter measurements to meters (1 cm = 0.01 m) and then apply Equation 10-6 as follows:

$$\mathfrak{R} = \frac{l}{\mu A}$$

$$= \frac{0.25 \text{ m}}{2.5 \times 10^{-3} \text{ Wb/A} \cdot \text{t} \cdot \text{m} \times 2.5 \times 10^{-3} \text{ m}^2}$$

$$= 40,000 \text{ A} \cdot \text{t/Wb}$$

We know that magnetomotive force, reluctance, and flux are comparable to voltage, resistance, and current, respectively, in an electrical circuit. This leads to the "Ohm's Law" equation for magnetic circuits expressed as Equation 10-7.

$$\Phi = \frac{\mathfrak{F}}{\mathfrak{R}} \qquad (10\text{-}7)$$

EXAMPLE SOLUTION

What is the total flux in a magnetic circuit that has a magnetomotive force of 100 A · t and a reluctance of 500×10^3 A · t/Wb?

EXAMPLE **SOLUTION**

We apply Equation 10-7 as follows:

$$\Phi = \frac{\mathfrak{F}}{\mathfrak{R}}$$

$$= \frac{100 \text{ A} \cdot \text{t}}{500 \times 10^3 \text{ A} \cdot \text{t/Wb}}$$

$$= 200 \times 10^{-6} \text{ Wb} = 200 \text{ } \mu\text{Wb}$$

Practice Problems

1. How much total flux does a 250 A · t magnetomotive force produce if the magnetic circuit has a reluctance of 2×10^6 A · t/Wb?

Answers to Practice Problems

1. 125 μWb

We also know that permeance in a magnetic circuit is similar to conductance in an electrical circuit. This leads to the development of Equation 10-8.

$$\wp = \frac{1}{\mathfrak{R}} \qquad (10\text{-}8)$$

Practice Problems

1. Use Equation 10-8 to determine the permeance of a circuit that has a reluctance of 2×10^8 A · t/Wb.

Answers to Practice Problems

1. $5 \times 10^{-9} \, \text{Wb/A} \cdot \text{t}$

BH Curve

Equation 10-4 expresses permeability in terms of flux density and magnetizing force. We could easily rearrange this equation to produce an expression for flux density:

$$\mu = \frac{B}{H}, \text{ or}$$

$$B = \mu \times H$$

KEY POINTS

A *BH* curve is a graph showing the flux density (*B*) in a material that results from a given magnetizing force (*H*).

In this form, we see that flux density depends on the strength of the magnetizing force and the permeability of the material. This seems intuitively correct, but there is another factor that complicates the issue—the permeability of ferromagnetic materials is not a constant. In fact, the permeability varies nonlinearly as the magnetizing force changes. For this reason, manufacturers of magnetic materials provide graphs that relate magnetizing force (*H*), flux density (*B*), and permeability (μ). A graph of this type is called a *BH* curve. Figure 10-33 shows a representative *BH* curve.

KEY POINTS

When all of the domains in a magnetic material have aligned with the field of the magnetizing source, we say the material is saturated. Any further increase in magnetizing force will produce flux external to the material.

Figure 10-33 shows that for low values of magnetizing force, there is only a slight increase in flux density (the curve has a shallow slope). In this region, only a few domains have begun to realign with the magnetizing field. Once a substantial number of domains begin to rotate into alignment, we see a notable increase in flux density for modest changes in magnetizing force (i.e., the curve in Figure 10-33 is steeper). This region is identified as the *linear region*, since there is a linear relationship between magnetizing force and flux density. Finally, at higher values of magnetizing force, nearly all of the domains will have aligned with the magnetizing field. Once this occurs, the flux density in the material is reaching its upper limit. Further increases in magnetizing force produce more total flux, but the flux increases are external to the material. We call this area of the curve the *saturation region*, because the magnetic material is saturated with magnetic flux.

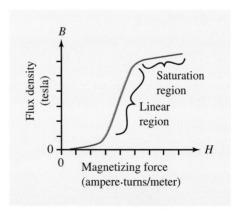

Figure 10-33. A *BH* curve relates magnetizing force, flux density, and permeability for a magnetic material.

Figure 10-34 illustrates how the various regions of the curve correspond to different values of permeability. Recall that permeability is the ratio of flux density to magnetizing force (i.e., $\mu = B/H$). In the lower regions of the curve, a given change in magnetizing force (H_1) produces only a modest change in flux density (B_1). This means the material is exhibiting a relatively low permeability in the lowest regions of the BH curve. In the linear region, as shown in Figure 10-34, a modest change in magnetizing force (H_2) produces a substantial change in flux density (B_2). This means the material is now exhibiting a higher permeability. Finally, as we enter the saturation region, we see that changes in magnetizing force (H_3) produce very small changes in flux density (B_3). This tells us that the permeability is very low in the saturation region of the BH curve. In fact, since the only flux increases after saturation are external to the material, the permeability in the saturation region is essentially the same as air (i.e., $\mu_r = 1$).

Figure 10-35 shows a test configuration that could be used to measure the values needed to plot a BH curve. Here, a known number of turns are wrapped around the material whose BH curve is to be plotted. The voltage source is adjusted to obtain different values of current, which of course causes varying amounts of magnetizing force ($H = \mathcal{F}/l = NI/l$). The technician performing the test would record the current and calculate the magnetizing force for a number of different points on the curve. The value of flux density would also be recorded for each of these points. Finally, the technician would use the recorded data to plot a BH curve that characterizes the material.

Figure 10-36 shows a BH curve provided by a manufacturer of ferromagnetic materials. The particular material characterized in Figure 10-36 is called *ferrite* and is a ferromagnetic ceramic. It is commonly used in power supplies, computers, and high-frequency circuits. The exact shape of a BH curve varies dramatically with different materials.

Hysteresis

The test configuration shown in Figure 10-35 can also be used to measure the data needed for another important magnetic graph. The graph is called a four-quadrant BH curve or,

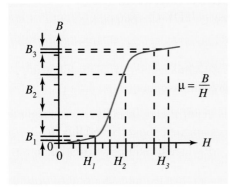

Figure 10-34. Permeability is the ratio of flux density (B) to magnetizing force (H) and varies as the magnetizing force changes.

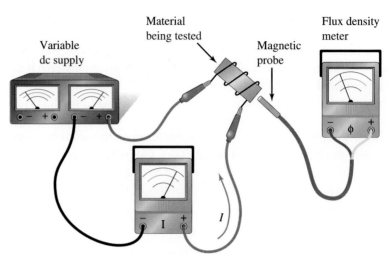

Figure 10-35. A technician's test setup that can be used to measure the values needed to plot a *BH* curve.

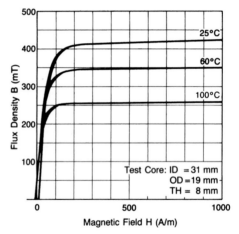

Figure 10-36. A typical *BH* curve. (Courtesy of TDK Corporation of America)

● **KEY POINTS**

A four-quadrant *BH* curve is called a hysteresis loop.

more commonly, a *hysteresis loop*. A technician performs the following procedural steps to obtain the data needed to plot a hysteresis loop:

1. The magnetizing force is slowly increased from zero to magnetic saturation. Several intermediate values of magnetizing force and flux density are recorded.

2. The magnetizing force is slowly reduced from saturation to zero. Several intermediate values are recorded.

3. The polarity of the dc supply is reversed, and steps 1 and 2 are repeated.

4. The original polarity of the dc source is restored and step 1 is repeated again.

A typical plot of the data obtained from this procedure is shown in Figure 10-37. The hysteresis loop shows us several important characteristics of the material. The "initial"

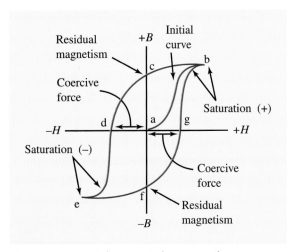

Figure 10-37. A hysteresis loop reveals many important characteristics about a magnetic material.

magnetization curve shown in Figure 10-37 is identical to the one developed in Figure 10-34. It begins at the origin of the graph with the material fully demagnetized (i.e., flux density = 0 and magnetizing force = 0). It progresses upward from point a to point b, where positive saturation is reached. This portion of the curve is generated during procedural step 1.

Procedural step 2 produces the portion of the curve that extends from point b to point c. At point c, the magnetizing force has been reduced to zero, but a definite amount of flux density remains! This is because many of the domains have remained in their aligned position; the material has been magnetized. The magnetism that remains after the magnetizing force has been removed is called **residual magnetism.** The property of the material that allows it to retain magnetism is called **retentivity.** We would expect permanent magnets to have a high degree of retentivity, and we would expect temporary magnets to have a low retentivity.

When the dc supply in Figure 10-35 is reversed during procedural step 3, and the magnetizing force is increased in the reverse direction, we move from point c to point d on the curve. At point d, the reverse magnetizing force is strong enough to overcome the fields of the aligned domains. The net flux density is zero. The magnetizing force required to overcome the residual magnetism is called the **coercive force.**

As we continue to increase the magnetizing force in the reverse polarity, the domains begin to align with the reversed field. We move from point d to point e as the domains align with the reverse field, and the material eventually saturates in the reverse polarity. If the magnetizing force is then reduced to zero, a certain level of residual magnetism will be left in the material. This is shown as point f on the curve in Figure 10-37.

We can complete the hysteresis loop by performing procedural step 4. It takes a definite amount of magnetizing force, called the coercive force, to overcome the internal fields of the aligned domains and produce a net flux density of zero. This is labeled as point g on the hysteresis loop in Figure 10-37. Finally, if we continue to increase the magnetizing force in the original polarity, the material will eventually saturate (point b). This completes the development of the hysteresis loop.

Different materials have different shaped hysteresis loops. Figure 10-38 shows two representative curves. The hysteresis loop shown in Figure 10-38(a) is fairly rectangular. Materials used in magnetic storage devices for computers have curves similar to this. Memory devices of this type store information digitally. That is, all data is stored as a series of two-state variables (e.g., yes/no, true/false, high/low). There are two characteristics of the curve that make this material well-suited to a magnetic memory device. First, when the magnetizing force is removed, there is a very high level of residual magnetism. This represents the stored information. One polarity of magnetism stores one digital state (e.g., yes or true), while the other polarity will represent the other digital state (e.g., no or false). Second, it takes a well-defined level of coercive force to cause the material to change polarity, and when it does switch, it makes a rapid transition. This is an important consideration for digital memory devices.

The material represented in Figure 10-38(b) shows a very low level of residual magnetism and a relatively large linear region. This material might make a good temporary magnet. It can be magnetized, but most of the magnetization goes away when the magnetizing force is removed.

It takes energy to cause the domains to change states. Some people view this as molecular friction. In any case, the energy consumption is evidenced as heat in the magnetic material, and is called **hysteresis loss.** The amount of hysteresis loss in a material is proportional to the area enclosed in the hysteresis loop.

KEY POINTS

The area of the loop is an indication of the amount of heat loss (called hysteresis loss) that will occur as the domains are rotated back and forth.

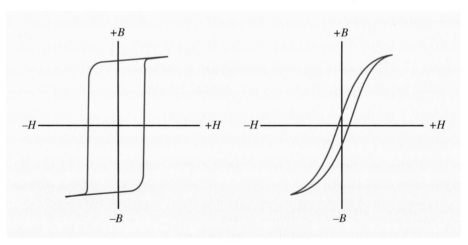

Figure 10-38. Representative hysteresis loops for different materials.

Exercise Problems 10.2

1. A magnetic field is considered to be composed of imaginary lines called _____.

2. A weber is the unit of measure for _____.

3. The symbol for flux is _____.

4. How many flux lines does it take to make 2.5 Wb?

5. Tesla is the unit of measure for _____ _____.

6. If 3 Wb of flux pass through a 0.75 m² area, what is the flux density?

7. What symbol is used to represent magnetomotive force (mmf)?

8. What is the unit of measure for mmf?

9. Name three factors that affect the value of magnetizing force in a magnetic circuit.

10. If a field intensity of 125 A · t/m is produced by a mmf of 3.25 A · t, what is the length of the magnetic path?

11. The symbol μ is used to represent the _____ of a magnetic material.

12. Wb/A · t · m is the unit of measure for what magnetic characteristic?

13. If a particular metal bar passes flux lines 2,000 times easier than air, what is its relative permeability?

14. What is the absolute permeability of a material that produces a flux density of 5 T when a magnetizing force of 350 A · t/m is applied?

15. What magnetic property opposes the passage of flux lines?

16. A *BH* curve is a graph of _____ versus _____.

17. Define the term residual magnetism.

18. What is the name given to describe the amount of magnetizing force required to produce zero flux density in a previously magnetized material?

19. What name is used to describe the condition in a material when flux density is at a maximum level?

20. A four-quadrant *BH* curve is called a _____ _____.

10.3 Electromagnetic Induction

When a magnetic field passes through a conductor, the flux lines interact with the electrons of the conductor such that a voltage is produced across the length of the wire. This voltage is called an **induced voltage** and has all the characteristics of any voltage source. If the conductor is part of a closed loop, then current will flow as a result of the induced voltage. We call the current an *induced current*. The process whereby a magnetic field causes an induced voltage or current in a conductor is called **electromagnetic induction.**

Electromagnetic induction is the basis for operation of many devices that a technician must understand. Motors, generators, d'Arsonval meter movements, transformers, and many other devices rely on electromagnetic induction for their operation. The underlying principle of induction is similar in all cases, however, so it is important to understand how induction works.

Figure 10-39(a) shows a conductor in the magnetic field created by a magnet. The flux lines, as previously described, form complete loops. Figure 10-39(a) only shows the portion of the magnetic field that is in the air gap between the poles. In Figure 10-39(a), neither the conductor nor the magnetic field is moving. Therefore, the flux does not cut through the conductor and no voltage is induced. In Figure 10-39(b), the conductor is moving, but it is moving parallel to the lines of flux. Again, no flux lines cut the conductor, so there is no induced voltage. Finally, in Figure 10-39(c), the conductor is moving

KEY POINTS

When a conductor is cut by a magnetic field, a voltage is produced in the conductor by a phenomenon called electromagnetic induction.

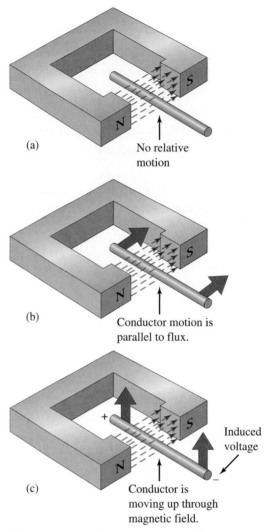

(a) No relative motion

(b) Conductor motion is parallel to flux.

(c) Conductor is moving up through magnetic field.

Induced voltage

Figure 10-39. Electromagnetic induction requires relative motion between a conductor and a magnetic field.

upward through the magnetic field. In this case, the flux lines do cut through the conductor and a voltage is induced in the conductor. It makes no difference whether the conductor moves through a stationary magnetic field, or a moving magnetic field cuts through a stationary conductor. It is simply necessary that we have relative motion of the field and the conductor for induction to occur.

Factors Affecting Induction

There are six basic factors that determine the magnitude and polarity of voltage induced into a conductor as it intersects a magnetic field:

- field strength
- relative speed

- relative angle
- number of turns
- direction of movement
- magnetic field polarity

These factors were originally identified by Michael Faraday and explained in Faraday's Laws. The first four factors listed affect the number of flux lines per second passing through the conductor (i.e., the rate of flux line intersection). These factors determine the magnitude of the induced voltage. The last two factors, direction of movement and polarity of the magnetic field, determine the polarity of induced voltage.

FIELD STRENGTH

If the magnetic field is stronger, there is a greater flux density between the poles of the magnet. As the conductor passes through this space, it intersects more lines per second, which produces a higher value of induced voltage. An analogy might be to imagine you are in a speedboat going perpendicular to a series of waves. If you go at a constant speed, then your rate of collision with the waves will depend on the density (how far apart are they?) of the waves. The magnitude of induced voltage in a moving conductor corresponds to the roughness of the ride in the boat.

RELATIVE SPEED

If the relative speed of motion between the conductor and the magnetic field is increased, then the conductor will intersect more lines per second. Again, the induced voltage will be higher. Going back to the speedboat analogy, if we increase the speed of the boat, then we will collide with the waves at a faster rate, and the ride will get rougher.

RELATIVE ANGLE

The angle between the conductor's motion and the flux lines affects the value of induced voltage. One extreme is a 90° angle, where the conductor is cutting through a maximum number of flux lines in a given time period. This causes maximum induced voltage if the other factors are held constant. The second extreme is 0°, where the conductor is moving parallel to the flux lines as shown in Figure 10-39(b). In this case, no flux lines are cut and the induced voltage is zero. Between 0° and 90°, the rate of flux cutting varies with the angle of cutting. This factor can also be illustrated with the speedboat analogy. If you run your boat parallel to the waves, then there will be no relative motion and you will have a smooth ride. As you begin to steer into the waves, the ride will get rougher until you are finally heading directly into the waves.

NUMBER OF TURNS

The conductors in Figure 10-39 remain straight as they pass through the magnetic field. Alternatively, we can form the wire into a coil. Now, each turn in the coil will intersect the magnetic field and will have an induced voltage. The voltages induced into each individual loop of the coil combine to form a higher composite voltage. If you increase the number of turns on the coil, you will increase the effective value of induced voltage.

KEY POINTS

Anything that affects the rate of flux cutting affects the magnitude of induced voltage.

DIRECTION OF MOVEMENT

Figure 10-40 shows a coil of wire being moved through a magnetic field. The ends of the coil are connected to a voltmeter. If the wire is moved in one direction, as shown in Figure 10-40(a), one polarity of voltage is induced in the coil. We arbitrarily label this as positive. In Figure 10-40(b), the coil is being moved in the opposite direction. There is relative motion between the conductor and the magnetic field, so a voltage is induced. The polarity of the voltage, however, is labeled as negative, since it is opposite from the polarity in Figure 10-40(a).

Figure 10-41 illustrates another left-hand rule used by technicians. This is called the left-hand rule for induction and can be used to determine the direction of induced current (or voltage) in a wire moving through a magnetic field. It works as follows:

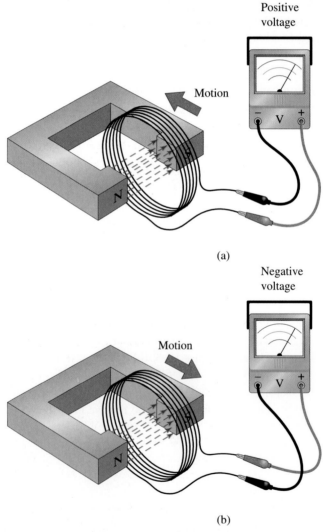

Figure 10-40. The direction of motion determines the polarity of induced voltage.

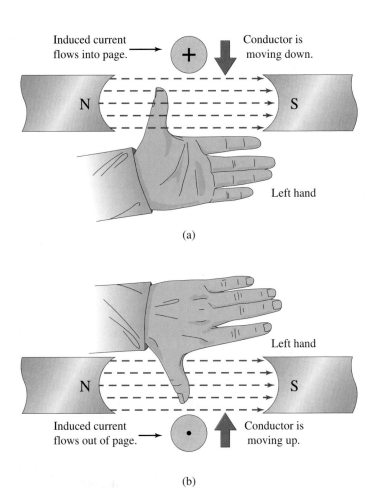

Figure 10-41. The left-hand rule for induction can be used to determine the polarity of induced voltage or current.

1. Point the fingers of your left hand in the direction of the flux lines (i.e., north to south).
2. Turn your palm such that it could catch the moving conductor.
3. Your thumb indicates the direction of induced electron current, or in the case of an open conductor, the negative side of the induced voltage.

The rule works regardless of whether it is the conductor or the field that is moving.

MAGNETIC FIELD POLARITY

As evidenced in the application of the left-hand rule for induction, the polarity of the induced voltage or current is also affected by the polarity of the magnetic field. For a given direction of conductor movement, the polarity of the induced voltage is directly determined by the polarity of the magnetic field.

● KEY POINTS

If the conductor is connected to a closed circuit, then the induced voltage causes a current to flow. This is often referred to as an induced current.

Practice Problems

1. Use the left-hand rule for induction to determine the direction of induced current in each of the conductors shown in Figure 10-42.

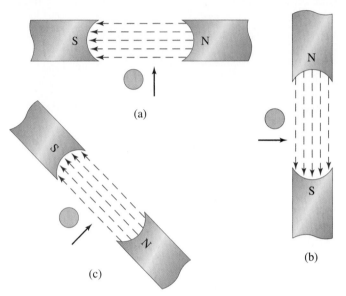

Figure 10-42. Determine the direction of induced current in each of these conductors.

2. Determine the direction of conductor movement for each of the cases shown in Figure 10-43.

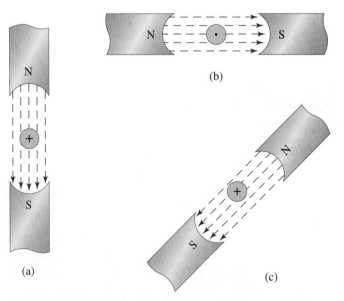

Figure 10-43. Which way are the conductors moving?

3. Use the left-hand rule for induction to correctly label the magnetic poles for each case in Figure 10-44.

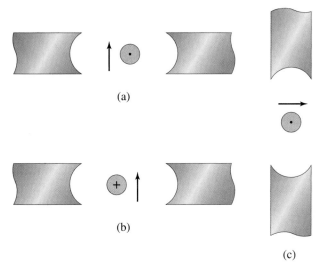

(a)

(b)

(c)

Figure 10-44. Label the magnetic poles.

Answers to Practice Problems

1. a. in (+) b. out (·) c. in (+)

2. a. right-to-left b. upward c. up and to the left

3. a. N–S b. S–N c. $\begin{array}{c}N\\S\end{array}$

MAGNETIC COUPLING BETWEEN COILS

Figure 10-45 illustrates yet another aspect of electromagnetic induction. Here, two coils are wound on a common ferromagnetic core. If the material used for the core has a high relative permeability, then most of the flux will be shared between the two coils. The flux lines find it easier to pass through the core than through the surrounding air.

The dc voltage source in Figure 10-45(a) is constant, which produces a constant magnetizing force. Although the core causes much of the flux to exist in the region of the second coil, the flux is not changing or moving. Recall that there must be relative motion between a magnetic field and a conductor in order to produce electromagnetic induction.

In Figure 10-45(b), the value of the dc supply is increasing, which causes an increasing magnetizing force and a corresponding increase in flux density in the core. As the field expands in the vicinity of the second coil, it will cut through the wires of the second winding and will induce a voltage of a certain polarity (positive in Figure 10-45b). Although this is a real and measurable voltage, it exists only so long as the magnetic field is changing. Once the dc supply reaches its maximum value and becomes steady, the induced voltage will drop to zero, regardless of the strength of the magnetic field.

⦿ KEY POINTS

If the magnetizing force in a given coil is changing, then its magnetic field will be expanding and contracting. If the moving field cuts through the turns of a second coil, a voltage will be induced in the second coil by electromagnetic induction.

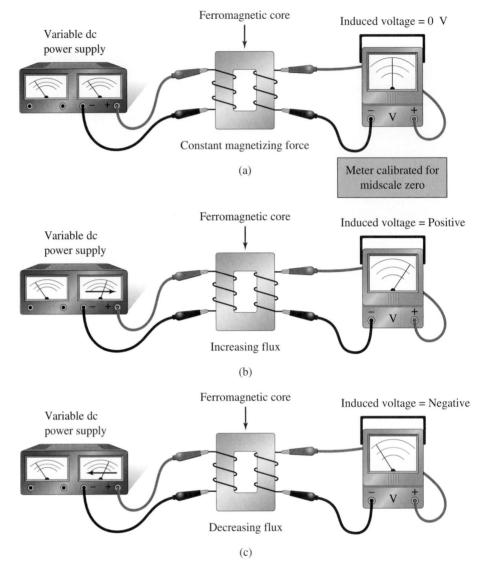

Ferromagnetic core

Induced voltage = 0 V

Variable dc
power supply

Constant magnetizing force

(a)

Meter calibrated for
midscale zero

Ferromagnetic core

Induced voltage = Positive

Variable dc
power supply

Increasing flux

(b)

Ferromagnetic core

Induced voltage = Negative

Variable dc
power supply

Decreasing flux

(c)

Figure 10-45. Flux changes created by one coil can induce a voltage in a second coil.

◉ KEY POINTS

The polarity of the
induced voltage
depends on whether the
field is expanding or
contracting (i.e., direc-
tion of motion). The
magnitude of the volt-
age depends on field
strength, rate of flux
change, and number of
turns in the coils.

Figure 10-45(c) shows that if the dc voltage source is then reduced from its maximum value, the magnetizing force and the flux density in the material will decrease proportionally. As the flux lines decrease, they will cut through the second coil and induce a voltage. The polarity of induced voltage will be opposite that produced by an expanding field. The polarity of the induced voltage in Figure 10-45(c) is labeled as negative.

ALTERNATIVE DEFINITION OF FLUX MEASUREMENT

Recall that total flux is measured in webers, where one weber corresponds to 1×10^8 flux lines. Some people object to this definition, since we are defining a unit of measurement for an imaginary quantity. Figure 10-46 shows an alternate way to define the amount of flux represented by one weber.

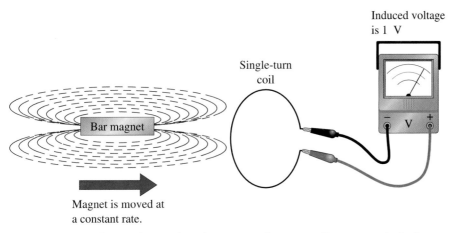

Induced voltage
is 1 V

Single-turn
coil

Bar magnet

Magnet is moved at
a constant rate.

Figure 10-46. If one volt is induced into a single-turn coil over a period of one second, then the amount of flux cutting the coil is exactly one weber.

Here, a magnetic field is moved past a conductor at a constant rate. The speed of movement is such that 1 V is induced into the conductor. Under these conditions, 1 Wb of flux will pass through the conductor every second. This same definition may be applied to a case where two coils are magnetically coupled via a common core (e.g., Figure 10-45). In this case, a sustained flux change of 1 Wb per second will induce a constant voltage of 1 V into a single-turn coil.

Basic Generator Action

A generator is a device that converts mechanical energy into electrical energy. A dc generator operates by spinning a coil inside a magnetic field. As the wires cut the flux lines, a voltage is induced. Figure 10-47 shows the basic configuration of a dc generator.

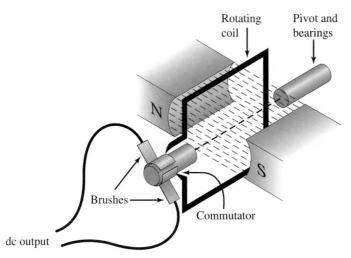

Rotating
coil

Pivot and
bearings

N

S

Brushes

Commutator

dc output

Figure 10-47. A dc generator rotates a coil in a magnetic field to produce a dc output voltage.

KEY POINTS

If the rate of change of flux is constant for a full second, and the voltage induced into a single turn on the second coil is 1 V, then we can say that the conductor was cut by an amount of flux equal to 1 Wb.

KEY POINTS

A generator is a device that converts mechanical energy into electrical energy. A dc generator spins a coil in a magnetic field to induce a voltage into the coil.

For simplicity, only a single-turn coil is shown. In a practical generator, the coil consists of many turns of wire, in order to produce a practical output voltage at a reasonable speed of rotation. As shown in Figure 10-47, the ends of the coil(s) connect to the segments of a commutator. The commutator segments are mounted to the pivotal shaft of the generator and provide a sliding contact to the external circuit via spring-loaded carbon brushes. Figure 10-48 shows a closer view of the commutator assembly. As the coil and commutator segments rotate in the magnetic field, the brushes slide along the surface of the commutator segments. This permits the external circuit to have a continuous electrical connection to the rotating coil. Note that a given brush will touch alternate segments as the coil/commutator assembly is rotated. The reason for this will be discussed shortly.

Figure 10-49 shows a dc generator as its coil passes through various points of its 360° rotation. In Figure 10-49(a), both sides of the loop are moving parallel to the flux lines, so no voltage is induced in the coil. By the time the coil has rotated 90°, as shown in Figure 10-49(b), both sides of the loop are moving perpendicular to the flux lines. This causes a maximum value of induced voltage. The left-hand rule for induction can be used to determine the polarity of the induced voltages. As indicated in Figure 10-49(b), the polarities of the two sides of the loop are additive. Be sure to note the overall polarity of voltage being generated.

As the loop rotates, it will move to the position shown in Figure 10-49(c). Here, the conductors are moving parallel to the flux lines, so no voltage will be induced. As the loop continues to rotate, it will pass through the position shown in Figure 10-49(d). Again, the conductors are moving perpendicular to the flux. A maximum value of induced voltage will be induced. However, note that the polarity is opposite from that shown in Figure 10-49(b). This means that the current in an external circuit would be reversed between the positions shown in Figure 10-49(b) and (d).

This represents a serious problem, since direct current is supposed to flow in one direction. The solution to the problem lies in the commutator. Close examination of the illustration shown in Figure 10-48 will reveal that the commutator effectively switches the coil connections. During half of the 360° rotation, a given brush is connected to one side of

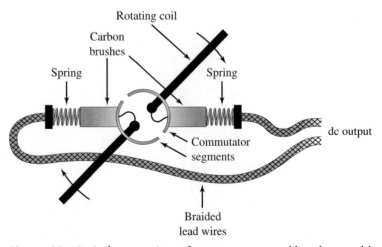

Figure 10-48. A close-up view of a commutator and brush assembly.

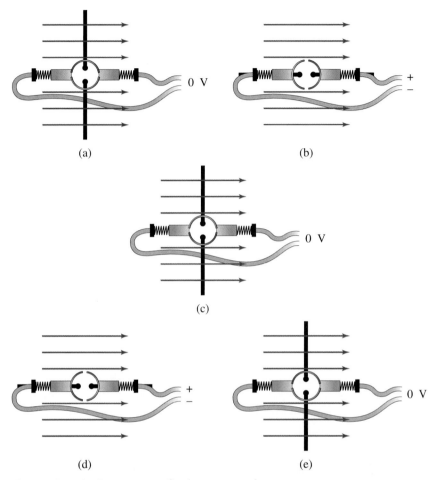

Figure 10-49. Generation of a dc output voltage.

the rotating coil. During this time (e.g., between the positions shown in Figure 10-49(a) and (c)), the polarity in the coils does not change. During the other half of the full 360° rotation, the given brush is connected to the opposite side of the rotating coil. So, both the coil polarity and the brush/coil connections have changed. The net result is that the polarity to the external circuit is the same for the entire 360° rotation of the coil.

As the coil passes through the positions shown in Figure 10-49(a) and (c), the switching of the brushes to the opposite side of the coil occurs. Recall that no voltage is being induced into the coil at these points.

With the addition of a commutator, we can supply current to an external circuit that flows in only one direction. It has one additional problem. Recall that the magnitude of induced voltage varied with the angle of flux intersection. As the coil in Figure 10-49 rotates, the angle of flux cutting is constantly changing; the output voltage also fluctuates continuously. This produces what is known as *pulsating dc* and is suitable for some purposes (e.g., battery charging). For many purposes, however, we need a dc source with a constant voltage output. This is easily obtained from a dc generator in either of two ways: adding more coils and commutator segments, or adding external filters (discussed in later chapters).

Basic Motor Action

Figure 10-50 shows how the interaction of magnetic fields causes a motor to rotate. The rotating shaft of a motor has coils that can move through a magnetic field. A single loop of the rotating coil is represented in Figure 10-50. Notice the construction is the same as that of a dc generator. In the case of a motor, however, we will connect the coil leads (via the commutator/brush assembly) to an external source of voltage. The external voltage forces current through the coil in the direction shown by the cross and dot symbols in Figure 10-50. The left-hand rule for conductors can be used to determine the polarity of the magnetic field around the coil wires as a direct result of the current flowing through them. This field is shown in Figure 10-50. Recall that like polarity flux lines repel, while opposite polarity fields attract. In Figure 10-50, we can see that the fields around the rotating coil wires interact with the stationary field from the magnets. The interaction causes one side of the coil to move downward, while the other side has an upward force. The combined force produces a rotational force called torque. With the polarities shown in Figure 10-50, the motor will rotate in a counterclockwise direction. In order to achieve continuous rotation, the commutator will have to switch the coil connections at just the right time.

Even though the diagram in Figure 10-50 represents a dc motor, it might be interesting for you to apply the left-hand rule for induction to the moving coils. You will find that a voltage is induced in the moving coils that is *exactly opposite* the polarity of the applied voltage. In effect, the induced voltage is in series with the external voltage. The result is that the current through the motor is much less than it would be without the induced voltage (called counter emf). You may have noticed the lights in your home or automobile getting momentarily dim when a motor first starts. This is because the motors have no counter emf until they start rotating; therefore, the initial starting current can be quite high.

d'Arsonval Meter Movements

Figure 10-51 shows how operation of a d'Arsonval meter movement relies on the interaction between magnetic fields. As the current passes through the meter coil, a magnetic field is formed around the coil wires. This field interacts with the permanent magnet field to produce a torque that moves the pointer upscale.

Electromagnetic induction can also be used to provide meter damping when the meter is not in use (e.g., during shipment). Damping is desired in these cases to keep the meter

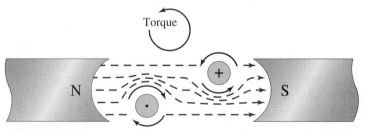

Figure 10-50. Interaction of magnetic fields causes a dc motor to rotate.

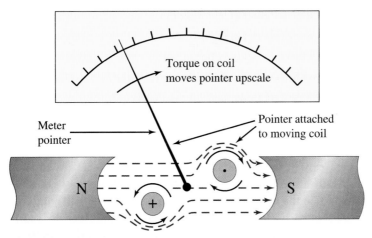

Figure 10-51. A basic d'Arsonval meter movement requires interaction of magnetic fields for its operation.

pointer from swinging wildly (and possibly bending) in response to a mechanical shock. Figure 10-52 illustrates magnetic meter damping.

First, the meter leads must be shorted together (with no series multipliers or other resistances). Many meters have a "transit" position on the mode switch, which shorts the meter movement internally. Alternatively, you can put the meter on its most sensitive current scale and simply short the external meter leads. Figure 10-52 shows that if the meter receives a mechanical shock that tries to swing the pointer upscale, an opposing magnetic field will be produced by the induced coil current. The faster the pointer tries to move, the stronger the opposing field. This is easily demonstrated in the laboratory and is an informative experiment.

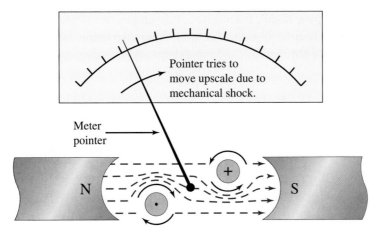

Induced coil current causes an opposing magnetic field, which dampens the pointer movement.

Figure 10-52. Magnetic meter damping can prevent damage caused by mechanical shocks.

Magnetic Force on an Electric Charge

KEY POINTS

There is no interaction between a magnetic field and a stationary electric charge.

Figure 10-53 shows a hypothetical experiment. Here, we have a free electron (a negative charge) surrounded by a magnetic field. Neither the electron nor the field is moving. Under these conditions, there is no interaction between the electron and the field.

Now consider the case represented in Figure 10-54. Here, the magnetic field is stationary, but the electron is moving (initially in a straight-line path). When the electron encounters the magnetic field, it is diverted from its previously straight flight.

KEY POINTS

When a particle that has either a positive or negative charge moves through a magnetic field, then its path will be altered by the force of the field.

If the moving charge moves parallel to the lines of flux, then there is no interaction. This is similar to a conductor moving parallel to the flux lines. If the charge is traveling at right angles to the field, then the field provides maximum force on the moving charge. The amount of force depends on the flux density, the velocity of the moving charge, the magnitude of the charge, and the angle of entry into the field. The resulting deflection can cause a slight bending of the charge's path, or it can cause the moving particle to enter a circular or spiral path. Television picture tubes work much like the cathode-ray tubes discussed in Chapter 9. However, deflection of the electron beam is accomplished with electromagnetic fields instead of static charges on deflection plates.

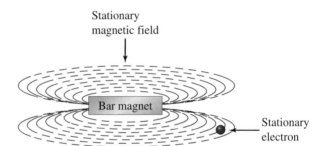

Figure 10-53. If a magnetic field and an electron have no relative motion, then there is no interaction between them.

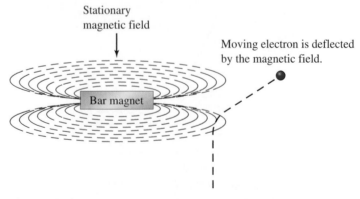

Figure 10-54. A magnetic field exerts a deflective force on a moving charge.

Exercise Problems 10.3

1. An induced voltage is only an imaginary concept and cannot perform any useful work. (True or False)

2. There must be a closed loop in order to have an induced current. (True or False)

3. If a conductor is moving parallel to the flux lines in a magnetic field, will a voltage be induced into the conductor?

4. List the four factors that affect the magnitude of the voltage induced into a conductor as it passes through a magnetic field.

5. What factors determine the polarity of voltage induced in a wire as it passes through a magnetic field?

6. Use the left-hand rule for induction to determine the direction of current flow in the conductor shown in Figure 10-55.

7. Determine the magnetic polarity in the induction diagram shown in Figure 10-56.

8. In Figure 10-57, the two coils share a common ferromagnetic core. There is a sizable magnetizing force in the left-most coil, yet the voltmeter on the right-most coil is reading 0 V. Explain why.

9. What is a commutator and what is its purpose?

10. Assume that the coil shown in Figure 10-58 is connected to a closed external circuit. Determine the direction of current through the coil.

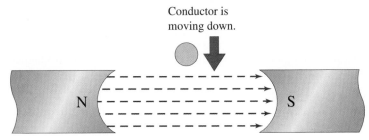

Figure 10-55. What is the direction of induced current?

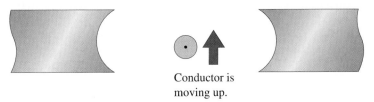

Figure 10-56. Label the poles of the magnet.

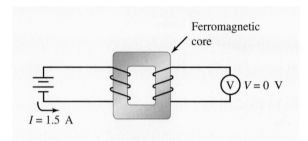

Figure 10-57. Why does the voltmeter read zero?

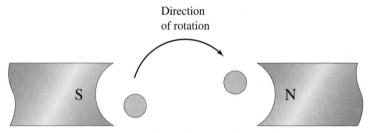

Figure 10-58. Determine the polarity of induced current.

10.4 Magnetic Circuit Analysis

KEY POINTS

The basic problem-solving methods used to analyze basic electrical circuits can be applied to the analysis of magnetic circuits.

A magnetic circuit can be related to a series electrical circuit; there are many analogies. These are shown in Table 10-1. We have already compared magnetomotive force in a magnetic circuit to electromotive force, or simply voltage, in an electrical circuit. We have also seen that permeance and reluctance in magnetic circuits are equivalent to conductance and resistance in an electrical circuit. Similarly, we can compare flux in a magnetic circuit to current in an electrical circuit. These analogies can help us remember the relationships in a magnetic circuit. Now, let us carry the analogy even further.

Figure 10-59 shows a magnetic circuit. Since the core material is ferromagnetic, most of the flux created by the coil is contained within the core material. In the case shown, the core material has an air gap. This is a common situation and is used in motors, generators, heads in tape recorders, and many other applications. As the flux travels around the complete magnetic loop, it encounters the opposition (reluctance) of the core and the reluctance of the air gap. These two reluctances are essentially in series and can be compared to resistances in a series electrical circuit. The total magnetomotive force is "dropped" (i.e., impressed) across the various circuit reluctances much like voltage in an electrical circuit is dropped across the series resistances. Further, the sum of the various mmf drops is equal to the total emf supplied to the coil. This is akin to Kirchhoff's Voltage Law for series circuits.

We can apply what we have learned to analyze the magnetic characteristics of the circuit shown in Figure 10-59. First, let's compute the magnetomotive force with Equation 10-2.

$$\mathscr{F} = NI$$
$$= 100 \text{ t} \times 2 \text{ A} = 200 \text{ A} \cdot \text{t}$$

MAGNETIC QUANTITY	SYMBOL	EQUATION	UNITS	ELECTRICAL QUANTITY	SYMBOL	EQUATION	UNITS
Magnetomotive force	$\mathscr{F}$	$\mathscr{F} = NI$	ampere · turn	Voltage	V	$V = IR$	volt
Reluctance	$\mathscr{R}$	$\mathscr{R} = l/\mu A$	ampere · turn per weber	Resistance	R	$R = V/I$	ohm
Flux	Φ	$\Phi = \mathscr{F}/\mathscr{R}$	weber	Current	I	$I = V/R$	ampere
Magnetizing force	H	$H = \mathscr{F}/l$	ampere · turn per meter	Electric field intensity	E	$E = V/l$	volts per meter
Flux density	B	$B = \Phi/A$	tesla	Current density	J	$J = I/A$	amperes per square meter
Permeability	μ	$\mu = B/H$	webers per ampere · turn · meter	Conductivity	σ	$\sigma = \rho$, where ρ = resistivity	siemens per meter
Permeance	$\wp$	$\wp = 1/\mathscr{R}$	webers per ampere · turn	Conductance	G	$G = 1/R$	siemens

Table 10-1. Comparison of Magnetic and Electric Quantities

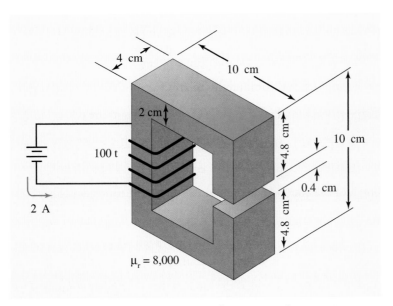

Figure 10-59. A magnetic circuit can be compared to a series electrical circuit.

Next, we can find the reluctance of the magnetic circuit. Total circuit reluctance consists of the "series" combination of the reluctance of the core and the reluctance of the air gap. Recall from Equation 10-6 that reluctance depends on the permeability of the material, the length of the magnetic path, and the cross-sectional area of the path. Let's find the reluctance of the air gap. The permeability of air is assumed to be the same as the permeability of a vacuum. That value of course is 1.257×10^{-6} Wb/A · t · m. The length of the

air gap portion of the magnetic path is given in Figure 10-59 as 0.4 cm or 0.004 m. Finally, the cross-sectional area of the gap (in meters) can be computed as follows:

$$A = width \times depth$$
$$= 4 \text{ cm} \times 2 \text{ cm}$$
$$= 8 \text{ cm}^2 = 800 \times 10^{-6} \text{ m}^2$$

We can now compute the reluctance of the air gap by applying Equation 10-6.

$$\Re_{air} = \frac{l}{\mu A}$$
$$= \frac{0.004 \text{ m}}{1.257 \times 10^{-6} \text{ Wb/A} \cdot t \cdot m \times 800 \times 10^{-6} \text{ m}^2}$$
$$= 3.98 \times 10^6 \text{ A} \cdot t/\text{Wb}$$

In order to compute the reluctance of the core, we must find the length of the magnetic path through the core. If the core were a straight bar, we would simply measure its length. In the present case, the distance around the core varies depending on whether you travel near the outer edges or more toward the inner edges of the core. We shall use the distance around the center of the core as the length in our computations. The length of the top, bottom, and left legs will be equal to 10 cm minus half of the width of the side legs. That is,

$$l_{top} = l_{bottom} = l_{left side} = 10 \text{ cm} - \frac{4 \text{ cm}}{2}$$
$$= 10 \text{ cm} - 2 \text{ cm} = 8 \text{ cm}$$

The length of the right side is computed similarly, except that we must subtract the length of the air gap:

$$l_{right side} = 10 \text{ cm} - \frac{4 \text{ cm}}{2} - 0.4 \text{ cm}$$
$$= 7.6 \text{ cm}$$

The total length of the core path is the simple sum of the four sides. That is,

$$l_{core} = l_{left side} + l_{right side} + l_{top} + l_{bottom}$$
$$= 8 \text{ cm} + 7.6 \text{ cm} + 8 \text{ cm} + 8 \text{ cm}$$
$$= 31.6 \text{ cm} = 0.3196 \text{ m}$$

The cross-sectional area of the core is the same as that previously computed for the air gap ($A = 800 \times 10^{-6}$ m²). The relative permeability of the core is given as 8,000. We compute its absolute permeability with Equation 10-5.

$$\mu_r = \frac{\mu_{core}}{\mu_v}, \text{ or}$$
$$\mu_{core} = \mu_r \times \mu_v$$
$$= 8,000 \times 1.257 \times 10^{-6} \text{ Wb/A} \cdot t \cdot m$$
$$= 10.06 \times 10^{-3} \text{ Wb/A} \cdot t \cdot m$$

We can now find the reluctance of the core path as follows:

$$\mathfrak{R}_{core} = \frac{l}{\mu A}$$

$$= \frac{0.32 \text{ m}}{10.06 \times 10^{-3} \text{ Wb/A} \cdot \text{t} \cdot \text{m} \times 800 \times 10^{-6} \text{m}^2}$$

$$= 39.76 \times 10^3 \text{ A} \cdot \text{t/Wb}$$

In a series electrical circuit, we found total resistance by summing the individual resistances. Similarly, to find total reluctance we add the individual reluctances.

$$\mathfrak{R}_T = \mathfrak{R}_{air} + \mathfrak{R}_{core}$$

$$= 3.98 \times 10^6 \text{ A} \cdot \text{t/Wb} + 39.7 \times 10^3 \text{ A} \cdot \text{t/Wb}$$

$$= 4.02 \times 10^6 \text{ A} \cdot \text{t/Wb}$$

Notice that the reluctance of the air gap was substantially larger than the reluctance of the core. We can now apply the "Ohm's Law" of magnetic circuits (Equation 10-7) to find total flux in the magnetic circuit.

$$\Phi = \frac{\mathcal{F}}{\mathfrak{R}}$$

$$= \frac{200 \text{ A} \cdot \text{t}}{4.02 \times 10^6 \text{ A} \cdot \text{t/Wb}}$$

$$= 49.75 \text{ } \mu\text{Wb}$$

Figure 10-60 illustrates a phenomenon called **fringing.** The flux lines in the air gap tend to repel each other. This causes them to extend beyond the edges of the core. Another way to view the cause of fringing is to remember that the air gap has a much lower permeability than the core. Consequently, we would expect a lower flux density (i.e., lines of flux farther apart). For very short air gaps (as shown in Figure 10-59), fringing can generally be ignored. This greatly simplifies the calculations, and means that we assume the flux density is the same in both the core and the air gap. This is a valid assumption for many electromagnetic applications.

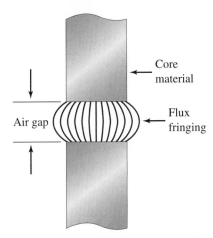

Figure 10-60. The flux lines tend to bulge at the air gap and cause fringing.

Based on the assumption that fringing can be neglected, let us compute flux density by applying Equation 10-1.

$$B = \frac{\Phi}{A}$$

$$= \frac{49.75 \ \mu\text{Wb}}{800 \times 10^{-6} \ \text{m}^2}$$

$$= 62.19 \times 10^{-3} \ \text{T}$$

We can apply Equation 10-7 to determine the magnetomotive force that is dropped across the air gap. This is similar to the voltage drop across an individual resistor in a series circuit.

$$\Phi = \frac{\mathcal{F}}{\mathcal{R}}, \text{ or}$$

$$\mathcal{F} = \Phi \times \mathcal{R}$$

$$= 49.75 \ \mu\text{Wb} \times 3.98 \times 10^6 \ \text{A} \cdot \text{t/Wb}$$

$$= 198 \ \text{A} \cdot \text{t}$$

In a similar manner, we could find the mmf across the core portion of the magnetic circuit.

$$\mathcal{F} = \Phi \times \mathcal{R}$$

$$= 49.75 \ \mu\text{Wb} \times 39.7 \times 10^3 \ \text{A} \cdot \text{t/Wb}$$

$$= 2 \ \text{A} \cdot \text{t}$$

Kirchhoff's Voltage Law also has an equivalent in magnetic circuits; the sum of the mmf drops around the magnetic circuit must equal the applied mmf. We express this relationship in Equation 10-9.

$$\mathcal{F}_{\text{total}} = \mathcal{F}_1 + \mathcal{F}_2 + \ldots + \mathcal{F}_N \qquad \textbf{(10-9)}$$

In the present case, we have

$$\mathcal{F}_{\text{total}} = \mathcal{F}_{\text{air}} + \mathcal{F}_{\text{core}}$$

$$= 198 \ \text{A} \cdot \text{t} + 2 \ \text{A} \cdot \text{t} = 200 \ \text{A} \cdot \text{t}$$

The magnetizing force in the air gap is a frequently needed quantity. We can compute this by applying Equation 10-3.

$$H_{\text{air}} = \frac{\mathcal{F}_{\text{air}}}{l_{\text{air}}}$$

$$= \frac{198 \ \text{A} \cdot \text{t}}{0.004 \ \text{m}}$$

$$= 49.5 \times 10^3 \ \text{A} \cdot \text{t/m}$$

Our comparison of magnetic and electrical circuits can be extended to include the analysis of series-parallel magnetic circuits. Figure 10-61 shows a circuit of this type. Here, the flux produced in the leftmost leg of the core will have two parallel branches in which to travel. Part of the flux will pass through the center leg; the remaining flux will pass through the leg with the air gap (comparable to Kirchhoff's Current Law). Since the air-gap leg is

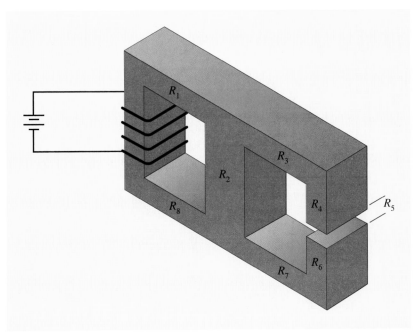

Figure 10-61. A series-parallel magnetic circuit.

bound to have more reluctance (i.e., lower permeability of air), we expect that less flux will pass through the air-gap leg than the center leg. The comparisons can be continued, but in general, we can combine the problem-solving methods that we learned for electrical circuits with the equations presented in this chapter to provide a method for analyzing magnetic circuits.

Exercise Problems 10.4

1. What is the magnetomotive force produced by a 50-t coil with 150 mA of current?
2. Compute the reluctance of an air gap in a ferromagnetic core if the length of the gap is 0.2 cm and it is 5 cm on each side.
3. Define the term fringing as it applies to magnetic circuits.
4. Compute the flux density of the air gap described in Problem 2, if the total flux in the gap is 100 µWb.
5. Compute the magnetizing force in an air gap if the flux is 200 µWb, the reluctance is 500×10^3 A · t/Wb, and the length is 0.1 cm.

   Refer to Figure 10-62 for Problems 6 through 10.

6. What is the magnetomotive force applied to the magnetic circuit?
7. What is the reluctance of the air gap?
8. What is the total reluctance of the circuit?
9. How much flux flows through the circuit?
10. What is the magnetizing force within the air gap?

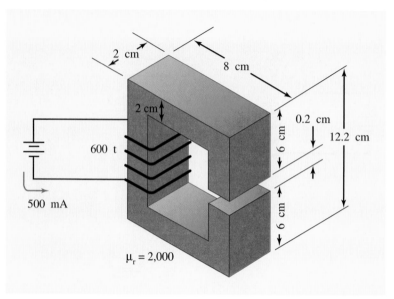

Figure 10-62. A magnetic circuit for analysis.

10.5 Applied Technology: Magnetic and Electromagnetic Devices

As an electronics technician, you will encounter a wide range of applications that make direct use of the electromagnetic principles presented in the preceding sections of this chapter. In the following sections we will examine several magnetic/electromagnetic applications. We will focus on how the devices operate.

Security Sensors

Figure 10-63 shows the cutaway view of a magnetically operated switch. One of the switch contacts is mounted on a movable arm made of a ferromagnetic material. The other contact is mounted on the plastic housing. A spring keeps the two contacts separated when there are no strong magnetic fields present. Two screws on the exterior of the plastic case provide a means for connecting the external circuit to the switch. As shown in Figure 10-63(a), the magnetic field from the plastic-housed magnet is too weak to attract the movable switch arm; the switch remains open.

In Figure 10-63(b), the magnet has been moved physically close to the magnetically operated switch. The field of the magnet attracts the movable switch arm and overcomes the tension of the spring. This brings the switch contacts together and closes the external circuit. Thus, the state of the switch (i.e., opened or closed) is determined by the location of the magnet. If the magnet is physically close to the switch, then the switch is closed; otherwise, it is open.

The switch pictured in Figure 10-63 would be classified as a single-pole single-throw (SPST) normally closed switch, since the switch is closed under static (nonalarm) conditions. Other forms such as normally open and single-pole double-throw (SPDT) are readily available.

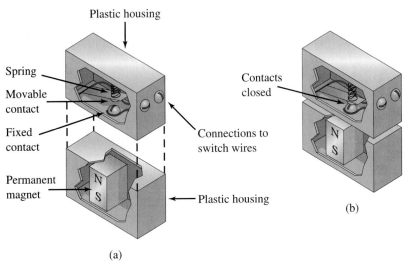

Figure 10-63. A cutaway view of a magnetically operated switch.

Figure 10-64 illustrates how the switch could be used in a burglar alarm application. Figure 10-64(a) shows the location of the magnetic switches throughout the house. Figure 10-64(b) shows a close up of how the switch/magnet assembly would be mounted on a door or window. The magnet is fastened on the door or movable part of the window; the switch itself is normally mounted on the door frame or stationary part of a sliding window. They are positioned such that the magnet and switch are in close proximity when the door or window is closed. The wires from the magnetically operated switch are routed through the house to other switches and, ultimately, to an electronic alarm controller.

As pictured in Figure 10-64(a), the various switches are all connected as part of a large series loop terminating at the controller. Once the circuit has been activated, the alarm will sound if any of the switches become open (as a result of its associated door or window opening).

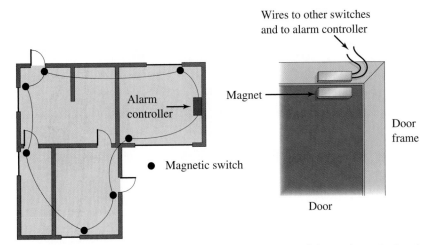

Figure 10-64. Magnetically operated switches are widely used as window/door sensors in alarm systems.

Magnetic Tachometer

Figure 10-65 shows how a magnetic tachometer works by sensing the presence or absence of the teeth on a ferromagnetic sprocket. If the sprocket is stationary, then the amount of flux flowing in the magnetic circuit is stable and no voltage is induced into the coil. The amount of flux that flows in the circuit at any given time is determined by the presence or absence of the teeth on the sprocket.

The presence of a ferromagnetic tooth between the two poles of the magnet causes the reluctance of that portion of the flux path to decrease. A reduced reluctance, of course, translates to a higher flux flow. If the sprocket is positioned so that the magnetic poles are between two adjacent teeth on the sprocket, then the reluctance in the magnetic path will increase, since the flux must flow through air (low permeability) rather than metal (high permeability).

As the sprocket rotates, then, the reluctance of the magnetic path continually changes from a high (no tooth between the poles) to low (tooth present). This causes the flux in the entire magnetic path to increase and decrease at a rate determined by the speed of rotation of the sprocket. As the flux increases and decreases (i.e., changes) in the region of the core internal to the coil, the moving flux will induce a voltage into the coil. It will not be a steady dc voltage, but rather a series of impulses or surges in voltage. These impulses are readily detected and counted by an electronic circuit. If we know the spacing of the teeth and the diameter of the sprocket, and we can measure the number of voltage impulses produced during a given time, then we can easily (electronically) compute and display the rotational speed of the sprocket in terms of revolutions per minute (rpm).

Magnetic Recording Devices

Many electronic devices rely on the use of magnetic storage for information. A cassette player/recorder stores voice and music information in the form of magnetized regions on

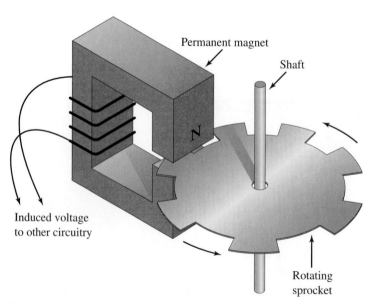

Figure 10-65. A magnetic tachometer measures the rpm of a sprocket with no physical contact.

the tape. Computer disks store huge amounts of information by magnetizing tiny regions of the disk in coded formats that represent numbers, letters, and computer commands. Most charge cards and many ID cards have a magnetic strip on the back side that holds personal information as tiny magnetized spots. Now let's briefly examine how the various magnetic recording devices store their information and how the information is subsequently retrieved.

Figure 10-66 illustrates a magnetic tape and a recording head. The magnetic tape consists of a tough, flexible mylar tape with a coating of ferromagnetic material called iron oxide. The iron oxide has a high level of retentivity. A motor drive mechanism causes the tape (or disk) to move across the surface of the recording head at a fixed rate.

Figure 10-67 shows a more detailed view of the recording head. It consists of a ferromagnetic core with a coil wrapped around one leg. There is an air gap on the portion of the head that rubs against the tape surface.

The music or digital computer information to be recorded is applied to the coil of the recording head. The information is represented electrically by a fluctuating voltage. When this varying voltage is applied to the coil of the recording head, it produces corresponding changes in flux. The flux changes are felt at all points in the magnetic circuit. As discussed in a previous section, the flux lines produce fringing at the air gap due to the lower permeability of the air gap. In this application, fringing is not only desired, it is essential. As the flux penetrates the iron-oxide coating, it causes the domains to align with the instantaneous polarity of the flux. As the flux continuously changes with the recording signal and the tape moves continuously across the recording head, the domains along the length of the recorded portion of the tape are left in a pattern that corresponds to the fluctuations of the recorded signal. The retentivity of the iron-oxide coating keeps the domains from rotating back to their random (unmagnetized) pattern.

Figure 10-68 illustrates how a recorded signal can be recovered from a magnetic tape. As the recorded tape moves past the playback head (which is constructed just like the recording head), the magnetic fields of the tape's domains cause flux variations in the magnetic circuit of the playback head. The flux changes cause a voltage to be induced into the coil. The induced voltage (which is subsequently amplified and fed to a speaker) corresponds to the flux changes, which correspond to the domain patterns, which correspond to the variations in the original recording signal. Thus, we have a means to record information magnetically, store it indefinitely, and then reproduce it anytime we like.

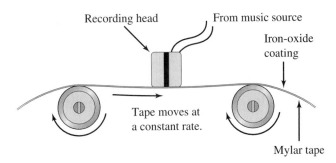

Figure 10-66. A magnetic tape recorder stores information in the form of tiny magnetized regions.

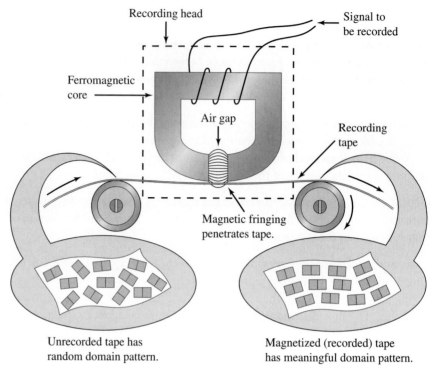

Figure 10-67. A magnetic recording head is an electromagnet with an air gap.

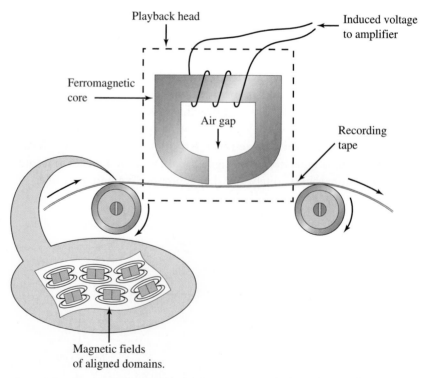

Figure 10-68. A playback head is used to sense the recorded information.

If you have the opportunity to examine the heads on a magnetic recorder, they will appear to be smooth, shiny, and solid (i.e., no apparent air gap). But, if you look very closely, you will see that the surface of the head that actually touches the tape is laminated (has layers of different materials). This laminated area of the head corresponds to the air gap pictured in Figures 10-67 and 10-68. In order to make the head smoother to produce less friction and wear on the tape, the air gap is generally filled in with a nonmagnetic material. So, it still functions as an air gap with regard to magnetic behavior, but it provides a mechanically smooth surface for the tape to slide over.

Speakers

The speakers in your home or car stereo rely on electromagnetism for their operation. Figure 10-69 shows the basic components of a permanent magnet (PM) speaker. The sound from a speaker is produced when the cone physically moves in and out (i.e., vibrates) at the frequency of the sounds being reproduced. When the cone moves forward, the air is compressed; when the cone moves backward, a low-pressure region is formed. The regions of high and low pressure travel through the air and strike your eardrum. The pressure changes on your eardrum vibrate the eardrum and produce a sensation called sound.

The center of the speaker cone is fastened to a coil (called a voice coil). The coil is physically positioned over a protruding portion of a magnet. The coil is free to move over the magnetic core without touching it.

The voice or music is applied to the voice coil as a changing voltage (current). As the current through the voice coil changes, the resultant magnetic field around it changes. The fluctuating field of the voice coil interacts with the field of the permanent magnet. The

KEY POINTS

The magnetic fields of the magnetized iron oxide induce a voltage into the coil of the playback head. The induced voltage is then amplified and interpreted by other electronic circuitry.

KEY POINTS

As the cone moves, it creates high- and low-pressure areas in the air in front of the speaker. These varying pressure areas create the sensation called sound when they strike the human eardrum.

KEY POINTS

The center of the speaker cone is fastened to the moving voice coil.

KEY POINTS

In a permanent magnet speaker, voltage changes that represent the sound are applied to the voice coil of the speaker. The applied signal causes current changes in the coil, which produce corresponding flux changes around the coil.

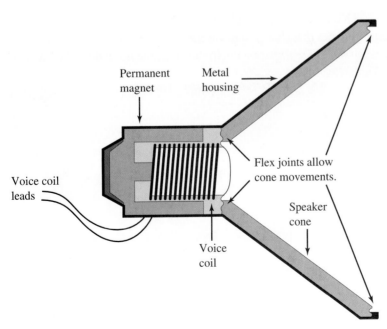

Figure 10-69. A permanent magnet speaker relies on electromagnetism.

force of attraction and repulsion between the two magnetic fields causes the voice coil (and the attached speaker cone) to move in and out in response to the applied voltage.

Larger changes in the voltage applied to the speaker coil move the cone further in and out, producing a louder sound. Faster changes in the applied voltage cause the cone to move in and out more quickly, producing a higher-frequency sound.

Relays

A relay is an electromagnetically operated switch. It is a fundamental component that is used in a tremendous range of electronic products. Several representative relays are shown in Figure 10-70. Applications for relays include motor control circuits for industrial machines, circuits to protect workers, switching circuits in automobiles, and power-switching circuits in notebook computers. Although relays are simple in principle, they can be used in very complex circuits, and pose a new set of troubleshooting challenges for technicians.

RELAY CONSTRUCTION

Figure 10-71 illustrates the basic component parts of a relay. Although relays vary greatly in size and appearance, they all have the components called out in Figure 10-71. As we examine the operation of the relay, it is important to remember that the relay is essentially a switch, a simple on-off/open-close switch. The relay differs from a standard switch in the way the switch is moved from one position to the other. A basic switch changes state when its mechanism is moved by the force from a human finger. The switching mechanism of a relay, by contrast, is moved between states by the force of attraction provided by an electromagnetic field.

First, locate the switch contacts on the relay illustration in Figure 10-71(a). The particular relay illustrated has double-pole double-throw (DPDT) contacts. It is effectively two electrically separate switches that change states at the same time. The relay in Figure 10-71(a) is shown in its *normal* or *deenergized* state. In this state, the normally closed switch contacts are

Figure 10-70. Several representative relays. (Courtesy of Potter & Brumfield, Inc.)

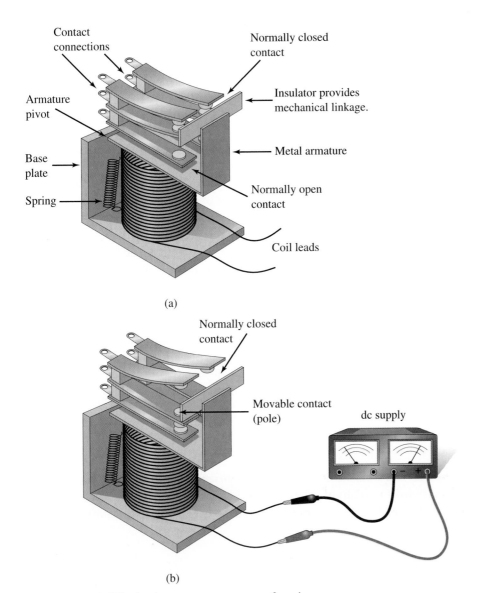

Figure 10-71. The basic component parts of a relay.

closed (shorted together), while the normally open contacts are open (physically separated). Spring tension on the movable contact forces the contacts to stay in the normal position.

Figure 10-71(b) shows the relay in its *energized* position. Here, current has been applied to the relay coil, producing a magnetic field in the vicinity of the metal armature. The movable metal armature is attracted by the magnetic field and moves into the position shown in Figure 10-71(b). The "pole" (movable contact) portion of the switches is mechanically linked to the metal armature. When the armature is attracted to the coil, the switch contacts change state; the normally closed contacts open and the normally open contacts close.

As long as sufficient current flows through the coil of the relay, the contacts will remain in the energized position. If the coil current is interrupted or allowed to fall below a certain value, then the force of the spring will snap the switch contacts back into the normal state

KEY POINTS

When current is applied to the relay coil, a magnetic field is created, which attracts a metal armature. The movable portion of the switch contacts is attached to the armature and changes the state of the switch when the relay coil is activated.

KEY POINTS

When current is removed from the relay coil, the spring returns the armature and switch contacts to their normal position.

(shown in Figure 10-71(a)). It is important to understand that the coil circuit and the switch circuit are electrically isolated from each other. Thus, for example, we might use a 5-V control circuit to activate the coil of a relay whose contacts switch a 2,000-V circuit.

RELAY SPECIFICATIONS

It takes a certain amount of current through the relay's coil to produce a magnetic field that is strong enough to move the armature and change the switch contacts. This value of current is called the **pull-in current** of the relay. It also takes a definite value of current through the coil to keep the relay in its energized state. If the coil current falls below a value called the **drop-out current,** then the force of the spring returns the relay to its deenergized state. Figure 10-72 shows that the values of pull-in and drop-out currents are very different. It takes more current to energize the relay initially than it does to keep it energized. This is because the armature is farther away from the coil initially. Once the relay has energized, the armature is close to the coil and does not require as much field intensity to keep it in position. Both pull-in and drop-out currents are specifications that are provided by the manufacturer. Pull-in current is the maximum coil current that is required to energize the relay. It may energize with less, but it is guaranteed to energize at the rated value of pull-in current. Drop-out current is the minimum amount of current required to keep the relay energized. It may stay energized with less current, but it is only guaranteed down to the rated level of drop-out current. Some manufacturers specify the energize/deenergize factors in terms of voltage, but the same meanings apply.

The wire used to wind the relay coil has a definite amount of resistance. The manufacturer labels this resistance as the coil resistance. Typical values range from a few ohms to several tens of thousands of ohms.

Since the relay coil is characterized in terms of current and resistance, it follows that we can also characterize it in terms of voltage. So, although it is the value of current that is more appropriately used to define the exact points of relay operation, it is common to charac-

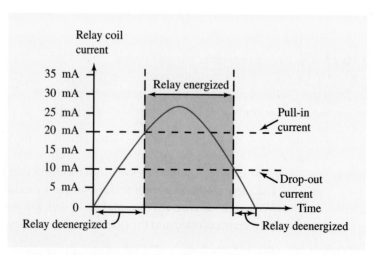

Figure 10-72. It takes more current to energize a relay than it does to keep it energized.

terize the general voltage range of the coil. For example, knowing that a coil is rated for 12-V operation quickly distinguishes it from a coil rated for 5-V operation. The intended application would immediately dictate the choice of coil voltages. As mentioned previously, some manufacturers specify the maximum **pick-up voltage** and the minimum **drop-out voltage.** These correspond functionally with the alternate specifications for pull-in and drop-out currents, respectively.

The contacts of the relays have specifications similar to those associated with discrete switches. The manufacturer normally supplies the following ratings for the contacts of a relay:

- maximum current
- maximum voltage between open contacts
- resistance of closed contacts
- maximum switched power
- number of operations (life expectancy)
- contact configuration (e.g., SPDT, DPDT, SPST, and so on)
- energize time
- deenergize time
- bounce time

Contact bounce occurs when a set of contacts changes states. Due mostly to surface imperfections, the contacts electrically bounce (i.e., repeatedly open and short) for a short time (typically hundreds of microseconds) as the contacts switch. For many circuits (e.g., lamp and motor controls), contact bounce is of no consequence. In high-speed digital circuits, however, contact bounce can cause serious problems.

Figure 10-73 shows a representative relay along with the manufacturer's specification sheet. Not all manufacturers supply all relay ratings, but a technician must be able to interpret the available specifications as they apply to a particular application.

Figure 10-74 shows four alternative schematic symbols for a relay. Many companies identify relays on their schematic diagrams with a "K" or a "CR" label. One of the most confusing aspects of the symbols shown in Figure 10-74(a) through 10-74(c), is that the contacts are not necessarily drawn in the same vicinity as the relay coil. If the coil is labeled K_5 for example, then the contacts might be labeled K_{5A}, K_{5B}, and K_{5C} to identify them as being associated with the K_5 coil, but they could be drawn on any part of the schematic. A technician analyzing the circuit must remember that all contacts associated with a particular relay coil will change states simultaneously.

Solenoids

In a previous section, we defined a solenoid as a wire that had been wound into a spiral coil. As such, it exhibits well-defined north and south magnetic poles when current passes through the coil. There is also a commercially available device called a solenoid that uses electromagnetism to create mechanical force. Figure 10-75 illustrates the component parts of a solenoid.

◗ KEY POINTS

A solenoid is a device that uses electromagnetism to cause mechanical motion that pushes, pulls, or twists an external mechanical system.

RN/RP SERIES ℝℓ
Dry Reed Relays
1 Form A, B, C and 2 Form A Contacts
Molded DIP and SIP Packages
Logic Compatible, Low-level Switching
Diode and Electrostatic Shielded Options

SPECIFICATIONS
Terminal Arrangement: Dual-In-Line; Single-In-Line
Contact Configurations: 1 Form A, B, C; 2 Form A
Contact Material: Ruthenium (1A, 1B, 2A); Rhodium (1C)
Contact Resistance: 200mΩ, max. (inital value) (1A, 1B, 2A)
          250mΩ, max. (inital value) (1C)
Operate Time: 1msec, max (1A, 1B, 2A); 3msec, max. (1C)
(Includes bounce)
Release Time: 1msec, max., (1A, 1B, 2A); 3msec, max. (1C)
Insulation Resistance: $10^{10}\Omega$ (1A, 1B, 2A); $10^{9}\Omega$ (1C)
Dielectric Strength: Coil to contact: 1,400VDC
          Across contacts: 250VDC (1A, 1B, 2A);
              150VDC (1C)
Electrical Life: 100×10^{6} (1A, 1B, 2A); 20×10^{6} (1C);
        200×10^{6} (1A/SIP)

TYPE LIST

Basic Type No.	Current Rating	Contact Config- uration	Rated Voltage (VDC)	Coil Resist. (Ω)	Nom. Power (mW)	Circuit Diagram (Bottom View)
RN1A	0.50A	1 Form A (DIP)	5	500	50	
			12	1000	144	
			24	2150	268	
RN1B	0.50A	1 Form B (DIP)	5	500	50	
			12	1000	144	
			24	2150	268	
RN1C	0.25A	1 Form C (DIP)	5	200	125	
			12	500	288	
			24	2150	268	
RN2A	0.50A	2 Form A (DIP)	5	140	179	
			12	500	288	
			24	2150	268	
RP1A	0.50A	1 Form A (SIP)	5	500	50	
			12	1000	144	
			24	2000	288	

P.C. BOARD LAYOUT
Dimensions in inches (mm).

RN1A, RN1B, RN1C, RN2A   RN1P

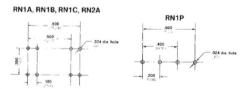

Figure 10-73. A relay specification sheet.
(Courtesy of IDEC Corporation)

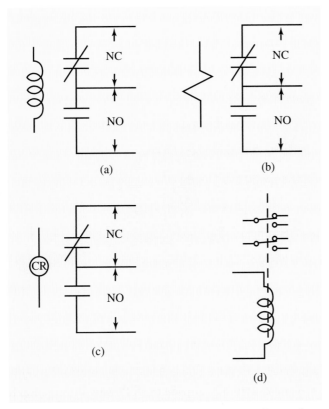

Figure 10-74. Alternative schematic symbols for a relay.

The plunger is an iron rod that is free to move inside the coil. The plunger acts as the core of the coil and concentrates the flux. The plungers in many solenoids have an internal spring that holds the plunger in the deenergized position.

Now, if we apply current to the coil, a magnetic field will be produced, causing the coil to act as an electromagnet, which will attract the iron plunger. Since the plunger is free to move, it quickly moves further inside the coil. When the coil current is interrupted, the internal (or sometimes external) spring returns the plunger to its original position.

The plunger can be constructed and positioned such that the attraction of the magnetic field causes the plunger to move in or to move out of the coil. Solenoids whose plungers move inward when coil current is applied are called *pull-type* solenoids. Those whose plungers move out when activated are called *push-type* solenoids. Some manufacturers also make a solenoid that produces a rotary or twisting force.

Solenoids are widely used for industrial and commercial electronics applications. Following are just a few examples of applications:

- to move an arm that diverts products moving down a conveyor belt, so they go into a different box or bin
- to activate the mechanical latch on an electrically operated door or gate
- to move the gear on the starting motor of a car while starting the engine, so that the teeth of the starting motor mesh with the gears on the flywheel

KEY POINTS

The solenoid consists of an electromagnet with a movable iron core. The movable core, called a plunger, is held in position by a spring.

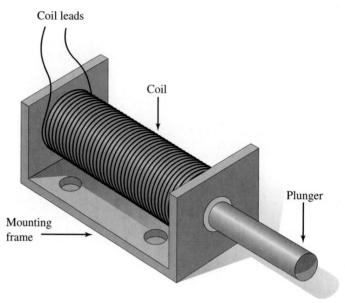

Figure 10-75. A solenoid uses electromagnetism to create mechanical force.

Manufacturers of solenoids typically provide the following information for technicians and designers:

- nominal coil voltage
- coil resistance
- coil power
- plunger motion (pull or push)
- intermittent or continuous operation
- force that can be produced by the plunger
- stroke length (how much the plunger moves)

Figure 10-76 shows some representative solenoids.

Magnetic Shielding

There are times when we want to shield a circuit or a device from a nearby magnetic field. At first, we might think that we would surround the area to be protected with a diamagnetic material that would prevent the passage of magnetic flux. While this may be a theoretically sound idea, in practice there are no materials that offer any substantial opposition to the passage of magnetic flux. Figure 10-77 shows the alternative that is used to keep magnetic fields away from certain areas.

The shield is made from a very high-permeability material and surrounds the area to be protected. When magnetic flux enters the region of the shield, it takes the easier (i.e., low-reluctance) path through the shield, rather than traveling through the protected (higher-reluctance) area. This type of shield is used to make a mechanical watch into an antimagnetic watch.

Figure 10-76. Representative solenoids.
(Courtesy of Deltrol Controls)

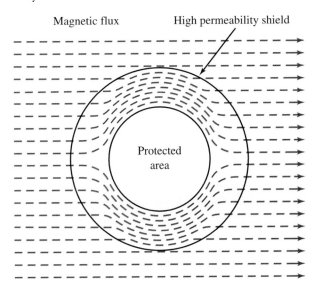

Figure 10-77. A magnetic shield keeps magnetic fields away from protected areas.

Exercise Problems 10.5

1. A magnetically operated switch responds to the position of a _____.

2. When the tooth of a ferromagnetic sprocket passes through the air gap of a magnetic tachometer sensor, the _____ of the magnetic circuit is decreased.

3. The sensor in a magnetic tachometer generates a voltage. (True or False)

4. Iron oxide is a diamagnetic material. (True or False)

5. A magnetic recording head has an air gap. (True or False)

6. The speaker cone in a permanent magnet speaker is fastened to the permanent magnet. (True or False)

7. The contacts of a relay are in the "normal" position when current is flowing through the relay coil. (True or False)

8. It takes more current to energize a relay than it does to keep it energized. (True or False)

9. A solenoid can push, pull, or twist the mechanical linkage of a machine. (True or False)

10. What does the stroke-length specification of a solenoid tell a technician?

11. Explain the meaning of the term *drop-out current* as it applies to relays.

12. What is meant by the term *contact bounce?*

10.6 Relay Circuit Analysis

A technician must be able to read the schematic diagram of a circuit that uses relays and determine how the circuit works. Analysis of relay circuits must be accomplished in two sequential stages:

1. Analyze the coil circuit to determine whether or not the relay is energized.

2. Analyze the switch circuit in its present state.

EXAMPLE SOLUTION

Determine whether the lamp in Figure 10-78 is lit.

EXAMPLE SOLUTION

The first step is to analyze the coil circuit to determine whether the relay is energized. By inspection, we can see that the relay coil circuit is completed through S_1 and V_1, so we know the relay will be energized.

Next, we examine the contact circuitry. In the case of the circuit shown in Figure 10-78, there is a complete circuit through L_1 and V_1 anytime the K_{1A} contacts are closed. Recall that contacts (like switches) are always drawn in their normal or deenergized position. So, in the case of the K_{1A} contacts, we expect them to be open. That is, we know the relay is energized, which means the normally closed contacts are open. The lamp, of course, is off, due to the open contacts that are part of its current loop.

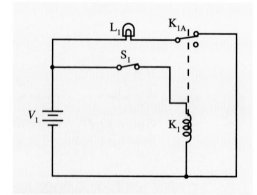

Figure 10-78. Is the lamp lit?

EXAMPLE SOLUTION

Determine which lamp or lamps are lit in Figure 10-79.

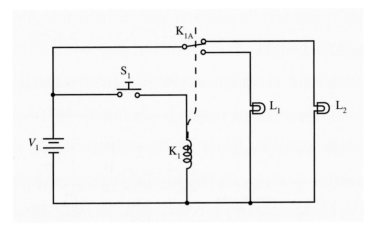

Figure 10-79. Which lamps are lit in this circuit?

EXAMPLE **SOLUTION**

Since S_1 is normally open (and not shown as being pressed), we can conclude that the coil circuit of K_1 is deenergized. This means that the relay contacts are in the normal or relaxed position as shown in the schematic diagram. By inspection, we can see that L_1 has no complete path for current and is not lit. L_2 does have a complete current path and is lit.

EXAMPLE SOLUTION

Determine which lamps are on for each position of switch S_1 in Figure 10-80.

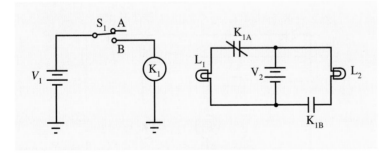

Figure 10-80. Which lamps are illuminated for each position of S_1?

EXAMPLE **SOLUTION**

When the switch is in position A (as shown in Figure 10-80), the relay is deenergized. Thus, the contacts will be in the state drawn on the schematic. As shown in Figure 10-80, lamp L_1 will be lit due to the complete current path through K_{1A} and V_2. Lamp L_2 will not be lit, because of the open contacts K_{1B}.

When switch S_1 is moved to position B, the relay coil will have a complete path for current and will energize. This causes the contacts K_{1A} and K_{1B} to change to their opposite state; K_{1A} contacts open and K_{1B} contacts close. This opens the current path for L_1 and provides a current path for L_2.

Note that the circuit shown in Figure 10-80 uses different voltage sources for the coil and contact circuits. This is an important feature of a relay—the ability to control one circuit from a totally isolated second circuit.

Practice Problems

1. Determine the state of the lamp for each position of the switch shown in Figure 10-81.

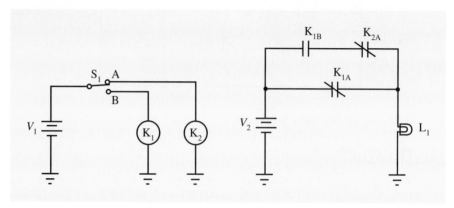

Figure 10-81. Which position(s) of the switch turns on the lamp?

2. What positions must switches S_1 and S_2 be in to illuminate the lamp in Figure 10-82?

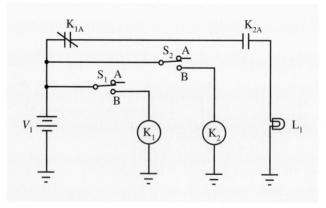

Figure 10-82. What must be done to cause the lamp to come on?

Answers to Practice Problems

1. Position A: L_1 on
    Position B: L_1 on

2. S_1 in position A and S_2 in position B

Self-Holding Contacts

Figure 10-83 shows a very common configuration for relay contacts. In its initial state, S_1 is closed, S_2 is open, and K_1 is deenergized. The lamp is off. When S_2 (a momentary contact switch) is pressed, K_1 energizes and closes contacts K_{1A}. The lamp is lit.

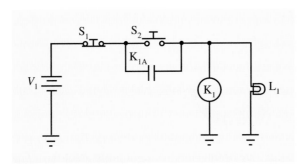

Figure 10-83. Relay K_1 has self-holding contacts.

Now, when S_2 is released, you might expect the relay to deenergize, but it does not. Closed contacts K_{1A} are in parallel with the S_2 contacts and continue to provide a current path for the relay coil. As long as the relay coil has a current path, it will stay energized, which keeps the K_{1A} contacts closed. The K_{1A} contacts are called **self-holding** contacts.

In order to return the circuit to its original condition, the coil current must be momentarily interrupted. This is the purpose of normally-closed switch S_1. If S_1 is pressed, even momentarily, then the relay current will be interrupted, and the coil will deenergize. This allows the K_{1A} contacts to open. Now, if S_1 is released (returns to its closed position), there will be no current path for the relay coil, since both S_2 and K_{1A} are open.

Chattering Contacts

Figure 10-84 illustrates another configuration for relay contacts called **chattering** or self-interrupting contacts. Although this configuration may be intentional in some specific application, it normally occurs accidentally during assembly of the circuit. It is important for a technician to quickly recognize the symptoms of this configuration.

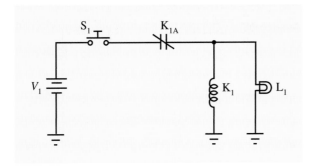

Figure 10-84. An example of chattering contacts.

Whenever S_1 is closed, the relay coil will have a current path through V_1, S_1, and K_{1A}. It will energize. But as soon as the relay energizes, the normally closed K_{1A} contacts open. This interrupts the current path for the relay coil, which allows it to deenergize. As soon as it deenergizes, however, the normally closed contacts K_{1A} close, which reestablishes a current path for the relay coil. The process continues as long as S_1 is closed. The relay contacts continually open and close (a mechanical operation). The rate of opening and closing varies with the type of relay, but it is nearly always fast enough to produce an audible clicking or buzzing sound. When a technician hears a relay buzz, it is a sure sign that a set of its own normally-closed contacts are in series with its coil.

Exercise Problems 10.6

1. Determine the state of the lamp for each position of S_1 in Figure 10-85.

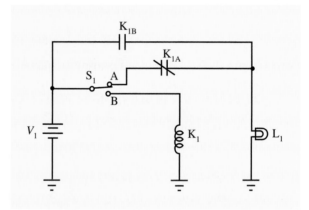

Figure 10-85. When is the lamp on?

2. What positions are required for switches S_1 and S_2 in Figure 10-86, in order to light the lamp?

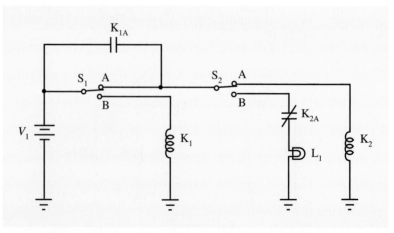

Figure 10-86. What must be done to light the lamp?

3. The K_{1A} contacts in Figure 10-87 are called _____ contacts.

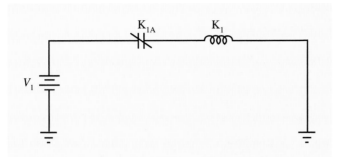

Figure 10-87. How are the K_{1A} contacts configured?

4. The K_{2A} contacts in Figure 10-88 are called _____ contacts.

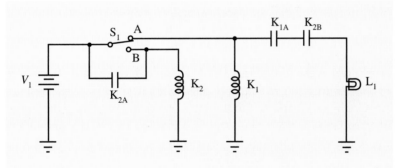

Figure 10-88. How are the K_{2A} contacts configured?

Chapter Summary

• An invisible, but measurable, region near a magnet that exhibits magnetic characteristics is called a magnetic field. We visualize the magnetic field as being composed of invisible lines called flux lines. Each flux line is continuous, is repelled by other flux lines, and cannot touch or intersect other flux lines. A magnetic field may be distorted by the effects of a nearby material.

• Ferromagnetic materials pass lines of flux easily and can distort a magnetic field. Paramagnetic and diamagnetic materials have no effect and a slight repelling effect, respectively, on magnetic fields.

• The north and south poles of a magnet are regions where the external flux lines are most concentrated. The direction of the flux lines is from the north pole to the south pole external to the magnet and from the south pole to the north pole internal to the magnet. Like poles of magnets will repel each other, while unlike poles attract. The magnitude of the attraction is directly proportional to the field strengths and inversely proportional to the distance between the poles.

- A material is magnetized by aligning its domains with an external magnetic field. Once the external field is removed, some of the domains may remain in their aligned orientation. This produces residual magnetism. The amount of residual magnetism is determined by the retentivity of the material.

- When current flows through a conductor, a circular magnetic field is produced around the wire. The circular field will repel or attract other magnetic fields and may cause physical movement of the conductor if the force is sufficient. The direction of the magnetic field can be determined with the left-hand rule for conductors. If the wire is wrapped into a spiral coil—called a solenoid—then the solenoid will have magnetic poles at its ends. The magnetic polarity can be determined with the left-hand rule for solenoids. The strength of the field is determined by the amount of current, the number of turns on the coil, the length of the coil, and the permeability of the core.

- Magnetic flux (Φ) is measured in webers (Wb) and represents 100,000,000 flux lines. Flux density (B) is measured in teslas (T). Magnetomotive force ($\mathcal{F}$) is measured in ampere $\cdot$ turns (A $\cdot$ t). Magnetizing force (H) is measured in ampere $\cdot$ turns/meter (A $\cdot$ t/m). Permeability (μ) expresses how easily a material will pass flux. It can be expressed as absolute or relative to a vacuum. Reluctance ($\mathcal{R}$) of a material determines the opposition to the flow of flux. It is measured in ampere $\cdot$ turns/weber (A $\cdot$ t/Wb).

- A *BH* curve relates the flux density (B) in a material to the magnetizing force (H) applied. A four-quadrant *BH* curve is called a hysteresis loop. A hysteresis loop reveals the permeability of a material, the amount of residual magnetism, the amount of coercive force required to overcome the residual magnetism, and the amount of energy loss required to switch the domains (hysteresis loss).

- Electromagnetic induction causes a voltage to be produced in a wire as it passes through a magnetic field. If the wire is part of a closed loop, then the induced voltage will cause a current (induced current) to flow through the circuit. The value of induced voltage is determined by field strength, relative speed, relative cutting angle, and number of turns of the conductor. The polarity of the induced voltage is determined by the direction of movement and the polarity of the magnetic field. The principle of induction is used in a dc generator. The left-hand rule for induction can be used to determine the polarity of induced voltage or current. If the changing magnetic field from one conductor cuts through another conductor, a voltage will be induced. Transformers (discussed in a later chapter) rely on this principle.

- Many of the same analytical skills required for the analysis of electrical circuits can be applied to the solution of magnetic circuits. Magnetomotive force is similar to voltage, permeance is equivalent to conductance, reluctance can be compared to resistance, and flux is comparable to current. With these analogies and a knowledge of the equations for magnetic circuits, a technician can analyze many magnetic circuits.

- A relay is a switch whose contacts are moved by the magnetic field from an electromagnet. The amount of current required to energize a relay is called pull-in current. A lesser current, called drop-out current, is required to keep the relay energized. These two critical points are sometimes identified by pick-up voltage and drop-out voltage specifications, respectively.

- Relay circuits are analyzed in two steps: first analyze the coil circuit, then analyze the contact circuit. If a relay's own normally-closed (NC) contacts are in series with its coil, they are called self-interrupting or chattering contacts. If the normally-open (NO) contacts of a relay provide an alternate current path for the relay coil when the relay is energized, they are called self-holding or latching contacts.

- A solenoid is a device that uses the magnetic field of a coil to attract a movable iron core or plunger. The plunger can be attached to the mechanical linkage of an external mechanical system to gain electrical control over a mechanical device. When coil current is interrupted, the plunger is returned to its normal position by an internal or external spring. Solenoids can produce pulling, pushing, or twisting force.

Review Questions

Section 10.1: Magnetic Fields

1. Invisible lines called flux lines form a region around a magnet called a _____ _____.

2. Which type of material passes flux lines more easily? (*air, iron*)

3. Air has a relative permeability of one. (True or False)

4. Two north poles will _____ each other.

5. Two south poles will _____ each other.

6. A north pole will _____ a south pole.

7. A paramagnetic material rejects the passage of flux lines. (True or False)

8. A ferromagnetic material would have a higher permeability than a diamagnetic material. (True or False)

9. If many of the domains in a ferromagnetic material are aligned, we say the material is _____. If all of the domains are aligned, we say the material is magnetically _____.

10. Determine the direction of the magnetic field around the conductor shown in Figure 10-89. (*cw, ccw*)

11. Which way does current flow in the conductor shown in Figure 10-90? (*left-to-right, right-to-left*)

12. The conductors shown in Figure 10-91 will (*attract, repel*) each other.

13. Determine the magnetic polarity of the solenoid shown in Figure 10-92.

14. Determine the magnetic polarity of the solenoid shown in Figure 10-93.

15. What factors affect the strength of the magnetic field of a solenoid?

16. Name two factors that determine the magnetic polarity of a solenoid.

Figure 10-89.
Determine the magnetic field direction.

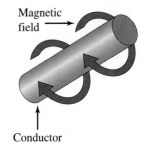

Figure 10-90.
Which way does current flow?

Figure 10-91.
Will these conductors repel or attract each other?

Section 10.2: Units of Measurement

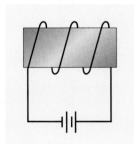

Figure 10-93.
Label the magnetic poles.

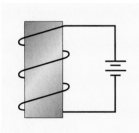

Figure 10-92.
Label the magnetic poles.

17. Φ is the symbol for _____.

18. A weber is the unit of measurement for _____.

19. B is the symbol for flux density. (True or False)

20. Flux density is measured in _____.

21. If a certain magnet produces 5 Wb of flux in a 1-m² area, what is the flux density?

22. If the flux density in a 5-m² area is 2.5 T, how much flux passes through the area?

23. What is the flux density in a material that has 10 mWb of flux passing through a 25-cm² area?

24. $\mathscr{F}$ is the symbol for _____.

25. Ampere · turns is the unit of measurement for _____.

26. What is the mmf produced by 1.5 A of current flowing through a 150-t coil?

27. If 275 mA flows through a 60-turn coil, what is the resulting mmf?

28. How much current must pass through a 250-t coil to produce an mmf of 25 A · t?

29. What is the symbol for magnetizing force?

30. How much magnetizing force is in a magnetic circuit that is 10 cm long and has a mmf of 100 A · t?

31. If a field intensity of 125 A · t/m is produced by a mmf of 2.5 A · t, what is the length of the magnetic circuit?

32. Permeance is a measure of the ease with which a magnetic field can be established in a material. (True or False)

33. Permeance of a magnetic circuit is similar to _____ in an electrical circuit.

34. The symbol for permeability is _____.

35. Relative permeability has no unit of measure. (True or False)

36. What is the absolute permeability of a material that produces a flux density of 5 T when a magnetizing force of 80 A · t/m is applied?

37. What is the relative permeability of the material described in the previous question?

38. What is the relative permeability of a material that has a flux density of 250 T with a magnetizing force of 75 A · t/m?

39. If the core material for a certain solenoid has a relative permeability of 5,000, what magnetizing force is required to produce a flux density of 2.5 T?

40. In a magnetic circuit, _____ is comparable to resistance in an electrical circuit.

41. For a given magnetizing force, a material with greater reluctance will have more flux. (True or False)

42. What is the reluctance of a bar that measures 2.5 cm × 5 cm × 50 cm and has a permeability of 20 mWb/A · t · m?

43. How much total flux does a 50 A · t mmf produce, if the magnetic circuit has a reluctance of 5 MA · t/Wb?

44. A *BH* curve is a graph of _____ _____ vs. _____ _____.

45. A four-quadrant *BH* curve is called a _____ _____.

46. If a portion of a *BH* curve is very steep (almost vertical), does this correspond to a region of higher or lower permeability?

47. What is the approximate permeability of a material that has entered magnetic saturation?

48. The magnetism that remains after the magnetizing force has been removed is called _____ _____.

49. A material with a high value of retentivity will have a high value of residual magnetism. (True or False)

50. What name is given to describe the amount of magnetizing force required to overcome the residual magnetism?

51. The amount of hysteresis loss in a magnetic material is proportional to the area of the hysteresis loop. (True or False)

52. What name is used to describe a material whose flux density is at a maximum level?

Section 10.3: Electromagnetic Induction

53. In order to have electromagnetic induction, there must be relative motion between a magnetic field and a conductor. (True or False)

54. The magnitude of an induced voltage is proportional to the strength of the magnetic field. (True or False)

55. The polarity of an induced voltage is determined by the number of turns in the coil. (True or False)

56. The polarity of the magnetic field and the direction of conductor movement determine the polarity of induced voltage when a conductor moves through a magnetic field. (True or False)

57. Use the left-hand rule to determine the direction of induced current in Figure 10-94.

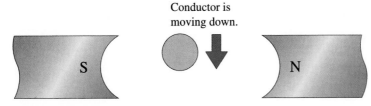

Figure 10-94. Which way does current flow?

58. Determine the direction of conductor motion in Figure 10-95.

Figure 10-95. Which way is the conductor moving?

59. Label the magnetic poles in Figure 10-96.

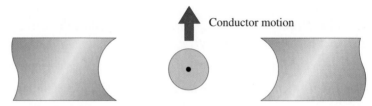

Figure 10-96. Label the poles.

60. In order for current in one coil to induce a voltage in a second coil, the two coils must be linked magnetically and the current in the first coil must be _____.

61. How much flux must cut a single-turn coil over a 1-s period to produce a sustained voltage of 1 V in the coil?

62. Figure 10-97 shows a loop of wire that is free to rotate in a magnetic field. If a current is passed through the wire in the indicated direction, which way will the loop of wire rotate?

Figure 10-97. Which way does the loop of wire rotate?

63. A magnetic field can deflect an electron if the electron is _____.

Section 10.4: Magnetic Circuit Analysis

64. Match the magnetic terms in the left column with the similar electrical terms in the right column:

Magnetic Terms	Electrical Terms
_____ Permeance	a. Conductance
_____ Flux	b. Current density
_____ Flux density	c. Electric field intensity
_____ Magnetizing force	d. Current
_____ Permeability	e. Resistance
_____ Magnetomotive force	f. Voltage
_____ Reluctance	g. Conductivity

65. What magnetomotive force is produced by a 175-t coil with 275 mA of current?

66. Compute the flux density in an air gap that is 0.5 cm long and measures 0.5 cm $\times$ 0.5 cm on a side if the total flux in the gap is 275 μWb.

67. Compute the magnetizing force in an air gap if the flux is 25 μWb, the reluctance is 200,000 A $\cdot$ t/Wb, and the length is 0.25 cm.

68. What is the magnetomotive force applied to the magnetic circuit shown in Figure 10-98?

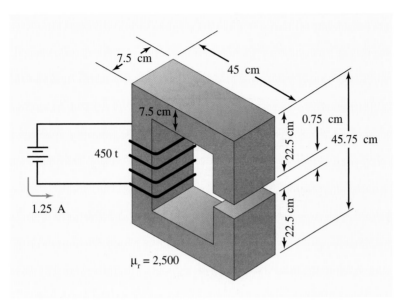

Figure 10-98. A magnetic circuit.

69. What is the reluctance of the air gap in Figure 10-98?
70. What is the total reluctance of the magnetic circuit shown in Figure 10-98?
71. How much flux flows in the magnetic circuit shown in Figure 10-98?
72. What is the magnetizing force within the air gap pictured in Figure 10-98?

Section 10.5: Applied Technology: Magnetic and Electromagnetic Devices

73. A magnetic tachometer must physically touch the shaft being monitored. (True or False)
74. The reluctance of the magnetic circuit in a magnetic tachometer increases when the ferromagnetic teeth enter the air gap. (True or False)
75. Describe how magnetic fringing is a necessary phenomenon in the recording head of a cassette recorder.
76. Magnetic storage tape is coated with a ferromagnetic material. (True or False)
77. A magnetic recording head requires an air gap. (True or False)
78. A material that is used as the storage medium in a magnetic storage system should have a high level of retentivity. (True or False)
79. A loud sound from a permanent magnet speaker requires less cone movement than a softer sound. (True or False)
80. What is the name given to the current required to energize a relay?
81. What is the name given to the voltage that allows the relay to deenergize?
82. Explain the meaning of the stroke-length specification for a solenoid.
83. Explain the meaning of bounce time on a relay specification sheet.

Section 10.6: Relay Circuit Analysis

84. Determine whether the lamp in Figure 10-99 is lit when the switch is not being pressed.

85. Is the lamp in Figure 10-99 lit, when the switch is pressed?

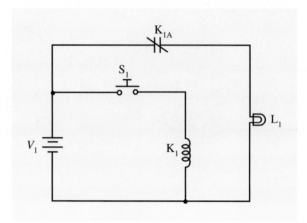

Figure 10-99. Is the lamp lit?

86. Refer to Figure 10-100. Complete Table 10-2 by entering ON or OFF for each of the lamps in each listed switch state.

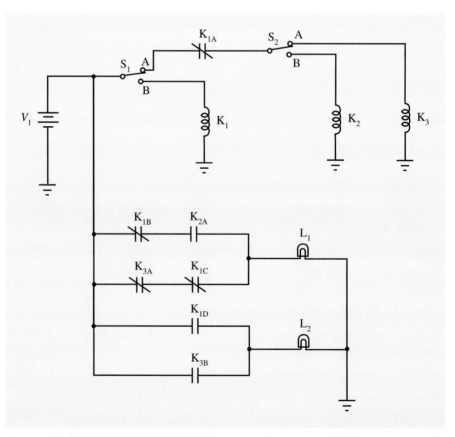

Figure 10-100. Analyze this circuit.

SWITCH STATES	L_1	L_2
$S_1 = A$; $S_2 = A$		
$S_1 = A$; $S_2 = B$		
$S_1 = B$; $S_2 = A$		
$S_1 = B$; $S_2 = B$		

Table 10-2.

87. Refer to the circuit shown in Figure 10-101. Assume that the switches are in the position shown when power is first applied. Describe what must be done to get L_1 to light.

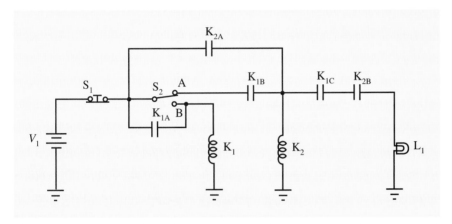

Figure 10-101. What must be done to light L_1?

88. Explain how to extinguish the lamp in Figure 10-101 once it comes on.

89. Suppose a technician applies power to a relay circuit and the relay begins to make a buzzing sound. Explain the most probable cause for this problem.

90. Explain how self-holding contacts work in a relay circuit.

TECHNICIAN CHALLENGE

The purpose of this Technician Challenge is to design a working burglar alarm system that utilizes many of the magnetic/electromagnetic principles and devices discussed in this chapter. Your design must utilize a minimum of the following devices:

- five magnetic switches for doors and windows
- one hidden manual switch
- one flashing lamp that requires 120 Vac to operate
- one chattering relay
- one standard relay
- one solenoid

The alarm must sense the opening of any one of the five doors/windows. When an opening is sensed, the chattering relay will begin to buzz and the lamp will begin to flash. Additionally, a solenoid will be activated, which releases the latch on the cage for your pet python. The buzzing and flashing will continue even if the door or window is shut again. The buzzing and flashing will continue until the alarm is reset by a hidden switch.

Draw the schematic diagram for your design. If practical, test your design (except for the python) in the laboratory with your instructor's guidance.

Equation List

(10-1) $\quad B = \dfrac{\Phi}{A}$

(10-2) $\quad \mathscr{F} = NI$

(10-3) $\quad H = \dfrac{\mathscr{F}}{l}$

(10-4) $\quad \mu = \dfrac{B}{H}$

(10-5) $\quad \mu_r = \dfrac{\mu}{\mu_v}$

(10-6) $\quad \mathfrak{R} = \dfrac{l}{\mu A}$

(10-7) $\quad \Phi = \dfrac{\mathscr{F}}{\mathfrak{R}}$

(10-8) $\quad \wp = \dfrac{1}{\mathfrak{R}}$

(10-9) $\quad \mathscr{F}_{total} = \mathscr{F}_1 + \mathscr{F}_2 + \ldots + \mathscr{F}_N$

Alternating Voltage and Current

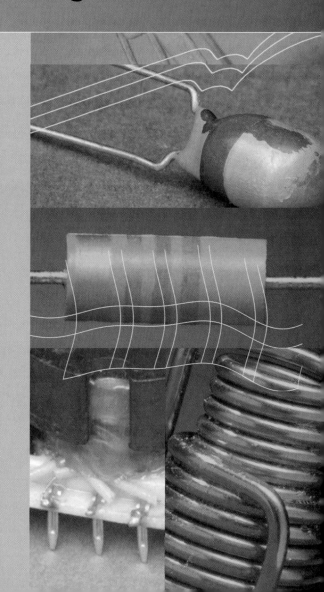

Alternating current and voltage is critical to the operation of nearly all practical electronic devices. The 120 volts supplied to your home by the power company, the signal voltages that cause music from your stereo, the radio waves that allow radio and television transmission, the radar waves used in speed detectors, the ultrasonic waves used in burglar alarms, and the telemetry signals used to communicate with a satellite are all examples of alternating current and voltage.

11.1 Generation of Alternating Voltage

A direct voltage or current (dc) has a certain value and a certain polarity. It remains relatively constant indefinitely. By contrast, an alternating voltage or current (ac) varies in both amplitude and polarity. To better appreciate these important characteristics of an alternating voltage or current, let us see how these voltages may be produced.

Rectangular Waveshapes

Figure 11-1 illustrates a very simple way to produce an alternating voltage. When the switch in Figure 11-1(a) is in position A, a positive voltage (with reference to ground) is applied to the resistor. V_{A_1} is essentially connected across the resistor. These times are labeled in Figure 11-1(b). When the switch is moved to position B, V_{A_2} is connected across the resistor, and a negative voltage (with reference to ground) can be measured. These times are also labeled in Figure 11-1(b). The horizontal axis of the graph is measured in time (e.g., seconds). Clearly, the rate at which the switch is moved between positions determines the duration of each voltage polarity interval on the graph.

Figure 11-2 shows the graph (called a **waveform**) of a voltage (or current, since $I = V/R$ and R is constant) that might be produced by the circuit shown in Figure 11-1. Several important characteristics of alternating voltages are labeled.

Note that the waveform in Figure 11-2 is repetitious and consists of a continuing series of **cycles.** The time required for one complete cycle is called the **period** of the waveform. A period consists of two opposite polarity **alternations.** The lengths of the positive and negative alternations are not necessarily the same, but the combined length is always equal to the period of the waveform. If the positive and negative alternations are equal, the waveform is said to be **symmetrical.** Waveforms with unequal alternations are called **asymmetrical.** If the cycles of the waveform are continuous (regardless of symmetry), the waveform is said to be **periodic.**

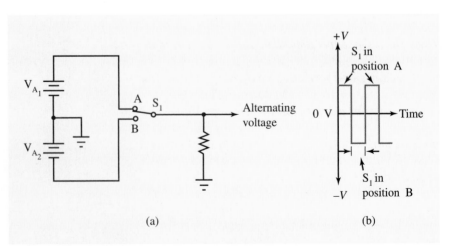

Figure 11-1. A simple way to produce alternating voltage.

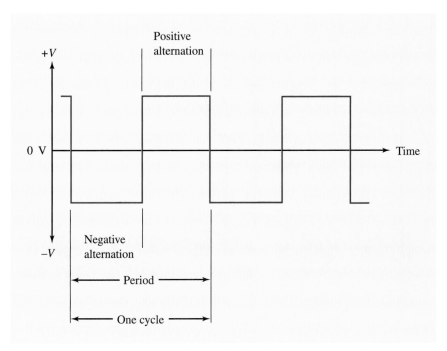

Figure 11-2. Some important characteristics of alternating waveforms.

A waveform like the one shown in Figure 11-2, which makes abrupt changes in polarity, is called a digital signal, a rectangular waveform, a square wave (if symmetrical), a pulse, or other similar classification. You will learn additional characteristics of this type waveform when you study digital electronics.

Sinusoidal Waveshapes

Figure 11-3 shows a loop of wire being rotated in a magnetic field. This is the basis of an ac generator or alternator. When we studied dc generators in Chapter 10, we saw how a segmented commutator and brush assembly is used to connect the rotating conductor to the external circuitry. In a similar manner, the **slip rings** and brushes in Figure 11-3 perform the task of making a continuous electrical connection to the rotating conductor. As shown in Figure 11-3, each end of the rotating loop is connected to a slip ring. The slip rings are mounted on the shaft that supports the rotating loop. Each of the slip rings is a continuous ring of smooth metal. The spring-loaded brushes slide along the surface of the slip rings as the conductor assembly rotates.

Figure 11-3 also shows a graph that plots the magnitude and polarity of the voltage that is produced as the conductor loop rotates. The first position shown in Figure 11-3(a) is labeled as 0°. Here, the conductors are moving parallel to the flux lines, so no voltage is generated. The graph shows 0 V at 0°. The sketch in Figure 11-3(b) shows the conductor as it passes directly across the pole faces of the magnet. In this case, the conductor is cutting a maximum number of flux lines and produces a maximum voltage. The left-hand rule for induction can be used to confirm the polarity of induced voltage. Since the conductor has moved one-fourth of a complete rotation, we label this as the 90° point. The graph shows that we have a maximum induced voltage at this point.

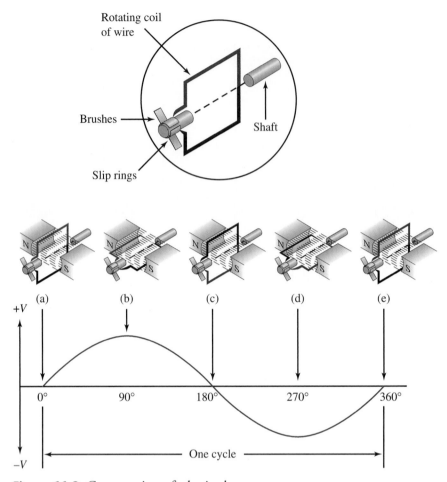

Figure 11-3. Construction of a basic alternator.

In Figure 11-3(c), the coil has reached the halfway point, and the conductors are once again moving parallel to the magnetic flux lines. No voltage is induced. This is shown graphically as the 180° point. Figure 11-3(d) shows the conductors moving across the pole faces and generating maximum voltage. Realize, however, that each of the conductors is moving across the opposite pole face from that shown in Figure 11-3(b). Thus, we would expect an opposite polarity of induced voltage. This expectation is labeled as the 270° point on the graph, and can easily be confirmed with the left-hand rule for induction.

Finally, Figure 11-3(e) shows the conductor back in its original position. The relative movement of the conductor and magnetic field is again zero, so no voltage is induced. This is identified as the 360° point on the graph. The graph clearly shows that the voltage varies continuously in amplitude and periodically changes polarity. Although the horizontal axis of the graph in Figure 11-3 is labeled in degrees of rotation, it could just as easily be labeled in units of time. That is, if the conductor is rotated at a given speed, then it will take a definite amount of time to make a full 360° rotation. A practical alternator, as with a practical dc generator, requires the combined output from many rotating loops of wire to produce usable levels of voltage.

Figure 11-4 shows two full cycles of alternating voltage with several important points labeled. A period is the time required for one full cycle. As with a rectangular wave, a

● KEY POINTS

The positive and negative alternations of a sinusoi-dal waveform are equal in time and amplitude.

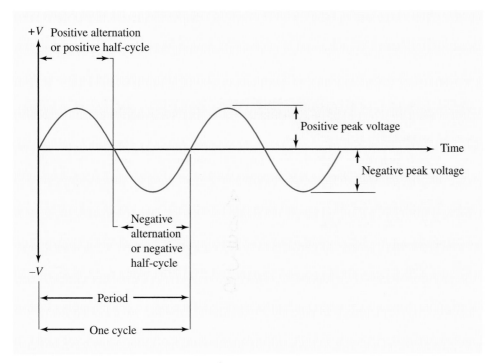

Figure 11-4. A basic sinusoidal waveform.

period consists of two opposite polarity alternations. Each alternation is also called a half-cycle. A waveform like that shown in Figure 11-4 is called a sine wave, or more generally a sinusoidal waveform. The two half-cycles of a sinusoidal waveform are identical except for their opposite polarities. Each half cycle has a maximum value called the peak value.

Function Generators

Figure 11-5 shows a function generator (also generically called a signal generator). This is a common item of test equipment used by electronic technicians. Nearly all function generators are capable of producing fundamental waveforms such as sine waves and rectangular waveforms.

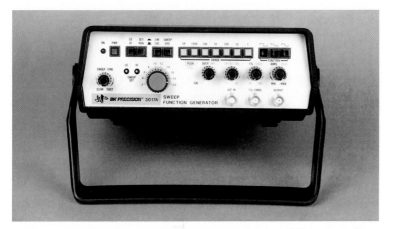

Figure 11-5. A representative function generator. (Courtesy of B&K Precision)

Some function generators can produce many other waveshapes, including arbitrary waveforms that can be programmed by the technician. In addition to selecting a particular waveshape, a technician can also adjust the voltage, symmetry, and time characteristics of the waveforms.

Exercise Problems 11.1

1. Match each of the labeled points on the waveform in Figure 11-6 to the following list of waveform characteristics:

   - positive alternation   A
   - negative alternation   D
   - period   E
   - time axis   C
   - voltage or current axis   B

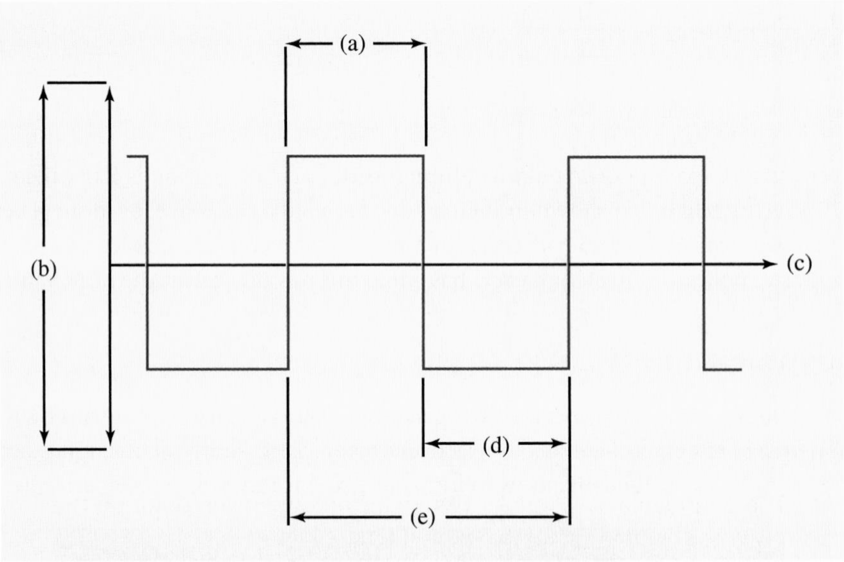

Figure 11-6. Identify the labeled areas.

2. Match each of the labeled points on the waveform in Figure 11-7 to the following list of waveform characteristics:

   - positive half-cycle   A
   - positive peak   F
   - negative peak   G
   - negative half-cycle   C
   - period   B
   - time axis   E
   - voltage or current axis   D

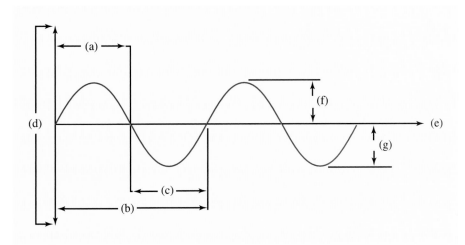

Figure 11-7. Identify the labeled areas.

11.2 Characteristics of Sine Waves

We examined some of the basic characteristics of sine waves in the previous section. We are now ready to examine these basic characteristics more closely and to identify other important characteristics that distinguish one sine wave from another. It is important to understand that the following characteristics apply to the sine waveshape itself. The sine wave may represent voltage, current, or another parameter.

Period

We have seen that the period of a sine wave is the time required for one full cycle of the waveform to occur. The measurement of a period does not have to begin at zero degrees. That is,

> The period of a sine wave is the time from any given point on the cycle to the same point on the following cycle.

The period is measured in time (t), and in most cases, it is measured in seconds (i.e., s, ms, μs, ns, ps).

EXAMPLE SOLUTION

Determine the period for the sine wave shown in Figure 11-8.

EXAMPLE **SOLUTION**

We measure the time from any point on one cycle to the same point on the following cycle. Suppose we started at the 0° point on the first cycle ($t = 0$). The 0° point on the next cycle starts at 2.0 ms. Therefore, the period is 2.0 ms – 0 ms = 2.0 ms.

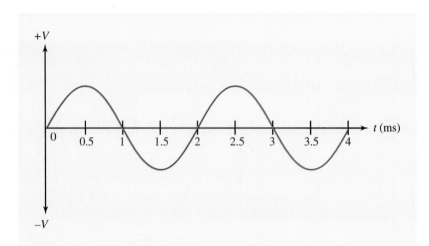

Figure 11-8. What is the period of this sine wave?

EXAMPLE SOLUTION

Determine the period for the sine wave shown in Figure 11-9.

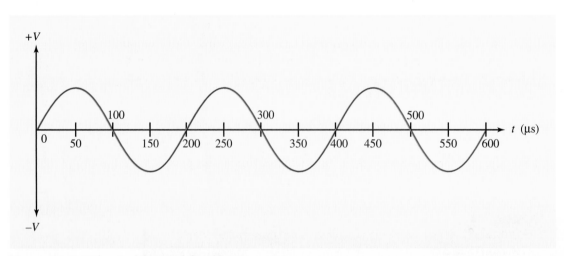

Figure 11-9. What is the period for this sine wave?

EXAMPLE **SOLUTION**

Let's measure the time between two consecutive 90° points. The first 90° point occurs at $t = 50$ μs. The 90° point on the next cycle occurs at $t = 250$ μs. The period is the difference between these two times, or $t = 250$ μs $- 50$ μs $= 200$ μs. Always subtract the smaller time value from the larger. A negative period has no meaning.

Practice Problems

1. What is the period for the sine wave shown in Figure 11-10?
2. Determine the period of the sine wave shown in Figure 11-11.

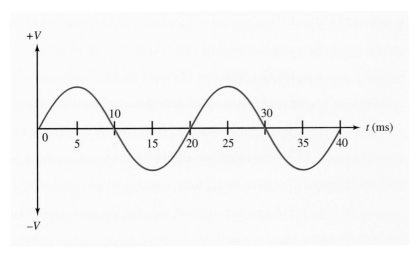

Figure 11-10. What is the period of this sine wave?

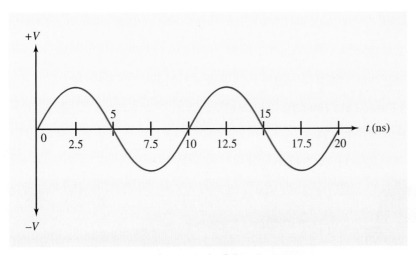

Figure 11-11. Determine the period of this sine wave.

Answers to Practice Problems

1. 20 ms   **2.** 10 ns

Frequency

The frequency of a sine wave (or any other periodic waveform) is the number of complete cycles in one second. In the case of the simple alternator (Figure 11-3), frequency (number of cycles per second) is equivalent to the speed of the alternator (revolutions per second). This is because each complete revolution produces one complete sine wave cycle.

Frequency (f) is measured in hertz (Hz). One hertz corresponds to one cycle per second. In fact, frequency used to be expressed in units of cycles per second. Usable frequencies range from less than 1 Hz (although 15 to 20 Hz is a more common low-frequency end) to frequencies over 50 billion hertz (50 GHz).

KEY POINTS

Frequency is measured in hertz and indicates the number of cycles that occur in one second.

Frequency and period have an inverse relationship. We use Equation 11-1 to determine the period of a known-frequency sine wave.

$$t = \frac{1}{f} \qquad (11\text{-}1)$$

Similarly, we use Equation 11-2 to find the frequency of a sine wave with a known period.

$$f = \frac{1}{t} \qquad (11\text{-}2)$$

EXAMPLE SOLUTION

If the period of a sine wave is 100 ms, what is its frequency?

EXAMPLE **SOLUTION**

We apply Equation 11-2 as follows:

$$f = \frac{1}{t}$$

$$= \frac{1}{100 \text{ ms}} = 10 \text{ Hz}$$

EXAMPLE SOLUTION

What is the period of a 250-MHz sine wave?

EXAMPLE **SOLUTION**

We apply Equation 11-1 as follows:

$$t = \frac{1}{f}$$

$$= \frac{1}{250 \text{ MHz}} = 4 \text{ ns}$$

Frequency-to-period and period-to-frequency conversions are especially easy with an electronic calculator. Simply enter the known quantity and press the reciprocal button (1/x).

Practice Problems

1. What is the period of a 10-kHz sine wave?
2. What is the frequency of a sine wave that has a period of 80 μs?
3. Find the period of a 4-GHz sine wave.
4. What frequency corresponds to a period of 175 ms?
5. One sine wave has a period of 3 ms, and another has a period of 6 ms. Which one has the higher frequency?

Answers to Practice Problems

1. 100 μs   2. 12.5 kHz

3. 250 ps   **4.** 5.7 Hz

5. The shorter period (3 ms) has the higher frequency (333.3 Hz).

Peak Value

We briefly discussed the peak sine wave value in an earlier section. It is the maximum voltage on a sine wave. By inspection of a sine wave graph, we can conclude several things about the peak value of a sine wave:

- Peak voltage occurs at two different points in the cycle.
- One peak point is positive, while the other is negative.
- The positive peak occurs at 90°.
- The negative peak occurs at 270°.
- The positive and negative peaks have equal magnitudes (opposite polarities).

EXAMPLE SOLUTION

If a sine wave has a value of +10 V (V_1) as it passes through the 90° point on the cycle, what voltage (V_2) will it have as it passes through the 270° point?

EXAMPLE **SOLUTION**

The 90° and 270° points correspond to the positive and negative peaks, respectively. These two points have equal magnitudes, but opposite polarities. Thus, we would expect the sine wave to have a value of $V_2 = -V_1 = -10$ volts as it passes through the 270° point.

Note the use of lowercase letters to represent instantaneous voltages. Instantaneous voltages are values that occur at a specific instant in time. This measurement is discussed more completely in a later section. Throughout the remainder of this text we will use lowercase letters to represent instantaneous values of alternating voltage and current, and uppercase letters for other values. This is an accepted practice in industry.

Practice Problems

1. If a certain sine wave of current has a negative peak of −1.5 A, what is its positive peak current value?

Answers to Practice Problems

1. 1.5 A

Average Value

The average value of any measured quantity is determined by summing all of the intermediate values and then dividing by the number of intermediate values. In the case of a sine wave, we need to consider two conditions: the average value for a full cycle, and the average value of a half-cycle.

AVERAGE VALUE OF A FULL SINE WAVE

The sine wave shown in Figure 11-12 has seventeen labeled points. If we measured the voltage at these seventeen points, added the values together, and then divided by 17, we could get an estimate of the average value for the full sine wave. It would only be an estimate, since we did not use an infinite number of points.

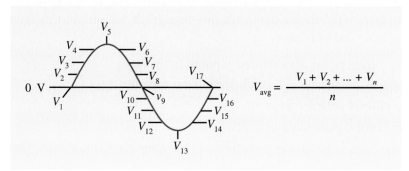

Figure 11-12. Determination of the average value of a sine wave.

Nevertheless, by inspection of Figure 11-12, we can see that no matter how many equally distributed points we choose, there will be as many positive values as there are negative values. Further, every positive value will have a corresponding negative value. For example, the positive values V_3 and V_5 in Figure 11-12 correspond to the negative values V_{11} and V_{13}, respectively. If we add equal positive and negative values together, we will always get zero. Therefore, we can conclude the following about the average value of a sine wave:

> A full cycle of any sine wave has an average value of zero.

This is an important consideration when troubleshooting amplifier circuits.

Practice Problems

1. What is the average value of a full sine wave that has a +100-V peak value?
2. What is the average value of a full sine wave with a negative peak of 3 V and a frequency of 10 MHz?
3. If the period of a certain sine wave is 5 ms, what is the average value for a full cycle?
4. What is the average value of a full sine wave if the peak value is 2.4 A?

Answers to Practice Problems

1. Zero
2. Zero
3. Zero
4. Zero

AVERAGE VALUE FOR ONE HALF-CYCLE

If we sum the voltage at a number (ideally an infinite number) of points in a single half-cycle of a sine wave and then divide by the number of sampled points, we can find the average value for one half-cycle. You can perform this calculation in the laboratory, by measuring a large number of points on a sine wave and then dividing by the number of points. However, since a sine wave has a consistent shape, higher mathematics can be used to prove the following relationships:

$$V_{avg} = 0.637\ V_P \qquad\qquad (11\text{-}3)$$

and

$$I_{avg} = 0.637\ I_P \qquad\qquad (11\text{-}4)$$

where V_p and I_p are the peak voltage and current values. Of course, current is still directly proportional to voltage in a given circuit as described by Ohm's Law.

EXAMPLE SOLUTION

What is the average value for one half-cycle of a sine wave that has a peak voltage of 100 V?

EXAMPLE **SOLUTION**

We apply Equation 11-3 as follows:

$$V_{avg} = 0.637\ V_P$$
$$= 100\ V \times 0.637 = 63.7\ V$$

Since the average value of a single half-cycle provides a more meaningful result in most cases than the average value of a full sine wave, it is customary to interpret the expression "average value of a sine wave" to mean average value of a half-cycle. While this may be technically incorrect, it is very common throughout industry. As a technician, it is important that you understand this interpretation. For the remainder of this text, will shall interpret V_{avg}, I_{avg}, and so on to mean the average of one half-cycle, unless otherwise stated.

Practice Problems

1. Compute the average value (V_{avg}) for a sine wave with a peak value of 20 V.
2. Find V_{avg} for a sine wave with a peak value of 150 mV.
3. What is the average voltage (V_{avg}) for a 1,000-V peak sine wave?
4. Compute I_{avg} for a sine wave with a 150 μA peak current.

Answers to Practice Problems

1. 12.74 V   2. 95.55 mV

3. 637 V   4. 95.55 μA

rms Value

One of the most important characteristics of a sine wave is its rms or effective value. The rms value describes the sine wave in terms that can be compared to an equivalent dc voltage. Figure 11-13(a) shows a 10-V_{rms} sine wave voltage source connected to a resistor. The resistor dissipates power in the form of heat. Figure 11-13(b) illustrates that a 10-V_{dc} source will produce the same amount of heat in the resistor. That is,

> The rms value of a sine wave produces the same heating effect in a resistance as an equal value of dc.

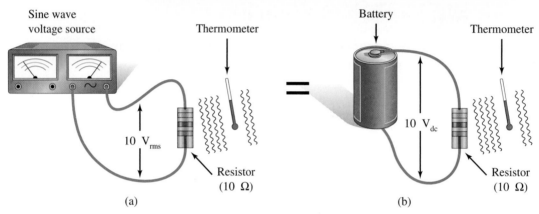

Figure 11-13. The rms voltage of a sine wave has the same heating effect as a similar dc voltage source.

The abbreviation rms stands for **root-mean-square**. For a given sine wave, we can take a large number (ideally an infinite number) of voltage values, square each value, sum our squared values together, divide by the number of points to get the average or mean squared value, and then extract the square root to get the rms value. Fortunately, higher mathematics can be used to derive Equation 11-5 and Equation 11-6, which describe the rms value of a sine wave in terms of its peak values.

$$V_{rms} = 0.707 \, V_P \qquad (11\text{-}5)$$

and

$$I_{rms} = 0.707 \, I_P \qquad (11\text{-}6)$$

EXAMPLE SOLUTION

What is the rms value of a sine wave that has a 25-V peak?

EXAMPLE SOLUTION

We apply Equation 11-5 as follows:

$$V_{rms} = 0707 \, V_P$$
$$= 0.707 \times 25 \text{ V} = 17.68 \text{ V}$$

EXAMPLE SOLUTION

What peak value of a sine wave is required to produce the same heating effect as a 48-V battery pack?

EXAMPLE **SOLUTION**

Since the rms value of a sine wave produces the same heating effect as dc, we know that our sine wave will have an rms value of 48 V. We compute the peak voltage by applying Equation 11-5.

$$V_{rms} = 0.707\ V_P, \text{ or}$$

$$V_P = \frac{V_{rms}}{0.707}$$

$$= \frac{48\ V}{0.707} = 67.89\ V$$

This latter calculation is used so often that it is generally remembered as separate equations.

$$V_P = \frac{V_{rms}}{0.707} = 1.414\ V_{rms} \qquad (11\text{-}7)$$

and

$$I_P = \frac{I_{rms}}{0.707} = 1.414\ I_{rms} \qquad (11\text{-}8)$$

It is important to note that most ac voltmeters—both digital and analog—are calibrated to display the rms value of the measured ac voltage. Further, most ac voltmeters only respond accurately to sinusoidal waveforms.

Peak-to-Peak Value

Figure 11-14 illustrates another measurement that can be used to describe a sine wave. The peak-to-peak voltage or current value of a sine wave is the difference between the two peak values. We can express this mathematically with Equations 11-9 and 11-10.

$$V_{PP} = 2\ V_P \qquad (11\text{-}9)$$

and

$$I_{PP} = 2\ I_P \qquad (11\text{-}10)$$

> **KEY POINTS**
>
> Peak-to-peak voltage is the difference between the positive and negative peak values. Its value is simply $2 \times V_p$.

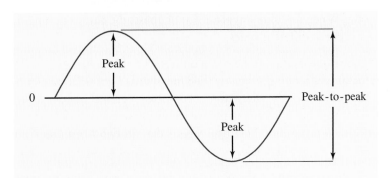

Figure 11-14. The peak-to-peak value of a sine wave is equal to the difference between the two peak values.

EXAMPLE SOLUTION

What is the peak-to-peak current of a sine wave that has a 750-mA peak current?

EXAMPLE SOLUTION

We apply Equation 11-10 as follows:

$$I_{PP} = 2\,I_P$$
$$= 2 \times 750 \text{ mA} = 1.5 \text{ A}$$

Practice Problems

1. If a sine wave has a peak voltage of 500 mV, what is its peak-to-peak value?
2. What is the peak-to-peak voltage of a sine wave that has a peak value of 12 V?
3. If the peak-to-peak voltage of a sine wave is 150 V, what is its peak voltage?

Answers to Practice Problems

1. 1.0 V   2. 24 V   3. 75 V

> **● KEY POINTS**
>
> The peak-to-peak measurement is useful to characterize a sine wave, but it is important to remember that the peak-to-peak voltage value never actually exists at any given instant in time.

It is important to note that the value of peak-to-peak voltage never really exists as a measurable voltage at some point in time. Recall that the horizontal axis of a sine wave graph can be marked off in units of time. This means that the positive and negative peaks of the sine wave do not occur at the same exact time. Therefore, as stated in an earlier section, the highest voltage or current of a sine wave is its peak value. The peak-to-peak value is mathematically twice the peak value, *but it does not exist as a measurable voltage at some instant in time*. It is, however, an important way to describe a sine wave. It is also the easiest value to interpret with an oscilloscope, since the scope display can show voltages that occur at different times.

Phase

> **● KEY POINTS**
>
> In-phase waveforms go through the corresponding points on the sine waves at the same time. Out-of-phase waveforms do not pass through corresponding points at the same time.

Phase is a relative term that is used to compare two or more sine waves that have the same frequency. When two sine waves are in phase, the various points on one sine wave (e.g., 0°, 90°, 180°, and so on) occur at exactly the same time as the corresponding points on the second sine wave. When two sine waves are out of phase, the corresponding points of the two sine waves occur at different times.

Figure 11-15 shows one method to help you understand phase relationships. Here, two identical alternators are rotating at exactly the same speed. They will produce two identical sine waves. However, one of the alternators is one-quarter turn ahead of the other one. Therefore, the corresponding points of the two resulting sine waves will occur at different times. This is illustrated in Figure 11-15(b). We say that the two sine waves are out of phase. In the case shown in Figure 11-15(b), we can say the two sine waves are 90° out of phase.

In order to describe the relative order of two out-of-phase waveforms, we use the terms leading and lagging. A waveform that has a leading phase of 45° is simply 45° ahead of another waveform. That is, any given point on the leading waveform occurs one-eighth

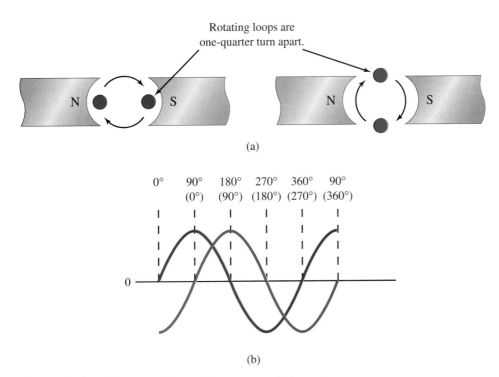

Figure 11-15. Corresponding points on out-of-phase sine waves occur at different times.

(45°/360°) of a cycle before the corresponding point of the second waveform. If we prefer, we can refer to the second waveform as lagging by 45°. We will further examine the use of the terms leading and lagging in Chapter 12.

ANGULAR MEASUREMENTS

There are two ways to express the angular measurements of a sine wave. We have already seen how degrees can be used to describe a particular point on a sine wave. A sine wave of voltage passes through 0 V at 0°, 180°, and 360°. Similarly, it has maximum positive and negative values at 90° and 270°, respectively. While it is mathematically correct to express angles larger than 360°, it is not generally done, since anything beyond 360° has a corresponding point on a previous cycle that is between 0° and 360°.

Angles can also be expressed in radians. One radian (rad) corresponds to 57.3°. As a technician, you should be able to perform calculations with either unit of measurement.

DEGREE-TO-RADIAN CONVERSION

We can express an angle given in degrees as an equivalent angle measured in radians by applying Equation 11-11.

$$radians = \frac{degrees}{57.3°} \qquad (11\text{-}11)$$

KEY POINTS

Phase is expressed as an angular measurement and can be measured in degrees or radians.

This can also be expressed as

$$\text{radians} = \text{degrees}\left(\frac{\pi}{180°}\right)$$

where π (the Greek letter pi) represents the approximate value of 3.1416. Scientific calculators have a π button. Pressing it automatically enters the value of 3.141... into your calculation.

EXAMPLE SOLUTION

Express the 30° point of a sine wave as an equivalent number of radians.

EXAMPLE **SOLUTION**

We apply Equation 11-11 as follows:

$$\text{radians} = \frac{\text{degrees}}{57.3°}$$

$$= \frac{30°}{57.3°} \approx 0.52 \text{ rad}$$

Practice Problems

1. 45° is the same as _____ radians.
2. 350° is the same as _____ radians.
3. 270° is the same as _____ radians.
4. 90° is the same as _____ radians.
5. 180° is the same as _____ radians.

Answers to Practice Problems

1. 0.785 or $\pi/4$ 2. 6.11   3. 4.7 or $3\pi/2$
4. 1.57 or $\pi/2$ 5. 3.14 or π

RADIAN-TO-DEGREE CONVERSION

If an angle is expressed in radians, we can convert it to an equivalent angle expressed in degrees by applying Equation 11-12.

$$\text{degrees} = \text{radians} \times 57.3° \qquad (11\text{-}12)$$

This can also be expressed as

$$\text{degrees} = \text{radians}\left(\frac{180°}{\pi}\right)$$

EXAMPLE SOLUTION

If a certain point on a sine wave is at 3.5 rad, express the same point in degrees.

EXAMPLE SOLUTION

We apply Equation 11-12 as follows:

$$\text{degrees} = \text{radians} \times 57.3°$$
$$= 3.5 \text{ rad } (57.3°) = 200.55°$$

Practice Problems

1. 2.5 rad is equivalent to _____ degrees.
2. 6.2827 rad is equivalent to _____ degrees.
3. 4 rad is equivalent to _____ degrees.
4. 1.75 rad is equivalent to _____ degrees.
5. 5.25 rad is equivalent to _____ degrees.

Answers to Practice Problems

1. 143.3   2. 360.2   3. 229.2

4. 100.3   5. 300.8

Instantaneous Value

As a sine wave progresses through its cycle, we can express the voltage (or current) at any instant in time as the instantaneous voltage (or current). Thus, for example, the instantaneous voltage of a sine wave is zero at 0°. Similarly, a sine wave with a 10-V peak amplitude, will have an instantaneous voltage of +10 V at 90° and −10 V at 270°. Equations 11-13 and 11-14 give us a way to determine the instantaneous value of a sine wave at any given angle.

$$v = V_P \sin\theta \qquad\qquad (11\text{-}13)$$

and

$$i = I_P \sin\theta \qquad\qquad (11\text{-}14)$$

where the lowercase v and i are used to represent instantaneous voltage and current values, θ is the angle, and sin is the trigonometric sine function.

The sine function is easily found on a scientific calculator by keying in the given angle (θ) and pressing the SIN button. Most scientific calculators can be configured to work in either degrees or radians. They will have a key labeled DRG, RAD, DEG, or other similar identification. Be sure your calculator is in the correct mode for the units of angular measurement you are using. A later section provides additional guidance in calculator operation.

KEY POINTS

The voltage at any specific instant in time is called the instantaneous voltage.

EXAMPLE SOLUTION

What is the instantaneous voltage at the 45° point of a sine wave with a peak voltage of 10 V?

EXAMPLE SOLUTION

We apply Equation 11-13 as follows:

$$v = V_P \sin\theta$$
$$= 10 \text{ V} \times \sin 45°$$
$$= 10 \text{ V} \times 0.7071 = 7.071 \text{ V}$$

EXAMPLE SOLUTION

If the peak current of a sine wave is 500 µA, what is the instantaneous current at 3.75 rad?

EXAMPLE **SOLUTION**

We apply Equation 11-14 as follows:

$$i = I_P \sin\theta$$
$$= 500 \text{ µA} \times \sin 3.75 \text{ rad}$$
$$= 500 \text{ µA} \times (-0.572) = -286 \text{ µA}$$

The negative current value, as with negative direct current values, simply means the current is going in the opposite direction from a positive current.

EXAMPLE SOLUTION

What is the peak voltage of a sine wave that has an instantaneous voltage of −25 V at 225°?

EXAMPLE **SOLUTION**

We apply Equation 11-13 as follows:

$$v = V_P \sin\theta, \text{ or}$$

$$V_P = \frac{v}{\sin\theta}$$

CONVERT FROM	CONVERT TO				
	PEAK	**PEAK TO PEAK**	**rms**	**INSTANTANEOUS**	**AVERAGE**
Peak	—	$V_{PP} = 2\,V_P$	$V_{rms} = 0.707\,V_P$	$v = V_P \sin\theta$	$V_{avg} = 0.637\,V_P$
Peak to Peak	$V_P = \dfrac{V_{PP}}{2}$	—	$V_{rms} = \dfrac{V_{PP}}{2.828}$	$v = 0.5\,V_{PP}\sin\theta$	$V_{avg} = 0.318\,V_{PP}$
rms	$V_P = 1.414\,V_{rms}$	$V_{PP} = 2.828\,V_{rms}$	—	$v = \dfrac{V_{rms}\sin\theta}{0.707}$	$V_{avg} = 0.9\,V_{rms}$
Instantaneous	$V_P = \dfrac{v}{\sin\theta}$	$V_{PP} = \dfrac{2v}{\sin\theta}$	$V_{rms} = \dfrac{0.707v}{\sin\theta}$	—	$V_{avg} = \dfrac{0.637\,v}{\sin\theta}$
Average	$V_P = \dfrac{V_{avg}}{0.637}$	$V_{PP} = 1.274\,V_{avg}$	$V_{rms} = 1.1\,V_{avg}$	$v = \dfrac{V_{avg}\sin\theta}{0.637}$	—

Table 11-1. Voltage Conversion Summary

$$= \frac{-25 \text{ V}}{\sin 225°}$$

$$= \frac{-25 \text{ V}}{-0.707} = 35.36 \text{ V}$$

Practice Problems

1. What is the instantaneous voltage of a sine wave at 30°, if its peak voltage is 500 V?
2. Compute the instantaneous voltage of sine wave at 2.8 rad, if its peak voltage is 200 mV.
3. Determine the maximum current value in a circuit that has a sinusoidal current with an instantaneous value of 740 mA at 1.6 rad.
4. What is the peak current in a circuit that has a sinusoidal current wave with an instantaneous current of 1.2 A at 50°?

Answers to Practice Problems

1. 250 V   2. 67 mV

3. 740.3 mA   4. 1.57 A

Time and Frequency Domains

All electronic signals can be viewed from either of two perspectives: time domain and frequency domain. It is useful to have a conceptual view of how these two domains interact.

TIME DOMAIN

We think of a time-domain signal as one whose instantaneous voltage changes over time. An oscilloscope, for example, displays the instantaneous values of a waveform as they change with time. We say an oscilloscope is a time-domain instrument.

FREQUENCY DOMAIN

We can also analyze an electronic signal in terms of its frequency content. That is, any given waveform can be shown to be composed of one or more sinusoidal signals at specific frequencies and amplitudes. A spectrum analyzer can be used to measure the frequency content of a waveform. We say that a spectrum analyzer is a frequency-domain instrument.

Converting between the time and frequency domains is well defined but requires advanced mathematics techniques. There are also many computer programs available that can make these conversions.

HARMONICS

Any repetitive, nonsinusoidal waveform in the time domain can be shown to be composed of a **fundamental frequency** and some combination of **harmonic frequencies.** The fundamental frequency is the basic frequency of the waveform as determined by the period

KEY POINTS

A technician must be able to view electronic signals in both time and frequency domains.

KEY POINTS

Periodic, nonsinusoidal waveforms in the time domain can be shown to consist of a mixture of the fundamental frequency and some combination of harmonic frequencies.

of the waveform. Harmonic frequencies are exact multiples of the fundamental frequency. Both fundamental and harmonic frequency components of the original waveform have sinusoidal characteristics in the time domain.

A frequency component of a waveform that is exactly twice the fundamental frequency is called the second harmonic. The third and fourth harmonic frequencies occur at three and four times the frequency of the fundamental, respectively. The first, third, fifth, and seventh harmonics, and so on are collectively called **odd harmonics.** Similarly, the second, fourth, sixth, and eighth harmonics are collectively called the **even harmonics.**

Now let us consider a specific example. Figure 11-16 illustrates that a square wave in the time domain corresponds to a mixture of the fundamental and the odd harmonic frequencies. Figure 11-16(a) shows the fundamental and the third and fifth harmonics. Figure 11-16(b) shows the algebraic sum of the three sine waves shown in Figure 11-16(a). Clearly, a rectangular waveform is starting to form. Figures 11-16(c) and (d) illustrate the effects of adding additional harmonic frequencies. The more harmonics we include, the more nearly the result resembles a square wave. A perfect square wave would have an infinite number of odd harmonics. The lower frequencies are primarily responsible for the flat, horizontal portions of the square wave. The higher harmonic frequencies contribute to the steepness of the rising and falling edges. Although the square wave is a selected example, any periodic nonsinusoidal waveform in the time domain can be decomposed into the fundamental and some combination of harmonics in the frequency domain.

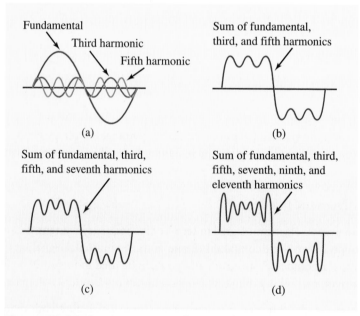

Figure 11-16. A square wave in the time domain corresponds to a mixture of the fundamental and odd harmonic frequencies in the frequency domain.

EXAMPLE SOLUTION

What is the third harmonic of a 1 kHz sine wave?

EXAMPLE SOLUTION

The third harmonic is three times the frequency of the fundamental, or

$$f_3 = 3 \times 1.0 \text{ kHz} = 3.0 \text{ kHz}$$

Practice Problems

1. What is the second harmonic frequency if the fundamental frequency is 2.5 MHz?
2. List the first four harmonics of a 25-MHz waveform.
3. List the first three odd harmonics of a 100-MHz sine wave.
4. List the first three even harmonics of a 250-MHz waveform.

Answers to Practice Problems

1. 5 MHz
2. 50 MHz, 75 MHz, 100 MHz, 125 MHz
3. 300 MHz, 500 MHz, 700 MHz
4. 500 MHz, 1,000 MHz, 1,500 MHz

Exercise Problems 11.2

1. If the frequency of a sine wave is 100 kHz, what is its period?
2. What is the period of the 60-Hz sinusoidal power supplied to your home by the power company?
3. If a sine wave has a period of 10 μs, what is its frequency?
4. A sine wave with a period of 2 ms has a frequency of _____ .
5. The positive peak of a sine wave occurs at _____ degrees.
6. The maximum negative voltage of a sine wave occurs at _____ radians.
7. A sine wave has a peak value of 150 A. Find its average value (I_{avg}).
8. If the average value (V_{avg}) of a sine wave is 35 V, what is its peak value?
9. What is the rms value for a sine wave with a 200-V peak?
10. What is the rms voltage for a sine wave with an average value of 12 V?
11. If a sine wave has a peak value of 25 V, what is its peak-to-peak value?
12. How much dc voltage is required to provide the same heating effect as a sine wave with a 200-V peak-to-peak amplitude?
13. What is the instantaneous voltage of a sine wave at 55° if its value at 90° is 100 V?
14. What is the peak voltage of a sine wave whose instantaneous voltage is –20 V at 300°?
15. A phase angle of 150° corresponds to _____ radians.
16. An angle of 2.5 rad is the same as an angle of _____ degrees.

11.3 Working with Phasors

Technicians frequently have to solve problems that involve phase relationships of one or more sinusoidal waveforms. It is very difficult to sketch sinusoidal curves accurately, but we need an accurate method of analyzing sinusoidal waveforms. Phasors provide us with a convenient way to represent sinusoidal waveforms and simplify many calculations.

Phasor Diagrams

KEY POINTS

The length of the phasor represents peak voltage, the angle of the phasor with respect to the horizontal axis corresponds to the instantaneous phase angle of the sine wave, and the distance from the end of the phasor to the horizontal axis represents the instantaneous voltage of the sine wave.

Figure 11-17 shows a phasor diagram. The phasor itself consists of a vector or "arrow" that can be rotated around a central point. The length of the phasor corresponds to the peak value of the sine wave. The angle of the phasor (θ) relative to the right-most horizontal axis corresponds to the angle of the sine wave. Finally, the distance from the point of the phasor to the horizontal axis represents the instantaneous voltage at the specified angle.

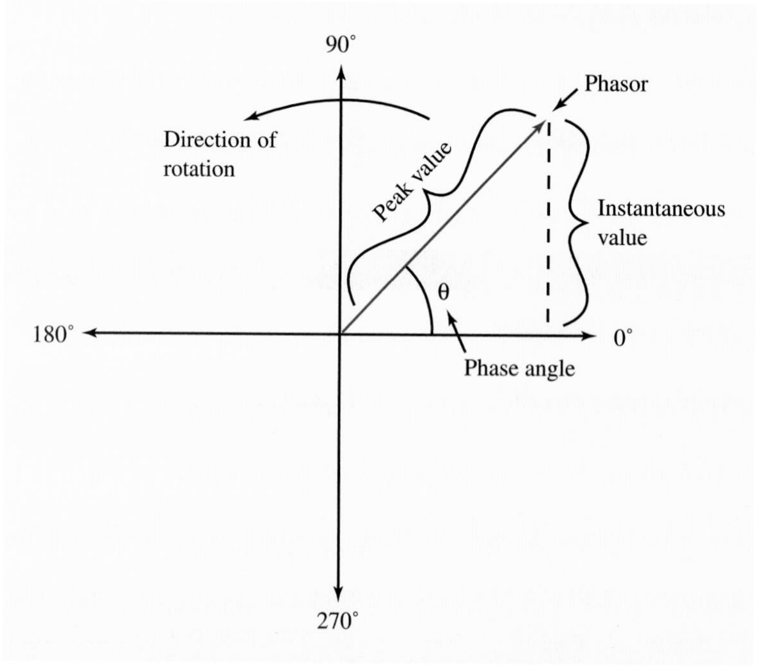

Figure 11-17. A phasor can be used to represent a sinusoidal quantity.

EXAMPLE SOLUTION

Figure 11-18(a) shows a phasor diagram for a particular sine wave. What is the peak value of the sine wave? What is the instantaneous voltage?

EXAMPLE SOLUTION

The peak voltage is represented on a phasor diagram by the length of the phasor. In the case shown in Figure 11-18, the peak voltage is 100 V. In Figure 11-18(b), we have drawn a horizontal dotted line from the point of the phasor to the vertical axis. The point of intersection is the instantaneous voltage. In this case, the instantaneous voltage is 50 V.

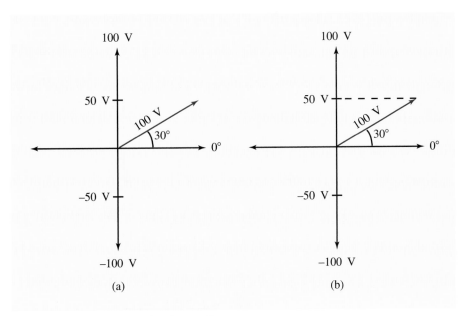

Figure 11-18. An example phasor diagram.

If the angle were not given, we could find it with a protractor. However, our accuracy would be limited by the preciseness of the phasor sketch. This would be tedious and inaccurate. Thus, we need yet a better way to calculate phasor values.

Right-Triangle Relationships

Figure 11-19 shows a phasor diagram. We have drawn a dotted line from the point of the phasor to the horizontal axis. This forms a triangle. Side "a" is a portion of the horizontal axis. Side "b" corresponds to the instantaneous voltage of the sine wave. Side "c" represents the peak voltage of the sine wave. The angle of the sine wave is represented by θ. A triangle drawn in this way will always have a right angle (90°), where the dotted line from the phasor point intersects the horizontal axis. Solution of problems involving triangles requires application of a branch of mathematics called trigonometry. Fortunately, the right-triangle calculations required to solve phasor problems are a well-defined subset of trigonometry. You do not have to understand all there is to know about trigonometry to solve these problems.

Figure 11-20(a) shows a right triangle with the sides labeled according to standard practice. The hypotenuse is always the longest side and is across from the right angle. Angle θ is used to represent the phase of a sine wave. The other two sides are named according to their relationship to θ. That is, the adjacent side is adjacent (i.e., forms one side of the angle) to θ, and the opposite side is opposite (i.e., across the triangle from) θ.

There are three basic equations that we must know in order to solve basic phasor problems.

$$\sin\theta = \frac{\text{opposite side}}{\text{hypotenuse}} \quad (11\text{-}15)$$

$$\cos\theta = \frac{\text{adjacent side}}{\text{hypotenuse}} \quad (11\text{-}16)$$

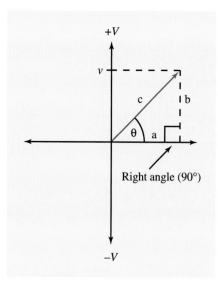

Figure 11-19. Phasor calculations are based on right-triangle calculations

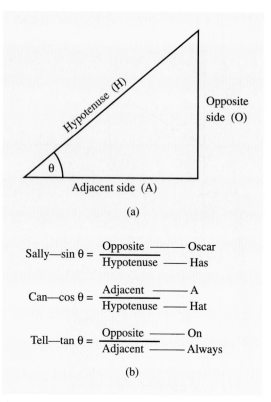

Figure 11-20. A right triangle with standard labels (a) and a memory aid (b).

$$\tan\theta = \frac{\text{opposite side}}{\text{adjacent side}} \qquad (11\text{-}17)$$

Sin, cos, and tan are basic functions (sine, cosine, and tangent) that are easily computed by your calculator. Equations 11-15, 11-16, and 11-17 are important relationships that will be used extensively in subsequent chapters. They should be committed to memory. Figure 11-20(b) shows a memory aid that is popular among technicians. Some technicians have different words, but basically the first letter in the expression "**S**ally **C**an **T**ell **O**scar **H**as **A** **H**at **O**n **A**lways" corresponds to the words **S**ine, **C**osine, **T**angent, **O**pposite, **H**ypotenuse, and **A**djacent. If you remember the phrase, then you can quickly construct the three equations.

Calculator Operation

There are several functions of your calculator that have not been required in the preceding chapters. We will, however, use them extensively in the following chapters, and you will certainly need to use them as a technician in industry. The trigonometric functions are of immediate interest to us. You might want to view the trigonometric function values as a code. Every angle has a corresponding numeric code called the sine (sin) of the angle. Each angle has a different code called the cosine (cos) of the angle. Finally, for every angle, there is a coded numerical value called the tangent (tan) of the angle. Scientific calculators have built-in look-up tables that allow you to determine the function values (or codes) for any given angle. Think of your calculator as your encoder/decoder device. You key in an angle.

Your calculator gives you a corresponding code. Alternatively, you can key in a coded value and the calculator will produce the value of the corresponding angle.

TRIGONOMETRIC FUNCTIONS

A scientific calculator will compute the sin, cos, and tan functions by simply keying in the value and pressing the appropriate button. The sine of an angle is computed when the SIN button is pressed. The cosine is found by pressing the COS button. And, finally, the tangent of an angle can be calculated by using the TAN button.

Many scientific calculators can work with angles expressed in either degrees or radians. Be certain that your calculator is in the degree mode, if your angles are expressed in degrees. Similarly, you must put your calculator in the radian mode, if you want to enter the angle values as radians.

EXAMPLE SOLUTION

Find the value of cos θ, when θ = 55°.

EXAMPLE **SOLUTION**

The key sequence required to obtain the value of a trigonometric function is the same for a standard engineering calculator or an RPN calculator. It consists of keying in the angle and pressing the appropriate function key. We must also ensure that the calculator is in the correct mode. Figure 11-21 shows a typical key sequence for the given problem. You should get a result of 0.5736. If your calculator produces an answer of 0.0221, then it is in the radian mode instead of the degree mode. Some calculators have the sin, cos, and tan functions as second, or shifted, functions. In these cases, you will need to press the "2nd" function key to invoke the cos function.

Figure 11-21.
A typical key sequence to compute cos 55°.

EXAMPLE SOLUTION

Find the value of tan θ, when θ = 2.75 rad.

EXAMPLE **SOLUTION**

Put your calculator into the radian mode. Look for a DRG or RAD button or consult your calculator manual. Figure 11-22 shows a typical key sequence. Your calculator should produce a result of –0.4129. If you get 0.048 for an answer, then your calculator is in the degree mode.

Figure 11-22.
A typical key sequence to compute tan 2.75 rad.

Practice Problems

Find the value of each of the following functions:

1. cos 28° = _____
2. sin 3.6 rad = _____
3. tan 150° = _____
4. sin 20° = _____
5. tan 175° = _____
6. tan 5 rad = _____
7. cos 1.8 rad = _____
8. sin 3.75 rad = _____
9. cos 200° = _____
10. sin 90° = _____

Answers to Practice Problems

1.	0.883	**6.**	−3.381
2.	−0.443	**7.**	−0.227
3.	−0.577	**8.**	−0.572
4.	0.342	**9.**	−0.94
5.	−0.087	**10.**	1

INVERSE TRIGONOMETRIC FUNCTIONS

KEY POINTS

Each of the basic functions has a corresponding inverse function: arcsin, arccos, and arctan.

There will be many times that we know the sine, cosine, or tangent of an angle, but we do not know the angle itself. That is, we known the "code," but we don't know the original angle. The sine, cosine, and tangent operations are called functions. The corresponding opposite (or decoding) operations are the arcsin, arccos, and arctan. These are called the inverse trigonometric functions. You can think of "arc" as meaning "angle whose." Thus, the expression arcsin 0.2 can be interpreted as "angle whose sine is 0.2."

EXAMPLE SOLUTION

We know the sine of an angle is 0.5. What is the size of the angle in degrees?

EXAMPLE SOLUTION

We need to find the angle whose sine is 0.5. That is, we need to find arcsin 0.5. First, we check to be sure our calculator is in the degree mode. Then, we key in the value of 0.5 and press the inverse sine button. In many cases, the inverse sine function is a second or shifted function of the sine key. You may have to press the "2nd" then the sine key to invoke the arcsin function. Some calculators label the inverse trigonometric functions as sin^{-1}, cos^{-1}, and tan^{-1}. Again, they are probably shifted functions. Another alternative labeling used on some calculators is ASIN, ACOS, and ATAN. Figure 11-23 shows a typical key sequence. You should get a result of 30°. If you get an answer of 0.524, then your calculator is in the radian mode.

·	5	2nd	SIN

Figure 11-23.
A typical calculator key sequence to compute arcsin 0.5.

Practice Problems

Find the value of the following inverse functions.

1.	arcsin 0.7 = _____		**6.**	arctan −0.577 = _____
2.	arccos −0.633 = _____		**7.**	arcsin −0.174 = _____
3.	arctan 0.466 = _____		**8.**	arccos 0.259 = _____
4.	arcsin 0.0872 = _____		**9.**	arctan −0.7 = _____
5.	arccos −1 = _____		**10.**	arcsin 0.2 = _____

Answers to Practice Problems

1.	44.4° or 0.78 rad	**3.**	24.99° or 0.44 rad
2.	129.27° or 2.26 rad	**4.**	5° or 0.087 rad

5. 180° or 3.14 rad   **8.** 74.99° or 1.31 rad

6. −29.98° or −0.52 rad   **9.** −34.99° or −0.611 rad

7. −10.02° or −0.175 rad   **10.** 11.54° or 0.2 rad

Phasor Calculations

In later chapters, we will use phasor calculations extensively to analyze circuits that contain inductors and/or capacitors. For now, however, let us concentrate on using the sine, cosine, and tangent functions and their inverse functions to solve right-triangle problems and basic sine wave phase problems.

EXAMPLE SOLUTION

Figure 11–24 shows a phasor diagram with known values for the peak and instantaneous voltages. Calculate the angle θ.

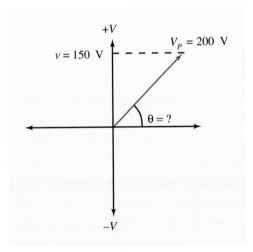

Figure 11-24. Find the angle θ.

EXAMPLE **SOLUTION**

In Figure 11-25, we have extended a vertical line from the phasor tip to the horizontal axis. This makes a right triangle. As long as we know the value of any two sides or at least one side and one of the smaller angles, we can find the rest by using Equations 11-15 through 11-17. In the present case, we know the hypotenuse and the opposite side. We can use Equation 11-15 to find the sine of the angle.

$$\sin \theta = \frac{\text{opposite side}}{\text{hypotenuse}}$$

$$\sin \theta = \frac{150 \text{ V}}{200 \text{ V}} = 0.75$$

If we know the sine of an angle, then the arcsin function will give us the value of the angle in degrees or radians.

$$\theta = \arcsin 0.75$$

$$\theta = 48.59° \text{ or } 0.848 \text{ rad}$$

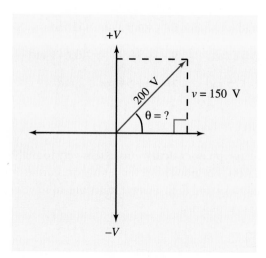

Figure 11-25. Form a right triangle and solve for θ.

EXAMPLE SOLUTION

Figure 11-26 shows a right triangle with the length of two sides given. Find the length of the remaining side, and find the value of the angle θ.

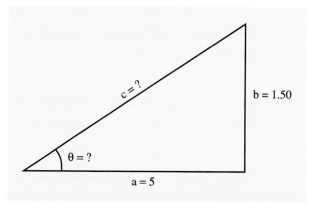

Figure 11-26. Find the angle θ and the length of side c.

EXAMPLE **SOLUTION**

Recall that if we know any two sides, we can find all other missing values. In this case, we know the adjacent side (a) and we know the opposite side (b). We can use Equation 11-17 to find the tangent of angle θ.

$$\tan\theta = \frac{\text{opposite side}}{\text{adjacent side}}$$

$$\tan\theta = \frac{1.5}{5} = 0.3$$

The arctan function will give us the actual value of the angle.

$$\theta = \arctan 0.3$$

$$\theta = 16.7° \text{ or } 0.29 \text{ rad}$$

We can use Equation 11-15 to determine side c (the hypotenuse).

$$\sin\theta = \frac{\text{opposite side}}{\text{hypotenuse}}$$

$$\text{hypotenuse} = \frac{\text{opposite side}}{\sin\theta} = \frac{1.5}{0.2873} = 5.22$$

Practice Problems

1. Find the length of side c and the value of angle θ in Figure 11-27.

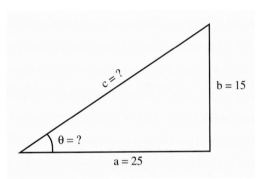

Figure 11-27. Find the missing values.

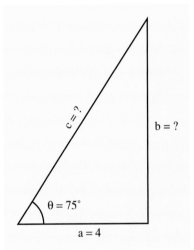

Figure 11-28.
Find the missing values.

2. Find the lengths of sides b and c in Figure 11-28.
3. Determine the lengths of sides a and c in Figure 11-29.

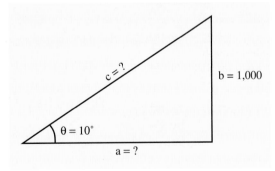

Figure 11-29. Find the missing values.

4. Calculate the value of θ in Figure 11-30.

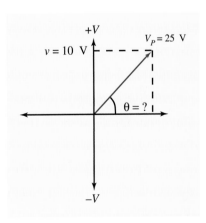

Figure 11-30.
What is the value of θ?

5. Find the value of θ in Figure 11-31.

Answers to Practice Problems

1. θ = 30.96° or 0.54 rad
   c = 29.16

2. b = 14.93
   c = 15.46

3. a = 5,671.28
   c = 5,758.77

4. 23.58° or 0.412 rad

5. −26.74° or −0.467 rad

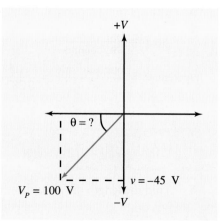

Figure 11-31. What is the value of θ?

Exercise Problems 11.3

1. Figure 11-32 shows a phasor diagram for a sine wave. What is the peak value of the sine wave?

2. What is the instantaneous voltage of the sine wave represented by the phasor diagram in Figure 11-32?

3. The longest side of a right triangle is called the _____ .

4. A right triangle always has a _____ -degree angle.

5. The sine of an angle can be found by dividing the _____ by the _____ .

6. Compute the following values:

    a. cos 100°   **b.** tan 27°   **c.** sin 278°
    d. cos 1.2 rad   **e.** tan 4.8 rad **f.** sin 2.9 rad

7. Determine the following angles and express in degrees:

    a. arccos 0.6   **b.** arcsin 0.09 **c.** arctan 11
    d. arcsin 0.91 **e.** arccos 0.91 **f.** arctan 30

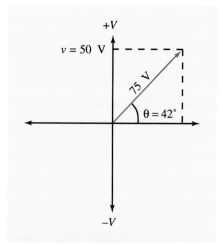

Figure 11-32. A phasor diagram.

8. Determine the following angles and express in radians:

    a. arccos 0.54   **b.** arcsin 0.04   **c.** arctan 9
    d. arcsin 0.95   **e.** arccos 0.95   **f.** arctan 26

9. What is the value of θ (in degrees) in Figure 11-33?

10. Find the length of side a in Figure 11-33.

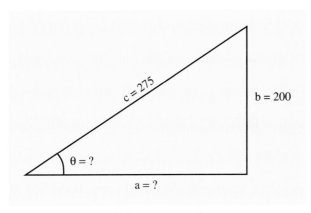

Figure 11-33. Find the missing values.

11.4 Resistive ac Circuit Analysis

Circuits composed of resistors and supplied by an ac voltage source can be analyzed with Ohm's and Kirchhoff's Laws. You must be careful, however, that you use the correct values. For example, if you want to find peak current, then use peak voltage in your calculations. If you want to find the rms voltage drop across a resistor, then use the rms value of current. For most purposes, power dissipation is computed using rms values of voltage and/or current. Although peak and average power are also useful computations for some applications (e.g., power supply circuits, antenna calculations, speaker design), unless specifically noted, power calculations in the remainder of the text will refer to rms values.

KEY POINTS

Analysis problems for ac resistive circuits may be simple or complex, but the method of solution is similar to that used to analyze dc resistive networks.

EXAMPLE SOLUTION

Figure 11-34 shows a simple series circuit composed of an ac source and a single resistor. Find the peak current in the circuit and the power dissipated in the resistor.

EXAMPLE SOLUTION

Since the value of the voltage source is given as a peak voltage, we can apply Ohm's Law directly to determine peak current.

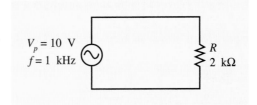

Figure 11-34. An ac circuit for analysis.

KEY POINTS

Although peak and average power calculations are important for some applications, the remainder of this text will concentrate on power dissipations using rms values of voltage and current.

$$I_P = \frac{V_P}{R}$$

$$= \frac{10\ \text{V}}{2\ \text{k}\Omega} = 5\ \text{mA}$$

We can use any of the power formulas to compute power, but we will need to use rms values of voltage and/or current. Let's use the $P = V^2/R$ formula. First, we will need to find the rms value of voltage.

$$V_{rms} = 0.707\ V_P$$

$$= 0.707 \times 10\ \text{V} = 7.07\ \text{V}$$

We can now compute power dissipation.

$$P = \frac{V^2}{R}$$

$$= \frac{(7.07\ \text{V})^2}{2\ \text{k}\Omega} = 24.99\ \text{mW}$$

EXAMPLE SOLUTION

Figure 11-35 shows a series-parallel circuit with an ac source. Complete Table 11-2 with reference to this circuit.

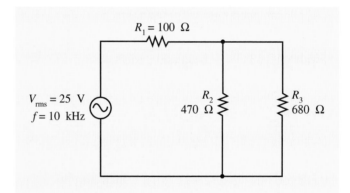

Figure 11-35. Analyze this circuit.

	VOLTAGE				CURRENT			
	V_P	V_{PP}	V_{rms}	V_{avg}	I_P	I_{PP}	I_{rms}	I_{avg}
R_1								
R_2								
R_3								
Total								

Table 11-2. Values for Figure 11-35

EXAMPLE SOLUTION

As with most circuit analysis problems, there are many ways to solve the problem. Let's begin by simplifying the circuit and sketching intermediate schematics. First, we note that R_2 and R_3 are in parallel. We can replace them with an equivalent resistance having a value of

$$R_{2,3} = \frac{R_2 R_3}{R_2 + R_3}$$

$$= \frac{470 \ \Omega \times 680 \ \Omega}{470 \ \Omega + 680 \ \Omega} = 277.9 \ \Omega$$

This step is shown in Figure 11-36. We can now find the total circuit resistance by combining the series resistances of R_1 and $R_{2,3}$.

$$R_T = R_1 + R_{2,3}$$

$$= 100 \ \Omega + 277.9 \ \Omega = 377.9 \ \Omega$$

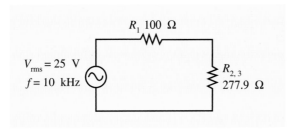

Figure 11-36. Simplification of the circuit shown in Figure 11-35.

The result at this point is shown in Figure 11-37.

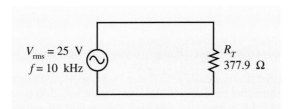

Figure 11-37. An equivalent circuit for Figure 11-35.

As a next step, let's compute the various voltage values for the applied voltage (V_T). V_{rms} is given. From this, we can find the others. We find V_p by using Equation 11-7.

$$V_{T(P)} = 1.414 V_{T(rms)}$$

$$= 1.414 \times 25 \ V = 35.35 \ V$$

We use Equation 11-9 to compute V_{pp}.

$$V_{T(PP)} = 2 V_{T(P)}$$

$$= 2 \times 35.35 \ V = 70.7 \ V$$

Finally, we can compute the average value by applying Equation 11-3.

$$V_{T(avg)} = 0.637 V_{T(P)}$$

$$= 0.637 \times 35.35 \ V = 22.52 \ V$$

This completes the first four boxes on the bottom row of Table 11-2. Since we know all forms of the applied voltage, and we know the value of total resistance, we can complete the lower row of Table 11-2 by using Ohm's Law. We find I_p as follows:

$$I_{T(P)} = \frac{V_{T(P)}}{R_T}$$

$$= \frac{35.35 \text{ V}}{377.9 \text{ }\Omega} = 93.54 \text{ mA}$$

Similarly, we compute I_{pp} as

$$I_{T(PP)} = \frac{V_{T(PP)}}{R_T}$$

$$= \frac{70.7 \text{ V}}{377.9 \text{ }\Omega} = 187.1 \text{ mA}$$

I_{rms} is computed as

$$I_{T(rms)} = \frac{V_{T(rms)}}{R_T}$$

$$= \frac{25 \text{ V}}{377.9 \text{ }\Omega} = 66.16 \text{ mA}$$

And we compute the total average value as

$$I_{T(avg)} = \frac{V_{T(avg)}}{R_T}$$

$$= \frac{22.52 \text{ V}}{377.9 \text{ }\Omega} = 59.59 \text{ mA}$$

If we wanted to compute total power dissipation, we could use any of the basic power formulas along with the relevant value of total current, voltage, or resistance.

We have computed everything we can with Figure 11-37, so let us move back to Figure 11-36 and continue. First, we note that total current also flows through R_1. This observation immediately gives us the values for all the R_1 currents in Table 11-2. We can find the voltages across R_1 as follows:

$$V_{1(P)} = I_{1(P)}R_1 = 93.54 \text{ mA} \times 100 \text{ }\Omega = 9.354 \text{ V}$$
$$V_{1(PP)} = I_{1(PP)}R_1 = 187.1 \text{ mA} \times 100 \text{ }\Omega = 18.71 \text{ V}$$
$$V_{1(rms)} = I_{1(rms)}R_1 = 66.16 \text{ mA} \times 100 \text{ }\Omega = 6.616 \text{ V}$$
$$V_{1(avg)} = I_{1(avg)}R_1 = 59.59 \text{ mA} \times 100 \text{ }\Omega = 5.959 \text{ V}$$

This completes the first row of answers in Table 11-2.

Kirchhoff's Law can be used to determine the various voltages across the $R_{2,3}$ resistance as follows:

$$V_{2,3(P)} = V_{T(P)} - V_{1(P)} = 35.35 \text{ V} - 9.354 \text{ V} = 26 \text{ V}$$
$$V_{2,3(PP)} = V_{T(PP)} - V_{1(PP)} = 70.7 \text{ V} - 18.71 \text{ V} = 51.99 \text{ V}$$
$$V_{2,3(rms)} = V_{T(rms)} - V_{1(rms)} = 25 \text{ V} - 6.616 \text{ V} = 18.38 \text{ V}$$
$$V_{2,3(avg)} = V_{T(avg)} - V_{1(avg)} = 22.52 \text{ V} - 5.959 \text{ V} = 16.56 \text{ V}$$

Because resistors R_2 and R_3 are in parallel, they will have the same voltages across them. Therefore, we can use the values for $V_{2,3}$ in both R_2 and R_3 rows of Table 11-2. This leaves only the current for R_2 and R_3 to be calculated. For this, we must move to Figure 11-35. There are many ways to calculate these currents. Let us use Ohm's Law for the R_2 calculations and Kirchhoff's Current Law for the R_3 calculations. The calculation for the R_2 currents are

$$I_{2(P)} = \frac{V_{2(P)}}{R_2} = \frac{26 \text{ V}}{470 \text{ }\Omega} = 55.32 \text{ mA}$$

$$I_{2(PP)} = \frac{V_{2(PP)}}{R_2} = \frac{51.99 \text{ V}}{470 \text{ }\Omega} = 110.6 \text{ mA}$$

$$I_{2(rms)} = \frac{V_{2(rms)}}{R_2} = \frac{18.38 \text{ V}}{470 \text{ }\Omega} = 39.11 \text{ mA}$$

$$I_{2(avg)} = \frac{V_{2(avg)}}{R_2} = \frac{16.56 \text{ V}}{470 \text{ }\Omega} = 35.23 \text{ mA}$$

Using Kirchhoff's Current Law for the R_3 current gives us the following results:

$$I_{3(P)} = I_{T(P)} - I_{2(P)} = 93.54 \text{ mA} - 55.32 \text{ mA} = 38.22 \text{ mA}$$

$$I_{3(PP)} = I_{T(PP)} - I_{2(PP)} = 187.1 \text{ mA} - 110.6 \text{ mA} = 76.5 \text{ mA}$$

$$I_{3(rms)} = I_{T(rms)} - I_{2(rms)} = 66.16 \text{ mA} - 39.11 \text{ mA} = 27.05 \text{ mA}$$

$$I_{3(avg)} = I_{T(avg)} - I_{2(avg)} = 59.59 \text{ mA} - 35.23 \text{ mA} = 24.36 \text{ mA}$$

The completed solution matrix is shown in Table 11-3.

	VOLTAGE (V)				CURRENT (mA)			
	V_P	V_{PP}	V_{rms}	V_{avg}	I_P	I_{PP}	I_{rms}	I_{avg}
R_1	9.354	18.71	6.616	5.959	93.54	187.1	66.16	59.59
R_2	26	51.99	18.38	16.56	55.32	110.6	39.11	35.23
R_3	26	51.99	18.38	16.56	38.22	76.5	27.05	24.36
Total	35.35	70.7	25	22.52	93.54	187.1	66.16	59.59

Table 11-3. Completed Solution Matrix for Figure 11-35

Exercise Problems 11.4

1. Complete a solution matrix similar to Table 11-3 for the circuit shown in Figure 11-38.

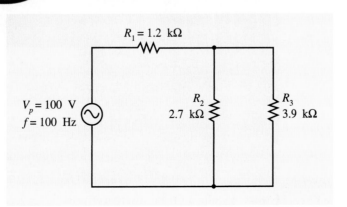

Figure 11-38. Complete a solution matrix for this circuit.

2. Refer to Figure 11-39 and complete a solution matrix similar to Table 11-3.

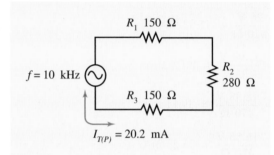

Figure 11-39. Complete a solution matrix for this circuit.

11.5 Applied Technology: Alternating Voltage Applications

There is an endless array of applications that rely on alternating voltage and current. The following sections provide a brief sampling of applications in several frequency ranges.

60-Hz Power Distribution

Undoubtedly, the most common application of alternating voltage is the 60-Hz power distribution that is supplied by power companies throughout the United States. The voltage is initially produced by alternators at such places as Hoover Dam near Las Vegas, Nevada. In the case of Hoover Dam, the alternators are rotated by the force of water as the Colorado River flows past the dam. Other power-generating plants use coal or nuclear power to produce steam, which ultimately provides the mechanical energy needed to spin the alternators. Figure 11-40 shows an alternator used to generate commercial power.

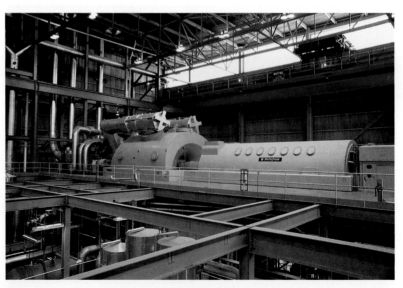

Figure 11-40. A 60-Hz alternator used for generation of commercial power. (Courtesy of Tampa Electric Company)

The voltage is initially generated at a level of several thousand volts. Near the power station, it is increased to several hundred thousand volts for long-distance transmission. Once the power has reached your neighborhood, it is again reduced to several thousand volts for local distribution. Finally, at a point just outside your home, the voltage is lowered to 220 V and routed into your house. At the service entrance (fuse or breaker box) in your house, the voltage is tapped off at a 120-V level for most appliances and lights, while the full 220 V is available for larger equipment such as air conditioners and electric dryers.

It may surprise you to learn that 120 V is an ideal rms value. (The peak voltage is nearly 170 V.) The actual rms voltage may change significantly throughout the day. It may be less than 100 V or higher than 130 V at certain times depending on the load demands and the capabilities of your power company.

The frequency is nominally 60 Hz and is established by the speed of the alternators. The frequency does vary slightly throughout the day. However, it is carefully compensated, so that the long-term average is very nearly 60 Hz. Many electrical and electronic devices (e.g., clocks and turntables) use the 60-Hz frequency as a time reference.

In most cases, transmission of 60-Hz energy is restricted to conductive wires and cables. No substantial energy is intentionally radiated at 60 Hz. Additional information regarding the transmission of 60-Hz power is presented in Chapter 17.

Sound Waves

Sound waves are made up of sinusoidal pressure changes in the air. We use sound waves in electronics for things like microphones and speakers. Sound also plays an important role in computers. Nearly all computers emit a "beep" to alert the user to an error. Some computers create and play music. Many industrial applications utilize sound as an audible annunciator to attract the operator's attention.

The frequency range for audible sound is from about 15 to 20 Hz on the low-frequency end to about 15 kHz to 20 kHz on the high-frequency end of the range. Sound waves travel through air at about 1,130 ft per second. This is roughly 1 ms per foot.

Ultrasonic Waves

Ultrasonic waves, like sound waves, are pressure changes in the air or other material. Ultrasonic applications include burglar alarms, range finders, and nondestructive inspection (NDI) equipment. In these cases, the ultrasonic sound is emitted by a transducer (equivalent of a speaker). It travels outward, reflects off objects (e.g., a burglar), and returns to a receiving transducer (like a microphone). Figure 11-41 illustrates this echo principle.

Some applications (e.g., a range finder on a camera) simply measure the amount of time that it takes the ultrasonic wave to travel from the "speaker" to the object and back to the "microphone." Since the waves travel at 1,130 feet per second, we can easily compute the distance to the object.

Other applications rely on a detected frequency difference between the transmitted and received signal. When the reflecting object is stationary, the received frequency is the same as the transmitted frequency. If the object is moving toward or away from the transmitter/receiver, then the received signal will be higher or lower, respectively, than the transmitted

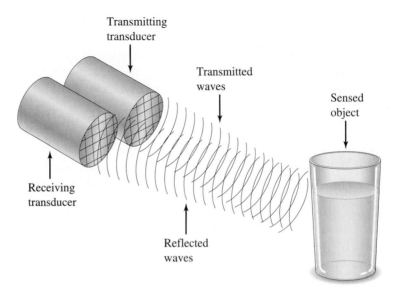

Figure 11-41. An ultrasonic receiver can detect an echo bounced off a sensed object.

signal. This shift in frequency is known as the doppler effect. This is the same phenomenon that causes the sound of a rapidly approaching train or race car to sound like a higher frequency. As the train or car passes you, the frequency drops to a much lower value.

Many medical electronics applications use ultrasonic frequencies. Ultrasound imaging systems can effectively "see" inside the body without the dangers associated with x rays. Ultrasonics are also used by oral hygienists to clean your teeth. Figure 11-42 shows a medical application for ultrasonics.

Ultrasonic energy travels through air at about 1,130 feet per second. The frequency range is from about 25 kHz to several hundreds of kilohertz.

Figure 11-42. An ultrasound machine used to view inside a human body. (Courtesy of GE Medical Systems)

Radio Waves

Radio, or electromagnetic, waves are used by a wide array of electronic devices. The most obvious applications include radio and television broadcasts. Other common applications include microwave communications links, satellite communications, and a variety of radar applications. Radio frequencies are also used for navigation. A small handheld receiver can intercept signals from a satellite and locate any spot in the world within a few feet.

Microwave ovens use radio frequencies. Other radio-frequency (rf) devices are used in industry to heat metal, melt plastic, and dry plywood. A similar device can be used to induce an artificial fever in a human.

Radio waves are transmitted from an antenna and travel through the air at very nearly the speed of light. The waves travel at about 300×10^6 meters per second, or 186,000 miles per second. Unlike sound waves, which are high- and low-pressure regions moving through the air, rf waves are electromagnetic. The moving wave has characteristics of both magnetic fields and electrostatic fields. The frequency range for electromagnetic radio waves begins below 20 Hz and extends into the hundreds of gigahertz. Figure 11-43 shows an antenna used for transmission of electromagnetic waves.

Figure 11-43. A radio-frequency antenna used to transmit electromagnetic energy. (Courtesy of Andrew Corporation)

1. Would a 10-Hz signal probably be referred to as ultrasonic?
2. The highest rf wave is about 100 MHz. (True or False)
3. Electromagnetic waves travel through the air at 1,130 feet per second. (True or False)

11.6 Oscilloscope Fundamentals

A previous chapter discussed oscilloscope operation as it applied to dc measurements. In this section, we will discuss the use of an oscilloscope for ac measurements. In particular, we will see how to use a scope to measure period, frequency, and voltage of a sine wave.

Voltage Measurements

Measurement of ac voltage on an oscilloscope consists of observing the amount of vertical deflection and multiplying by the setting of the vertical sensitivity control. Most modern scopes have internal microprocessors that automatically interpret the average, peak, peak-to-peak, and rms values of a measured wave. The values are displayed on the oscilloscope screen as text. Figure 11-44 shows a scope presentation with on-screen text of measured values.

Some less sophisticated scopes require the technician to measure and interpret the voltage manually. As a technician, you should be capable of operating any type of scope. It is best to learn on a basic scope with few automatic operations. If you later encounter a more sophisticated scope that has automatic measurements, then it will be easy to operate. On the other hand, if you only know how to operate a scope with automatic measurement capability, then you will be unable to operate a more basic scope.

○ **KEY POINTS**

An oscilloscope can directly measure peak-to-peak voltage and period. Once these two values are known, a technician can calculate the peak, rms, and average voltages and the frequency by applying basic sine wave equations.

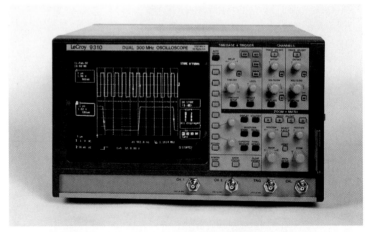

Figure 11-44. An oscilloscope with on-screen display of measured values. (Courtesy of LeCroy Corporation, Chestnut Ridge, NY)

Figure 11-45 illustrates how to measure the peak-to-peak voltage of a sine wave. The setting of the horizontal and triggering controls of the scope are not critical, but they should be adjusted to provide a stable display that includes at least one full cycle of the measured wave. If you prefer, you can slow the horizontal sweep down to the point where there are so many cycles on the screen that they appear as a solid band. It doesn't matter, so long as you can accurately determine the amplitude of the peaks.

The vertical gain control should be adjusted to provide the maximum deflection that still allows the entire sine wave to be viewed on the screen. The more deflection you have, the more accurate your measurement can be. The measurement itself consists of counting the number of divisions between the extreme negative peak and the extreme positive peak. The vertical position control can be used to move the displayed waveform into a more convenient position. In most cases, if the negative peak is placed directly on one of the horizontal graticule lines, then it is easier to determine the amount of deflection. Otherwise (as pictured in Figure 11-45), both peaks will appear in the middle portion of a division, making it more difficult to count divisions.

Once you have determined the number of divisions of vertical deflection, you simply multiply this number by the setting of the vertical gain (or sensitivity) control. This is expressed by Equation 11-18.

> **● KEY POINTS**
>
> In general, the accuracy of interpretation can be increased by expanding the displayed waveform as much as possible.

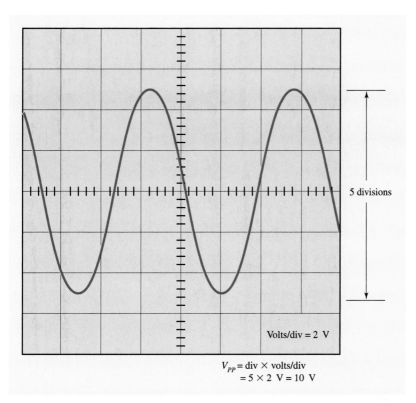

$$V_{PP} = \text{div} \times \text{volts/div}$$
$$= 5 \times 2 \text{ V} = 10 \text{ V}$$

Figure 11-45. Measurement of the peak-to-peak voltage of a sine wave.

$$V_{PP} = \text{No. of vertical divisions} \times \text{volts per division} \quad (11\text{-}18)$$

In the case pictured in Figure 11-45, we have five divisions of deflection, and the vertical gain control is set for 2 volts/division. The peak-to-peak voltage of the sine wave is computed as

$$V_{PP} = \text{No. of vertical divisions} \times \text{volts per division}$$
$$= 5 \text{ divisions} \times 2 \text{ V/division} = 10 \text{ V}$$

Practice Problems

1. Measure the peak-to-peak voltage of the sine wave in Figure 11-46.
2. What is the peak-to-peak voltage of the sine wave in Figure 11-47?
3. Determine the peak-to-peak voltage of the sine wave in Figure 11-48.

Answers to Practice Problems

1. 20 V

2. 60 mV

3. 0.36 V

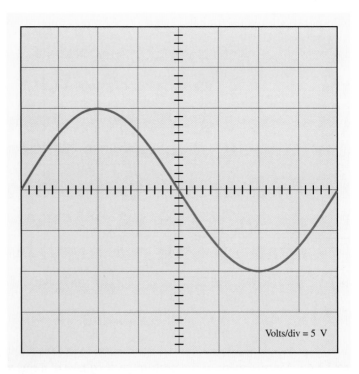

Figure 11-46. Measure peak-to-peak voltage.

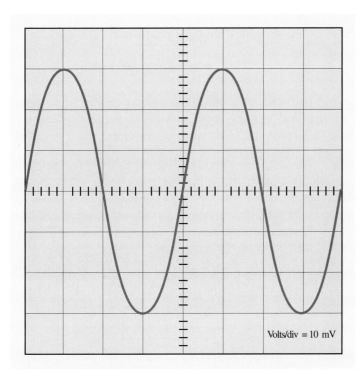

Figure 11-47. What is the peak-to-peak voltage of this sine wave?

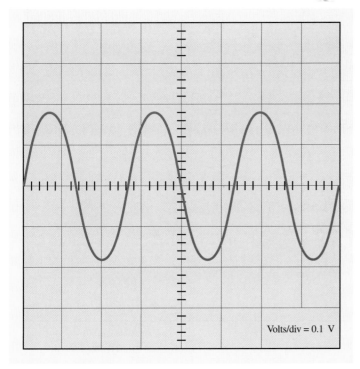

Figure 11-48. Determine the peak-to-peak voltage.

Once the peak-to-peak voltage of a sine wave has been measured, a technician can readily determine the peak, rms, or average value by applying Equations 11-3, 11-5, and 11-9.

Period Measurements

To measure the period of a sine wave, we essentially count the number of divisions between any point on a cycle and the corresponding point on the next cycle. It is usually most convenient to measure between the zero-crossing points, or between two consecutive peaks of the same polarity.

The vertical controls should be adjusted so that the sine wave nearly fills the screen. The triggering controls should be adjusted to provide a stable display. The sweep rate should be adjusted to provide as few cycles as possible on the screen as long as there is at least one full cycle visible. Figure 11-49 illustrates the measurement of period.

In Figure 11-49, the technician has adjusted the vertical positioning control, so that the sine wave is centered around the middle line on the screen. This is essential if you plan to

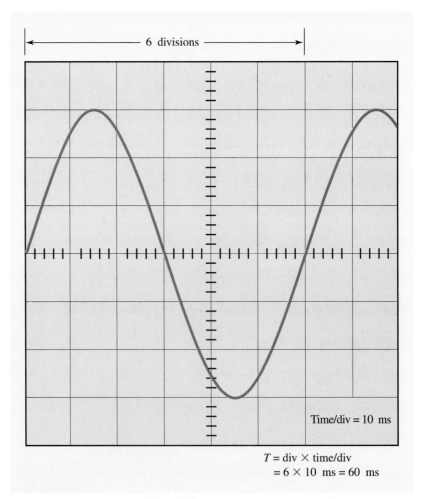

$$T = \text{div} \times \text{time/div}$$
$$= 6 \times 10 \ \text{ms} = 60 \ \text{ms}$$

Figure 11-49. Measuring the period of a sine wave.

measure between the zero-crossing points. The period is determined by counting the number of divisions between corresponding zero-crossing points on two consecutive half-cycles, and multiplying this number by the setting of the sweep rate control. This is expressed by Equation 11-19.

$$t = \text{No. of horizontal divisions} \times \text{sweep rate} \qquad \textbf{(11-19)}$$

In the case shown in Figure 11-49, the horizontal distance between the two corresponding zero-crossing points is 6 divisions. The sweep rate is set for 10 ms per division. The period is determined with Equation 11-19.

$$t = \text{No. of horizontal divisions} \times \text{sweep rate}$$
$$= 6 \text{ divisions} \times 10 \text{ ms/division} = 60 \text{ ms}$$

Figure 11-50 shows an alternate way to measure period. Here, the technician is measuring the time between two consecutive positive peaks. In this case, the waveform does not have to be centered on the screen. The period of the sine wave shown in Figure 11-50 is determined by applying Equation 11-19.

$$t = \text{No. of horizontal divisions} \times \text{sweep rate}$$
$$= 4.5 \text{ divisions} \times 100 \text{ μs/division} = 450 \text{ μs}$$

Either of the preceding methods provides better results when the vertical gain is set to provide a large vertical deflection. In the case of measuring between zero-crossing points,

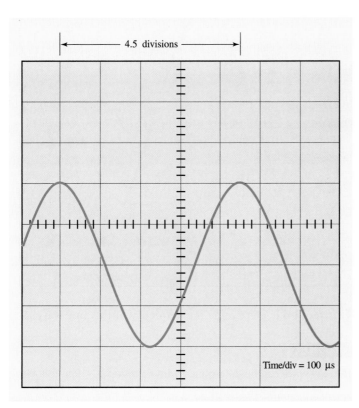

Figure 11-50. An alternate way to measure period.

maximum accuracy can be obtained when the vertical deflection is adjusted to extend beyond the limits of the screen. This makes the zero-crossing points on the wave appear more vertical and well-defined. The technician must be very certain that the trace is centered when this method is used. When measuring between two peaks, increased vertical amplitude makes the peaks appear sharper. It is easier to pinpoint the exact peak. If you are measuring between the positive peaks, it is all right for the negative peaks to extend off the screen.

Frequency Measurements

An oscilloscope does not measure frequency directly, unless it has a built-in microprocessor to perform automatic measurements. Without automatic measurements, the technician must measure period, as described in the preceding section. Then Equation 11-2 must be applied to determine the frequency.

EXAMPLE SOLUTION

What is the frequency of the sine wave shown in Figure 11-51?

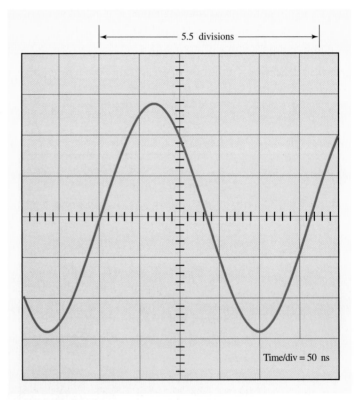

Figure 11-51. Determine the frequency of this sine wave.

EXAMPLE SOLUTION

The period of the sine wave can be determined with Equation 11-19 as follows:

t = No. of horizontal divisions × sweep rate

= 5.5 div × 50 ns/div = 275 ns

The frequency can now be computed by applying Equation 11-2.

$$f = \frac{1}{t}$$

$$= \frac{1}{275 \text{ ns}} = 3.636 \text{ MHz}$$

Exercise Problems 11.6

1. Determine the following values for the sine wave pictured in Figure 11-52.

    a. peak-to-peak voltage   **b.** average voltage   **c.** rms voltage
    d. frequency   **e.** period

2. Measure the following values on the waveform displayed in Figure 11-53.

    a. peak-to-peak voltage   **b.** average voltage   **c.** rms voltage
    d. frequency   **e.** period

3. Determine the following values for the sine wave pictured in Figure 11-54.

    a. peak-to-peak voltage   **b.** average voltage   **c.** rms voltage
    d. frequency   **e.** period

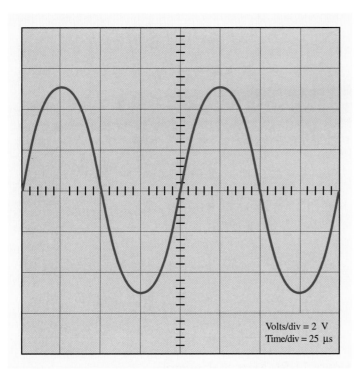

Figure 11-52. Measure this signal.

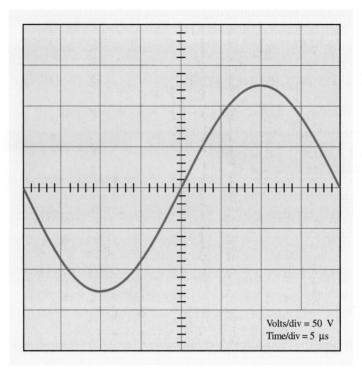

Figure 11-53. Measure this waveform.

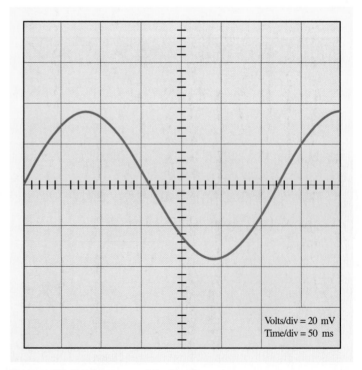

Figure 11-54. Measure this signal.

Chapter Summary

- Alternating voltage and current is at the heart of a vast majority of electronic applications. A sinusoidal voltage continuously varies in amplitude and periodically reverses polarity. A sinusoidal current continuously varies in amplitude and periodically changes direction. Sinusoidal waveforms consist of a series of repetitive cycles. The time for one cycle is called the period of the waveform. Frequency describes how many full cycles occur in a one-second interval. Each cycle is composed of two half-cycles with opposite polarities. Sine waves can be generated by an alternator or electronically (e.g., a function generator). If the durations of the two half-cycles are equal (as in a sine wave), the waveform is said to be symmetrical. An asymmetrical waveform has unequal half-cycles.

- The highest instantaneous voltage or current point on a sine wave is called the peak. Each sine wave has a positive and a negative peak. The average value of a full sine wave is zero, since it is positive as long as it is negative. As a general practice, the term average, when used to describe sine waves, refers to the average of one half-cycle. The average voltage is $0.637 \times V_P$. The root-mean-square (rms) voltage of a sine wave describes the heating effect of the waveform. It takes an amount of dc equal to the rms value of a sine wave to produce the same amount of heat in a resistance. The rms voltage is equal to $0.707 \times V_P$. The peak-to-peak voltage does not actually exist as an instantaneous value, but it is useful to describe the sine wave. It is always equal to twice the peak value.

- The phase of a sine wave describes an angular point on the wave at a specific time. Zero and 180 degrees coincide to the zero-crossing points of the sine wave. The peak values occur at 90° and 270°. If two sine waves are in phase, then the corresponding points of the two waveforms occur at the same time. Out-of-phase sine waves have corresponding points occurring at different times.

- Phase angle can be expressed in degrees or radians. One radian is equal to 57.3°. Either unit of measure is easy to use with a scientific calculator.

- The instantaneous voltage of a sine wave varies with each instant in time. For a given peak amplitude, the instantaneous voltage depends on the instantaneous phase of the sine wave. More specifically, $v = V_P \sin \theta$.

- A phasor diagram is a pictorial representation of the instantaneous phase in a sine wave. It can also show the phase relationships between multiple sine waves. The length of a phasor arrow represents the peak voltage of a sine wave. The distance of the arrowhead from the horizontal axis represents the instantaneous voltage of the sine wave. Finally, the angle formed by the phasor and the horizontal axis corresponds to the instantaneous phase of the sine wave.

- Phasor calculations are based on right-triangle calculations. A technician should be able to compute the sine, cosine, and tangent functions of an angle. Each of these functions also has a corresponding inverse function (arcsin, arccos, and arctan).

- Resistive circuits with an ac source are analyzed in a manner consistent with the dc analysis presented in previous chapters. Care must be exercised to ensure that corresponding voltages and currents are used (e.g., rms voltage with rms current).

- Applications that utilize alternating voltage and current are virtually endless. New applications are being created continuously. The applications can be grossly categorized according to frequency of operation. The frequency of alternating voltages and currents used in electronics extends from subaudio to well over 100 GHz.

- An oscilloscope can be used to measure period. The technician must compute frequency. The oscilloscope can be used to measure peak-to-peak voltage. The technician generally computes peak, average, and rms values. Many scopes have an internal microprocessor that automatically computes sine wave values and displays them on the scope screen.

Review Questions

Section 11.1: Generation of Alternating Voltage

1. The time required to complete a full cycle of alternating voltage is called the _____ .

2. How many alternations are required to make one full cycle?

3. If the positive and negative half-cycles of an alternating voltage are equal, the waveform is (*symmetrical, asymmetrical*).

4. If the cycles of a waveform are continuous, the waveform is said to be _____ .

5. What is another name for an ac generator?

6. Slip rings are segmented like a commutator on a dc generator. (True or False)

7. Why are the brushes on an alternator spring loaded?

8. When a looped conductor is rotated at a constant speed in a magnetic field, a sinusoidal wave is produced. What conductor position corresponds to zero degrees on the sine wave?

9. The maximum instantaneous value of a sine wave is called the _____ voltage.

10. What type of test equipment can generate sine waves, rectangular waves, and waves of other shapes?

Section 11.2: Characteristics of Sine Waves

11. What is the basic unit of measurement for the period of a sine wave?

12. What sine wave quantity describes how many cycles occur in a one-second interval?

13. What is the basic unit of measurement for the frequency of a sine wave?

14. If the period of a certain sine wave is 2.5 ms, what is its frequency?

15. What is the frequency of a sine wave that has a period of 75 ns?

16. A sine wave with a period of 100 ns has a higher frequency than a sine wave with a period of 100 ms. (True or False)

17. What is the period of a 5-MHz sine wave?

18. If the frequency of a certain sine wave is 2.5 GHz, what is its period?

19. What is the period of a 60-Hz sine wave?

20. What is the period of a 550-kHz sine wave?

21. How many times does a peak occur during one cycle of a sine wave?

22. The positive peak of a sine wave occurs at 180°. (True or False)

23. The negative peak of a sine wave occurs at 270°. (True or False)

24. If a sine wave has a positive peak current of 2.75 A, what is its negative peak value?

25. What is the average value of a *full cycle* of a sine wave, if the peak voltage is 10 V?

26. What is the average value of a *full sine wave,* if the peak voltage is 1,000 V?

27. What is the average value (V_{avg}) of a sine wave with a peak voltage of 25 V?

28. If the average value of a sine wave is 108 V, what is its peak value?

29. It takes an amount of dc equal to the average value of a sine wave to cause the same heat in a resistance. (True or False)

30. If a sine wave has a peak-to-peak current value of 10 A, what is its rms value?

31. What is the rms value of a sine wave that has a 150-mV average value?

32. What is the peak voltage of a sine wave that has a 200-μV rms value?

33. What is the peak-to-peak voltage of a 10-V peak sine wave?

34. Which sine wave would cause the most heat in a resistor?
     a) 10 V$_{PP}$ b) 10 V$_{rms}$ c) 10 V$_{avg}$ d) 10 V$_P$

35. Angular measurement of a sine wave can be expressed in _____ or _____ .

36. When the corresponding points on two similar frequency sine waves occur at different times, we say the sine waves are (*in, out*) of _____ .

37. If the instantaneous phase of a certain sine wave is 45° and the peak voltage is 100 V, what is the instantaneous voltage?

38. A phase angle of 2 rad is the same as a phase angle of _____ degrees.

39. A phase angle of 35° is the same as a phase angle of _____ radians.

40. A 100° angle can be expressed as an angle of _____ radians.

41. A 5-rad angle can be expressed as a _____ -degree angle.

42. What is the instantaneous voltage of a sine wave at 25° if its peak voltage is 30 V?

43. What is the instantaneous voltage of a sine wave at 105° if the rms voltage is 100 V?

44. What is the peak voltage of a sine wave that has an instantaneous voltage of 10 V at 1.2 rad?

45. If a sine wave at 3 rad has an instantaneous voltage of 120 V, what is its peak-to-peak value?

46. What is the frequency of the fifth harmonic of a 75-MHz waveform?

47. Seventy-five megahertz is the _____ harmonic frequency of a 25-MHz fundamental.

Section 11.3: Working with Phasors

48. The length of a phasor represents the _____ voltage of a sine wave.

49. The phase of a sine wave is represented on a phasor diagram by the angle between the phasor and the vertical axis. (True or False)

50. The distance of the phasor tip from the horizontal axis represents the _____ voltage of a sine wave.

51. The longest side of a right triangle is called the _____ .

52. The cosine of an angle in a right triangle is obtained by dividing the _____ by the _____ .

53. If the _____ in a right triangle is divided by the _____ , the result is a number that is the tangent of the angle.

54. If you know the tangent of an angle, how do you find the angle itself?

55. If you know the cosine of an angle, how do you find the angle itself?

56. What is the sine of a 35° angle?

57. What is the sine of a 3.2-rad angle?

58. What is the tangent of a 23° angle?

59. What is the cosine of a 1.6-rad angle?

60. If the tangent of an angle is 5.6, what is the angle in degrees?

61. If the tangent of an angle is 4.25, what is the size of the angle in radians?

62. Refer to Figure 11-55. What is the length of side a?

63. What is the length of side b in Figure 11-55?

64. What is the value of θ in Figure 11-56?

65. What is the length of side c in Figure 11-56?

66. What is the peak voltage of the sine wave represented in Figure 11-57?

67. What is the instantaneous voltage of the sine wave represented in Figure 11-57?

68. What is the instantaneous voltage of the sine wave represented in Figure 11-58?

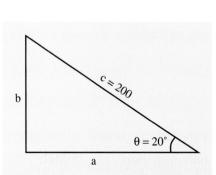

Figure 11-55. A right-triangle problem.

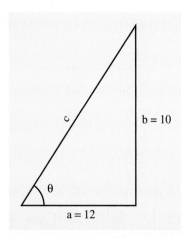

Figure 11-56. A right-triangle problem.

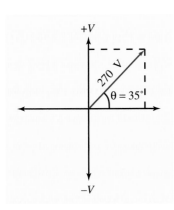

Figure 11-57. A phasor diagram.

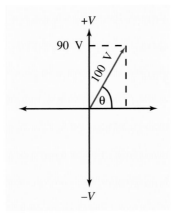

Figure 11-58. A phasor diagram.

69. What is the phase angle of the sine wave represented in Figure 11-58?

70. Compute the value of arccos 0.6 and express in radians.

71. Express the value of arcsin 0.8 in degrees.

72. The value of arctan 20 corresponds to _____ degrees.

Section 11.4: Resistive ac Circuit Analysis

Refer to Figure 11-59 for Questions 73–77.

73. What is the peak current?

74. What is the power dissipation in the resistor?

75. What is the average value of input voltage?

76. What is the peak-to-peak input voltage?

77. What is the instantaneous current at 50°?

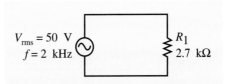

Figure 11-59. Analyze this circuit.

Refer to Figure 11-60 for Questions 78–87.

78. What is the total resistance?

79. What is the total rms current?

80. What is the peak voltage across R_1?

81. What is the average current through R_4?

82. What is the peak-to-peak voltage across R_2?

83. What is the rms current through R_3?

84. What is the total average current?

85. What is the total average current over a full cycle?

86. What is the total power dissipation?

87. What is the rms current through R_4?

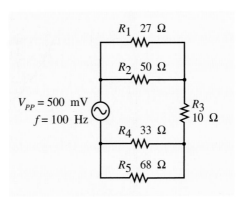

Figure 11-60. Analyze this circuit.

Section 11.5: Applied Technology: Alternating Voltage Applications

88. What is the frequency of the power distribution network in the United States?

89. When the power used in your home is first generated, it is a much lower voltage. (True or False)

90. The nominal value of voltage available at the wall sockets in your home is about 108 V average.

91. Sound and ultrasonic waves both travel through air at about 1,130 feet per second. (True or False)

92. What is the name of the effect that explains the frequency shift in an ultrasonic system when the reflecting object is moving?

93. An ultrasonic transmitter transducer is like a _____ in an audio system.

94. An ultrasonic receiver transducer is like a _____ in an audio system.

95. Radio frequency waves travel through air at approximately _____ meters per second.

96. Radio waves move through the air as high- and low-pressure changes. (True or False)

97. Radio waves are called electromagnetic waves. (True or False)

98. It is possible to have radio waves at least as high as 2,000 MHz. (True or False)

Section 11.6: Oscilloscope Fundamentals

99. What is the period of the sine wave shown in Figure 11-61?

100. What is the rms voltage of the sine wave shown in Figure 11-61?

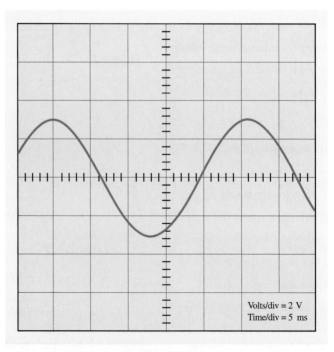

Volts/div = 2 V
Time/div = 5 ms

Figure 11-61. Interpret this display.

101. What is the frequency of the sine wave shown in Figure 11-62?

102. What is the peak voltage of the sine wave shown in Figure 11-62?

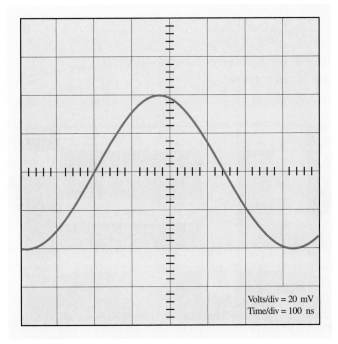

Volts/div = 20 mV
Time/div = 100 ns

Figure 11-62. Interpret this display.

TECHNICIAN CHALLENGE

Following is a crossword puzzle with entries that are related to the material presented in this chapter. It will give you one more opportunity to recall the facts and terms learned in this chapter.

ACROSS:

1. A unit of angular measurement.
8. One of the factors that determine the instantaneous voltage of a sine wave.
11. A type of current that is steady and flows in one direction.
12. The name of a 90° angle.
13. $0.637\ V_p$.
14. The longest side of a right triangle.
15. A point in an ac circuit where the current splits.
16. $0.707\ V_p$.

DOWN:

1. The "r" in rms.
2. A unit of angular measurement.
3. Alternating current.

4. Arccos is the inverse function of _____ .
5. A way to produce alternating voltage.
6. One positive alternation + one negative half-cycle.
7. Arctan is the inverse function of _____ .
8. Maximum voltage is induced when the conductor moves past the _____ in an alternator.
9. The alternator in your car is mechanically linked by _____ .
10. In the phrase "sin θ," θ is an _____ .

Equation List

(11-1) $t = \dfrac{1}{f}$

(11-2) $f = \dfrac{1}{t}$

(11-3) $V_{avg} = 0.637\, V_P$

(11-4) $I_{avg} = 0.637\, I_P$

(11-5) $V_{rms} = 0.707\, V_P$

(11-6) $I_{rms} = 0.707\, I_P$

(11-7) $V_P = \dfrac{V_{rms}}{0.707} = 1.414\, V_{rms}$

(11-8) $I_P = \dfrac{I_{rms}}{0.707} = 1.414\, I_{rms}$

(11-9) $V_{PP} = 2\, V_P$

(11-10) $I_{PP} = 2\, I_P$

(11-11) $radians = \dfrac{degrees}{57.3°}$

(11-12) $degrees = radians \times 57.3°$

(11-13) $v = V_P \sin\theta$

(11-14) $i = I_P \sin\theta$

(11-15) $\sin\theta = \dfrac{opposite\ side}{hypotenuse}$

(11-16) $\cos\theta = \dfrac{adjacent\ side}{hypotenuse}$

(11-17) $\tan\theta = \dfrac{opposite\ side}{adjacent\ side}$

(11-18) V_{PP} = No. of vertical divisions $\times$ volts per division

(11-19) t = No. of horizontal divisions $\times$ sweep rate

key terms

choke

copper loss

eddy currents

effective series resistance (*ESR*)

inductive reactance

Q

skin effect

objectives

After completing this chapter, you should be able to:

1. List six factors that affect electromagnetic induction.

2. State Faraday's and Lenz's Laws regarding electromagnetic induction.

3. List the factors that affect the value of an inductor.

4. Name and describe at least six types of inductors.

5. Determine the total inductance of a circuit having inductors connected in series, parallel, series-parallel, or in a complex configuration.

6. Contrast the effects of an inductor in an ac versus a dc circuit.

7. State the factors that determine the inductive reactance of an inductor.

8. Calculate the inductive reactance of an inductor in a given circuit.

9. Calculate the total inductive reactance of a circuit having multiple inductive reactances connected in series, parallel, series-parallel, or in a complex configuration.

10. Calculate the *Q* of a coil.

11. State the phase relationship between current and voltage in an inductive circuit.

12. Explain the power dissipation of an ideal inductance.

13. Identify possible reasons for power loss in a nonideal inductance.

14. Explain how to identify defects in an inductor.

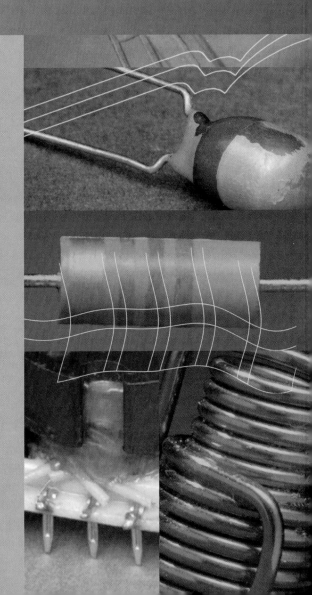

Inductance and Inductive Reactance

Inductors, like resistors, are a fundamental building block of electronic circuits. Like resistors, inductors tend to impede current flow. Unlike resistors, however, the opposition offered by inductors is different for dc and ac circuits. Inductors are also called coils or **chokes.** *The latter term comes from the use of some inductors to suppress (i.e., choke out) undesired frequencies. The basis for understanding the effects of an inductor in an electronic circuit lies in the understanding of electromagnetic induction.*

12.1 Electromagnetic Induction

Figure 12-1 illustrates the basic principle of electromagnetic induction. Here, two coils are wound on a common core. The core is a material with a high permeability, so the flux lines are more concentrated. One of the coils is connected to an alternating current source. As the current changes continuously, so does the magnetic flux within the core. Since a significant percentage of this flux is common to the second coil, the turns of the second coil will be intercepted by the changing flux lines. You will recall that when a conductor is cut by moving magnetic flux, a voltage is induced. Figure 12-1 shows that a voltage can be measured in the second coil, even though there is only a magnetic connection to the first coil. The process of inducing a voltage into a conductor with a changing magnetic field is called electromagnetic induction.

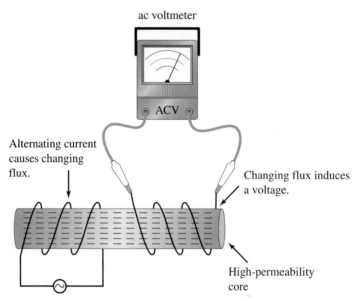

Figure 12-1. A demonstration of electromagnetic induction.

Review of Factors Affecting Induction

In a previous chapter, we discussed the factors that determine the amount and polarity of induced voltage. The factors are:

- field strength
- rate of relative motion
- angle of relative motion
- number of turns on the conductor
- direction of relative motion
- polarity of the magnetic field

These factors are illustrated in Figure 12-2.

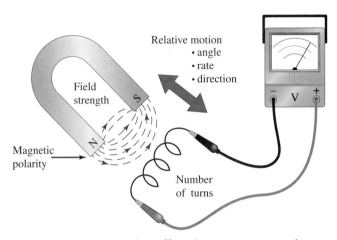

Figure 12-2. Factors that affect electromagnetic induction.

FIELD STRENGTH

The amount of induced voltage is directly proportional to the strength of the magnetic field. A stronger field produces more voltage.

RATE OF RELATIVE MOTION

In order to induce a voltage, there must be relative motion between the magnetic flux and the conductor. The conductor can move through the flux (as in a generator), or the changing flux can cut through the conductor (as illustrated in Figure 12-1). In either case, the amount of induced voltage is directly proportional to the rate of relative motion. If the conductor and flux are moving toward or away from each other faster, then a higher voltage will be induced.

ANGLE OF RELATIVE MOTION

If the conductor moves parallel to the flux lines, then it will not be intercepted by the flux and no induction will occur. If the conductor moves at right angles to the flux (cuts directly through it), then maximum interception of flux occurs, and the maximum voltage will be induced.

NUMBER OF TURNS IN THE CONDUCTOR

We can form the conductor into a coil to increase the amount of induced voltage. As the coiled inductor moves through the magnetic flux, each turn of the coil will have an induced voltage. The individual voltages are additive. Thus, the amount of induced voltage is directly proportional to the number of turns on the coil.

DIRECTION OF RELATIVE MOTION

The direction of flux interception determines the polarity of induced voltage. If the flux periodically increases and decreases (i.e., moves in both directions), then it will induce both polarities of voltage into a nearby conductor.

Polarity of Magnetic Field

The polarity of the magnetic field is the second factor that determines the polarity of induced voltage. If the polarity of the field is changed, then the polarity of the induced voltage will change.

Faraday's Law

Michael Faraday found that moving a magnet through a coil of wire produced a voltage in the coil. In 1831, he formally documented what we call Faraday's Law. In short, Faraday's Law states the following:

> When a coil of wire is intercepted by a magnetic field, a voltage is induced into the coil that is directly proportional to the number of turns on the coil and directly proportional to the rate of change of flux relative to the coil.

Lenz's Law

When current flows in a wire, it causes a magnetic field to be formed around the wire. If the wire is formed into a coil, then the magnetic field becomes concentrated and magnetic poles are formed. We already know that the polarity of induced voltage (or current in a closed circuit) is determined by the polarity of the magnetic field and the relative direction of motion. Lenz's Law describes the relationship of induced current polarity to the polarity of the moving magnetic field that caused the current. Lenz's Law can be summarized as follows:

> When a current is induced in a coil by a changing magnetic field, the current creates a second magnetic field. The magnetic field produced by the current has a polarity that opposes the changes in the original magnetic field.

This important principle is illustrated in Figure 12-3. In Figure 12-3(a), the north pole of a magnet is approaching a coil. An induced current flows in the coil such that a magnetic field that opposes the motion of the moving field is produced. In the case shown in Figure 12-3(a), the coil forms a north pole on the end nearest the approaching north pole. The two fields now oppose the relative motion.

Figure 12-3(b) illustrates the effects when the magnet is moved away from the coil. Since the field is moving in the opposite direction, we would expect an opposite polarity to be induced in the coil. As shown in Figure 12-3(b), the resulting current flow causes a magnetic field that opposes the motion of the original magnetic field. In the case shown in Figure 12-3(b), the induced current causes a south pole to form nearest the departing north pole. The unlike poles tend to attract, which retards the movement of the departing field.

Inductors

An inductor is essentially a coil of wire. When current flows through the turns of an inductor, a magnetic field is produced around each turn. If the current is changing, then the magnetic

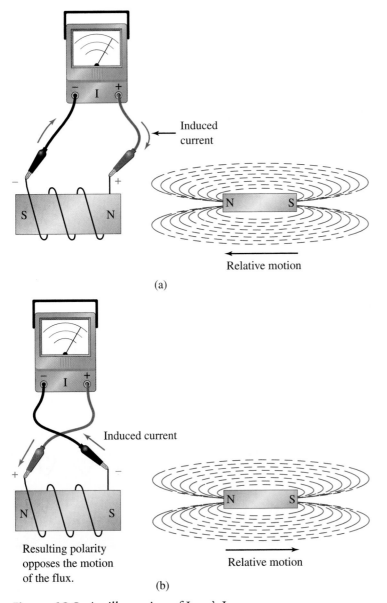

(a)

(b)

Figure 12-3. An illustration of Lenz's Law.

field will change. Any change in the magnetic field associated with one turn in the coil will produce a voltage (induced voltage) in the other turns of the coil as the flux lines intersect the wire. Thus, any change in current produces a change in flux which, according to Faraday's Law, causes an induced voltage in each turn of the coil. Further, Lenz's Law tells us that the polarity of induced voltage in each turn will be such that it opposes the original change in current. This is a very important concept that describes the basic behavior of an inductor.

When current changes in an inductor, a voltage is induced in the turns of the inductor that opposes the initial change in the current.

The key word here is *change*. If, for example, a certain inductor had a very high—but steady—current, then there would be a very strong magnetic field around the coil. But there would be no induced voltage, since the field is stationary (i.e., not changing). If, by contrast, a current changes in an inductor, then there can be a very high value of induced voltage. The value of induced voltage is, according to Faraday's Law, determined by the number of turns and by how quickly the current (and therefore the flux) changes. Many technicians prefer to state the behavior of an inductor yet another way:

> An inductor opposes a change in current.

This is a simple but important concept. An inductor does not oppose the actual flow of current. It opposes any changes in current. Thus, if the current tries to increase or decrease, the inductor will oppose the changing current and will try to keep it constant.

Figure 12-4 illustrates the net effect of the self-induced voltage in an inductor that has a changing current flowing through it. In Figure 12-4(a), the voltage source is increasing, which causes an increase in current through the coil. However, the increasing current causes a changing flux. As the magnetic field builds out around the coil, it induces a voltage into the turns of the coil. As illustrated in Figure 12-4(a), the polarity of the induced voltage opposes the current change. Thus, the net increase in current will be less, since the effective voltage in the circuit is less (remember series-opposing voltages subtract).

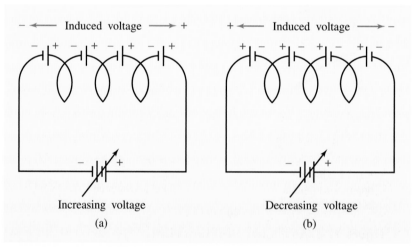

Figure 12-4. The self-induced voltage in an inductor opposes any changes in current.

Figure 12-4(b) shows the results when the current (same polarity) is decreased. As the current decreases, the magnetic field begins to collapse. As it collapses, a voltage is induced into the turns of the coil. But, the field is moving in the opposite direction from that shown in Figure 12-4(a), so the induced voltage polarity will also be opposite. As you can see from Figure 12-4(b), the induced voltage is now series-aiding with the applied voltage. This tends to compensate for or oppose or retard the decrease in current.

The property of an inductor that allows it to exhibit the characteristics just described is called inductance (L). Inductance is measured in henries (H). More specifically, if the current through an inductor changes at the rate of one ampere per second and causes one volt of induced voltage, then the inductor has a value of one henry.

Factors Affecting Inductance

The inductance of a coil is determined by the physical characteristics of the coil and the magnetic characteristics of the core. Let's examine each of these individually.

Number of Turns

Each turn of the coil has an induced voltage. Intuitively, we know that more turns must produce more inductance. In fact, the amount of inductance is directly proportional to the square of the number of turns on the coil. If, for example, we increased the number of turns on a coil by a factor of 4, then the inductance would increase by 4^2 or 16 times. This is illustrated in Figure 12-5.

Figure 12-5(a) shows a coil with 10 turns and an inductance of 100 µH. In Figure 12-5(b), the number of turns has been increased to 50. This factor-of-5 increase produces an increase of 5^2 or 25 times in the inductance. Thus, the inductance of the coil shown in Figure 12-5(b) is 2,500 µH. This computation assumes that no other factors were changed.

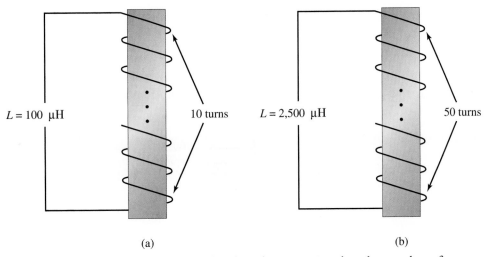

$L = 100$ µH   10 turns $L = 2,500$ µH   50 turns

(a)   (b)

Figure 12-5. The inductance of a coil is directly proportional to the number of turns squared.

Practice Problems

1. If the number of turns on a 10-mH coil is tripled, what will be the new value of inductance?
2. If the turns on a coil are increased from 25 to 65 turns, how much does the inductance increase?

3. If a technician removes half of the turns from a 100-turn coil, what happens to the value of inductance in the coil?

Answers to Practice Problems

1. 90 mH   **2.** Inductance increases 6.76 times.

3. Inductance reduces to one-fourth its original value.

LENGTH OF THE MAGNETIC CIRCUIT

This characteristic of a coil is easy to discuss, but very complex to calculate. In general, the longer the magnetic path, the lower the inductance of a coil. This is illustrated in Figure 12-6.

In Figure 12-6(a), the average length of the magnetic circuit is 26 cm. In Figure 12-6(b), the magnetic path length has been increased to 56 cm. Since the path length is longer in Figure 12-6(b), the inductance will be proportionally less. In particular, the magnetic path length has increased by 56 cm/26 cm, or a factor of 2.15. This will produce a reduction in the inductance of the coil by a factor of 1/2.15 or 46.5%.

Accurate computation of the magnetic path length for a solenoid is much more complex. The portion of the path within the coil is easily computed as the length of the core, or in the case of an air-core solenoid, the length of the coil itself. However, the portion of the complete magnetic path outside of the coil is not clearly defined. Its computation requires advanced mathematics.

CROSS-SECTIONAL AREA OF THE COIL

The size of the turns on a coil has a direct effect on the amount of inductance exhibited by the coil. Figure 12-7 illustrates the effects on inductance caused by a change in the cross-sectional area of the coil.

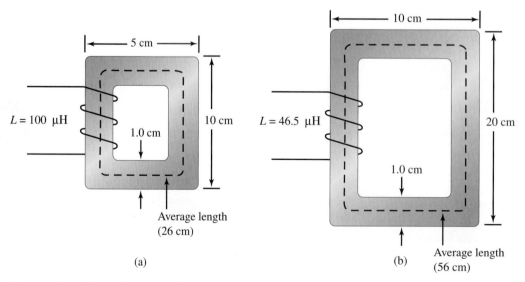

Figure 12-6. The inductance of a coil is inversely proportional to the length of the magnetic circuit.

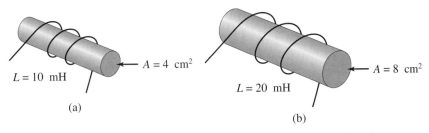

$L = 10$ mH $\quad\leftarrow A = 4$ cm^2

$L = 20$ mH $\quad\leftarrow A = 8$ cm^2

(a)

(b)

Figure 12-7. The inductance of a coil is directly proportional to the cross-sectional area of the coil.

In Figure 12-7(a), the coil has a cross-sectional area of 4 cm^2 and has an inductance of 10 mH. In Figure 12-7(b), the cross-sectional area of the coil is doubled (with all other factors constant). This increased area causes the inductance to double to 20 mH.

Practice Problems

1. If the cross-sectional area of a coil is increased by a factor of five, what happens to the value of inductance?
2. If the cross-sectional area of a coil is reduced to one-tenth its original size, what effect does this change have on the value of inductance for the coil?
3. The inductance of a coil is (*directly, inversely*) proportional to the cross-sectional area of the coil.

Answers to Practice Problems

1. Inductance increases to five times the original inductance.
2. Inductance decreases to one-tenth the original inductance.
3. directly

PERMEABILITY OF THE CORE

Recall that flux density is directly proportional to the permeability of a material. If we increase the permeability of the core material for a coil, we will produce a corresponding increase in flux density. Thus, a given change in current will now produce a larger change in flux. The increased flux changes create higher values of self-induced voltage in the turns of the coil. This, of course, translates to increased inductance in the coil. This is illustrated in Figure 12-8.

Figure 12-8(a) shows an air-core coil with an inductance of 1.0 μH. Figure 12-8(b) shows the same coil with a ferrite core inserted. The ferrite shown has a relative permeability of 2,000. Thus, the inductance will be increased by a factor of 2,000. The inductance in Figure 12-8(b) is 2,000 μH. Most technicians know intuitively that iron- or ferrite-core coils have a much higher inductance than air-core coils.

Air core

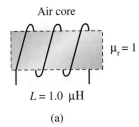

$\mu_r = 1$

$L = 1.0$ μH

(a)

Ferrite core

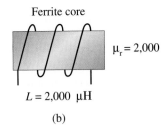

$\mu_r = 2,000$

$L = 2,000$ μH

(b)

Figure 12-8. The inductance of a coil is directly proportional to the permeability of the core material.

1. The core of a 50-μH coil has a relative permeability of 500. What happens to the inductance of the coil if it is wound on a core having a relative permeability of 5,000?

2. If the core of a 10-mH coil is changed from a material with a relative permeability of 4,000 to one with a relative permeability of 1,000, what will happen to the inductance of the coil?

Answers to Practice Problems

1. Inductance increases to 500 μH.   2. Inductance decreases to 2.5 mH.

All of the factors discussed that affect inductance can be used in an equation to compute inductance. Equation 12-1 describes their relationships to inductance.

$$L = \mu \frac{N^2 A}{l} \qquad (12\text{-}1)$$

where L is the inductance in henries, μ is the absolute permeability of the core, N is the number of turns, A is the cross-sectional area of the coil in square meters, and l is the length of the magnetic circuit in meters.

Since the permeability of most materials used by technicians is given as a relative permeability, this expression can also be written as Equation 12-2, by applying Equation 10-5.

$$L = \mu_r \times 1.257 \times 10^{-6} \times \frac{N^2 A}{l} \qquad (12\text{-}2)$$

where L is the inductance in henries, μ_r is the relative permeability of the core, N is the number of turns, A is the cross-sectional area of the coil in square meters, and l is the length of the magnetic circuit in meters.

Equations 12-1 and 12-2 show some very important relationships, but they produce only approximations of the true inductance of a coil. There are many nonideal effects, plus the poorly defined characteristics of the complete magnetic circuit, that make an accurate prediction of inductance very difficult to achieve. However, Equations 12-1 and 12-2 are fairly accurate for single-layer, air-core coils whose diameters are no more than one-tenth the coil length. For many technicians, the real value of Equation 12-1 or Equation 12-2 lies in the simple expression of relationships.

KEY POINTS

The inductance of an inductor or coil is determined by the:

- Physical dimensions of the coil
- Number of turns on the coil
- Permeability of the core material

The inductance is:

- Inversely proportional to the length of the magnetic path
- Directly proportional to the cross-sectional area of the coil, the permeability of the core material, and the square of the number of turns on the coil

1. A changing current in one coil can produce a voltage in a nearby coil through a process called _____ _____.

2. The amount of induced voltage in a conductor is _____ proportional to the strength of the moving magnetic field.

3. The polarity of induced voltage is determined by the number of turns on the coil. (True or False)

4. The angle of relative motion between a conductor and a magnetic field affects the amount of induced voltage. (True or False)

5. An inductor opposes any changes in current. (True or False)

6. Inductance is measured in _____.

7. The symbol for inductance is _____.

8. H is the abbreviation for _____.

9. If the number of turns on an inductor is increased from 100 to 300, the inductance of the coil will (*increase, decrease*) by a factor of _____.

10. If 25 turns are removed from a 100-turn coil, what is the effect on inductance?

11. If the cross-sectional area of a coil is made five times larger, what will happen to the inductance of the coil?

12. If the magnetic path length of a torroidal inductor is made shorter, what happens to the inductance of the coil?

13. If the permeability of the core material for a particular coil is changed from a relative permeability of 500 to a relative permeability of 15,000, what is the effect on inductance?

12.2 Self-Induction

In the preceding section we briefly discussed self-induction. When a changing current flows through the windings of a coil, a changing magnetic field is produced around the turns of the coil. This changing magnetic field intercepts the other turns on the coil and induces a voltage in them. This voltage is induced by a process called self-induction. Lenz's Law tells us that the polarity of self-induced voltage will be such that it opposes the initial current change. Faraday's Law tells us that the magnitude of the self-induced voltage will be determined by the number of turns on the coil and by the rate of change of flux. The rate of change of flux is determined by the rate of change of current. Now let us examine how various waveshapes of current affect the value of induced voltage in an inductor.

Effect of a Ramp Current

Figure 12-9(a) shows an inductor being driven by a current source. The amplitude of the current varies continuously. First, it slowly ramps from −1 mA to +1 mA during times t_1 and t_2. Then, it changes steadily, but more quickly, from +1 mA to −1 mA during times t_3 and t_4. The cycle then repeats.

As shown in Figure 12-9(b), the inductor voltage is constant as long as the rate of change of current is constant. The polarity of the self-induced voltage is not dependent on the polarity (i.e., direction) of current flow, but rather the direction of change. During the time period t_1, the value of current is steadily decreasing from −1 mA to zero. The polarity of the inductor voltage opposes this decrease. During time interval t_2, the current is increasing in the opposite direction. Note the polarity of the inductor is still the same in order to oppose this increase.

◯ KEY POINTS

If a current ramp is applied to an inductor, the value of self-induced voltage is constant, since the rate of change of current is constant.

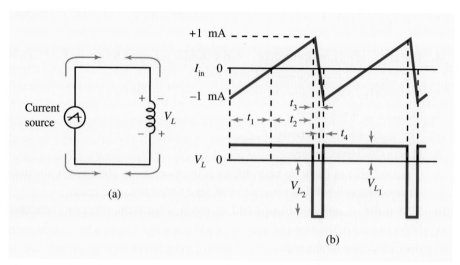

Figure 12-9. An inductor driven by a ramp current source generates a rectangular waveform of self-induced voltage.

A similar situation occurs on the other portion of the cycle. At the beginning of time t_3, the current stops increasing and begins to decrease toward zero. The inductor opposes this change. As you can see in Figure 12-9(b), the polarity of the self-induced voltage changes at the beginning of time t_3. During time t_3, the current decays steadily toward zero. The inductor voltage remains constant, since the rate of change of current is constant. During t_4, the current is building up in the opposite direction. The polarity of the inductor again opposes this increase.

The rate of change of current is greater during times t_3 and t_4 in Figure 12-9(b). This causes a greater value of self-induced voltage. Compare the magnitudes of V_{L_1} and V_{L_2} in Figure 12-9(b). You are encouraged to study Figure 12-9 and the preceding discussion carefully until you full appreciate what is being demonstrated. The very heart of inductor behavior is being shown.

Effect of a Step Current

Figure 12-10 shows an inductor connected to a current source capable of making sudden changes in value. The current changes from zero to +1 mA in a very short (but not zero) time period. Since this represents a very high rate of change of current, the inductor develops a substantial value of self-induced voltage that opposes the increasing current. This is shown in Figure 12-10(b) at time t_1.

Once the current reaches its 1-mA level, it is constant. Since the associated flux is no longer changing, there can be no induced voltage. So, between times t_1 and t_2 in Figure 12-10(b), the inductor has no self-induced voltage.

At time t_2 in Figure 12-10(b), the current changes rapidly from 1 mA to zero. During this brief period, the inductor experiences a high rate of change of flux, and therefore produces a relatively high value of self-induced voltage. Also note the polarity of induced voltage is opposite, since it is opposing a decreasing current.

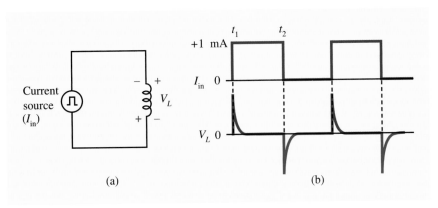

Figure 12-10. An inductor driven by a current source that makes sudden current changes.

There is no scale on the voltage graph in Figure 12-10(b), but it is important to note the value of self-induced voltage can easily reach thousands of volts for a modest current change. We will learn to compute specific values of induced voltage in a later section.

Effect of a Sine Wave Current

When we apply a sine wave of current to an inductive circuit, the coil behaves in the same manner as previously described. That is, it generates a self-induced voltage whose polarity opposes any change in current, and whose magnitude is determined by the rate of change of flux.

First, let us examine a sine wave of current more closely, by studying Figure 12-11. Recall that the magnitude of self-induced voltage in a coil is proportional to the rate of change in current. When the sine wave of current reaches its 90° and 270° points, the instantaneous rate of change of current is zero. For example, between the 0° and 90° points on the sine wave, the current is increasing, but its rate of increase slows as it approaches 90°. As the current wave passes through 90°, the rate of change passes through zero. Between 90° and 180° on the sine wave, the current is decreasing toward zero. As it approaches zero, it is decreasing, but it is changing the most rapidly. As the current wave passes through zero (at the 180° point), it changes direction and begins increasing between the 180° and 270° points.

Since induced voltage is directly proportional to the rate of change of voltage, we would expect maximum induced voltage at the 0° and 180° points on the current wave, since the current is changing most rapidly at these points. Similarly, we would expect no induced voltage at the 90° and 270° points on the current wave, since the rate of change of current is zero at these points.

Now let us consider the polarity of induced voltage. The direction of change (i.e., increasing or decreasing) changes as we pass through 90° and 270°. Thus, we would expect the polarity of induced voltage to change at these same points. The direction of current changes as it passes through zero, but so does the direction of change, so we will expect no change in the polarity of induced voltage at the 0° and 180° points on the current wave.

KEY POINTS

When a sine wave of current is applied to an inductor, the self-induced voltage is proportional to the rate of change of current.

KEY POINTS

The maximum levels of self-induced voltage occur at the 0° and 180° points of the current wave, since these are the points where the sine wave has the greatest rate of change.

KEY POINTS

The rate of change of current is zero at the 90° and 270° points, so no voltage is induced at these points.

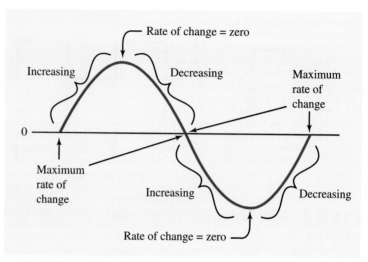

Figure 12-11. Analysis of a sine wave.

Figure 12-12(a) shows a coil driven by a sinusoidal current source. Figure 12-12(b) shows the relationship between current and voltage in the circuit. Note that the induced voltage is maximum at the points where the sine wave of current has the highest rates of change (0° and 180°). The induced voltage is zero when the rate of change of current is zero (90° and 270° on the current wave). Also note that the induced voltage changes polarity as the current wave passes through 90° and 270°. It is at these points that the current direction remains the same, but the direction of change is opposite.

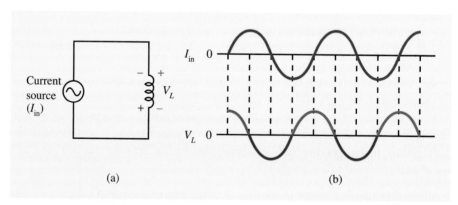

Figure 12-12. Current and voltage are 90° out of phase in an inductor.

General Equation for Self-Induced Voltage

Regardless of the specific waveshape of current through an inductor, the self-induced voltage will be directly proportional to the value of inductance and the rate of change of current flow through the inductor. This is summarized by Equation 12-3.

$$V_L = L\frac{\Delta i}{\Delta t} = L\frac{di}{dt} \qquad (12\text{-}3)$$

where the lowercase *d* or the Greek letter delta (Δ) represents the phrase "a change in," *L* is the value of inductance, *i* is the amount of current change, and *t* is the time during which the current change occurred.

EXAMPLE SOLUTION

If the current through a 100-μH inductor changes by 100 mA in a 10-ms period, what is the value of voltage across the inductor?

EXAMPLE **SOLUTION**

We apply Equation 12-3 as follows:

$$V_L = L\frac{di}{dt}$$

$$= 100 \ \mu H \times \frac{100 \ mA}{10 \ ms}$$

$$= 1.0 \ mV$$

Practice Problems

1. Determine the voltage developed across a 10-mH coil, if the current through it changes by 500 mA in a 100-μs period.
2. What is the value of self-induced voltage when the current changes at the rate of 100 A/s through a 250-mH inductor?
3. What rate of change of current is required to produce 500 mV across a 5-H inductor?

Answers to Practice Problems

1. 50 V   2. 25 V   3. 100 mA/s

Exercise Problems 12.2

1. Refer to Figure 12-13. Draw the waveshape of self-induced voltage that will be developed by the inductor.

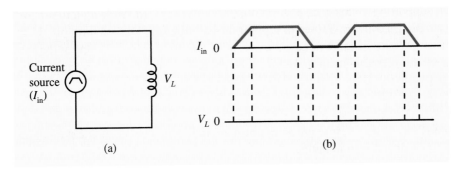

(a)   (b)

Figure 12-13.
What is the wave-shape of V_L?

2. The maximum rate of change on a sine wave occurs at the _____ and _____ points.

3. The minimum rate of change on a sine wave occurs at the _____ and _____ points.

12.3 Types of Inductors

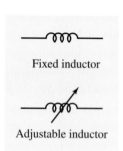

Figure 12-14.
Generic schematic symbols for a fixed and an adjustable inductor.

All inductors are similar in certain respects. They all possess inductance. They all generate a self-induced voltage in direct proportion to the rate of change of current. And the polarity of self-induced voltage will always oppose any changes in current. There are many different types of inductors, however, and they vary dramatically in value, primary application, and their physical appearance.

Inductors can be categorized into two classes: fixed and adjustable (variable). Figure 12-14 shows the schematic symbols for these two general classes of inductors.

Inductors can also be classified by the type of core material used. Some, but not all, manufacturers indicate the type of core material by using a slightly modified schematic symbol. We will examine several types of inductors in the following paragraphs. Variations to the generic schematic symbol will be shown where appropriate.

Air Core

⬤ **KEY POINTS**

An inductor can be made to be fixed or adjustable.

Probably the simplest type of inductor is an air-core coil. Figure 12-15(a) shows a picture of a commercial air-core inductor. An air-core coil is represented by the standard schematic symbol, Figure 12-15(b).

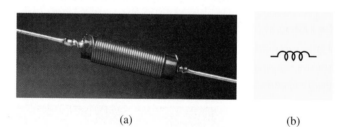

(a)   (b)

Figure 12-15. (a) An air-core coil and (b) its schematic symbol.

⬤ **KEY POINTS**

Inductors can be classed according to their core material: iron, ferrite, powdered iron, and air.

If the wire used to wind the coil is very stiff, then no support is needed for the coil, and it can be self-supporting. In many cases, however, the wire is not sufficiently rigid to support itself. In these cases, a coil form is used to support the coil winding. The coil form may be made of cardboard, plastic, or other nonconductive, nonmagnetic material. As long as the material has a relative permeability of near unity, then the coil is considered to have an air core.

Iron Core

Figure 12-16(a) shows a picture of an iron-core inductor. The schematic symbol for an iron-core coil is shown in Figure 12-16(b).

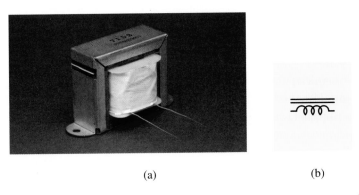

(a)   (b)

Figure 12-16. (a) An iron-core inductor and (b) its schematic symbol.

In most cases, the iron core is made of several laminated layers of thin iron sheets. Each metal sheet is electrically insulated from the other sheets by a thin coating of varnish and iron oxide. The layers of insulation have no significant effect on the overall magnetic characteristics of the core, but they greatly increase the electrical resistance of the core. This reduces core losses and is described in more detail in a later section.

Iron-core inductors are generally used for low-frequency applications in the audio range. They are easily recognized by their physical size, weight, and general appearance.

Powdered-Iron Core

Individual granules of iron can be individually coated with an insulative material and then pressed into a solid pellet that can be used as the core of a small inductor. The iron granules provide the high permeability needed to obtain the desired values of inductance. The insulative material makes the electrical resistance of the core very high to reduce power losses in the core. Many powdered-iron core inductors are adjustable. The powdered-iron slug is threaded and can be screwed in and out of the coil. As the core is inserted further, the inductance of the coil increases. This type of coil is generally used for radio frequencies.

Ferrite Core

A ferrite is a ceramic material that provides very high values of permeability, but has a high electrical resistance. From a user's point of view, there is little physical difference between a ferrite core and a powdered-iron core inductor. Figure 12-17 shows a typical ferrite core (or powdered-iron core) inductor along with the schematic symbol. Ferrite core inductors are used from frequencies in the low kilohertz to frequencies in the hundreds of megahertz.

Ferrite Beads as Inductors

Ferrite beads are also available in a wide array of sizes and material types. A ferrite bead has a hole (or holes) through which a wire or leaded component may be inserted. (Note: Some beads are sold with the wire already inserted.) The presence of the bead adds series inductance to the wire. The bead also adds resistance to the circuit.

◗ KEY POINTS

Ferrite beads can also be used as inductors at some frequencies. The ferrite bead can be slipped over either a wire or the lead of another component. It acts like a series resistance and a series inductance.

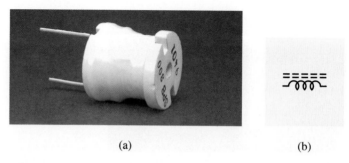

(a)   (b)

Figure 12-17. (a) A ferrite (or powdered-iron) core inductor and (b) its schematic symbol.

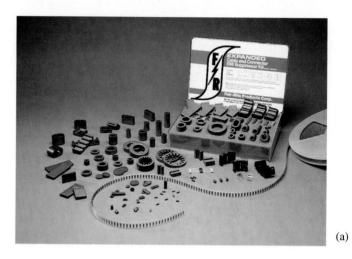

(a)

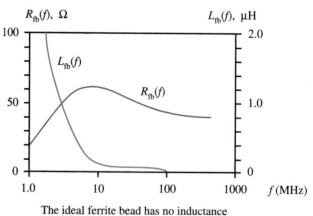

The ideal ferrite bead has no inductance
and substantial resistance at high frequencies.

(b)

Figure 12-18. (a) Ferrite beads are available in many different styles. (Courtesy of Fair-Rite Products Corporation) (b) The inductance and resistance of the ferrite material varies with frequency. (© TKC, Pinellas Park, FL)

Figure 12-18(a) shows an assortment of ferrite beads. Surface-mount ferrite beads are also shown. The exact behavior of the ferrite bead is largely dependent on frequency. The graph in Figure 12-18(b) shows that the inductance and resistance of the bead vary greatly with frequency. It is important to note that although the bead is slipped over the wire (even an insulated wire in some cases), the bead's inductance and resistance characteristics appear as if they were in series with the inserted wire. The frequency characteristics shown in Figure 12-18(b) are only representative. Exact performance varies considerably with the specific type of ferrite material.

Ferrite beads find extensive use in high-frequency circuits. They are also widely used in personal and handheld computers to suppress unwanted emissions that might otherwise interfere with nearby radio and television reception.

Molded Inductors

Figure 12-19 shows a molded inductor. Some molded inductors are easily confused with a resistor, since they may have a similar appearance and both have color-coded values. Manufacturers generally put a double-width band on inductors to distinguish them from resistors.

KEY POINTS

Molded inductors often look like color-coded resistors. The presence of a double-width band identifies the component as an inductor.

Figure 12-19. A molded inductor.

The color-coded bands (or dots in some cases) for a molded inductor are interpreted in much the same way as resistor color codes. There are, however, three major differences. First, the double-width band is not included as part of the color-coded value. Second, the value of the inductor is given in microhenries. Third, a gold band can serve one of two different purposes. If it appears as the last band, it indicates a 5% tolerance, similar to a resistor tolerance. If, however, the gold band appears in a position other than the tolerance position, it represents a decimal point.

EXAMPLE SOLUTION

What is the value of the molded inductor shown in Figure 12-20?

EXAMPLE **SOLUTION**

The wide silver band identifies the component as an inductor. The value is decoded as follows:

Figure 12-20. What is the value of this inductor?

red—2; gold—decimal point; violet—7;
tolerance—10%. The component is a 2.7-µH inductor with a 10% tolerance.

EXAMPLE SOLUTION

Determine the value of the component shown in Figure 12-21.

EXAMPLE SOLUTION

The wide band identifies the component as a molded inductor. Its value is decoded as follows:

orange—3; white—9; gold—5%. The component is a 39-μH inductor with a 5% tolerance.

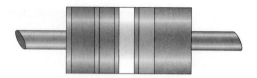

Figure 12-21. What is the value of this component?

Practice Problems

1. Determine the values for each of the components shown in Figure 12-22.

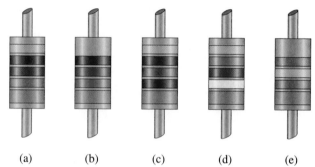

(a)   (b)   (c)   (d)   (e)

Figure 12-22. Determine the value of each of these components.

Answers to Practice Problems

1. **a.** 270 μH ± 5%
    b. 560 μH
    c. 180 μH ± 5%
    d. 475 μH ± 5%
    e. 2.2 μH

Surface-Mount

Figure 12-23 shows a picture of a surface-mount inductor. Except for the small physical size and the absence of leads, a surface-mount inductor functions like any other inductor. Because of their small size, they require care when soldering and desoldering them on a printed circuit board. Surface-mount inductors are often too small to print or color code the value. If the inductor has no apparent markings, then the technician must use an inductance meter to measure the value of inductance, or locate the value on a schematic diagram.

Figure 12-23. A surface-mount inductor is commonly used in high-density printed circuit boards.

Exercise Problems 12.3

1. Iron-core inductors are generally used for frequencies in excess of 100 MHz. (True or False)
2. How is a decimal point indicated on a molded inductor?
3. What is the purpose of the double-width color band on a molded inductor?
4. What is the value of a molded inductor that has the following color bands: a wide brown band, followed by blue, gray, and silver bands?
5. What is the value of a molded inductor that has the following color bands: a wide silver band, followed by red, gold, red, and silver bands?

12.4 Multiple Circuit Inductances

Any time two or more inductances are used near each other, there exists a possibility that the expanding and contracting magnetic field from one inductor will intercept the windings of another inductor. This complicates analysis, since the voltage induced by the external inductor may be in or out of phase with the self-induced voltage of the inductor. In either case, however, the effective value of inductance is altered.

For the purposes of this chapter, we shall assume that there is no interaction of magnetic fields between multiple inductors. This is a valid assumption in most cases where two or more physical coils are being considered. Chapter 17 discusses the case where there is magnetic interaction between two or more inductors.

Series Inductors

Figure 12-24 shows three inductors connected in series. Each inductor responds to changes in the common series current. The self-induced voltages all act to oppose any changes in current. The combined inductance of series-connected inductors is the simple sum of the individual inductors. This is expressed in Equation 12-4.

KEY POINTS

As long as the magnetic fields from the various inductors do not interact, series and parallel combinations of inductors can be combined using the same basic procedures used with resistors.

KEY POINTS

The total inductance of series-connected coils is found by summing the individual inductances.

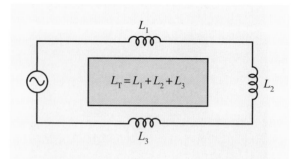

Figure 12-24. Series-connected inductors combine like series resistors.

$$L_T = L_1 + L_2 + L_3 + \ldots + L_N \qquad (12\text{-}4)$$

This equation is not valid if any of the coils are magnetically linked.

EXAMPLE SOLUTION

If $L_1 = 10\ \mu H$, $L_2 = 50\ \mu H$ and $L_3 = 25\ \mu H$ in Figure 12-24, what is the total circuit inductance?

EXAMPLE **SOLUTION**

We apply Equation 12-4 as follows:

$$
\begin{aligned}
L_T &= L_1 + L_2 + L_3 \\
&= 10\ \mu H + 50\ \mu H + 25\ \mu H \\
&= 85\ \mu H
\end{aligned}
$$

Practice Problems

1. What is the combined inductance if four 10-μH coils are connected in series?
2. If a 150-μH and a 250-μH inductor are connected in series, what is the total inductance?
3. How much inductance must be connected in series with a 20-mH inductor to have a total inductance of 45 mH?

Answers to Practice Problems

1. $40\ \mu H$ 2. $400\ \mu H$ 3. $25\ mH$

KEY POINTS

The total inductance of several parallel-connected coils is found using the reciprocal formula.

Parallel Inductors

Figure 12-25 shows three parallel-connected inductors. Since the total current divides among the various inductors, any given inductor experiences less than the total change in current. The self-induced voltage for a given inductor will be correspondingly smaller than

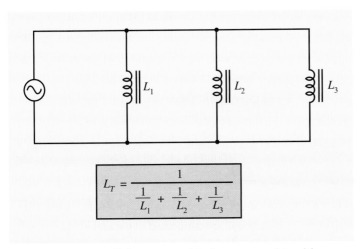

Figure 12-25. Parallel-connected inductors combine like parallel resistors.

if the total current change were applied to the coil. As with parallel-connected resistors, the total inductance is found by using a reciprocal formula:

$$L_T = \frac{1}{\dfrac{1}{L_1} + \dfrac{1}{L_2} + \dfrac{1}{L_3} + \ldots + \dfrac{1}{L_N}} \qquad \text{(12-5)}$$

As you might expect, the combined inductance of parallel-connected inductors is always less than the smallest individual inductance.

EXAMPLE SOLUTION

If L_1 = 80 mH, L_2 = 50 mH and L_3 = 35 mH in Figure 12-25, what is the total circuit inductance?

EXAMPLE **SOLUTION**

We apply Equation 12-5 as follows:

$$L_T = \frac{1}{\dfrac{1}{L_1} + \dfrac{1}{L_2} + \dfrac{1}{L_3}}$$

$$= \frac{1}{\dfrac{1}{80 \text{ mH}} + \dfrac{1}{50 \text{ mH}} + \dfrac{1}{35 \text{ mH}}}$$

$$= 16.37 \text{ mH}$$

It would be helpful to realize that the shortcut equations for parallel resistors can also be applied to parallel inductances. For example, Equation 12-6 can be used to find the combined inductance of several (N) parallel-connected inductors that all have the same (L) value.

$$L_T = \frac{L}{N} \qquad \text{(12-6)}$$

where L is the common inductance value and N is the number of parallel-connected inductances. Equation 12-7 is another useful relationship that computes the total inductance of two coils in parallel.

$$L_T = \frac{L_1 L_2}{L_1 + L_2} \qquad (12\text{-}7)$$

Practice Problems

1. What is the total inductance if five 25-mH chokes are connected in parallel?
2. If a 10-µH and a 45-µH coil are connected in parallel, what is the total inductance?
3. What is the total inductance if a 5-mH, a 25-mH, and a 10-mH inductor are connected in parallel?

Answers to Practice Problems

1. 5 mH   2. 8.18 µH   3. 2.94 mH

Other Network Configurations

Although not required for most practical applications, inductors can be connected in series-parallel or complex configurations. The total inductance for these networks can be found using the same basic procedures that are used for series-parallel and complex combinations of resistors. This method of analysis is not valid if any of the coils are magnetically coupled to each other.

Exercise Problems 12.4

1. What is the total inductance of the circuit shown in Figure 12-26?

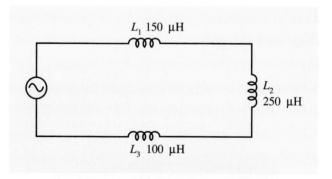

Figure 12-26. Find total inductance.

2. What is the value of L_3 in Figure 12-27?

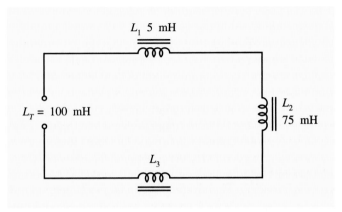

Figure 12-27. What is the inductance of L_3?

3. Find the total inductance of the circuit in Figure 12-28.

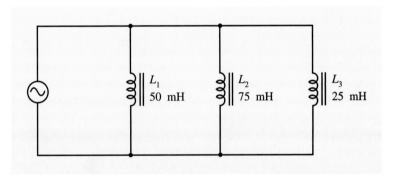

Figure 12-28. Find the total inductance of this circuit.

4. Determine the value of L_2 in Figure 12-29.

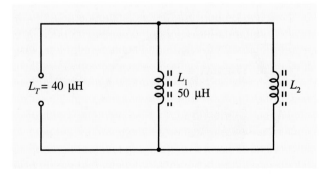

Figure 12-29. What is the value of L_2?

5. Compute the total inductance for the circuit shown in Figure 12-30.

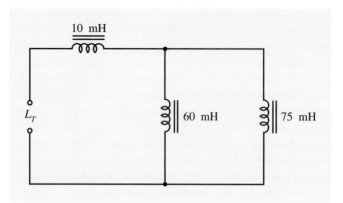

Figure 12-30. Find the total inductance of this circuit.

12.5 Inductance in DC Circuits

Now let us examine the behavior of an inductor in a dc circuit. Figure 12-31(a) shows an inductor in series with a switch, a resistor, and (in position A) a battery. When the switch is in position B, as shown in Figure 12-31(a), the battery is removed from the circuit. There is no current, and there is no voltage across either the inductor or the resistor. The coil has no magnetic field.

Now, if we move the switch to position A, as shown in Figure 12-31(b), the battery is connected across the circuit. The current tries to increase suddenly from zero to some higher value. This represents a very high rate of change of current. The magnetic field will build rapidly around the coil. Since the rate of change of flux is high, the coil will have a high value of self-induced voltage. As shown in Figure 12-31(b), at the first instant the self-induced voltage in the coil will be 12 V. Additionally (according to Lenz's Law), the

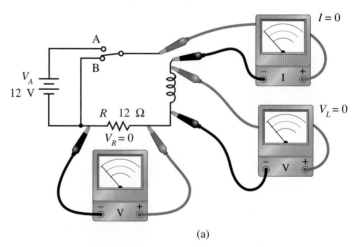

(a)

Figure 12-31. The response of an inductive circuit to a switched dc source.

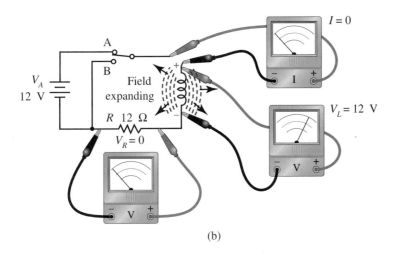

(b)

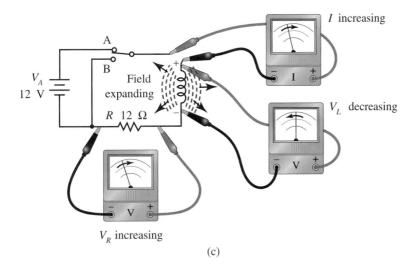

(c)

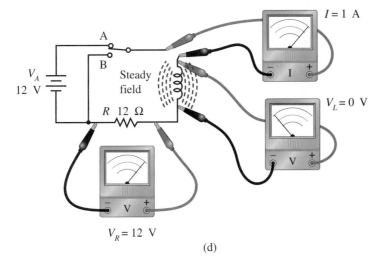

(d)

Figure 12-31. *Continued*

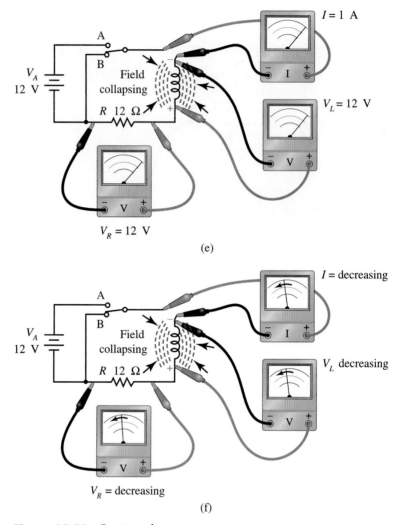

Figure 12-31. *Continued*

polarity will be such that it opposes the current increase. Now at the first instant, the 12 V of self-induced voltage completely opposes the 12 V of the battery to yield a net circuit voltage of zero. Thus, at this first instant, there is no current in the circuit and no voltage across the resistor.

Figure 12-31(c) shows the circuit a few moments after the battery was connected. The rate of expansion of the magnetic field is decreasing, so the self-induced voltage is also decreasing. Since the inductor voltage is offering less opposition to the battery voltage, the current continues to increase. Since the current flows through the resistor, its voltage will also be increasing.

Eventually, the current will stop increasing and the magnetic field will stop expanding. This condition is shown in Figure 12-31(d). The magnetic field is at its maximum level, but it is stationary. Therefore, no voltage is induced into the windings of the coil. At this point, the inductor is nothing more than a coiled piece of wire. It drops no voltage and offers no opposition to current. As shown in Figure 12-31(d), the full supply voltage is across the resistor, and the current is determined by Ohm's Law. This condition will exist as long as the switch remains in position A.

Figure 12-31(e) shows the circuit condition at the instant the switch is moved back to position B. With the power source disconnected, the field of the inductor will begin to collapse. As it does, it will induce a voltage into the windings of the coil. Because the flux is now moving in the opposite direction, the induced voltage will be of the opposite polarity. The coil is now acting as a voltage source, and it (momentarily) keeps the current flowing at the same 1-A value. As indicated in Figure 12-31(e), the current remains the same for the first instant, as does the resistor voltage. The inductor voltage, as shown in Figure 12-31(e), is –12 V. (Note that the negative sign merely indicates that the polarity is opposite from that previously shown).

Figure 12-31(f) shows the circuit conditions a few moments after the switch was moved back to position B. Here, the magnetic field is continuing to collapse, but its rate of collapse is decreasing as the field dissipates. Consequently, the voltage induced into the coil is also decreasing. Since the voltage produced by the coil is the only source of voltage in the circuit at this time, then the resistor voltage and the current will also be decaying at the same rate. Eventually, the magnetic field will completely collapse. At this time, there will be no voltage or current in the circuit, and we will return to the conditions depicted in Figure 12-31(a).

In Chapter 13, you will learn how to calculate the amount of time it takes for the magnetic field to build to maximum or to decay to zero. For now, it is important for you to understand the following points:

- When a direct current is switched through a coil, the self-induced voltage is maximum at the first instant and prevents any current flow.
- As the magnetic field builds out around the coil, the self-induced voltage slowly drops and the current through the circuit slowly rises.
- Once the magnetic field becomes stable (limited by the circuit resistance or the magnetic saturation of the core), then there is no self-induced voltage across the coil. The circuit current is strictly limited by the circuit resistance. The inductor has no effect on current flow.
- When an attempt is made to abruptly reduce the direct current through an inductor to zero, the magnetic field around the inductor begins to collapse. At the first instant, the induced voltage is adequate to keep the same value of current flowing through the circuit. The coil acts as a voltage source as it returns energy to the circuit. As the magnetic field decays, the self-induced voltage and the circuit current decrease toward zero.
- It takes a definite amount of time for the current to change from one value to another. The inductor prevents any abrupt current changes.

The fourth item in the preceding list describes an inductor characteristic that allows us to produce a high voltage from a lower battery voltage. Figure 12-32 shows a circuit similar to the one previously discussed, with the addition of R_2. Figure 12-32(a) shows the circuit condition after the magnetic field has had time to fully expand. The circuit is stable with a current of 1 A.

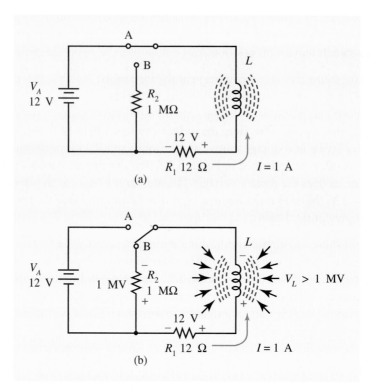

Figure 12-32. An inductor can create a high voltage from a low battery voltage.

Figure 12-32(b) shows the circuit at the first instant after the switch is moved to position B. You will recall that the collapsing inductor field generates enough self-induced voltage to keep the current constant for a moment. Note, however, that the resistance of the circuit is now in excess of 1 MΩ. When 1 A flows through R_2, it generates one million volts! So what is the source for this high voltage? Since R_2 is so large, the current tries to decrease abruptly when the switch is moved to position B. This high rate of change of current produces a high rate of change of flux, which causes a high self-induced voltage. In the present case, the self-induced voltage will be in excess of one million volts!

This same principle is used in the ignition system of an automobile to provide the high voltage needed to arc across the spark gap in the spark plugs. The automobile only has a 12-V dc source, but the spark plugs require several thousand volts to arc across the gap and ultimately ignite the fuel vapor.

Exercise Problems 12.5

1. What determines the amount of self-induced voltage across an inductor when used in a dc circuit?
2. Explain how it is possible to obtain a high voltage from a low-voltage battery.

12.6 Inductance in ac Circuits

When an inductor is used in a circuit that has a sinusoidal current, the self-induced voltage is also sinusoidal. The self-induced voltage always has a polarity that opposes any changes in current. The magnitude of the self-induced voltage is determined by the rate of change of current through the inductor. Since the maximum rate of change for a sine wave of current occurs at the 0° and 180° points, these are the points that correspond to maximum inductor voltage.

Inductive Reactance

The self-induced voltage in an inductor is constantly changing if the current through the inductor is sinusoidal. However, the polarity of the induced voltage always opposes the changing current. This opposition results in a lower value of current than would otherwise flow. The opposition to a sinusoidal alternating current flow presented by an inductor is called **inductive reactance**. It is measured in ohms. The generic symbol for reactance is X. Inductive reactance is represented by the symbol X_L.

FACTORS AFFECTING INDUCTIVE REACTANCE

Anything that affects the value of self-induced voltage in the inductor will affect its inductive reactance. Thus, we would expect the number of turns on the coil, permeability of the core, rate of change of current, and other such factors to be related to inductive reactance. They are. However, we can group all the factors such as physical coil character-istics and magnetic core characteristics together, since they collectively determine the value of inductance. That is, by referring to the value of inductance, we are inherently referring to all of the physical and magnetic factors. We can account for the rate of change of current factor by referring to the frequency of the current or voltage in the inductor. Equation 12-8 provides us with a direct method for computing the inductive reactance of an inductor:

$$X_L = 2\pi f L \qquad (12\text{-}8)$$

EXAMPLE SOLUTION

How much inductive reactance is provided by a 10-mH coil when operated at a frequency of 25 kHz?

EXAMPLE SOLUTION

We apply Equation 12-8 as follows:

$$X_L = 2\pi f L$$
$$= 6.28 \times 25 \text{ kHz} \times 10 \text{ mH}$$
$$= 1.57 \text{ k}\Omega$$

EXAMPLE SOLUTION

What frequency causes a 100-µH inductor to present an inductive reactance of 5 kΩ?

EXAMPLE SOLUTION

We apply Equation 12-8 as follows:

$$X_L = 2\pi f L, \text{ or}$$

$$f = \frac{X_L}{2\pi L}$$

$$= \frac{5 \text{ k}\Omega}{6.28 \times 100 \text{ }\mu\text{H}}$$

$$= 7.96 \text{ MHz}$$

Practice Problems

1. Calculate the inductive reactance provided by a 33-µH coil at a frequency of 2.5 MHz.
2. Determine the value of inductance needed to produce an inductive reactance of 1.5 kΩ when operated at a frequency of 250 kHz.
3. Compute the inductive reactance of a 200-mH inductor, if it is operated at a frequency of 1.2 kHz.
4. What frequency causes a 500-mH inductor to have an inductive reactance of 470 Ω?

Answers to Practice Problems

1. 518.1 Ω   2. 955.4 µH   3. 1.51 kΩ   4. 149.7 Hz

OHM'S LAW WITH X_L

Inductive reactance is measured in ohms, and it offers opposition to current flow much like a resistance does. All of the relationships between voltage, current, and resistance described by Ohm's Law are equally applicable to sinusoidal inductive circuits. That is,

$$V_L = I_L X_L \qquad (12\text{-}9)$$

$$I_L = \frac{V_L}{X_L} \qquad (12\text{-}10)$$

$$X_L = \frac{V_L}{I_L} \qquad (12\text{-}11)$$

EXAMPLE SOLUTION

Refer to Figure 12-33. What is the rms value of current flowing in the circuit?

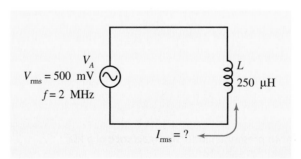

Figure 12-33. How much current flows in this circuit?

EXAMPLE SOLUTION

First, we need to determine the inductive reactance in the circuit. We can apply Equation 12-8.

$$X_L = 2\pi f L$$
$$= 6.28 \times 2 \text{ MHz} \times 250 \text{ }\mu\text{H}$$
$$= 3.14 \text{ k}\Omega$$

Now, we can apply Ohm's Law (Equation 12-10) to find the value of current.

$$I = \frac{V_L}{X_L}$$
$$= \frac{500 \text{ mV}}{3.14 \text{ k}\Omega} = 159.2 \text{ }\mu\text{A}$$

Since the voltage was given as an rms value, the computed current is also an rms value.

Practice Problems

1. Calculate the value of rms current for the circuit shown in Figure 12-34.

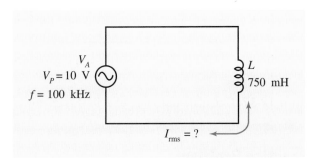

Figure 12-34. Find the rms current in this circuit.

2. What is the value of peak-to-peak current for the circuit shown in Figure 12-35?

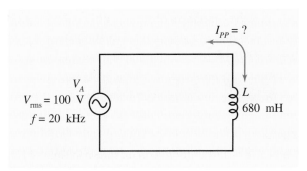

Figure 12-35. Find the value of I_{PP}.

3. What is the peak value of voltage applied to the circuit shown in Figure 12-36?

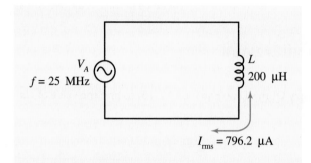

Figure 12-36. Find the peak voltage in this circuit.

4. What is the value of the inductor in the circuit shown in Figure 12-37?

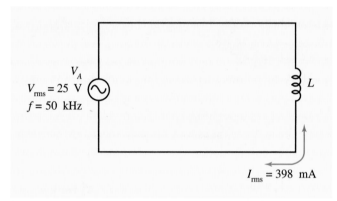

Figure 12-37. Calculate the value of L in this circuit.

Answers to Practice Problems

1. 15.01 μA   **2.** 3.31 mA   **3.** 35.35 V   **4.** 200 μH

INDUCTIVE REACTANCES IN SERIES

When the individual inductive reactances of two or more series-connected inductors are known, the total inductive reactance can be found by summing the individual reactances. This is expressed by Equation 12-12.

$$X_{L_T} = X_{L_1} + X_{L_2} + X_{L_3} + \ldots + X_{L_N} \quad (12\text{-}12)$$

EXAMPLE SOLUTION

Determine the total inductive reactance for the circuit shown in Figure 12-38.

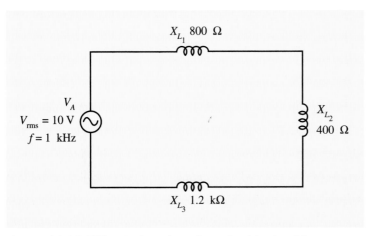

Figure 12-38. What is the value of X_{L_T} in this circuit?

EXAMPLE **SOLUTION**

We apply Equation 12-12 as follows:

$$X_{L_T} = X_{L_1} + X_{L_2} + X_{L_3}$$
$$= 800\ \Omega + 400\ \Omega + 1.2\ k\Omega = 2.4\ k\Omega$$

Practice Problems

1. What is the total inductive reactance (X_{L_T}) in the circuit shown in Figure 12-39?
2. What is the value of X_{L_2} in Figure 12-40?
3. What is the total inductive reactance if 5 coils are connected in series and each coil has a 2.5-kΩ reactance?
4. What happens to the value of inductive reactance in a circuit if the frequency is increased?

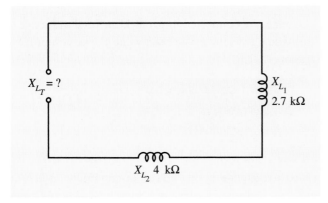

Figure 12-39. Find the value of X_{L_T}.

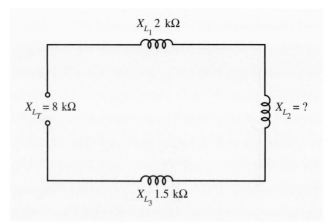

Figure 12-40. Find the value of X_{L_2}.

5. What happens to the value of inductive reactance in a circuit if the applied voltage is increased?

6. What happens to the value of inductive reactance in a circuit if the value of inductance is increased?

Answers to Practice Problems

1. $6.7 \text{ k}\Omega$

2. $4.5 \text{ k}\Omega$

3. $12.5 \text{ k}\Omega$

4. Inductive reactance increases.

5. Inductive reactance stays the same.

6. Inductive reactance increases.

INDUCTIVE REACTANCES IN PARALLEL

When inductive reactances are connected in parallel, the total inductive reactance is less than the smallest individual reactance. The exact value can be computed with the reciprocal formula (Equation 12-13).

$$X_{L_T} = \cfrac{1}{\cfrac{1}{X_{L_1}} + \cfrac{1}{X_{L_2}} + \cfrac{1}{X_{L_3}} + \ldots + \cfrac{1}{X_{L_N}}} \quad (12\text{-}13)$$

EXAMPLE SOLUTION

What is the total inductive reactance for the circuit shown in Figure 12-41?

EXAMPLE SOLUTION

We apply the reciprocal formula (Equation 12-13) as follows:

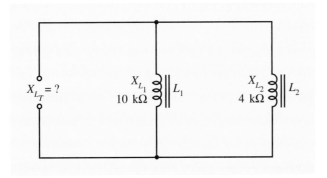

Figure 12-41. Find the total inductive reactance in this circuit.

$$X_{L_T} = \cfrac{1}{\cfrac{1}{X_{L_1}} + \cfrac{1}{X_{L_2}}}$$

$$= \cfrac{1}{\cfrac{1}{10\ k\Omega} + \cfrac{1}{4\ k\Omega}} = 2.86\ k\Omega$$

Practice Problems

1. What is the total inductive reactance if 2 coils having reactances of 27 kΩ and 68 kΩ are connected in parallel?
2. What is the total inductive reactance if 4 coils are connected in parallel and each coil has a reactance of 100 Ω?
3. What value of inductive reactance must be used to parallel a 10-kΩ inductive reactance, if the total reactance must be 2,000 Ω.

Answers to Practice Problems

1. 19.33 kΩ 2. 25 Ω 3. 2.5 kΩ

Phase Relationships

You will recall from a previous discussion that the voltage across an inductor is proportional to the rate of change of current through the inductor. If the current waveform is a sine wave, then it follows that the points of highest self-induced voltage will coincide with the 0° and 180° points of the current waveform, since these points have the highest rate of change. Similarly, there will be no self-induced voltage as the current wave passes through the 90° and 270° points, since the rate of change of current at these points is zero. These relationships lead to the waveforms shown in Figure 12-42.

As you can see from the waveforms in Figure 12-42(b), the voltage peaks coincide with the points of maximum rate of change of current. Similarly, the zero-crossing points of

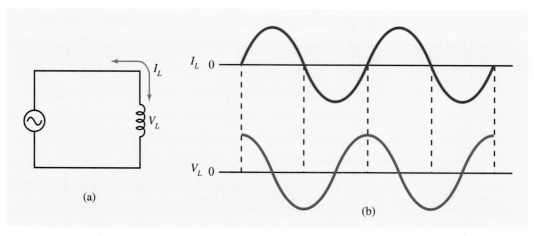

Figure 12-42. The current in an inductor lags the voltage by 90°.

the voltage waveform coincide with the peaks (minimum rate of change) of the current waveform.

Both current and voltage waveforms are sinusoidal, but the voltage waveform is 90° out of phase with the current waveform. That is, the current and voltage peaks do not occur at the same time. More specifically, the voltage waveform leads (i.e., is ahead of) the current waveform by 90°. Alternatively, we may say the current waveform lags (i.e., is behind) the voltage waveform.

This is a very important characteristic of inductors. Let's state it as a rule.

> The sinusoidal voltage across an inductor will always lead the sinusoidal current by 90°.

Exercise Problems 12.6

1. What is the inductive reactance offered by a 10-mH coil when the applied frequency is 175 kHz?
2. Calculate the frequency at which a 250-μH coil has an inductive reactance of 100 Ω.
3. Which coil will have the highest reactance at a particular frequency: 100 μH or 100 mH?
4. Inductive reactance is (*directly, inversely*) proportional to frequency.
5. Inductive reactance is (*directly, inversely*) proportional to inductance.
6. If 10 V$_{rms}$ at 25 MHz is applied to a 500-μH coil, how much rms current will flow?
7. What is the peak voltage across a 33-mH inductor if 100 mA of current flow at a frequency of 200 kHz?
8. What size inductor is needed to provide 2.5 kΩ of inductive reactance when supplied by a 100-MHz source?

9. If the sine wave of current in an inductor is passing through 150°, what angle is the inductor voltage passing through at this same time?

10. If the voltage across an inductor is at its 45° point, at what point is the current wave?

12.7 Q of an Inductor

Although the primary purpose of an inductor is to provide inductance (or inductive reactance), a practical inductor also adds resistance to the circuit. The ratio of inductive reactance (the desired quantity) to the resistance (the undesired quantity) is called the quality of the coil. The symbol for quality is Q. Since Q is a simple ratio, it has no units of measurement. The computation of a coil's Q is accomplished with Equation 12-14.

$$Q = \frac{X_L}{ESR} \qquad (12\text{-}14)$$

where X_L is the inductive reactance of the coil and ESR is the effective series resistance of the coil. The ESR of the coil is more complex than the simple dc resistance of the wire itself. It is actually a summation of several coil losses that collectively dissipate heat. This combined apparent resistance is called the coil's effective series resistance, or ESR. We will discuss the various components that make up the ESR in a subsequent section. Figure 12-43 illustrates the way the ESR appears to the external circuitry.

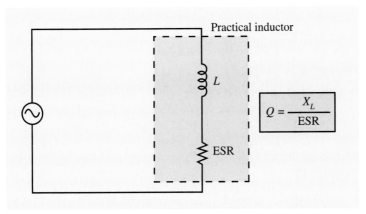

Figure 12-43. A practical inductor also adds resistance to the circuit in the form of an effective series resistance, or ESR.

EXAMPLE SOLUTION

What is the Q of a 250-mH coil that has an ESR of 100 Ω when operated at a frequency of 3 kHz?

EXAMPLE **SOLUTION**

First, we compute the inductive reactance of the coil with Equation 12-8.

$X_L = 2\pi fL$

$\quad = 6.28 \times 3 \text{ kHz} \times 250 \text{ mH}$

$\quad = 4.71 \text{ k}\Omega$

Next, we compute the Q of the coil by applying Equation 12-14.

$$Q = \frac{X_L}{ESR}$$

$$= \frac{4.71 \text{ k}\Omega}{100 \text{ }\Omega} = 47.1$$

Although the Q of a coil is a calculable numeric quantity, technicians often speak of Q in relative terms. Generally, a coil with a Q of less than ten is called a low-Q coil. Similarly, if a coil has a Q of over 100, it is usually called a high-Q coil.

Both factors (X_L and ESR) of Q are frequency-dependent. Therefore, Q is frequency dependent. The Q of a coil must be specified at a particular operating or test frequency. At frequencies well below the optimum operating range, the Q decreases due to a falling X_L. At frequencies substantially higher than the optimum operating range, the Q decreases due to the increase in ESR.

Practice Problems

1. What is the Q of a coil that has an inductive reactance of 600 Ω and an ESR of 27 Ω?

2. Would a coil that has an ESR of 5 Ω and an inductive reactance of 100 Ω, generally be classified as a high-Q coil?

3. What is the most ESR a 33-mH coil can have, if it is to have a Q of at least 40 at a frequency of 100 kHz?

Answers to Practice Problems

1. 22.22   2. No   3. 518.1 Ω

Power Dissipation

A perfect inductor dissipates no power (i.e., generates no heat). Figure 12-44 helps explain why an ideal inductor dissipates no power.

Figure 12-44(a) shows an ideal (i.e., ESR = 0) inductor connected to an ac source. Figure 12-44(b) illustrates the resulting current, voltage, and power dissipation waveforms. The current and voltage waveforms are 90° out of phase as discussed in a previous section. The power dissipation waveform (P_L) was constructed by using the power formula $P = VI$. That is, if we multiply the instantaneous voltage and current values at many points on the voltage and current curves, we will obtain the results shown in Figure 12-44(b) as P_L. Clearly, the power dissipation curve has a frequency that is twice the applied frequency. Of more immediate interest, however, are the polarities of the power dissipation curve and the average value.

During times when the power dissipation curve is positive, the inductor is taking energy from the circuit. This occurs while the magnetic field is expanding around the coil (i.e., the current is increasing). During times when the current is decreasing (between 90° and

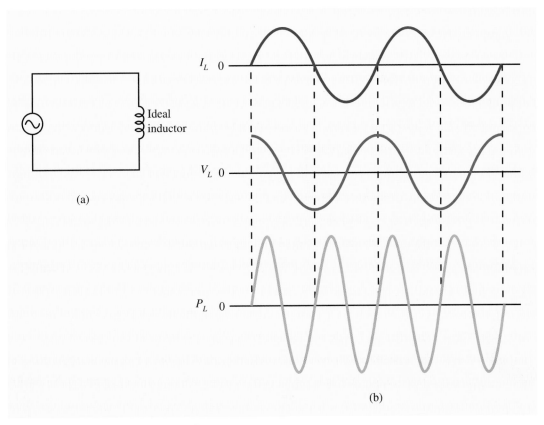

(a)

(b)

Figure 12-44. An ideal inductor dissipates no power.

180° and between 270° and 360° on the current wave), the magnetic field is collapsing. This actually generates a voltage and returns power to the circuit. These times are indicated in Figure 12-44(b) as a negative power curve. So, half of the time the inductor is taking energy from the circuit where it is stored in the magnetic field. The other half of the time, the energy stored in the magnetic field is returned to the circuit. The net power loss is zero.

Inductor Losses

A practical inductor has a nonzero ESR, which causes a positive power dissipation. The ESR of a coil is made up of several losses. Some of the losses occur in the coil winding, while others occur in the core of the coil. Let's examine each of them individually.

Loss Due to Wire Resistance

The copper wire used to wind the inductor has a definite amount of ohmic resistance. You will recall from our discussion of conductors in Chapter 8, that the resistance of a wire is directly proportional to its length and inversely proportional to its cross-sectional area. Both of these factors affect the resistance of the coil wire and, therefore, the Q of the coil. Loss due to the resistance of the wire is called **copper loss** or I^2R loss. It is not frequency dependent.

KEY POINTS

A practical coil has a number of power losses that appear as a resistance in series with the coil winding. This apparent resistance is called the effective series resistance, or *ESR*, of the coil.

KEY POINTS

Winding losses include the power dissipated by the ohmic resistance of the wire used to wind the coil and the increase in effective wire resistance due to skin effect.

Each moving electron has a surrounding magnetic field.

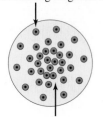

Flux density is highest in the center of the conductor.

(a)

Current flows near the surface.

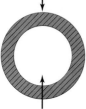

There is no substantial current flowing in the central regions of the conductor.

(b)

Figure 12-45.
Skin effect reduces the effective cross-sectional area of a wire.

LOSS DUE TO SKIN EFFECT

When alternating current flows through an inductor, each moving electron has an associated changing magnetic field. The fields from the various electrons interact with neighboring electrons and affect their movement. Figure 12-45(a) illustrates that the cumulative flux density of the individual electron fields increases toward the center of the wire. When electrons try to flow in these central regions, they are impeded. The result is shown in Figure 12-45(b). The electrons tend to flow near the surface of the wire. This phenomenon is called the **skin effect.** As the frequency of the current increases, so does its tendency to flow near the surface. As indicated in Figure 12-45(b), the effective cross-sectional area of the wire is reduced due to skin effect. This means that the effective resistance of the wire will increase as the applied frequency increases.

Since the resistance increase due to skin effect only occurs at higher frequencies, you cannot measure it with an ohmmeter. However, it does increase the ESR and power dissipation of the coil. It also decreases the Q of the coil.

HYSTERESIS LOSS

You will recall from our study of basic magnetic theory that a magnetic material has domains that align with external magnetic fields. In the case of a magnetic-core inductor, the domains in the core must realign themselves every half cycle. This requires energy and the energy must come from the ac source. As the frequency is increased, the hysteresis loss also increases, since the domain-switching energy is being consumed more often.

As discussed in Chapter 10, the degree of hysteresis loss for a particular material is indicated by the area of its hysteresis loop. Cores for inductors are generally made from a material that has a high permeability, but a narrow hysteresis loop.

EDDY-CURRENT LOSS

When a changing magnetic field intersects a conductor, it induces a voltage in the conductor. The induced voltage can cause current to flow. We know from the basic power formula ($P = VI$) that there must be an associated power dissipation.

When the changing magnetic field of an inductor cuts through the magnetic core, it can induce a voltage in the core, if the core is conductive. Many materials that are highly magnetic are also conductive. Since the core is relatively large, different potentials are induced into different regions of the core at any instant in time. This causes current to flow from one point in the core to another. However, since the flux pattern is continually changing, so does the specific current paths within the core. Currents induced into the core of an inductor are called **eddy currents.** They are essentially circulating currents whose paths are dynamic, but they do draw energy from the power source. Figure 12-46 illustrates the formation of eddy currents in the core of an inductor.

We can reduce the power dissipation due to eddy currents by using a high-resistance core. This is one advantage of a ferrite or powdered-iron core. These materials provide high permeability, but they also have a high ohmic resistance. This limits the magnitude of the eddy currents. Eddy currents increase with higher frequencies, since the voltage induced into the core material increases.

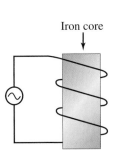

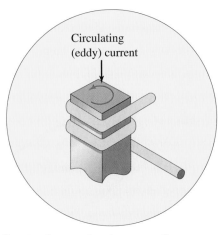

Iron core

Circulating (eddy) current

Figure 12-46. Eddy currents flow in the conductive core of an inductor.

You will recall from an earlier section that iron-core coils normally use a laminated core. The core consists of thin sheets of core material separated with an insulative coating. This allows the core to provide the high permeability needed to obtain high values of inductance, but the insulative barriers keep the eddy current losses to a minimum.

Exercise Problems 12.7

1. A coil with a Q of 8 could be classified as a (*high, low*)-Q coil.
2. If the ESR of a coil increases, the Q _____.
3. Eddy currents are unaffected by frequency. (True or False)
4. Hysteresis loss is unaffected by frequency. (True or False)
5. Skin effect is unaffected by frequency. (True or False)
6. Copper loss is unaffected by frequency. (True or False)
7. What is the Q of a 10-mH coil that has an ESR of 75 Ω when it is operated at a frequency of 25 kHz?
8. Q is unaffected by frequency. (True or False)
9. If a certain coil has an inductive reactance of 22 kΩ and an ESR of 1.5 kΩ, what is the Q of the coil?
10. What can be done to minimize losses due to eddy currents?

12.8 Troubleshooting Inductors

Since an inductor is nothing more than a wire wrapped around a core, there are only three defects that are probable. An open in the coil winding is probably the most common malfunction. Figure 12-47 illustrates the use of an ohmmeter to detect the open winding. If the winding is open, the ohmmeter will indicate infinite (∞) resistance.

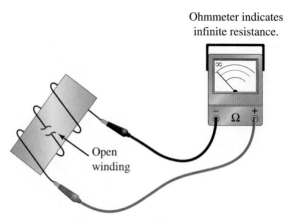

Ohmmeter indicates
infinite resistance.

Open
winding

Figure 12-47. An open coil has infinite resistance.

When an ohmmeter is used to check the condition of a good coil, the meter will indicate the resistance of the wire. This varies with the type of coil being considered. It can be less than one ohm or as high as several thousand ohms. As a technician, you should measure and record the normal resistances of coils used in circuits you are expected to troubleshoot. That way, you can contrast the resistance values of suspected coils with those of known good coils.

Figure 12-48 shows another possible defect that can occur in a multilayer coil. Here, the insulation between two adjacent turns has disintegrated, allowing the windings to short together. This effectively shorts out or bypasses a portion of the coil. Depending on the location of the short and the method of winding the coil, the short may only bypass a single turn, or it may bypass a significant portion of the inductor. A shorted coil is sometimes very difficult to detect with an ohmmeter. The ohmmeter will read a value of coil resistance that is less than the normal value for the coil. However, since the normal value may only be one or two ohms, a typical ohmmeter may not absolutely identify shorted turns.

Windings usually become shorted as a direct result of overheating or internal arcing. In either case, the short may be accompanied by physical clues such as discoloration of the coil, visible melting of the insulation, or a characteristic, pungent odor.

Figure 12-49 shows a test instrument that technicians use to measure the value of inductance in a coil. If a significant number of turns in the coil are shorted, or if the coil is open,

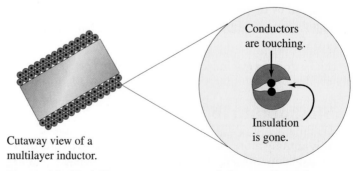

Conductors
are touching.

Cutaway view of a
multilayer inductor.

Insulation
is gone.

Figure 12-48. Adjacent turns on a multilayer coil can become
shorted if the insulation breaks down.

Figure 12-49. An inductance meter measures the actual value of an inductor. It can be used to detect opens or shorts in the coil winding. (Courtesy of Leader Instruments Corporation)

then the inductance meter will easily and quickly identify the defect. For accurate measurements in most circuits, at least one lead of the inductor must be removed from the circuit while it is being measured.

The final defect that is likely to occur in coils is similar to the case of shorted windings. If the coil overheats, the insulation that separates the winding from the metallic core can break down. This allows the coil winding to contact the conductive core material. In the case of iron-core coils, this is generally a catastrophic failure, since the core is often connected directly to the chassis ground of the equipment. This means that the current that would ordinarily flow through the coil is bypassed directly to ground. Additionally, the short to ground effectively bypasses the inductive reactance of the coil, which may allow a substantial current to flow.

An ohmmeter can be used to detect a winding that is shorted to the core. Figure 12-50 illustrates this technique. Figure 12-50(a) shows that a good coil will have an infinite

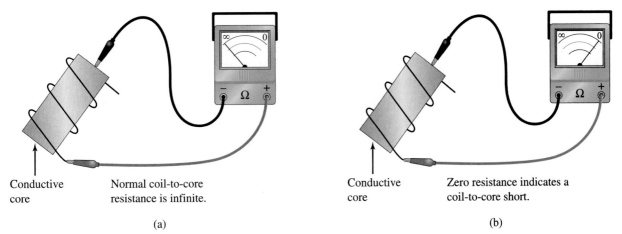

Conductive core   Normal coil-to-core resistance is infinite.

Conductive core   Zero resistance indicates a coil-to-core short.

(a)   (b)

Figure 12-50. (a) An ohmmeter can distinguish between a normal coil winding and (b) a winding that is shorted to the core.

resistance (at least higher than a typical ohmmeter will read) between the coil winding and the metallic core. If the winding shorts to the core, as shown in Figure 12-50(b), then an ohmmeter check will easily detect this condition by indicating zero ohms.

Exercise Problems 12.8

1. If the resistance of a coil winding measures infinity, and the normal value is 500 Ω, what is the most likely defect?
2. If the normal resistance of a coil is 500 Ω, but it measures 300 Ω, what is the most likely defect?
3. What is the normal ohmmeter indication for a good coil when measuring the resistance between the winding and the core?

Chapter Summary

• When current flows through a conductor, it creates a magnetic field around the conductor. If the current is changing, then the magnetic field will be changing. If a changing magnetic field intercepts a conductor, a voltage will be induced into the inductor by a process known as electromagnetic induction.

• When the magnetic field of a coil intercepts the turns of the same coil, the induced voltage is called a self-induced voltage. The value of the self-induced voltage is determined by the strength of the magnetic field, the rate of change of flux, the angle of flux cutting, and the number of turns in the coil. These factors are summarized in Faraday's Law. The polarity of the induced voltage is determined by the direction of the changing magnetic field (i.e., expanding or contracting) and the polarity of the field. According to Lenz's Law, the polarity of the induced voltage will always oppose the change in current. That is, the inductor works to keep the current from changing (i.e., increasing or decreasing).

• Inductance (L) is measured in henries (H). One henry will produce a self-induced voltage of one volt when its current changes at the rate of one ampere per second. The value of inductance for a particular coil is determined by its physical characteristics: number of turns, length of the magnetic circuit, cross-sectional area, and permeability of the core. The inductance of a coil is inversely proportional to the length of the magnetic circuit. It is directly proportional to all other factors.

• Since the value of self-induced voltage in a coil is dependent on the rate of change of current through the coil, the waveshape of self-induced voltage may differ from the waveshape of current. That is, the value of self-induced voltage is dependent on the rate of change of current, but it is unaffected by the absolute value of current. When a sine wave is applied to an inductor, the waveshapes of both voltage and current are sinusoidal, but they are 90° out of phase with each other. More specifically, the voltage waveform leads the current waveform by 90° in a pure inductance.

- There are many different kinds of inductors. They are generally classified as fixed or variable. They can also be classified by the type of material used for the core. Examples include air-core, iron-core, and ferrite-core. A ferrite bead can be added to a wire or the lead of a component to add inductance and resistance to the circuit. The resistance and inductance of the bead appear as series elements to the circuit. Both inductors and ferrite beads are available as surface-mount devices.

- When inductors are combined in series, parallel, or another configuration and not linked magnetically, the total inductance can be found with the same general procedures used to combine resistances. Series inductances add together to form the total inductance. The total inductance of parallel inductances can be computed with the reciprocal formula.

- When inductors are used in dc circuits, there is no effect except when the current value changes. The inductor tends to oppose any changes in current, and may develop substantial voltages, if the current changes quickly.

- When inductors are used in ac circuits, they also oppose changes in current. However, since the current is changing continuously, we express this opposition as inductive reactance. Inductive reactance (X_L) is measured in ohms. It varies directly with the applied frequency and the value of inductance.

- Multiple values of inductive reactance connected in series, parallel, or other configurations, may be combined in much the same manner as resistances, provided there is no magnetic coupling between the various inductors. Additionally, Ohm's Law can be applied to inductive circuits in much the same way as it applies to resistive circuits. Inductive reactance is substituted for resistance in the Ohm's Law equations.

- An ideal inductor has no resistance. It offers only inductance to the circuit. A practical inductor has several forms of power loss, which manifest as a series resistance. This apparent resistance is called the effective series resistance, or ESR of the coil. A portion of the ESR is contributed by the winding itself: copper loss and skin effect. The remaining portion of the ESR is due to losses in the core of the inductor: hysteresis loss and eddy currents. Copper loss is unaffected by frequency. The other factors that make up the ESR are frequency dependent.

- The quality, or Q, of a coil indicates how ideal a coil is. Q is computed as the ratio of inductive reactance to ESR. Since both of these factors are frequency dependent, the Q of a coil is specified at a particular frequency.

- Since an ideal coil has no resistance, it dissipates no power. Energy that is taken from the circuit during one portion of the cycle is returned to the circuit during a subsequent time period. A practical inductor dissipates power in the ESR of the coil.

- A technician can use an inductance meter or an ohmmeter to locate defects in a coil. The normal resistance of a coil winding should be known in advance by the technician. If the resistance of a suspected coil is substantially less than the normal value, then one or more turns of the coil could be shorted. If the coil winding measures infinity, then the coil has an open winding. Finally, the windings of a coil may short to the core. An ohmmeter can detect this condition.

Review Questions

Section 12.1: Electromagnetic Induction

1. When a conductor passes through a magnetic field, the induced voltage is (*directly, inversely*) proportional to field strength.

2. If the relative rate of motion between a conductor and a magnetic field is increased, the induced voltage will (*increase, decrease*).

3. When a conductor moves parallel to the magnetic lines of force, (*minimum, maximum*) voltage is induced.

4. When a conductor moves perpendicular to the magnetic lines of force, (*minimum, maximum*) voltage is induced.

5. If the number of turns on a coil that is being moved through a magnetic field is increased, what happens to the value of induced voltage?

6. Name two factors that determine the polarity of the voltage that will be induced into a conductor that is moving through a magnetic field.

7. According to Faraday's Law, the amount of induced voltage is directly proportional to the rate of change of flux. (True or False)

8. According to Faraday's Law, the amount of induced voltage is directly proportional to the number of turns on a coil. (True or False)

9. The polarity of self-induced voltage in a coil always (*aids, opposes*) the initial current change.

10. Which current would receive the greatest opposition from a 100-mH inductor: a 10-μA current at a frequency of 25 MHz or a 10-A current from a dc source?

11. Inductors oppose a _____ in current.

12. The value of an inductor is inversely proportional to the number of turns on the inductor. (True or False)

13. If the number of turns on an inductor is doubled, what happens to the value of inductance?

14. If the length of the magnetic circuit in a coil is increased, what happens to the value of inductance?

15. If the cross-sectional area of a coil is increased, what happens to the value of inductance?

16. The inductance of a coil is directly proportional to the cross-sectional area of the winding. (True or False)

17. If the number of turns on an inductor is increased by three, but the cross-sectional area of the coil is reduced to one-third of its original size, what happens to the value of inductance?

18. If the relative permeability of the core material in a coil is increased by a factor of ten, what will happen to the value of inductance?

19. The inductance of a coil is directly proportional to the permeability of the core. (True or False)

Section 12.2: Self-Induction

20. Briefly define the term *self-induction*.

21. Describe the voltage waveform across an inductor during times when the rate of change of current through the coil is constant.

22. Explain why a sudden change in current values through a coil produces high voltage.

23. When a sine wave of current flows through an inductor, the maximum value of induced voltage occurs at the _____ and _____-degree points on the current wave.

24. What points on a sine wave correspond to the minimum rate of change?

Section 12.3: Types of Inductors

25. Draw the schematic symbol for a fixed inductor.

26. Draw the schematic symbol for a variable inductor.

27. A coil wound on a nonmagnetic core is generally called a(n) _____-core inductor.

28. Iron-core transformers are generally used at frequencies in excess of 25 MHz. (True or False)

29. The permeability of a powdered-iron core is (*higher, lower*) than the permeability of an air core.

30. The ohmic resistance of a powdered-iron core is (*low, high*).

31. Powdered-iron core coils are generally used for frequencies below 100 Hz. (True or False)

32. A ferrite bead adds both _____ and _____ to a circuit.

33. The opposition to current flow offered by a ferrite bead varies with frequency. (True or False)

34. Molded inductors are often color coded. Their values are assumed to be in _____.

35. What is the value of a molded inductor that has the following color bands: silver (wide), gray, red, silver?

36. What is the value of a molded inductor that has the following color bands: silver, red, gold, violet, gold?

37. What is the primary advantage of a surface-mount inductor over a leaded inductor?

Section 12.4: Multiple Circuit Inductances

38. What is the total inductance of three 100-mH coils that are series-connected?

39. If a 25-μH coil is connected in series with a 50-μH coil, what is the total inductance in the circuit?

40. If the total inductance of three series-connected coils is 250 μH, and two of the coils have values of 50 μH and 75 μH, what is the value of the third coil?

41. How much inductance must be connected in series with 830 mH to obtain a total inductance of 1.2 H?

42. The total inductance of series-connected coils is always (*smaller, larger*) than the largest individual coil.

43. The total inductance of parallel-connected coils is always (*smaller, larger*) than the smallest individual coil.

44. What is the total inductance of four parallel-connected 100-µH coils?

45. If L_1, L_2, and L_3 have values of 40 mH, 50 mH, and 200 mH, respectively, and are parallel-connected, what is the total circuit inductance?

46. What value of inductance must be connected in parallel with a 50-µH inductor to obtain a total circuit inductance of 15 µH?

47. What is the total inductance for the circuit shown in Figure 12-51?

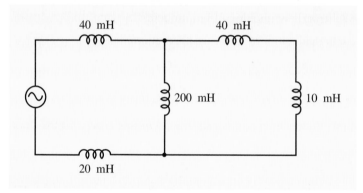

Figure 12-51. What is the total inductance in this circuit?

Section 12.5: Inductance in DC Circuits

48. When direct current is switched into a coil, the self-induced voltage is (*minimum, maximum*) at the first instant.

49. When direct current passes through a coil and the circuit has had time to stabilize, the self-induced voltage across the coil is zero. (True or False)

50. An inductor has no effect on direct current unless it changes values. (True or False)

51. An inductor prevents abrupt changes in the current in a dc circuit. (True or False)

Section 12.6: Inductance in AC Circuits

52. If an inductor has a sinusoidal waveform of current, what is the waveform of voltage?

53. Inductive reactance is unaffected by frequency of operation. (True or False)

54. Inductive reactance is directly proportional to the value of inductance. (True or False)

55. What is the inductive reactance of a 75-mH coil when operated at a frequency of 150 kHz?

56. What frequency causes a 2.5-mH coil to present an inductive reactance of 130 Ω?

57. What is the inductive reactance of a 20-mH coil at a frequency of 5.6 MHz?

58. Inductive reactance is measured in _____.

59. What is the value of rms current in the circuit shown in Figure 12-52?

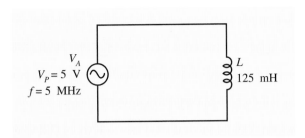

Figure 12-52. Find the rms current in this circuit.

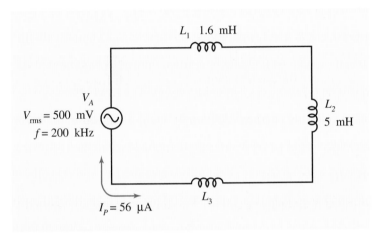

Figure 12-53. Determine the value of L_3.

60. What is the value of L_3 in Figure 12-53?

61. What is the value of X_{L_3} in Figure 12-53?

62. What is the value of peak-to-peak voltage across L_2 in Figure 12-53?

63. What is the value of rms voltage across L_1 in Figure 12-53?

64. If three inductive reactances of 30 kΩ each are connected in series, what is the total inductive reactance?

65. If three inductive reactances of 30 kΩ each are connected in parallel, what is the total inductive reactance?

66. What is the total inductive reactance if 47 kΩ of inductive reactance are connected in parallel with 20 kΩ of inductive reactance?

67. If a sine wave of current flowing through an inductor is passing through its 90° point, at what point is the voltage wave across the inductor?

68. Current through an inductor (*leads, lags*) the voltage across the inductor by _____ degrees.

69. The voltage across an inductor (*leads, lags*) the current through the inductor by _____ degrees.

70. The maximum sinusoidal current through an inductor does not occur at the same time as its maximum voltage. (True or False)

Section 12.7: Q of an Inductor

71. A practical inductor adds both inductance and _____ to the circuit.

72. What is the unit of measurement for Q?

73. What does the abbreviation ESR represent when used to describe an inductor?

74. A technician can measure ESR with an ohmmeter. (True or False)

75. What is the Q of a 100-mH inductor that has an ESR of 100 Ω when operated at a frequency of 12 kHz?

76. If a 10-μH coil has an ESR of 7 Ω at a frequency of 2.5 MHz, what is its Q?

77. Would a coil with an inductive reactance of 25 kΩ and an ESR of 200 be considered a low-Q coil?

78. Briefly explain in your own words why an ideal inductor dissipates no power.

79. What is the name of the phenomenon that causes high-frequency current to flow near the surface of a conductor?

80. Name two ways to reduce the eddy-current loss in an inductor.

81. The Q of a coil changes with frequency. (True or False)

Section 12.8: Troubleshooting Inductors

82. If the measured resistance of a coil is 1.5 Ω and the normal value is 4.5 Ω, what is a possible defect?

83. If an ohmmeter check of the winding in an iron-core inductor indicates infinite resistance, is the coil good?

84. What is the normal value of resistance that should be measured between the winding of a coil and its metallic core?

TECHNICIAN CHALLENGE

Figure 12-54 shows the schematic diagram of a circuit that can be used as an electric fence charger. Electric fences are often used to keep farm animals and pets restricted to a particular area. If an animal touches the electric fence, it receives a harmless, but memorable, electric shock. The circuit operates from 12 Vdc, but it provides high-voltage to the fence.

Study the circuit shown in Figure 12-54 carefully to determine how it works. Then, accomplish the following:

- Write a complete theory of operation to describe how the circuit works. This description will be used in the service manual for the charger.
- State the purpose of every component.
- If practical, build the circuit in the laboratory to verify its operation. You can replace the one-second timer with a manually operated switch. Inductor L_1 should be as large as practical. (You might use the coil from another relay or a winding on a transformer, if no large inductors are available.)

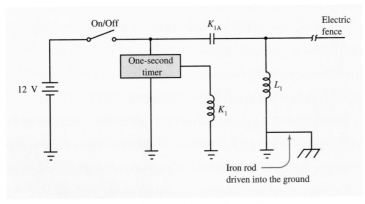

Figure 12-54. A circuit for an electric fence charger.

Equation List

(12-1) $\quad L = \mu \dfrac{N^2 A}{l}$

(12-2) $\quad L = \mu_r \times 1.257 \times 10^{-6} \times \dfrac{N^2 A}{l}$

(12-3) $\quad V_L = L\dfrac{\Delta i}{\Delta t} = L\dfrac{di}{dt}$

(12-4) $\quad L_T = L_1 + L_2 + L_3 + \ldots + L_N$

(12-5) $\quad L_T = \dfrac{1}{\dfrac{1}{L_1} + \dfrac{1}{L_2} + \dfrac{1}{L_3} + \ldots + \dfrac{1}{L_N}}$

(12-6) $\quad L_T = \dfrac{L}{N}$

(12-7) $\quad L_T = \dfrac{L_1 L_2}{L_1 + L_2}$

(12-8) $\quad X_L = 2\pi f L$

(12-9) $\quad V_L = I_L X_L$

(12-10) $\quad I_L = \dfrac{V_L}{X_L}$

(12-11) $\quad X_L = \dfrac{V_L}{I_L}$

(12-12) $\quad X_{L_T} = X_{L_1} + X_{L_2} + X_{L_3} + \ldots + X_{L_N}$

(12-13) $\quad X_{L_T} = \dfrac{1}{\dfrac{1}{X_{L_1}} + \dfrac{1}{X_{L_2}} + \dfrac{1}{X_{L_3}} + \ldots + \dfrac{1}{X_{L_N}}}$

(12-14) $\quad Q = \dfrac{X_L}{\text{ESR}}$

objectives

After completing this chapter, you should be able to:

1. Use both polar and rectangular notation to calculate the following quantities in a series, parallel, or series-parallel *RL* circuit:

   admittance
   apparent power
   component currents
   component voltages
   phase angle between total
       voltage and total current

   power factor
   reactive power
   total current
   total impedance
   true power

   The calculations will include both polar and rectangular notation.

2. Draw a phasor diagram to represent a given series or parallel *RL* circuit.

3. Express circuit quantities in either polar or rectangular form.

4. State and apply the Pythagorean theorem.

5. Calculate the *RL* time constant of an *RL* circuit.

6. State the general requirements for *RL* differentiator and integrator circuits.

7. Determine the approximate pulse response of an *RL* circuit.

8. Name at least two applications that utilize *RL* circuits.

9. Explain the relationship between and calculate the values of the following quantities: true power, apparent power, reactive power, power factor.

Resistive-Inductive Circuit Analysis

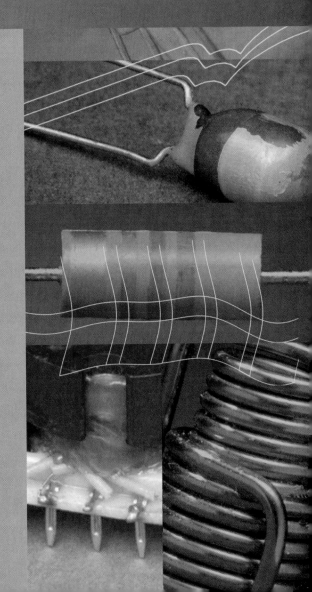

When resistance (R) and inductance (L) are combined in the same circuit, we refer to it as an RL circuit. As you might expect, the characteristics of the circuit lie between those of a pure resistive or a pure inductive circuit. The various components that make up the circuit may be connected as series, parallel, series-parallel, or complex. The basic principles that have been emphasized in previous chapters such as Ohm's Law and Kirchhoff's Laws will still apply, but we need to include the concept of phase relationships.

13.1 Series *RL* Circuits with Sinusoidal Currents

Let us begin by learning some of the basic characteristics of series *RL* circuits. We will consider a few mathematical relationships, but will defer most computations until a later section. Figure 13-1 shows a series *RL* circuit consisting of one resistor and one inductor. Let's examine some basic characteristics of this circuit configuration.

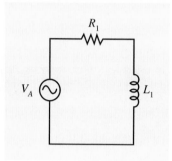

Figure 13-1. A simple series *RL* circuit.

Series *RL* Circuit Characteristics

CURRENT

KEY POINTS

The current in a series *RL* circuit is the same in all parts of the circuit.

Since we are considering a series circuit, we know that the current will be the same in all parts of the circuit. If we knew, for example, that the current through L_1 was 150 mA, then we would immediately know the value of current through R_1 and through the source. This important principle is true in any series circuit. It is illustrated in Figure 13-2.

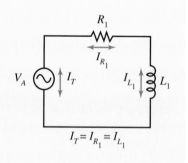

$$I_T = I_{R_1} = I_{L_1}$$

Figure 13-2. The current is the same through all components in a series *RL* circuit.

KEY POINTS

The voltage drops across each component in a series *RL* circuit are directly proportional to the resistance or reactance values.

VOLTAGE DROPS

We recall from Ohm's Law that the voltage across a component is determined by its resistance (or reactance) and the value of current flowing through it. Since current is the same in all parts of a series *RL* circuit, the various components will have voltage drops that are proportional to the resistance or reactance values. Those with higher resistances or reactances will drop more voltage than components with smaller resistances or reactances. This is illustrated in Figure 13-3.

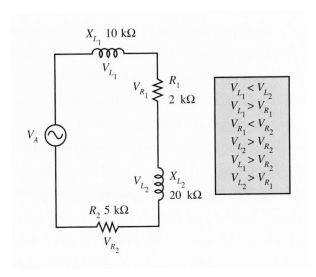

Figure 13-3. Component voltage drops in a series *RL* circuit are proportional to their resistance or reactance values.

PHASE RELATIONSHIPS

We can easily predict the phase relationships in a series *RL* circuit by applying what we already know. Consider these three facts:

- Current is the same in all parts of the circuit at all times.
- Current and voltage are in phase in a resistance.
- Current lags voltage by 90° in an inductance.

As shown in Figure 13-4, the voltage waveforms across the two inductors will have the same phase (they lead the common current by 90°). The inductor voltage waveforms will be 90° out of phase with the resistor voltage waveforms. Specifically, the inductor voltages will lead the resistor voltages by 90°. Finally, we would expect the two resistor voltage waveforms to be in phase with each other.

POWER FACTOR

The **power factor** of a circuit is a dimensionless (no units of measure) quantity that describes the phase relationship between total current and total voltage. A later section will show that the power factor is numerically equal to the cosine of the phase angle. This means that it has numeric values that range between zero and one.

A power factor of one or unity indicates a zero-degree phase relationship between current and voltage. That is, the circuit is purely resistive. Under these conditions, all power that

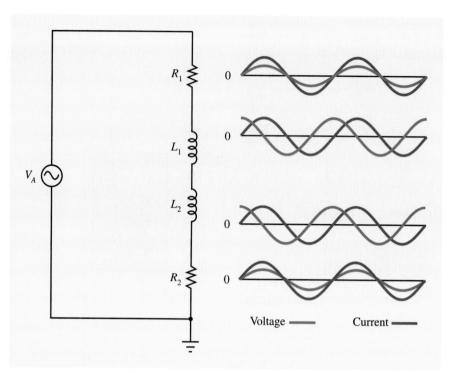

Figure 13-4. Phase relationships in a series *RL* circuit.

leaves the source is transferred to the load and converted to another form of energy (e.g., heat in a resistor, mechanical energy in a motor, heat and light energy in a lamp, and so on). A power factor of zero indicates a purely reactive circuit. In this case, energy leaves the source and is temporarily stored in the reactive component. During a later portion of the ac cycle, energy is returned from the reactive component to the source. Thus, energy moves back and forth between the source and the external circuit, but it is not converted to another more useful form. In most cases, we want the energy to be transformed to provide useful work. The power factor gives us a numerical measure of how much of the energy is actually being transformed and how much is being returned to the source.

IMPEDANCE

We know that the total opposition to current flow in a series resistive circuit is found by adding the individual resistances in the circuit. Similarly, we determine total inductive reactance in a series circuit by adding together the individual inductive reactance values. This same concept extends to include series *RL* circuits, but we have to consider one additional point—phase.

First, the total opposition to current flow in a circuit consisting of both resistance and reactance is called **impedance.** As you would expect, it is measured in ohms. We use the letter Z to represent impedance. Thus, we might describe the total opposition to current (impedance) of a particular circuit by saying $Z = 12.5 \text{ k}\Omega$.

Impedance in a series *RL* circuit is computed by adding the individual resistance and inductive reactance values using phasor addition. This is the same concept as simple addition, but it accounts for the fact that the instantaneous opposition to current flow depends on the phase of the current (i.e., the instantaneous rate of change of current). We will examine phasor addition in a subsequent section.

SUSCEPTANCE

Recall that the reciprocal of a pure resistance is conductance (G). This is a measurement of the ease with which current flows through a resistance. **Susceptance** (B) is a comparable quantity of a pure reactance. We can express it formally with Equation 13-1.

$$B = \frac{1}{X_L} \qquad (13\text{-}1)$$

Susceptance, like conductance, is measured in siemens (S).

KEY POINTS

Just as conductance is the reciprocal of resistance, susceptance is the reciprocal of reactance.

EXAMPLE SOLUTION

What is the susceptance of a 10-mH inductor when it is operating at a frequency of 100 kHz?

EXAMPLE **SOLUTION**

We will apply Equation 13-1, but first we need to compute the value of inductive reactance.

$$X_L = 2\pi f L$$
$$= 6.28 \times 100 \text{ kHz} \times 10 \text{ mH} = 6.28 \text{ k}\Omega$$

Now we can apply Equation 13-1.

$$B = \frac{1}{X_L}$$
$$= \frac{1}{6.28 \text{ k}\Omega} = 159.2 \text{ }\mu\text{S}$$

Practice Problems

1. What is the susceptance of a coil that has a 25-kΩ reactance?
2. How much susceptance will a 100-μH coil have, if it is operated at a frequency of 2.5 MHz?
3. At what frequency must a 75-μH inductor be operated in order to have a susceptance of 5 μS?

Answers to Practice Problems

1. 40 μS   2. 636.9 μS   3. 424.6 MHz

ADMITTANCE

Conductance (G) and susceptance (B) are the reciprocals of pure resistance and pure reactance. Both conductance and susceptance are measured in siemens. When the circuit has a

KEY POINTS

Conductance (G), susceptance (B), and admittance (Y) are all measured in siemens (S).

combination of resistance and reactance, we express its total opposition to current flow as impedance. The reciprocal of impedance is called **admittance** (Y). It is also measured in siemens. We can express it formally as Equation 13-2.

$$Y = \frac{1}{Z} \qquad\qquad (13\text{-}2)$$

EXAMPLE SOLUTION

What is the admittance of a series RL circuit that has a 10-kΩ impedance?

EXAMPLE **SOLUTION**

We apply Equation 13-2 as follows:

$$Y = \frac{1}{Z} = \frac{1}{10\ \text{k}\Omega} = 100\ \mu\text{S}$$

Circuit Simplification

We can simplify a series RL circuit such as the one shown in Figure 13-5 by combining the various resistances into an equivalent resistance and by combining the various inductances (assuming no magnetic linkage) into a single equivalent inductance. We use our previously discussed equations for each of these simplification steps.

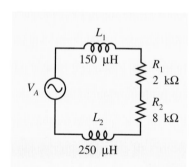

Figure 13-5. An unsimplified RL circuit.

EXAMPLE SOLUTION

Simplify the RL circuit shown in Figure 13-5.

EXAMPLE **SOLUTION**

First, we combine the resistances as follows:

$$R_T = R_1 + R_2$$
$$= 2\ \text{k}\Omega + 8\ \text{k}\Omega = 10\ \text{k}\Omega$$

Next we combine the series inductances in a similar manner:

$$L_T = L_1 + L_2$$
$$= 150\ \mu\text{H} + 250\ \mu\text{H} = 400\ \mu\text{H}$$

The simplified circuit is shown in Figure 13-6. It consists of a single equivalent resistance and a single equivalent inductance. It should be noted that we can further simplify this circuit by adding (phasor addition) the equivalent resistance and the equivalent inductance to obtain the overall impedance of the circuit. We shall discuss this final simplification step momentarily.

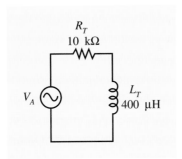

Figure 13-6. The equivalent circuit for the original circuit shown in Figure 13-5.

Practice Problems

1. Draw the equivalent circuit for the *RL* circuit shown in Figure 13-7.

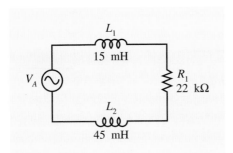

Figure 13-7. Simplify this circuit.

2. Draw the equivalent circuit for the *RL* circuit shown in Figure 13-8.

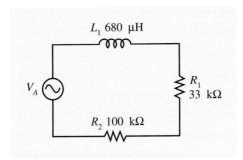

Figure 13-8. Simplify this circuit.

3. Simplify the circuit shown in Figure 13-9.

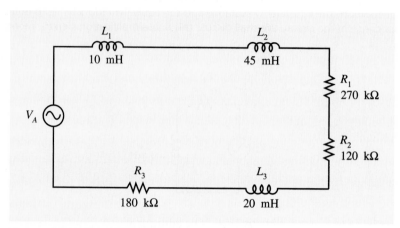

Figure 13-9. Simplify this circuit.

Answers to Practice Problems

1.

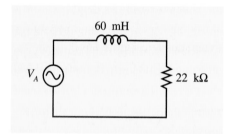

2.

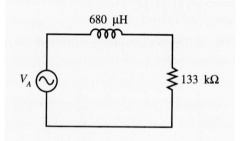

3.

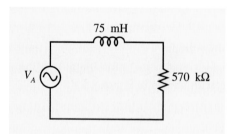

Phasor Representation

Let us now take a closer look at the phase relationships in a series *RL* circuit. As we proceed, it is important to remember that the term *phase* refers to time. That is, if two sinusoidal waveforms are out of phase, then the corresponding points occur at different times. Technicians use phasor diagrams to help visualize the phase relationships in a circuit.

Voltage Phasors

As previously noted, the voltage waveforms across the components in a series *RL* circuit are not all in phase. The current waveforms, however, must be identical, since the instantaneous current is the same in all parts of a series circuit. For this reason, we shall use current as the reference phasor when sketching phasor diagrams for series circuits. The voltage phasors will be drawn relative to the reference phasor.

KEY POINTS

We begin to draw a voltage phasor diagram for a series *RL* circuit by sketching the current phasor. Current is used as a reference because it is common to all components. The reference phasor on a phasor diagram is normally drawn in the 0° position.

EXAMPLE SOLUTION

Draw a phasor diagram to represent the circuit shown in Figure 13-10.

Figure 13-10. Draw the phasors for this circuit.

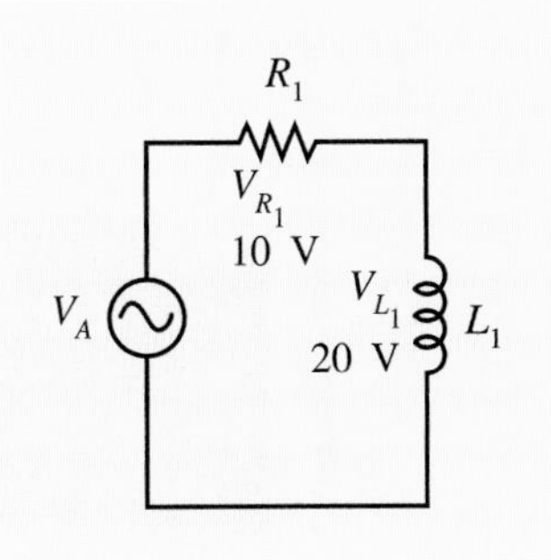

EXAMPLE SOLUTION

First, we sketch the reference phasor (current). As shown in Figure 13-11, it is customary to draw the reference phasor extending horizontally to the right. The relative lengths of the current and voltage phasors are not critical. It is, for example, impossible to discuss the relative magnitudes of 100 V and 100 A. They have different units and cannot be compared. For clarity, however, we normally make the reference phasor longer than any others.

Next, we can add a phasor representing the voltage drop across the series resistance. Since voltage and current are in phase through a resistance, we will sketch the resistor voltage phasor (V_{R_1}) in the same direction as our current phasor (I). As previously mentioned, its length relative to the reference phasor is not critical.

KEY POINTS

The phasor representing resistive voltage is drawn in phase with the current; the phasor representing inductive voltage is drawn 90° ahead (counterclockwise) of the current phasor.

Finally, we add the phasor to represent the inductive voltage (V_{L_1}). We know that inductor voltage leads current by 90°, so we will draw the inductive voltage phasor 90° ahead (more counterclockwise) of the current phasor. The resistive voltage and inductive voltage phasors have common units (volts). Therefore, we can indicate their relative magnitudes on the phasor diagram. In the present case, we shall make the V_{L_1} phasor (20 V) twice as long as the V_{R_1} phasor (10 V).

We can also indicate the phase angle and magnitude of the applied voltage. We can do this graphically by simply completing a parallelogram where V_{R_1} and V_{L_1} are two of the sides and the axes of the graph are the other two sides. The phasor sum is represented by the diagonal of the parallelogram. As shown in Figure 13-11, the diagonal is drawn from the origin to the opposite side of the parallelogram.

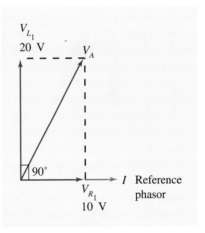

Figure 13-11. A phasor diagram for the circuit in Figure 13-10.

KEY POINTS

The relative lengths of the voltage phasors correspond to the relative values of the resistive and inductive voltages. Their phasor sum is the applied voltage.

KEY POINTS

The voltage drops in a series *RL* circuit may be summed to determine the total voltage (Kirchhoff's Voltage Law). However, because the inductive and resistive voltage drops are 90° out of phase, they must be summed using phasor addition.

Phasor Addition

When you first learned how to add numbers, you restricted yourself to positive integers. Later you extended your skills to include fractions and decimals. Later studies further extended your scope to include algebraic addition, which allowed you to sum positive and negative numbers. We will now extend your addition skills even more to include phasor addition. This will allow us to add two quantities that are out of phase.

Figure 13-12 illustrates the three basic levels of addition as they might apply to a summation of mechanical forces. In Figure 13-12(a), the two forces are in the same direction. They can be combined with simple arithmetic addition. Figure 13-12(b) is more complex. Here, the forces are acting in opposite directions. Summation of the forces requires algebraic addition to account for the signed numbers. Finally, Figure 13-12(c) shows the two forces pulling at a right angle to each other. Summation of these right-angle forces requires vector addition. This is numerically equivalent to phasor addition.

Arithmetic addition: Total force = force 1 + force 2

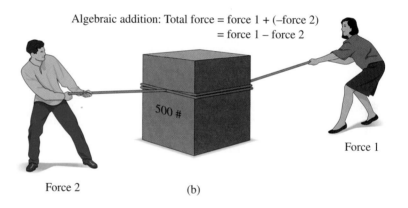

Force 2

Force 1

(a)

Algebraic addition: Total force = force 1 + (−force 2)
= force 1 − force 2

Force 1

Force 2   (b)

Vector (phasor) addition: Total force = $\sqrt{(\text{force }1)^2 + (\text{force }2)^2}$

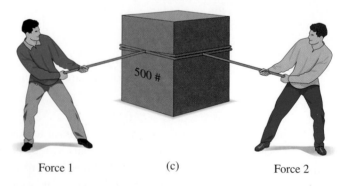

Force 1   (c)   Force 2

Figure 13-12. A mechanical analogy showing addition of forces.

As shown in Figure 13-12(c), right-angle forces are combined by finding the square root of the sum of the squares of the individual forces. This stems from a basic theorem in right-angle trigonometry called the Pythagorean theorem.

In the case of a series RL circuit, we know the inductive and resistive voltages are 90° out of phase. We also know that the total voltage in a series circuit is equal to the sum of the voltage drops. This leads us to Equation 13-3 for determining the applied voltage in a series RL circuit.

$$V_T = V_A = \sqrt{V_R^2 + V_L^2} \qquad \text{(13-3)}$$

Note that V_T and V_A are both commonly used to represent total or applied voltage.

EXAMPLE SOLUTION

Find the value of applied voltage for the circuit shown in Figure 13-13.

EXAMPLE **SOLUTION**

The phasor diagram for this circuit is shown in Figure 13-14. We compute the value of applied voltage by applying Equation 13-3 as follows:

$$V_A = \sqrt{(V_L)^2 + (V_R)^2}$$
$$= \sqrt{(25 \text{ V})^2 + (10 \text{ V})^2}$$
$$= \sqrt{625 + 100} = \sqrt{725} \approx 26.93 \text{ V}$$

You can check your calculations by applying the following rule:

> The result of phasor addition will always be greater than either individual phasor but less than the arithmetic sum of the two phasors.

EXAMPLE SOLUTION

Determine the value of resistor voltage for the circuit shown in Figure 13-15.

EXAMPLE **SOLUTION**

The phasor diagram for this circuit is shown in Figure 13-16. We find V_R by applying Equation 13-3 as follows:

$$V_A = \sqrt{(V_L)^2 + (V_R)^2}, \text{ or}$$
$$V_A^2 = ((V_L)^2 + (V_R)^2), \text{ or}$$
$$(V_R)^2 = (V_A)^2 - (V_L)^2; \text{ therefore}$$
$$V_R = \sqrt{(V_A)^2 - (V_L)^2}$$
$$= \sqrt{(10 \text{ V})^2 - (5 \text{ V})^2}$$
$$= \sqrt{100 - 25} = \sqrt{75} \approx 8.66 \text{ V}$$

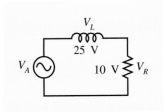

Figure 13-13. Find the value of applied voltage in this circuit.

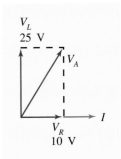

Figure 13-14. Phasor diagram for the circuit shown in Figure 13-13.

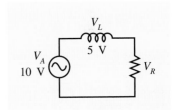

Figure 13-15. Find the voltage across the resistor in this circuit.

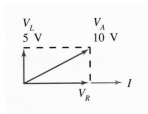

Figure 13-16. Phasor diagram for the circuit shown in Figure 13-15.

CALCULATOR SEQUENCES

One possible set of calculator button sequences for phasor addition is shown in Figures 13-17 and 13-18 for standard and RPN engineering calculators, respectively.

$$a = 5$$
$$b = 7$$
$$c = \sqrt{5^2 + 7^2} \cong 8.6$$

Figure 13-17. Calculator sequence for phasor addition on a standard engineering calculator.

$$a = 5$$
$$b = 7$$
$$c = \sqrt{5^2 + 7^2} \cong 8.6$$

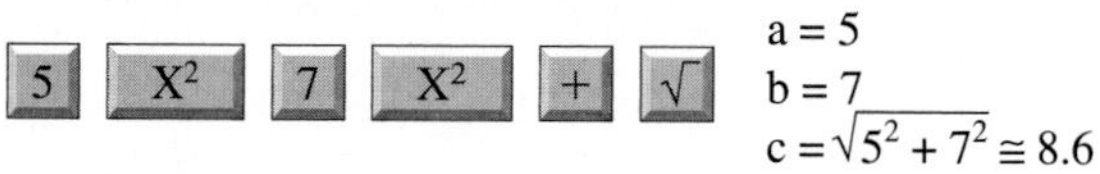

Figure 13-18. Calculator sequence for phasor addition on a RPN engineering calculator.

Practice Problems

1. Determine the value of applied voltage for the circuit shown in Figure 13-19.
2. What is the value of V_A in Figure 13-20?
3. Find the resistor voltage in Figure 13-21.

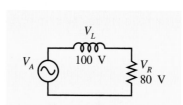

Figure 13-19. Find the applied voltage.

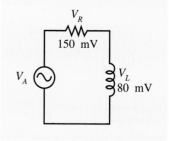

Figure 13-20. Determine V_A in this circuit.

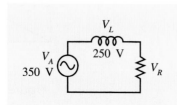

Figure 13-21. What is the resistor voltage?

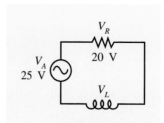

Figure 13-22. Determine the inductor voltage.

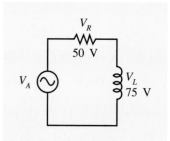

Figure 13-23. Find the total applied voltage in this circuit.

4. What is the voltage drop across the inductor in Figure 13-22?

5. What is the total (i.e., applied) voltage in Figure 13-23?

Answers to Practice Problems

1. 128.1 V   2. 244.9 V   3. 170 mV

4. 15 V   5. 90.14 V

IMPEDANCE PHASORS

The voltage drops across the components in a series circuit are directly proportional to their relative resistances and reactances. Therefore, the impedance phasor diagram is identical (except for labels) to the voltage phasor diagram. Additionally, we can compute impedance of the series RL circuit by adding (phasor addition) the individual resistance and reactance. This is reflected in Equation 13-4.

$$Z = \sqrt{R^2 + X_L^2} \qquad (13\text{-}4)$$

EXAMPLE SOLUTION

Draw the impedance phasor diagram and determine the total impedance for the circuit shown in Figure 13-24.

EXAMPLE **SOLUTION**

The phasor diagram is shown in Figure 13-25. We compute impedance with Equation 13-4 as follows:

$$Z = \sqrt{R^2 + X_L^2}$$
$$= \sqrt{(10\ \text{k}\Omega)^2 + (20\ \text{k}\Omega)^2}$$
$$= \sqrt{500 \times 10^6} \approx 22.36\ \text{k}\Omega$$

Since our result is larger than either individual phasor and less than their arithmetic sum, we have some assurance that our calculation is correct.

Practice Problems

1. Draw the impedance phasor diagram and compute total impedance for the series RL circuit shown in Figure 13-26.

2. What is the total impedance in a series RL circuit that has 82 kΩ of inductive reactance and 47 kΩ of resistance?

3. How much resistance must be connected in series with 10 kΩ of inductive reactance to produce a total impedance of 15 kΩ?

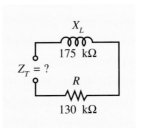

Figure 13-26. Find the impedance in this circuit.

KEY POINTS

Impedance phasors are drawn by sketching the resistance phasor and the inductive reactance phasor 90° apart. Their phasor sum is the total circuit impedance.

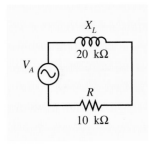

Figure 13-24. Draw the impedance phasor diagram for this circuit.

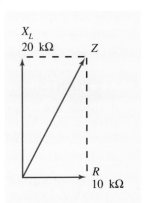

Figure 13-25. Phasor diagram for the circuit shown in Figure 13-24.

Answers to Practice Problems

1.

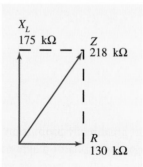

2. 94.5 kΩ   **3.** 11.18 kΩ

KEY POINTS

We can draw power phasors to describe quantities in a series *RL* circuit.

KEY POINTS

Power dissipated by the circuit resistance is called true power and is measured in watts.

KEY POINTS

The power that is stored in the magnetic field of the inductance is called reactive power and is measured in volt-amperes-reactive (VAR).

KEY POINTS

The phasor sum of true and reactive powers is called apparent power and is measured in volt-amperes (VA).

POWER PHASORS

We know from discussions in Chapter 12 that a purely inductive circuit dissipates no power. During one portion of the cycle, power is stored in the inductor's magnetic field. The power is returned to the circuit during the remaining portion of the cycle.

In a series *RL* circuit then, we must be concerned with two different types of power. The most obvious power is that dissipated by the resistance in the circuit. This can be computed with any one of the basic power formulas (e.g., $P = I^2R$) and is measured in watts. We refer to this as **real power** or **true power.**

The power that is taken from and subsequently returned to the circuit by the inductance is called **reactive power.** Reactive power is measured in volt-amperes-reactive (VAR).

Total power in a series *RL* circuit can be found by summing (phasor addition) the reactive and true power components. Total power in an *RL* circuit is called **apparent power.** Apparent power is measured in volt-amperes (VA). Figure 13-27 shows a power phasor diagram for a series *RL* circuit. Note the phasor positions.

We shall consider additional power relationships in a later section, but we can already express total or apparent power as the phasor sum of the true power and reactive power components. This phasor sum is given by Equation 13-5.

$$P_A = \sqrt{P_T^2 + P_R^2} \qquad (13\text{-}5)$$

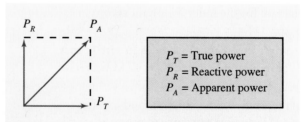

Figure 13-27. A power phasor diagram.

EXAMPLE SOLUTION

A series *RL* circuit has 10 VAR of reactive power and 10 W of true power. What is the value of apparent power in the circuit?

EXAMPLE **SOLUTION**

We apply Equation 13-5 as follows:

$$P_A = \sqrt{P_T{}^2 + P_R{}^2}$$
$$= \sqrt{(10 \text{ W})^2 + (10 \text{ VAR})^2}$$
$$= \sqrt{200} = 14.14 \text{ VA}$$

Note the phasor sum is larger than either individual phasor but is less than the arithmetic sum.

Practice Problems

1. What is the apparent power in a series *RL* circuit if the true power is 120 W and the reactive power is 95 VAR?
2. What is the apparent power in a series *RL* circuit that has a resistive power dissipation of 250 mW and a reactive power of 100 mVAR?
3. How much power is dissipated by the resistance in a series *RL* circuit if the apparent power is 20 VA and the reactive power is 15 VAR?
4. What is the reactive power in a series *RL* circuit if the apparent power is 9 mVA and the resistive power is 5 mW?

Answers to Practice Problems

1. 153.1 VA **2.** 269.3 mVA **3.** 13.23 W **4.** 7.48 mVAR

Exercise Problems 13.1

1. What can be said about the relative current through the various components in a series *RL* circuit?
2. If the current through the resistance in a series *RL* circuit is 100 mA, how much current flows through the inductance?
3. If the total current in a series *RL* circuit is 2.75 A, how much current flows through the circuit inductance?
4. Refer to Problem 3. How much current flows through the circuit resistance?
5. The voltage drops in a series *RL* circuit must be summed using _____ _____ to compute the total circuit voltage.
6. The voltage across the resistance in a series *RL* circuit is (*in, out*) of phase with the current.

7. The voltage across the inductance in a series *RL* circuit is (*in, out*) of phase with the current.

8. The total current in a series *RL* circuit is out of phase with the applied voltage by an angle that is between zero and ninety degrees. (True or False)

9. If a certain series *RL* circuit has an 8.2-kΩ resistor and a 12-kΩ inductive reactance, what is the value of the circuit impedance?

10. How much resistance must be connected in series with a 250-Ω inductive reactance to produce a total circuit impedance of 400 Ω?

11. A circuit consists of a 10-kΩ, a 5-kΩ, and a 25-kΩ resistor connected in series with two 25-mH coils. Draw the simplified circuit.

12. Draw the simplified circuit for a series *RL* circuit that consists of three 150-kΩ resistors in series with four 100-μH inductors.

13. Draw a voltage phasor diagram to represent the circuit shown in Figure 13-28.

14. Refer to Figure 13-28. What is the value of applied voltage?

15. Draw an impedance phasor diagram to represent the circuit shown in Figure 13-29.

16. Compute the total circuit impedance for the circuit shown in Figure 13-29.

17. If the voltage across the resistance in a series *RL* circuit is 18 V and the applied voltage is 25 V, what is the voltage drop across the inductance?

18. The unit of measure for apparent power is _____ .

19. The unit of measure for true power is _____ .

20. Apparent power in a series *RL* circuit can be found by summing the true power and the _____ power using phasor addition.

21. The unit of measure for admittance is _____ .

22. The unit of measure for susceptance is _____ .

23. What is the susceptance of a coil that has 25 kΩ of reactance?

24. What is the admittance of a circuit that has an impedance of 275 kΩ?

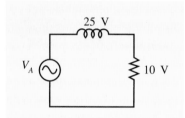

Figure 13-28. A series *RL* circuit.

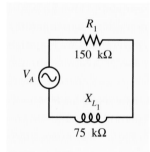

Figure 13-29.
A series *RL* circuit.

13.2 Parallel *RL* Circuits with Sinusoidal Currents

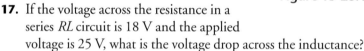

Now let us learn some of the basic characteristics and relationships found in a parallel *RL* circuit. Figure 13-30 shows two resistors that are in parallel with two inductors. Since all of the components are in parallel, we can classify this circuit as a parallel circuit.

Parallel *RL* Circuit Characteristics
VOLTAGE DROPS

⦿ **KEY POINTS**

Every component in a parallel *RL* circuit has exactly the same value of voltage.

Since every component in a parallel *RL* circuit is connected directly across the voltage source, we know that all components will have identical voltage drops. This is consistent with what we know about resistive parallel circuits.

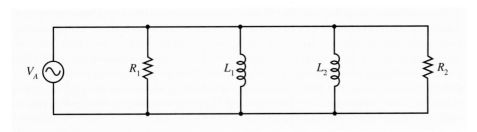

Figure 13-30. A basic parallel *RL* circuit.

PHASE RELATIONSHIPS

We can easily predict the phase relationships in a parallel *RL* circuit by applying what we already know. Consider these three facts:

- Voltage is the same across all components in the circuit.
- Current and voltage are in phase in a resistance.
- Current lags voltage by 90° in an inductance.

As shown in Figure 13-31, the voltage waveforms across all components will have the same phase. The current waveforms through the inductors will be 90° out of phase with the circuit voltage waveforms. Specifically, the inductor currents will lag the circuit voltage by

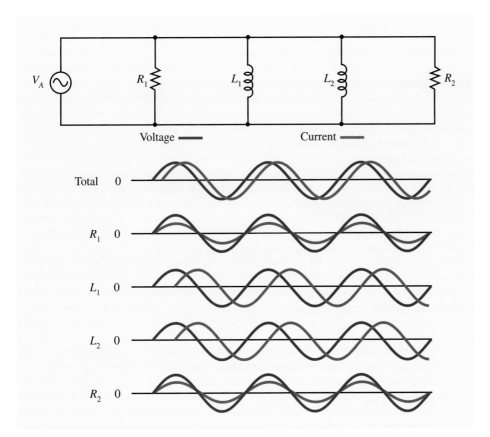

Figure 13-31. Phase relationships in a parallel *RL* circuit.

90°. Finally, we would expect the resistor current waveforms to be in phase with the circuit voltage. We can also conclude that the resistive currents and the inductive currents will be 90° out of phase with each other.

CURRENT

Each branch of a parallel *RL* circuit draws a value of current that is determined by the applied voltage and the resistance or reactance of the branch (Ohm's Law). The current in a given branch is unaffected by the value of currents in other branches.

As with a resistive parallel circuit, the individual branch currents in a parallel *RL* circuit are summed to find total current (Kirchhoff's Current Law). However, since the currents are 90° out of phase with each other, we must sum them using phasor addition. We know, then, that the total current will be greater than any individual branch current, but it will be less than the arithmetic sum of the branch currents. Equation 13-6 expresses this summation more formally.

$$I_T = \sqrt{I_R{}^2 + I_L{}^2} \qquad (13\text{-}6)$$

EXAMPLE SOLUTION

Compute the total current in a parallel *RL* circuit if the resistive branch has 3.5 A of current and the inductive branch has 5.25 A.

EXAMPLE **SOLUTION**

We apply Equation 13-6 as follows:

$$I_T = \sqrt{I_R{}^2 + I_L{}^2}$$
$$= \sqrt{(3.5 \text{ A})^2 + (5.25 \text{ A})^2}$$
$$\approx 6.31 \text{ A}$$

Practice Problems

1. The resistive current in a parallel *RL* circuit is 100 mA and the inductive current is 65 mA. What is the value of total current in the circuit?
2. If the resistive branch of a parallel *RL* circuit has 375 µA and the inductive branch has 275 µA, how much current flows through the source?
3. How much resistive current must flow in a parallel *RL* circuit to produce 200 mA when combined with 150 mA of inductive current?
4. If the total current in a parallel *RL* circuit is 2.9 A, and the inductive branch has 1.8 A, how much current flows through the resistive branch of the circuit?

Answers to Practice Problems

1. 119.3 mA   2. 465 µA   3. 132.3 mA   4. 2.27 A

IMPEDANCE

Since the current in a parallel *RL* circuit is greater than any individual branch current, it follows that total circuit impedance must be less than any individual branch resistance or reactance. This is consistent with what we learned about resistive parallel circuits. Computation of the value of circuit impedance is somewhat different, however, since the currents for the resistive and reactive branches are 90° out of phase. We will learn to compute impedance in a later paragraph.

SUSCEPTANCE AND ADMITTANCE

The characteristics of a parallel *RL* circuit have the same definitions as discussed for series circuits.

Circuit Simplification

The various resistances in a multiresistor parallel *RL* circuit can be combined as in a simple resistive parallel circuit. Similarly, the various branch inductances (assuming no magnetic linkage) can be combined as we did with a purely inductive parallel circuit. The result of these two simplifications produces an equivalent circuit having a single resistance and a single inductance in parallel with the voltage source.

EXAMPLE SOLUTION

Draw the equivalent circuit for the parallel *RL* circuit shown in Figure 13-32.

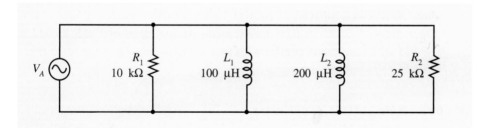

Figure 13-32. Simplify this circuit.

EXAMPLE SOLUTION

First, we combine the parallel resistors using any of the methods discussed for parallel resistive circuits. Let's select the product-over-the-sum method.

$$R_T = \frac{R_1 R_2}{R_1 + R_2}$$

$$= \frac{10\ k\Omega \times 25\ k\Omega}{10\ k\Omega + 25\ k\Omega}$$

$$\approx 7.14\ k\Omega$$

Similarly, we combine the parallel inductors as we did with purely inductive parallel circuits. For variety, let's select the reciprocal method.

$$L_T = \cfrac{1}{\cfrac{1}{L_1} + \cfrac{1}{L_2}}$$

$$= \cfrac{1}{\cfrac{1}{100\ \mu H} + \cfrac{1}{200\ \mu H}}$$

$$\approx 66.7\ \mu H$$

These results are summarized in Figure 13-33.

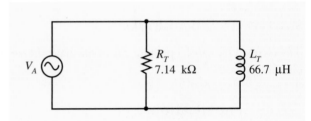

Figure 13-33. An equivalent circuit for the original circuit shown in Figure 13-32.

Practice Problems

1. Draw the equivalent circuit if three resistors having values of 10 kΩ, 22 kΩ, and 27 kΩ are connected in parallel with a 50-mH inductor.

2. Draw the equivalent circuit if three inductors having values of 275 μH, 100 μH, and 330 μH are connected in parallel with a 3.9-kΩ resistor.

3. Draw the equivalent circuit if three resistors having values of 5 kΩ, 2.7 kΩ, and 4.7 kΩ are connected in parallel with two inductors having values of 25 mH and 75 mH.

Answers to Practice Problems

1.   **2.**   **3.**

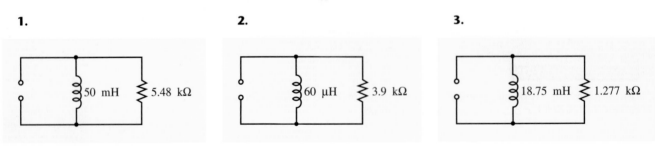

Phasor Representation

There are two types of phasor diagrams that are particularly useful when analyzing parallel *RL* circuits: current phasors and power phasors.

CURRENT PHASORS

The current phasor diagram for a parallel *RL* circuit will seem sensible, if we consider the following facts that we already know:

- The branch currents in a parallel circuit must be summed to find the total current (Kirchhoff's Current Law).
- The resistive branch currents and the inductive branch currents are 90° out of phase.
- Voltage is the same for all components in a parallel circuit.

We will always use voltage as the reference phasor when working with parallel circuits, because voltage is common to all components. Refer to Figure 13-34 as we determine the positions of the remaining phasors.

We know that current and voltage are always in phase in a resistive circuit. Therefore, the phasor representing resistive current will be drawn in the same relative position as voltage on our phasor diagram. Inductive current, by contrast, is 90° out of phase with the voltage. In particular, the current lags the voltage by 90°. Our phasor diagram must show this relationship, so we sketch the inductive current phasor 90° behind (clockwise) the voltage or reference phasor.

Finally, as shown in Figure 13-34, we can complete the parallelogram and sketch the diagonal, which represents the phasor sum of the two current phasors. Since the current phasors represent branch currents, the phasor sum will represent the total current in the circuit (Kirchhoff's Current Law).

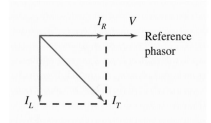

Figure 13-34. Current phasor diagrams for parallel circuits use voltage as the reference phasor.

EXAMPLE SOLUTION

Draw a current phasor diagram to represent the parallel *RL* circuit shown in Figure 13-35.

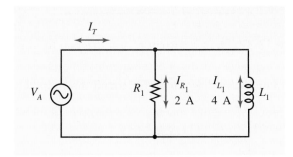

Figure 13-35. Draw a current phasor diagram for this circuit.

EXAMPLE **SOLUTION**

First, we draw our reference phasor representing circuit voltage. Next, we sketch the resistive current phasor (2 A) in the same relative position as the reference phasor, since current and

KEY POINTS

Current phasors are drawn using voltage as the reference phasor. The phasor representing resistive current is aligned with the voltage or reference phasor. The phasor representing inductive current is drawn 90° behind (clockwise) the voltage or reference phasor. Completion of the phasor parallelogram for the inductive and resistive current phasors produces a phasor (diagonal of the parallelogram) that represents total current.

voltage are in phase through a resistive circuit. Next, we sketch the inductive current phasor (4 A), such that it lags 90° behind the reference phasor, because current lags voltage by 90° in an inductive circuit. Finally, we complete the parallelogram, as shown in Figure 13-36, and compute the value of I_T using Equation 13-6.

$$I_T = \sqrt{I_R^2 + I_L^2}$$
$$= \sqrt{(2\ \text{A})^2 + (4\ \text{A})^2}$$
$$\approx 4.47\ \text{A}$$

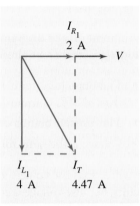

Figure 13-36. A current phasor diagram for the circuit shown in Figure 13-35.

Practice Problems

1. Draw a current phasor diagram for the circuit shown in Figure 13-37. Be sure to compute the value of total current.

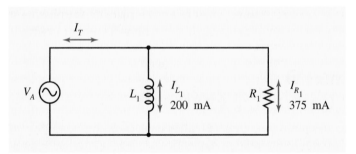

Figure 13-37. Draw a current phasor diagram for this circuit.

2. Draw a current phasor diagram for the circuit shown in Figure 13-38.

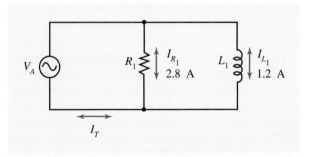

Figure 13-38. Draw a current phasor diagram for this circuit.

3. Simplify the circuit shown in Figure 13-39 by combining similar currents. Then draw the current phasor diagram and compute total current.

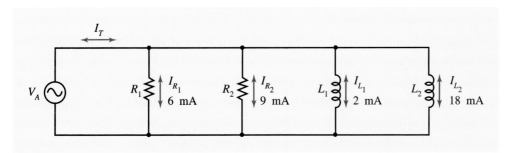

Figure 13-39. Simplify this circuit and draw a current phasor diagram.

Answers to Practice Problems

1.

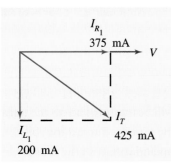

2.

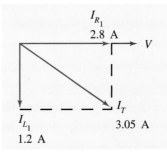

3.

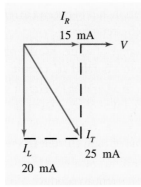

POWER PHASORS

You will recall from our previous study of resistive circuits, that power calculations are the same whether the circuit is series, parallel, or more complex. This same relationship holds true for *RL* circuits.

In a parallel *RL* circuit, then, we are interested in the values of true power, reactive power, and apparent power. The meaning and relationships of these powers were discussed with reference to series *RL* circuits and will not be repeated here.

The power phasors for a parallel *RL* circuit are drawn in the same relative positions as the current phasors. Figure 13-40 shows a representative power phasor diagram for a parallel *RL* circuit.

KEY POINTS

Power phasors can be used to represent the values of true power, reactive power, and apparent power in a parallel *RL* circuit. The definitions and relationships of these powers is similar to those discussed for series *RL* circuits.

KEY POINTS

Power phasors for parallel *RL* circuits are drawn in the same relative positions as the current phasors.

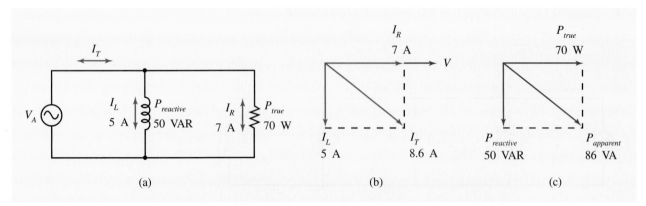

Figure 13-40. A parallel *RL* circuit with corresponding current and power phasor diagrams.

Exercise Problems 13.2

1. If the voltage drop across the resistance in a parallel *RL* circuit is 10 V, how much voltage will be felt across the inductance?

2. A certain parallel *RL* circuit consists of a 100-Ω resistor, a 350-mH coil, and a 25-V sinusoidal power source. How much voltage will be measured across the resistor?

3. Refer to Problem 2. How much voltage will be measured across the coil?

4. What is the phase relationship between current and voltage in the resistive branch of a parallel *RL* circuit?

5. What is the phase relationship between current and voltage in the inductive branch of a parallel *RL* circuit?

6. What is the possible range of values for the phase relationship between total current and total voltage in a parallel *RL* circuit?

7. If the resistive current in a parallel *RL* circuit is 120 mA and the inductive current is 225 mA, what is the value of total current in the circuit?

8. If a parallel *RL* circuit has a resistive current of 3.7 A and an inductive current of 1.8 A, what is the value of total current in the circuit?

9. How much resistive current must flow in a parallel *RL* circuit in order to combine with 3 A of inductive current to produce 5 A of total current?

10. The total current in a parallel *RL* circuit is 390 mA. The resistive branch current is 175 mA. Calculate the value of inductive branch current.

11. The total impedance in a parallel *RL* circuit is (*smaller, larger*) than any individual branch resistance or reactance.

12. Simplify the circuit shown in Figure 13-41.

13. Combine similar values in Figure 13-42 to simplify the circuit.

14. What quantity is used as the reference phasor for a current phasor diagram representing a parallel *RL* circuit?

15. The total current in a parallel *RL* circuit can be computed by simple arithmetic addition of the individual branch currents. (True or False)

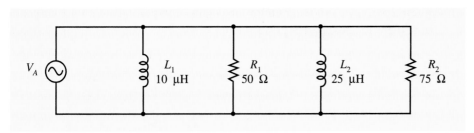

Figure 13-41. Simplify this circuit.

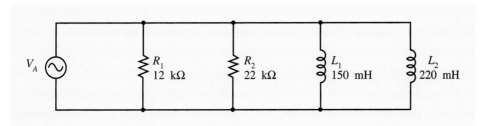

Figure 13-42. Simplify this circuit.

16. Kirchhoff's Current Law applies to parallel *RL* circuits, but the currents must be combined using _____ _____ .

17. Draw a current phasor diagram for the circuit shown in Figure 13-43.

18. Calculate the value of total current in Figure 13-43.

19. Draw a current phasor diagram for the circuit shown in Figure 13-44.

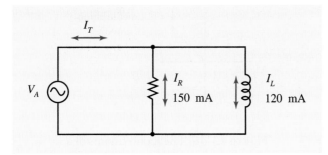

Figure 13-43. Draw a current phasor diagram for this circuit.

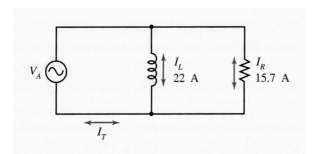

Figure 13-44. Draw a current phasor diagram for this circuit.

20. Draw a power phasor diagram for the circuit shown in Figure 13-45, and compute the value of apparent power.

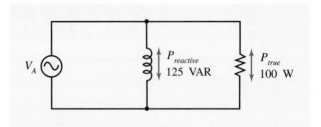

Figure 13-45. Draw a power phasor diagram for this circuit.

13.3 Representing Circuit Quantities

There are several ways that alternating circuit quantities can be expressed. We shall employ two common methods that are used by technicians: polar and rectangular notation. Further, we will concentrate on developing a knowledge of these notations that can be directly applied to the solution of circuit analysis problems.

Regardless of the notation used, we must have a way to represent the value of a phasor. Figure 13-46 illustrates the general requirement of a notation system.

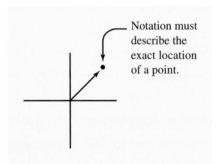

Figure 13-46. The requirements of a notation system for representing phasors.

Rectangular Notation

The rectangular method of notation considers the point to be located on a two-dimensional graph. Location of a particular point can be specified by listing the two corresponding coordinates. Figure 13-47 illustrates rectangular notation.

The two-dimensional graph consists of a real axis and an imaginary axis (sometimes called the reactance axis). It is customary to label the imaginary axis with the letter j. Mathematicians label the imaginary axis with the letter i, but technicians prefer to use j to avoid

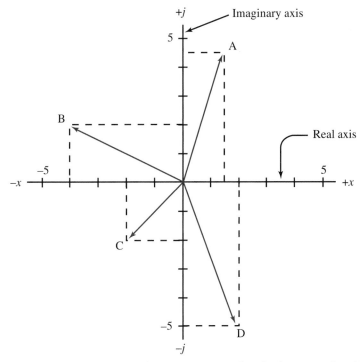

Figure 13-47. Rectangular notation specifies the horizontal and vertical coordinates of a given point.

confusion with the letter i, which is used to represent current. In either case, each axis has a positive and negative region. The horizontal scale is positive on the right half, and the vertical scale is positive on the upper half. A point is identified by specifying the coordinates of the point. We always list the horizontal coordinate first, and we preface the vertical coordinate with a lowercase j. The letter j is often called the j-operator and implies a 90° phase relationship.

EXAMPLE SOLUTION

Write the rectangular notation to describe each of the points shown in Figure 13-47.

EXAMPLE **SOLUTION**

Point A in Figure 13-47 can be described by the horizontal coordinate of 1.5 and the vertical coordinate of 4.5. We write this as 1.5 + j4.5. Similarly, point B can be described as –4 + j2, since it is 4 units to the left and 2 units up. Point C is 2 units to the left (negative) and 2 units down (negative), so we describe it as –2 – j2. Finally, point D is 2 units to the right (positive) and 5 units down (negative). We describe it as 2 – j5.

It is important to note that rectangular notation is used to represent a point on a graph. For our purposes, this point corresponds to the tip of a phasor. The phasor may represent current, voltage, impedance, or power. Thus, we might have a voltage (V_1) expressed as V_1 = +25 + j42 V, or perhaps a current (I_3) written as I_3 = –6 + j2.5 A.

Practice Problems

1. Write the rectangular notation for each of the current phasors shown in Figure 13-48.

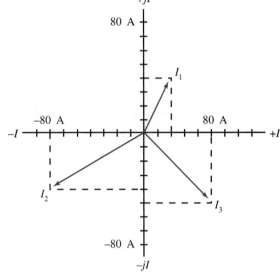

Figure 13-48. Write the rectangular notation to describe each of these phasors.

2. Write the rectangular notation to describe each of the voltage phasors shown in Figure 13-49.

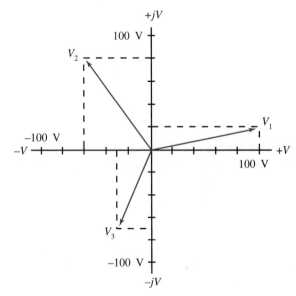

Figure 13-49. Express these phasors using rectangular notation.

3. Express each of the power phasors shown in Figure 13-50 in their rectangular form.

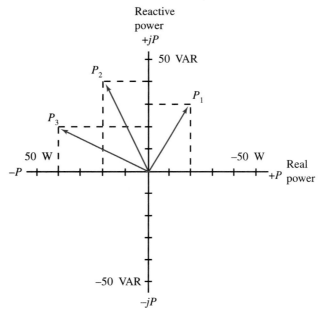

Figure 13-50. Write the rectangular notation for each of these phasors.

Answers to Practice Problems

1.
$I_1 = 20 + j40$ A
$I_2 = -70 - j40$ A
$I_3 = 50 - j50$ A

2.
$V_1 = 100 + j20$ V
$V_2 = -60 + j80$ V
$V_3 = -30 - j70$ V

3.
$P_1 = 20 + j30$ VA
$P_2 = -20 + j40$ VA
$P_3 = -40 + j20$ VA

Polar Notation

Figure 13-51 shows an alternate method of describing the location of points (or phasors). A point described in the polar coordinate system consists of two parts: the magnitude (length of the phasor) and the angle (relative to the rightmost horizontal axis). Thus, for example, a voltage phasor that is 25 units (volts) in length and is positioned at an angle of 45° is described as 25 V ∠45°. The length of the phasor is always represented with a positive number, regardless of its position. The angle may be expressed in either degrees or radians.

EXAMPLE SOLUTION

Write the polar notation for each of the phasors shown in Figure 13-52.

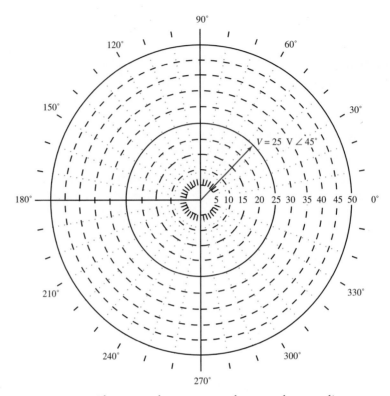

Figure 13-51. Phasor can be represented on a polar coordinate system.

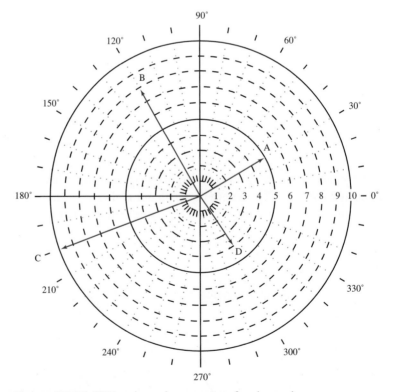

Figure 13-52. Write the polar notation for these phasors.

Phasor A is 5 units long. Its angle is 30°. We can describe the phasor as A = 5 ∠30°. Phasor B is 8 units long and is positioned at 120°. We write it as B = 8 ∠120°. Similarly, phasors C and D can be described as 10 ∠200° and 4 ∠305°.

As with rectangular notation, the units are determined by the quantity being represented. Thus, for example, we may have a voltage that is described as 125 V ∠45° or perhaps a current written as 6.5 mA ∠175°.

Practice Problems

1. Write the polar notation for each of the current phasors shown in Figure 13-53.

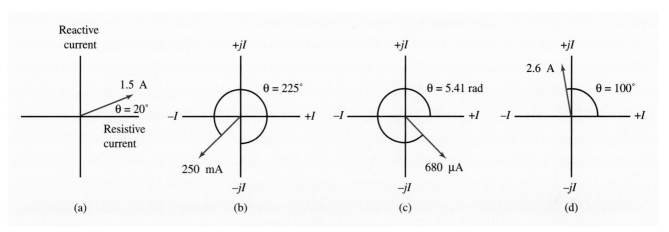

Figure 13-53. Write the polar notation for these current phasors.

2. Write the polar notation for the impedance phasors shown in Figure 13-54.

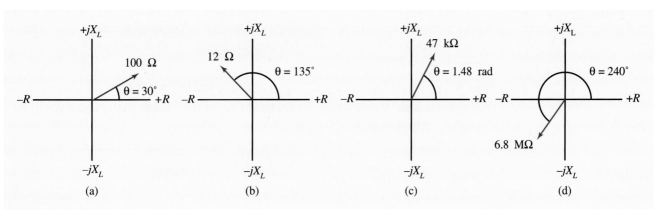

Figure 13-54. Express these impedance phasors in polar form.

3. Write the polar notation for each of the power phasors shown in Figure 13-55.

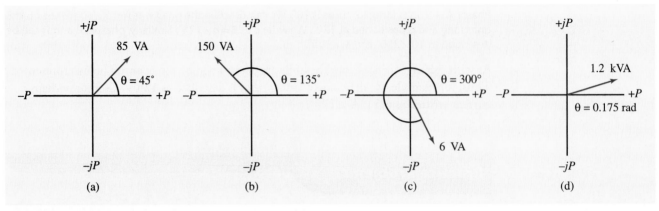

Figure 13-55. Express these power phasors in polar form.

Answers to Practice Problems

1. a. 1.5 A ∠20°
    b. 250 mA ∠225°
    c. 680 μA ∠5.41 rad
    d. 2.6 A ∠100°

2. a. 100 Ω ∠30°
    b. 12 Ω ∠135°
    c. 47 kΩ ∠1.48 rad
    d. 6.8 MΩ ∠240°

3. a. 85 VA ∠45°
    b. 150 VA ∠135°
    c. 6 VA ∠300°
    d. 1.2 kVA ∠0.175 rad

Polar-to-Rectangular Conversions

Certain calculations (e.g., multiplication and division) are more easily accomplished with phasors expressed in polar form. Other calculations (e.g., addition and subtraction) are easier when the phasors are expressed in rectangular form. If you have an engineering calculator, it makes little difference which form you use. But as a technician, you will want to be familiar with both, and you will need to convert between the two notations.

Conversion of polar notation to rectangular notation is just an application of the right-angle trigonometry we discussed in Chapter 11. Figure 13-56 shows how to make the conversion.

The phasor V_A is initially specified in polar form as 100 V ∠55°. This is the hypotenuse of our right triangle. The opposite and adjacent sides of the triangle (relative to angle θ) are V_L and V_R, respectively. We can solve for V_L as follows:

$$\sin\theta = \frac{\text{opposite}}{\text{hypotenuse}} = \frac{V_L}{V_A}$$

Transposing for V_L gives us Equation 13-7.

$$V_L = V_A \sin\theta \qquad (13\text{-}7)$$

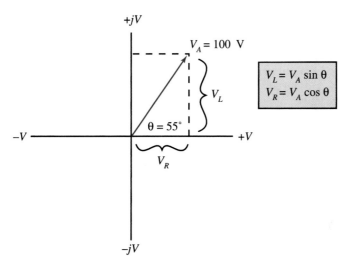

Figure 13-56. Conversion of this voltage phasor from polar to rectangular notation is an application of right-angle trigonometry.

In a similar manner, we can find an expression for V_R.

$$\cos\theta = \frac{\text{adjacent}}{\text{hypotenuse}} = \frac{V_R}{V_A}$$

Transposing for V_R gives us Equation 13-8.

$$V_R = V_A\cos\theta \qquad\qquad (13\text{-}8)$$

Substituting our known values into Equation 13-7 gives us the following results:

$$V_L = V_A\sin\theta$$
$$= 100\text{ V} \times \sin 55°$$
$$= 100\text{ V} \times 0.8192 = 81.92\text{ V}$$

Similarly, we find V_R by applying Equation 13-8.

$$V_R = V_A\cos\theta$$
$$= 100\text{ V} \times \cos 55°$$
$$= 100\text{ V} \times 0.5736 = 57.36\text{ V}$$

We can now express our phasor in rectangular form.

$$V_A = 100\text{ V }\angle 55° = 57.36 + j81.92\text{ V}$$

Conversion of impedance, power, or current phasors from polar form to rectangular form is approached in the same way. These conversions are summarized in Table 13-1 in the next section.

EXAMPLE SOLUTION

Convert the series impedance phasor 250 Ω ∠30° to an equivalent rectangular form.

EXAMPLE SOLUTION

Figure 13-57 shows a sketch of the given phasor. The known value of impedance is the hypotenuse of our right triangle. The opposite side is X_L and the adjacent side is the value of resistance.

The real or resistive part of our rectangular form will be computed as

$$\cos\theta = \frac{\text{adjacent}}{\text{hypotenuse}}$$

$$\cos\theta = \frac{R}{Z}$$

$$R = Z\cos\theta$$

The imaginary or reactive part of the rectangular form can be found as

$$\sin\theta = \frac{\text{opposite}}{\text{hypotenuse}}$$

$$\sin\theta = \frac{X_L}{Z}$$

$$X_L = Z\sin\theta$$

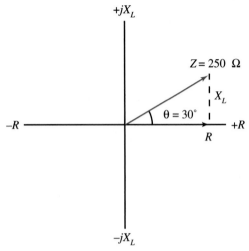

Figure 13-57. Express this phasor in rectangular notation.

Of course, we will preface this with the letter j to identify it as the reactive portion of the impedance phasor. We can combine these two parts into a single expression and obtain Equation 13-9.

$$Z\angle\theta = Z\cos\theta + jZ\sin\theta \qquad (13\text{-}9)$$

Substituting values into Equation 13-9 yields our complete rectangular notation for the given phasor.

$$
\begin{aligned}
Z\angle\theta &= Z\cos\theta + jZ\sin\theta \\
&= 250\ \Omega\cos 30° + j250\ \Omega\sin 30° \\
&= 250\ \Omega \times 0.866 + j250\ \Omega \times 0.5 \\
&= 216.5\ \Omega + j125\ \Omega
\end{aligned}
$$

Practice Problems

1. Convert the phasor shown in Figure 13-58 to rectangular notation.
2. Express the phasor shown in Figure 13-59 in rectangular notation.
3. The impedance phasor for a series circuit is 1,800 Ω $\angle 50°$. Express the circuit impedance using rectangular notation.
4. If the total current phasor for a parallel circuit is described as 2.75 A $\angle 125°$, how will the current be expressed in rectangular notation?

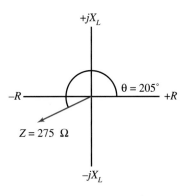

Figure 13-58. Express this phasor in rectangular form.

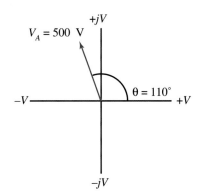

Figure 13-59. Convert this phasor to rectangular form.

Answers to Practice Problems

1. $Z = -249.2 - j116.2\ \Omega$ **2.** $V_A = -171 + j469.8\ V$

3. $Z = 1.16\ k\Omega + j1.38\ k\Omega$ **4.** $I_T = -1.58 + j2.25\ A$

Rectangular-to-Polar Conversions

Conversion of a phasor expressed in rectangular notation to an equivalent phasor expressed in polar notation is another application of basic right-triangle mathematics. Consider the phasor shown in Figure 13-60 and expressed as $Z = 25 + j50\ \Omega$.

We have already learned to compute the length of the phasor with Equation 13-4, or simply the Pythagorean theorem. In this case, we can compute the magnitude of Z as follows:

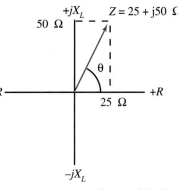

Figure 13-60. Convert this phasor to polar form.

$$Z = \sqrt{R^2 + X_L{}^2}$$
$$= \sqrt{25^2 + 50^2} = 55.9\ \Omega$$

The opposite and adjacent sides of our right triangle are known. The opposite side is X_L (50 Ω), and the adjacent side is R (25 Ω). If we know any two sides of a right triangle, we can find the angle θ by applying one of the basic trigonometric equations presented in Chapter 11. In the present case, we will use the tangent function, since we know the opposite and adjacent sides of the triangle.

$$\tan\theta = \frac{\text{opposite}}{\text{adjacent}}$$
$$= \frac{X_L}{R} = \frac{50\ \Omega}{25\ \Omega} = 2;\ \text{therefore}$$
$$\theta = \arctan 2 = 63.4°$$

So, we write $Z = 55.9\ \Omega\ \angle 63.4°$.

EXAMPLE SOLUTION

If the total voltage in a series *RL* circuit is expressed as $V_A = 10 + j7$ V, write this voltage using polar notation.

EXAMPLE SOLUTION

It is always a good idea to sketch the problem on a phasor diagram. Figure 13-61 shows a sketch of the given phasor.

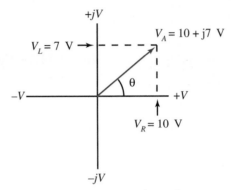

Figure 13-61. Express this voltage in polar notation.

We can use phasor addition (i.e., Equation 13-3) to determine the length of the polar phasor.

$$V_A = \sqrt{V_R^2 + V_L^2}$$
$$= \sqrt{10^2 + 7^2}$$
$$= \sqrt{149} = 12.21 \text{ V}$$

Again, we know the opposite and adjacent sides of a right triangle, so we will want to use the tangent function to find the angle θ.

$$\tan\theta = \frac{\text{opposite}}{\text{adjacent}}$$
$$= \frac{V_L}{V_R} = \frac{7 \text{ V}}{10 \text{ V}} = 0.7; \text{ therefore}$$
$$\theta = \arctan 0.7 \approx 35°$$

So, we write $V_A = 12.21$ V $\angle 35°$.

Practice Problems

1. If the total current in a parallel *RL* circuit is expressed as $I_T = 40 + j10$ mA, write this current using polar notation.

2. Express the impedance of a series circuit in polar notation, if the rectangular form is $Z = 2$ k $+ j3$ kΩ.

3. What is the polar form of the total voltage in a series circuit, if it is expressed in rectangular notation as $V_A = 5 + j2$ V.

Answers to Practice Problems

1. $I_T = 41.23$ mA $\angle 14.04°$ **2.** $Z = 3.6$ kΩ $\angle 56.3°$

3. $V_A = 5.39$ V $\angle 21.8°$

Although conversion from polar to rectangular notation and conversion from rectangular to polar notation is based on previously learned methods, Table 13-1 is provided as a way of summarizing the various conversions.

CIRCUIT TYPE	QUANTITY	POLAR TO RECTANGULAR	RECTANGULAR TO POLAR
Series	Voltage	$V_A \angle \theta = V_A \cos\theta + j\, V_A \sin\theta$	$V_A = \sqrt{V_R^2 + V_L^2}$ $\theta = \arctan \dfrac{V_L}{V_R}$
	Impedance	$Z \angle \theta = Z \cos\theta + j\, Z \sin\theta$	$Z = \sqrt{R^2 + X_L^2}$ $\theta = \arctan \dfrac{X_L}{R}$
	Power	$P_A \angle \theta = P_A \cos\theta + j\, P_A \sin\theta$	$P_A = \sqrt{P_T^2 + P_R^2}$ $\theta = \arctan \dfrac{P_R}{P_T}$
Parallel	Current	$I_T \angle \theta = I_T \cos\theta + j\, I_T \sin\theta$	$I_T = \sqrt{I_R^2 + I_L^2}$ $\theta = \arctan \dfrac{I_L}{I_R}$
	Power	$P_A \angle \theta = P_A \cos\theta + j\, P_A \sin\theta$	$P_A = \sqrt{P_T^2 + P_R^2}$ $\theta = \arctan \dfrac{P_R}{P_T}$

Table 13-1. Phasor Notation Conversion Chart

Calculator Sequences

Many engineering calculators provide direct conversions between polar and rectangular notation. Typical operation involves keying in the polar form and pressing a key labeled P→R to get the rectangular form. Similarly, the R→P key will convert a rectangular phasor into its polar equivalent. Other calculators use a key labeled POLAR to toggle between the polar and rectangular modes. The phasors may be entered using either type of notation, but they will be displayed in the selected mode.

If your calculator does not have the capability of direct phasor conversion, then you can use one of the pairs of sequences shown in Figures 13-62 and 13-63.

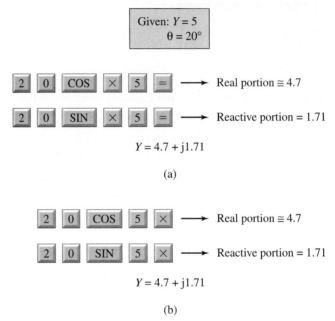

Figure 13-62. Calculator sequences for polar-to-rectangular conversion. Sequences are shown for (a) a standard engineering calculator and (b) an RPN engineering calculator.

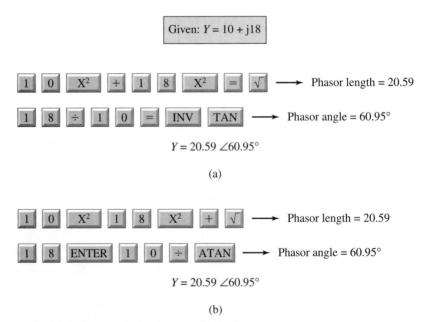

Figure 13-63. Calculator sequences for rectangular-to-polar conversion. Sequences are shown for (a) a standard engineering calculator and (b) an RPN engineering calculator.

13.4 *RL* Circuit Calculations

We are now in a position to completely analyze *RL* circuits. We shall consider series, parallel, series-parallel, and complex circuit configurations. In all cases, it will be assumed that there is no magnetic linkage between the various inductors. Most of our discussion will be simple application of previously learned rules, principles, and procedures. We should be able to compute any of the following circuit quantities in an *RL* circuit: voltage, current, power, phase relationships, impedance, or admittance. Additionally, we should be able to work with sinusoidal quantities expressed in any of the following ways: peak, peak to peak, or rms.

It is not practical to illustrate every possible combination of knowns and unknowns even for a simple *RL* circuit. Therefore, we will concentrate on developing a method of analysis that relies heavily on previously learned techniques. Thus, analysis of *RL* circuits should be viewed as a simple extension of prior knowledge rather than as a totally new subject.

Series *RL* Circuit Computations

We now have all the tools necessary to analyze a series *RL* circuit. We will formally identify some additional equations as we work through the examples, but it is important to remember that the equations are not really new. They are simply applications of previously studied principles. If you view them in this way, then you will be reinforcing previously learned material instead of trying to learn new material.

As with other circuit analysis problems we have done, there are many paths that lead to the correct solution of all quantities. The more experience you gain, the easier it will be to identify an optimum path. In all cases involving single equations, we must know all but one of the parameters in order to solve the equation. You will also remember that it is generally helpful to find currents early in the solution of series circuit problems and voltages early in the solution of parallel circuit problems, since these quantities are common to all components. Finally, if you are unable to compute a particular quantity, then calculate some other value. The more circuit values that become known, the easier it will be to find an equation that can be solved for a particular quantity.

> **◗ KEY POINTS**
>
> When analyzing series circuits, we try to find current early on. For parallel circuits, we generally strive to find the common voltage early in the problem.

EXAMPLE SOLUTION

Analyze the *RL* circuit shown in Figure 13-64, and complete the matrix shown in Table 13-2.

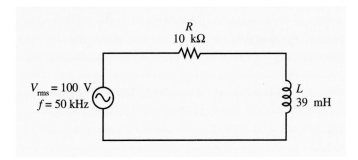

Figure 13-64. Analyze this *RL* circuit.

CIRCUIT QUANTITY		RESISTOR	INDUCTOR	TOTAL
Current (rms)				
Voltage	V_P			
	V_{PP}			
	V_{rms}			100 V
Resistance/Reactance/Impedance		10 kΩ		
Conductance/Susceptance/Admittance				
Power				
Phase Angle (overall)				
Power Factor (overall)				

Table 13-2. Solution Matrix for the Circuit in Figure 13-64

EXAMPLE SOLUTION

Let's compute the peak and the peak-to-peak values for the applied voltage as a first step.

$$V_P = 1.414 V_{rms}$$
$$= 1.414 \times 100 \text{ V} = 141.4 \text{ V}$$

We can double this to get the peak-to-peak value.

$$V_{PP} = 2V_P$$
$$= 2 \times 141.4 \text{ V} = 282.8 \text{ V}$$

Since we know the value of inductance and the applied frequency, we can determine the value of inductive reactance.

$$X_L = 2\pi f L$$
$$= 6.28 \times 50 \text{ kHz} \times 39 \text{ mH} = 12.25 \text{ kΩ}$$

We can now add (phasor addition) the resistance and reactance values to determine total impedance.

$$Z = \sqrt{R^2 + X_L^2}$$
$$= \sqrt{(10 \text{ kΩ})^2 + (12.25 \text{ kΩ})^2}$$
$$= \sqrt{250.1 \times 10^6} = 15.81 \text{ kΩ}$$

We can find the conductance, susceptance, and admittance of the circuit by finding the reciprocals of resistance, reactance, and impedance, respectively. Let's begin with conductance.

$$G = \frac{1}{R} = \frac{1}{10 \text{ kΩ}} = 100 \text{ μS}$$

We find susceptance in a similar manner.

$$B = \frac{1}{X_L} = \frac{1}{12.25 \text{ k}\Omega} = 81.63 \text{ μS}$$

The reciprocal of impedance will give us the value of admittance.

$$Y = \frac{1}{Z} = \frac{1}{15.81 \text{ k}\Omega} = 63.25 \text{ μS}$$

Now that we know total voltage and total impedance in a series circuit, we can apply Ohm's Law to determine the total current. We can express this formally as Equation 13-10.

$$I_T = \frac{V_T}{Z} \qquad\qquad (13\text{-}10)$$

As with any Ohm's Law problem, we may use voltage expressed in any form (i.e., peak, peak-to-peak, or rms), but the computed value of current will be in the same form. In this case, let us compute the rms value of current by using the rms value of applied voltage.

$$I_T = \frac{V_T}{Z} = \frac{100 \text{ V}}{15.81 \text{ k}\Omega} = 6.325 \text{ mA}$$

We can use Ohm's Law to determine the individual voltage drops across the resistance and reactance in the circuit. Let's use rms current, which will produce rms voltage as an answer.

$$V_R = I_R R = 6.325 \text{ mA} \times 10 \text{ k}\Omega = 63.25 \text{ V}$$

Equation 13-11 expresses a similar application of Ohm's Law to determine the voltage across the inductor.

$$V_L = I_T X_L \qquad\qquad (13\text{-}11)$$

For the circuit shown in Figure 13-64, we compute the rms value for V_L as follows:

$$V_L = I_T X_L = 6.325 \text{ mA} \times 12.25 \text{ k}\Omega = 77.48 \text{ V}$$

Determination of peak and peak-to-peak values for the resistive and inductive voltages requires straightforward application of the basic sine wave relationships.

$$V_{R_{(P)}} = 1.414 V_{R_{(rms)}} = 1.414 \times 63.25 \text{ V} = 89.44 \text{ V}$$
$$V_{R_{(PP)}} = 2 V_{R_{(P)}} = 2 \times 89.44 \text{ V} = 178.9 \text{ V}$$

Similarly,

$$V_{L_{(P)}} = 1.414 V_{L_{(rms)}} = 1.414 \times 77.48 \text{ V} = 109.6 \text{ V}$$
$$V_{L_{(PP)}} = 2 V_{L_{(P)}} = 2 \times 109.6 \text{ V} = 219.2 \text{ V}$$

Now let us calculate the power in the circuit. Although some applications (e.g., heat calculations in transistors and radiated power from antennas) require computation of peak and average power, we will concentrate on the more common expression of power that utilizes rms values. Recall that rms values of current and voltage produce the same power (heating effect) as an equivalent amount of dc. Let's first compute the power dissipated in the resistance (i.e., real or true power).

$$P_T = I_{R_{(rms)}} \times V_{R_{(rms)}} = 6.325 \text{ mA} \times 63.25 \text{ V} = 400.1 \text{ mW}$$

We can use a similar calculation to determine the reactive power for the circuit inductance.

$$P_R = I_{L_{(rms)}} V_{R_{(rms)}} = 6.325 \text{ mA} \times 77.48 \text{ V} = 490.1 \text{ mVAR}$$

Finally, we can find apparent power by applying the same procedure using total current and voltage values, or by summing (phasor addition) the values of true power and reactive power. Let's choose the addition method.

$$P_A = \sqrt{P_T^2 + P_R^2}$$

$$= \sqrt{(400.1 \text{ mW})^2 + (490.1 \text{ mVAR})^2}$$

$$= \sqrt{400.3 \times 10^{-3}} \approx 632.7 \text{ mVA}$$

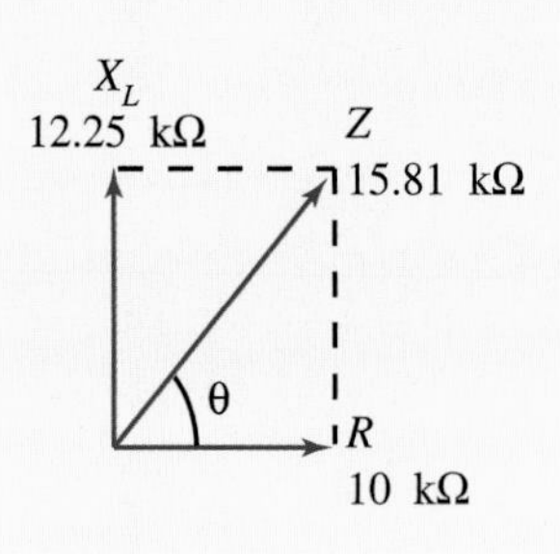

We can obtain the phase angle by using any one of the voltage, impedance, or power phasor diagrams. Figure 13-65 shows the impedance phasor diagram. We shall use it to compute the phase angle (θ).

$$\tan\theta = \frac{\text{opposite}}{\text{adjacent}} = \frac{X_L}{R}$$

$$= \frac{12.25 \text{ k}\Omega}{10 \text{ k}\Omega} = 1.225$$

$$\theta = \arctan 1.225 = 50.77°$$

Figure 13-65.
An impedance phasor diagram for the circuit shown in Figure 13-64.

The power factor of a circuit is another way to describe the phase relation between current and voltage in a circuit. It is simply the cosine of the phase angle as expressed in Equation 13-12.

$$\text{Power factor} = pf = \cos\theta \qquad (13\text{-}12)$$

We additionally describe the power factor to be leading or lagging. Current is always assumed to be the reference. In the case of *RL* circuits, the current will always lag the voltage, so we will always have a lagging power factor. In a later chapter, we will study capacitive circuits that have a leading power factor. For the present example, we find the power factor as follows:

$$pf = \cos\theta = \cos 50.77° = 0.632 \text{ (lagging)}$$

This completes our analysis of the circuit shown in Figure 13-64. The completed solution matrix is shown in Table 13-3.

CIRCUIT QUANTITY		RESISTOR	INDUCTOR	TOTAL
Current (rms)		6.325 mA	6.325 mA	6.325 mA
Voltage	V_P	89.44 V	109.6 V	141.4 V
	V_{PP}	178.9 V	219.2 V	282.8 V
	V_{rms}	63.25 V	77.48 V	100 V
Resistance/Reactance/Impedance		10 kΩ	12.25 kΩ	15.81 kΩ
Conductance/Susceptance/Admittance		100 μS	81.63 μS	63.25 μS
Power		400.1 mW	490.1 mVAR	632.7 mVA
Phase Angle (overall)		50.77°		
Power Factor (overall)		0.632 (lagging)		

Table 13-3. Completed Solution Matrix for the Circuit in Figure 13-64

It is possible to write and subsequently memorize literally hundreds of equations related to the solution of series *RL* circuit problems. This approach is not recommended. First, it is a difficult task. Second, it is very unlikely that you will be able to remember hundreds of equations for each type of circuit.

A more practical approach requires you to apply previously learned principles. In the case of *RL* circuit analysis, you can still rely on Ohm's Law, Kirchhoff's Laws, the general power equations, basic sine wave formulas, and the right-triangle relationships previously studied. The only "new" material involves recognition that summation of resistive and reactive quantities requires phasor addition. So in general, it is recommended that you utilize your prior knowledge as much as possible. Not only is this an easier approach to problem solving, but it strengthens your understanding of the fundamental principles that represent the heart of electronics.

Practice Problems

1. An *RL* circuit consists of an 18-kΩ resistor, a 4.0-mH inductor, and an ac voltage source that maintains 500 mV rms at a frequency of 1.2 MHz. Analyze the circuit and complete a solution matrix similar to Table 13-2.

2. A 25-V peak voltage source produces 150-MHz sine waves. It is connected in a series *RL* circuit consisting of a 100-µH coil and a 91-kΩ resistor. Analyze the circuit and complete a solution matrix similar to Table 13-2.

3. What is the value of peak current that flows in the circuit shown in Figure 13-66?

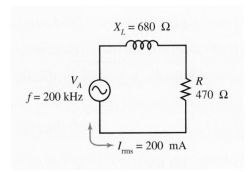

$X_L = 680$ Ω

V_A
$f = 200$ kHz

R
470 Ω

$I_{rms} = 200$ mA

Figure 13-66. A series *RL* circuit.

4. What is the peak-to-peak voltage across the inductor in Figure 13-66?
5. What is the value of impedance for the circuit in Figure 13-66?
6. What is the value of inductance for the coil shown in Figure 13-66?
7. What is the reactive power in Figure 13-66?
8. How much true power is dissipated by the circuit in Figure 13-66?
9. What is the phase angle for the circuit shown in Figure 13-66?
10. What is the rms value of applied voltage (V_A) in Figure 13-66?

Answers to Practice Problems

1.

CIRCUIT QUANTITY		RESISTOR	INDUCTOR	TOTAL
Current (rms)		14.24 μA	14.24 μA	14.24 μA
Voltage	V_P	362.4 mV	606.9 mV	707 mV
	V_{PP}	724.8 mV	1.214 V	1.414 V
	V_{rms}	256.3 mV	429.2 mV	500 mV
Resistance/Reactance/Impedance		18 kΩ	30.14 kΩ	35.11 kΩ
Conductance/Susceptance/Admittance		55.56 μS	33.18 μS	28.48 μS
Power		3.65 μW	6.11 μVAR	7.12 μVA
Phase Angle (overall)		59.16°		
Power Factor (overall)		0.513 (lagging)		

2.

CIRCUIT QUANTITY		RESISTOR	INDUCTOR	TOTAL
Current (rms)		134.96 μA	134.96 μA	134.96 μA
Voltage	V_P	17.37 V	17.98 V	25 V
	V_{PP}	34.74 V	35.95 V	50 V
	V_{rms}	12.28 V	12.71 V	17.68 V
Resistance/Reactance/Impedance		91 kΩ	94.2 kΩ	131 kΩ
Conductance/Susceptance/Admittance		10.99 μS	10.62 μS	7.63 μS
Power		1.66 mW	1.72 mVAR	2.39 mVA
Phase Angle (overall)		45.99°		
Power Factor (overall)		0.695 (lagging)		

3. 282.8 mA **4.** 384.6 V **5.** 826.6 Ω

6. 541.4 μH **7.** 27.2 VAR **8.** 18.8 W

9. 55.35° **10.** 165.3 V

Parallel *RL* Circuit Computations

We shall approach the analysis of parallel *RL* circuits as an application of previously learned techniques and concepts. We will highlight some "new" equations and state them formally for completeness. It is important, however, for you to understand that these added equations are simple extensions of basic circuit analysis methods that you already know. Concentrate on understanding how the various methods tie in with previously mastered material, rather than trying to memorize the equations.

EXAMPLE SOLUTION

Analyze the parallel *RL* circuit shown in Figure 13-67, and complete the solution matrix in Table 13-4.

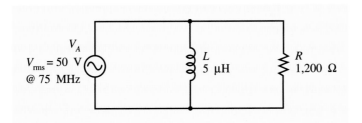

Figure 13-67. Analyze this parallel *RL* circuit.

CIRCUIT QUANTITY		RESISTOR	INDUCTOR	TOTAL
Current (rms)				
Voltage	V_P			
	V_{PP}			
	V_{rms}			50 V
Resistance/Reactance/Impedance		1200 Ω		
Conductance/Susceptance/Admittance				
Power				
Phase Angle (overall)				
Power Factor (overall)				

Table 13-4. Solution Matrix for the Circuit in Figure 13-67

EXAMPLE **SOLUTION**

The applied voltage (50 V rms) is common to all components. We can use the sine wave equations to determine the peak and peak-to-peak values.

$$V_P = 1.414 V_{rms} = 1.414 \times 50 \text{ V} = 70.7 \text{ V}$$
$$V_{PP} = 2V_P = 2 \times 70.7 \text{ V} = 141.4 \text{ V}$$

Now let's compute the inductive reactance.

$$X_L = 2\pi fL = 6.28 \times 75 \text{ MHz} \times 5 \text{ μH} = 2.36 \text{ k}\Omega$$

We can apply Ohm's Law to determine the current flow through the two parallel branches.

$$I_R = \frac{V_R}{R} = \frac{50 \text{ V}}{1,200 \text{ Ω}} = 41.67 \text{ mA, and}$$

$$I_L = \frac{V_L}{X_L} = \frac{50 \text{ V}}{2.36 \text{ k}\Omega} = 21.19 \text{ mA}$$

As with previously studied parallel circuits, we can add the branch currents to determine the total current (Kirchhoff's Current Law). Since our currents are 90° out of phase, we will sum the currents with phasor addition. This can be expressed formally as Equation 13-13.

$$I_T = \sqrt{I_R^2 + I_L^2} \qquad \text{(13-13)}$$

In the present case, we have

$$
\begin{aligned}
I_T &= \sqrt{I_R^2 + I_L^2} \\
&= \sqrt{(41.67 \text{ mA})^2 + (21.19 \text{ mA})^2} \\
&= \sqrt{2.185 \times 10^{-3}} = 46.75 \text{ mA}
\end{aligned}
$$

As with other parallel circuits, we can apply Ohm's Law to find the total opposition to current flow (impedance).

$$Z = \frac{V_A}{I_T} = \frac{50 \text{ V}}{46.75 \text{ mA}} = 1.07 \text{ k}\Omega$$

We can compute conductance, susceptance, and admittance by finding the reciprocals of resistance, reactance, and impedance. We compute conductance as

$$G = \frac{1}{R} = \frac{1}{1,200 \ \Omega} = 833.33 \ \mu\text{S}$$

Susceptance is the reciprocal of reactance.

$$B = \frac{1}{X_L} = \frac{1}{2.36 \text{ k}\Omega} = 423.7 \ \mu\text{S}$$

In a similar manner, admittance is the reciprocal of impedance.

$$Y = \frac{1}{Z} = \frac{1}{1.07 \text{ k}\Omega} = 934.58 \ \mu\text{S}$$

The true power dissipated in the resistor can be computed with any one of the basic power formulas.

$$P_T = \frac{V_R^2}{R} = \frac{(50 \text{ V})^2}{1,200 \ \Omega} = 2.08 \text{ W}$$

The reactive power is found in a similar manner.

$$P_R = I_L V_L = 21.19 \text{ mA} \times 50 \text{ V} = 1.06 \text{ VAR}$$

Finally, apparent power can be found as the phasor sum of true power and reactive power, or we can simply apply one of the power equations to the circuit using total values. Let's choose the latter method using total voltage and total current.

$$P_A = I_T V_A = 46.75 \text{ mA} \times 50 \text{ V} = 2.34 \text{ VA}$$

Figure 13-68 shows a phasor diagram for the circuit shown in Figure 13-67. With the help of this diagram, we can readily determine the phase angle and power factor.

Since we know all three sides of the right triangle shown in Figure 13-68, we can use any one of the sine, cosine, or tangent functions. Let's choose the cosine function, since the cosine will also provide the value of the power factor.

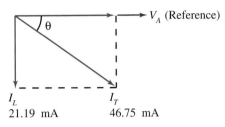

Figure 13-68. A phasor diagram for the circuit shown in Figure 13-67.

$$\cos\theta = \frac{\text{adjacent}}{\text{hypotenuse}} = \frac{I_R}{I_T}$$

$$= \frac{41.67 \text{ mA}}{46.75 \text{ mA}} = 0.891$$

$$\theta = \arccos 0.891 \approx 27°$$

This completes the analysis of the circuit shown in Figure 13-67. The completed solution matrix is shown in Table 13-5.

CIRCUIT QUANTITY		RESISTOR	INDUCTOR	TOTAL
Current (rms)		41.67 mA	21.19 mA	46.75 mA
Voltage	V_P	70.7 V	70.7 V	70.7 V
	V_{PP}	141.4 V	141.4 V	141.4 V
	V_{rms}	50 V	50 V	50 V
Resistance/Reactance/Impedance		1,200 Ω	2.36 kΩ	1.07 kΩ
Conductance/Susceptance/Admittance		833.33 μS	423.7 μS	934.58 μS
Power		2.08 W	1.06 VAR	2.34 VA
Phase Angle (overall)		27°		
Power Factor (overall)		0.891 (lagging)		

Table 13-5. Completed Solution Matrix for the Circuit in Figure 13-67

Practice Problems

1. What is the inductive reactance of the coil shown in Figure 13-69?
2. What is the rms voltage across the inductor in Figure 13-69?
3. How much peak voltage is across the inductor in Figure 13-69?

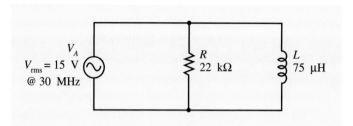

Figure 13-69. A parallel *RL* circuit.

4. The value of resistor voltage in Figure 13-69 is greater than the value of coil voltage. (True or False)
5. The resistor voltage in Figure 13-69 is in phase with the coil voltage. (True or False)
6. How much peak current flows through the resistor in Figure 13-69?
7. What is the rms current through the coil in Figure 13-69?
8. What is the total rms current for the circuit in Figure 13-69?
9. What is the impedance of the circuit shown in Figure 13-69?
10. What is the power factor of the circuit shown in Figure 13-69?
11. Calculate the apparent power for Figure 13-69.
12. Calculate the true power for Figure 13-69.
13. What is the value of reactive power in Figure 13-69?
14. What is the phase angle between voltage and current in Figure 13-69?
15. Calculate the peak value of total current for the circuit in Figure 13-69.
16. Compute the susceptance of the coil in Figure 13-69.
17. What is the admittance of the circuit shown in Figure 13-69?

Answers to Practice Problems

1. 14.13 kΩ	**2.** 15 V	**3.** 21.21 V
4. False	**5.** True	**6.** 964.1 µA
7. 1.06 mA	**8.** 1.26 mA	**9.** 11.89 kΩ
10. 0.541	**11.** 18.9 mVA	**12.** 10.23 mW
13. 15.92 mVAR	**14.** 57.29°	**15.** 1.78 mA
16. 70.77 µS	**17.** 84.1 µS	

Series-Parallel *RL* Circuits with Sinusoidal Currents

For the most part, analysis of series-parallel *RL* circuits consists of applying techniques and equations previously learned. Try to consider this section as an extension of what you already know. Do not think of it as an entirely new subject.

CIRCUIT SIMPLIFICATION

The first step toward analyzing a series-parallel *RL* circuit is to reduce the circuit complexity by combining all sets of series or parallel resistances and inductances. Sets of series resistances or inductances are combined into an equivalent resistance or inductance by adding the individual component values. Sets of parallel resistances or inductances are combined into an equivalent resistance or inductance by using your choice of a parallel resistor/inductor equation.

EXAMPLE SOLUTION

Simplify the *RL* circuit shown in Figure 13-70 by combining sets of series and parallel resistances and inductances.

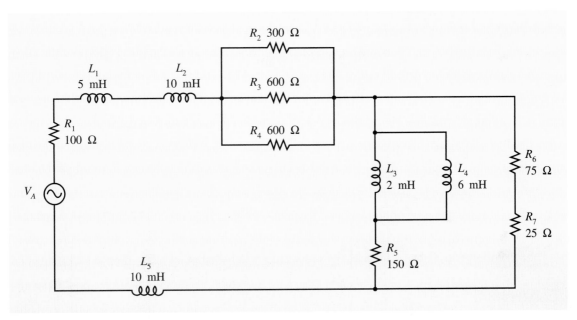

Figure 13-70. Simplify this circuit.

EXAMPLE SOLUTION

We can start by noting that resistors R_2, R_3, and R_4 are in parallel. We can combine them with the reciprocal formula for parallel resistors.

$$R_{2,3,4} = \cfrac{1}{\cfrac{1}{R_2} + \cfrac{1}{R_3} + \cfrac{1}{R_4}}$$

$$= \cfrac{1}{\cfrac{1}{300} + \cfrac{1}{600} + \cfrac{1}{600}} = 150 \ \Omega$$

This result is in series with R_1. We compute this equivalent as follows:

$$R_{1,2,3,4} = R_1 + R_{2,3,4}$$
$$= 100 \ \Omega + 150 \ \Omega = 250 \ \Omega$$

KEY POINTS

For series-parallel circuits, we combine sets of series and parallel components through a series of progressively simpler sketches, until we reach a fully simplified circuit that reflects the total impedance. We then work our way back through the intermediate sketches computing voltage and current values for each component.

In a similar manner, we can combine series resistors R_6 and R_7.

$$R_{6,7} = R_6 + R_7$$
$$= 75\ \Omega + 25\ \Omega = 100\ \Omega$$

There are no more series or parallel combinations of resistors, so let's combine the inductances. First, we note that L_1, L_2, and L_5 are in series. We combine them as follows:

$$L_{1,2,5} = L_1 + L_2 + L_5$$
$$= 5\ \text{mH} + 10\ \text{mH} + 10\ \text{mH} = 25\ \text{mH}$$

Inductors L_3 and L_4 are in parallel. We can use the reciprocal or the product-over-the-sum equation. Let's choose the latter.

$$L_{3,4} = \frac{L_3 L_4}{L_3 + L_4}$$
$$= \frac{2\ \text{mH} \times 6\ \text{mH}}{2\ \text{mH} + 6\ \text{mH}} = 1.5\ \text{mH}$$

There are no more combinations of series or parallel resistances. Neither are there any more combinations of series or parallel inductances. The simplified circuit is shown in Figure 13-71.

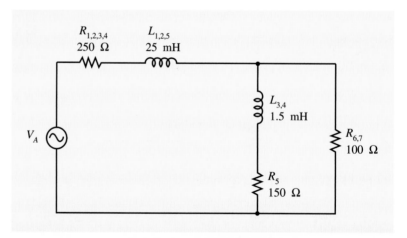

Figure 13-71. The simplified circuit for Figure 13-70.

Practice Problems

1. Simplify the series-parallel *RL* circuit shown in Figure 13-72.
2. Combine sets of series and parallel resistances and inductances in Figure 13-73 to produce a simplified circuit.
3. Simplify the *RL* circuit shown in Figure 13-74.

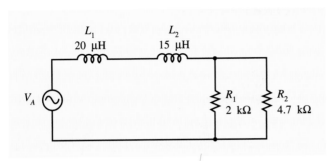

Figure 13-72. Simplify this series-parallel *RL* circuit.

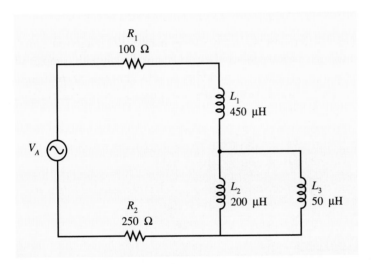

Figure 13-73. Simplify this *RL* circuit.

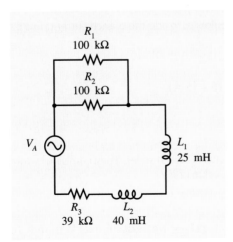

Figure 13-74. Simplify this series-parallel *RL* circuit.

Answers to Practice Problems

1.

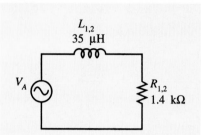

2.

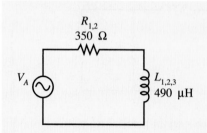

3.

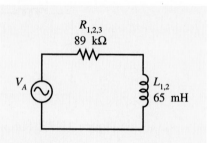

CALCULATIONS WITH POLAR AND RECTANGULAR NOTATION

Although it is possible to thoroughly analyze series-parallel *RL* circuits using methods previously discussed, it is much simpler to perform the calculations directly with complex circuit quantities expressed in polar and rectangular form. Addition and subtraction of complex quantities are most easily accomplished when the quantities are expressed in rectangular notation. The rule for addition and subtraction of complex quantities in rectangular form is stated as follows:

> Combine (add or subtract as required) the resistive terms to obtain the resistive term in the result.
>
> Combine (add or subtract as required) the reactive terms to obtain the reactive term in the result.

EXAMPLE SOLUTION

A parallel *RL* circuit has branch currents of 25 + j15 A (I_1) and 10 + j5 A (I_2). What is the total current?

EXAMPLE **SOLUTION**

Kirchhoff's Current Law requires us to add the branch currents in a parallel circuit to find the total current.

$$I_T = I_1 + I_2$$
$$= (25 + \text{j}15 \text{ A}) + (10 + \text{j}5 \text{ A})$$
$$= 35 + \text{j}20 \text{ A}$$

EXAMPLE SOLUTION

Two complex voltages (V_1 and V_2) are in series. They combine to form V_A. If V_A is 100 + j50 V and V_1 is 20 + j10 V, what is the value of V_2?

EXAMPLE **SOLUTION**

According to Kirchhoff's Voltage Law, we must subtract V_1 from V_A (i.e., $V_2 = V_A - V_1$).

$$V_2 = V_A - V_1$$
$$= (100 + \text{j}50 \text{ V}) - (20 + \text{j}10 \text{ V})$$
$$= 80 + \text{j}40 \text{ V}$$

Multiplication and division of complex circuit quantities is most easily accomplished if the values are expressed in polar notation. The rule for multiplication and division of quantities expressed in polar notation is as follows:

> Multiply or divide (as required) the two magnitudes to obtain the magnitude portion of the result, and
>
> Add the angles to obtain the angle in the result for multiplication,
>
> or
>
> Subtract the divisor (bottom number) angle from the dividend (top number) angle to obtain the angle in the result for division.

EXAMPLE SOLUTION

Use Ohm's Law to compute voltage if the current is 25 mA $\angle 40°$ and the impedance is 35 Ω $\angle 25°$.

EXAMPLE **SOLUTION**

We apply the multiplication rule for polar notation as follows:

$V = IZ$

$\quad = 25 \text{ mA } \angle 40° \times 35 \text{ Ω } \angle 25°$

$\quad = 0.875 \text{ V } \angle 65°$

EXAMPLE SOLUTION

Use Ohm's Law to compute current flow in a circuit, if the voltage is 50 V $\angle 75°$ and the impedance is 10 Ω $\angle 20°$.

EXAMPLE **SOLUTION**

$I = \dfrac{V}{Z}$

$\quad = \dfrac{50 \text{ V } \angle 75°}{10 \text{ Ω } \angle 20°} = 5 \text{ A } \angle 55°$

Practice Problems

1. Use Ohm's Law to find the voltage in a circuit, where the current is 5 mA $\angle 20°$ and the impedance is 2 kΩ $\angle 5°$.
2. Use Ohm's Law to calculate the impedance in a circuit, where the voltage is 75 V $\angle 80°$ and the current is 5 mA $\angle 10°$.
3. How much current flows in a circuit that has a voltage of 50 V $\angle 25°$ and an impedance of 10 Ω $\angle 20°$?

Answers to Practice Problems

1. 10 V $\angle 25°$ 2. 15 kΩ $\angle 70°$ 3. 5 A $\angle 5°$

SERIES-PARALLEL *RL* CIRCUIT COMPUTATIONS

We now have all the tools necessary to fully analyze a series-parallel *RL* circuit. Remember that there are many alternative routes to a solution. Just have confidence in Ohm's and Kirchhoff's Laws, and rely on your knowledge of series-parallel circuits to evolve a solution.

EXAMPLE SOLUTION

Analyze the series-parallel *RL* circuit shown in Figure 13-75, and complete the solution matrix shown in Table 13-6.

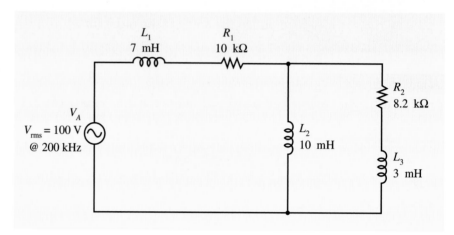

Figure 13-75. A series-parallel *RL* circuit.

CIRCUIT QUANTITY	R_1	R_2	L_1	L_2	L_3	TOTAL
Voltage (rms)						100 V
Current (rms)						
Resistance/Reactance/Impedance	10 kΩ	8.2 kΩ				
Conductance/Susceptance/Admittance						
Power						
Phase Angle (overall)						
Power Factor (overall)						

Table 13-6. A Solution Matrix for the Circuit in Figure 13-75

EXAMPLE **SOLUTION**

The first step is to simplify the circuit by combining sets of series and parallel resistors and inductors. In the case of Figure 13-75, no direct simplifications are possible. For our next step, let's compute the reactance of the three inductors.

$$X_{L_1} = 2\pi f L = 6.28 \times 200 \text{ kHz} \times 7 \text{ mH} = 8.79 \text{ k}\Omega$$

$$X_{L_2} = 2\pi f L = 6.28 \times 200 \text{ kHz} \times 10 \text{ mH} = 12.56 \text{ k}\Omega$$

$$X_{L_3} = 2\pi f L = 6.28 \times 200 \text{ kHz} \times 3 \text{ mH} = 3.77 \text{ k}\Omega$$

Now let's work toward finding the impedance of the circuit. We would take this same general approach if the circuit were a purely resistive circuit. We begin by finding the impedance presented by the parallel branches consisting of L_2 and the series combination of R_2 and L_3. Since there are only two branches, we can use the product-over-the-sum approach. We will use the references shown in Figure 13-76. We shall use polar notation when multiplication and division are required. In cases involving addition and subtraction, we will use rectangular notation. Impedances Z_B and Z_C can be written as

$$Z_B = 0 + j12.56 \text{ k}\Omega$$
$$Z_C = 8.2 \text{ k}\Omega + j3.77 \text{ k}\Omega$$

We can also express each of these in polar form.

$$Z_B = \sqrt{R^2 + X_{L_2}^2}$$
$$= \sqrt{0^2 + (12.56 \text{ k}\Omega)^2} = 12.56 \text{ k}\Omega$$
$$\theta_B = 90° \text{ (purely inductive); therefore}$$
$$Z_B = 12.56 \text{ k}\Omega \angle 90°$$

We find the polar form of Z_C in a similar manner.

$$Z_C = \sqrt{R_2^2 + X_{L_3}^2}$$
$$= \sqrt{(8.2 \text{ k}\Omega)^2 + (3.77 \text{ k}\Omega)^2} = 9.03 \text{ k}\Omega$$
$$\theta_C = \arctan\frac{X_{L_3}}{R_2} = \arctan\frac{3.77 \text{ k}\Omega}{8.2 \text{ k}\Omega}$$
$$= \arctan 0.46 = 24.7°; \text{ therefore}$$
$$Z_C = 9.03 \text{ k}\Omega \angle 24.7°$$

We can now combine Z_B and Z_C using the product-over-the-sum equation.

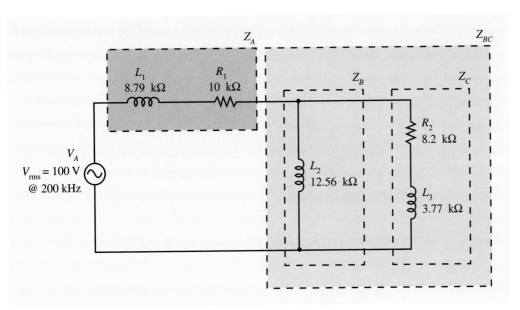

Figure 13-76. Network reference designation for the circuit shown in Figure 13-75.

$$Z_{BC} = \frac{Z_B Z_C}{Z_B + Z_C}$$

$$= \frac{(12.56\ \text{k}\Omega\ \angle 90°)(9.03\ \text{k}\Omega\ \angle 24.7°)}{(0 + j12.56\ \text{k}\Omega) + (8.2\ \text{k}\Omega + j3.77\ \text{k}\Omega)}$$

$$= \frac{(113.4 \times 10^6)\ \angle 114.7°}{8.2\ \text{k}\Omega + j16.33\ \text{k}\Omega}$$

$$= \frac{(113.4 \times 10^6)\ \angle 114.7°}{18.27\ \text{k}\Omega\ \angle 63.34°} = 6.21\ \text{k}\Omega\ \angle 51.36°$$

$$= 3.88\ \text{k}\Omega + j4.85\ \text{k}\Omega$$

Our progress to this point is shown in Figure 13-77.

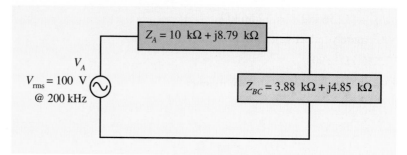

Figure 13-77. Result of combining Z_B and Z_C in Figure 13-76.

We now have a simple series circuit. To find total impedance, we simply add the two series impedances.

$$Z_T = Z_A + Z_{BC}$$

$$= (10\ \text{k}\Omega + j8.79\ \text{k}\Omega) + (3.88\ \text{k}\Omega + j4.85\ \text{k}\Omega)$$

$$= 13.88\ \text{k}\Omega + j13.64\ \text{k}\Omega = 19.46\ \text{k}\Omega\ \angle 44.5°$$

This calculation yields the total impedance and the phase angle of the circuit. Figure 13-78 shows the fully simplified circuit.

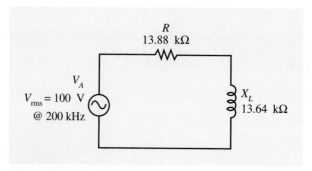

Figure 13-78. An equivalent circuit for the original circuit shown in Figure 13-75.

Just as we would do with a resistive circuit analysis, we now find the total current and work our way back through the partially simplified circuit sketches. We find total current by applying Ohm's Law.

$$I_T = \frac{V_A}{Z_T}$$

$$= \frac{100 \text{ V}}{19.46 \text{ k}\Omega \ \angle 44.5°}$$

$$= 5.14 \text{ mA} \ \angle -44.5°$$

By inspection of the original circuit, we can see that this same current flows through L_1 and R_1. We can use Ohm's Law to compute the voltage drop across these two components.

$$V_{L_1} = I_T X_{L_1}$$

$$= (5.14 \text{ mA} \ \angle -44.5°) \times (8.79 \text{ k}\Omega \ \angle 90°)$$

$$= 45.18 \text{ V} \ \angle 45.5°, \text{ and}$$

$$V_{R_1} = I_T R_1$$

$$= (5.14 \text{ mA} \ \angle -44.5°) \times (10 \text{ k}\Omega \ \angle 0°)$$

$$= 51.4 \text{ V} \ \angle -44.5°$$

The total current divides (Figure 13-76) through branch impedances Z_B and Z_C. We can determine the values of current in each branch with the current divider formula, or by applying Ohm's and Kirchhoff's Laws. Let's choose the latter approach. First, we find the voltage dropped across Z_{BC} with Ohm's Law.

$$V_{BC} = I_T Z_{BC}$$

$$= (5.14 \text{ mA} \ \angle -44.5°) \times (6.21 \text{ k}\Omega \ \angle 51.36°)$$

$$= 31.92 \text{ V} \ \angle 6.86°$$

This voltage will be felt across the parallel combination of L_2 and the R_2, L_3 network. We can find the current through either branch with Ohm's Law.

$$I_B = \frac{V_B}{Z_B}$$

$$= \frac{31.92 \text{ V} \ \angle 6.86°}{12.56 \text{ k}\Omega \ \angle 90°}$$

$$= 2.54 \text{ mA} \ \angle -83.14°$$

Applying Kirchhoff's Current Law will give us the value of current through the Z_C branch.

$$I_C = I_T - I_B$$

$$= (3.67 \text{ mA} - j3.6 \text{ mA}) - (303.4 \ \mu\text{A} - j2.52 \text{ mA})$$

$$= 3.36 \text{ mA} - j1.08 \text{ mA} = 3.53 \text{ mA} \ \angle -17.82°$$

We can see from Figure 13-76, that the I_C current flows through both R_2 and L_3. We can use Ohm's Law to find the voltage drops across these two components.

$$V_{R_2} = I_C R_2$$

$$= (3.53 \text{ mA} \ \angle -17.82°) \times (8.2 \text{ k}\Omega \ \angle 0°)$$

$$= 28.95 \text{ V} \ \angle -17.82°, \text{ and}$$

$$V_{L_3} = I_C X_{L_3}$$

$$= (3.53 \text{ mA} \ \angle -17.82°) \times (3.77 \text{ k}\Omega \ \angle 90°)$$

$$= 13.31 \ \angle 72.18°$$

This completes the voltage and current calculations for the circuit. We can complete the conductance/susceptance/admittance row in Table 13-6 by finding the reciprocal of the corresponding resistance, reactance, or impedance.

$$G_1 = \frac{1}{R_1} = \frac{1}{10 \text{ k}\Omega} = 100 \text{ } \mu\text{S}$$

$$G_2 = \frac{1}{R_2} = \frac{1}{8.2 \text{ k}\Omega} = 122 \text{ } \mu\text{S}$$

$$B_1 = \frac{1}{X_{L_1}} = \frac{1}{8.79 \text{ k}\Omega} = 113.8 \text{ } \mu\text{S}$$

$$B_2 = \frac{1}{X_{L_2}} = \frac{1}{12.56 \text{ k}\Omega} = 79.62 \text{ } \mu\text{S}$$

$$B_3 = \frac{1}{X_{L_3}} = \frac{1}{3.77 \text{ k}\Omega} = 265.3 \text{ } \mu\text{S}$$

$$Y_T = \frac{1}{Z_T} = \frac{1}{19.46 \text{ k}\Omega} = 51.39 \text{ } \mu\text{S}$$

We can find the various power values with any of the power formulas. Let's use the $P = VI$ form.

$$P_{R_1} = V_{R_1}I_T = 51.4 \text{ V} \times 5.14 \text{ mA} = 264.2 \text{ mW}$$

$$P_{R_2} = V_{R_2}I_C = 28.95 \text{ V} \times 3.53 \text{ mA} = 102.2 \text{ mW}$$

$$P_{L_1} = V_{L_1}I_T = 45.18 \text{ V} \times 5.14 \text{ mA} = 232.2 \text{ mVAR}$$

$$P_{L_2} = V_{L_2}I_B = 31.92 \text{ V} \times 2.54 \text{ mA} = 81.08 \text{ mVAR}$$

$$P_{L_3} = V_{L_3}I_C = 13.31 \text{ V} \times 3.53 \text{ mA} = 46.98 \text{ mVAR}$$

$$P_T = V_A I_T = 100 \text{ V} \times 5.14 \text{ mA} = 514 \text{ mVA}$$

KEY POINTS

Series-parallel seems more complicated since it requires us to use polar or rectangular notation to keep track of the various phase angles in the circuit. The basic procedure, however, is consistent with the analysis of resistive series-parallel circuits.

Finally, we can compute the overall circuit power factor by finding the cosine of the overall phase angle.

$$\text{Power factor} = \cos\theta = \cos 44.5° = 0.713(\text{lagging})$$

This completes our analysis of the circuit shown in Figure 13-75. The completed solution matrix is shown in Table 13-7.

Although the calculations may seem rather extensive, it is important to note that they are all based on circuit analysis principles that you have already studied. Confidence will come through practice.

CIRCUIT QUANTITY	R_1	R_2	L_1	L_2	L_3	TOTAL
Voltage (rms)	51.4 V	28.95 V	45.18 V	31.92 V	13.31 V	100 V
Current (rms)	5.14 mA	3.53 mA	5.14 mA	2.54 mA	3.53 mA	5.14 mA
Resistance/Reactance/ Impedance	10 kΩ	8.2 kΩ	8.79 kΩ	12.56 kΩ	3.77 kΩ	19.46 kΩ
Conductance/Susceptance/Admittance	100 μS	122 μS	113.8 μS	79.62 μS	265.3 μS	51.39 μS
Power	264.2 mW	102.2 mW	232.2 mVAR	81.08 mVAR	46.98 mVAR	514 mVA
Phase Angle (overall)	44.5°					
Power Factor (overall)	0.713 (lagging)					

Table 13-7. A Completed Solution Matrix for the Circuit in Figure 13-75

1. Analyze the *RL* circuit shown in Figure 13-79 and complete the solution matrix shown in Table 13-8.

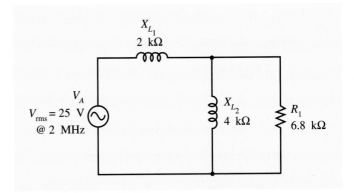

Figure 13-79. Analyze this *RL* circuit.

CIRCUIT QUANTITY	R_1	L_1	L_2	TOTAL
Voltage (rms)				25 V
Current (rms)				
Resistance/Reactance/Impedance	6.8 kΩ	2 kΩ	4 kΩ	
Conductance/Susceptance/Admittance				
Power				
Phase Angle (overall)				
Power Factor (overall)				

Table 13-8. A Solution Matrix for the Circuit in Figure 13-79

2. Analyze the *RL* circuit shown in Figure 13-80 and complete the solution matrix shown in Table 13-9.

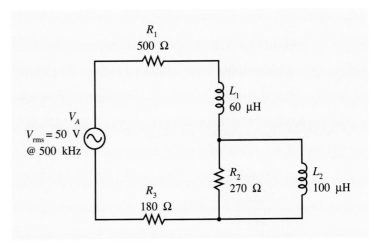

Figure 13-80. Analyze this *RL* circuit.

CIRCUIT QUANTITY	R_1	R_2	R_3	L_1	L_2	TOTAL
Voltage (rms)						50 V
Current (rms)						
Resistance/Reactance/Impedance	500 Ω	270 Ω	180 Ω			
Conductance/Susceptance/Admittance						
Power						
Phase Angle (overall)						
Power Factor (overall)						

Table 13-9. A Solution Matrix for the Circuit in Figure 13-80

Answers to Practice Problems

1.

CIRCUIT QUANTITY	R_1	L_1	L_2	TOTAL
Voltage (rms)	16.36 V	9.49 V	16.36 V	25 V
Current (rms)	2.4 mA	4.74 mA	4.1 mA	4.74 mA
Resistance/Reactance/Impedance	6.8 kΩ	2 kΩ	4 kΩ	5.27 kΩ
Conductance/Susceptance/Admittance	147.1 µS	500 µS	250 µS	189.8 µS
Power	39.3 mW	45 mVAR	67.1 mVAR	118.5 mVA
Phase Angle (overall)	70.63°			
Power Factor (overall)	0.332 (lagging)			

2.

CIRCUIT QUANTITY	R_1	R_2	R_3	L_1	L_2	TOTAL
Voltage (rms)	27.95 V	11.43 V	10.1 V	10.5 V	11.43 V	50 V
Current (rms)	55.9 mA	42.3 mA	55.9 mA	55.9 mA	36.4 mA	55.9 mA
Resistance/Reactance/Impedance	500 Ω	270 Ω	180 Ω	188.4 Ω	314 Ω	895.1 Ω
Conductance/Susceptance/Admittance	2 mS	3.7 mS	5.6 mS	5.3 mS	3.2 mS	1.1 mS
Power	1.56 W	484 mW	565 mW	587 mVAR	416 mVAR	2.79 VA
Phase Angle (overall)	21.1°					
Power Factor (overall)	0.933 (lagging)					

Complex *RL* Circuit Computations

Although it will not be formally presented here, it is important for you to realize that analysis of complex *RL* circuits can be accomplished using the same network theorems presented for analysis of resistive circuits. That is, Thevenin's Theorem, Norton's Theorem, tee-to-pi conversions, mesh analysis, and so on can be applied to the solution of complex *RL* circuits. The only significant difference is that all calculations must account for the presence of a nonzero phase angle. That is, your circuit quantities must be expressed in polar or rectangular notation.

KEY POINTS

Even complex *RL* networks can be analyzed in a manner similar to complex resistive networks, provided the circuit quantities are expressed in polar or rectangular notation.

Exercise Problems 13.4

Refer to Figure 13-81 for Problems 1–10.

1. What is the reactance of L_1?
2. What is the impedance of the circuit?
3. What is the admittance of the circuit?
4. How much current flows through R_1?
5. What is the voltage drop across L_1?
6. What is the power dissipated by R_2?
7. What is the phase angle between total voltage and total current in the circuit?
8. What is the power factor of the circuit?
9. How much current flows through L_1?
10. What is the apparent power in the circuit?

Refer to Figure 13-82 for Problems 11–20.

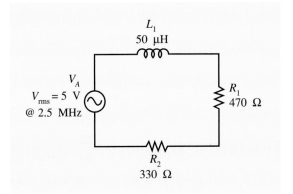

Figure 13-81. A series *RL* circuit.

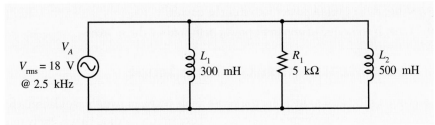

Figure 13-82. A parallel *RL* circuit.

11. What is the voltage across L_1?
12. What is the voltage across L_2?
13. How much current flows through R_1?
14. How much current flows through L_1?
15. What is the total current flow in the circuit?
16. What is the total impedance of the circuit?
17. What is the susceptance of L_2?

18. How much true power is dissipated in the circuit?

19. What is the power factor of the circuit?

20. What is the phase angle between total current and total voltage in the circuit? Refer to Figure 13-83 for Problems 21–30.

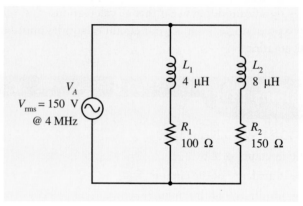

Figure 13-83. A series-parallel *RL* circuit.

21. What is the series branch impedance of L_1 and R_1?

22. What is the series branch impedance of L_2 and R_2?

23. What is the total impedance in the circuit?

24. How much total current flows in the circuit?

25. What is the total admittance in the circuit?

26. What is the voltage drop across R_1?

27. What is the voltage drop across L_2?

28. What is the phase angle between total voltage and total current in the circuit?

29. What is the overall power factor of the circuit?

30. What is the value of apparent power delivered by the source?

13.5 Phase Measurements with a Scope

KEY POINTS

Technicians must be able to measure phase relationships with an oscilloscope.

A technician must be able to measure the phase angle between two sinusoidal waveforms. There are many ways to accomplish this. We will examine a method that is easily understood and produces excellent results.

Most oscilloscopes in use today have at least two independent vertical channels (many have more). Phase measurements with a dual-channel scope consist of the following steps:

1. Connect the two waveforms to be measured to the vertical input jacks of the oscilloscope.

2. Connect the ground jack of the scope—or, preferably, the ground clip on one of the probes—to the circuit common.

3. Adjust the vertical deflection (i.e., vertical gain) of each channel such that the waveforms fill the whole screen. The zero-reference of the sine waves should be in the center of the screen. Note that the vertical channels do not have to remain calibrated for phase measurements.

4.   Adjust the sweep rate and horizontal position controls until one half-cycle of the waveforms spans nine divisions on the scope's graticule.

Figure 13-84(a) shows a simple *RL* circuit. Figure 13-84(b) shows a scope screen after the preceding sequence has been accomplished. Since one half-cycle covers nine divisions, this means that each major division on the scope graticule represents 20°. Further, since each major division consists of five smaller divisions, each smaller graticule mark represents 20°/5 or 4°.

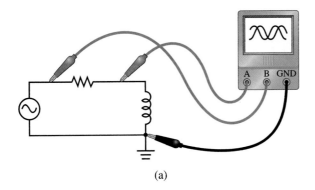

(a)

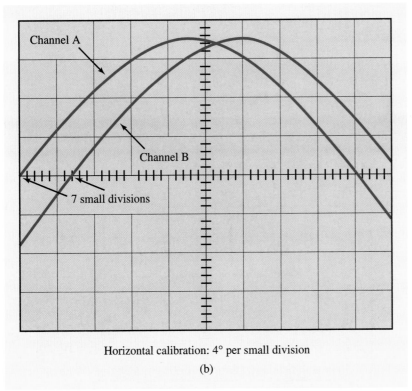

Horizontal calibration: 4° per small division

(b)

Figure 13-84. Phase measurements with a dual-channel oscilloscope.

The phase difference between the two waveforms is determined by simply counting the number of small divisions between corresponding points on the two waveforms. It is most convenient and accurate to measure between two points on the center graticule line of the scope. For the case shown in Figure 13-84, the two waveforms are seven small divisions apart or

$$\text{Phase angle} = \theta = \text{number of divisions} \times \text{degrees per division}$$

$$= 7 \text{ divisions} \times \frac{4^\circ}{\text{division}} = 28^\circ$$

There are other combinations of vertical and horizontal settings that can be used similarly. For example, the horizontal controls can be adjusted so that one full cycle spans six major divisions. In this case, each major division represents 360°/6 or 60°. Each minor division now represents 60°/5 or 12°. In general, greater reading accuracy can be achieved when the waveform is expanded as much as possible. In the case of the horizontal axis, nine major divisions for one half-cycle works well for most oscilloscopes. The vertical channel may be set for any desired gain as long as the zero reference remains in the center of the screen. For optimum reading accuracy, the vertical gain should be as high as practical. Even if the waveform extends beyond the vertical limits of the screen, accurate measurements in the center of the screen can be made. The large vertical deflection makes it easier to identify the exact point of intersection on the zero-reference line.

Practice Problems

1. Determine the phase angle between the waveforms shown in Figure 13-85.

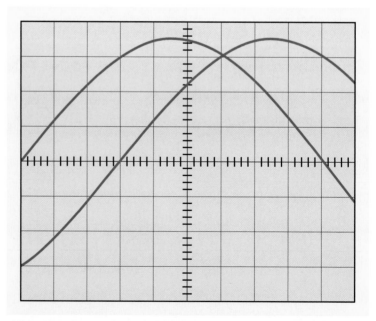

Figure 13-85. Determine the phase angle between the two sine waves.

2. What phase angle is being displayed in Figure 13-86?

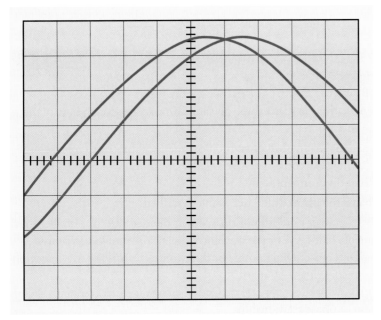

Figure 13-86. Measure the phase angle being displayed.

3. Measure the phase angle between the two waveforms in Figure 13-87.

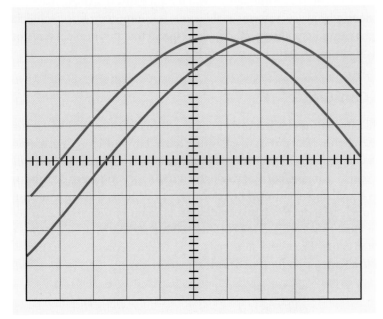

Figure 13-87. Determine the phase angle being displayed.

Answers to Practice Problems

1. 60°   **2.** 24°   **3.** 26°

Exercise Problems 13.5

1. If one half-cycle spans three divisions on a scope screen, how many degrees does each division represent?

2. If each major division of an oscilloscope is calibrated to represent 30° and each major division consists of five smaller divisions, how many degrees does each smaller division represent?

3. What is the phase difference being described?

   a. Each major division of the scope is divided into five smaller divisions.

   b. One half-cycle of the waveforms spans nine major divisions.

   c. The zero-crossing points of the two waveforms are two major plus two smaller divisions apart.

4. Explain why larger vertical deflections make it easier to obtain accurate interpretation of phase relationships.

13.6 Pulse Response of *RL* Circuits

KEY POINTS

Current cannot change abruptly in an inductor. If the voltage across an inductor is changed suddenly, then the current will also change, but more gradually.

We have thoroughly discussed the analysis of *RL* circuits that have sinusoidal inputs. Nonsinusoidal input voltages cause a radically different circuit response and require substantially different analytical methods. Although we will not present a thorough analysis of *RL* circuit response to nonsinusoidal waveforms at this time, it is important for a technician to understand *L/R* time constants and how *RL* circuits respond to nonsinusoidal waveforms. We will concentrate on pulse or rectangular waveforms.

L/R Time Constant

KEY POINTS

It takes five time constants (5τ) for the current to make its full change. During the first time constant, the current will make about 63% of its total change.

Figure 13-88 shows a simple circuit that will produce nonsinusoidal waveforms (pulses) for a series *RL* circuit. When the switch is first moved to position A, current will try to increase from zero to some nonzero value (limited by the value of *R*). You will recall from Chapter 12 that a changing current through an inductor produces a self-induced voltage that opposes the changing current (Lenz's Law). The result is that current cannot build up instantly through an inductor. It takes a definite amount of time for the current to increase. The total time required for current buildup is divided into five time periods called time constants. The Greek letter tau (τ) is used to represent one time constant. In the case of *RL* circuits, τ is computed with Equation 13-14.

$$\tau = \frac{L}{R} \qquad (13\text{-}14)$$

where τ is measured in seconds, *L* in henrys, and *R* in ohms.

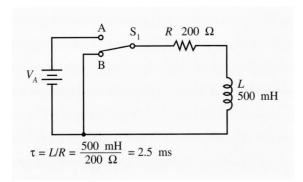

$$\tau = L/R = \frac{500\ \text{mH}}{200\ \Omega} = 2.5\ \text{ms}$$

Figure 13-88. An *RL* circuit with a pulse input.

Figure 13-89(a) shows how the current builds up through the series circuit. At *t* = 0 (the first instant), the self-induced voltage in the coil prevents any current flow. At time *t* = τ, the inductor current has risen to 63.2% of its final value. During each additional time period of τ seconds, the current builds to 63.2% of the remaining distance. For example, at the end of the first time interval (τ) the current has another 36.8% to go. During the time between τ and 2τ, it will increase 63.2% of this 36.8% or 23.3% for a total of 86.5% at time *t* = 2τ.

Once 5τ seconds have elapsed, the current will have climbed to more than 99% of its final value. In theory, it will continue increasing forever, but in practice, we consider it as having reached its final value in time period 5τ. The final value of current can be computed with Ohm's Law. It is determined by the value of dc input voltage and the value of the series resistance.

When the switch is returned to position B in Figure 13-88, the current will decay toward zero. But, again, the inductor will oppose the changing current. In this case, the voltage induced by the collapsing magnetic field of the inductor acts as an energy source to keep current flowing. As the field decays, so does the inductor current. Figure 13-89(b) shows that the decaying current also takes five time constants (τ) to reach its final value (zero).

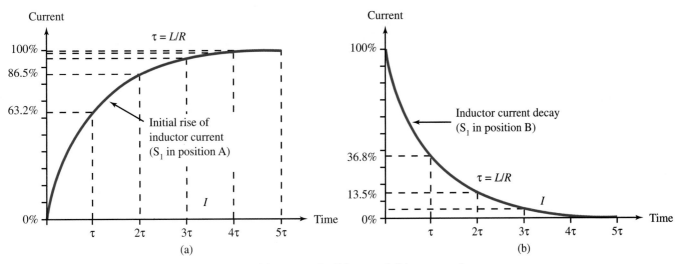

Figure 13-89. An *L/R* time constant curve: (a) current buildup, and (b) current decay.

Circuit Waveforms

Since current changes require a specific amount of time (based on the values of L and R), then a given current change may not be possible in the time allowed. That is, if the input voltage is removed before the current has had time to build to maximum, then it will begin decaying from some lower value. Figure 13-90 illustrates the response of an RL circuit to a pulse of three different frequencies.

In Figure 13-90(b), each polarity of the input pulse is present for a period of time equal to one time constant. Thus, current does not have sufficient time to reach maximum before the input voltage is removed. Similarly, it is not given time to decay to zero between pulses.

Figure 13-90(c) illustrates a short time constant. Here, the input pulse duration is present for an amount of time equal to ten time constants. Clearly, the current (and therefore resistor voltage) is allowed to reach maximum and to decay to zero. The inductor voltage appears as short-duration voltage spikes. Notice that the inductor waveform has a dual

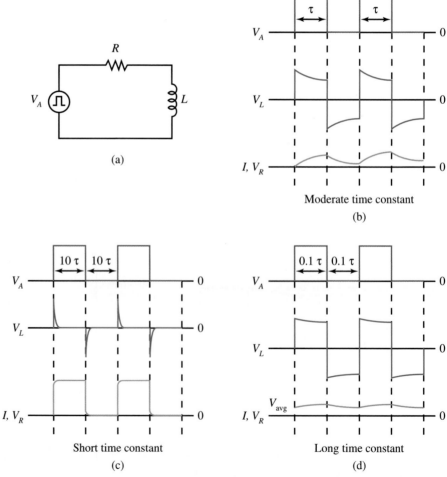

Figure 13-90. (a) An RL circuit and the effect of (b) moderate, (c) short, and (d) long time constants on RL circuit response.

polarity. The expanding and collapsing magnetic field causes opposite polarities of induced voltage in the inductor. The waveform across the inductor is called a differentiated waveform. The entire circuit is often called a **differentiator** circuit. To qualify as a differentiator, the circuit must have a short time constant and the output must be taken across the inductor.

Finally, Figure 13-90(d) illustrates the effects of a long time constant, where the input pulse duration is one-tenth of one time constant. Here, the current is barely allowed to increase or decrease. That is, the long time constant produces a nearly steady current flow. This effect is useful for filtering circuits in electronic power supplies. If the output is taken across the resistor, we call the circuit an **integrator** circuit. An integrator circuit requires a long time constant.

Magnitude of Self-Induced Voltage

We know from Chapter 12 that the magnitude of induced voltage in an inductor is determined by the rate of change of current. This can be expressed formally as Equation 13-15.

$$V_L = L\frac{di}{dt} = L\frac{\Delta i}{\Delta t} \qquad (13\text{-}15)$$

The di or Δi stands for change in current and dt or Δt stands for change in time. Inductance (L) is measured in henrys.

EXAMPLE SOLUTION

If the current through a 10-mH inductor changes steadily by 2 A over a period of 500 ms, what is the value of self-induced voltage?

EXAMPLE SOLUTION

We apply Equation 13-15 as follows:

$$V_L = L\frac{di}{dt}$$

$$= 10 \text{ mH}\left(\frac{2 \text{ A}}{500 \text{ ms}}\right) = 40 \text{ mV}$$

EXAMPLE SOLUTION

At a particular instant, the current through a 100-µH inductor is changing at the rate of 5 A/ms. What is the value of induced voltage?

EXAMPLE SOLUTION

$$V_L = L\frac{di}{dt}$$

$$= 100 \text{ µH}\left(\frac{5 \text{ A}}{1.0 \text{ ms}}\right) = 500 \text{ mV}$$

KEY POINTS

If the output is taken across the inductor in an *RL* circuit with a short time constant, the circuit is called a differentiator circuit.

KEY POINTS

If the *RL* circuit has a long time constant and the output is taken across the resistor, then the circuit is called an integrator circuit.

KEY POINTS

The value of induced voltage in an inductor is directly proportional to the inductance. It is also directly proportional to the rate of change of current.

Exercise Problems 13.6

1. What is the L/R time constant for a series circuit consisting of a 250-mH coil and a 1.5-kΩ resistor?

2. If the current through an inductor tries to change from 0 to 5 A, what will be the value of current after 1 time constant?

3. If the current through an inductor tries to decrease from 400 mA to 0, what will be the value of current after 2 time constants?

4. When current tries to change in a series RL circuit consisting of a 10-mH coil and a 10-kΩ resistor, how long does it take for the current to reach its final value?

5. If the current through a 175-μH inductor changes at the rate of 10 mA/ns, what is the value of induced voltage?

13.7 Applied Technology: *RL* Circuits

There are numerous applications for RL circuits. Most applications are an integral part of a more extensive circuit (e.g., a palm-top computer). Let's examine the basic operation of two common RL circuit applications: power supply filters and frequency selective circuits.

Power Supply Filter Circuits

Figure 13-91 illustrates the basic operation of a filter circuit for an electronic power supply. Most electronic circuits require dc voltage for operation, but it is often desirable to operate the system from the standard 120 Vac power line. An electronic power supply circuit is used to convert the 120-Vac, 60-Hz power distributed by the power company to a pulsating dc voltage. This pulsating or surging voltage can then be applied to an RL circuit as shown in Figure 13-91. Since the inductor opposes any changes in current, the current is held fairly constant. If the current through the inductor remains constant, then

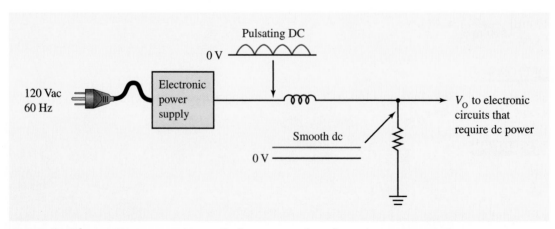

Figure 13-91. An *RL* circuit can smooth the output of an electronic power supply.

the voltage across the resistor and the voltage to the electronic circuitry will be held constant. The result is a smooth dc voltage similar to that which would be produced by a battery.

Frequency Selective Circuits

Electronic circuits often need to distinguish between waveforms that differ in frequency. Figure 13-92 shows how an *RL* circuit can accept a mixture of 1-kHz and 100-kHz sine waves at its input but produce separate outputs for the two differing frequencies.

Let's consider what happens to a 100-kHz waveform as it passes through the circuit. First, we see that the series combination of R_1, L_1 is in parallel with the series combination of R_2, L_2. Both *RL* circuits will receive the 100-kHz waveform. In the case of the R_1, L_1 branch, nearly all of the voltage will be dropped across L_1, since its reactance is ten times greater than the resistance of R_1. So, if we measure the voltage across L_1, we will get nearly 100% of the full amplitude of the 100-kHz voltage.

When the 100-kHz waveform is applied to the R_2, L_2 network, nearly all of the voltage is dropped across L_2 for reasons previously listed. However, in this case, the output signal is taken across R_2. Therefore, R_2 will have only a very small percentage of the original 100-kHz input voltage.

If we consider the circuit effects on the 1-kHz voltage input, we get an opposite result. The reactance of L_1 and L_2 is only one-tenth the value of R_1 and R_2 at a frequency of 1 kHz. This means that R_1 will drop nearly all of the voltage at 1 kHz, leaving very little to be measured across L_1. Similarly, L_2 drops only a very small percentage of the 1-kHz voltage, which means that most of it is developed across R_2. Thus, the 1-kHz input voltage appears at the lower output, but it is prevented from passing through to the upper output. This general principle is used for tone controls and crossover networks in audio equipment.

KEY POINTS

RL circuits can also be used to distinguish between two or more signals on the basis of their frequencies.

KEY POINTS

High frequencies can be allowed to pass or can be rejected depending on the circuit configuration. Similarly, lower frequencies can be either passed or rejected.

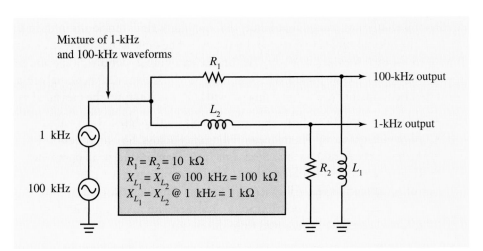

Figure 13-92. A frequency selective *RL* filter circuit.

1. In your own words, explain how a pulsating current can be smoothed with an *RL* circuit.
2. Refer to Figure 13-92. If a 100-Hz sine wave were applied to the circuit, describe the circuit behavior.
3. Describe the response of the circuit in Figure 13-92, if a 2-MHz sine wave were connected to the input.

13.8 Troubleshooting *RL* Circuits

It is generally not difficult to troubleshoot *RL* circuits. They are essentially combinations of the purely resistive and purely inductive circuits previously studied. The following represent the most probable defects in an *RL* circuit:

*   open coil
*   open resistor
*   coil winding-to-core short
*   shorted turns on the coil
*   shorted resistor (usually solder bridge on a printed circuit board)
*   resistor value out of tolerance

All of these defects for series, parallel, or more complex *RL* networks can be located using the same troubleshooting logic presented for purely resistive circuits coupled with the coil verification checks presented in Chapter 12.

Following are several troubleshooting problems using the Pshooter method. Recall that each Pshooter circuit has an associated table that lists the various testpoints in the circuit along with the normal value of voltage and resistance. The table also provides a bracketed ([ ]) number for each measurable entry. The bracketed number identifies a specific entry in the Pshooter lookup table provided in Appendix A, which gives you the actual measured value of the circuit quantity. The following list will help you interpret the measured values:

*   All testpoint (TPxx) measurements are made with reference to ground.
*   All nontestpoint values are measured directly across the component.
*   All resistance measurements reflect the approximate reading of an ohmmeter.
*   Resistance values for all testpoints (TPxx) are measured with the applied voltage disconnected. All other components remain connected.
*   Resistance tests for all non-testpoint measurements are made with the specified component removed from the circuit.

1. Refer to Figure 13-93 and the Pshooter chart shown in Table 13-10. Locate the defective component in as few measurements as possible by applying a logical, systematic troubleshooting method.

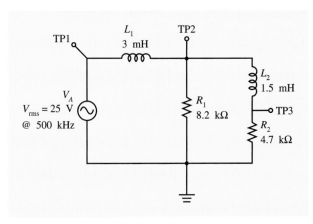

Figure 13-93. Troubleshoot this circuit

TEST-POINT	VOLTAGE (rms)		RESISTANCE	
	NORMAL	ACTUAL	NORMAL	ACTUAL
V_A	25 V	[78]	—	—
R_1	8.51 V	[200]	8.2 kΩ	[164]
R_2	6.02 V	[72]	4.7 kΩ	[12]
TP1	25 V	[163]	2.99 kΩ	[196]
TP2	8.51 V	[41]	2.99 kΩ	[185]
TP3	6.02 V	[137]	2.99 kΩ	[42]
L_1	20.19 V	[70]	1.2 Ω	[36]
L_2	6.02 V	[105]	0.7 Ω	[94]

Table 13-10. PShooter Data for Figure 13-93

2. Use the PShooter data in Table 13-11 to troubleshoot the circuit shown in Figure 13-93.

TEST-POINT	VOLTAGE (rms)		RESISTANCE	
	NORMAL	ACTUAL	NORMAL	ACTUAL
V_A	25 V	[154]	—	—
R_1	8.51 V	[152]	8.2 kΩ	[143]
R_2	6.02 V	[209]	4.7 kΩ	[36]
TP1	25 V	[190]	2.99 kΩ	[159]
TP2	8.51 V	[69]	2.99 kΩ	[96]
TP3	6.02 V	[85]	2.99 kΩ	[101]
L_1	20.19 V	[216]	1.2 Ω	[213]
L_2	6.02 V	[144]	0.7 Ω	[203]

Table 13-11. PShooter Data for Figure 13-93

3. Use the PShooter data in Table 13-12 to troubleshoot the circuit shown in Figure 13-93.

TEST-POINT	VOLTAGE (rms)		RESISTANCE	
	NORMAL	ACTUAL	NORMAL	ACTUAL
V_A	25 V	[21]	—	—
R_1	8.51 V	[178]	8.2 kΩ	[17]
R_2	6.02 V	[20]	4.7 kΩ	[100]
TP1	25 V	[180]	2.99 kΩ	[171]
TP2	8.51 V	[37]	2.99 kΩ	[23]
TP3	6.02 V	[168]	2.99 kΩ	[108]
L_1	20.19 V	[129]	1.2 Ω	[10]
L_2	6.02 V	[27]	0.7 Ω	[61]

Table 13-12. PShooter Data for Figure 13-93

Chapter Summary

- Series, parallel, series-parallel, and complex RL circuits share many basic characteristics with purely resistive circuits. The current is the same for all components in a series circuit. The voltage is the same for all components in a parallel circuit. Ohm's and Kirchhoff's Laws may be applied to RL circuits as long as we account for the effects of out-of-phase quantities.

- Total opposition to current flow in an RL circuit is called impedance (Z) and is measured in ohms. The reciprocal of impedance is called admittance (Y) and is measured in siemens.

- Sets of series or parallel inductors may be combined to simplify the circuit. Simplification equations are similar to those used for resistive circuits, as long as there is no magnetic linkage between the coils.

- Summing of quantities that are 90° out of phase requires a process called phasor addition. The result of phasor addition is larger than either of the two original numbers but smaller than their arithmetic sum.

- There are three power measurements that are important in an RL circuit. True power is dissipated by the resistance in the circuit and measured in watts. Reactive power is acquired by the inductance in the circuit and measured in volt-amperes-reactive (VAR). The product of total voltage and total current (also the phasor sum of true and reactive power) is called apparent power and is measured in volt-amperes (VA). The relative magnitudes of reactive and true power determine the power factor of the circuit. Power factor is equal to the cosine of the phase angle between true power and apparent power.

- Out-of-phase circuit quantities (e.g., current, voltage, impedance, and power) can be represented in either of two ways: polar or rectangular notation. Polar notation specifies the magnitude of the phasor and its angle (e.g., V_3 = 150 V $\angle 35°$). Rectangular notation describes a phasor in terms of its real (horizontal axis) and reactive (vertical axis) components (e.g., Z_T = 25 + j75 Ω). Conversion from one form to the other can be done with basic right-triangle relationships. Engineering calculators make the conversion process very easy.

- Series and parallel *RL* circuit analysis can be accomplished using the same basic techniques applicable to resistive circuits provided phase relationships are included. When circuit quantities that are 90° out of phase are summed (e.g., the voltage drops around a series circuit or the branch currents in a parallel circuit), then they must be summed with phasor addition.

- Analysis techniques used with resistive circuits are also applicable to series-parallel (or even complex) *RL* circuits, but the out-of-phase circuit quantities must be represented with polar or rectangular notation. Addition and subtraction is generally easier when the numbers are expressed in rectangular form. Polar form makes multiplication and division simpler.

- When a pulse waveform is applied to an *RL* circuit, it takes a definite amount of time for the current to increase through the coil. One time constant (τ) is equal to the inductance divided by the resistance ($\tau = L/R$). The current can increase (or decrease) 63.2% of the total change in one time constant. For practical purposes, it takes five time constants for the current to stabilize after a pulse input.

Review Questions

Section 13.1: Series *RL* Circuits with Sinusoidal Currents

1. The current is the same through all components in a series *RL* circuit. (True or False)

2. In a series *RL* circuit, the highest component voltage drop will be across the component with the (*highest, lowest*) resistance or reactance.

3. If a 10-kΩ resistor is in series with a coil having an inductive reactance of 27 kΩ, which will drop the most voltage?

4. What is the phase relationship between voltage and current through the resistance of a series *RL* circuit?

5. What is the phase relationship between voltage and current through the inductance of a series *RL* circuit?

6. The current waveform through the resistance and the current waveform through the inductance in a series *RL* circuit are in phase. (True or False)

7. The voltage waveform across the resistance and the voltage waveform across the inductance in a series *RL* circuit are in phase. (True or False)

8. Total opposition to current flow in a series *RL* circuit is called _____ .

9. Susceptance is measured in _____ .

10. What is the unit of measurement for admittance?

11. Impedance of a series *RL* circuit is measured in _____ .

12. Draw an equivalent circuit for the circuit shown in Figure 13-94.

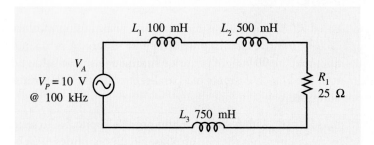

Figure 13-94. Draw an equivalent circuit.

13. A series *RL* circuit consists of the following: a 10-kΩ resistor, 2.5-kΩ resistor, 5-mH coil, 2-mH coil, and an ac source. Draw the simplified circuit.

14. What circuit quantity is used as the reference phasor in a series *RL* circuit?

15. Draw an impedance phasor diagram to represent the circuit shown in Figure 13-95. Include the value of impedance.

16. Draw a voltage phasor diagram to represent the circuit shown in Figure 13-95. Include the value of applied voltage.

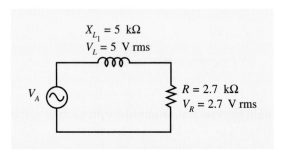

Figure 13-95. A series *RL* circuit.

17. True power is a measure of the power dissipated in the (*resistance, inductance*) of an *RL* circuit.

18. Reactive power is measured in _____ .

19. In any given *RL* circuit, which of the following—true power, reactive power, or apparent power—is always the largest? Why?

20. What is the susceptance of a coil that has an inductive reactance of 2.9 kΩ?

Section 13.2: Parallel *RL* Circuits with Sinusoidal Currents

21. In a parallel *RL* circuit, larger resistances or reactances have larger voltage drops. (True or False)

22. In a parallel *RL* circuit, smaller resistances or reactances have larger currents. (True or False)

23. The voltage and current through the resistance in a parallel *RL* circuit are in phase. (True or False)

24. The voltage and current through the inductance in a parallel *RL* circuit are in phase. (True or False)

25. If a two-branch parallel *RL* circuit has currents of $I_R = 1.5$ A and $I_L = 2.5$ A, what is the total current in the circuit?

26. If the total current in a two-branch parallel circuit is 500 mA and the resistive current is 250 mA, what is the value of inductive current?

27. How much resistive current must flow in a parallel *RL* circuit to produce 500 μA when combined with 350 μA of inductive current?

28. Draw the simplified circuit for the *RL* circuit shown in Figure 13-96.

29. What is the peak-to-peak voltage across L_2 in Figure 13-96?

30. How much peak current flows through R_1 in Figure 13-96?

31. What is the rms current through R_1 in Figure 13-96?

32. Calculate the total rms current in Figure 13-96.

33. What is the total impedance of the circuit shown in Figure 13-96?

34. Draw a current phasor diagram to represent the circuit shown in Figure 13-96.

35. Draw a power phasor diagram to represent the circuit shown in Figure 13-96.

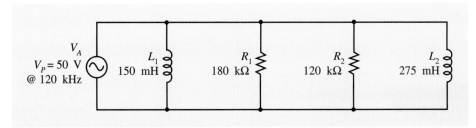

Figure 13-96. A parallel *RL* circuit.

Section 13.3: Representing Circuit Quantities

36. Write the rectangular notation to describe each of the points shown in Figure 13-97.

37. Write the rectangular notation for each of the current phasors shown in Figure 13-98.

38. Write the rectangular notation for each of the voltage phasors shown in Figure 13-99.

39. Convert the series impedance phasor 100 Ω ∠20° to rectangular notation.

40. Convert the series impedance phasor 2,700 Ω ∠45° to rectangular notation.

41. If the total current phasor for a parallel circuit were described as 150 mA ∠75°, how would the current be expressed in rectangular notation?

42. If the total voltage in a series *RL* circuit is expressed as 125 + j55 V, write this voltage using polar notation.

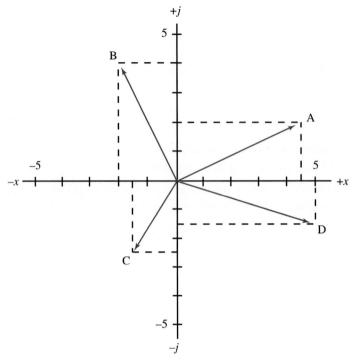

Figure 13-97. Describe the indicated points.

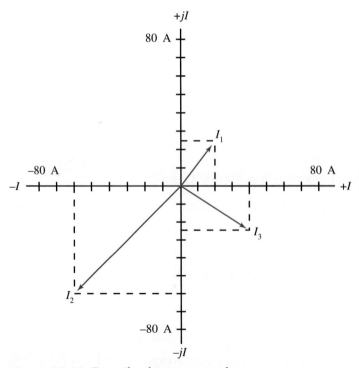

Figure 13-98. Describe these current phasors.

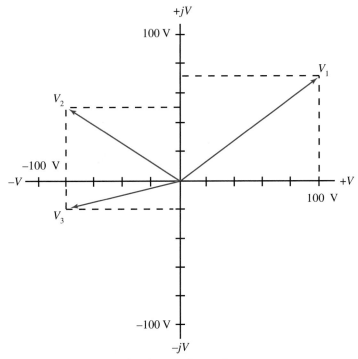

Figure 13-99. Describe these voltage phasors.

43. The total current in a parallel circuit is described as 27 + j25 mA. Express this current in polar notation.

44. What is the polar form of a voltage that is described as 250 + j100 V?

Section 13.4: *RL* Circuit Calculations

Refer to Figure 13-100 for Questions 45–55.

45. What is the value of inductive reactance?

46. Compute the total circuit impedance.

47. How much current flows through the coil?

48. How much current flows through the resistor?

49. What is the voltage drop across the coil?

50. How much voltage is dropped across the resistor?

51. How much power is dissipated by the resistor?

52. What is the reactive power in the circuit?

53. Calculate the apparent power in the circuit.

54. What is the phase angle between total voltage and total current in the circuit?

55. What is the power factor of the circuit?

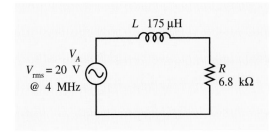

Figure 13-100. A series *RL* circuit for analysis.

Refer to Figure 13-101 for Questions 56–65.

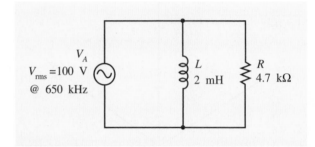

Figure 13-101. A parallel circuit for analysis.

56. What is the value of inductive reactance?

57. What is the voltage across the coil?

58. What is the voltage drop across the resistor?

59. How much current flows through the coil?

60. What is the total current in the circuit?

61. What is the total impedance of the circuit?

62. Compute the total admittance for the circuit.

63. What is the value of true power in the circuit?

64. What is the phase angle between total voltage and total current in the circuit?

65. What is the power factor for the circuit?

Refer to Figure 13-102 for Questions 66–75.

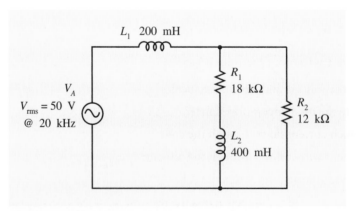

Figure 13-102. A series-parallel circuit for analysis.

66. What is the impedance of the series combination of R_1 and L_2?

67. What is the combined impedance of R_1, R_2, and L_2?

68. What is the total impedance of the circuit?

69. How much current flows through L_1?

70. What is the voltage drop across L_1?

71. What is the voltage drop across R_2?

72. How much current flows through R_2?

73. What is the true power in the circuit?

74. What is the phase angle between total voltage and total current in the circuit?

75. Calculate the circuit's power factor.

Section 13.5: Phase Measurements with a Scope

76. When measuring the phase angle between two sine waves with a dual-trace oscillo-scope, why is it helpful to adjust the controls until the waveforms fill the major part of the screen?

77. What phase relationship is being displayed in Figure 13-103?

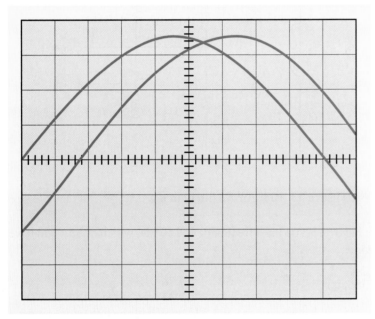

Figure 13-103. Measure the phase angle.

Section 13.6: Pulse Response of *RL* Circuits

78. A time constant is measured in _____ .

79. The Greek letter _____ is used to represent a time constant.

80. What is the time constant of a series *RL* circuit consisting of a 470-Ω resistor and a 375-mH coil?

81. Current in a series *RL* circuit can rise to _____ percent of its final value in one time constant when it is driven by a pulse voltage.

82. It takes _____ time constants for current in a series *RL* circuit to reach its approximate final value when driven by a voltage pulse.

83. If a pulse is applied to a series *RL* circuit consisting of a 200-Ω resistor and a 750-μH inductance, how long does it take for the current to reach its maximum value?

84. If the current through a 250-mH inductor changes at the rate 2.75 A/s, how much voltage is developed across the coil?

85. How much voltage is developed across a 120-mH coil if the current through it changes at the rate of 600 mA/μs?

Section 13.7: Applied Technology: *RL* Circuits

86. If V_A in Figure 13-104 has a frequency of 100 kHz, which output will have the greatest amplitude?

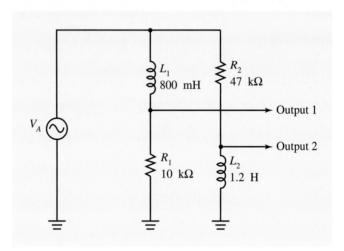

Figure 13-104. An *RL* filter circuit.

87. If V_A in Figure 13-104 has a frequency of 150 Hz, which output will have the greatest amplitude?

Section 13.8: Troubleshooting *RL* Circuits

88. If a 10-mH coil has a resistance of 2.5 Ω, which of the following is most likely correct?
    a. open coil   b. shorted coil   c. normal coil

89. If the coil in a series *RL* circuit develops an open, describe the voltage that will be measured across the coil.

90. If the resistor in a series *RL* circuit develops an open, describe the voltage that will be measured across the coil.

91. If the coil in a series *RL* circuit is found to be open, which of the following may have caused the problem originally?
    a. series resistor open   b. series resistor shorted   c. power supply voltage too low

It is important for a technician to be able to apply previously learned knowledge and skills to the solution of unfamiliar problems. Figure 13-105 gives you an opportunity to apply your circuit analysis skills to a complex *RL* circuit.

We have not formally discussed the solution of complex *RL* networks, but we have discussed all of the individual tools needed to analyze the problem. Apply what you have learned in this and previous chapters and meet the technician challenge!

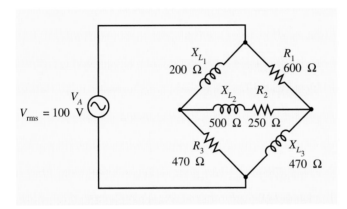

Figure 13-105. Analyze this complex *RL* circuit.

Here's the challenge. Analyze the circuit shown in Figure 13-105 to determine the following values:

- total current
- phase angle between total current and total voltage
- voltage across X_{L_2}
- current through R_1
- total impedance of the circuit

Equation List

(13-1) $B = \dfrac{1}{X_L}$

(13-2) $Y = \dfrac{1}{Z}$

(13-3) $V_T = V_A = \sqrt{V_R{}^2 + V_L{}^2}$

(13-4) $Z = \sqrt{R^2 + X_L{}^2}$

(13-5) $P_A = \sqrt{P_T{}^2 + P_R{}^2}$

(13-6) $I_T = \sqrt{I_R{}^2 + I_L{}^2}$

(13-7) $V_L = V_A \sin\theta$

(13-8) $V_R = V_A \cos\theta$

(13-9) $Z \angle\theta = Z\cos\theta + jZ\sin\theta$

(13-10) $I_T = \dfrac{V_T}{Z}$

(13-11) $V_L = I_T X_L$

(13-12)   Power factor $= pf = \cos\theta$

(13-13) $I_T = \sqrt{I_R{}^2 + I_L{}^2}$

(13-14) $\tau = \dfrac{L}{R}$

(13-15) $V_L = L\dfrac{di}{dt} = L\dfrac{\Delta i}{\Delta t}$

key terms

capacitive reactance
dielectric
dielectric constant
dissipation factor
ESL
ESR

leakage current
permittivity
power factor
quality factor
temperature coefficient

objectives

After completing this chapter, you should be able to:

1. List the basic requirements of a capacitor.

2. List the factors that affect capacitance.

3. Describe the current flow in an ac or dc capacitive circuit.

4. Name at least five types of capacitors and describe their unique features.

5. Interpret the value of a capacitor from the manufacturer's markings.

6. Calculate the total capacitance of multiple capacitors connected in a series, parallel, series-parallel, or complex network.

7. Test a capacitor to verify its condition or identify a defect.

8. Identify the following capacitor lead styles: axial, radial, SMT, integrated package.

Capacitance and Capacitive Reactance

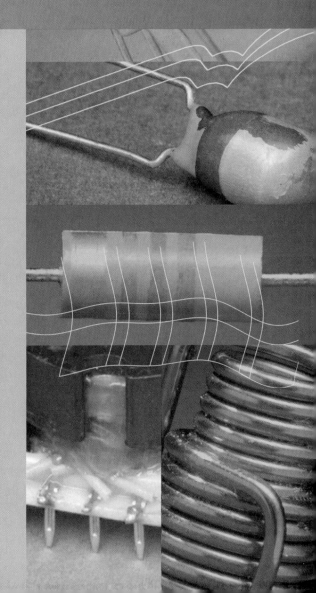

A capacitor is another fundamental electronic component like resistors and inductors. Certain characteristics of the capacitor are similar to characteristics of resistors or inductors. Other characteristics are unique to capacitors. Since capacitors are used in nearly every nontrivial electronic system, you will want to learn all you can about how they work.

14.1 Capacitance Fundamentals

Capacitance can be defined as the ability to store electrical energy in an electrostatic field. This is somewhat comparable to inductance, which is the ability to store electrical energy in an electromagnetic field. A capacitor is simply a physical device designed to have a certain amount of capacitance.

KEY POINTS

Capacitors are essentially two conductors physically and electrically separated by an insulator or dielectric.

Figure 14-1(a) shows the essential parts of a simple parallel-plate capacitor. It consists of two conductors—generally called the plates of the capacitor—separated by an insulator. The insulator material is called the **dielectric**. Figure 14-1(b) shows the generic schematic symbol for a capacitor. We will look at other, more specific, symbols in a later section.

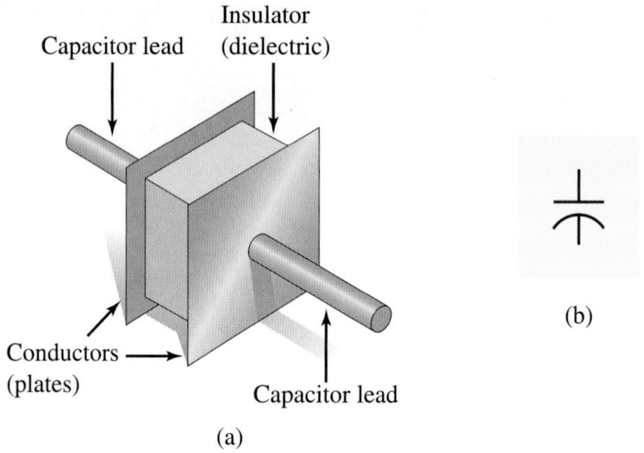

Figure 14-1. (a) Basic construction of a capacitor and (b) the generic symbol for a capacitor.

Charges and Electric Fields

Figure 14-2 shows a capacitor connected through a switch to a battery and a series resistor. In Figure 14-2(a), the switch is open and no current is flowing in the circuit. The capacitor has 0 V across it, as you might anticipate.

In Figure 14-2(b), the switch has been closed. The potential on the negative battery terminal causes electrons to leave the battery and move to the lower plate of the capacitor. Since the dielectric of the capacitor is an insulator, the electrons cannot continue through the capacitor. However, a similar process is occurring on the upper capacitor plate. Here, the positive potential of the battery terminal attracts electrons from the upper capacitor plate. As electrons leave the upper plate and travel to the battery terminal, the upper plate of the capacitor is left with a deficiency of electrons (i.e., a positive charge).

KEY POINTS

A capacitor has the ability to store energy in an electrostatic field within its dielectric.

You will recall from Chapter 2, that an electric field exists between charged bodies. In the case of the capacitor in Figure 14-2(b), there is an electrostatic field set up within the dielectric that extends between the positive and negative plates of the capacitor. We refer to the accumulation of charges on the capacitor plates as "charging the capacitor." A voltmeter connected across the capacitor (as shown in Figure 14-2) will increase as the capacitor is charged.

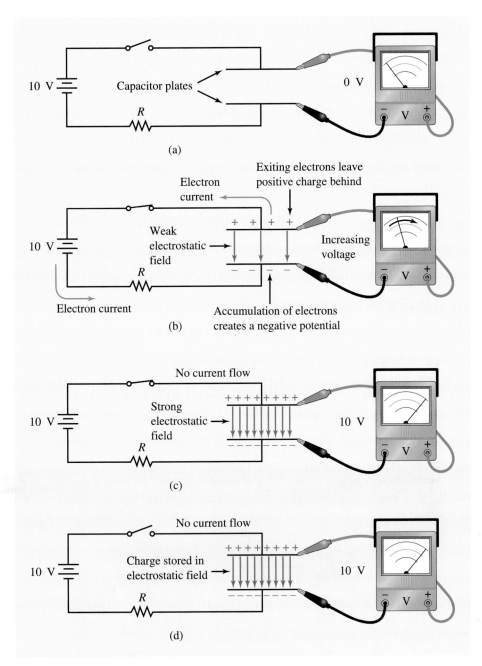

Figure 14-2. Charging a capacitor.

If you closely examine the polarity of the charge that is accumulating on the capacitor plates in Figure 14-2(b), you will see that the increasing capacitor voltage is opposing the battery voltage. Each time an electron moves to the lower capacitor plate and/or leaves the upper plate, the charge on the capacitor and the opposition to the battery voltage increases. The increased opposition results in reduced current flow.

KEY POINTS

When a capacitor is first connected to a dc source, a charging current flows through the external circuit. The charging current decays as the capacitor accumulates a charge. The charging current stops when the capacitor voltage reaches the value of supply voltage.

Eventually, enough electrons will have moved around the circuit to produce a capacitor charge that is equal (but opposite) to the battery voltage. This condition is pictured in Figure 14-2(c). Since the battery and capacitor voltages are equal and opposite, there will be no current flow in the circuit. This is a stable condition and will remain as long as the battery voltage is available.

In Figure 14-2(d), the switch has been opened. The electrons that have accumulated on the lower plate of the capacitor cannot move around the circuit to neutralize the positive charge on the upper plate. The charge is trapped on the capacitor. In theory, the voltage (charge) will stay on the capacitor indefinitely. In practice, it will eventually leak off primarily due to imperfections in the dielectric.

Figure 14-3 shows the results of connecting a charged capacitor across a circuit. In Figure 14-3(a), the switch is open, so the charge remains on the capacitor. No current flows and there is no voltage dropped across the resistor.

In Figure 14-3(b), the switch has closed, providing a path for the accumulated electrons on the lower plate to move around the circuit to the positive charge on the upper plate. As the electrons move around the circuit, they represent current flow through the resistor. The electron movement (amount of current) is maximum when the switch is first closed. As each electron makes its trip around the circuit, both positive and negative plates become more neutral. As the plates lose their charge, there is less potential available to cause current flow. The voltage across the capacitor and across the resistor will decay as the capacitor discharges. The time required to fully discharge the capacitor depends on several circuit variables, but it can range from fractions of a picosecond to literally months.

Unit of Measurement for Capacitance

You may recall from Chapter 1 that the unit of measurement for capacitance (C) is the farad (F). When each plate of a capacitor has a one-coulomb charge (one positive and one negative) and the resulting capacitor voltage is one volt, we say the capacitance is one farad. We can express this formally with Equation 14-1.

$$C = \frac{Q}{V} \qquad\qquad (14\text{-}1)$$

where C is expressed in farads, Q is the charge in coulombs, and V is the voltage across the capacitor. This is a relationship that is worth remembering, since it is used as the heart of the derivation of many other circuit analysis equations.

EXAMPLE SOLUTION

If 250 µC of charge produce a 6-V potential across a capacitor, what is the value of the capacitor?

EXAMPLE **SOLUTION**

We apply Equation 14-1 as follows:

$$C = \frac{Q}{V}$$

$$= \frac{250 \ \mu C}{6 \ V} = 41.67 \ \mu F$$

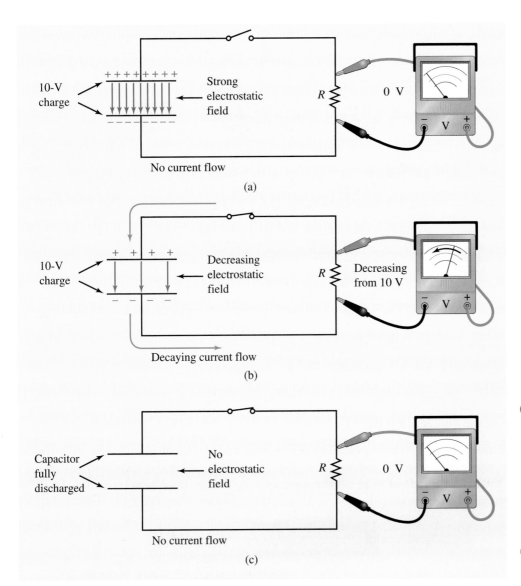

Figure 14-3. Discharging a capacitor.

Values for practical capacitors are generally in the subfarad range. Values of 10 pF, 0.01 μF, and 10,000 μF are representative values for typical capacitors.

Factors Affecting Capacitance

It is the physical characteristics of a capacitor that determine its capacitance. There are three primary factors to consider: area of the plates, distance between the plates, and the type of material used for the dielectric. Figure 14-4 illustrates these factors.

PLATE AREA

The greater the plate area, the more charge the capacitor can hold. If all other factors remain constant and we double the area of the plates, then we can store twice as much

KEY POINTS

The farad is too large a measurement for most purposes; practical capacitors are generally sized in the 5-pF to 30,000-μF range.

KEY POINTS

The primary factors that affect the value of a capacitor are its physical characteristics:

- area of its plates
- the distance between its plates
- the type of material used for the dielectric.

KEY POINTS

Capacitance is directly proportional to the area of the plates and the permittivity of the dielectric.

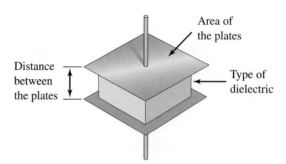

Figure 14-4. The physical characteristics of a capacitor determine its value.

charge for a given voltage. The value of capacitance is directly proportional to the area of the plates.

PLATE SEPARATION

By contrast, if all other factors remain the same and we double the distance between the plates of a capacitor, then we will only have half as much electric field intensity for a given voltage. The value of capacitance is inversely proportional to the distance between the plates.

DIELECTRIC MATERIAL

Permittivity is a measure of a material's ability to concentrate an electrostatic field. This is similar to permeability with reference to magnetic fields. Relative permittivity (ϵ_r) of a material is the ratio of its absolute permittivity (ϵ_m) to the absolute permittivity of a vacuum (ϵ_0). That is,

$$\epsilon_r = \frac{\epsilon_m}{\epsilon_0} \qquad (14\text{-}2)$$

The relative permittivity of a dielectric is generally called its **dielectric constant** (k). The value of capacitance is directly proportional to the permittivity of the dielectric material. Table 14-1 lists the dielectric constants for several materials used as insulators for capacitors.

Since capacitance is directly proportional to the area of the plates (A) and the permittivity of the dielectric (ϵ_m), but is inversely proportional to the distance between the plates (d), we may write the following equation:

$$C = \frac{\epsilon_m A}{d}$$

But from Equation 14-2, we can write

$$\epsilon_r = \frac{\epsilon_m}{\epsilon_0} = \text{dielectric constant} = k, \text{ or}$$

$$\epsilon_m = k\epsilon_0$$

TYPE OF MATERIAL	APPROXIMATE DIELECTRIC CONSTANT
Air	1.0
Aluminum oxide	8
Ceramic	>1,000
Glass	7.0
Mica	6.0
Mylar	3.0
Polystyrene	2.5
Tantalum oxide	24
Teflon®	2.0
Waxed paper	3.0

Table 14-1. Dielectric Materials Used in Capacitors and Their Approximate Dielectric Constants

Substituting this result into the prior equation gives us

$$C = \frac{k\epsilon_0 A}{d}$$

The absolute permittivity of a vacuum (ϵ_0) is equal to 8.85×10^{-12}. Replacing ϵ_0 in the prior equation with this constant gives us an equation for capacitance in terms of its physical characteristics.

$$C = \frac{8.85 \times 10^{-12} kA}{d}$$

Rearranging this equation gives us Equation 14-3, which is a form that technicians generally prefer.

$$C = \frac{8.85 kA}{10^{12} d} \qquad (14\text{-}3)$$

where C is in farads, d is in meters, and A is in square meters. Table 14-2 summarizes these same relationships.

FACTOR	PROPORTIONALITY	EXAMPLE
Plate area	Direct	Area ↑ Capacitance ↑
Distance between plates	Inverse	Distance ↑ Capacitance ↓
Dielectric constant (k)	Direct	k ↑ Capacitance ↑

Table 14-2. Relationship of Physical Factors Affecting Capacitance

EXAMPLE SOLUTION

Calculate the capacitance of a parallel-plate capacitor with a plate area of 0.05 m², a plate separation of 0.01 m, and a mylar dielectric.

EXAMPLE **SOLUTION**

From Table 14-1, we know the dielectric constant of mylar is 3.0. Substituting this into **Equation 14-3** gives us the following results.

$$C = \frac{8.85kA}{10^{12}d}$$

$$= \frac{8.85 \times 3 \times 0.05}{10^{12} \times 0.01} = 132.8 \text{ pF}$$

Multiplate Capacitors

Manufacturers often use multiple sets of capacitor plates to obtain higher values of capacitance. Figure 14-5 illustrates multiplate construction.

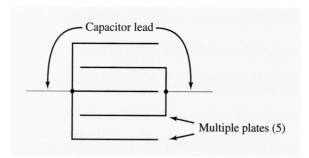

Figure 14-5. Multiple sets of plates produce higher values of capacitance.

The total capacitance for a multiplate capacitor is computed in basically the same way as for a two-plate capacitor, but we need to include the effects of the added plates. Equation 14-4 includes this added factor.

$$C = \frac{8.85kA}{10^{12}d}(n-1) \qquad (14\text{-}4)$$

where n is the total number of conductive plates in the capacitor.

EXAMPLE SOLUTION

Compute the value of the capacitor illustrated in Figure 14-6.

EXAMPLE **SOLUTION**

We apply Equation 14-4 as follows:

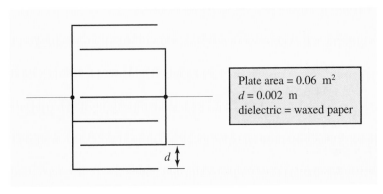

Figure 14-6. Compute the value of this capacitor.

$$C = \frac{8.85 kA}{10^{12} d}(n-1)$$

$$= \frac{8.85 \times 3 \times 0.06}{10^{12} \times 0.002}(7-1)$$

$$= 0.0048 \ \mu F$$

Note that technically, the answer could be expressed as 4.8 nF, but it is industry practice to express capacitance in terms of picofarad (pF) or microfarad (μF) units.

Current Flow in a Capacitive Circuit

It is interesting that a capacitor has an internal insulator (open circuit), and yet current can flow through the external circuit. It is important for you to be able to visualize this action. The discussion relevant to Figure 14-2 illustrated current flow in a dc capacitive circuit. In the circuit external to the capacitor, there will be current flow (i.e., movement of electrons) as long as the capacitor is either charging or discharging. Once the capacitor has fully charged or fully discharged, there is no current in the circuit. The length of time that the current flows is directly proportional to the size of the capacitor. The greater the capacitance, the longer it takes to charge or discharge it through a given circuit.

One interesting application of capacitance—although not intentional—is illustrated in Figure 14-7. You already know that an open in a series circuit will drop the entire applied voltage. An open circuit is essentially a capacitor, since it is two conductors (the wires on either side of the open) separated by an insulator (the air between the two conductors). As you might expect, the value of capacitance is very small, so it charges almost instantly to the applied voltage. Nevertheless, it is a capacitor, and now you have yet another way to view the effects of an open circuit.

Figure 14-8 illustrates current flow in an ac circuit. In this case the source voltage is changing continuously—in both polarity and amplitude—so the capacitor is charging or discharging continuously. This means there will always be current in the external circuit. It appears as if the current were actually flowing through the capacitor, but it is important to remember that the dielectric is an insulator, so no current can literally flow through the capacitor. The value of current in a capacitive circuit with an alternating source is directly proportional to the value of the capacitor. Greater capacitance corresponds to increased current flow.

KEY POINTS

When a capacitor is connected to an ac source, the applied voltage is continuously changing polarities and amplitudes. This causes the capacitor to be in a continuous state of charge or discharge. This means there will be current through the external circuit as long as the ac source is connected.

KEY POINTS

No current flows through the capacitor with either dc or ac supplies because the dielectric is an insulator.

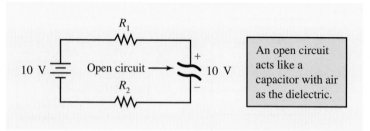

Figure 14-7. An open circuit forms an accidental capacitance that charges to the value of the applied voltage.

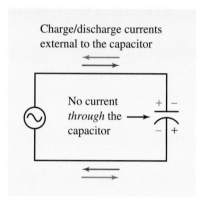

Figure 14-8. Current flow in an ac circuit.

Exercise Problems 14.1

1. Capacitors store energy in a(n) _____ field.
2. A capacitor is basically two _____ separated by a(n) _____.
3. The dielectric in a capacitor is a good conductor. (True or False)
4. In a dc circuit, electrons flow through the capacitor. (True or False)
5. In an ac circuit, electrons flow through the capacitor. (True or False)
6. In a capacitive circuit with a dc source voltage, electrons flow continuously through the external circuit. (True or False)
7. Capacitance is measured in _____.
8. What is the value of a capacitor that has 25 V across it, when it has a charge of 6.5 mC?
9. Name the three primary factors that affect the value of a capacitor.
10. If the dielectric of a capacitor were changed from air to glass with no other physical changes, what would happen to the value of capacitance?
11. Calculate the capacitance of a parallel-plate capacitor with a plate area of 0.04 m^2, a plate separation of 0.003 m, and a mylar dielectric.
12. Calculate the capacitance of a parallel-plate capacitor with a total of nine plates, a plate area of 0.035 m^2, a plate separation of 0.0015 m, and a mica dielectric.

14.2 Capacitor Construction

There are many different kinds of capacitors. They vary dramatically in size, shape, value, reliability, stability, and so on. Nevertheless, they all share the same fundamental characteristic: two conductors separated by an insulator.

Capacitor Types

It is important for a technician to be able to identify the major capacitor types by their physical appearance. Additionally, you should know any outstanding advantages or disadvantages of a particular technology. This latter subject will be discussed in a later section.

There are many characteristics that can be used to classify capacitors. Some are physical and some are electrical. Electrical factors include such things as capacitance, operating voltage, temperature performance, frequency of operation, and either fixed or variable capacitance. Physical factors include size, weight, and lead arrangement.

Lead Styles

Although there are many subtle variations, there are four major arrangements for lead placement. Figure 14-9 shows capacitors with the leads protruding from opposite ends of a cylindrical body. This type of lead pattern is called axial leads.

Figure 14-10 shows radial-lead capacitors. A capacitor with radial leads can be identified by both leads protruding from the same end of the body. The body may be cylindrical, square, flat, or another shape.

Surface-mount technology (SMT) is rapidly becoming the standard for many products (e.g., computers, consumer electronics, and satellite communications). Figure 14-11 shows some typical surface-mount capacitors. These are very small and delicate. Products that utilize surface-mount components are generally assembled and soldered by automatic machines or robots. However, component-level repair and replacement is generally the responsibility of an electronics technician.

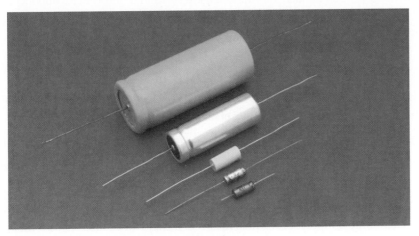

Figure 14-9. Capacitors with axial leads.

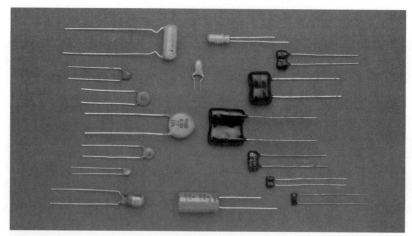

Figure 14-10. Capacitors with radial leads.

Figure 14-11. Surface-mount capacitors. The head of a common straight pin is shown for a size comparison.

Figure 14-12 shows some integrated capacitors. This type of device provides several capacitors (and sometimes other components) within a single package. The individual components may be isolated from each other or they may be connected as a network. There are many package styles and interconnect options.

FIXED AND VARIABLE CAPACITORS

All of the capacitors pictured in Figures 14-9 through 14-12 are fixed-value capacitors. Their capacitance value is determined at the time of manufacture and cannot be altered.

Figure 14-13 shows some variable capacitors. The basic range of adjustment is established during manufacture of the capacitor, but the exact value can be adjusted in the field by a technician.

POLARIZED AND NONPOLARIZED CAPACITORS

Most capacitors are nonpolarized, which means that they can be inserted into the circuit in either direction. Some types of capacitors, however, are polarized. Polarized capacitors

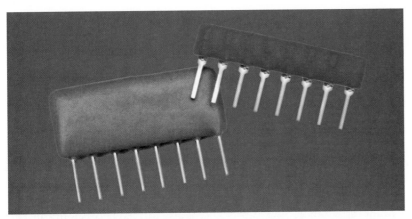

Figure 14-12. Integrated capacitors have multiple capacitors inside a single, multipin package.

Figure 14-13. The capacitance value of variable capacitors can be adjusted.

always have markings to indicate the polarity of the capacitor. One common marking method is to put a plus (+) sign near the positive terminal. Alternatively, some manufacturers place a series of minus signs (–) along one side of the capacitor with an arrow pointing to the negative terminal. When the technician inserts a polarized capacitor into a circuit, it is very important that the positive side of the capacitor be connected to the more positive terminal in the circuit. If a polarized capacitor is inserted backward, then it is very likely to explode violently. At a minimum, you will be startled when the capacitor explodes with the loudness of a firecracker. Worse, you may receive eye damage or skin cuts from flying debris or have chemicals splattered on your face. You should not be afraid of polarized capacitors, but it certainly makes sense to be very certain that they are connected properly before applying power. Since a polarized capacitor must always have the correct voltage polarity across it, you should never connect a polarized capacitor to an ac source.

Accidental Capacitors

Not all capacitance in a practical circuit is put there in the form of a physical capacitor. Recall the basic description of a capacitor is two conductors separated by an insulator. This definition extends to include all sets of conductors that are separated with a nonconductive material. Figure 14-14 illustrates an accidental or stray (also called parasitic) capacitance formed between two insulated wires connected between two circuits. The wires (two conductors) are separated by an insulator (air and wire insulation), which satisfies the

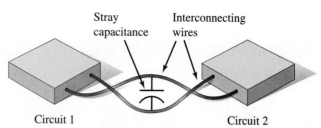

Figure 14-14. Stray capacitance is formed between any two conductors that are separated by a nonconducting material.

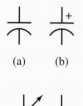

Figure 14-15.
Schematic symbols for (a) a generic capacitor, (b) a polarized capacitor, and (c) and (d) variable capacitors.

requirement for capacitance. Stray capacitance is also formed between nearby copper traces on a printed circuit board.

For many applications, stray capacitance can be safely ignored because it is usually quite small (a few picofarads). However, in high-frequency circuits, stray capacitance can interfere with the desired operation of the circuit. In these cases, technicians use very short wires, coaxial cable, and well-planned cable routing to minimize the effects of stray capacitance.

SCHEMATIC SYMBOLS

There are several schematic symbols for capacitors that a technician must be able to interpret. Figure 14-15(a) shows the general symbol for capacitance. This symbol is used unless a more specific symbol is needed. Figure 14-15(b) shows the schematic representation for a polarized capacitor. The plus (+) sign, of course, indicates the side of the capacitor that is connected to the more positive potential. Figure 14-15(c) shows the schematic symbol for a variable capacitor. The symbol shown in Figure 14-15(d) also represents a variable capacitor, but generally implies a very limited range of adjustment.

All of the schematic symbols shown in Figure 14-15 have one straight and one curved line to represent the capacitor. These are the most widely used symbols. A capacitor symbol is sometimes drawn with two straight lines. This is less common, but it is accepted practice in some industries.

Capacitor Ratings and Marking

Capacitors have numerous ratings, but the five that are most important to a technician are capacitance value, voltage rating, temperature coefficient, tolerance, and leakage.

CAPACITANCE VALUE

This rating needs minimal explanation, since the very purpose of the capacitor is to provide a certain amount of capacitance. Capacitance values are nearly always specified in either picofarads or microfarads.

In older style capacitors, the values were often indicated by colored dots or bands similar to a resistor color code. Several different color codes were used on capacitors, but most are now obsolete. Values for today's capacitors are generally indicated by printing directly on the capacitor. The value is expressed is picofarads or microfarads. Figure 14-16 shows how the popular picofarad code is interpreted.

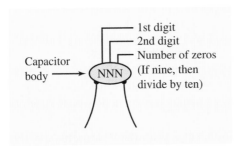

Figure 14-16. Interpretation of the picofarad code for capacitors.

EXAMPLE SOLUTION

A ceramic capacitor has a value code of 102. What is its value?

EXAMPLE **SOLUTION**

The code is interpreted as shown in Figure 14-16 as 10 plus 2 more zeros, or 1,000 pF.

Practice Problems

1. Determine the value of capacitance for each of the following capacitor markings:
   a. 103   **b.** 225   **c.** 105

Answers to Practice Problems

1. **a.** $0.01\ \mu F$ **2.** **b.** $2.2\ \mu F$ **3.** **c.** $1.0\ \mu F$

An alternative to the picofarad code is to simply print the value directly on the capacitor. For example, a 100-pF capacitor might be marked with 100. Similarly, a 0.05-μF capacitor might be marked as .05. Since both of these are simply numbers with no indication of the units (picofarad or microfarad), interpretation can be confusing at first. With practice, however, a technician can readily determine the correct value. Correct interpretation relies on the technician's knowledge of physical capacitor sizes. A 0.05-μF capacitor, for example, might be roughly the same diameter as your index finger. A 0.05-pF capacitor, by contrast, would be so small you could hardly see it. Similarly, a 1,000-pF capacitor might be about 1/8 inch across, whereas a 1,000-μF capacitor might be 1 to 3 inches long. Thus, with a little experience, there is rarely any confusion. Additionally, by the time the capacitance value is 10 μF or so, the capacitors are usually physically large enough to include a μ, μF, mfd, or even MFD to indicate microfarads.

Surface-mount capacitors generally have no printed markings to indicate value. Here, a technician must measure the capacitance, or, in most cases, simply replace a suspected component.

VOLTAGE RATING

Every capacitor has a voltage rating. This specifies the peak ac voltage or the maximum dc voltage that can be applied to the capacitor. If this voltage is exceeded, the dielectric can break down and allow current to flow through the capacitor. This is a destructive process for most types of capacitors. Figure 14-17 illustrates voltage breakdown in a capacitor.

Figure 14-17(a) illustrates a capacitor with no charge. Notice the undistorted orbits of the atoms making up the dielectric. Figure 14-17(b) shows a capacitor that has been charged to a voltage less than the breakdown voltage. The charge on the positive plate attracts the orbiting electrons, while the charge on the negative plate repels the electrons. Since the atoms are relatively immobile and the electrons are tightly held in their orbits, the orbits are distorted. The electrons strain toward the positive plate as a direct result of the electrostatic field that passes through the dielectric.

Figure 14-17(c) shows what happens if the voltage across the capacitor exceeds the breakdown voltage rating. Here, the electrostatic field has become so intense that it has actually ripped some of the electrons out of their orbits. Once an electron is free, it moves quickly toward the positive plate. On the way to the plate, the electron may collide with other particles and free even more electrons. Electrons moving toward the positive plate constitute current flow. Because the dielectric has a relatively high voltage across it, a substantial amount of power (heat) is produced (i.e., $P = VI$). This releases even more electrons and contributes to the breakdown of the dielectric. In an instant, a current path has been established between the plates of the capacitor. The applied voltage must be interrupted to stop the current flow.

⬤ KEY POINTS

If the voltage rating on a capacitor is exceeded, then the dielectric may be punctured. As soon as the dielectric breaks down, current through the dielectric may be quite high. Breakdown is generally, but not always, a destructive process.

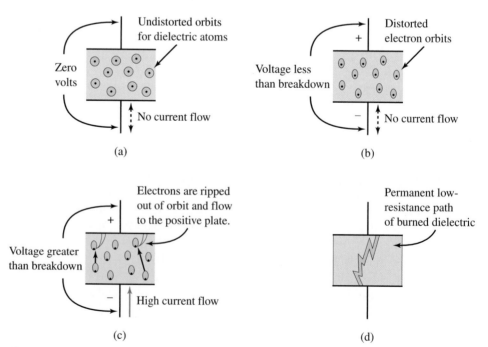

Figure 14-17. If the voltage across a capacitor exceeds its voltage rating, then the dielectric will puncture or break down.

Once the applied voltage has been removed, the dielectric may or may not "heal" itself. In most cases, the high current flow and intense, localized heat causes a conductive path of burned material to form between the plates. The capacitor is shorted or, at least, resistive. This is illustrated in Figure 14-17(d). Certain types of capacitors (e.g., oil dielectric) are self-healing. Once the excessive voltage has been removed, the dielectric regains its high resistance properties.

The value of voltage breakdown is determined by the dielectric strength of the dielectric and the spacing between the plates of the capacitor. Dielectric strength is specified as volts per meter. Do not confuse dielectric strength with dielectric constant (previously discussed). Table 14-3 lists the approximate dielectric strengths of some materials commonly used as capacitor dielectrics.

TYPE OF MATERIAL	APPROXIMATE DIELECTRIC STRENGTH (VOLTS/METER)
Air	3×10^6
Ceramic	39×10^6
Glass	50×10^6
Mica	200×10^6
Polystyrene	24×10^6
Tantalum oxide	4×10^6
Teflon®	60×10^6
Waxed paper	47×10^6

Table 14-3. Dielectric Materials Used in Capacitors and Their Approximate Dielectric Strengths

TEMPERATURE COEFFICIENT

The **temperature coefficient** of a capacitor describes how the value of capacitance varies with changes in temperature. The temperature coefficient specification consists of two parts: polarity and magnitude.

The polarity of the temperature coefficient tells whether the capacitance increases or decreases with increasing temperature. The capacitance of a capacitor with a positive temperature coefficient increases as temperature increases. The capacitance of a capacitor with a negative temperature coefficient decreases as the temperature increases. Some capacitors are designed to have a zero temperature coefficient, which means their capacitance value is relatively unaffected by changes in temperature.

The magnitude portion of the temperature coefficient specification tells how much the capacitance value changes for a given change in temperature. It is specified in parts per million per degree Centigrade (ppm/°C).

KEY POINTS

The temperature coefficient of a capacitor describes how the capacitance tracks with temperature. It can be positive, negative, or zero and is specified as parts per million per degree Centigrade (ppm/°C).

The complete temperature coefficient is always specified in the manufacturer's specification sheets, but it is not always printed on the physical capacitor. When it is printed on the capacitor, the polarity is listed first as N, P, or NP for negative, positive, and zero temperature coefficients, respectively. The magnitude is listed next.

EXAMPLE SOLUTION

Three capacitors have the markings N750, P350, and NPO. What do these marks mean?

EXAMPLE **SOLUTION**

These are the temperature coefficients for the capacitors. The N750 mark indicates a capacitor with a negative temperature coefficient of 750 ppm/°C. The P350 mark identifies a positive temperature coefficient of 350 ppm/°C. The NPO marking specifies a capacitor with a zero temperature coefficient. Practical capacitors with NPO ratings still vary slightly with temperature, but their coefficients are generally less than 30 ppm/°C.

The value of a particular capacitor for a given temperature change and a given temperature coefficient can be found with Equation 14-5.

$$C_{actual} = C_{rated} + \left((T_{actual} - T_{rated}) \frac{T_C}{10^6} \right) C_{rated} \quad (14\text{-}5)$$

where C_{rated} is the nominal capacitance at the rated temperature (T_{rated}), and T_C is the capacitor's temperature coefficient. The sign of T_C is positive (+) or negative (–) for positive or negative temperature coefficients, respectively.

EXAMPLE SOLUTION

If a capacitor is rated for 1,000 pF at 25°C, and is marked as N250, how much capacitance will it have at 60°C?

EXAMPLE **SOLUTION**

First, we compute the magnitude of the change as follows:

$$C_{actual} = C_{rated} + \left((T_{actual} - T_{rated}) \frac{T_C}{10^6} \right) C_{rated}$$

$$= 1{,}000 \text{ pF} + \left((60\,^{\circ}\text{C} - 25\,^{\circ}\text{C}) \frac{-250}{10^6} \right) 1{,}000 \text{ pF}$$

$$= 1{,}000 \text{ pF} - 8.75 \text{ pF} = 991.25 \text{ pF}$$

Practice Problems

1. Compute the value of a capacitor rated for 25 μF at 25°C and a P350 temperature coefficient if the capacitor is operated at 75°C.

2. Find the value of a capacitor rated for 800 pF at 25°C and a temperature coefficient of N650 if the capacitor is operated at 5°C.

3. What is the value of a 470-pF capacitor (rated at 25°C) if it is marked as NPO and is operated at 45°C?

Answers to Practice Problems

1. 25.44 µF   **2.** 810.4 pF   **3.** 470 pF

TOLERANCE

The tolerance of a capacitor is interpreted the same way as the tolerance rating on a resistor. That is, although a particular capacitor is intended to have a specific value of capacitance, manufacturing tolerances may introduce some error. Thus, a 680-pF capacitor with a ±10% tolerance may have actual capacitance values in the following range.

$$\text{maximum deviation} = \text{tolerance} \times \text{marked value}$$
$$= 0.1 \times 680 \text{ pF} = 68 \text{ pF, and}$$
$$\text{capacitance range} = \text{marked value} \pm \text{maximum deviation}$$
$$= 680 \text{ pF} \pm 68 \text{ pF, so}$$
$$C_{\text{minimum}} = 612 \text{ pF, and}$$
$$C_{\text{maximum}} = 748 \text{ pF}$$

Capacitor tolerances may be high relative to many resistors. Additionally, they are often asymmetrical. Typical symmetrical tolerances for capacitors include ±5%, ±10%, and ±20%, although tighter tolerances are available. Typical asymmetrical tolerances include −10%/+150%, −20%/+80%, and −0%/+50%.

LEAKAGE CURRENT

Ideally, the dielectric of a capacitor is a perfect insulator and allows no current flow. A practical dielectric, however, does permit a small current to flow. The current that flows through the dielectric is called **leakage current.** Do not confuse leakage current with the current that flows through the dielectric after breakdown occurs. Breakdown current can be quite large and destructive. Leakage current, by contrast, flows at normal operating voltages and is very small.

Some manufacturers specify the leakage of a capacitor in terms of insulation resistance rather than leakage current. This is the value of dc resistance measured between the terminals of the capacitor. As you would expect, this is a very high value (often beyond the range of ordinary ohmmeters).

In general, we want the leakage current to be as small as possible. This is equivalent to saying we want the dielectric resistance to be as high as possible. In either case, the leakage rating must be specified at a particular voltage, temperature, and humidity, since each of these factors affects the absolute value of leakage.

Capacitor Technologies

Many different types of materials are used to manufacture capacitors. A technician should know the major characteristics of a given capacitor technology. We will take a brief look at several of the more common types.

ALUMINUM ELECTROLYTIC

Probably the two most outstanding characteristics of aluminum electrolytic capacitors are their high capacities and the fact that they are polarized. The basic construction of an aluminum electrolytic capacitor is illustrated in Figure 14-18. The plates of the capacitor are made from aluminum sheets or ribbons that have been etched to create a rough surface. A rough surface provides much greater surface area (i.e., higher capacitance) than a smooth polished surface. The plates are separated with a thin paper or gauze that is saturated with a conductive material called the electrolyte. The sandwich shown in Figure 14-18(a) is rolled tightly and inserted into an aluminum can. The negative plate is generally connected to the metal can. A lead is welded to the positive plate and brought out through an insulative seal on the top of the aluminum can as shown in Figure 14-18(b).

At this point in the manufacture of the aluminum capacitor, its positive and negative plates are shorted by the conductive electrolyte. The dielectric is formed chemically by connecting a voltage to the capacitor. As current flows through the newly constructed capacitor, a thin coating (about 1×10^{-8} inch thick) of aluminum oxide is formed on the surface of the positive plate. This oxide is an insulator and forms the dielectric for the capacitor. As the dielectric forms, the effective dc resistance of the capacitor increases until

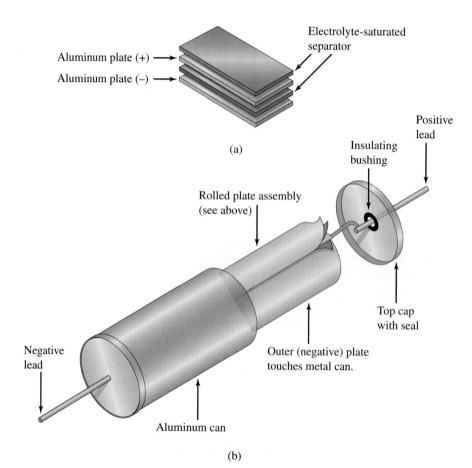

Figure 14-18. Construction of an aluminum electrolytic capacitor.

it reaches the prescribed levels. It should be noted, however, that leakage currents for electrolytic capacitors are much higher than for other types of capacitors. The paper separator material in the capacitor roll prevents direct electrical contact between the two plates.

Figure 14-19 shows some representative aluminum electrolytic capacitors. Typical values range from a few microfarads to tens of thousands of microfarads. Breakdown voltages range from a few volts to several hundred volts. Reverse voltages of as little as 1 or 2 V can destroy the dielectric. Aluminum electrolytics are only useful at low frequencies. Higher frequencies (as low as hundreds of kilohertz) cause the capacitor to exhibit high impedances due to the inadvertent series inductance (called equivalent series inductance, or **ESL**) associated with the construction of the capacitor.

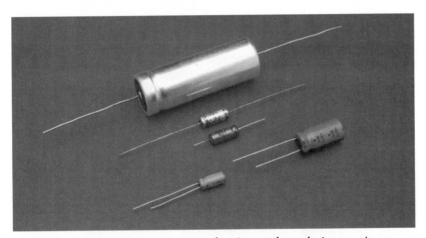

Figure 14-19. Some representative aluminum electrolytic capacitors.

Aluminum electrolytics have a somewhat limited shelf life. The dielectric can deteriorate through lack of use. This condition can sometimes be detected with an ohmmeter before inserting the capacitor into a circuit. Although most aluminum electrolytics are polarized, a few are made nonpolar by creating an oxide coating on both of the aluminum plates.

CERAMIC

Figure 14-20 illustrates the basic construction of a ceramic capacitor. It consists of two metal disks or plates separated by a thin ceramic sheet (dielectric). The entire assembly is covered with an insulative coating. Practical values of capacitance can be obtained in relatively small packages due to the high dielectric constant (>1,000) exhibited by ceramic materials.

A multilayer ceramic capacitor has a similar construction, but it uses several alternating layers of metal and dielectric to obtain a greater total surface area, and, therefore, increased capacitance. Multilayer ceramic capacitors are also called monolithic ceramic capacitors.

Ceramic capacitor values range from as small as 1 pF (disk) to at least 10 μF (multilayer). Common voltage ratings extend from as low as 3 V to as high as several thousand volts. Ceramic capacitors are well-suited for high-frequency applications due to their low series inductance (ESL). Figure 14-21 shows some representative ceramic capacitors.

Figure 14-22 shows an array of multilayer ceramic chip capacitors.

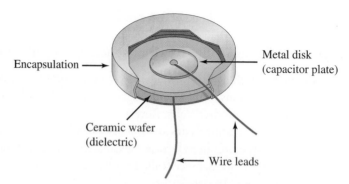

Encapsulation

Metal disk
(capacitor plate)

Ceramic wafer
(dielectric)

Wire leads

Figure 14-20. Construction of a ceramic disk capacitor.

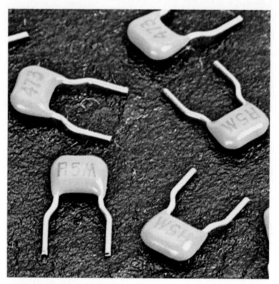

Figure 14-21. Ceramic capacitors. (Courtesy of
AVX Corporation)

Figure 14-22. Multilayer ceramic chip capacitors. (Courtesy of AVX
Corporation)

PLASTIC FILM

Figure 14-23 shows the basic construction of a film capacitor. A four-layer sandwich made of alternating layers of metal foil and insulative film (dielectric) is rolled into a compact cylinder. There is a wide array of materials used for the dielectric film, including polystyrene, mylar, polypropylene, and polycarbonate.

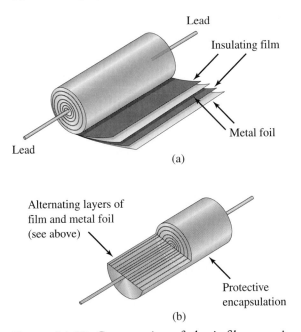

Lead

Insulating film

Metal foil

Lead

(a)

Alternating layers of
film and metal foil
(see above)

Protective
encapsulation

(b)

Figure 14-23. Construction of plastic film capacitors.

Film capacitor values range from as low as 47 pF to as high as 25 µF. Voltage ratings are available from 50 to over 1,000 V. The electrical performance of the capacitor varies with the specific type of plastic used for the dielectric film. Figure 14-24 shows some representative plastic film capacitors.

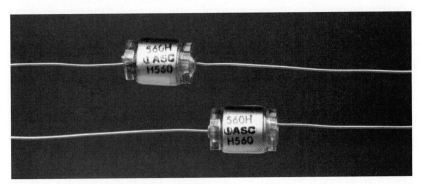

Figure 14-24. Plastic film capacitors.

METALLIZED FILM

Figure 14-25 illustrates the construction of a metallized film (also called metal-film) capacitor. It is made by depositing a thin metal coating on the surface of two plastic (e.g., polycarbonate,

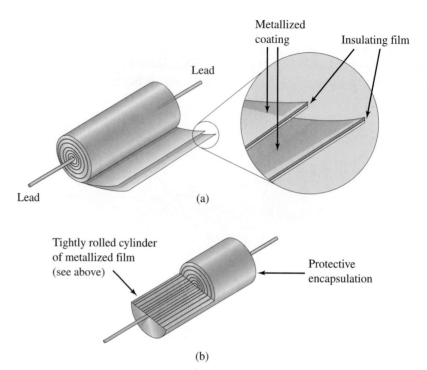

Figure 14-25. Construction of a metallized film capacitor.

polypropylene) sheets or films. The two films are then stacked and rolled into a compact cylinder. Leads are attached, and the entire assembly is sealed with a protective coating.

Values for metallized film capacitors begin as low as 0.0033 μF and range as high as 75 μF. Breakdown voltages of 50 to several hundred volts are typical. One interesting characteristic of a metallized film capacitor is that it tends to be self-healing after it has experienced dielectric breakdown. This is because the metal coatings (capacitor plates) are so thin that they vaporize around the point where the breakdown occurred. Even though a portion of the plate is essentially gone after a breakdown, the two plates are prevented from shorting together by the absence of conductive material in the vicinity of the pinhole that was punched through the dielectric by the high voltage. As long as the breakdown was confined to a small pinhole in the dielectric, the capacitor can continue to perform satisfactorily after a breakdown. Figure 14-26 shows some representative metallized film capacitors.

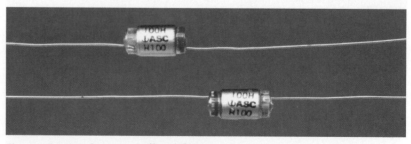

Figure 14-26. Some metallized film capacitors.

MICA

Figure 14-27 shows some mica capacitors. Their construction is similar to a multilayer ceramic capacitor, except that the dielectric material is a thin sheet of mica. The plates of the capacitor can be actual ribbons of metal foil, or they may take the form of a deposited coating like that used in metallized film capacitors. Silver is the metal that is deposited on the mica sheets. This type of capacitor is also called a silver mica capacitor. The completed assembly is generally dipped to provide a durable, insulative seal around the capacitor.

Mica capacitors are available with values as low as 2 pF and as high as several tens of thousands of picofarads. Breakdown voltages range from 100 V to several thousand volts.

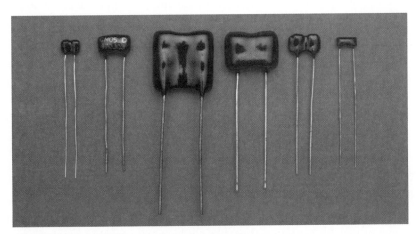

Figure 14-27. Representative mica capacitors.

PAPER

Paper capacitors are constructed essentially the same way as plastic film capacitors illustrated in Figure 14-23. The dielectric material, however, is a thin ribbon of paper instead of the plastic sheets used in plastic film capacitors. The paper is generally impregnated with oil, which increases the overall dielectric strength of the insulating material.

Paper capacitor technology is one of the oldest. It has been replaced in many applications by other newer technologies. Paper capacitor values range from as low as 0.01 μF to as high as 10 μF. Very high breakdown voltages are available, including values as high as several tens of thousands of volts.

TANTALUM

Tantalum capacitors are polarized electrolytics, similar in many ways to the aluminum electrolytics. They are characterized by high values of capacitance for a given physical size. This space savings is caused by two factors. First, the dielectric is a very thin layer of tantalum oxide (comparable to the aluminum oxide dielectric in aluminum electrolytics). Second, the dielectric constant of tantalum oxide is relatively high (24).

There are three types of tantalum capacitor technologies: wet foil, wet slug, and solid slug. The wet foil construction is comparable to the construction of the aluminum electrolytic, but a thin tantalum foil is used instead of aluminum foil. The wet slug uses a sintered slug of tantalum as the positive electrode. The electrolyte is a gel, which surrounds the tantalum slug. A layer of tantalum oxide is formed between the electrolyte and the tantalum, which serves as the dielectric.

Figure 14-28 illustrates the construction of a solid slug tantalum capacitor. The positive electrode is a porous pellet of tantalum. As with other electrolytic capacitors, a thin insulative film (tantalum pentoxide) is formed through electrochemical action. This is the dielectric of the capacitor. The electrolyte is dry and is pressed between the tantalum core and an outer electrode made of silver and coated with graphite. The silver electrode is the negative lead of the capacitor. Finally, the entire assembly is dipped in epoxy to provide a protective seal for the capacitor.

Tantalum capacitors have a much longer (indefinite) shelf life than aluminum electrolytics, and they exhibit substantially lower leakage currents. Most tantalum capacitors are designed for low voltage (i.e., <100 V) operation. Under certain conditions (e.g., high-impedance circuits), the tantalum oxide dielectric is self-healing after a voltage breakdown. Typical values for tantalum electrolytics range from 0.1 μF to as high as 2,200 μF. Figure 14-29 shows some representative tantalum capacitors.

VARIABLE CAPACITORS

KEY POINTS

Variable capacitors may be adjusted for a specific value by a technician.

A variable capacitor is manufactured to have a nominal range of capacitance. The exact value of the capacitor is then set by the technician or operator in the field. There are many different types of variable capacitors, but they all share one common characteristic—their specific capacitance can be adjusted.

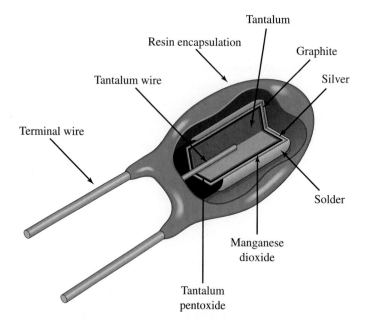

Figure 14-28. Construction of a tantalum capacitor.

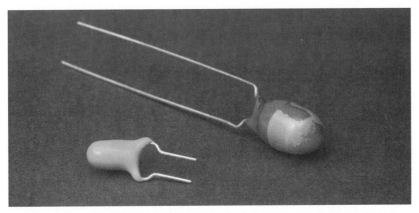

Figure 14-29. Tantalum capacitors offer a very high capacitance-to-volume ratio.

Exercise Problems 14.2

1. When a capacitor's leads protrude from opposite ends of its body, the leads are called _____ leads.
2. When both leads protrude from the same end of the capacitor's body, they are called _____ leads.
3. Describe the lead arrangement on a surface-mount capacitor.
4. Describe why it is important for a technician to distinguish between polarized and nonpolarized capacitors.
5. Explain how stray capacitance is formed in a circuit.
6. Draw the schematic symbol for a variable capacitor.
7. What is the ideal value of current that flows through the dielectric of a capacitor?
8. What is the name of the current that flows through the dielectric of a capacitor that is being operated below its rated voltage value?
9. What happens to the value of current through a capacitor's dielectric if the voltage rating on the capacitor is exceeded?
10. Explain the meaning of the marking N750 on a capacitor.
11. If a capacitor is rated for 2,500 μF at 25°C and is marked as P500, how much capacitance will it have at 75°C?
12. What is the range of allowable values that a 25-μF capacitor can have if its tolerance is listed as −20%/+80%?
13. Capacitors are either fixed or _____.

14.3 Multiple Circuit Capacitances

Two or more capacitors may be connected in series, parallel, or other configuration to produce an equivalent value of capacitance. The equations for combining capacitors are different than the equations for resistors, but they are very easy to remember if you learn them in a logical way (i.e., avoid casual memorization).

Series-Connected Capacitors

Figure 14-30(a) shows two series-connected capacitors. Figure 14-30(b) illustrates a way to help you remember the relative (i.e., larger or smaller) value of the equivalent capacitance. As shown in Figure 14-30(b), when the capacitors are connected in series, we are essentially increasing the thickness of the dielectric. That is, we are increasing the distance between the plates. You already know that increased plate separation results in decreased capacitance. So, we can make the following conclusion:

> When capacitors are connected in series, the total capacitance is less than the value of any of the individual capacitances.

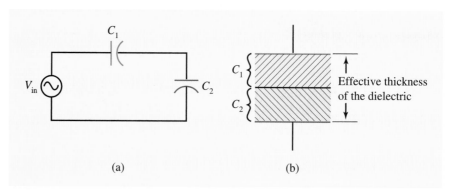

(a) (b)

Figure 14-30. (a) Series-connected capacitors. (b) The distance between the plates is effectively increased when capacitors are connected in series.

TOTAL CAPACITANCE

KEY POINTS

Series-connected capacitances combine like parallel-connected resistors (i.e., reciprocal equation). Thus, the total capacitance of series-connected capacitors is always less than the smallest individual capacitance.

Equation 14-6 can be used to compute the equivalent capacitance for series-connected capacitors. Note that it is the same basic form as the reciprocal equation used to compute the equivalent resistance of parallel-connected resistors.

$$C_T = \cfrac{1}{\cfrac{1}{C_1} + \cfrac{1}{C_2} + \dots + \cfrac{1}{C_N}} \qquad (14\text{-}6)$$

EXAMPLE SOLUTION

A 4.7-μF and a 6.8-μF capacitor are connected in series. What is the total capacitance?

EXAMPLE **SOLUTION**

We apply Equation 14-6 as follows:

$$C_T = \cfrac{1}{\cfrac{1}{C_1} + \cfrac{1}{C_2}}$$

$$= \cfrac{1}{\cfrac{1}{4.7\ \mu F} + \cfrac{1}{6.8\ \mu F}}$$

$$= \cfrac{1}{359.82 \times 10^3} = 2.78\ \mu F$$

Practice Problems

1. Calculate the value of a 4.7-μF and a 10-μF capacitor in series.
2. What is the total capacitance when a 200-pF capacitor is connected in series with a 680-pF capacitor?
3. What is the total capacitance of four series-connected capacitors that have the following values: 330 μF, 470 μF, 680 μF, and 1,000 μF?

Answers to Practice Problems

1. 3.2 μF   2. 154.6 pF   3. 131.1 μF

The shortcut equations that were used for computing the resistance of parallel resistors can also be applied to the solution of series-connected capacitors. For example, if several equal-value capacitors are connected in series, their combined capacitance can be computed with Equation 14-7.

$$C_T = \frac{C_X}{N} \qquad (14\text{-}7)$$

where C_X is the common capacitor value and N is the number of equal-valued capacitors.

Similarly, the product-over-the-sum formula used to find the combined resistance of two parallel-connected resistors can be used to compute the total capacitance of two series-connected capacitors as represented by Equation 14-8.

$$C_T = \frac{C_1 C_2}{C_1 + C_2} \qquad (14\text{-}8)$$

Practice Problems

1. What is the combined value of four series-connected 100-μF capacitors?
2. What is the value of five 20-μF capacitors connected in series?
3. What is the value of a 20-μF and a 5-μF capacitor connected in series?

Answers to Practice Problems

1. 25 μF **2.** 4 μF **3.** 4 μF

SERIES CAPACITORS WITH A dc SOURCE

If two capacitors are connected in series across a dc source, a charging current will flow until the total charge on the capacitors is equal to the value of the applied voltage. Since the capacitors are in series, the current—and therefore the accumulated coulombs of charge—will be the same. However, rearrangement of Equation 14-1 shows us that the voltage on a given capacitor is inversely proportional to the value of capacitance. That is,

$$C = \frac{Q}{V}, \text{ or}$$

$$V = \frac{Q}{C}$$

This means that when two capacitors are connected in series across a dc source, the one with the smallest capacitance will have the most voltage. Of course, the sum of the two voltage drops must still be equal to the applied voltage according to Kirchhoff's Voltage Law.

Parallel-Connected Capacitors

KEY POINTS

Parallel-connected capacitances combine like resistors in series. The total capacitance of parallel-connected capacitors is equal to the sum of the individual capacitances.

Figure 14-31(a) shows two capacitors connected in parallel. Figure 14-31(b) illustrates a way to help you remember the relative (i.e., larger or smaller) value of the equivalent capacitance. As capacitors are added in parallel, the effective plate area, and therefore the total capacitance, increases.

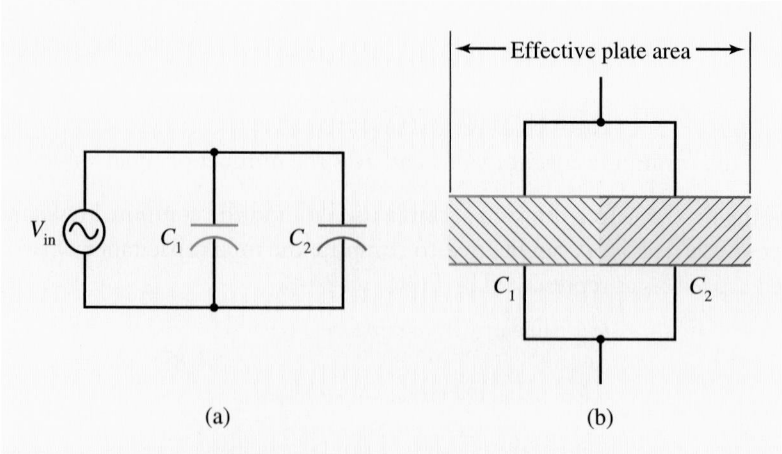

Figure 14-31. (a) Parallel-connected capacitors. (b) The effective area of the plates is increased when capacitors are connected in parallel.

We can make the following conclusion:

> When capacitors are connected in parallel, the total capacitance is equal to the sum of the individual capacitances.

Although the reasons are totally different, you should recognize this general statement as being similar to the rule for series-connected resistors. Equation 14-9 can be used to compute the equivalent capacitance for parallel-connected capacitors.

$$C_T = C_1 + C_2 + \ldots + C_N \qquad \text{(14-9)}$$

EXAMPLE SOLUTION

A 33-μF and a 10-μF capacitor are connected in parallel. What is the total capacitance?

EXAMPLE **SOLUTION**

We apply Equation 14-9 as follows:

$$C_T = C_1 + C_2$$
$$= 33\ \mu F + 10\ \mu F = 43\ \mu F$$

Practice Problems

1. Calculate the value of a 27-μF and a 47-μF capacitor in parallel.
2. What is the total capacitance when a 270-pF capacitor is connected in parallel with a 470-pF capacitor?
3. What is the total capacitance of four parallel-connected capacitors that have the following values: 200 μF, 390 μF, 430 μF, and 1,000 μF?

Answers to Practice Problems

1. 74 μF   2. 740 pF   3. 2,020 μF

Series-Parallel Capacitor Networks

The total capacitance of several capacitors connected in a series-parallel circuit is found in a way similar to that used for series-parallel resistive circuits. That is, we replace sets of series-connected capacitors and sets of parallel-connected capacitors with a single equivalent capacitor. When replacing a set of series capacitors, we use the series capacitor equation (Equation 14-6). When a set of parallel capacitors is replaced, we use the parallel capacitor equation (Equation 14-9). The replacement process continues until we have a single equivalent capacitor whose value is equal to the total capacitance of the original circuit.

EXAMPLE SOLUTION

What is the total capacitance of the circuit shown in Figure 14-32?

EXAMPLE **SOLUTION**

First, we combine parallel-connected capacitors C_2 and C_3 by applying Equation 14-9 as follows:

$$C_{2,3} = C_2 + C_3$$
$$= 20\ \mu F + 80\ \mu F = 100\ \mu F$$

KEY POINTS

The basic analysis method for a series-parallel capacitive circuit is similar to that used to analyze series-parallel resistive circuits:

- Sets of pure series or pure parallel capacitances are replaced with an equivalent capacitance.
- The equivalent capacitance is found by applying the appropriate series or parallel capacitor equation.
- Substitution continues until only a single capacitance (total capacitance) remains.

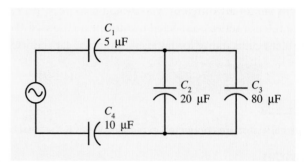

Figure 14-32. What is the total capacitance of this series-parallel circuit?

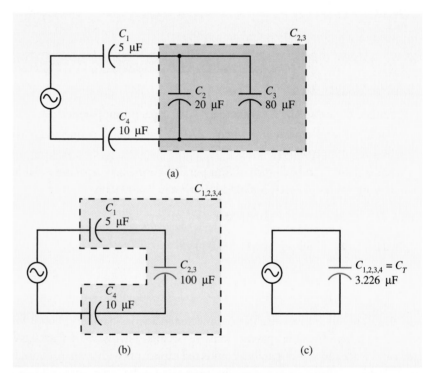

Figure 14-33. Steps in the simplification of the circuit shown in Figure 14-33.

Figures 14-33(a) and (b) show our progress so far.

Next, we can replace series-connected capacitors C_1, $C_{2,3}$, and C_4 with an equivalent capacitance by applying Equation 14-6.

$$C_T = \cfrac{1}{\cfrac{1}{C_1} + \cfrac{1}{C_{2,3}} + \cfrac{1}{C_4}}$$

$$= \cfrac{1}{\cfrac{1}{5 \ \mu F} + \cfrac{1}{100 \ \mu F} + \cfrac{1}{10 \ \mu F}} = 3.226 \ \mu F$$

This completes the simplification of our circuit. The result is shown in Figure 14-33(c).

Practice Problems

1. Compute the total capacitance for the circuit shown in Figure 14-34.

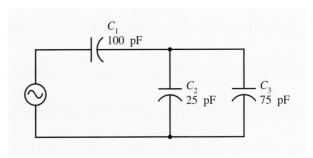

Figure 14-34.
Determine the total capacitance in this circuit.

2. What is the total capacitance of the circuit shown in Figure 14-35?

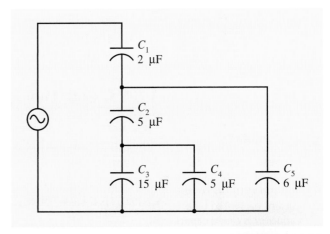

Figure 14-35.
Find the total capacitance.

3. What is the value of capacitor C_1 in Figure 14-36?

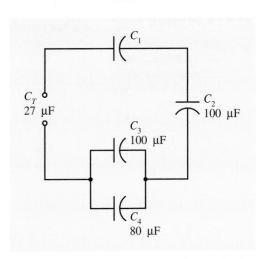

Answers to Practice Problems

1. 50 pF 2. 1.67 µF 3. 46.55 µF

Figure 14-36.
Find the value of C_1.

1. Connecting capacitors in series (*decreases, increases*) the total capacitance.
2. Connecting capacitors in parallel (*decreases, increases*) the total capacitance.
3. What is the combined capacitance of a 15-μF and a 10-μF capacitor connected in series?
4. What value of capacitance must be connected in series with a 680-pF capacitor to obtain a total capacitance of 300 pF?
5. What is the total capacitance when a 470-pF capacitor is connected in parallel with a 330-pF capacitor?
6. What is the total capacitance if five 100-pF capacitors are connected in parallel?
7. What is the total capacitance if five 100-pF capacitors are connected in series?
8. If a technician has three capacitors and wants to combine them to produce the highest possible capacitance, they should be connected in (*series, parallel, series-parallel*).
9. If a 10-μF capacitor is connected in series with a parallel network consisting of 5-μF and 20-μF capacitors, what is the total capacitance?

14.4 Capacitive Reactance

Capacitors offer opposition to current flow in a manner similar to resistors and inductors, although the mechanism is different. The opposition to sinusoidal current flow presented by a capacitor is called **capacitive reactance.** As you might expect, it is measured in ohms.

CAPACITIVE REACTANCE VARIES WITH FREQUENCY

Capacitive reactance, like inductive reactance, is affected by frequency. Capacitive reactance is inversely proportional to frequency. Thus, the higher the frequency of operation for a given capacitor size, the lower the value of capacitive reactance. Figure 14-37 illustrates this important relationship. In each of the three sketches in Figure 14-37, the applied voltage is held constant at 10 V. Only the frequency is changed.

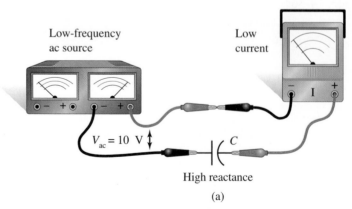

Figure 14-37. Higher frequencies produce lower capacitive reactances.

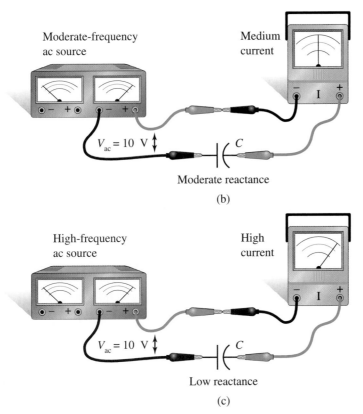

Moderate-frequency ac source

Medium current

$V_{ac} = 10$ V

Moderate reactance

(b)

High-frequency ac source

High current

$V_{ac} = 10$ V

Low reactance

(c)

Figure 14-37. *(Continued)*

Let's examine this relationship yet another way. To charge a given capacitor to a given voltage requires a definite amount of charge (Equation 14-1). As the frequency is increased, the same amount of charge (i.e., same number of coulombs) must move past a given point in the circuit in less time. With more coulombs of charge flowing past a point every second, we must have a higher current. Recall that current is a measure of coulombs per second flowing past a given point (Equation 2-3). Now, if the voltage is constant and the current has increased, then Ohm's Law will tell us that resistance (in this case reactance) must have decreased.

EXAMPLE SOLUTION

If the frequency in a capacitive circuit is reduced to one-half of its original value, what happens to the capacitive reactance in the circuit?

EXAMPLE **SOLUTION**

The capacitive reactance will increase, since capacitive reactance is inversely proportional to frequency. In the present example, the capacitive reactance would double.

CAPACITIVE REACTANCE VARIES WITH CAPACITANCE

Capacitive reactance is inversely proportional to capacitance. If the capacitance in a circuit is increased, then the capacitive reactance will decrease (for a fixed frequency). This is shown numerically in the next section. However, we know intuitively that it takes less current flow to charge a smaller capacitor to a given voltage than it does to charge a larger capacitor to the same voltage. Having less current in a circuit for a given voltage is equivalent to saying there is more opposition—capacitive reactance in this case—to current flow.

CAPACITORS IN DC CIRCUITS

If a dc source (i.e., $f = 0$ Hz) is connected to a capacitor, then current will only flow long enough to charge the capacitor. Once the capacitor is charged to the applied voltage, no more current flows, and the dc circuit is essentially open-circuited by the capacitor. The term capacitive reactance has no meaning in the case of dc circuits. Capacitive reactance applies only to sinusoidal circuits.

Calculating Capacitive Reactance

We can determine the value of capacitive reactance for a given capacitor at a given frequency by applying Equation 14-10.

$$X_C = \frac{1}{2\pi fC} \tag{14-10}$$

Many technicians prefer to divide the 2π to obtain an alternate form of the equation. We express it here as Equation 14-11.

$$X_C = \frac{0.159}{fC} \tag{14-11}$$

In either case, we can see that capacitive reactance is inversely proportional to both frequency and capacitance.

EXAMPLE SOLUTION

Compute the reactance of a 560-pF capacitor when operated at 150 kHz. Also determine the amount of current that will flow if the 150-kHz source supplies 120 V rms.

EXAMPLE **SOLUTION**

We apply Equation 14-10 (or Equation 14-11) to compute the value of capacitive reactance.

$$X_C = \frac{1}{2\pi fC}$$

$$= \frac{1}{6.28 \times 150 \times 10^3 \times 560 \times 10^{-12}}$$

$$= 1.896 \text{ k}\Omega$$

Now we can apply Ohm's Law to determine the current flow.

$$I_C = \frac{V_C}{X_C}$$

$$= \frac{120 \text{ V}}{1.896 \text{ k}\Omega} = 63.29 \text{ mA rms}$$

Practice Problems

1. What is the reactance of a 100-pF capacitor operated at 10 MHz?
2. How much reactance is provided by a 25-μF capacitor when powered by a 350-Hz source?
3. At what frequency does a 0.01-μF capacitor have a reactance of 5 kΩ?
4. If a 470-pF capacitor must provide 12 kΩ of capacitive reactance, what must be the frequency of operation?
5. What value of capacitor will provide 100 kΩ of reactance at a frequency of 500 Hz?

Answers to Practice Problems

1. 159.2 Ω   **2.** 18.2 Ω   **3.** 3.18 kHz

4. 28.2 kHz   **5.** 0.0032 μF

Capacitive Reactance in Series

When several values of capacitive reactance are connected in series, they can be combined just like series resistances. That is, the total capacitive reactance is simply the sum of the individual reactances. We express this formally with Equation 14-12.

$$X_{C_T} = X_{C_1} + X_{C_2} + \ldots + X_{C_N} \qquad \text{(14-12)}$$

Do not confuse capacitive reactance with total capacitance, which requires use of the reciprocal equation (Equation 14-6).

EXAMPLE SOLUTION

Determine the total capacitive reactance for the circuit shown in Figure 14-38.

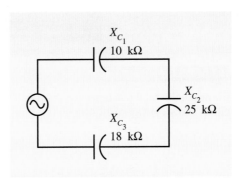

Figure 14-38. What is the total capacitive reactance of this circuit?

EXAMPLE **SOLUTION**

We apply Equation 14-12 as follows:

$$X_{C_T} = X_{C_1} + X_{C_2} + X_{C_3}$$
$$= 10 \text{ k}\Omega + 25 \text{ k}\Omega + 18 \text{ k}\Omega = 53 \text{ k}\Omega$$

Practice Problems

1. Find the total reactance of the series circuit shown in Figure 14-39.

2. What is the total reactance if three identical capacitors are connected in series and each has a reactance of 40 kΩ?

3. How much capacitive reactance must be connected in series with a capacitive reactance of 600 Ω to produce a total capacitive reactance of 2 kΩ?

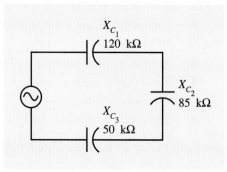

Figure 14-39. Find the total reactance in this circuit.

Answers to Practice Problems

1. 255 kΩ   **2.** 120 kΩ   **3.** 1.4 kΩ

Capacitive Reactance in Parallel

When capacitive reactances are connected in parallel, they combine like resistances in parallel. We can express this formally with Equation 14-13.

$$X_{C_T} = \cfrac{1}{\cfrac{1}{X_{C_1}} + \cfrac{1}{X_{C_2}} + \dots + \cfrac{1}{X_{C_N}}} \qquad (14\text{-}13)$$

EXAMPLE SOLUTION

Compute the total reactance for the parallel circuit shown in Figure 14-40.

EXAMPLE **SOLUTION**

We apply Equation 14-13 as follows:

$$X_{C_T} = \cfrac{1}{\cfrac{1}{X_{C_1}} + \cfrac{1}{X_{C_2}}}$$

$$= \cfrac{1}{\cfrac{1}{100\ \Omega} + \cfrac{1}{400\ \Omega}} = 80\ \Omega$$

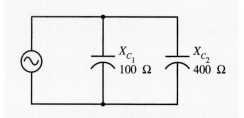

Figure 14-40. What is the total capacitive reactance in this circuit?

Practice Problems

1. What is the total capacitive reactance if three capacitors with reactances of 680 Ω, 820 Ω, and 1,000 Ω are connected in parallel?

2. What is the total capacitive reactance if three identical capacitors having individual reactances of 100 Ω are connected in parallel?

3. How much capacitive reactance must be connected in parallel with a 12-kΩ capacitive reactance to produce a total capacitive reactance of 4,000 Ω?

Answers to Practice Problems

1. 271 Ω   2. 33.33 Ω   3. 6 kΩ

Series-Parallel Capacitive Reactance

Series-parallel connections of capacitive reactances are simplified in exactly the same way as their resistive counterparts. That is, sets of series or parallel reactances are replaced with an equivalent reactance. The value of the equivalent reactance is found by applying the relevant series (Equation 14-12) or parallel (Equation 14-13) reactance equation.

Analysis of Complex Capacitive Circuits

It is unlikely that you will have a real-world need to analyze purely capacitive circuits connected in a complex circuit configuration. You may, however, encounter such a scheme as part of a theoretical study. In any case, capacitive reactance provides you with a convenient tool for determining the total capacitance of a complex capacitive circuit. The sequential procedure follows:

1. Assume an applied frequency unless one is given.
2. Compute the values of capacitive reactance for each capacitor in the network.
3. Apply your favorite network theorem to the reactance values using the exact procedures presented for use with resistive circuits.
4. Compute total capacitance from the resulting value of total reactance.

KEY POINTS

Series-parallel connections of capacitive reactance can be simplified in the same way series-parallel resistive circuits are simplified—sets of series or parallel reactances are replaced with an equivalent reactance.

KEY POINTS

The total capacitance of a complex capacitive network can be found by determining the reactances of the individual capacitors, simplifying the reactances with a network theorem, and then computing total capacitance from the value of total reactance provided by the network simplification.

Exercise Problems 14.4

1. _____ _____ is the opposition to current flow offered by a capacitor.
2. If the frequency applied to a capacitive circuit decreases, what happens to the value of capacitance?
3. If the frequency applied to a capacitive circuit decreases, what happens to the value of capacitive reactance?
4. Capacitive reactance is (*directly, inversely*) proportional to frequency.

5. Capacitive reactance is (*directly, inversely*) proportional to capacitance.

6. What is the reactance of a 15-µF capacitor operating at 200 Hz?

7. What value of capacitor provides 1,000 Ω of reactance at a frequency of 2.5 MHz?

8. If the following capacitive reactances—100 Ω, 250 Ω, 680 Ω—are connected in series, what is the total capacitive reactance of the circuit?

9. If the following capacitive reactances— 270 Ω, 390 Ω, 820 Ω—are connected in parallel, what is the total capacitive reactance of the circuit?

10. If a capacitor having a reactance of 1,000 Ω is connected in series with the parallel reactance combination of 2,500 Ω and 1,500 Ω, what is the total reactance of the circuit?

14.5 Power Dissipation

KEY POINTS

Ideal capacitors dissipate no heat.

An ideal capacitor, like an ideal inductor, dissipates no power. It takes energy from the circuit while being charged, but subsequently returns that power to the circuit when it is discharged. The net consumption of power by the capacitor is, therefore, zero.

Power Losses in a Practical Capacitor

KEY POINTS

A practical capacitor dissipates heat.

There are losses in a practical capacitor that convert some of the stored energy to heat. Basically there are three causes of power loss or power dissipation in a capacitor:

1. nonzero resistance of the capacitor plates and the connecting wires,
2. leakage resistance of the dielectric, and
3. dielectric dissipation (also called dielectric hysteresis).

These losses are represented in Figure 14-41.

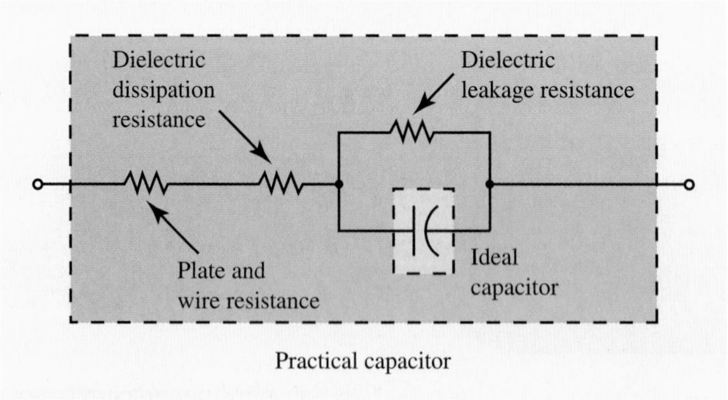

Figure 14-41. Losses in a practical capacitor can be viewed as resistances.

PLATE AND LEAD WIRE RESISTANCE

Although the capacitor plates and connecting lead wires are conductors, they are not ideal conductors. They have some resistance. The amount of resistance added by the plates and lead wires is quite small and varies with temperature and frequency. As shown in Figure 14-41, this resistive loss appears as a series resistance. For many noncritical applications, the effects of plate and lead wire resistance are negligible.

DIELECTRIC LEAKAGE RESISTANCE

The capacitor's dielectric material is an insulator, but it is not a perfect insulator. Its resistance is less than infinity. As shown in Figure 14-41, the dielectric leakage acts like a high-value resistance in parallel with an ideal capacitor. The effects of dielectric leakage loss can be disregarded for many noncritical applications.

DIELECTRIC DISSIPATION RESISTANCE

In a previous discussion on capacitor voltage ratings (refer to Figure 14-17), we noted that the orbits of the atoms in the dielectric material were distorted by the charge on the capacitor. When the charge decreases to zero, the distorted orbits return to their normal shape. When the orbits change from normal to distorted, energy is taken from the circuit. As the orbits return to their normal shape, some energy is converted to heat in the dielectric. Although it is not technically accurate, many technicians like to think of dielectric dissipation as heat produced by friction as the immobile dielectric atoms are pulled one way and then the other each half-cycle.

The two series resistances shown in Figure 14-41 are generally grouped into a single series resistance called the **equivalent series resistance** or ESR. This value is provided by the capacitor manufacturer. In most cases, we want the ESR of the capacitor to be as low as possible.

Quality of a Capacitor

The ratio of energy stored to energy converted to heat is called the **quality factor** or Q of the capacitor. The quality factor is a simple ratio (i.e., no units of measure). It is interpreted much the same way as Q for inductors. It is defined as a ratio of reactance to resistance as we did with inductors. In the case of capacitors, Q can be expressed with Equation 14-14.

$$Q = \frac{X_C}{\text{ESR}} \qquad (14\text{-}14)$$

where ESR is the equivalent series resistance of the capacitor. Since X_C varies with frequency, we know that Q will also vary with frequency.

The **dissipation factor** is the inverse of the quality factor. Thus, the higher the Q of a capacitor, the lower the dissipation factor (i.e., less energy converted to heat). The dissipation factor can be computed with Equation 14-15.

$$\text{DF} = \frac{1}{Q} \qquad (14\text{-}15)$$

The energy loss in a capacitor can also be described by the capacitor's **power factor.** The power factor of a capacitor can be computed with Equation 14-16.

KEY POINTS

The ESR is composed of two losses:

• resistance of capacitor plates and lead wires and

• dielectric dissipation.

KEY POINTS

The ratio of stored energy to energy lost in the ESR is called the quality factor of the capacitor. It is more often described as the ratio of reactance to resistance, where the resistance is the ESR of the capacitance.

$$pf = \frac{\text{ESR}}{\sqrt{\text{ESR}^2 + X_C^2}} \qquad (14\text{-}16)$$

Chapter 15 presents additional material regarding calculation of power factors in an *RC* circuit. A nonideal capacitor is itself an *RC* circuit (refer to Figure 14-41).

Exercise Problems 14.5

1. How much power is dissipated as heat in an ideal capacitor?
2. List and describe the three causes of heat loss in a practical capacitor.
3. Plate and lead wire resistance in a capacitor is generally very (*low, high*).
4. Dielectric leakage resistance in a capacitor is generally very (*low, high*).
5. Dielectric dissipation resistance in a capacitor is generally very (*low, high*).
6. If the reactance of a particular capacitor is 1,000 Ω, and its ESR is 10, what is the *Q* of the capacitor?
7. If the reactance of a particular capacitor is 1,000 Ω, and its ESR is 10, what is the dissipation factor of the capacitor?
8. If the dissipation factor of a particular capacitor is 0.009 (0.9%), what is its *Q*?
9. What is the dissipation factor of a capacitor that has a *Q* of 200?
10. If the reactance of a capacitor is 250 Ω and the ESR is 5 Ω, what is the power factor of the capacitor?

14.6 Troubleshooting Capacitors

Capacitor malfunctions may be divided into three general classes of failures: open, short, and increased leakage. A technician must be able to test capacitors and determine their condition. Three methods that can be used to diagnose capacitor malfunctions are substitution, use of a capacitance tester, and ohmmeter tests.

Substitution

The condition of a suspected capacitor can easily be determined by replacing it in the circuit with another capacitor known to be good. At first this appears to be a poor technique, but it is actually wise in many cases. First, most capacitors are very inexpensive and readily available. You can remove and replace a capacitor faster than you can remove, test, and replace it. Generally, the labor savings far outweighs the cost of the capacitor. Second, the act of removing the capacitor for testing and the subsequent resoldering can stress an otherwise good capacitor, which may introduce even more trouble. In the case of surface-mount capacitors (quickly becoming the most popular type), substitution is the recommended method. You should always keep a supply of common capacitor values on hand for this purpose.

Substitution for purposes of troubleshooting is of no value unless the substituted component is *known to be good*. Many technicians and repair shops eventually accumulate a wide assortment of new and used parts mixed together in a box or drawer. Beware! If you accidentally substitute a defective capacitor, then your results will be misleading.

Capacitance Testers

Second only to substitution, the use of a capacitor tester is the preferred way to determine the condition of a capacitor. Although some capacitor testers are restricted to the testing of capacitors only, many technicians use an LCR tester, which can be used to test inductors, capacitors, and resistors. A representative LCR tester is shown in Figure 14-42.

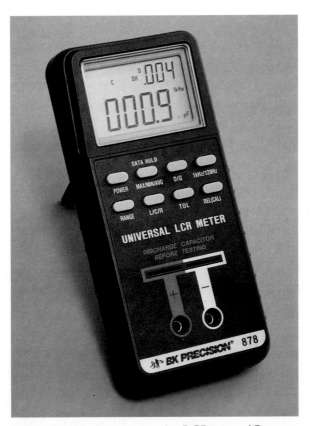

Figure 14-42. A representative LCR tester. (Courtesy B&K Precision)

The use of the capacitor (or LCR) tester is very straightforward. The technician connects the capacitor to be tested, and the display indicates the capacitance value. The leakage resistance can also be measured in some cases. Finally, if the capacitor is open or shorted, it is indicated by the tester.

Ohmmeter Tests

In some cases, a technician can get an estimate of the condition of a capacitor by using an analog ohmmeter. The method is illustrated in Figure 14-43. First, the capacitor to be tested is fully discharged by shorting its leads as shown in Figure 14-43(a). Next, the ohmmeter is switched to one of the higher ranges (e.g., R × 10k, R × 100k, or R × 1M) and connected across the capacitor (observe the polarity of electrolytics). As soon as the leads make contact, the meter pointer will swing to near 0 Ω on the scale as indicated in

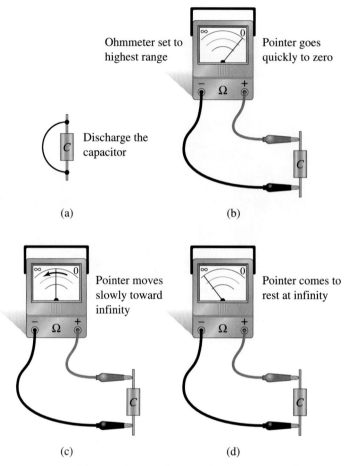

Ohmmeter set to highest range

Pointer goes quickly to zero

Discharge the capacitor

(a)

(b)

Pointer moves slowly toward infinity

Pointer comes to rest at infinity

(c)

(d)

Figure 14-43. Testing the condition of a capacitor with an ohmmeter.

KEY POINTS

Shorts can always be located with an ohmmeter.

Figure 14-43(b). It will then move slowly toward infinity on the scale, as shown in Figure 14-43(c), as the capacitor charges to the internal battery voltage of the ohmmeter. Finally, the pointer will come to rest near infinite ohms as shown in Figure 14-43(d). Now that we have discussed the general method, let us be more specific.

First, the time required for the meter to move between zero and infinite ohms (Figure 14-43[b] through Figure 14-43[d]) varies dramatically from several minutes to a few microseconds. The exact time depends on the meter (and range) used as well as the size of the capacitor. If the time span is too short, the pointer movement may be barely perceptible or even undetectable. In this case, use the highest range on the ohmmeter. For larger capacitor sizes (e.g., 0.1 µF and up), the pointer movement is easily detected. Generally, the larger the value of capacitance, the lower the range on the ohmmeter that gives a usable indication.

KEY POINTS

Opens can only be located with an ohmmeter when measuring relatively large (>1,000 pF) capacitance values.

Regardless of the capacitor size, a shorted capacitor will cause the ohmmeter to deflect to zero ohms and remain there. This means the dielectric has been damaged and the capacitor plates are electrically connected. The capacitor must be replaced.

If the capacitor is open, there will be no ohmmeter indication. This is a useful test for large capacitors, but provides no detectable information for smaller capacitors.

Finally, a capacitor that has a low dielectric leakage resistance (called a leaky capacitor) will cause the pointer to come to rest at some resistance lower than infinity. The classification of the measured value as good or bad must be made by the technician based on a comparison with a good capacitor of the same type. Ceramic capacitors, for example, have normal leakage resistances so high that most ohmmeters indicate infinite ohms. Aluminum electrolytics, by contrast, may have normal leakage resistances as low as 1 mΩ.

In general, the ohmmeter test for capacitors should be interpreted as follows. If the ohmmeter test indicates a defective capacitor (i.e., open, shorted, or leaky), then it is definitely defective. If, however, no defective indication can be observed on the meter, the capacitor might be good. You cannot be certain with an ohmmeter test.

Increased ESR

The ESR of a capacitor—particularly electrolytic capacitors—can increase as the capacitor ages. This causes circuit symptoms that may be the same as if a capacitor with too small of a value were used. This effect is often seen when troubleshooting power supply circuits. Detection of this type of defect is strongly dependent on the technician's knowledge of the circuit and the normal behavior of the capacitor in the given circuit.

> **KEY POINTS**
>
> Leaky capacitors can sometimes be located with an ohmmeter, particularly when measuring high values of capacitance.

Exercise Problems 14.6

1. Explain why capacitor substitution is used by many technicians as a cost-effective test method.
2. An LCR tester can be used to detect shorted or open capacitors. (True or False)
3. An ohmmeter has the following indications while testing a capacitor: When the meter is first connected, the pointer moves to zero and remains there. What is the condition of the capacitor?

Chapter Summary

• A capacitor is formed when two conductors are separated by an insulator. Capacitance is measured in farads. The value of a capacitor is determined by its physical characteristics: plate area, plate separation, and type of dielectric. Capacitance is directly proportional to the total area of the plates and the dielectric constant of the material separating the plates. It is inversely proportional to the distance between the plates.

• No current (except a tiny leakage current) flows through a capacitor. However, any change in circuit voltage causes a charging or discharging current through the circuit external to the capacitor. The dc circuits have only momentary current in the external circuit. Sinusoidal circuits have sinusoidal current flowing in the external circuits.

• Capacitors are available in several lead styles (e.g., axial, radial, SMT). They can be fixed or variable, polarized or nonpolarized, and they may be intentional (as a physical capacitor) or accidental (stray capacitance). All capacitors have capacitance, breakdown voltage, tolerance, leakage current, and temperature coefficient ratings. There are numerous technologies used to manufacture capacitors. Each technology has relative advantages and disadvantages.

- Series-connected capacitors combine like parallel-connected resistors (i.e., reciprocal equation). Parallel-connected capacitors combine like series-connected resistors (i.e., they add directly). Series-parallel combinations of capacitance can be simplified by replacing sets of series or parallel capacitors with an equivalent capacitance.

- Capacitive reactance is the opposition to sinusoidal current flow provided by a capacitor. It is measured in ohms and is inversely proportional to both frequency and capacitance. Connections of multiple capacitive reactances can be simplified using the same procedures used for resistive circuits.

- Ideal capacitors do not dissipate power. Practical capacitors have internal losses that appear as resistances and dissipate power in the form of heat. The resistance of the plates and leads and the dielectric dissipation are collectively called the equivalent series resistance (ESR). The ESR causes power dissipation in a capacitor. The quality (Q) of a capacitor is specified by the ratio of reactance to ESR at a particular frequency.

- Technicians must be able to test and classify capacitors as good or defective. A defective capacitor may be shorted, open, or leaky. Identification of a defective capacitor can be achieved by substitution or by using an LCR tester, a capacitance tester, or an ohmmeter. Substitution provides the most reliable indication, while ohmmeter tests are generally the most unreliable.

Review Questions

Section 14.1: Capacitance Fundamentals

1. A capacitor stores energy in an electromagnetic field. (True or False)

2. A capacitor consists of two _____ separated by a(n) _____.

3. The insulator material used in a capacitor is called the _____.

4. When a capacitor is connected to a dc source, substantial dc current flows through the dielectric. (True or False)

5. Capacitance is measured in _____.

6. If a charge of 100 µC produces a 10-V potential across a capacitor, what is the value of the capacitor?

7. A 680-pF capacitor has a voltage of 25 V across it. What is its charge in coulombs?

8. The ability of a material to concentrate an electrostatic field is called its

   _____.

9. Relative permittivity of a material is more commonly called its _____ _____.

10. Rank the following materials based on their approximate dielectric constant: air, ceramic, tantalum oxide.

11. Capacitance is (*directly, inversely*) proportional to the area of the plates.

12. Capacitance is (*directly, inversely*) proportional to the separation of the plates.

13. Capacitance is (*directly, inversely*) proportional to the dielectric constant.

14. Manufacturers often use multiple plates in a capacitor to increase the value of capacitance. (True or False)

Section 14.2: Capacitor Construction

15. In which style of capacitor do the leads extend from opposite ends of a cylindrical body?

16. In which style of capacitor do the leads extend from the same end of the capacitor?

17. What style of capacitor has no wire leads?

18. What type of capacitor package provides several capacitors within a single physical package?

19. Why is it important to connect a polarized capacitor correctly?

20. Explain the term stray capacitance.

21. What happens if the voltage rating of a capacitor is exceeded?

22. What is a self-healing capacitor?

23. A ceramic capacitor is marked as N750. What does this indicate?

24. What do the letters NPO on a capacitor indicate?

25. If a capacitor is rated for 560 pF at 25°C and is marked as P750, how much capacitance will it have at 50°C?

26. What is the value of a 1,000-pF capacitor (rated at 25°C) if it is marked as NPO and is operated at 35°C?

27. Compute the value of a capacitor that is rated for 680 pF at 25°C, has an N750 temperature coefficient, and is operated at 80°C.

28. A practical capacitor permits a small current to flow through the dielectric. This current is called _____ _____.

29. Capacitors with the largest values are electrolytics. (True or False)

30. Name the two types of electrolytic capacitors.

31. How can a technician identify an electrolytic capacitor from its physical markings?

32. What type of capacitor technology provides the highest capacitance for a given physical size?

Section 14.3: Multiple Circuit Capacitances

33. When capacitors are connected in series, total capacitance is (*increased, decreased*).

34. Paralleling capacitors (*increases, decreases*) total capacitance.

35. Multiple capacitances in series combine like resistances connected in _____.

36. Multiple capacitances in parallel combine like resistances in _____.

37. What is the total capacitance of a 2,700-pF and a 1,800-pF capacitor in series?

38. If four 1,000-pF capacitors are connected in series, what is the total capacitance?

39. What size capacitor must be connected in series with a 0.01-µF capacitor to produce a total capacitance of 0.002 µF?

40. What is the combined capacitance of the following capacitors—680 pF, 1,000 pF and 1,500 pF—if they are connected in series?

41. The total capacitance of a parallel capacitive circuit is less than the smallest branch capacitance. (True or False)

42. What is the total capacitance if 100 µF and 250 µF are in parallel?

43. Calculate the total capacitance when the following capacitors—270 pF, 330 pF, and 470 pF—are connected in parallel.

44. What value capacitor must be connected in parallel with 2.2 µF to produce a total capacitance of 5.5 µF?

45. What is the total capacitance of the circuit shown in Figure 14-44?

46. Find the total capacitance for the circuit shown in Figure 14-45.

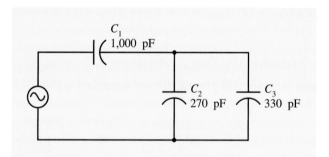

Figure 14-44. Calculate total capacitance.

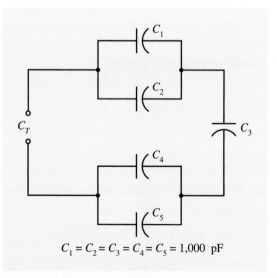

$C_1 = C_2 = C_3 = C_4 = C_5 = 1,000 \ pF$

Figure 14-45. Find total capacitance.

Section 14.4: Capacitive Reactance

47. Capacitive reactance opposes sinusoidal current flow. (True or False)

48. Capacitive reactance opposes dc current flow. (True or False)

49. As frequency increases for a given capacitance, what happens to the value of capacitive reactance?

50. Capacitive reactance is directly proportional to frequency. (True or False)

51. What is the value of capacitive reactance for a 2,700-pF capacitor when operated at 250 MHz?

52. How much capacitive reactance is presented by a 0.005-μF capacitor at a frequency of 750 Hz?

53. At what frequency does a 0.05-μF capacitor have a reactance of 53 kΩ?

54. What value of capacitor is required to produce 10 kΩ of reactance at a frequency of 50 MHz?

55. What is the total reactance if reactances of 10 kΩ and 25 kΩ are connected in series?

56. Compute the total reactance of the following series-connected capacitive reactances: 150 kΩ, 220 kΩ, 270 kΩ, and 330 kΩ.

57. Capacitive reactances in series combine like resistors in _____.

58. Capacitive reactances in parallel combine like resistances in _____.

59. What is the total capacitive reactance if capacitive reactances of 500 Ω and 1,000 Ω are connected in parallel?

60. If three capacitive reactances of 100 Ω each are connected in parallel, what is the total reactance?

61. Calculate the total capacitive reactance for the circuit shown in Figure 14-46.

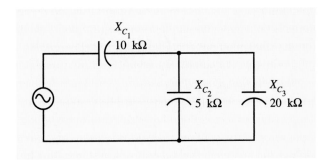

Figure 14-46. Find total capacitive reactance.

62. What is the total capacitive reactance for the circuit shown in Figure 14-47?

63. Find the total capacitance for the bridge circuit shown in Figure 14-48.

64. Compute the total capacitance for the circuit shown in Figure 14-49.

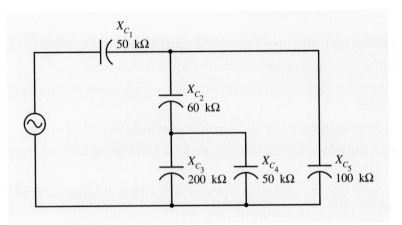

Figure 14-47. Compute total capacitive reactance.

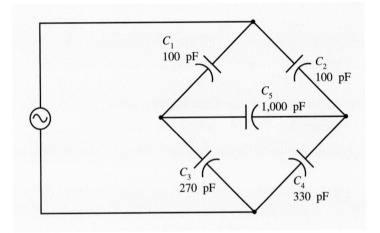

Figure 14-48. Compute total capacitance.

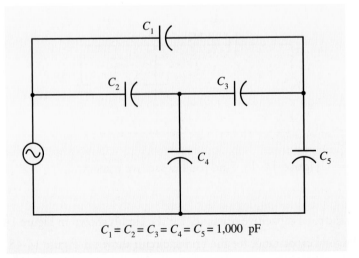

$C_1 = C_2 = C_3 = C_4 = C_5 = 1,000$ pF

Figure 14-49. Find total capacitance.

Section 14.5: Power Dissipation

65. An ideal capacitor dissipates no power. (True or False)

66. A practical capacitor converts some energy to heat. (True or False)

67. What is the ideal value of resistance for the capacitor's plates and lead wires?

68. What is the ideal value of resistance for the capacitor's dielectric?

69. Name two factors comprising the ESR of a capacitor.

70. If the reactance of a capacitor remained constant but the ESR could be reduced, what would happen to the Q of the capacitor?

71. If the Q of a capacitor is increased, what happens to the dissipation factor?

72. If the reactance of a capacitor is 2,500 Ω and its ESR is 5 Ω, what is the Q of the capacitor?

73. What is the dissipation factor of a 1,000-pF capacitor with an ESR of 5 Ω, if it is operated at 320 kHz?

74. What is the power factor of a capacitor with a reactance of 1,000 Ω and an ESR of 12 Ω?

Section 14.6: Troubleshooting Capacitors

75. Explain what is meant by the substitution method of capacitor checking.

76. Explain why an LCR tester or a capacitor tester is a better diagnostic tool for checking capacitors than a simple ohmmeter.

77. When testing a 10-pF capacitor with an ohmmeter, it measures 0 Ω at all times. Is the capacitor definitely defective?

78. When testing a 10-pF capacitor with an ohmmeter, it measures infinite ohms at all times. Is the capacitor definitely defective?

79. When testing a 1,000-μF capacitor with an ohmmeter, it first measures 0 Ω, but then slowly increases to 6.5 MΩ. Is the capacitor probably good or probably defective?

TECHNICIAN CHALLENGE

Let us assume that you are the senior technician on a remote radar site in northern Alaska. You have diagnosed a malfunction in the system's transmitter and have identified three defective capacitors. The capacitors are nonstandard values installed at the factory during final alignment of the transmitter. The values are 1,270 pF, 1,930 pF, and 1,333 pF.

The only capacitors that you have in your supply room are 270 pF, 390 pF, and 1,000 pF. Your challenge is to develop circuit configurations using any number of the available capacitors to produce the three values of needed capacitance.

Equation List

$$(14\text{-}1) \quad C = \frac{Q}{V}$$

$$(14\text{-}2) \quad \epsilon_r = \frac{\epsilon_m}{\epsilon_0}$$

$$(14\text{-}3) \quad C = \frac{8.85kA}{10^{12}d}$$

$$(14\text{-}4) \quad C = \frac{8.85kA}{10^{12}d}(n-1)$$

$$(14\text{-}5) \quad C_{\text{actual}} = C_{\text{rated}} + \left((T_{\text{actual}} - T_{\text{rated}}) \frac{T_C}{10^6} \right) C_{\text{rated}}$$

$$(14\text{-}6) \quad C_T = \frac{1}{\dfrac{1}{C_1} + \dfrac{1}{C_2} + \dots + \dfrac{1}{C_N}}$$

$$(14\text{-}7) \quad C_T = \frac{C_X}{N}$$

$$(14\text{-}8) \quad C_T = \frac{C_1 C_2}{C_1 + C_2}$$

$$(14\text{-}9) \quad C_T = C_1 + C_2 + \dots + C_N$$

$$(14\text{-}10) \quad X_C = \frac{1}{2\pi fC}$$

$$(14\text{-}11) \quad X_C = \frac{0.159}{fC}$$

$$(14\text{-}12) \quad X_{C_T} = X_{C_1} + X_{C_2} + \dots + X_{C_N}$$

$$(14\text{-}13) \quad X_{C_T} = \frac{1}{\dfrac{1}{X_{C_1}} + \dfrac{1}{X_{C_2}} + \dots + \dfrac{1}{X_{C_N}}}$$

$$(14\text{-}14) \quad Q = \frac{X_C}{\text{ESR}}$$

$$(14\text{-}15) \quad \text{DF} = \frac{1}{Q}$$

$$(14\text{-}16) \quad pf = \frac{\text{ESR}}{\sqrt{\text{ESR}^2 + X_C^2}}$$

Resistive-Capacitive Circuit Analysis

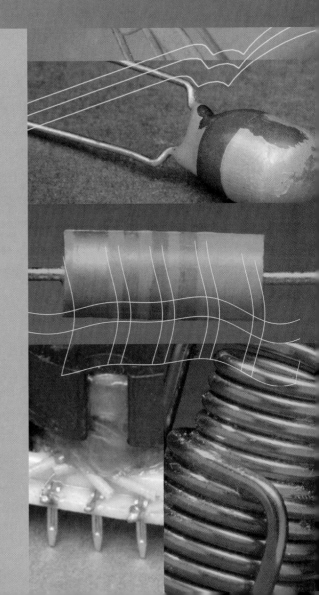

When resistance (R) and capacitance (C) are combined in the same circuit, we refer to it as an RC circuit. The characteristics of an RC circuit lie somewhere between those of a purely resistive circuit and a purely capacitive circuit. RC circuits may be configured as series, parallel, series-parallel, or complex circuits. The basic circuit analysis methods presented in previous chapters are applicable to RC circuits, but we need to consider phase relationships much in the same way as we did with RL circuits.

15.1 Series *RC* Circuits with Sinusoidal Current

As we work through the basic characteristics of series *RC* circuits, you should try to relate the information to previously studied material. There is a high degree of similarity between series *RC* circuits and other series circuits, in particular series *RL* circuits.

Series *RC* Circuit Characteristics

Figure 15-1 shows a series *RC* circuit consisting of one resistor and one capacitor. Let's examine the characteristics of this circuit configuration.

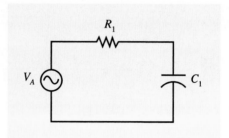

Figure 15-1. A series *RC* circuit.

CURRENT

Because the circuit in Figure 15-1 can be identified as a series circuit, we immediately know that the current must be the same in all parts of the circuit. If, for example, we knew that the current through C_1 was 575 µA, then we would also know that 575 µA was flowing through R_1 and the source. This is an important circuit characteristic. It is illustrated in Figure 15-2.

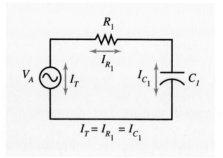

Figure 15-2. The current is the same through all components in a series *RC* circuit.

VOLTAGE DROPS

Ohm's Law tells us that the voltage across a component is proportional to its resistance (or reactance) and the value of current flowing through it. In the case of a series *RC* circuit, the current is the same through all components. Therefore, the various components will have voltage drops that are proportional to the resistance or reactance values. Those with higher resistances or reactances will have correspondingly higher voltage drops than components with less resistance or reactance. This circuit principle is illustrated in Figure 15-3.

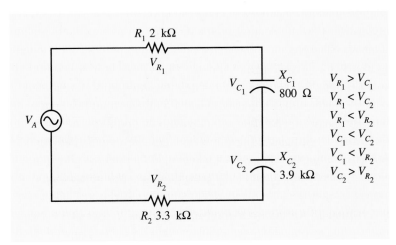

Figure 15-3. Component voltage drops in a series *RC* circuit are proportional to the resistance or reactance values.

Additionally, the voltage drops in a series circuit can be summed to find total voltage (Kirchhoff's Voltage Law). In the case of series *RC* circuits, the voltages must be summed using phasor addition, but the basic law is still valid.

PHASE RELATIONSHIPS

Our understanding of resistors and capacitors as isolated components will allow us to determine the phase relationships in a series *RC* circuit. We already know these things:

- Current is the same in all parts of the circuit at all times.
- Current and voltage are in phase in a resistance.
- Current leads voltage by 90° in a capacitor.

Figure 15-4 illustrates the phase relationships in the various components of a series *RC* circuit. Note that the current waveform is the same for all components. Also note that the current and voltage are in phase in the resistances. In the case of the capacitors, the current leads the voltage by 90°. Since the two capacitors have identical currents and the respective voltages both lag by 90°, it follows that the voltage waveforms for C_1 and C_2 are in phase with each other.

POWER FACTOR

The power factor in an *RC* circuit has the same meaning as the power factor in an *RL* circuit. That is, the power factor indicates the relative phase relationship between total current and total voltage. Since current is always used as a reference when speaking of power factors, we know that an *RC* circuit must have a leading power factor, since current leads voltage.

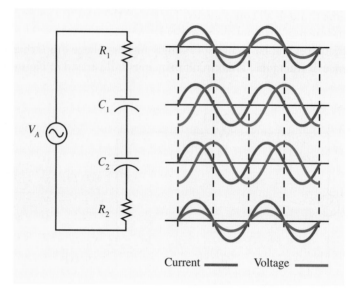

Figure 15-4. Phase relationships in a series *RC* circuit.

The power factor for an *RC* circuit is computed the same way as for an *RL* circuit (power factor = cos θ).

IMPEDANCE

As with *RL* circuits, impedance is the total opposition to current flow. It includes the combined effects of the resistance and capacitive reactance in the circuit. You will recall from your knowledge of series circuits that total opposition to current flow is found by adding the individual component resistances or reactances. In the case of *RC* circuits, we must use phasor addition, since the resistive and reactive components are 90° out of phase. Impedance in a series *RC* circuit can be found with Equation 15-1:

$$Z = \sqrt{R^2 + X_C^2} \qquad (15\text{-}1)$$

You will recognize this as a phasor addition problem similar to the impedance calculation for series *RL* circuits.

EXAMPLE SOLUTION

Calculate the impedance for the circuit shown in Figure 15-5.

EXAMPLE SOLUTION

We apply Equation 15-1 as follows:

$$Z = \sqrt{R^2 + X_C^2}$$
$$= \sqrt{(10\ \text{k}\Omega)^2 + (20\ \text{k}\Omega)^2} = 22.36\ \text{k}\Omega$$

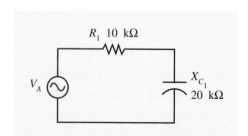

Figure 15-5. What is the impedance of this circuit?

Practice Problems

1. What is the impedance of a series circuit consisting of a 4.7-kΩ resistor and a 2.2-kΩ capacitive reactance?

2. Calculate the impedance of a series circuit consisting of a 680-Ω resistor and a 1.5-kΩ capacitive reactance.

3. What is the impedance of a series circuit consisting of a 2.7-kΩ resistor and a 1,000-pF capacitor? The circuit is operating at 40 kHz.

Answers to Practice Problems

1. 5.19 kΩ 2. 1.65 kΩ 3. 4.81 kΩ

SUSCEPTANCE AND ADMITTANCE

Susceptance and admittance have essentially the same meaning for RC circuits as they do for RL circuits. In the case of RC circuits, susceptance (B) is the reciprocal of capacitive reactance, as reflected in Equation 15-2.

$$B = \frac{1}{X_C} \qquad (15\text{-}2)$$

Admittance is defined as the reciprocal of impedance, just as it is in RL circuits. Recall that both susceptance and admittance are measured in siemens (S).

EXAMPLE SOLUTION

Find the susceptance and the admittance for the circuit shown in Figure 15-6.

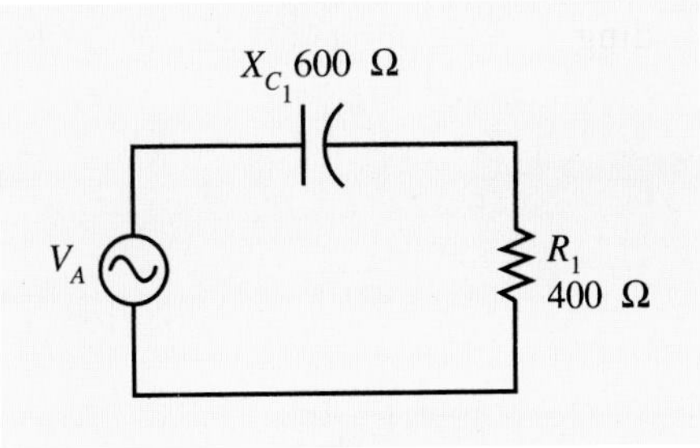

EXAMPLE SOLUTION

We find the susceptance by direct application of Equation 15-2.

$$B = \frac{1}{X_C}$$

$$= \frac{1}{600 \ \Omega} = 1.67 \text{ mS}$$

Next, we find the impedance of the circuit by applying Equation 15-1.

$$Z = \sqrt{R^2 + X_C{}^2}$$

$$= \sqrt{(400 \ \Omega)^2 + (600 \ \Omega)^2} = 721.11 \ \Omega$$

The reciprocal of impedance gives us the value of admittance:

$$Y = \frac{1}{Z} = \frac{1}{721.11 \ \Omega} = 1.39 \text{ mS}$$

Figure 15-6. What is the susceptance and admittance of this circuit?

1. What is the susceptance of a 0.01-μF capacitor when it is operating at a frequency of 2.5 MHz?

2. If the capacitive reactance of a capacitor is 2.9 kΩ, what is its susceptance?

3. What is the admittance of a series RC circuit whose impedance is 18 kΩ?

Answers to Practice Problems

1. 157 mS   **2.** 344.8 μS   **3.** 55.56 μS

Circuit Simplification

Simplification of a series RC circuit is a straightforward process that combines the various resistive elements into a single resistance and all capacitive elements into a single capacitance. Recall that series resistances are combined by simple addition, but series capacitances require use of the reciprocal equation.

EXAMPLE SOLUTION

Simplify the circuit shown in Figure 15-7.

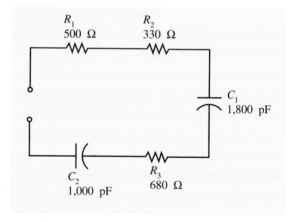

Figure 15-7. Simplify this circuit.

EXAMPLE **SOLUTION**

We combine the resistors as we would in a simple resistive circuit.

$$R_T = R_1 + R_2 + R_3$$

$$= 500 \ \Omega + 330 \ \Omega + 680 \ \Omega = 1.51 \ k\Omega$$

Next, we combine C_1 and C_2 using the reciprocal equation presented in Chapter 13.

$$C_T = \frac{1}{\dfrac{1}{C_1} + \dfrac{1}{C_2}}$$

$$= \frac{1}{\dfrac{1}{1,800 \ pF} + \dfrac{1}{1,000 \ pF}} = 642.86 \ pF$$

The simplified circuit is shown in Figure 15-8. It could be further simplified by combining the total resistance and total capacitance into an equivalent impedance.

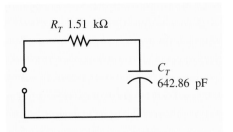

Figure 15-8. A simplified version of the circuit shown in Figure 15-7.

Practice Problems

1. Simplify the *RC* circuit shown in Figure 15-9.

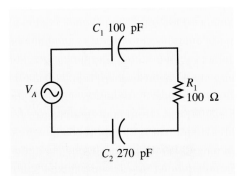

Figure 15-9. Simplify this circuit.

2. Reduce the circuit shown in Figure 15-10 to a circuit consisting of one resistance and one capacitance.

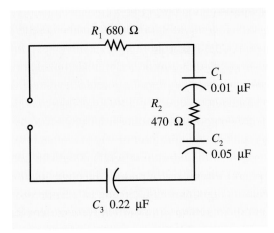

Figure 15-10. Reduce this circuit.

3. Simplify the *RC* circuit shown in Figure 15-11.

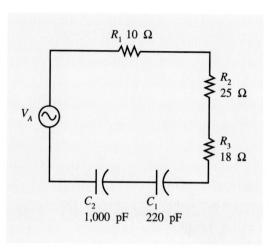

Figure 15-11. Simplify this circuit.

Answers to Practice Problems

1.

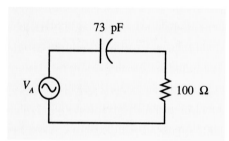

2.

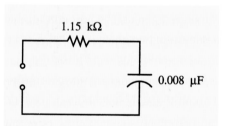

3.

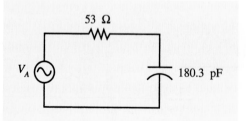

Phasor Representation

Technicians often use phasor diagrams to represent the phase relationships in a series *RC* circuit. In principle, phasor diagrams for *RC* circuits are identical to the ones used to explain *RL* circuits. They differ somewhat, however, since current leads voltage in a capacitive circuit, but lags in an inductive circuit.

VOLTAGE PHASORS

Current is the same for all components in a series *RC* circuit. For this reason, we will use current as the reference phasor. The various voltage phasors can then be properly drawn with reference to the common current phasor. Also recall that the total voltage in a series circuit is found by adding the individual component voltage drops. In the case of an *RC* circuit, we must use phasor addition as reflected in Equation 15-3:

$$V_A = \sqrt{V_R^2 + V_C^2} \qquad \text{(15-3)}$$

Recall that the resistive and capacitive voltage drops are 90° out of phase with each other.

EXAMPLE SOLUTION

Draw a voltage phasor diagram to represent the circuit shown in Figure 15-12.

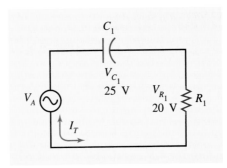

Figure 15-12. Draw a voltage phasor diagram for this circuit.

EXAMPLE **SOLUTION**

First, we draw a horizontal phasor extending to the right as our reference phasor. This represents the current in our series circuit. This phasor is shown in Figure 15-13.

Next, we add our resistive voltage phasor. Since voltage and current are in phase in a resistance, our resistive voltage phasor has the same position as the current phasor. This is shown in Figure 15-13. Recall that the length of a voltage phasor with respect to a current phasor is arbitrary.

We can now sketch the phasor to represent the capacitive voltage. Since voltage in a capacitor lags the current by 90°, we will draw our V_{C_1} phasor downward (i.e., 90° behind the current phasor). Recall that the phasors are assumed to rotate in a counterclockwise direction. Figure 15-13 shows the position of the capacitive voltage phasor. Also note the relative lengths of the two voltage phasors.

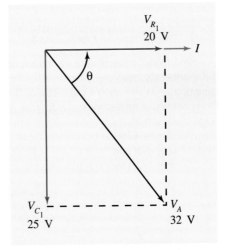

Figure 15-13. A voltage phasor diagram for the circuit shown in Figure 15-12.

We can add (phasor addition) the values of V_{R_1} and V_{C_1} to find total voltage. You will recall that we can perform this addition graphically, by simply completing a parallelogram between the two phasors being added. In Figure 15-13, we complete a parallelogram between V_{R_1} and V_{C_1}. Their phasor sum corresponds to total voltage. It is represented by the diagonal of the parallelogram. The value of the applied voltage can be found by performing the numerical phasor addition process with Equation 15-3.

$$V_A = \sqrt{V_{R_1}^2 + V_{C_1}^2}$$

$$= \sqrt{(20 \text{ V})^2 + (25 \text{ V})^2} = 32.02 \text{ V}$$

Finally, we label the angle between total current and total voltage as θ. Figure 15-13 illustrates the completed phasor diagram.

IMPEDANCE PHASORS

Impedance phasors show how the resistance and capacitive reactance in a series RC circuit combine to form circuit impedance. Because the component voltage drops in a series circuit are directly proportional to the relative resistances and reactances, the impedance phasor diagram has the same basic relationships as the voltage phasor diagram.

EXAMPLE SOLUTION

Draw an impedance phasor diagram to represent the circuit shown in Figure 15-14.

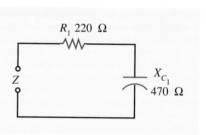

Figure 15-14. Draw an impedance phasor diagram for this circuit.

EXAMPLE SOLUTION

It is customary to draw the resistance phasor in the reference position (i.e., extending horizontally to the right). This is shown in Figure 15-15.

Next, we add the phasor to represent capacitive reactance. We draw this phasor 90° behind the resistive phasor as shown in Figure 15-15. By completing the parallelogram between the R_1 and X_{C_1} phasors, we can sketch the position of circuit impedance. Recall that impedance in a series circuit is equal to the sum (phasor sum in this case) of the individual component resistances or reactances.

Finally, as shown in Figure 15-15, we label the circuit phase angle as θ. This is always the angle formed by the resistance and impedance phasors.

POWER PHASORS

The power phasor diagram for a series RC circuit is similar to the power phasor diagram for a series RL circuit, except the reactive power lags true power in a series RC circuit. Figure 15-16 shows a basic power phasor diagram for a series RC circuit.

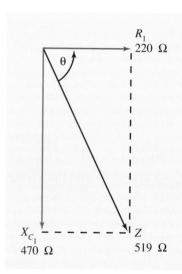

Figure 15-15. An impedance phasor diagram to represent the circuit shown in Figure 15-14.

The phase angle (θ) is formed between the true power and the apparent power phasors. The basic relationships between true power, reactive power, and apparent power were discussed in Chapter 13 and are not repeated here.

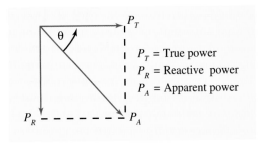

P_T = True power
P_R = Reactive power
P_A = Apparent power

Figure 15-16. A power phasor diagram for a series *RC* circuit.

Exercise Problems 15.1

1. If the current through the resistance in a series *RC* circuit is 225 mA, what is the current through the series capacitance?

2. If the current through the resistance in a series *RC* circuit increases, what must happen to the current through the series capacitance?

3. What is the relationship between the voltage drops in a series *RC* circuit and the applied voltage?

4. The current through a series *RC* circuit leads the resistive voltage drop by 90°. (True or False)

5. The current through a series *RC* circuit leads the capacitive voltage drop by 90°. (True or False)

6. The voltage drop across the resistance in a series *RC* circuit leads the capacitive voltage drop by 90°. (True or False)

7. Total opposition to current flow in a series *RC* circuit is called _____.

8. What is the admittance of a series *RC* circuit whose impedance is 150 kΩ?

9. What is the impedance of a 120-kΩ resistor in series with a capacitive reactance of 85 kΩ?

10. What is the susceptance of a capacitor that has a reactance of 250 Ω?

11. Draw a voltage phasor diagram to represent the circuit shown in Figure 15-17. Be sure to include the value of V_A.

12. Draw an impedance phasor diagram for the circuit shown in Figure 15-17. Be sure to include the value of impedance.

13. In a series *RC* circuit, apparent power is the phasor sum of true power and reactive power. (True or False)

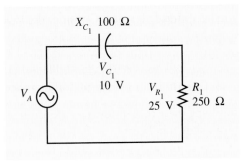

Figure 15-17. A series *RC* circuit.

15.2 Parallel *RC* Circuits with Sinusoidal Current

Now let's examine parallel *RC* circuits and identify some of their important characteristics. You will find much similarity between parallel *RC* circuits and other parallel circuits studied previously (i.e., resistive and inductive circuits).

Parallel *RC* Circuit Characteristics

VOLTAGE DROPS

The voltage drop is the same across all components in a parallel *RC* circuit, just as it is the same for every parallel circuit. Figure 15-18 illustrates this point. If we know the voltage across any one component in a parallel circuit, we immediately know the voltage across all components.

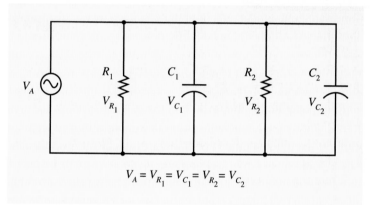

Figure 15-18. Every component in a parallel *RC* circuit has the same voltage drop.

PHASE RELATIONSHIPS

It is easy to predict the phase relationships in a parallel *RC* circuit if you simply apply what you have already learned. Consider the following:

- Voltage is the same across all components in the circuit.
- Current and voltage are in phase in a resistance.
- Current leads voltage by 90° in a capacitance.

Applying these three known relationships allows us to sketch the current and voltage waveforms shown in Figure 15-19. First, note that the voltage waveforms across all components are identical. Next, see that the current and voltage waveforms for the resistances are in phase. The current waveform through the capacitors, by contrast, leads the capacitive voltage waveforms by 90°. Finally, we see that the total current waveform leads the total voltage waveform by some angle less than 90°. We would expect the total current and voltage to have a relationship somewhere between a purely resistive circuit ($\theta = 0°$) and a purely capacitive circuit ($\theta = 90°$).

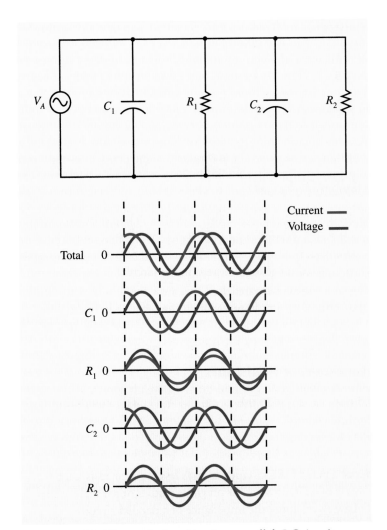

Figure 15-19. Phase relationships in a parallel *RC* circuit.

CURRENT

As with all parallel circuits, the current through each branch of a parallel *RC* circuit is determined by the applied voltage and the resistance or reactance of the particular branch. The current in a given branch is unaffected by the currents in other branches.

We know that the individual branch currents in a parallel circuit combine to form total current. This is also true for a parallel *RC* circuit, but the resistive and capacitive currents must be combined with phasor addition as indicated by Equation 15-4.

$$I_T = \sqrt{I_R^2 + I_C^2} \qquad (15\text{-}4)$$

KEY POINTS

The branch currents combine (phasor addition) to form total current.

EXAMPLE SOLUTION

Compute the total current in a parallel *RC* circuit if the resistive branch has 2.6 A of current and the capacitive branch has 1.8 A.

EXAMPLE **SOLUTION**

We apply Equation 15-4 as follows:

$$I_T = \sqrt{I_R^2 + I_C^2}$$
$$= \sqrt{(2.6 \text{ A})^2 + (1.8 \text{ A})^2} = 3.16 \text{ A}$$

Practice Problems

1. If the resistive branch of a parallel RC circuit has 125 mA and the capacitive branch has 250 mA, how much current flows through the source?
2. The resistive current in a parallel RC circuit is 500 μA and the capacitive current is 375 μA. What is the total current in the circuit?
3. If the total current in a parallel RC circuit is 5 A and the capacitive branch has 3 A, how much current flows through the resistive branch?
4. How much capacitive current must flow in a parallel RC circuit to produce 180 mA when combined with 100 mA of resistive current?

Answers to Practice Problems

1. 279.5 mA   **2.** 625 μA   **3.** 4 A   **4.** 149.7 mA

IMPEDANCE

The impedance of a parallel RC circuit is always less than the resistance or reactance of any one branch. This is consistent with our previous studies of parallel circuits. As before, we can use Ohm's Law to compute the impedance of a parallel circuit.

EXAMPLE SOLUTION

If the resistive current in a parallel RC circuit is 100 mA and the capacitive current is 50 mA when 100 V are applied to the circuit, what is the impedance of the circuit?

EXAMPLE **SOLUTION**

In order to apply Ohm's Law to find total impedance, we need to find total voltage and total current. We use Equation 15-4 to compute total current.

$$I_T = \sqrt{I_R^2 + I_C^2}$$
$$= \sqrt{(100 \text{ mA})^2 + (50 \text{ mA})^2} = 111.8 \text{ mA}$$

Since total voltage is given, we may now use Ohm's Law to compute impedance.

$$Z = \frac{V_A}{I_T}$$
$$= \frac{100 \text{ V}}{111.8 \text{ mA}} = 894.45 \text{ }\Omega$$

SUSCEPTANCE AND ADMITTANCE

Definitions and relationships for susceptance and admittance in parallel RC circuits are similar to those discussed with reference to series RC circuits. They are not repeated here.

Circuit Simplification

A parallel RC circuit consisting of multiple resistances and/or multiple capacitances can be simplified to a circuit consisting of a single resistance and a single capacitance. Resistive branches are combined according to the rules for parallel resistive circuits. Capacitive branches are combined according to the rules for capacitive circuits.

EXAMPLE SOLUTION

Simplify the parallel RC circuit shown in Figure 15-20.

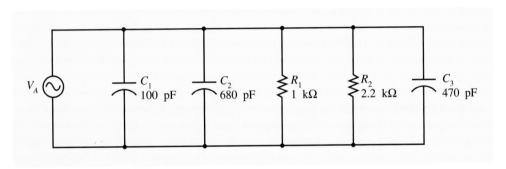

Figure 15-20. Simplify this circuit.

EXAMPLE **SOLUTION**

First, we combine the resistive branches.

$$R_T = \frac{R_1 R_2}{R_1 + R_2}$$

$$= \frac{1.0 \text{ k}\Omega \times 2.2 \text{ k}\Omega}{1.0 \text{ k}\Omega + 2.2 \text{ k}\Omega} = 687.5 \ \Omega$$

Next, we combine the capacitive branches.

$$C_T = C_1 + C_2 + C_3$$

$$= 100 \text{ pF} + 680 \text{ pF} + 470 \text{ pF} = 1{,}250 \text{ pF}$$

The simplified circuit is shown in Figure 15-21.

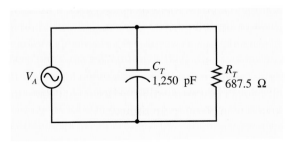

Figure 15-21. The simplified form of the circuit shown in Figure 15-20.

> **KEY POINTS**
>
> Susceptance and admittance have definitions and relationships similar to those in series RC circuits.

> **KEY POINTS**
>
> A parallel RC circuit consisting of multiple resistive branches and/or multiple capacitive branches can be simplified by combining similar branches.

> **KEY POINTS**
>
> Resistive branches are combined with the reciprocal equation, whereas capacitive branches are additive.

Practice Problems

1. Draw the equivalent circuit if three resistors having values of 330 Ω, 470 Ω, and 680 Ω are connected in parallel with two capacitors having values of 25 μF and 80 μF.

2. Draw the equivalent circuit if three capacitors having values of 2,200 pF, 1,000 pF, and 1,800 pF are connected in parallel with three resistors having values of 10 kΩ, 22 kΩ, and 18 kΩ.

3. Draw the simplified circuit if a 120-kΩ resistor is connected in parallel with two capacitors having values of 5 μF and 25 μF.

Answers to Practice Problems

1.

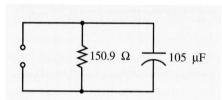

2.

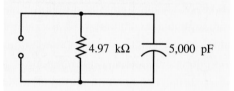

3.

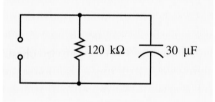

Phasor Representation

The two types of phasor diagrams that are most useful to a technician when analyzing parallel *RC* circuits are current phasors and power phasors.

CURRENT PHASORS

KEY POINTS

Current and power phasors are used by technicians to help visualize phase relationships in a parallel *RC* circuit.

If we apply what we already know about resistors, capacitors, and basic circuit theory, we should be able to determine the phase relationships between current and voltage in a parallel *RC* circuit. Consider the following:

- The branch currents must be summed to find total current (Kirchhoff's Current Law).
- Voltage is the same across all components in a parallel circuit.
- Current and voltage are in phase in a resistance.
- Current leads voltage by 90° in a capacitance.

EXAMPLE SOLUTION

Draw a current phasor diagram for the circuit shown in Figure 15-22.

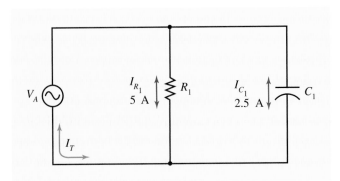

Figure 15-22. Draw a current phasor diagram for this circuit.

EXAMPLE SOLUTION

Since voltage is common to all components in a parallel circuit, we will use voltage as the reference phasor. Figure 15-23 shows the voltage phasor V_A drawn in the reference position.

Next, we draw the resistive current phasor at the same angle as our reference phasor, since current and voltage are in phase in a resistance. The phasor representing capacitive current must be shown leading the voltage by 90°. Since the phasors rotate counterclockwise, we will have to sketch the capacitive current phasor in the upward direction. This phasor is labeled as I_{C_1} in Figure 15-23.

We can determine the location of the total current phasor by completing the parallelogram between the capacitive and resistive current phasors. We can determine the numerical value of this phasor by applying Equation 15-4.

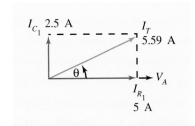

Figure 15-23. A current phasor diagram for the circuit shown in Figure 15-22.

$$I_T = \sqrt{I_R^2 + I_C^2}$$

$$= \sqrt{(5\ A)^2 + (2.5\ A)^2} = 5.59\ A$$

Finally, as shown in Figure 15-23, we label the phase angle (θ) between total current and total voltage.

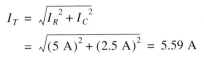

Practice Problems

1. Draw a current phasor diagram for the circuit shown in Figure 15-24. Be sure to compute the value of total current.

2. Simplify the circuit shown in Figure 15-25 by combining similar currents, then draw the current phasor diagram and compute total current.

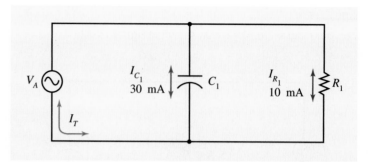

Figure 15-24. Draw the current phasor diagram for this circuit.

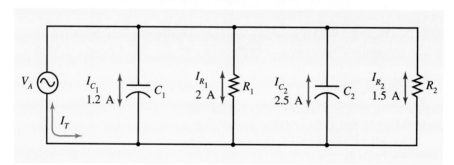

Figure 15-25. Simplify this circuit and draw a current phasor diagram.

Answers to Practice Problems

1.

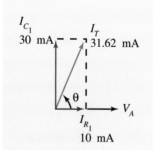

2.

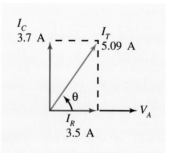

KEY POINTS

The power phasors have the same relative relationships as with RL circuits and series RC circuits. The positions of the power phasors correspond to the positions of the current phasors.

POWER PHASORS

The relationships between true power, reactive power, and apparent power are similar for both series and parallel RC circuits. The power phasor diagram for a parallel RC circuit is drawn in the same relative position as the current phasors, where true power, reactive power, and apparent power correspond to the phasor positions of resistive current, capacitive current, and total current, respectively. Figure 15-26 shows the relationships in a power phasor diagram for a parallel RC circuit.

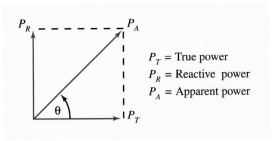

Figure 15-26. A generic power phasor diagram for a parallel RC circuit.

Exercise Problems 15.2

1. The total voltage in a parallel RC circuit is equal to the phasor sum of the individual branch voltages. (True or False)

2. What is the phase relationship between current and voltage in a resistive branch of a parallel RC circuit?

3. What is the phase relationship between current and voltage in a capacitive branch of a parallel RC circuit?

4. What is the phase relationship between total current and total voltage in a parallel RC circuit?

5. The total current in a parallel RC circuit can be found by summing (phasor addition) the individual branch currents. (True or False)

6. If the resistive branch current of a parallel RC circuit is 125 mA and the capacitive current is 75 mA, how much current flows through the source?

7. If the total current in a parallel RC circuit is 200 μA and the resistive current is 100 μA, what is the value of capacitive current?

8. What is the impedance of a parallel RC circuit if the applied voltage is 12 V rms and the total current is 1.6 A rms?

9. Simplify the parallel RC circuit shown in Figure 15-27.

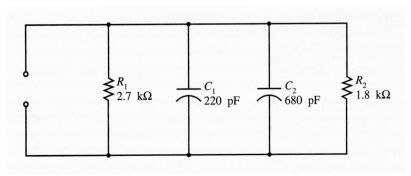

Figure 15-27. Simplify this circuit.

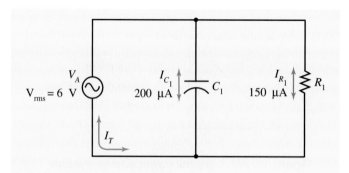

Figure 15-28. A parallel *RC* circuit.

10. Draw the current phasor diagram to represent the circuit shown in Figure 15-28. Be sure to compute the value of I_T.

11. Refer to Figure 15-28. What is the impedance of the circuit?

12. What is the susceptance of the capacitive branch of the circuit in Figure 15-28?

13. Compute the admittance of the circuit shown in Figure 15-28.

15.3 RC Circuit Calculations

We are now ready to thoroughly analyze all types of circuit configurations for *RC* circuits. We shall explore series, parallel, series-parallel, and complex circuits. In all cases, we will rely heavily on previously learned equations, relationships, and techniques. As much as practical, we will apply our prior knowledge to the logical solution of unfamiliar problems. This method is much better than rote memorization of lists of equations suited to a specific kind of problem. Once you gain confidence in applying the basic circuit analysis techniques to new circuit types, you will have some very powerful skills that will serve you the rest of your career.

Series *RC* Circuit Computations

When analyzing series *RC* circuits, as with any series circuit, it is generally helpful to compute current as soon as practical. Once you know current, then you have one factor that is common to all components. Remember, if you know any two factors (e.g., current, voltage, power, resistance, reactance, and so on), then you can calculate most of the remaining factors. There are a few exceptions involving frequency.

EXAMPLE SOLUTION

Analyze the series *RC* circuit shown in Figure 15-29, and complete the solution matrix shown in Table 15-1.

EXAMPLE **SOLUTION**

There are many paths that lead to the solution of this type of problem. Experience will help you find the most direct one, but for now, compute any unknown you can. The more values you compute, the more options you will

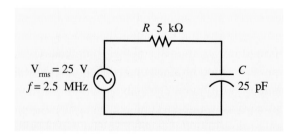

Figure 15-29. Analyze this series *RC* circuit.

CIRCUIT QUANTITY		RESISTOR	CAPACITOR	TOTAL
Current (rms)				
Voltage	V_P			
	V_{PP}			
	V_{rms}			25 V
Resistance/Reactance/Impedance		5 kΩ		
Conductance/Susceptance/Admittance				
Power				
Phase Angle (overall)				
Power Factor (overall)				

Table 15-1. Solution Matrix for the Circuit in Figure 15-29

have for subsequent calculations. Let's begin by computing the peak and peak-to-peak values of applied voltage.

$$V_P = 1.414 V_{rms}$$
$$= 1.414 \times 25 \text{ V} = 35.35 \text{ V}$$

Doubling the peak voltage will give us the peak-to-peak value.

$$V_{PP} = 2 V_P$$
$$= 2 \times 35.35 \text{ V} = 70.7 \text{ V}$$

We can now find the value of capacitive reactance, since we know the capacitance and the frequency.

$$X_C = \frac{1}{2\pi f C}$$

$$= \frac{1}{6.28 \times 2.5 \text{ MHz} \times 25 \text{ pF}} = 2.55 \text{ k}\Omega$$

We now have enough information to find the circuit impedance by applying Equation 15-1.

$$Z = \sqrt{R^2 + X_C^2}$$
$$= \sqrt{(5 \text{ k}\Omega)^2 + (2.55 \text{ k}\Omega)^2} = 5.61 \text{ k}\Omega$$

Since we know the applied voltage and the circuit impedance, we can compute current flow in the circuit. We normally use rms values unless we have a specific reason to do otherwise.

$$I_T = \frac{V_A}{Z}$$

$$= \frac{25 \text{ V}}{5.61 \text{ k}\Omega} = 4.46 \text{ mA}$$

Of course, if we want to know the peak or peak-to-peak current, we can repeat this calculation, or simply convert the rms current value directly. In any case, the value of current that we have computed will flow through the entire circuit.

We can now find the individual component voltage drops by applying Ohm's Law.

$$V_R = I_R R$$
$$= 4.46 \text{ mA} \times 5 \text{ k}\Omega = 22.3 \text{ V}$$

A similar calculation gives us the capacitor voltage drop.

$$V_C = I_C X_C$$
$$= 4.46 \text{ mA} \times 2.55 \text{ k}\Omega = 11.37 \text{ V}$$

We should always be alert for opportunities to catch any errors. We know that Kirchhoff's Voltage Law says the sum of the component voltage drops in a series circuit must equal the supply voltage. Do they in this case? Well, probably so, if we remember that the resistor and capacitor voltages are out of phase, so they must be added with phasor addition. Let's check our work at this point with Kirchhoff's Law.

$$V_A = \sqrt{V_R^2 + V_C^2}$$
$$= \sqrt{(22.3 \text{ V})^2 + (11.37 \text{ V})^2} = 25 \text{ V}$$

We now have the confidence to proceed. Let's use our knowledge of basic sine wave relationships and convert the rms component voltage drops to peak and peak-to-peak values.

$$V_{R_P} = 1.414 V_{R_{rms}} = 1.414 \times 22.3 \text{ V} = 31.53 \text{ V}$$
$$V_{R_{PP}} = 2 V_{R_P} = 2 \times 31.53 \text{ V} = 63.06 \text{ V}$$

Similarly,

$$V_{C_P} = 1.414 V_{C_{rms}} = 1.414 \times 11.37 \text{ V} = 16.08 \text{ V}$$
$$V_{C_{PP}} = 2 V_{C_P} = 2 \times 16.08 \text{ V} = 32.16 \text{ V}$$

We can find the values of conductance, susceptance, and admittance by simply taking the reciprocal of resistance, reactance, and impedance, respectively.

$$G = \frac{1}{R} = \frac{1}{5 \text{ k}\Omega} = 200 \text{ μS}$$

We find susceptance in a similar manner.

$$B = \frac{1}{X_C} = \frac{1}{2.55 \text{ k}\Omega} = 392.2 \text{ μS}$$

The reciprocal of impedance will give us the value of admittance.

$$Y = \frac{1}{Z} = \frac{1}{5.61 \text{ k}\Omega} = 178.25 \text{ μS}$$

We can use any of the basic power relationships to compute power in the circuit. We normally use rms values, unless we have some specific reason to compute another quantity (e.g., peak power). Let's begin by finding true power (i.e., that which is dissipated by the resistance as heat).

$$P_T = I_{R_{rms}} \times V_{R_{rms}} = 4.46 \text{ mA} \times 22.3 \text{ V} = 99.5 \text{ mW}$$

A similar calculation will give us the value of reactive power due to the capacitance.

$$P_R = I_{C_{rms}} \times V_{C_{rms}} = 4.46 \text{ mA} \times 11.37 \text{ V} = 50.7 \text{ mVAR}$$

Finally, we can sum (phasor addition) the true and reactive powers to find total or apparent power.

$$P_A = \sqrt{P_T^2 + P_R^2}$$
$$= \sqrt{(99.5 \text{ mW})^2 + (50.7 \text{ mVAR})^2} = 111.7 \text{ mVA}$$

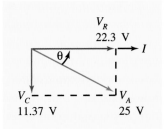

Figure 15-30.
Voltage phasor diagram for the circuit in Figure 15-29.

Let's use the voltage phasor diagram to help us identify the phase angle. Figure 15-30 shows the voltage phasor diagram for the circuit in Figure 15-29. The phase angle (θ) is always the angle between total voltage and total current. Since we know all three sides of the right triangle, we can use any of the basic trigonometric functions to find the phase angle. Let's choose the cosine function, so that we will also be finding the power factor of the circuit (i.e., power factor = cos θ).

$$\cos \theta = \frac{\text{adjacent}}{\text{hypotenuse}} = \frac{V_R}{V_A} = \frac{22.3 \text{ V}}{25 \text{ V}} = 0.892$$
$$\theta = \arccos 0.892 = 26.9°$$

The power factor can be further described as leading, since current is always assumed to be the reference when describing power factors as leading or lagging.

This completes our analysis of the circuit shown in Figure 15-29. The completed solution matrix is shown in Table 15-2.

CIRCUIT QUANTITY		RESISTOR	CAPACITOR	TOTAL
Current (rms)		4.46 mA	4.46 mA	4.46 mA
Voltage	V_P	31.53 V	16.08 V	35.35 V
	V_{PP}	63.06 V	32.16 V	70.7 V
	V_{rms}	22.3 V	11.37 V	25 V
Resistance/Reactance/Impedance		5 kΩ	2.55 kΩ	5.61 kΩ
Conductance/Susceptance/Admittance		200 μS	392.2 μS	178.25 μS
Power		99.5 mW	50.7 mVAR	111.7 mVA
Phase Angle (overall)		26.9°		
Power Factor (overall)		0.892 (leading)		

Table 15-2. Solution Matrix for the Circuit in Figure 15-29

We could also have solved this problem using rectangular and polar notation to describe the various circuit quantities. When representing capacitive reactance with rectangular notation we use the $-j$ prefix. Use of polar and rectangular notation can reduce your calculations, depending on the type of engineering calculator you are using. For example, let us recompute current using complex notation.

$$I_T = \frac{V_A}{Z}$$

$$= \frac{25 \text{ V}}{5 \text{ k}\Omega - j2.55 \text{ k}\Omega}$$

$$= 4.46 \text{ mA } \angle 26.9°$$

We get the value of current and the phase angle in a single calculation.

Practice Problems

1. Analyze the series RC circuit shown in Figure 15-31, and complete a solution matrix similar to Table 15-1.

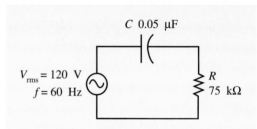

Figure 15-31. Analyze this circuit.

2. Analyze the circuit shown in Figure 15-32 and complete a solution matrix like Table 15-1.

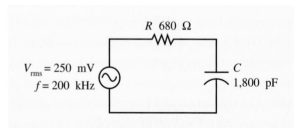

Figure 15-32. Analyze this circuit.

3. Analyze the circuit shown in Figure 15-33, and complete a solution matrix similar to Table 15-1. Be sure to include an entry in the matrix for frequency.

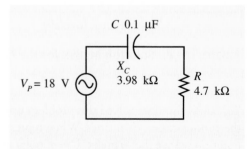

Figure 15-33. Analyze this circuit.

Answers to Practice Problems

1.

CIRCUIT QUANTITY		RESISTOR	CAPACITOR	TOTAL
Current (rms)		1.31 mA	1.31 mA	1.31 mA
Voltage	V_P	139 V	98.4 V	169.7 V
	V_{PP}	278 V	196.8 V	339.4 V
	V_{rms}	98.3 V	69.6 V	120 V
Resistance/Reactance/Impedance		75 kΩ	53.1 kΩ	91.9 kΩ
Conductance/Susceptance/Admittance		13.3 μS	18.8 μS	10.9 μS
Power		128.8 mW	91.2 mVAR	157.2 mVA
Phase Angle (overall)		35.3°		
Power Factor (overall)		0.816 (leading)		

2.

CIRCUIT QUANTITY		RESISTOR	CAPACITOR	TOTAL
Current (rms)		308.2 μA	308.2 μA	308.2 μA
Voltage	V_P	0.296 V	0.193 V	0.35 V
	V_{PP}	0.593 V	0.386 V	0.7 V
	V_{rms}	0.210 V	0.136 V	0.25 V
Resistance/Reactance/Impedance		680 Ω	442.3 Ω	811.2 Ω
Conductance/Susceptance/Admittance		1.47 mS	2.26 mS	1.23 mS
Power		64.7 μW	41.9 μVAR	77.1 μVA
Phase Angle (overall)		33°		
Power Factor (overall)		0.839 (leading)		

3.

CIRCUIT QUANTITY		RESISTOR	CAPACITOR	TOTAL
Current (rms)		2.07 mA	2.07 mA	2.07 mA
Voltage	V_P	13.8 V	11.7 V	18 V
	V_{PP}	27.5 V	22.3 V	36 V
	V_{rms}	9.7 V	8.2 V	12.7 V
Resistance/Reactance/Impedance		4.7 kΩ	3.98 kΩ	6.16 kΩ
Conductance/Susceptance/Admittance		212.8 µS	251.3 µS	162.3 µS
Power		20.1 mW	17 mVAR	26.3 mVA
Phase Angle (overall)		40.3°		
Power Factor (overall)		0.763 (leading)		
Frequency (overall)		400.1 Hz		

Parallel *RC* Circuit Computations

KEY POINTS

When analyzing a parallel circuit, voltage should be found early in the problem, since it is common to all components.

Analysis of parallel *RC* circuits should seem like a logical progression that represents more of an application of previously learned material than a series of new concepts. There is no need to memorize a set of new, specialized equations. Ohm's and Kirchhoff's Laws and basic circuit theory are all you need.

EXAMPLE SOLUTION

Complete the solution matrix given in Table 15-3 for the parallel *RC* circuit shown in Figure 15-34.

CIRCUIT QUANTITY		RESISTOR	CAPACITOR	TOTAL
Current (rms)				
Voltage	V_P			
	V_{PP}			
	V_{rms}			75 V
Resistance/Reactance/Impedance		2.2 kΩ		
Conductance/Susceptance/Admittance				
Power				
Phase Angle (overall)				
Power Factor (overall)				

Table 15-3. Solution Matrix for the Circuit in Figure 15-34

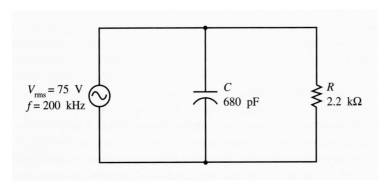

Figure 15-34. Analyze this parallel RC circuit.

EXAMPLE SOLUTION

In a parallel circuit, the voltage is common to all components. Let's begin by applying the basic sine wave equations to convert the rms voltage to peak and peak-to-peak values.

$$V_{A_P} = 1.414 V_{A_{rms}} = 1.414 \times 75 \text{ V} = 106.05 \text{ V}$$

$$V_{A_{PP}} = 2 V_{A_P} = 2 \times 106.05 \text{ V} = 212.1 \text{ V}$$

Of course, these values are the same for the applied voltage, the resistor voltage, and the capacitor voltage, since they are all in parallel.

Now, let's compute the capacitive reactance of the circuit.

$$X_C = \frac{1}{2\pi f C} = \frac{1}{6.28 \times 200 \text{ kHz} \times 680 \text{ pF}} = 1.17 \text{ k}\Omega$$

Since we know the voltage and the resistance or reactance for each branch, we can now compute the current flow in each branch with Ohm's Law. First, let's compute the resistive current.

$$I_R = \frac{V_R}{R} = \frac{75 \text{ V}}{2.2 \text{ k}\Omega} = 34.1 \text{ mA}$$

We have computed rms current, but we could easily convert it to peak or peak-to-peak with the basic sine wave equations. We find current through the capacitive branch in a similar way.

$$I_C = \frac{V_C}{X_C} = \frac{75 \text{ V}}{1.17 \text{ k}\Omega} = 64.1 \text{ mA}$$

Total current in a parallel circuit is equal to the sum (phasor sum in the case of RC circuits) of the individual branch currents. We use Equation 15-4 to compute the phasor sum.

$$I_T = \sqrt{I_R^2 + I_C^2}$$

$$= \sqrt{(34.1 \text{ mA})^2 + (64.1 \text{ mA})^2} = 72.61 \text{ mA}$$

With known values for total current and total voltage, we can apply Ohm's Law to find circuit impedance.

$$Z = \frac{V_A}{I_T} = \frac{75 \text{ V}}{72.61 \text{ mA}} = 1.03 \text{ k}\Omega$$

We can compute conductance, susceptance, and admittance by finding the reciprocals of resistance, reactance, and impedance, respectively. We compute conductance as

$$G = \frac{1}{R} = \frac{1}{2{,}200 \ \Omega} = 454.5 \ \mu S$$

Susceptance is the reciprocal of reactance.

$$B = \frac{1}{X_C} = \frac{1}{1.17 \ k\Omega} = 854.7 \ \mu S$$

In a similar manner, admittance is the reciprocal of impedance.

$$Y = \frac{1}{Z} = \frac{1}{1.03 \ k\Omega} = 970.9 \ \mu S$$

We can use the basic power equations to compute the various powers in the circuit. Let's begin with the true power dissipated in the resistance.

$$P_T = V_R I_R = 75 \ V \times 34.1 \ mA = 2.56 \ W$$

The reactive power is found in a similar manner.

$$P_R = V_C I_C = 75 \ V \times 64.1 \ mA = 4.81 \ VAR$$

Total power in any circuit is the sum of the individual powers. In this case, we have true power and reactive power, which are 90° out of phase, so we must sum them with phasor addition to get total, or apparent power.

$$P_A = \sqrt{P_T{}^2 + P_R{}^2}$$
$$= \sqrt{(2.56 \ W)^2 + (4.81 \ VAR)^2} = 5.45 \ VA$$

Of course, we could also have used one of the basic power equations to compute total (apparent) power.

Figure 15-35 shows a phasor diagram for the circuit shown in Figure 15-34. With the aid of this diagram, we can easily find the phase angle and power factor.

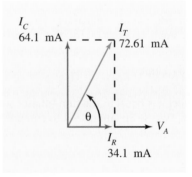

Figure 15-35. A phasor diagram for the circuit shown in Figure 15-34.

Since we now know all three sides of the right triangle shown in Figure 15-35, we can use any one of the sine, cosine, or tangent functions to find the phase angle (θ). If we choose the cosine function, it will also provide us with the power factor of the circuit (power factor = cos θ).

$$\cos\theta = \frac{\text{adjacent}}{\text{hypotenuse}} = \frac{I_R}{I_T}$$

$$= \frac{34.1 \text{ mA}}{72.61 \text{ mA}} = 0.47$$

$$\theta = \arccos 0.47 = 61.97°$$

Since current is always the reference when describing power factors, we will list the power factor as leading.

This completes our analysis of the circuit shown in Figure 15-34. The completed solution matrix is given in Table 15-4.

CIRCUIT QUANTITY		RESISTOR	CAPACITOR	TOTAL
Current (rms)		34.1 mA	64.1 mA	72.61 mA
Voltage	V_P	106.05 V	106.05 V	106.05 V
	V_{PP}	212.1 V	212.1 V	212.1 V
	V_{rms}	75 V	75 V	75 V
Resistance/Reactance/Impedance		2.2 kΩ	1.17 kΩ	1.03 kΩ
Conductance/Susceptance/Admittance		454.5 µS	854.7 µS	970.9 µS
Power		2.56 W	4.81 VAR	5.45 VA
Phase Angle (overall)		61.97°		
Power Factor (overall)		0.47 (leading)		

Table 15-4. Completed Solution Matrix for the Circuit in Figure 15-34.

Practice Problems

1. Complete a solution matrix similar to Table 15-3 for the circuit shown in Figure 15-36.

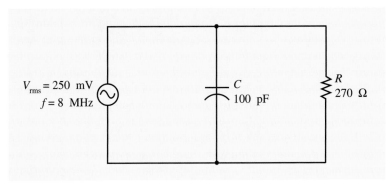

$V_{rms} = 250$ mV
$f = 8$ MHz

C
100 pF

R
270 Ω

Figure 15-36. Analyze this circuit.

2. Analyze the circuit shown in Figure 15-37, and complete a solution matrix similar to Table 15-3.

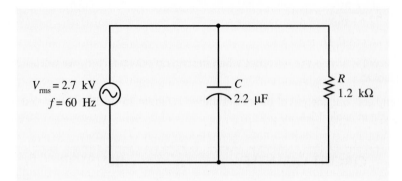

Figure 15-37. Analyze this circuit.

3. Complete a solution matrix similar to Table 15-3 for the circuit shown in Figure 15-38.

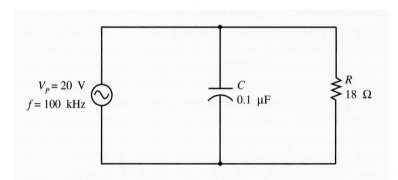

Figure 15-38. Analyze this circuit.

Answers to Practice Problems

1.

CIRCUIT QUANTITY		RESISTOR	CAPACITOR	TOTAL
Current (rms)		925.9 μA	1.26 mA	1.56 mA
Voltage	V_P	0.354 V	0.354 V	0.354 V
	V_{PP}	0.707 V	0.707 V	0.707 V
	V_{rms}	0.25 V	0.25 V	0.25 V
Resistance/Reactance/Impedance		270 Ω	199 Ω	159.9 Ω
Conductance/Susceptance/Admittance		3.7 mS	5.03 mS	6.25 mS
Power		231.5 μW	315 μVAR	390 μVA
Phase Angle (overall)			53.6°	
Power Factor (overall)			0.594 (leading)	

2.

CIRCUIT QUANTITY		RESISTOR	CAPACITOR	TOTAL
Current (rms)		2.25 A	2.23 A	3.17 A
Voltage	V_P	3.82 kV	3.82 kV	3.82 kV
	V_{PP}	7.64 kV	7.64 kV	7.64 kV
	V_{rms}	2.7 kV	2.7 kV	2.7 kV
Resistance/Reactance/Impedance		1.2 kΩ	1.21 kΩ	852.3 Ω
Conductance/Susceptance/Admittance		833 μS	826 μS	1.17 mS
Power		6.08 kW	6.02 kVAR	8.56 kVA
Phase Angle (overall)		44.8°		
Power Factor (overall)		0.71 (leading)		

3.

CIRCUIT QUANTITY		RESISTOR	CAPACITOR	TOTAL
Current (rms)		785.7 mA	888 mA	1.19 A
Voltage	V_P	20 V	20 V	20 V
	V_{PP}	40 V	40 V	40 V
	V_{rms}	14.14 V	14.14 V	14.14 V
Resistance/Reactance/Impedance		18 Ω	15.92 Ω	11.88 Ω
Conductance/Susceptance/Admittance		55.56 mS	62.81 mS	84.16 mS
Power		11.11 W	12.56 VAR	16.83 VA
Phase Angle (overall)		48.7°		
Power Factor (overall)		0.66 (leading)		

Series-Parallel *RC* Circuits with Sinusoidal Current

The solution of series-parallel *RC* problems is very similar to the simplification of resistive or *RL* series-parallel circuit problems. First we simplify the circuit, then we perform computations to determine total current, impedance, and so on. Finally, we work to expand the circuit to compute the individual component voltages, currents, and so forth.

CIRCUIT SIMPLIFICATION

To simplify a series-parallel *RC* circuit, we replace sets of series or parallel resistances with an equivalent resistance. Similarly, we replace sets of series or parallel capacitances with an equivalent capacitance. In each case, we use the basic equations for combining series or parallel resistances or capacitances.

EXAMPLE SOLUTION

Simplify the circuit shown in Figure 15-39 by combining sets of series and parallel resistances and capacitances.

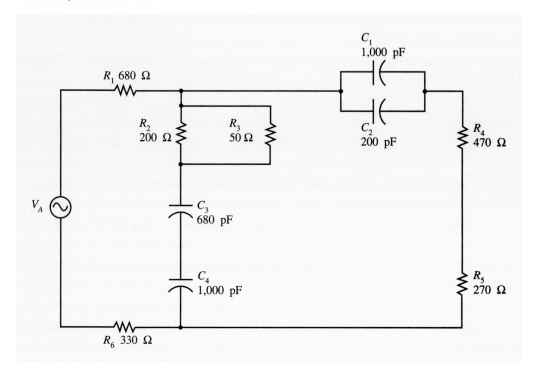

Figure 15-39. Simplify this circuit.

EXAMPLE SOLUTION

Let's start by noting that capacitors C_1 and C_2 are in parallel. We combine them as follows:

$$C_{1,2} = C_1 + C_2 = 1,000 \text{ pF} + 200 \text{ pF} = 1,200 \text{ pF}$$

Resistors R_4 and R_5 are in series and can be combined.

$$R_{4,5} = R_4 + R_5 = 470 \ \Omega + 270 \ \Omega = 740 \ \Omega$$

Resistors R_1 and R_6 are in series and combine as follows.

$$R_{1,6} = R_1 + R_6 = 680 \ \Omega + 330 \ \Omega = 1,010 \ \Omega$$

Resistors R_2 and R_3 are directly in parallel and can be replaced with an equivalent value.

$$R_{2,3} = \frac{R_2 R_3}{R_2 + R_3} = \frac{200 \ \Omega \times 50 \ \Omega}{200 \ \Omega + 50 \ \Omega} = 40 \ \Omega$$

Capacitors C_3 and C_4 are in series and can be combined with the reciprocal equation.

$$C_{3,4} = \frac{1}{\dfrac{1}{C_3} + \dfrac{1}{C_4}}$$

$$= \frac{1}{\dfrac{1}{680 \text{ pF}} + \dfrac{1}{1,000 \text{ pF}}} = 404.8 \text{ pF}$$

There are no more straightforward series or parallel combinations. Our circuit simplification at this point is shown in Figure 15-40. We can further simplify the circuit, but we will need to use rectangular or polar notation to represent the various branch impedances. We will perform this type of calculation in the next section.

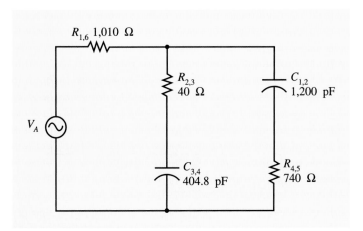

Figure 15-40. A partially simplified version of the circuit shown in Figure 15-39.

Practice Problems

1. Simplify the circuit shown in Figure 15-41 by replacing sets of series or parallel components.

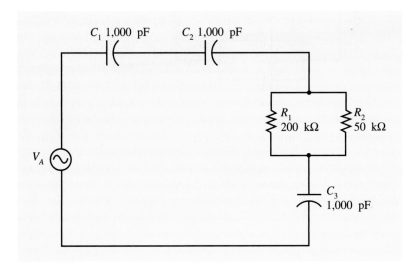

Figure 15-41. Simplify this circuit.

2. Simplify the circuit shown in Figure 15-42 by replacing sets of series or parallel components.

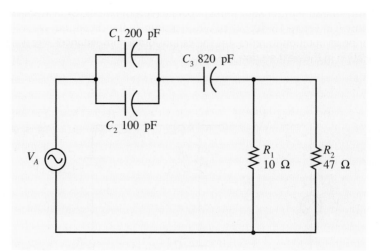

Figure 15-42. Simplify this circuit.

3. Simplify the circuit shown in Figure 15-43 by replacing sets of series or parallel components.

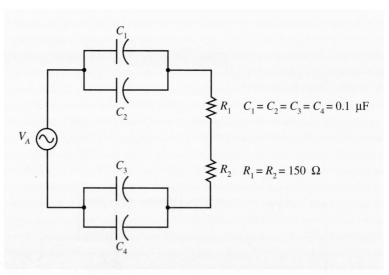

Figure 15-43. Simplify this circuit.

Answers to Practice Problems

1.

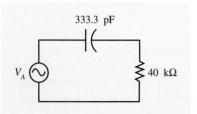

2.

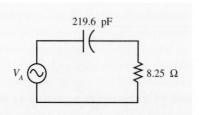

3.

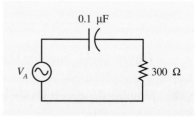

CIRCUIT COMPUTATIONS

We now have all the tools necessary to fully analyze a series-parallel *RC* circuit. You must remember that there are many paths that lead to a valid solution. At this point, calculate any circuit value you can find. The more values you have, the more options you will have for subsequent calculations. As always, use Ohm's and Kirchhoff's Laws and your basic knowledge of circuit configurations as the foundation of your analysis.

EXAMPLE SOLUTION

Analyze the series-parallel *RC* circuit shown in Figure 15-44, and complete the solution matrix given in Table 15-5.

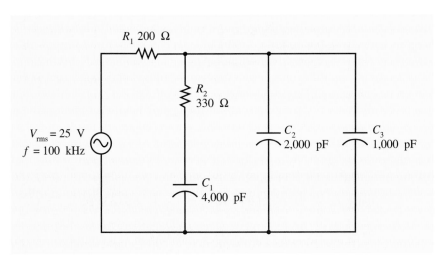

Figure 15-44. A series-parallel *RC* circuit.

CIRCUIT QUANTITY	R_1	R_2	C_1	C_2	C_3	TOTAL
Voltage (rms)						25 V
Current (rms)						
Resistance/Reactance/Impedance	200 Ω	330 Ω				
Conductance/Susceptance/Admittance						
Power						
Phase Angle (overall)						
Power Factor (overall)						

Table 15-5. A Solution Matrix for the Circuit in Figure 15-44

EXAMPLE **SOLUTION**

Our first step will be to combine any sets of series or parallel resistances or capacitances. By inspection, we see that capacitors C_2 and C_3 are in parallel and can be combined.

$$C_{2,3} = C_2 + C_3 = 2{,}000 \text{ pF} + 1{,}000 \text{ pF} = 3{,}000 \text{ pF}$$

There are no more resistances or capacitances that are directly in series or parallel. Our partially simplified circuit is shown in Figure 15-45.

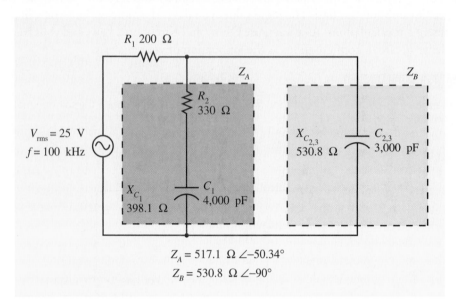

$$Z_A = 517.1 \ \Omega \ \angle{-50.34°}$$
$$Z_B = 530.8 \ \Omega \ \angle{-90°}$$

Figure 15-45. A partially simplified version of the circuit shown in Figure 15-44.

Now let us compute the reactances of the two capacitances shown in Figure 15-45.

$$X_{C_1} = \frac{1}{2\pi f C_1} = \frac{1}{6.28 \times 100 \text{ kHz} \times 4{,}000 \text{ pF}} = 398.1 \ \Omega$$

Similarly, we find the reactance of equivalent capacitor $C_{2,3}$.

$$X_{C_{2,3}} = \frac{1}{2\pi f C_{2,3}} = \frac{1}{6.28 \times 100 \text{ kHz} \times 3{,}000 \text{ pF}} = 530.8 \ \Omega$$

The reactances for C_1 and $C_{2,3}$ are labeled on Figure 15-45. The two parallel branches are also labeled as Z_A and Z_B to simplify their reference in equations.

The next step toward fully simplifying this circuit is to combine parallel branches Z_A and Z_B. These impedances can be expressed individually as follows:

$$Z_A = 330 \ \Omega - j398.1 \ \Omega$$
$$Z_B = 0 - j530.8 \ \Omega$$

It might be beneficial (depending on your specific calculator) to also determine the polar form of these impedances. Note the minus sign associated with the capacitive reactance, since it is drawn downward on the impedance phasor diagram.

$$Z_A = \sqrt{R_2{}^2 + X_{C_1}{}^2}$$

$$= \sqrt{(330 \ \Omega)^2 + (398.1 \ \Omega)^2} = 517.1 \ \Omega$$

$$\theta_A = \arctan \frac{X_{C_1}}{R_2} = \arctan \frac{-398.1 \ \Omega}{330 \ \Omega}$$

$$= \arctan -1.206 = -50.34°; \text{ therefore}$$

$Z_A = 517.1\ \Omega\ \angle{-50.34}°$

Conversion of Z_B to polar form is rather straightforward.

$Z_B = 530.8\ \Omega\ \angle{-90}°$

Since impedances Z_A and Z_B are in parallel, we can combine them with the reciprocal equation, or simply the product-over-the-sum approach.

$$Z_{A,B} = \frac{Z_A Z_B}{Z_A + Z_B}$$

$$= \frac{(517.1\ \Omega\ \angle{-50.34}°)\ (530.8\ \Omega\ \angle{-90}°)}{(330\ \Omega - j398.1\ \Omega) + (0 - j530.8\ \Omega)}$$

$$= \frac{(274.48 \times 10^3)\ \angle{-140.34}°}{330\ \Omega - j928.9\ \Omega}$$

$$= 278.44\ \Omega\ \angle{-69.9}° \ = 95.7\ \Omega - j261.5\ \Omega$$

Figure 15-46 shows the results of our simplification to this point.

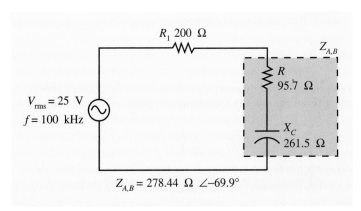

$$R_1\ 200\ \Omega$$

$$V_{rms} = 25\ \text{V}$$
$$f = 100\ \text{kHz}$$

$$Z_{A,B}$$
$$R\ 95.7\ \Omega$$
$$X_C\ 261.5\ \Omega$$

$$Z_{A,B} = 278.44\ \Omega\ \angle{-69.9}°$$

Figure 15-46. A partially simplified version of the circuit shown in Figure 15-44.

In the equivalent series circuit shown in Figure 15-46, we can see that total impedance can be found by summing R_1 and $Z_{A,B}$. We perform this calculation as follows:

$$Z_T = R_1 + Z_{A,B}$$

$$= 200\ \Omega + (95.7\ \Omega - j261.5\ \Omega)$$

$$= 295.7\ \Omega - j261.5\ \Omega \ = 394.7\ \Omega\ \angle{-41.5}°$$

By converting our final impedance value to polar form, we also find the phase angle of the circuit. The fully simplified circuit is shown in Figure 15-47. Of course, we could determine the value of equivalent capacitance, if desired. That is,

$$X_C = \frac{1}{2\pi f C},\ \text{or}$$

$$C = \frac{1}{2\pi f X_C}$$

$$= \frac{1}{6.28 \times 100\ \text{kHz} \times 261.5\ \Omega} = 6{,}089\ \text{pF}$$

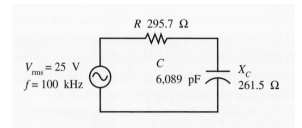

Figure 15-47. A fully simplified equivalent circuit for the circuit originally shown in Figure 15-44.

KEY POINTS

Once total impedance has been computed, rectangular and polar notation may be avoided if you do not need to know the phase angles associated with each component.

Now, just as we would do in a resistive circuit analysis problem, we want to find total current and then work our way back through the partially simplified sketches as we compute individual component voltages and currents. If desired, we could use rectangular or polar notation for the current, voltages, and impedances in the following calculations. While this complicates the calculations, it does give us the phase angle at every point in the circuit. For purposes of this example, we will use the magnitude portion of the circuit quantities only. This will give us the correct values of current, voltage, and so on, but we will not have the phase angles at the intermediate circuit points. Let's compute total current with Ohm's Law.

$$I_T = \frac{V_A}{Z_T} = \frac{25 \text{ V}}{394.7 \text{ } \Omega} = 63.34 \text{ mA}$$

Let's transfer this total current value to the partially simplified sketch shown in Figure 15-46. Clearly, total current flows through R_1. We can apply Ohm's Law to compute the voltage drop across R_1.

$$V_{R_1} = I_{R_1} R_1 = 63.34 \text{ mA} \times 200 \text{ } \Omega = 12.67 \text{ V}$$

The remaining voltage is dropped across $Z_{A,B}$ in Figure 15-46. We compute its voltage drop as follows:

$$V_{Z_{A,B}} = I_{Z_{A,B}} Z_{A,B} = 63.34 \text{ mA} \times 278.44 \text{ } \Omega = 17.64 \text{ V}$$

By inspection of Figures 15-45 and 15-46, we can see that this same voltage will be felt across both branches Z_A and Z_B.

Now, let's apply Ohm's Law to branches Z_A and Z_B to find the current flow.

$$I_{Z_A} = \frac{V_{Z_A}}{Z_A} = \frac{17.64 \text{ V}}{517.1 \text{ } \Omega} = 34.11 \text{ mA}$$

$$I_{Z_B} = \frac{V_{Z_B}}{Z_B} = \frac{17.64 \text{ V}}{530.8 \text{ } \Omega} = 33.23 \text{ mA}$$

We can see from the circuit diagram in Figure 15-45, that the I_{Z_A} current flows through R_2 and C_1. Similarly, the I_{Z_B} current flows through $C_{2,3}$. To determine the amount of current through C_2 and C_3 individually, we can use the current divider method or simply apply Ohm's Law. Let's choose the latter approach. First, we need to compute the individual reactances of C_2 and C_3.

$$X_{C_2} = \frac{1}{2\pi f C_2} = \frac{1}{6.28 \times 100 \text{ kHz} \times 2,000 \text{ pF}} = 796.18 \text{ } \Omega$$

$$X_{C_3} = \frac{1}{2\pi f C_3} = \frac{1}{6.28 \times 100 \text{ kHz} \times 1,000 \text{ pF}} = 1.59 \text{ k}\Omega$$

Now applying Ohm's Law, we get

$$I_{C_2} = \frac{V_{C_2}}{X_{C_2}} = \frac{17.64 \text{ V}}{796.18 \text{ } \Omega} = 22.16 \text{ mA}$$

$$I_{C_3} = \frac{V_{C_3}}{X_{C_3}} = \frac{17.64 \text{ V}}{1.59 \text{ k}\Omega} = 11.09 \text{ mA}$$

The individual voltage drops across R_2 and C_1 can be found by applying Ohm's Law.

$$V_{R_2} = I_{R_2}R_2 = 34.11 \text{ mA} \times 330 \text{ } \Omega = 11.26 \text{ V}$$

$$V_{C_1} = I_{C_1}X_{C_1} = 34.11 \text{ mA} \times 398.1 \text{ } \Omega = 13.58 \text{ V}$$

This completes the voltage and current calculations for the circuit. We can complete the Conductance/Susceptance/Admittance row in Table 15-5 by finding the reciprocal of the corresponding resistance, reactance, or impedance.

$$G_1 = \frac{1}{R_1} = \frac{1}{200 \text{ } \Omega} = 5 \text{ mS}$$

$$G_2 = \frac{1}{R_2} = \frac{1}{330 \text{ } \Omega} = 3.03 \text{ mS}$$

$$B_1 = \frac{1}{X_{C_1}} = \frac{1}{398.1 \text{ } \Omega} = 2.51 \text{ mS}$$

$$B_2 = \frac{1}{X_{C_2}} = \frac{1}{796.18 \text{ } \Omega} = 1.26 \text{ mS}$$

$$B_3 = \frac{1}{X_{C_3}} = \frac{1}{1.59 \text{ k}\Omega} = 628.9 \text{ } \mu\text{S}$$

$$Y_T = \frac{1}{Z_T} = \frac{1}{394.7 \text{ } \Omega} = 2.53 \text{ mS}$$

We can compute the various powers with any of the power equations. Let's choose to use the $P = VI$ form in all cases.

$$P_{R_1} = V_{R_1}I_{R_1} = 12.67 \text{ V} \times 63.34 \text{ mA} = 802.5 \text{ mW}$$

$$P_{R_2} = V_{R_2}I_{R_2} = 11.26 \text{ V} \times 34.11 \text{ mA} = 384.1 \text{ mW}$$

$$P_{C_1} = V_{C_1}I_{C_1} = 13.58 \text{ V} \times 34.11 \text{ mA} = 463.2 \text{ mVAR}$$

$$P_{C_2} = V_{C_2}I_{C_2} = 17.64 \text{ V} \times 22.16 \text{ mA} = 390.9 \text{ mVAR}$$

$$P_{C_3} = V_{C_3}I_{C_3} = 17.64 \text{ V} \times 11.09 \text{ mA} = 195.6 \text{ mVAR}$$

$$P_T = V_T I_T = 25 \text{ V} \times 63.34 \text{ mA} = 1.58 \text{ VA}$$

Of course, we could have computed total power as a sum (phasor addition) of the individual powers.

The last entry in our solution matrix is power factor. Power factor, you will recall, is simply the cosine of the phase angle (θ). In the present case, we have

$$\text{power factor} = \cos\theta = \cos -41.5° = 0.749$$

It is a leading power factor, since we have a capacitive circuit.

This completes our analysis of the circuit shown in Figure 15-44. The completed solution matrix is shown in Table 15-6.

CIRCUIT QUANTITY	R_1	R_2	C_1	C_2	C_3	TOTAL
Voltage (rms)	12.67 V	11.26 V	13.58 V	17.64 V	17.64 V	25 V
Current (rms)	63.34 mA	34.11 mA	34.11 mA	22.16 mA	11.09 mA	63.34 mA
Resistance/Reactance/Impedance	200 Ω	330 Ω	398.1 Ω	796.18 Ω	1.59 kΩ	394.7 Ω
Conductance/Susceptance/Admittance	5 mS	3.03 mS	2.51 mS	1.26 mS	628.9 μS	2.53 mS
Power	802.5 mW	384.1 mW	463.2 mVAR	390.9 mVAR	195.6 mVAR	1.58 VA
Phase Angle (overall)			−41.5°			
Power Factor (overall)			0.749 (leading)			

Table 15-6. A Completed Solution Matrix for the Circuit in Figure 15-44

Practice Problems

1. Analyze the circuit shown in Figure 15-48, and complete a solution matrix similar to the one in Table 15-5.
2. Analyze the circuit shown in Figure 15-49, and complete a solution matrix similar to the one in Table 15-5.

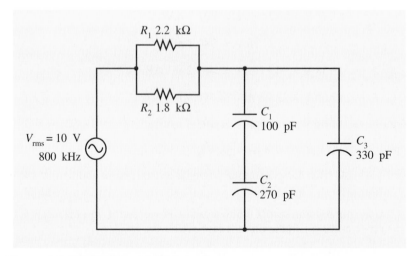

Figure 15-48. Analyze this RC circuit.

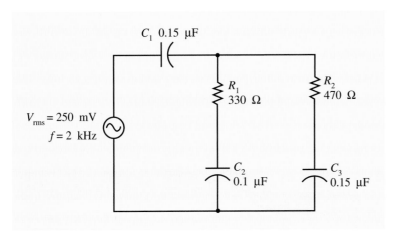

Figure 15-49. Analyze this *RC* circuit.

Answers to Practice Problems

1.

CIRCUIT QUANTITY	R_1	R_2	C_1	C_2	C_3	TOTAL
Voltage (rms)	8.95 V	8.95 V	3.26 V	1.21 V	4.47 V	10 V
Current (rms)	4.07 mA	4.97 mA	1.64 mA	1.64 mA	7.4 mA	9.04 mA
Resistance/Reactance/Impedance	2.2 kΩ	1.8 kΩ	1.99 kΩ	737.2 Ω	603.2 Ω	1.11 kΩ
Conductance/Susceptance/Admittance	455 µS	556 µS	503 µS	1.36 mS	1.66 mS	901 µS
Power	36.4 mW	44.5 mW	5.35 mVAR	1.98 mVAR	33.1 mVAR	90.4 mVA
Phase Angle (overall)	−26.5°					
Power Factor (overall)	0.895 (leading)					

2.

CIRCUIT QUANTITY	R_1	R_2	C_1	C_2	C_3	TOTAL
Voltage (rms)	42.5 mV	73.6 mV	149.4 mV	102.6 mV	83.1 mV	0.25 V
Current (rms)	128.8 µA	156.6 µA	281.5 µA	128.8 µA	156.6 µA	281.5 µA
Resistance/Reactance/Impedance	330 Ω	470 Ω	530.8 Ω	796.2 Ω	530.8 Ω	888 Ω
Conductance/Susceptance/Admittance	3.03 mS	2.13 mS	1.88 mS	1.26 mS	1.88 mS	1.13 mS
Power	5.47 µW	11.53 µW	42.1 µVAR	13.2 µVAR	13 µVAR	70.4 µVA
Phase Angle (overall)	−76°					
Power Factor (overall)	0.242 (leading)					

Complex *RC* Circuit Computations

The solution of complex *RC* circuits can be realized using the same network theorems used for the solution of complex resistive circuits. When simplification methods such as Thevenin's and Norton's Theorems, mesh analysis, and so on are applied to a complex *RC* network, the circuit quantities must be expressed in polar or rectangular notation to account for the nonzero phase angle. The basic method of simplification, however, remains identical.

Exercise Problems 15.3

Refer to Figure 15-50 for Problems 1–10.

1. What is the reactance of *C*?

2. What is the impedance of the circuit?

3. What is the admittance of the circuit?

4. How much current flows through *R*?

5. What is the total current in the circuit?

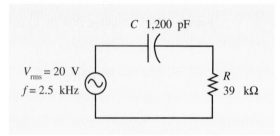

Figure 15-50. A series *RC* circuit.

6. What is the voltage drop across *R*?

7. What is the voltage drop across *C*?

8. What is the phase angle (θ) between total voltage and total current in the circuit?

9. What is the power factor of the circuit?

10. What is the apparent power in the circuit?

Refer to Figure 15-51 for Problems 11–20.

11. What is the peak voltage across C_1?

12. What is the rms voltage across C_2?

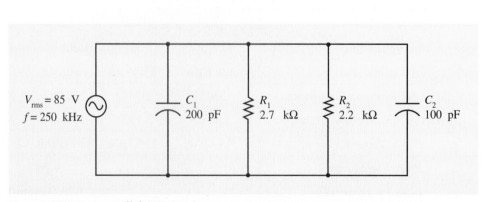

Figure 15-51. A parallel *RC* circuit.

13. How much current flows through R_1?
14. How much current flows through C_2?
15. What is the total current in the circuit?
16. What is the total impedance of the circuit?
17. What is the susceptance of C_1?
18. How much true power is dissipated by the circuit?
19. What is the phase angle between total voltage and total current in the circuit?
20. What is the apparent power in the circuit?

Refer to Figure 15-52 for Problems 21–30.

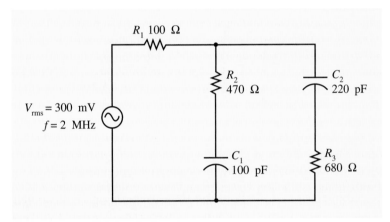

Figure 15-52. A series-parallel RC circuit.

21. What is the branch impedance of R_2 and C_1?
22. What is the branch impedance of R_3 and C_2?
23. What is the total impedance of the circuit?
24. How much current flows in the circuit?
25. What is the total admittance of the circuit?
26. What is the voltage drop across R_1?
27. What is the voltage drop across the entire R_2C_1 branch?
28. How much voltage is dropped across R_3?
29. What is the overall power factor of the circuit?
30. How much apparent power is delivered by the source?

15.4 Pulse Response of RC Circuits

We have thoroughly discussed the analysis of RC circuits with sinusoidal inputs. Nonsinu-soidal input waveforms cause a radically different circuit response and require different analytical methods. It is not necessary for you to be able to thoroughly analyze the nonsi-nusoidal response of an RC circuit at this time, but it is important for you to understand RC time constants and how an RC circuit responds to a pulse or rectangular waveform.

RC Time Constant

Figure 15-53 shows a simple circuit that will manually produce pulse waveforms to an *RC* circuit as the switch is moved between positions A and B. When the switch is first moved to position A (Figure 15-53a), current will flow through the *RC* circuit and begin to charge the capacitor. The initial value of current is limited only by the resistor. Charging current continues until the capacitor voltage is equal to the battery voltage.

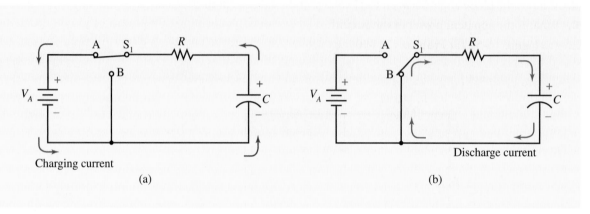

(a)   (b)

Figure 15-53. An *RC* circuit with a pulse input.

Now we know from previous discussions that for a given size capacitor to charge to a particular voltage requires a definite amount of charge (i.e., a definite number of electrons). You will recall the expression for charge is $Q = CV$. It takes time for the electrons to move around the circuit and accumulate on the capacitor. How long does it take for enough electrons to move around the circuit to charge the capacitor to a value equal to the supply voltage? Well, that depends on the value of charging current. Remember, current is a measure of how many electrons move past a given point in one second. As stated previously, the resistor limits the value of current in the circuit. The more resistance we have in the resistor, the lower the current and the longer it takes the capacitor to reach full charge.

It is also important to note that the value of charging current decays as the capacitor becomes charged. In essence, the accumulating charge on the capacitor provides more and more opposition to the applied voltage (like series-opposing voltage sources). When the capacitor is fully charged, the opposing voltages are equal and no more current can flow.

The total time required for the capacitor to fully charge is divided into five time periods called time constants. The Greek letter tau (τ) is used to represent one time constant. In the case of *RC* circuits, τ is computed with Equation 15-5.

$$\tau = RC \qquad (15\text{-}5)$$

where τ is measured in seconds, *R* is measured in ohms, and *C* is measured in farads.

Figure 15-54(a) illustrates the timing of the capacitor voltage and the charging current. The graph indicates that after a time interval of $t = \tau$, the capacitor voltage has increased to 63.2% of its full-charge value. At that same time, charging current has decreased to 36.8% of its initial value. During each additional time period of τ seconds, the capacitor voltage increases by 63.2% of the remaining voltage. For example, at the end of the first time interval (τ), the capacitor voltage has another 36.8% to go. During the time between τ and 2τ it will increase 63.2% of this 36.8% or 23.3% for a total of 86.5% at time $t = 2\tau$.

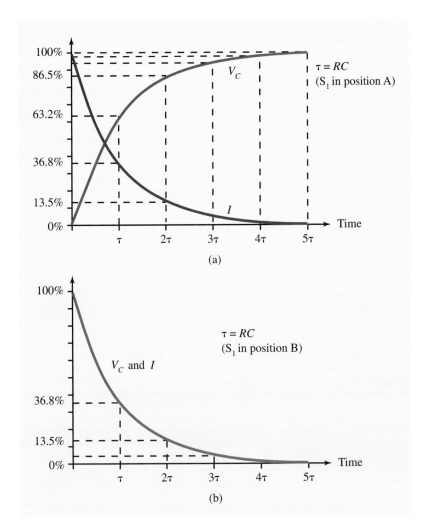

Figure 15-54. An *RC* time constant graph: (a) charging and (b) discharging.

Once 5τ seconds have elapsed, the capacitor voltage will have increased to more than 99% of its full-charge value. Likewise, the charging current will have decreased to less than 1% of its initial value. For practical purposes, we consider the capacitor to have reached full charge in five time constants (5τ).

If the capacitor in Figure 15-53(a) has reached full charge with the switch in position A, there will be no additional current flow. The charge on the capacitor is equal, but opposite, to the battery potential. Now, if we move the switch to position B, the charge on the capacitor acts like a voltage source and causes a discharge current to flow. This action is

 **KEY POINTS**

It takes five time constants for a capacitor to fully charge or discharge regardless of the amount of voltage involved.

illustrated in Figure 15-53(b). Again, the amount of initial current is limited by the resistor. The higher the discharge current (i.e., the lower the value of the resistor), the quicker the capacitor can discharge its accumulated voltage. The discharge current will continue until the charge on the capacitor has decreased to zero.

Figure 15-54(b) illustrates the timing of the current and capacitor voltage during the discharging period. At time $t = 0$, the capacitor is at full charge and the discharge current is maximum. At time $t = \tau$, the capacitor voltage and discharge current have decayed to 36.8% of their initial values. The current and voltage continue to decrease by 63.2% of the remaining voltage each time constant. Again, for practical purposes, we consider the capacitor to be fully discharged after five time constants (5τ).

It is important to realize that the time required to fully charge or discharge a capacitor (5τ) is determined only by the value of R and C. It is unaffected by the value of voltage in the circuit. If either R or C is made larger, then it takes longer to charge or discharge the capacitor.

EXAMPLE SOLUTION

If a 10-kΩ resistor and a 2.5-μF capacitor are connected in series across a 10-V battery, how long does it take for the capacitor to have a 10-V charge?

EXAMPLE **SOLUTION**

We know it takes five time constants to charge or discharge a capacitor. So, let's compute the time constant with Equation 15-5.

$$\tau = RC = 10\ k\Omega \times 2.5\ \mu F = 25\ ms$$

We multiply by five to find the total charge (or discharge) time.

$$charge\ time = 5\tau = 5 \times 25\ ms = 125\ ms$$

Practice Problems

1. What is the RC time constant for the circuit shown in Figure 15-55?
2. How many time constants are required for the capacitor in Figure 15-55 to reach full charge, once the switch has been closed?

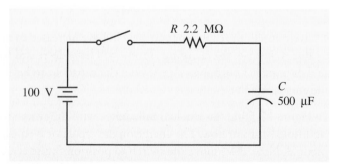

Figure 15-55. An RC circuit.

3. What will be the approximate value of current in the circuit shown in Figure 15-55, after five time constants?

4. Refer to Figure 15-56. Capacitor C_1 is charged to 1.5 V. Capacitor C_2 is charged to 2.0 kV. The two switches are mechanically linked and move together. When the switches are closed, which capacitor will discharge to 0 V first? Explain.

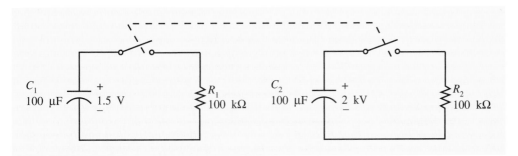

Figure 15-56. An RC time constant problem.

Answers to Practice Problems

1. 1,100 s or 18.33 min   2. 5   3. zero

4. Both capacitors will discharge in five time constants. Since the time constants are the same, the discharge time will be the same.

Precise RC Calculations

If we need to know the voltage or current in an RC circuit at some specific instant in time while it is charging or discharging to some given dc level, then the preceding calculations are insufficient. A more accurate method requires application of Equation 15-6 (for voltage) and Equation 15-7 (for current):

$$v_C = V - (V - V_0)\epsilon^{-t/\tau} \qquad (15\text{-}6)$$

$$i_C = I - (I - I_0)\epsilon^{-t/\tau} \qquad (15\text{-}7)$$

where

v_C and i_C are the capacitor voltage and current at a specific time t;

V and I are the circuit values that will exist after five time constants;

V_0 and I_0 are the initial values of voltage and current;

ϵ is a constant equal to 2.718 (natural logarithm base);

t is the specific time when voltage or current is to be computed;

τ is the RC time constant.

EXAMPLE SOLUTION

Assume that the switch in Figure 15-57 was left in position B until the capacitor had fully discharged. What is the voltage on the capacitor 0.5 s after moving the switch to position A?

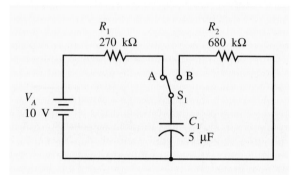

Figure 15-57. An *RC* charge/discharge circuit.

EXAMPLE **SOLUTION**

Equation 15-6 provides the tool for this type of problem. Here, the initial capacitor voltage (V_0) is zero. The final voltage after five time constants (V) will be 10 V. Since the switch will be in position A, the time constant is computed with Equation 15-5.

$$\tau = R_1 C_1$$
$$= 270 \text{ k}\Omega \times 5 \text{ }\mu\text{F} = 1.35 \text{ s}$$

We can now apply Equation 15-6 to determine the capacitor voltage after 0.5 s.

$$v_C = V - (V - V_0)\epsilon^{-t/\tau}$$
$$= 10 - (10 - 0)\epsilon^{-0.5/1.35}$$
$$= 10 - 10 \times 0.6905 = 3.095 \text{ V}$$

Practice Problems

Refer to Figure 15-57 for the following practice problems.

1. If the capacitor has been fully discharged in position B, calculate the capacitor voltage 1.9 s after the switch is moved to position A.
2. If the capacitor has been fully charged in position A, calculate the capacitor voltage 200 ms after moving the switch to position B.
3. The capacitor is fully discharged in position B. It is moved to position A for 1.5 s, then switched back to position B. What is the capacitor voltage 750 ms after returning to position B?

Answers to Practice Problems

1. 7.55 V   2. 9.43 V   3. 5.38 V

Sometimes we are interested in how long it takes a capacitor to charge or discharge to a particular voltage.

$$t = \tau ln\left(\frac{V - V_0}{V - v_C}\right) \qquad (15\text{-}8)$$

EXAMPLE SOLUTION

If the capacitor in Figure 15-57 has been fully charged in position A, how long does it take to reach 4 V after the switch is moved to position B?

EXAMPLE **SOLUTION**

We apply Equation 15-8 directly.

$$t = \tau ln\left(\frac{V - V_0}{V - v_c}\right)$$

$$= 680 \text{ k}\Omega \times 5 \text{ }\mu F \times ln\left(\frac{0 - 10}{0 - 4}\right)$$

$$= 3.4 \times ln\,2.5 = 3.4 \times 0.9163 = 3.115 \text{ s}$$

Practice Problems

Refer to Figure 15-57 for the following problems.

1. If the capacitor has been fully discharged in position B, how long does it take to reach 2 V after the switch is moved to position A?

2. If the capacitor has been fully charged in position A, how long does it take to reach 8 V after the switch is moved to position B?

3. The capacitor is fully charged in position A. The switch is moved to position B for 4 s and then returned to position A. How long does it take for the capacitor to reach 8 V after the switch has been returned to position A?

Answers to Practice Problems

1. 301.2 ms   2. 758.7 ms   3. 1.68 s

Circuit Waveforms

Charging and discharging of a capacitor in an *RC* circuit requires a definite amount of time (determined by the values of resistance and capacitance). If the input voltage level changes too quickly, the capacitor may not have enough time to fully charge or discharge. Figure 15-58 illustrates the response of an *RC* circuit to a pulse of three different frequencies.

In Figure 15-58(b), each polarity of the input pulse is present for a period of time equal to one time constant. Thus, the capacitor does not have sufficient time to reach full charge before the input voltage is removed. Similarly, it is not given time to decay to zero between pulses.

KEY POINTS

When an *RC* circuit is supplied by a pulsed voltage waveform, its response is dependent on the length of the *RC* time constant relative to the duration of the pulse voltages.

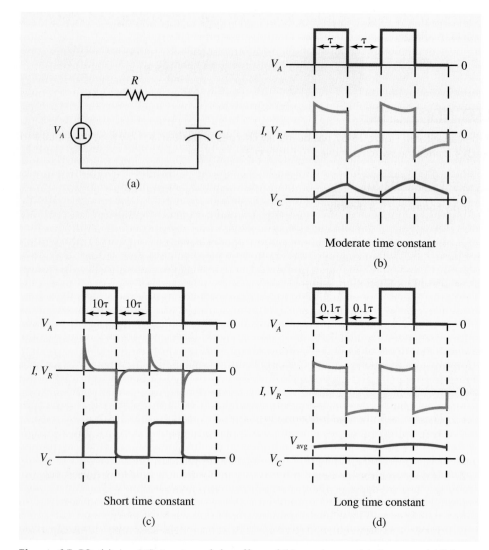

Figure 15-58. (a) An *RC* circuit and the effect of (b) moderate, (c) short, and (d) long time constant on *RC* circuit response.

Figure 15-58(c) illustrates a short time constant. Here, the input pulse duration is present for an amount of time equal to ten time constants. Clearly, the capacitor voltage is allowed time to reach maximum and to decay to zero. The resistor voltage appears as short duration voltage spikes. Notice that the resistor waveform has a dual polarity caused by the opposite direction of current flow during the charge and discharge periods. The waveform across the resistor is called a differentiated waveform. The entire circuit is often called a **differentiator** circuit. To qualify as a differentiator, the circuit must have a short time constant and the output must be taken across the resistor.

Finally, Figure 15-58(d) illustrates the effects of a long time constant, where the input pulse duration is one-tenth of one time constant. Here, the capacitor voltage is barely allowed time to increase or decrease. The resistor voltage has a waveform that is nearly identical to the input waveform. This effect is useful for coupling signals between subsequent stages in an amplifier circuit. When the output is taken across the resistor, the circuit is called an *RC* **coupling** circuit. If the output is taken across the capacitor, we call the circuit an **integrator** circuit. Both coupling and integrator circuits require long time constants.

Magnitude of Capacitor Current

The value of charging (or discharging) current in a capacitor is directly related to the rate of change of voltage across the capacitor. The higher the current, the faster the voltage can increase or decrease. In the same way, if the voltage across a capacitor changes, then a charging or discharging current must flow. The value of that current is proportional to the rate of change of capacitor voltage. This is an important concept and is expressed more formally by Equation 15-9.

$$i_c = C\frac{dv}{dt} = C\frac{\Delta v}{\Delta t} \qquad (15\text{-}9)$$

where i_c is the instantaneous capacitor current, C is the value of the capacitor in farads, dv or Δv stands for change in voltage, and dt or Δt stands for change in time.

EXAMPLE SOLUTION

If the voltage across a 1,000-pF capacitor increases steadily by 5 V over a period of 250 μs, what is the value of capacitor current?

EXAMPLE **SOLUTION**

First, it is interesting to note that you may encounter this type of problem when working on analog-to-digital converter circuits used in computers. We apply Equation 15-9 as follows:

$$i_c = C\frac{dv}{dt}$$

$$= 1{,}000 \text{ pF} \times \frac{5 \text{ V}}{250 \text{ μs}} = 20 \text{ μA}$$

EXAMPLE SOLUTION

At a particular instant, the voltage across a 25-μF capacitor is changing at the rate of 10 V/ms. What is the value of capacitor current at that time?

EXAMPLE **SOLUTION**

We apply Equation 15-9 as follows.

$$i_c = C\frac{dv}{dt}$$

$$= 25 \text{ μF} \times \frac{10 \text{ V}}{1 \text{ ms}} = 250 \text{ mA}$$

KEY POINTS

A series *RC* circuit can be any of the following:

• Differentiator: short time constant with the output taken across the resistor.

• Integrator: long time constant with the output taken across the capacitor.

• Coupling circuit: long time constant with the output taken across the resistor.

KEY POINTS

The value of charging or discharging current in a capacitor at any instant in time is directly proportional to the value of the capacitor and the rate of change of voltage across the capacitor.

Practice Problems

1. What is the current through a 150-µF capacitor if its voltage is changing by 4 V every 2 µs?
2. What value of current must be provided to a 1,500-pF capacitor in order to cause its voltage to increase by 5 V in 100 ms?
3. The voltage across a capacitor increases by 25 V in 1.5 µs when it is supplied from a 2.75-A current source. What is the value of the capacitor?

Answers to Practice Problems

1. 300 A   2. 75 nA   3. 0.165 µF

Exercise Problems 15.4

1. What is the RC time constant of an 1,800-pF capacitor in series with a 1.5-MΩ resistor?
2. What is the RC time constant of a 25-µF capacitor and a 200-kΩ resistor?
3. What size capacitor must be connected in series with a 75-kΩ resistor to produce a circuit with a 165-ms time constant?
4. If the voltage waveform across the resistor in a series RC circuit with a pulse input is nearly identical to the input pulse shape, the circuit must have a (*short, long*) time constant.
5. If the pulse duration of an input waveform is 100 times longer than the RC time constant of a series RC circuit, the circuit has a (*short, long*) time constant.
6. If the voltage across a 2.2-µF capacitor increases by 15 V in 3 ms, what is the value of current flow?

15.5 Applied Technology: *RC* Circuits

RC circuits are used in nearly every practical electronic device. This includes such things as computers, microwave ovens, telephone systems, satellite communications equipment, industrial control equipment, automotive electronics, and so on. We will briefly examine three common *RC* circuit applications: power supply filters, frequency selective circuits, and time delay circuits.

Power Supply Filter Circuits

Figure 15-59 illustrates the basic operation of a filter circuit for an electronic power supply. Most electronic circuits require dc voltage for operation, but it is often desirable to operate the system from the standard 120-Vac power line. A power supply circuit is used to convert the 120-Vac, 60-Hz power distributed by the power company to a pulsating dc voltage. This pulsating or surging voltage can then be applied to an *RC* circuit as shown in

---

○ KEY POINTS

Electronic power supplies nearly always use an *RC* circuit to filter or smooth the dc output voltage so that it more closely approximates the steady voltage from a battery.

Figure 15-59. The *RC* network is designed to have a long time constant. This means the voltage across the capacitor cannot follow the quick increases and decreases in the pulsating voltage from the electronic power supply. Rather, the capacitor voltage is a relatively smooth dc. The voltage across the capacitor is used to supply power to the subsequent electronic circuitry. The overall result is a smooth dc voltage similar to that produced by a battery.

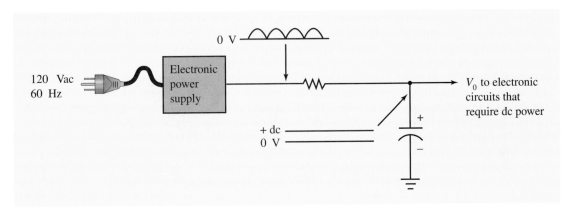

Figure 15-59. An *RC* circuit can smooth the output of an electronic power supply.

Frequency Selective Circuits

RC circuits can be used for frequency selective circuits in the same way as described in Chapter 13 for *RL* circuits. This is a very common application. Figure 15-60 shows a closely related application involving a public address system.

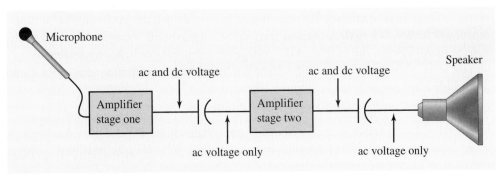

Figure 15-60. A public address (PA) system.

Here, a weak microphone signal is amplified by two sequential amplifier stages and used to drive a speaker. The electronic amplifier stages require dc voltages in order to operate, but the ac signals representing the sound into the microphone are the only voltages that should be allowed to pass through the system. By connecting the stages together with a capacitor, we can isolate the dc voltages of the two stages and yet couple the desired ac signal. Remember that capacitors act like open circuits to dc voltages, but can have low reactances to ac if they are sized correctly. Another capacitor is used to couple the signal from the last amplifier stage to the speaker. Here again, we only want the ac signals to reach the speaker. If dc is allowed to pass through the speaker, the sound may be distorted and/or damage may result.

Time Delay Circuits

RC circuits are widely used to provide time delay action in electronic circuits. Figure 15-61 illustrates one particular method.

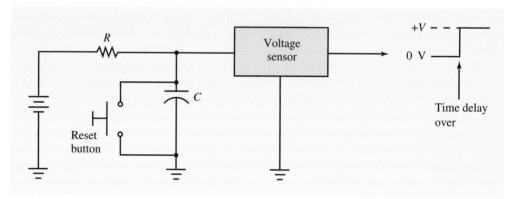

Figure 15-61. *RC* circuits can be used to provide electronic time delays.

When the reset button in Figure 15-61 is pressed, the capacitor voltage is quickly discharged via the low resistance of the switch contacts. When the button is released, the capacitor begins to charge via resistor *R*. As the capacitor charges, its voltage is monitored by an electronic circuit called a voltage sensor. As long as the capacitor voltage is below a certain voltage level (established by the voltage sensor circuit), the output of the voltage sensor will be 0 V. When the capacitor reaches the trigger-point voltage established by the voltage sensor circuit, the output of the voltage sensor quickly switches to a different voltage level. This change in output voltage can be used to drive an indicator lamp, sound an audible alarm, or activate a computer circuit.

So what determines the length of the time delay? The time delay is determined by how long it takes the capacitor to charge to the trigger voltage. If we increase the value of either *R* or *C*, the time delay will be increased. In practical time delay circuits, resistor *R* is often a rheostat, which allows a variable time delay.

Exercise Problems 15.5

1. In your own words, explain how a pulsating waveform can be smoothed with an *RC* circuit that has a long time constant.
2. In your own words, explain why a capacitor appears to block dc in an application like that shown in Figure 15-60.
3. Refer to Figure 15-61. During the time delay, the capacitor is (*charging, discharging*).

15.6 Troubleshooting *RC* Circuits

RC circuits are relatively easy to troubleshoot. The possible malfunctions are essentially combinations of the troubles found in the purely resistive and purely capacitive circuits previously studied. The following represent the most probable defects in an *RC* circuit:

- open capacitor
- open resistor
- shorted capacitor
- shorted resistor (usually solder bridge on a printed circuit board)
- leaky capacitor
- resistor value out of tolerance

All of these defects for series, parallel, or more complex *RC* networks can be located using the same troubleshooting logic presented for purely resistive circuits coupled with the capacitor verification checks presented in Chapter 14.

Exercise Problems 15.6

Following are several troubleshooting problems using the PShooter method. Recall that each PShooter circuit has an associated table that lists the various testpoints in the circuit along with the normal values of voltage and resistance. The table also provides a bracketed ([ ]) number for each measurable entry. The bracketed number identifies a specific entry in the PShooter lookup table provided in Appendix A, which gives you the actual measured value of the circuit quantity. The following list will help you interpret the measured values:

- All testpoint (TPxx) measurements are made with reference to ground.
- All non-testpoint values are measured directly across the component.
- All resistance measurements reflect the approximate reading of an ohmmeter.
- Resistance values for all testpoints (TPxx) are measured with the applied voltage disconnected. All other components remain connected.
- Resistance tests for all non-testpoint measurements are made with the specified component removed from the circuit.
- Capacitance measurements are made with the component removed from the circuit.
- The entry 0→∞ is used to indicate the normal movement of an ohmmeter pointer when testing a capacitor (i.e., it starts at zero and moves toward infinity).

1. Refer to Figure 15-62 and the PShooter chart shown in Table 15-7. Locate the defective component in as few measurements as possible by applying a logical, systematic troubleshooting method.

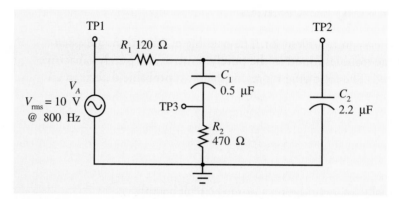

Figure 15-62. Troubleshoot this circuit.

TESTPOINT	rms VOLTAGE		CAPACITANCE		RESISTANCE	
	NORMAL	ACTUAL	NORMAL	ACTUAL	NORMAL	ACTUAL
V_A	10 V	[35]	—	—	—	—
C_1	3.49 V	[29]	0.5 µF	[118]	0→∞	[191]
C_2	5.4 V	[6]	2.2 µF	[115]	0→∞	[5]
R_1	7.88 V	[44]	—	—	120 Ω	[66]
R_2	4.12 V	[126]	—	—	470 Ω	[187]
TP1	10 V	[104]	—	—	∞	[196]
TP2	5.4 V	[148]	—	—	∞	[113]
TP3	4.12 V	[54]	—	—	470 Ω	[80]

Table 15-7. PShooter Data for Figure 15-62

TESTPOINT	rms VOLTAGE		CAPACITANCE		RESISTANCE	
	NORMAL	ACTUAL	NORMAL	ACTUAL	NORMAL	ACTUAL
V_A	10 V	[49]	—	—	—	—
C_1	3.49 V	[105]	0.5 µF	[88]	0→∞	[22]
C_2	5.4 V	[179]	2.2 µF	[30]	0→∞	[32]
R_1	7.88 V	[71]	—	—	120 Ω	[150]
R_2	4.12 V	[25]	—	—	470 Ω	[62]
TP1	10 V	[208]	—	—	∞	[64]
TP2	5.4 V	[40]	—	—	∞	[4]
TP3	4.12 V	[155]	—	—	470 Ω	[112]

Table 15-8. PShooter Data for Figure 15-62

2. Use the PShooter data in Table 15-8 to troubleshoot the circuit shown in Figure 15-62.

3. Use the PShooter data in Table 15-9 to troubleshoot the circuit shown in Figure 15-62.

TESTPOINT	rms VOLTAGE		CAPACITANCE		RESISTANCE	
	NORMAL	ACTUAL	NORMAL	ACTUAL	NORMAL	ACTUAL
V_A	10 V	[146]	—	—	—	—
C_1	3.49 V	[200]	0.5 µF	[68]	0→∞	[82]
C_2	5.4 V	[41]	2.2 µF	[194]	0→∞	[140]
R_1	7.88 V	[104]	—	—	120 Ω	[113]
R_2	4.12 V	[188]	—	—	470 Ω	[134]
TP1	10 V	[184]	—	—	∞	[36]
TP2	5.4 V	[67]	—	—	∞	[196]
TP3	4.12 V	[176]	—	—	470 Ω	[121]

Table 15-9. PShooter Data for Figure 15-62

Chapter Summary

- Series, parallel, series-parallel, and complex RC circuits are similar to simple resistive circuits in many ways. The current is the same in all parts of a series circuit, whereas the voltage is the same across all components in a parallel circuit. Simplification of series-parallel and complex RC circuits is accomplished using the same methods as used with resistive circuits. The difference is that circuit quantities for RC circuits must be represented in polar or rectangular notation to account for the effects of phase differences. Basic circuit theory, Ohm's Law, and Kirchhoff's Laws still apply and still form the basis of all RC circuit analyses with sinusoidal input voltages.

- Total opposition to current flow in an RC circuit is called impedance (Z) and is measured in ohms. Its reciprocal is called admittance (Y) and is measured in siemens. True power is dissipated in the resistance of an RC circuit. Reactive power is caused by the capacitance in the circuit. As with RL circuits, total power is called apparent power and is the phasor sum of the true and reactive power components.

- When a pulse waveform is applied to an RC circuit, it takes a definite amount of time for the capacitor to charge or discharge. One time constant (τ) is equal to the resistance in ohms times the capacitance in farads ($\tau = RC$). The capacitor voltage can increase (or decrease) 63.2% of its total charge in one time constant. For practical purposes, it takes five time constants for a capacitor to fully charge or discharge after an abrupt voltage change at the input of the RC circuit.

- The relative lengths of the RC time constant and the duration of an input pulse voltage determine how an RC circuit will respond to a pulse. If the time constant is ten or more times as long as the input pulse duration, we say it is a long time constant. If the time constant is one-tenth of the pulse duration or less, then it is classed as a short time constant.

- Troubleshooting of *RC* circuits requires application of the troubleshooting methods used for comparable resistive circuit and capacitive circuit configurations. As with any troubleshooting problem, a technician should make smart measurements, but make as few as practical to locate the defect.

Review Questions

Section 15.1: Series *RC* Circuits with Sinusoidal Current

1. The current is the same magnitude through all components in a series *RC* circuit. (True or False)

2. The current through every component in a series *RC* circuit is in phase with the current through every other component. (True or False)

3. A series *RC* circuit consists of a 10-kΩ resistance and a 5-kΩ capacitive reactance. If total current is 10 mA, how much current flows through the resistance? How much through the capacitance?

4. The total voltage in a series *RC* circuit is equal to the phasor sum of the individual component voltage drops. (True or False)

5. Each resistance or reactance in a series *RC* circuit drops a voltage that is inversely proportional to the value of resistance or reactance. (True or False)

6. What is the phase relationship between voltage and current in a capacitor that is part of a series *RC* circuit?

7. What is the phase relationship between voltage and current in a resistor that is part of a series *RC* circuit?

8. What can be said about the range of possible values for the overall phase angle in a series *RC* circuit?

9. The impedance in a series *RC* circuit is equal to the phasor sum of the individual resistances and reactances. (True or False)

10. Calculate the impedance of a series *RC* circuit consisting of a 1,500-pF capacitor and a 5-Ω resistor if they are operated at 30 MHz.

11. What is the impedance of a series circuit consisting of a 3.9-kΩ resistor and a 680-pF capacitor if the operating frequency is 50 kHz?

12. The total impedance of a series *RC* circuit decreases as frequency increases. (True or False)

13. What size capacitor must be connected in series with a 22-kΩ resistor to produce an impedance of 27 kΩ if the operating frequency is 4.6 MHz?

14. If a series *RC* circuit has a resistance of 300 Ω, a capacitive reactance of 400 Ω, and an impedance of 500 Ω, what is the susceptance of the circuit? What is the admittance of the circuit?

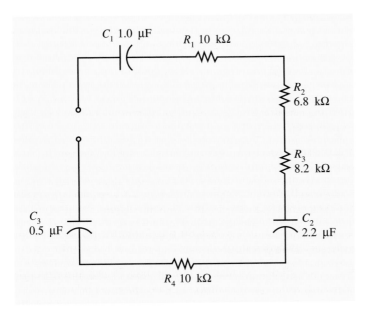

Figure 15-63. Simplify this circuit.

15. Draw an equivalent simplified circuit for the circuit shown in Figure 15-63.

16. Draw the voltage phasor diagram for the circuit shown in Figure 15-64. Include the value of V_A.

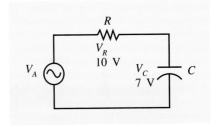

Figure 15-64. A series RC circuit.

17. Draw the impedance phasor diagram for the circuit shown in Figure 15-65. Include the value of Z.

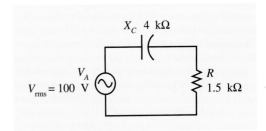

Figure 15-65. A series RC circuit.

18. Draw the power phasor diagram for the circuit shown in Figure 15-65.

19. In a series RC circuit, reactive power and true power are always the same value. (True or False)

20. In a series RC circuit, apparent power is the phasor sum of the capacitive (reactive) and resistive (true) powers. (True or False)

Section 15.2: Parallel *RC* Circuits with Sinusoidal Current

21. The current is the same through every branch of a parallel *RC* circuit. (True or False)

22. The voltage drops of the individual components in a parallel *RC* circuit may be summed with phasor addition to find total voltage. (True or False)

23. What is the phase relationship between the current and voltage in a capacitor that is part of a parallel *RC* circuit?

24. What is the phase relationship between the current and voltage in a resistor that is part of a parallel *RC* circuit?

25. The current in each branch of a parallel *RC* circuit is inversely proportional to the resistance or reactance in that branch. (True or False)

26. The total current in a parallel *RC* circuit is equal to the phasor sum of the individual branch currents. (True or False)

27. If the resistive branch of a parallel *RC* circuit has 275 mA and the capacitive branch has 325 mA, how much current flows through the source?

28. If the total current in a parallel *RC* circuit is 1.5 A and the resistive branch has 0.75 A, how much current flows through the capacitive branch?

29. The impedance of a parallel *RC* circuit is always less than the resistance or reactance of a given branch. (True or False)

30. If the resistive current in a parallel *RC* circuit is 22 mA and the capacitive current is 35 mA when the input voltage is 75 V, what is the impedance of the circuit?

31. Simplify the circuit shown in Figure 15-66.

32. Draw the equivalent, simplified circuit if three capacitors having values of 100 pF, 200 pF, and 300 pF are connected in parallel with three resistors having values of 100 Ω, 200 Ω, and 300 Ω.

33. Draw the current phasor diagram for the circuit shown in Figure 15-67. Include the value of total current.

34. Draw the power phasor diagram for the circuit shown in Figure 15-67.

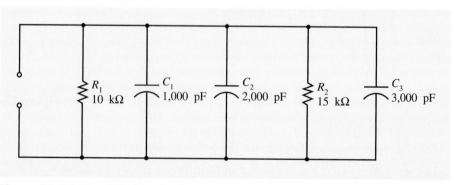

Figure 15-66. Simplify this circuit.

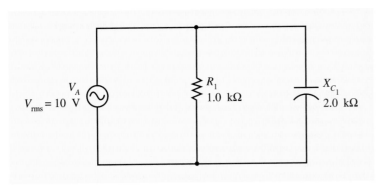

Figure 15-67. A parallel RC circuit.

Section 15.3: *RC* Circuit Calculations

Refer to Figure 15-68 for Questions 35–44.

35. What is the reactance of C?

36. What is the impedance of the circuit?

37. How much current flows through the resistor?

38. What is the voltage drop across C?

39. What is the voltage drop across R?

40. What is the power factor of the circuit?

41. What is the phase angle (θ) of the circuit?

42. What is the apparent power in the circuit?

43. How much true power is dissipated by the circuit?

44. What is the susceptance in the circuit?

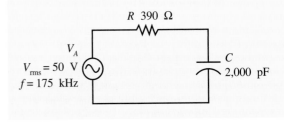

Figure 15-68. Analyze this RC circuit.

45. Complete the solution matrix in Table 15-10 for the circuit shown in Figure 15-69.

46. Complete a solution matrix like the one in Table 15-10 for the circuit shown in Figure 15-70.

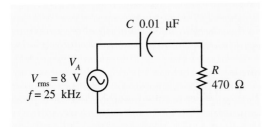

Figure 15-69. Analyze this circuit.

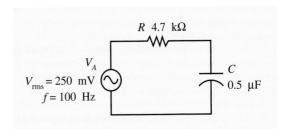

Figure 15-70. Analyze this circuit.

CIRCUIT QUANTITY		RESISTOR	CAPACITOR	TOTAL
Current (rms)				
Voltage	V_P			
	V_{PP}			
	V_{rms}			
Resistance/Reactance/Impedance				
Conductance/Susceptance/Admittance				
Power				
Phase Angle (overall)				
Power Factor (overall)				

Table 15-10. A Solution Matrix

Refer to Figure 15-71 for Questions 47–58.

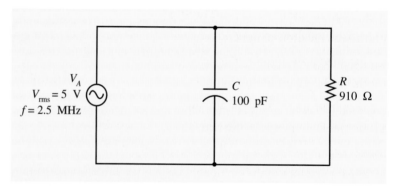

Figure 15-71. A parallel RC circuit.

47. What is the value of capacitive reactance?
48. What is the value of susceptance in the circuit?
49. How much current flows through C?
50. How much current flows through R?
51. What is the value of total current in the circuit?
52. What is the impedance of the circuit?
53. What is the admittance of the circuit?
54. What is the phase angle between total voltage and total current in the circuit?
55. What is the power factor of the circuit?
56. How much reactive power is in the circuit?
57. How much true power is dissipated in the resistance?

58. What is the value of apparent power?
59. Complete a solution matrix like Table 15-10 for the circuit shown in Figure 15-72.

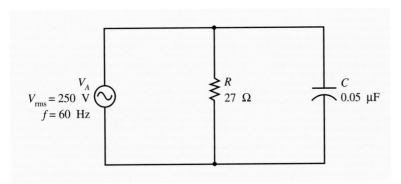

Figure 15-72. Analyze this circuit.

60. Complete a solution matrix like Table 15-10 for the circuit shown in Figure 15-73.

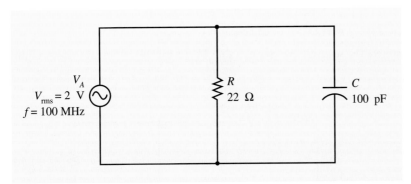

Figure 15-73. Analyze this circuit.

Refer to Figure 15-74 for Questions 61–75.

61. What is the reactance of C_1?
62. What is the reactance of C_2?
63. What is the impedance of the R_1, C_2 branch?
64. What is the impedance of the R_2, C_3 branch?
65. What is the total impedance of the circuit?
66. How much total current flows in the circuit?
67. What is the voltage drop across C_1?
68. What is the voltage drop across R_1?
69. What is the current through C_3?
70. What is the voltage drop across R_2?
71. What is the value of apparent power in the circuit?

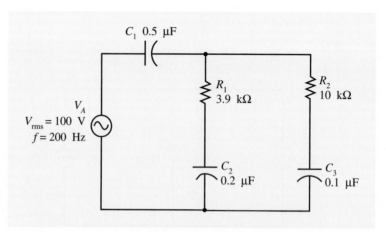

Figure 15-74. A series-parallel *RC* circuit.

72. What is the phase angle between current and voltage in C_3?
73. What is the phase angle between current and voltage in R_1?
74. What is the phase angle between total current and total voltage in the circuit?
75. What is the power factor of the circuit?
76. Complete the solution matrix in Table 15-11 for the circuit shown in Figure 15-75.

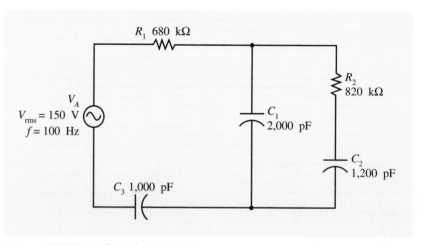

Figure 15-75. Analyze this circuit.

CIRCUIT QUANTITY	R_1	R_2	C_1	C_2	C_3	TOTAL
Voltage (rms)						
Current (rms)						
Resistance/Reactance/Impedance						
Conductance/Susceptance/Admittance						
Power						
Phase Angle (overall)						
Power Factor (overall)						

Table 15-11. A Solution Matrix

Section 15.4: Pulse Response of *RC* Circuits

77. A capacitor will charge to 63.2% of the applied voltage in one time constant. (True or False)

78. If a capacitor that is charged to 100 V is connected across a resistor, how many time constants does it take for the capacitor voltage to decrease to about 36.8 V?

79. If the *RC* time constant of a circuit is 2.5 S, how long does it take to fully charge the capacitor?

80. What is the *RC* time constant of a series circuit consisting of a 100-μF capacitor and a 2.7-kΩ resistor?

81. What value of resistance must be used with a 2,200-pF capacitor to produce a time constant of 220 μs?

82. It takes more total time to discharge a capacitor with 100 V than it does if the capacitor only had 10 V. (True or False)

83. If the pulse duration applied to an *RC* circuit is much longer than the *RC* time constant, we can classify the time constant as a _____ time constant.

84. If the time constant of an *RC* circuit is much longer than the input pulse duration, we can classify the time constant as a _____ time constant.

85. At a particular instant, the voltage across a 0.005-μF capacitor is changing at the rate of 100 V/μs. What is the value of capacitor current at that time?

86. If the rate of change of voltage across a capacitor is increased, what happens to the charging or discharging current?

87. When the voltage across a capacitor changes from 5 V to 10 V in 1 ms, the current is 10 μA. How much current would flow if the voltage changed from 1,000 V to 1,005 V in 1 ms?

88. A 100-kΩ resistor and a 1,000-pF capacitor are connected in series. What is the voltage across the capacitor 50 μs after a 25-V source is connected to the *RC* circuit?

89. A 10-μF capacitor is charged to 25 V. If it is then connected across a 150-kΩ resistor, how long does it take to discharge to 2.5 V?

Section 15.5: Applied Technology: *RC* Circuits

90. List two applications that use *RC* circuits.

91. If an *RC* circuit were providing the delay for a time delay circuit, what would happen to the delay time if the capacitor was increased? What if the resistor was increased?

92. Explain how an *RC* circuit can be used to distinguish between ac and dc voltages.

Section 15.6: Troubleshooting *RC* Circuits

93. Name three possible defects in a capacitor.

94. Name three possible defects in a resistor.

95. Use the PShooter data provided in Table 15-12 to troubleshoot the circuit shown in Figure 15-76.

TESTPOINT	rms VOLTAGE		CAPACITANCE		RESISTANCE	
	NORMAL	ACTUAL	NORMAL	ACTUAL	NORMAL	ACTUAL
V_A	75 V	[28]	—	—	—	—
C_1	18.1 V	[117]	0.5 μF	[34]	$0 \rightarrow \infty$	[82]
C_2	33.6 V	[72]	0.33 μF	[141]	$0 \rightarrow \infty$	[105]
R_1	18.1 V	[189]	—	—	1.8 kΩ	[59]
R_2	47 V	[125]	—	—	2.7 kΩ	[91]
TP1	75 V	[136]	—	—	∞	[123]
TP2	57.7 V	[75]	—	—	∞	[91]
TP3	33.6 V	[22]	—	—	∞	[67]

Table 15-12. PShooter Data for Figure 15-76

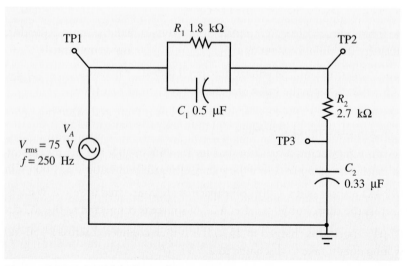

Figure 15-76. Troubleshoot this circuit.

A technician must be able to apply previously learned knowledge and skills to the solution of unfamiliar problems. Figure 15-77 gives you an opportunity to apply your circuit analysis skills to a complex RC circuit.

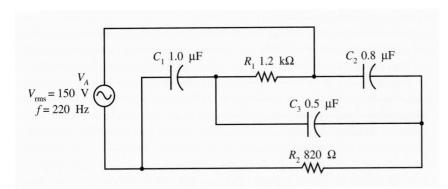

Figure 15-77. Analyze this complex RC circuit.

We have not formally discussed the solution of complex RC circuits, but we have discussed all of the individual tools needed to analyze the problem. Apply what you have learned in this and preceding chapters and meet the technician challenge!

Here's the challenge. Analyze the circuit shown in Figure 15-77 to determine the following values:

- total current
- phase angle between total current and total voltage
- voltage across C_1
- current through R_2
- total impedance of the circuit (express in polar form)

Equation List

(15-1) $Z = \sqrt{R^2 + X_C^2}$

(15-2) $B = \dfrac{1}{X_C}$

(15-3) $V_A = \sqrt{V_R^2 + V_C^2}$

(15-4) $I_T = \sqrt{I_R^2 + I_C^2}$

(15-5) $\tau = RC$

(15-6) $v_C = V - (V - V_0)\epsilon^{-t/\tau}$

(15-7) $i_C = I - (I - I_0)\epsilon^{-t/\tau}$

(15-8) $t = \tau \ln\!\left(\dfrac{V - V_0}{V - v_C}\right)$

(15-9) $i_c = C\dfrac{dv}{dt} = C\dfrac{\Delta v}{\Delta t}$

key terms

bandpass
bandwidth
cutoff frequency
half-power point
negative feedback

positive feedback
Q of an RLC circuit
resonant frequency
selectivity
self-resonant frequency

objectives

After completing this chapter, you should be able to:

1. Draw an equivalent RC or RL circuit for a given RLC circuit.

2. Calculate the following quantities in a series, parallel, or series-parallel RLC circuit:

   admittance
   apparent power
   component currents
   component voltages
   phase angle between total
      voltage and total current

   power factor
   reactive power
   total current
   total impedance
   true power
   susceptance

3. State the following characteristics of series and parallel RLC circuits at resonance, above resonance, and below resonance: a. phase relationships; b. relative component voltages and currents; c. equivalent circuit.

4. Explain the causes and effects of self-resonance for inductors and capacitors.

5. Explain how to tune an RLC circuit to a given resonant frequency.

6. Troubleshoot an RLC circuit.

RLC Circuit Analysis

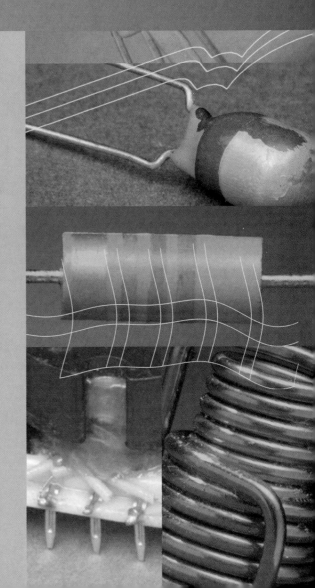

Because we have emphasized the application of Ohm's and Kirch-hoff's Laws for all types of circuit analysis problems presented in previous chapters, the analysis of RLC circuits should seem very logical to you. We will continue to apply our previously mastered skills and knowledge to the solution of unfamiliar circuit problems. This approach drastically reduces your effort (e.g., very few new equations and concepts), and it gives you ongoing practice in the basic techniques, which will increase your skills and your long-term memory of the important circuit analysis methods.

16.1 Introduction to *RLC* Circuits

When inductance, capacitance, and resistance are all part of the same circuit, we call it an *RLC* circuit. As you might expect, the characteristics of an *RLC* circuit can appear capacitive, resistive, or inductive, depending on the relative component values. Before we begin a rigorous numerical analysis of *RLC* circuits, let's examine the basic characteristics of series and parallel *RLC* networks.

Series *RLC* Circuits

Figure 16-1 shows a series *RLC* circuit. Most of the characteristics of series *RLC* circuits are similar to previously studied series circuits.

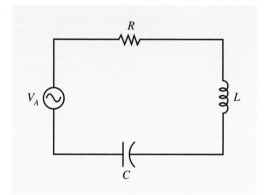

Figure 16-1. A series *RLC* circuit.

CURRENT

The current in a series *RLC* circuit is the same through every component. The current has exactly the same value at the same time in all parts of the circuit. This is equivalent to saying that the current through any given component is in phase with the current through every other component. This important relationship is illustrated in Figure 16-2.

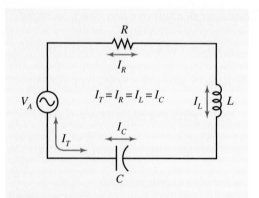

Figure 16-2. The current is the same in all parts of a series *RLC* circuit.

VOLTAGE DROPS

Since the current is the same through all components in a series *RLC* circuit, it follows that the voltage drops across the various components will be determined by the relative resistance or reactance of the component. Higher resistances or reactances will drop more voltage (Ohm's Law). This is illustrated in Figure 16-3.

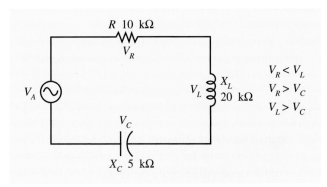

Figure 16-3. Components in an *RLC* circuit drop voltage in proportion to the resistance or reactance.

As with every other series circuit, the sum of the voltage drops must equal the applied voltage (Kirchhoff's Voltage Law). As you might expect, however, the voltages in a series *RLC* circuit must be summed with phasor addition to include the effects of the relative phases. Since inductive voltages lead the common current by 90° and capacitive voltages lag the common current by 90°, it follows that inductive and capacitive voltages are 180° out of phase in a series *RLC* circuit. This means that they tend to cancel each other. We can express this more formally with Equation 16-1.

$$V_{REACTIVE} = V_L + V_C \qquad (16\text{-}1)$$

where V_L is always positive and V_C is always negative. This assignment of polarities follows from the orientation of the capacitive and inductive voltage phasors. If the result is positive, then the net reactive voltage is inductive (i.e., voltage will be leading current). If the result is negative, then the net reactive voltage is capacitive (i.e., current will be leading voltage).

EXAMPLE SOLUTION

If the inductive voltage in a series *RLC* circuit is 10 V and the capacitive voltage is 8 V, what is the net effect of these two voltages?

EXAMPLE **SOLUTION**

Since they are 180° out of phase, they act like two opposing voltage sources. The result is found with Equation 16-1.

$$V_{REACTIVE} = V_L + V_C = 10 \text{ V} + (-8 \text{ V}) = 2 \text{ V}$$

This means that the inductive voltage has cancelled the capacitive voltage and still has a 2-V net potential. The circuit will act inductively (i.e., voltage will lead current).

EXAMPLE SOLUTION

What is the net reactive voltage in a series *RLC* circuit that has 25 V across the inductor and 45 V across the capacitor?

EXAMPLE **SOLUTION**

We apply Equation 16-1 as follows:

$$V_{REACTIVE} = V_L + V_C = 25 \text{ V} + (-45 \text{ V}) = -20 \text{ V}$$

This means that the capacitive voltage has completely cancelled the inductive voltage and still has a 20-V potential. The circuit will act capacitively (i.e., current will lead voltage).

○ **KEY POINTS**

Phasor addition is used to combine the net reactive voltage with the resistive voltage since they are 90° out of phase.

In order to apply Kirchhoff's Voltage Law to a series *RLC* circuit, we find the phasor sum of the resistive voltage and the net reactive voltage as computed by Equation 16-1. We can combine these two steps into Equation 16-2.

$$V_A = \sqrt{(V_L - V_C)^2 + V_R^2} \text{ , or}$$
$$V_A = \sqrt{V_R^2 + (V_L - V_C)^2} \qquad (16\text{-}2)$$

Clearly, these two alternatives are equivalent, but some technicians have a strong preference for one form or the other. In either case, the calculations have been combined into one equation. Additionally, both reactive voltages can be entered as positive numbers, so there is less opportunity for error.

EXAMPLE SOLUTION

What is the value of input voltage for the circuit shown in Figure 16-4?

EXAMPLE **SOLUTION**

We apply Equation 16-2 as follows:

$$V_A = \sqrt{(V_L - V_C)^2 + V_R^2}$$
$$= \sqrt{(20 \text{ V} - 15 \text{ V})^2 + (10 \text{ V})^2} = 11.18 \text{ V}$$

○ **KEY POINTS**

It is possible for the applied voltage to be less than one or more of the component voltages.

This value should seem peculiar to you at first, since you are not accustomed to seeing individual component voltage drops in a series circuit that are numerically larger than the input voltage. This is a very interesting characteristic of the *RLC* circuit and one that will be explained in more detail in a later section. For now, trust Kirchhoff's Voltage Law.

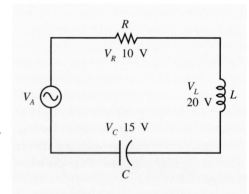

Figure 16-4. Find the value of V_A.

Practice Problems

1.   If a series *RLC* circuit has an inductive voltage drop of 18 V, a capacitive voltage drop of 15 V, and a resistive voltage drop of 6 V, what is the value of applied voltage?

2. What is the total voltage in a series *RLC* circuit if it has an inductive voltage drop of 175 V, a capacitive voltage drop of 150 V, and a resistive voltage drop of 10 V?

3. If a series *RLC* circuit has an inductive voltage drop of 200 mV, a capacitive voltage drop of 100 mV, and a resistive voltage drop of 100 mV, what is the value of applied voltage?

Answers to Practice Problems

1. 6.7 V   **2.** 26.9 V   **3.** 141.4 mV

IMPEDANCE

The impedance of any series circuit is equal to the sum of the individual resistances and reactances. In the case of a series *RLC* circuit, the summing operation must include the effects of differing phase relationships. You will recall from the phasor diagrams for series *RL* and series *RC* circuits, that X_L is shown at 90° and X_C is drawn at −90°. Since the two reactances are 180° out of phase they tend to cancel. Their net result and the resistance are then summed with phasor addition to yield the value of total impedance. We can express this process with Equation 16-3.

$$Z = \sqrt{(X_L - X_C)^2 + R^2}, \text{ or}$$
$$Z = \sqrt{R^2 + (X_L - X_C)^2}$$

(16-3)

EXAMPLE SOLUTION

What is the total impedance of the circuit shown in Figure 16-5?

EXAMPLE **SOLUTION**

We apply Equation 16-3 as follows:

$$Z = \sqrt{(X_L - X_C)^2 + R^2}$$
$$= \sqrt{(4 \text{ k}\Omega - 7 \text{ k}\Omega)^2 + (5 \text{ k}\Omega)^2} = 5.83 \text{ k}\Omega$$

Here again, your prior experience may alert you that this is a peculiar value. You have never seen a series circuit whose total impedance was less than the resistance or reactance of some of its series components. This interesting characteristic of a series *RLC* circuit is very important and will be addressed further in a later section. For now, accept that the total impedance is sometimes less than the reactance of the inductor or the capacitor.

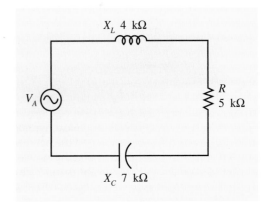

Figure 16-5. Find the impedance of this circuit.

PHASOR RELATIONSHIPS

The phase relationships and the related phasor diagram for a series *RLC* circuit is a logical extension of *RL* and *RC* circuits. It is essentially a combination of the two. The only new aspect is that the two reactive components (voltage, reactance, or reactive power) must be combined algebraically, since they are 180° out of phase, before you complete the parallelogram to locate total voltage, impedance, or apparent power.

EXAMPLE SOLUTION

Draw the voltage phasor diagram for the circuit shown in Figure 16-6.

EXAMPLE SOLUTION

Because it is a series circuit, we will use current as the reference phasor. The resistive voltage will be in phase with the current (as in all resistors). The inductive voltage will be leading the current by 90° (as in all inductors), and the capacitive voltage will be lagging the current by 90° (as in all capacitors). These relationships are shown in Figure 16-7.

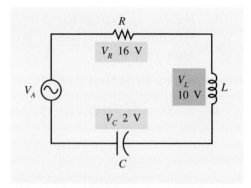

Figure 16-6. Draw the voltage phasor diagram for this circuit.

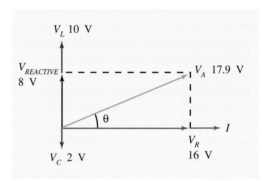

Figure 16-7. A voltage phasor diagram for the circuit shown in Figure 16-6.

Next, we combine the inductive voltage (V_L) and the capacitive voltage (V_C) algebraically to find the net reactive voltage ($V_{REACTIVE}$). By inspection, we can see that the difference is 8 V and that the inductive voltage is larger. So, we can plot the $V_{REACTIVE}$ phasor as shown in Figure 16-7. If the magnitude and direction of the $V_{REACTIVE}$ phasor is not obvious, then we can always apply Equation 16-1 as follows:

$$V_{REACTIVE} = V_L + V_C = 10 \text{ V} + (-2 \text{ V}) = +8 \text{ V}$$

We can now complete the parallelogram to locate the position of the total voltage phasor (V_A). We can compute its value by finding the phasor sum of $V_{REACTIVE}$ and V_R by applying Equation 16-2.

$$V_A = \sqrt{(V_L - V_C)^2 + V_R^2}$$
$$= \sqrt{(10 \text{ V} - 2 \text{ V})^2 + (16 \text{ V})^2} = 17.9 \text{ V}$$

This value is shown on the completed phasor diagram in Figure 16-7. Also shown is the phase angle (θ) between total voltage and total current.

EXAMPLE SOLUTION

Draw an impedance phasor diagram for the circuit shown in Figure 16-8.

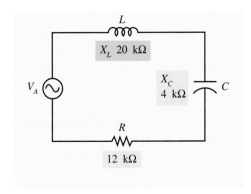

Figure 16-8. Draw the impedance phasor diagram for this circuit.

EXAMPLE **SOLUTION**

First, we draw the resistance phasor in the reference position. As with *RL* circuits, the inductive reactance phasor is drawn 90° ahead of the resistance phasor. The capacitive reactance phasor, as with *RC* circuits, is drawn 90° behind the resistive phasor. These are all shown in Figure 16-9.

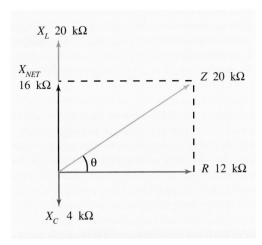

Figure 16-9. An impedance phasor diagram for the circuit shown in Figure 16-8.

Since the phasors representing inductive and capacitive reactance are 180° out of phase, we combine them algebraically to find the net reactance (X_{NET}). In most cases, the magnitude and polarity of X_{NET} is apparent from the phasor diagram. If necessary, we can consider X_L to be positive and X_C to be negative and then apply Equation 16-4.

$$X_{NET} = X_L + X_C \qquad (16\text{-}4)$$

In our present example, we have

$$X_{NET} = X_L + X_C = 20 \text{ k}\Omega + (-4 \text{ k}\Omega) = 16 \text{ k}\Omega$$

Finally, we combine the net reactance (X_{NET}) with the resistance (*R*) with phasor addition to find the circuit impedance. We can locate the phasor graphically by completing the parallelogram as shown in Figure 16-9. The numerical value is found with Equation 16-3.

$$Z = \sqrt{(X_L - X_C)^2 + R^2}$$
$$= \sqrt{(20 \text{ k}\Omega - 4 \text{ k}\Omega)^2 + (12 \text{ k}\Omega)^2} = 20 \text{ k}\Omega$$

This is shown on the completed phasor diagram in Figure 16-9. Also shown is the phase angle (θ) between the resistance and impedance phasors.

EXAMPLE SOLUTION

Draw the power phasor diagram for the circuit shown in Figure 16-10.

EXAMPLE SOLUTION

First, we draw true power in the reference position. Reactive power due to the inductive reactance (P_{R_L}) is drawn 90° ahead of the true power phasor. Reactive power due to the capacitive reactance (P_{R_C}) is drawn 90° behind the true power phasor. These are shown in Figure 16-11. They are consistent with the phasor diagrams for *RC* and *RL* circuits.

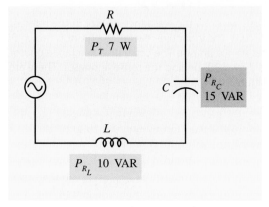

Figure 16-10. Draw the power phasor diagram for this circuit.

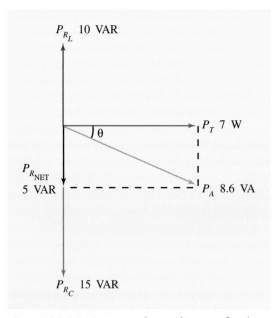

Figure 16-11. A power phasor diagram for the circuit shown in Figure 16-10.

Next, we find the net reactive power ($P_{R_{NET}}$) by algebraically combining the two reactive power phasors (P_{R_L} and P_{R_C}). In most cases, the magnitude and direction can be determined by inspection, but Equation 16-5 provides a more formal approach.

$$P_{R_{NET}} = P_{R_L} + P_{R_C} \qquad \text{(16-5)}$$

where P_{R_L} is assumed to be positive and P_{R_C} is assumed to be negative. For our present example, we have

$$P_{R_{NET}} = P_{R_L} + P_{R_C} = 10 \text{ VAR} + (-15 \text{ VAR}) = -5 \text{ VAR}$$

where the minus sign tells us that the net reactive power is capacitive.

Finally, we sum (phasor addition) the net reactive power with the true power to find the total or apparent power. Graphically, we can complete the parallelogram as shown in Figure 16-11. Numerically, we can apply Equation 16-6.

$$P_A = \sqrt{(P_{R_L} - P_{R_C})^2 + P_T^2} \text{, or} \qquad \text{(16-6)}$$
$$P_A = \sqrt{P_T^2 + (P_{R_L} - P_{R_C})^2}$$

In the case of Figure 16-11, we compute apparent power as

$$P_A = \sqrt{(P_{R_L} - P_{R_C})^2 + P_T^2}$$
$$= \sqrt{(10 \text{ VAR} - 15 \text{ VAR})^2 + (7 \text{ W})^2} = 8.6 \text{ VA}$$

This is shown on the completed phasor diagram in Figure 16-11. Also shown is the phase angle (θ), which is always drawn between the true power and apparent power phasors.

Parallel *RLC* Circuits

Figure 16-12 shows a basic parallel *RLC* circuit. These circuits possess many of the characteristics of other parallel circuits.

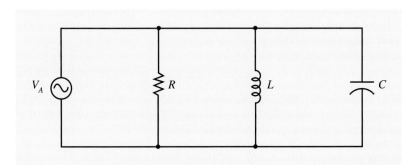

Figure 16-12. A parallel *RLC* circuit.

CURRENT

As with other parallel circuits, the total current in a parallel *RLC* circuit is equal to the sum of the individual branch currents. Since the branch currents are out of phase, however, they must be summed with phasor addition. More specifically, capacitive branch current and inductive branch current are 180° out of phase and must be combined algebraically to find the net reactive current (I_X). This is expressed more formally by Equation 16-7.

KEY POINTS

The voltage drop is the same across every component in a parallel *RLC* circuit.

$$I_X = I_L + I_C \qquad (16\text{-}7)$$

where I_L is assumed to be negative and I_C is assumed to be positive. These polarity assignments are based on the phasor diagrams for parallel *RL*, *RC*, and *RLC* circuits. Total current is then the phasor sum of resistive current and the net reactive current. All of this can be expressed simply as Equation 16-8.

$$I_T = \sqrt{(I_L - I_C)^2 + I_R^2}, \text{ or}$$
$$I_T = \sqrt{I_R^2 + (I_L - I_C)^2} \qquad (16\text{-}8)$$

KEY POINTS

The reactive currents are combined algebraically to produce a net reactive current. Phasor addition is then used to combine the net reactive current with the resistive current.

EXAMPLE SOLUTION

What is the total current flow for the circuit shown in Figure 16-13?

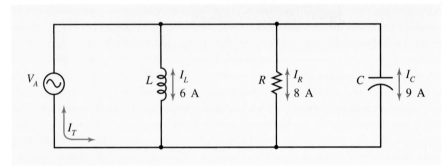

Figure 16-13. Find the total current flow in this circuit.

EXAMPLE **SOLUTION**

We apply Equation 16-8 directly.

$$I_T = \sqrt{(I_L - I_C)^2 + I_R^2}$$
$$= \sqrt{(6 \text{ A} - 9 \text{ A})^2 + (8 \text{ A})^2} = 8.54 \text{ A}$$

KEY POINTS

Some current circulates between the inductor and the capacitor and does not contribute to total current. This means that it is possible for total current to be less than the current through the reactive branches.

This illustrates an interesting characteristic of a parallel *RLC* circuit; one or more branch currents may be larger than total current. This seems contrary to common sense at first, but if you remember that the inductive and capacitive currents are 180° out of phase and tend to cancel, then it is more understandable.

Yet another way to explain this strange characteristic is to consider that because the two reactive currents are 180° out of phase, when current is flowing upward in the inductive branch, it will be flowing downward in the capacitive branch. For purposes of illustration, let's assume that the value of inductive current is larger than the value of capacitive current. If we examine the electron flow in the circuit at a specific instant in time and apply Kirchhoff's Current Law, it will appear as shown in Figure 16-14. Not all of I_C and I_L flow back to the source. Part of the current circulates between *L* and *C*. Only the difference between I_L and I_C flows to the source. This difference is, of course, the net reactive current (I_X).

KEY POINTS

The branch currents are inversely proportional to the branch resistance or reactance (Ohm's Law).

Another characteristic of a parallel *RLC* circuit is that the individual branch currents are inversely proportional to the branch resistance (Ohm's Law). The higher the resistance or

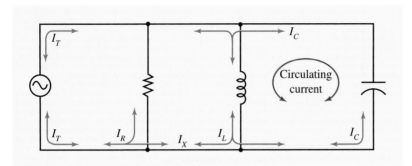

Figure 16-14. Some current circulates between the inductive and capacitive branches.

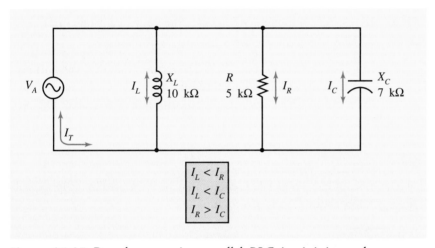

Figure 16-15. Branch current in a parallel *RLC* circuit is inversely proportional to the branch resistance or reactance.

reactance in a particular branch, the lower the current through that particular branch. Figure 16-15 illustrates this relationship.

VOLTAGE DROPS

The voltage is the same across every component in a parallel *RLC* circuit, just as with every other parallel circuit you have studied. Regardless of the relative values of resistance or reactance in the branches, the voltages across every component will be identical at all times. This also implies that the voltage waveform across a given component is in phase with the voltage waveform of every other component. Figure 16-16 illustrates the voltage relationships in a parallel *RLC* circuit.

IMPEDANCE

The impedance of a parallel circuit can be found with Ohm's Law ($Z = V_A/I_T$), just as we have done with previous parallel circuits. We can also compute impedance with the reciprocal equation provided we represent the circuit quantities in polar and/or rectangular form to account for the effects of differing phase angles.

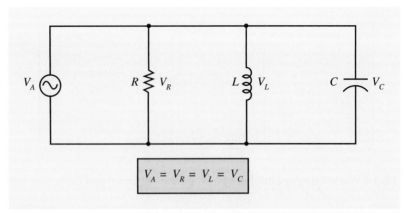

Figure 16-16. All component voltages are the same in a parallel *RLC* circuit.

One very interesting characteristic of a parallel *RLC* circuit is that the overall impedance is not necessarily lower than the smallest branch resistance or reactance. This follows from Ohm's Law and the previous observation that total current may be smaller than some of the branch currents.

PHASE RELATIONSHIPS

The phase relationships in a parallel *RLC* circuit may be illustrated by simply combining what we already know about *RC* and *RL* circuits.

EXAMPLE SOLUTION

Draw the current phasor diagram for the circuit shown in Figure 16-17.

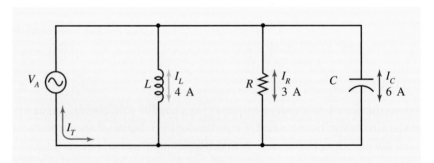

Figure 16-17. Draw a current phasor diagram for this circuit.

EXAMPLE SOLUTION

First, we draw the voltage phasor as a reference, since it is common to all components. The resistive current phasor (I_R) is drawn in phase with the voltage phasor. The inductive current phasor (I_L) lags the circuit voltage by 90°, and the capacitive current phasor (I_C) leads the circuit voltage by 90°. These phasor positions are all shown in Figure 16-18.

Next, we algebraically combine the inductive and capacitive currents to find the net reactive current (I_X). If its magnitude and direction are not obvious, we can use Equation 16-7 to determine its value.

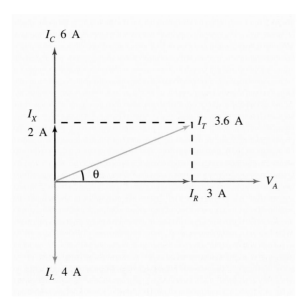

Figure 16-18. A current phasor diagram for the circuit shown in Figure 16-17.

$$I_X = I_L + I_C = -4\ \text{A} + 6\ \text{A} = 2\ \text{A}$$

Once we have located the net reactive current phasor (I_X), we can combine it with the resistive current phasor (I_R) to find total current (I_T). We do this graphically by completing the parallelogram as shown in Figure 16-18. We find its numerical value through phasor addition by applying Equation 16-8.

$$I_T = \sqrt{(I_L - I_C)^2 + I_R^2}$$
$$= \sqrt{(4\ \text{A} - 6\ \text{A})^2 + (3\ \text{A})^2} = 3.6\ \text{A}$$

This is shown on the completed phasor diagram in Figure 16-18. Also shown is the phase angle (θ), which is always located between the total voltage and total current phasors.

EXAMPLE SOLUTION

Draw the power phasor diagram for the circuit shown in Figure 16-19.

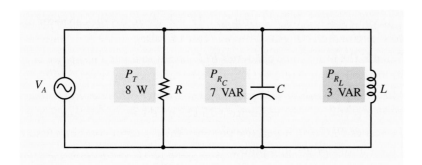

Figure 16-19. Draw the power phasor diagram for this circuit.

KEY POINTS

The power relationships in a series or parallel *RLC* circuit are essentially a combination of the relationships found in *RL* and *RC* circuits.

EXAMPLE SOLUTION

First, we sketch the true power phasor in the reference position, as with RC and RL circuits. The reactive power phasor for the inductive branch (P_{R_L}) is drawn 90° behind the true power phasor, as we did with RL circuits. Similarly, the reactive power phasor for the capacitive branch (P_{R_C}) is drawn leading true power by 90°, as with RC circuits. These phasors are all shown in Figure 16-20.

Next, we algebraically combine the two reactive powers to find the net reactive power ($P_{R_{NET}}$). In most cases, the magnitude and direction of the phasor can be determined by inspection. Otherwise, we can apply Equation 16-5.

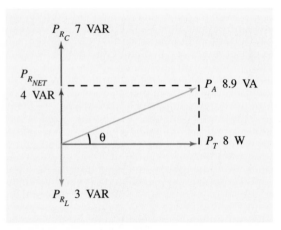

Figure 16-20. A power phasor diagram for the circuit shown in Figure 16-19.

$$P_{R_{NET}} = P_{R_L} + P_{R_C} = -3 \text{ VAR} + 7 \text{ VAR} = 4 \text{ VAR}$$

We can now combine the net reactive power phasor ($P_{R_{NET}}$) with the true power phasor (P_T) to find total or apparent power (P_A). We can locate the phasor position graphically by completing the parallelogram as shown in Figure 16-20. The numerical value of the phasor is found with Equation 16-6.

$$P_A = \sqrt{(P_{R_L} - P_{R_C})^2 + P_T^2}$$
$$= \sqrt{(3 \text{ VAR} - 7 \text{ VAR})^2 + (8 \text{ W})^2} = 8.9 \text{ VA}$$

This is shown on the completed phasor diagram in Figure 16-20. The phase angle (θ) is also labeled. It is always the angle between true power (P_T) and apparent power (P_A).

Exercise Problems 16.1

1. The voltage is the same across all components in a series *RLC* circuit. (True or False)

2. The voltage across the inductor and the voltage across the capacitor in a series *RLC* circuit are 180° out of phase. (True or False)

3. What is the impedance of a series *RLC* circuit that has a resistance of 22 kΩ, inductive reactance of 47 kΩ, and a capacitive reactance of 35 kΩ?

4. How much current flows through the capacitance of a series *RLC* circuit if the total current is 100 mA, the resistance is 25 Ω, and the inductive reactance is 50 Ω?

5. In a series *RLC* circuit, reactive power in the capacitor combines with reactive power in the inductor to produce _____ _____ _____.

6. Is it possible for the total voltage in a series *RLC* circuit to be less than the voltage across a single component?

7. The voltage across every component in a parallel circuit is in phase with the voltage across every other component. (True or False)

8. The current through every component in a parallel circuit is in phase with the current through every other component. (True or False)

9. It is possible for the total voltage in a parallel *RLC* circuit to be less than one of the branch voltage drops. (True or False)

10. How much total current flows through a parallel *RLC* circuit that has 10-mA resistive current, 25-mA capacitive current, and 30-mA inductive current?

16.2 *RLC* Circuit Calculations

We now have all the analytical tools needed to thoroughly analyze an *RLC* circuit with a sinusoidal input voltage. The analysis procedure should appear as a logical extension of *RC* and *RL* circuits. Rely on Ohm's and Kirchhoff's Laws and your basic understanding of circuit configurations to guide you through the analysis.

Series *RLC* Circuits

Analysis of series *RLC* circuits is similar to the analysis of *RC* or *RL* circuits. The specific steps, of course, vary with the problem, but the general sequence follows:

1. Compute all reactances.
2. Calculate impedance.
3. Find total current.
4. Compute component voltage drops.
5. Calculate other circuit parameters (e.g., power, phase angle, power factor).

EXAMPLE SOLUTION

Analyze the circuit shown in Figure 16-21, and complete the solution matrix in Table 16-1.

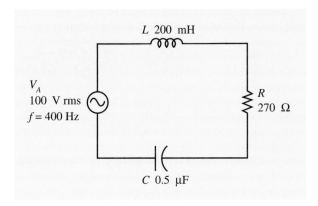

V_A
100 V rms
$f = 400$ Hz

L 200 mH

R 270 Ω

C 0.5 μF

Figure 16-21. Analyze this series *RLC* circuit.

CIRCUIT QUANTITY		RESISTOR	INDUCTOR	CAPACITOR	TOTAL
Current (rms)					
Voltage	V_P				
	V_{PP}				
	V_{rms}				100 V
Resistance/Reactance/Impedance		270 Ω			
Conductance/Susceptance/Admittance					
Power					
Phase Angle (overall)					
Power Factor (overall)					

Table 16-1. Solution Matrix for the Circuit in Figure 16-21

EXAMPLE SOLUTION

Let's first find the reactances in the circuit.

$$X_L = 2\pi f L = 6.28 \times 400 \text{ Hz} \times 200 \text{ mH} = 502 \text{ }\Omega$$

$$X_C = \frac{1}{2\pi f C} = \frac{1}{6.28 \times 400 \text{ Hz} \times 0.5 \text{ }\mu\text{F}} = 796 \text{ }\Omega$$

We can now apply Equation 16-3 to find the impedance of the circuit.

$$Z = \sqrt{(X_L - X_C)^2 + R^2}$$
$$= \sqrt{(502 \text{ }\Omega - 796 \text{ }\Omega)^2 + (270 \text{ }\Omega)^2} = 399 \text{ }\Omega$$

Now we use Ohm's Law to find total current in the circuit.

$$I_T = \frac{V_A}{Z} = \frac{100 \text{ V}}{399 \text{ }\Omega} = 250.6 \text{ mA}$$

Because it is a series circuit, this same current will flow through all components.

We now know the individual resistances/reactances, and we know the current through each component. We can use Ohm's Law to compute the individual voltage drops.

$$V_R = I_R R = 250.6 \text{ mA} \times 270 \text{ }\Omega = 67.7 \text{ V}$$
$$V_L = I_L X_L = 250.6 \text{ mA} \times 502 \text{ }\Omega = 125.8 \text{ V}$$
$$V_C = I_C X_C = 250.6 \text{ mA} \times 796 \text{ }\Omega = 199.5 \text{ V}$$

These can all be converted to peak and peak-to-peak values by applying the basic sine wave equations.

$$V_{R_p} = V_{R_{rms}} \times 1.414 = 67.7 \text{ V} \times 1.414 = 95.7 \text{ V}$$
$$V_{R_{PP}} = V_{R_p} \times 2 = 95.7 \text{ V} \times 2 = 191.4 \text{ V}$$

$$V_{L_p} = V_{L_{rms}} \times 1.414 = 125.8 \text{ V} \times 1.414 = 177.9 \text{ V}$$

$$V_{L_{PP}} = V_{L_p} \times 2 = 177.9 \text{ V} \times 2 = 355.8 \text{ V}$$

$$V_{C_p} = V_{C_{rms}} \times 1.414 = 199.5 \text{ V} \times 1.414 = 282.1 \text{ V}$$

$$V_{C_{PP}} = V_{C_L} \times 2 = 282.1 \text{ V} \times 2 = 564.2 \text{ V}$$

The rms value of input voltage can also be converted to peak and peak-to-peak values in the same way.

$$V_{A_p} = V_{A_{rms}} \times 1.414 = 100 \text{ V} \times 1.414 = 141.4 \text{ V}$$

$$V_{A_{PP}} = V_{A_p} \times 2 = 141.4 \text{ V} \times 2 = 282.8 \text{ V}$$

We can find conductance, susceptance, and admittance by simply taking the reciprocal of the resistance, reactance, and impedance values, respectively.

$$G = \frac{1}{R} = \frac{1}{270 \ \Omega} = 3.7 \text{ mS}$$

$$B_L = \frac{1}{X_L} = \frac{1}{502 \ \Omega} = 1.99 \text{ mS}$$

$$B_C = \frac{1}{X_C} = \frac{1}{796 \ \Omega} = 1.26 \text{ mS}$$

$$Y = \frac{1}{Z} = \frac{1}{399 \ \Omega} = 2.51 \text{ mS}$$

True power dissipated by the resistance can be found with any one of the basic power equations. Let's use the $P = VI$ relationship.

$$P_T = V_R I_R = 67.7 \text{ V} \times 250.6 \text{ mA} = 16.97 \text{ W}$$

Reactive power in each of the reactances is found in a similar way.

$$P_{R_L} = V_L I_L = 125.8 \text{ V} \times 250.6 \text{ mA} = 31.53 \text{ VAR}$$

$$P_{R_C} = V_C I_C = 199.5 \text{ V} \times 250.6 \text{ mA} = 49.99 \text{ VAR}$$

Total power can also be found with our choice of basic power equations, or we can use Equation 16-6 as a way to sum the individual component powers. Let's continue with the $P = VI$ method.

$$P_A = V_A I_T = 100 \text{ V} \times 250.6 \text{ mA} = 25.1 \text{ VA}$$

Let's take time out to draw the phasor diagram for the circuit. It will help ensure that we use the correct values to compute phase angle and power factor. We can use our choice of voltage, impedance, or power phasor diagrams. Let's choose to sketch a diagram for the voltage phasors. Because it is a series circuit, we will use current as the reference phasor. The resistive voltage phasor (V_R) is always in phase with current. Inductive voltage always leads current by 90°, and capacitive voltage always lags current by 90°. These phasors are all shown in Figure 16-22.

Next, we find the net reactive voltage ($V_{REACTIVE}$) by applying Equation 16-1.

$$V_{REACTIVE} = V_L + V_C = 125.8 \text{ V} + (-199.5 \text{ V}) = -73.7 \text{ V}$$

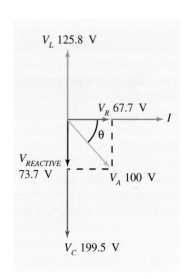

Figure 16-22.
Voltage phasor diagram for the circuit shown in Figure 16-21.

This allows us to locate and sketch the phasor representing total voltage (V_A) as shown in Figure 16-22.

The phase angle (θ) always appears between total current and total voltage. We can find its value by applying any of the basic right-triangle equations. Let's use the cosine equation so that we will find the power factor at the same time.

$$\text{power factor} = \cos\theta = \frac{\text{adjacent}}{\text{hypotenuse}} = \frac{V_R}{V_A}$$

$$= \frac{67.7\ \text{V}}{100\ \text{V}} = 0.677$$

$$\text{phase angle} = \theta = \arccos 0.677 = 47.39°$$

Since current leads the voltage in this circuit, we will label the power factor as leading. The completed solution matrix is given in Table 16-2.

The completed phasor diagram in Figure 16-22 clearly shows that total current is leading total voltage in the circuit. You will recall that this is a characteristic of a capacitive (*RC*) circuit. In effect, the high value of capacitive reactance (796 Ω) has cancelled the smaller value of inductive reactance (502 Ω). The remaining capacitive reactance (294 Ω) combines with the resistance (270 Ω) to produce the characteristics of a simple *RC* circuit (i.e., current leads voltage).

CIRCUIT QUANTITY		RESISTOR	INDUCTOR	CAPACITOR	TOTAL
Current (rms)		250.6 mA	250.6 mA	250.6 mA	250.6 mA
Voltage	V_P	95.7 V	177.9 V	282.1 V	141.4 V
	V_{PP}	191.4 V	355.8 V	564.2 V	282.8 V
	V_{rms}	67.7 V	125.8 V	199.5 V	100 V
Resistance/Reactance/Impedance		270 Ω	502 Ω	796 Ω	399 Ω
Conductance/Susceptance/Admittance		3.7 mS	1.99 mS	1.26 mS	2.51 mS
Power		16.97 W	31.53 VAR	49.99 VAR	25.1 VA
Phase Angle (overall)		47.39°			
Power Factor (overall)		0.677 (leading)			

Table 16-2. Completed Solution Matrix for the Circuit in Figure 16-21

Practice Problems

1. Analyze the circuit shown in Figure 16-23, and complete a solution matrix similar to Table 16-1.

2. Analyze the circuit shown in Figure 16-24, and complete a solution matrix similar to Table 16-1.

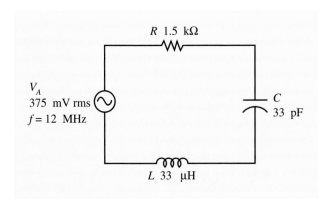

Figure 16-23. Analyze this *RLC* circuit.

Figure 16-24. Analyze this *RLC* circuit.

Answers to Practice Problems

1.

CIRCUIT QUANTITY		RESISTOR	INDUCTOR	CAPACITOR	TOTAL
Current (rms)		145.9 μA	145.9 μA	145.9 μA	145.9 μA
Voltage	V_P	0.31 V	0.514 V	82.95 mV	0.53 V
	V_{PP}	0.62 V	1.03 V	165.9 mV	1.06 V
	V_{rms}	0.219 V	0.363 V	58.67 mV	0.375 V
Resistance/Reactance/Impedance		1.5 kΩ	2.49 kΩ	402.1 Ω	2.57 kΩ
Conductance/Susceptance/Admittance		666.7 μS	401.6 μS	2.49 mS	389.1 μS
Power		31.95 μW	52.96 μVAR	8.56 μVAR	54.7 μVA
Phase Angle (overall)		54.3°			
Power Factor (overall)		0.583 (lagging)			

2.

CIRCUIT QUANTITY		RESISTOR	INDUCTOR	CAPACITOR	TOTAL
Current (rms)		7.61 mA	7.61 mA	7.61 mA	7.61 mA
Voltage	V_P	35.5 V	142 V	97.92 V	56.56 V
	V_{PP}	71 V	284 V	195.8 V	113.1 V
	V_{rms}	25.1 V	100.5 V	69.25 V	40 V
Resistance/Reactance/Impedance		3.3 kΩ	13.2 kΩ	9.1 kΩ	5.26 kΩ
Conductance/Susceptance/Admittance		303 μS	75.76 μS	109.9 μS	190.1 μS
Power		191 mW	764.8 mVAR	527 mVAR	304.4 mVA
Phase Angle (overall)		51.2°			
Power Factor (overall)		0.627 (lagging)			

Parallel *RLC* Circuits

Analysis of parallel *RLC* circuits is similar to the analysis of equivalent *RC* or *RL* circuits. The specific steps, of course, vary with the specific problem, but the general sequence follows:

1. Compute all reactances.
2. Calculate branch currents.
3. Find total current.
4. Compute impedance.
5. Calculate other circuit parameters (e.g., power, phase angle, power factor).

EXAMPLE SOLUTION

Analyze the parallel *RLC* circuit shown in Figure 16-25, and complete the solution matrix provided in Table 16-3.

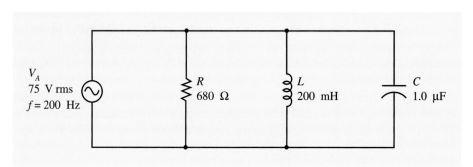

Figure 16-25. Analyze this parallel *RLC* circuit.

CIRCUIT QUANTITY		RESISTOR	INDUCTOR	CAPACITOR	TOTAL
Current (rms)					
Voltage	V_P				
	V_{PP}				
	V_{rms}				75 V
Resistance/Reactance/Impedance		680 Ω			
Conductance/Susceptance/Admittance					
Power					
Phase Angle (overall)					
Power Factor (overall)					

Table 16-3. Solution Matrix for the Circuit in Figure 16-25

EXAMPLE SOLUTION

First, we apply the basic reactance equations to compute inductive and capacitive reactance.

$$X_L = 2\pi f L = 6.28 \times 200 \text{ Hz} \times 200 \text{ mH} = 251.2 \ \Omega$$

$$X_C = \frac{1}{2\pi f C} = \frac{1}{6.28 \times 200 \text{ Hz} \times 1.0 \ \mu\text{F}} = 796 \ \Omega$$

We can now use Ohm's Law to compute the individual branch currents. The voltage is, of course, the same for all components.

$$I_R = \frac{V_R}{R} = \frac{75 \text{ V}}{680 \ \Omega} = 110.3 \text{ mA}$$

$$I_L = \frac{V_L}{X_L} = \frac{75 \text{ V}}{251.2 \ \Omega} = 298.6 \text{ mA}$$

$$I_C = \frac{V_C}{X_C} = \frac{75 \text{ V}}{796 \ \Omega} = 94.2 \text{ mA}$$

The branch currents can be combined to find total current. We use Equation 16-8 as follows:

$$I_T = \sqrt{(I_L - I_C)^2 + I_R^2}$$
$$= \sqrt{(298.6 \text{ mA} - 94.2 \text{ mA})^2 + (110.3 \text{ mA})^2} = 232.3 \text{ mA}$$

Application of Ohm's Law will give us the value of impedance in the circuit.

$$Z = \frac{V_A}{I_T} = \frac{75 \text{ V}}{232.3 \text{ mA}} \approx 322.9 \ \Omega$$

The remaining circuit quantities can be computed in most any sequence. Let's convert the rms voltages to equivalent peak and peak-to-peak values by applying the basic sine wave equations.

$$V_{R_P} = V_{R_{rms}} \times 1.414 = 75 \text{ V} \times 1.414 = 106.1 \text{ V}$$

$$V_{R_{PP}} = V_{R_P} \times 2 = 106.1 \text{ V} \times 2 = 212.2 \text{ V}$$

Since the voltage is the same across all components in a parallel circuit, this same value applies to V_A, V_C, and V_L.

The conductance, susceptance, and admittance is found by taking the reciprocal of resistance, reactance, and impedance, respectively.

$$G = \frac{1}{R} = \frac{1}{680 \ \Omega} = 1.47 \text{ mS}$$

$$B_L = \frac{1}{X_L} = \frac{1}{251.2 \ \Omega} = 3.98 \text{ mS}$$

$$B_C = \frac{1}{X_C} = \frac{1}{796 \ \Omega} = 1.26 \text{ mS}$$

$$Y = \frac{1}{Z} = \frac{1}{322.9 \ \Omega} = 3.1 \text{ mS}$$

We can use any of the basic power equations to compute power in the circuit. For purposes of illustration, let's use the basic form of $P = I^2 R$.

$$P_T = I_R^2 R = (110.3 \text{ mA})^2 \times 680 \ \Omega = 8.27 \text{ W}$$

$$P_{R_L} = I_L^2 X_L = (298.6 \text{ mA})^2 \times 251.2 \ \Omega = 22.4 \text{ VAR}$$

$$P_{R_C} = I_C^2 X_C = (94.2 \text{ mA})^2 \times 796 \ \Omega = 7.06 \text{ VAR}$$

$$P_A = I_T^2 Z = (232.3 \text{ mA})^2 \times 322.9 \ \Omega = 17.4 \text{ VA}$$

We could just as easily have used Equation 16-6 to compute total or apparent power. Sometimes, it's good to use both calculations as a way to catch errors.

Before we calculate phase angle and power factor, let's sketch the phasor diagram to lessen the chance for error. We can use either a current or power phasor diagram. Let's choose the current phasor diagram. Because it is a parallel circuit, we will use voltage as the reference phasor. The resistive current phasor is always drawn in phase with voltage. The inductive current always lags voltage by 90°, while capacitive current always leads voltage by 90°. These phasors are sketched on the diagram in Figure 16-26.

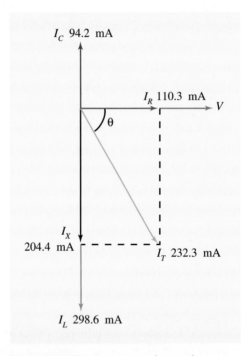

Figure 16-26. A current phasor diagram for the circuit in Figure 16-25.

Next, we algebraically combine the two reactive currents to find the net reactive current (I_X). We can use Equation 16-7.

$$I_X = I_L + I_C = -298.6 \text{ mA} + 94.2 \text{ mA} = -204.4 \text{ mA}$$

The minus sign tells us that the net current is inductive and should be drawn downward on the phasor diagram. If we complete the parallelogram between net reactive current (I_X) and resistive current (I_R), we can locate the phasor for total current (I_T). This is sketched in Figure 16-26.

The phase angle is always measured between total voltage and total current, as marked on the diagram in Figure 16-26. We can compute its value with any of the basic right-triangle equations. If we use the cosine function, then we will also find the value of the power factor.

$$pf = \cos \theta = \frac{\text{adjacent}}{\text{hypotenuse}} = \frac{I_R}{I_T}$$

$$= \frac{110.3 \text{ mA}}{232.2 \text{ mA}} = 0.475$$

$$\theta = \arccos 0.475 = 61.6°$$

The circuit has a lagging power factor, since total current is lagging total voltage. You will recall that an *RL* circuit exhibits a similar characteristic. In effect, the high value of inductive current (298.6 mA) has cancelled the lesser value of capacitive current (94.2 mA). The remaining reactive current ($I_X = 204.4$ mA) combines with resistive current (I_R) to produce a total current (I_T) that lags the total voltage, just as would occur in a simple *RL* circuit. The completed solution matrix is given in Table 16-4.

CIRCUIT QUANTITY		RESISTOR	INDUCTOR	CAPACITOR	TOTAL
Current (rms)		110.3 mA	298.6 mA	94.2 mA	232.3 mA
Voltage	V_P	106.1 V	106.1 V	106.1 V	106.1 V
	V_{PP}	212.2 V	212.2 V	212.2 V	212.2 V
	V_{rms}	75 V	75 V	75 V	75 V
Resistance/Reactance/Impedance		680 Ω	251.2 Ω	796 Ω	322.9 Ω
Conductance/Susceptance/Admittance		1.47 mS	3.98 mS	1.26 mS	3.1 mS
Power		8.27 W	22.4 VAR	7.06 VAR	17.4 VA
Phase Angle (overall)		61.6°			
Power Factor (overall)		0.475 (lagging)			

Table 16-4. Completed Solution Matrix for the Circuit in Figure 16-25

Practice Problems

1. Analyze the circuit shown in Figure 16-27, and complete a solution matrix similar to Table 16-3.

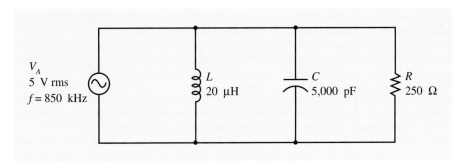

Figure 16-27. Analyze this circuit.

2. Analyze the circuit shown in Figure 16-28, and complete a solution matrix similar to Table 16-3.

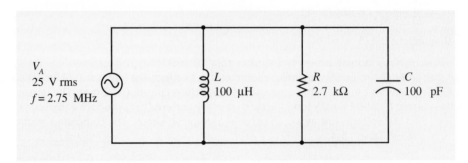

Figure 16-28. Analyze this circuit.

Answers to Practice Problems

1.

CIRCUIT QUANTITY		RESISTOR	INDUCTOR	CAPACITOR	TOTAL
Current (rms)		20 mA	46.82 mA	133.4 mA	88.9 mA
Voltage	V_P	7.07 V	7.07 V	7.07 V	7.07 V
	V_{PP}	14.14 V	14.14 V	14.14 V	14.14 V
	V_{rms}	5 V	5 V	5 V	5 V
Resistance/Reactance/Impedance		250 Ω	106.8 Ω	37.47 Ω	56.24 Ω
Conductance/Susceptance/Admittance		4 mS	9.37 mS	26.7 mS	17.8 mS
Power		100 mW	234.1 mVAR	667 mVAR	444.5 mVA
Phase Angle (overall)		77°			
Power Factor (overall)		0.225 (leading)			

2.

CIRCUIT QUANTITY		RESISTOR	INDUCTOR	CAPACITOR	TOTAL
Current (rms)		9.26 mA	14.5 mA	43.2 mA	30.16 mA
Voltage	V_P	35.35 V	35.35 V	35.35 V	35.35 V
	V_{PP}	70.7 V	70.7 V	70.7 V	70.7 V
	V_{rms}	25 V	25 V	25 V	25 V
Resistance/Reactance/Impedance		2.7 kΩ	1.73 kΩ	579 Ω	828.9 Ω
Conductance/Susceptance/Admittance		370.4 μS	578 μS	1.73 mS	1.21 mS
Power		231.5 mW	362.5 mVAR	1.08 VAR	754.0 mVA
Phase Angle (overall)		72.1°			
Power Factor (overall)		0.307 (leading)			

Series-Parallel *RLC* Circuits

The analysis of series-parallel *RLC* circuits follows the same basic procedure as analysis of resistive series-parallel circuits. We simplify the circuit by replacing sets of series or parallel components with equivalent components. If we encounter a series or parallel combination of similar components (i.e., all resistors, all inductors, or all capacitors), we replace them with an equivalent value. If the series or parallel components are dissimilar, then we represent the impedances with rectangular or polar notation and simplify them into an equivalent network. In general, rectangular and polar notation ensure that we account for the effects of phase differences. At a minimum, we will use rectangular or polar notation as we work toward finding total impedance. When we work back through the circuit to find individual currents and voltages, we may elect to omit the complex notation if we do not need to know the phase relationships at intermediate points in the circuit. Although it is impractical to describe a sequential procedure that is optimum for every type of series-parallel *RLC* circuit, the following will provide some general guidance.

1. Compute individual reactances.
2. Simplify the circuit to find total impedance.
3. Compute total current.
4. Compute individual voltages and currents.
5. Compute all other desired quantities.

DIRECT ANALYSIS OF SERIES-PARALLEL *RLC* CIRCUITS

We will emphasize the direct solution of *RLC* circuit analysis problems using techniques and procedures that are consistent with previously studied circuit analysis methods. Later, we will briefly examine an alternative method that can be used to simplify certain types of series-parallel problems.

EXAMPLE SOLUTION

Analyze the series-parallel *RLC* circuit shown in Figure 16-29, and complete the solution matrix shown in Table 16-5.

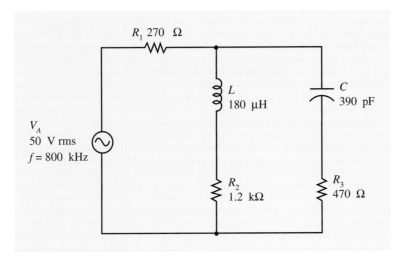

Figure 16-29. Analyze this circuit.

CIRCUIT QUANTITY	R_1	R_2	R_3	C	L	TOTAL
Voltage (rms)						50 V
Current (rms)						
Resistance/Reactance/Impedance	270 Ω	1.2 kΩ	470 Ω			
Conductance/Susceptance/Admittance						
Power						
Phase Angle (overall)						
Power Factor (overall)						

Table 16-5. A Solution Matrix for the Circuit in Figure 16-29

EXAMPLE SOLUTION

First, we find the reactance of the inductor and the capacitor by applying the basic reactance equations.

$$X_L = 2\pi fL = 6.28 \times 800 \text{ kHz} \times 180 \text{ μH} = 904 \text{ Ω}$$

$$X_C = \frac{1}{2\pi fC} = \frac{1}{6.28 \times 800 \text{ kHz} \times 390 \text{ pF}} = 510 \text{ Ω}$$

Next, let's combine the two parallel branches consisting of L, R_2, C, and R_3. For ease of notation, we will label these two branches as Z_A and Z_B as shown in Figure 16-30. We can express these two impedances as follows:

$$Z_A = 1{,}200 \text{ Ω} + j904 \text{ Ω, or}$$

$$Z_A = 1{,}502 \text{ Ω} \angle 37°$$

$$Z_B = 470 \text{ Ω} - j510 \text{ Ω, or}$$

$$Z_B = 693.5 \text{ Ω} \angle{-47.4°}$$

These two parallel branches (Z_A and Z_B) can be combined with the reciprocal formula for parallel circuits or the product-over-the-sum equation. Let's use the latter method to find the equivalent impedance $Z_{A,B}$.

$$Z_{A,B} = \frac{Z_A Z_B}{Z_A + Z_B}$$

$$= \frac{(1{,}502 \text{ Ω} \angle 37°)(693.5 \text{ Ω} \angle{-47.4°})}{(1{,}200 \text{ Ω} + j904 \text{ Ω}) + (470 \text{ Ω} - j510 \text{ Ω})}$$

$$= 556 \text{ Ω} - j243.8 \text{ Ω} = 607.1 \text{ Ω} \angle{-23.7°}$$

The simplified circuit, at this point, is shown in Figure 16-31.

We can now find total impedance by combining R_1 and the RC network labeled $Z_{A,B}$. Since it is a series circuit, we can sum the two portions to find total impedance. That is,

$$Z_T = R_1 + Z_{A,B}$$

$$= 270 \text{ Ω} + (556 \text{ Ω} - j243.8 \text{ Ω}) = 826 \text{ Ω} - j243.8 \text{ Ω}$$

$$= 861.2 \text{ Ω} \angle{-16.4°}$$

The −16.4° is the phase angle between resistance and impedence, which is the overall phase angle (θ) of the circuit. The resulting simplified circuit is shown in Figure 16-32.

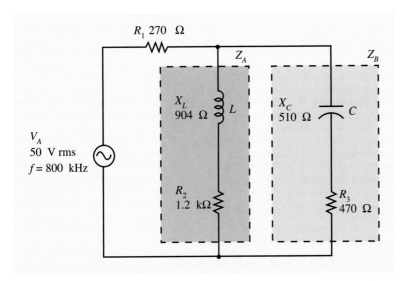

Figure 16-30. A simplification step in the analysis of the circuit shown in Figure 16-29.

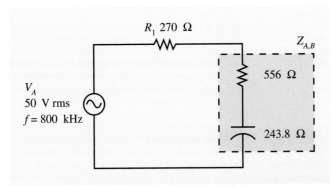

Figure 16-31. A partially simplified version of the circuit shown in Figure 16-29.

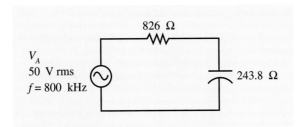

Figure 16-32. The simplified equivalent for the circuit shown in Figure 16-29.

Ohm's Law can now be used to compute total current flow.

$$I_T = \frac{V_A}{Z_T} = \frac{50 \text{ V}}{861.2 \text{ }\Omega \text{ } \angle{-16.4}°} = 58.1 \text{ mA } \angle{16.4}°$$

Because we have no apparent need to know the phase relationships between the individual component voltages and currents, we will no longer represent the circuit quantities as complex numbers. If we did want to know the intermediate phase angles, then we would simply perform all calculations with the circuit quantities expressed in rectangular and polar notation.

We are now ready to work our way back through the circuit to compute individual component voltages and currents. First, we know by inspection that total current flows through R_1. We compute its voltage drop with Ohm's Law.

$$V_{R_1} = I_{R_1} R_1 = 58.1 \text{ mA} \times 270 \text{ } \Omega = 15.69 \text{ V}$$

According to Kirchhoff's Voltage Law, the rest of the supply voltage must be dropped across $Z_{A,B}$ in Figure 16-31. Because of the phase relationships, however, we cannot do a simple arithmetic subtraction of voltages. We must either perform the subtraction with complex numbers, or simply compute the voltage drop across $Z_{A,B}$ with Ohm's Law. Let's use the latter approach.

$$V_{A,B} = I_{A,B} \times Z_{A,B} = 58.1 \text{ mA} \times 607.1 \text{ } \Omega = 35.27 \text{ V}$$

We can now transfer this voltage ($V_{A,B}$) back to Figure 16-30. This is redrawn in Figure 16-33.

The current through Z_A and Z_B can be found with Ohm's Law as follows.

$$I_A = \frac{V_{A,B}}{Z_A} = \frac{35.27 \text{ V}}{1{,}502 \text{ } \Omega} = 23.48 \text{ mA}$$

$$I_B = \frac{V_{A,B}}{Z_B} = \frac{35.27 \text{ V}}{693.5 \text{ } \Omega} = 50.86 \text{ mA}$$

Since L and R_2 are in series, they will both have the same current as Z_A. Similarly, the Z_B current will flow through both C and R_3. This completes our calculations for component currents. We can now use Ohm's Law to find the voltage drops across each component in branches Z_A and Z_B.

$$V_L = I_A X_L = 23.48 \text{ mA} \times 904 \text{ } \Omega = 21.23 \text{ V}$$

$$V_{R_2} = I_A R_2 = 23.48 \text{ mA} \times 1{,}200 \text{ } \Omega = 28.18 \text{ V}$$

$$V_C = I_B X_C = 50.86 \text{ mA} \times 510 \text{ } \Omega = 25.94 \text{ V}$$

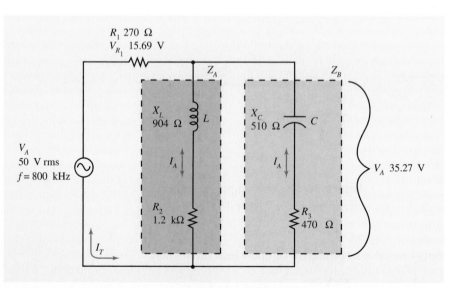

Figure 16-33. $V_{A,B}$ appears across parallel networks Z_A and Z_B.

$$V_{R_3} = I_B R_3 = 50.86 \text{ mA} \times 470 \ \Omega = 23.9 \text{ V}$$

The various conductances, susceptances, and admittances can be found by taking the reciprocal of the resistance, reactance, and impedance, respectively.

$$G_1 = \frac{1}{R_1} = \frac{1}{270 \ \Omega} = 3.7 \text{ mS}$$

$$G_2 = \frac{1}{R_2} = \frac{1}{1{,}200 \ \Omega} = 833.3 \ \mu\text{S}$$

$$G_3 = \frac{1}{R_3} = \frac{1}{470 \ \Omega} = 2.13 \text{ mS}$$

$$B_L = \frac{1}{X_L} = \frac{1}{904 \ \Omega} = 1.11 \text{ mS}$$

$$B_C = \frac{1}{X_C} = \frac{1}{510 \ \Omega} = 1.96 \text{ mS}$$

$$Y = \frac{1}{Z_T} = \frac{1}{861.2 \ \Omega} = 1.16 \text{ mS}$$

We can use our choice of the basic power equations to compute the circuit power. Let's choose the basic form of $P = VI$.

$$P_{T_1} = V_{R_1} I_{R_1} = 15.69 \text{ V} \times 58.1 \text{ mA} = 911.6 \text{ mW}$$

$$P_{T_2} = V_{R_2} I_{R_2} = 28.18 \text{ V} \times 23.48 \text{ mA} = 661.7 \text{ mW}$$

$$P_{T_3} = V_{R_3} I_{R_3} = 23.9 \text{ V} \times 50.86 \text{ mA} = 1.22 \text{ W}$$

$$P_{R_L} = V_L I_L = 21.23 \text{ V} \times 23.48 \text{ mA} = 498.5 \text{ mVAR}$$

$$P_{R_C} = V_C I_C = 25.94 \text{ V} \times 50.86 \text{ mA} = 1.32 \text{ VAR}$$

$$P_A = V_A I_T = 50 \text{ V} \times 58.1 \text{ mA} = 2.91 \text{ VA}$$

Power factor is the only remaining circuit quantity to be computed. Power factor is always equal to the cosine of the overall phase angle (θ).

$$pf = \cos\theta = \cos -16.4° = 0.959 \text{ (leading)}$$

We know the power factor is leading, since the overall circuit simplifies to an *RC* circuit (Figure 16-32), and we know that current always leads voltage in an *RC* circuit. The completed solution matrix is shown in Table 16-6.

CIRCUIT QUANTITY	R_1	R_2	R_3	C	L	TOTAL
Voltage (rms)	15.69 V	28.18 V	23.9 V	25.94 V	21.23 V	50 V
Current (rms)	58.1 mA	23.48 mA	50.86 mA	50.86 mA	23.48 mA	58.1 mA
Resistance/Reactance/Impedance	270 Ω	1.2 kΩ	470 Ω	510 Ω	904 Ω	861.2 Ω
Conductance/Susceptance/Admittance	3.7 mS	833.3 μS	2.13 mS	1.96 mS	1.11 mS	1.16 mS
Power	911.6 mW	661.7 mW	1.22 W	1.32 VAR	498.5 mVAR	2.91 VA
Phase Angle (overall)	−16.4°					
Power Factor (overall)	0.959 (leading)					

Table 16-6. A Completed Solution Matrix for the Circuit in Figure 16-29

1. Analyze the circuit shown in Figure 16-34, and complete a solution matrix similar to Table 16-5.

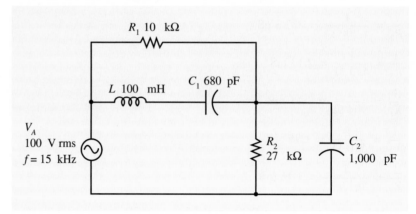

Figure 16-34. Analyze this circuit.

2. Analyze the circuit shown in Figure 16-35, and complete a solution matrix similar to Table 16-5.

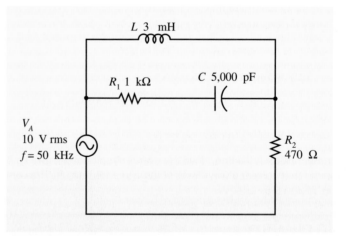

Figure 16-35. Analyze this circuit.

Answers to Practice Problems

1.

CIRCUIT QUANTITY	R_1	R_2	C_1	C_2	L	TOTAL
Voltage (rms)	34.88 V	65.52 V	87.96 V	65.52 V	53.08 V	100 V
Current (rms)	3.49 mA	2.43 mA	5.64 mA	6.17 mA	5.64 mA	6.63 mA
Resistance/Reactance/Impedance	10 kΩ	27 kΩ	15.61 kΩ	10.62 kΩ	9.42 kΩ	15.09 kΩ
Conductance/Susceptance/Admittance	100 μS	37 μS	64.1 μS	94.2 μS	106.2 μS	66.3 μS
Power	121.7 mW	159.2 mW	496.1 mVAR	404.3 mVAR	299.4 mVAR	663 mVA
Phase Angle (overall)	−64.96°					
Power Factor (overall)	0.423 (leading)					

2.

CIRCUIT QUANTITY	R_1	R_2	C	L	TOTAL
Voltage (rms)	6.17 V	3.22 V	3.93 V	7.32 V	10 V
Current (rms)	6.17 mA	6.85 mA	6.17 mA	7.77 mA	6.85 mA
Resistance/Reactance/Impedance	1 kΩ	470 Ω	636.9 Ω	942 Ω	1.46 kΩ
Conductance/Susceptance/Admittance	1 mS	2.13 mS	1.57 mS	1.06 mS	684.9 μS
Power	38.1 mW	22.1 mW	24.3 mVAR	56.9 mVAR	68.5 mVA
Phase Angle (overall)	28.4°				
Power Factor (overall)	0.880 (lagging)				

TRANSFORMATION OF SERIES-PARALLEL RLC CIRCUITS

The previously discussed method for analyzing series-parallel RLC circuits is applicable to all types of series-parallel RLC configurations. There is one common series-parallel config-uration that deserves special mention. Figure 16-36 shows a parallel LC circuit. The resis-tance of the coil appears as a resistance in series with the coil inductance. Thus, the apparently parallel circuit must be analyzed as a series-parallel circuit.

Some technicians prefer to convert the series-parallel configuration (shown in the left portion of Figure 16-36) to an equivalent parallel RLC circuit (shown in the right portion of Figure 16-36). The primary motivation for this conversion is to avoid the use of complex numbers. If your engineering calculator manipulates complex numbers directly, then there is little advantage to the conversion process.

As long as the Q of the coil (i.e., Equation 12-14 where $Q = X_L/\text{ESR}$) is greater than 10 (the usual case), then Equation 16-9 can be used to transform the series resistance (ESR) into an equivalent parallel resistance (R_P).

$$R_P = \text{ESR} \times Q^2 \qquad (16\text{-}9)$$

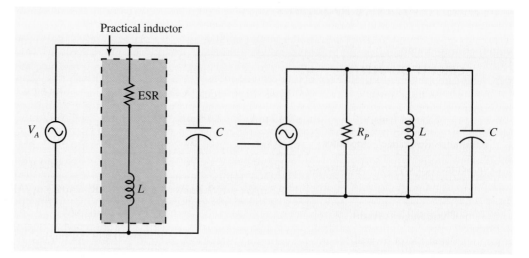

Figure 16-36. Transformation of a series-parallel *RLC* circuit into an equivalent parallel *RLC* circuit.

EXAMPLE SOLUTION

A parallel (ideally) *LC* circuit consists of a 1,000-pF capacitor and a 120-µH coil that has an equivalent series resistance of 18 Ω. The circuit is being operated at 600 kHz. Convert the circuit to an equivalent parallel circuit.

EXAMPLE **SOLUTION**

The basic conversion process is illustrated in Figure 16-36 and works for coils with Q factors greater than 10. Let's first determine the Q of the coil.

$$X_L = 2\pi fL = 6.28 \times 600 \text{ kHz} \times 120 \text{ µH} = 452.2 \text{ Ω}$$

$$Q_{(COIL)} = \frac{X_L}{ESR} = \frac{452.2 \text{ Ω}}{18 \text{ Ω}} = 25$$

Since this qualifies for our conversion method, we can apply Equation 16-9.

$$R_P = R_S Q^2 = 18 \text{ Ω} \times 25^2 = 11.3 \text{ kΩ}$$

Complex *RLC* Circuits

Complex *RLC* circuits can be analyzed by applying the same techniques and same network theorems that were used to analyze complex resistive circuits. The primary difference is that circuit quantities in the complex *RLC* circuit must be represented in polar or rectangular notation to account for the effects of differing phase angles. Ohm's and Kirchhoff's Laws remain your two most important analytical tools.

Exercise Problems 16.2

Refer to Figure 16-37 for Problems 1–10. All values are rms unless otherwise stated.

    1.   What is the value of inductive reactance?

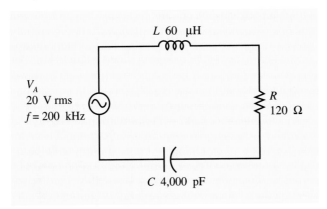

Figure 16-37. Analyze this circuit.

2. What is the value of capacitive reactance?
3. What is the impedance of the circuit?
4. How much total current flows through the circuit?
5. How much current flows through C?
6. What is the voltage drop across L?
7. What is the net reactive voltage?
8. What is the admittance of the circuit?
9. What is the overall phase angle of the circuit?
10. What is the circuit's power factor?

Refer to Figure 16-38 for Problems 11–20. All values are rms unless otherwise stated.

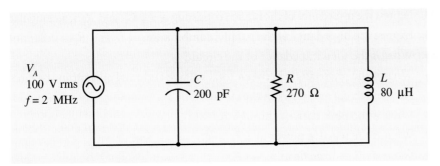

Figure 16-38. Analyze this circuit.

11. What is the value of inductive reactance in the circuit?
12. What is the rms voltage across C?
13. How much peak voltage is dropped across R?
14. How much current flows through L?
15. How much current flows through C?
16. What is the net reactive current?
17. What is the total current in the circuit?
18. What is the impedance of the circuit?

19. What is the phase angle of the circuit?

20. How much true power is dissipated by the circuit?

Refer to Figure 16-39 for Problems 21–30. All values are rms unless otherwise stated.

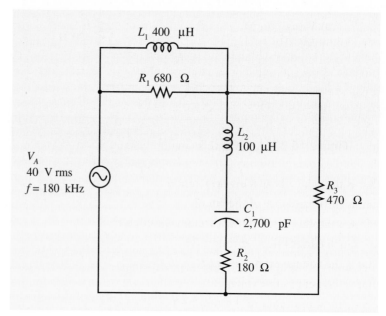

Figure 16-39. Analyze this circuit.

21. What is the reactance of L_2?

22. What is the reactance of C_1?

23. Express the impedance of the branch consisting of L_2, C_1, and R_2 in rectangular form.

24. Express the impedance of the parallel combination of L_1 and R_1 in polar notation.

25. What is the total impedance of the circuit?

26. What is the phase angle between total voltage and total current in the circuit?

27. How much current flows in the circuit?

28. What is the peak current through R_3?

29. What is the reactive power in L_2?

30. What is the voltage drop across R_2?

16.3 Resonance

The concept of resonance is very important to an electronics technician, since it plays such a central role in the operation of many electronic devices. Although we will address the subject of resonance as a separate section, it is very important for you to realize that resonance is nothing more than a description of the circuit characteristics of an *RLC* circuit at a particular operating frequency. That is, every *RLC* circuit that we have previously discussed and analyzed has a resonant frequency and exhibits the characteristics of resonance that are detailed in this section. A resonant *RLC* circuit is exactly like any other *RLC* circuit, except it is operating at a particular frequency called the resonant frequency.

Mechanical Resonance

Before we begin our discussion of resonant *RLC* circuits, let's briefly consider the more familiar aspects of mechanical resonance. Figure 16-40 shows the motion associated with a common pendulum. In Figure 16-40(a), the pendulum has been forcibly pulled to one side of its center or rest position. Once it is released, it accelerates toward its rest position as illustrated in Figure 16-40(b). When it reaches the center or rest position, as indicated in Figure 16-40(c), it has gained substantial momentum and continues to swing beyond its rest position. Its movement slows as it moves farther past the rest position. Eventually, as shown in Figure 16-40(d), it stops and begins to move back toward its rest position. Again, it accelerates toward the rest position, as shown in Figure 16-40(e), so that it has substantial momentum when it reaches the rest position, as shown in Figure 16-40(f). Finally, it begins to slow and eventually reaches its initial position, as represented in Figure 16-40(a).

Although the basic movement of the pendulum is familiar, there are other characteristics that are important. First, the rate (i.e., number of swings or cycles per second) at which the pendulum swings is solely determined by the length of the pendulum (assuming the force of gravity is a constant). That is, the frequency at which the pendulum oscillates is determined by the physical characteristic of length. Second, as the pendulum swings, it will lose energy as a result of air resistance, friction at the pivot point, bending of the pendulum arm, and so on. This means that each time the pendulum tries to return to its initial position, as represented in Figure 16-40(a), it will fall a little short. Unless additional energy is added to overcome the energy losses (e.g., we give it another push), the pendulum will continue to lose energy and will eventually stop in its rest position.

Now let's consider what would have to be done to keep the pendulum in motion indefinitely. Can you see that at the end of each cycle, we only need to add the amount of energy that was lost by the pendulum? If we add this tiny amount of energy each cycle, the pendulum will continue to make full swings.

Now consider when this external energy must be added. You know intuitively that for maximum efficiency, we would want to push the pendulum exactly as it reaches its maximum deflection. This is comparable to pushing a child in a swing. If we try to push the pendulum at the wrong time or in the wrong direction, we will actually interfere with the natural rhythm of the pendulum. If we try to make it swing at any frequency other than its natural frequency, we will have to add substantial energy. If, by contrast, we add

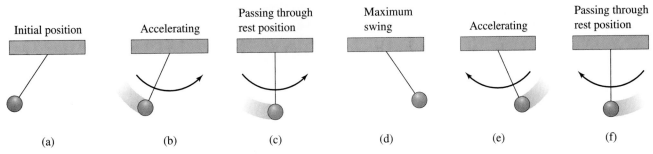

Figure 16-40. The movement of a simple pendulum.

the energy at the correct time, we aid the natural frequency and very minimal energy is required. Electrical resonance exhibits similar effects, as we will soon see.

Many readers will be familiar with yet another example of mechanical resonance that is illustrated in Figure 16-41. Here, we have a crystal goblet with only a small amount of liquid. If a finger is rubbed smoothly and lightly around the rim of the glass, the glass will begin to vibrate at a very definite frequency. Once it has started to oscillate, you can sustain the oscillation with only the slightest movement of your finger along the rim. Like the pendulum, the crystal goblet is exhibiting mechanical resonance. If we add energy at the right time each cycle, we need only add enough energy to compensate for losses. In Figure 16-41, we are adding the energy from a sliding finger. Some singers can add the energy with their voice (excessive energy can shatter the glass). The frequency of oscillation is determined by the physical characteristics of the glass (e.g., size, thickness, amount of liquid). Unless we change these physical characteristics, we cannot easily alter the frequency of oscillation. This natural frequency of oscillation that is exhibited by the pendulum and by the crystal goblet is called the resonant frequency.

Parallel Resonance

The similarities between electrical and mechanical resonance are easier to understand when considering a parallel *RLC* circuit, so let's begin by studying the characteristics of a parallel resonant circuit.

Figure 16-42 shows a parallel *RLC* circuit being operated at three frequencies. In Figure 16-42(a), the circuit is operated below the resonant frequency. At low frequencies, X_L, which decreases with decreasing frequencies, will be lower than X_C, which increases with decreasing frequencies. Since both *L* and *C* have the same voltage (i.e., they are in parallel), we know that I_L will be greater than I_C (Ohm's Law). The phasor diagram in Figure 16-42(a) illustrates the relationships in the circuit.

Rub moistened finger gently and smoothly around rim.

Glass vibrates at its resonant frequency.

Figure 16-41. An interesting demonstration of resonance.

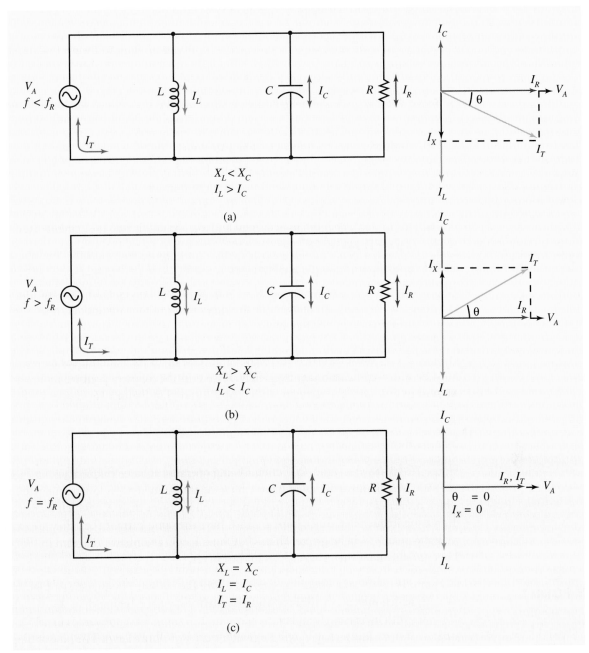

Figure 16-42. A parallel *RLC* circuit (a) below resonance, (b) above resonance, and (c) at resonance.

In Figure 16-42(b), the circuit is operated above the resonant frequency. At high frequencies, X_L, which increases with increasing frequencies, will be higher than X_C, which decreases with increasing frequencies. Since both L and C have the same voltage (i.e., they are in parallel), we know that I_C will be greater than I_L (Ohm's Law). The phasor diagram in Figure 16-42(b) illustrates the relationships in the circuit.

Now let's consider the circuit conditions if the input frequency is adjusted such that the inductive reactance and the capacitive reactance are equal. This is the **resonant frequency** of the circuit. Figure 16-42(c) illustrates the *RLC* circuit at the resonant frequency. Because the inductive and capacitive branches have equal reactances and equal voltages, we know the two branch currents will be equal in magnitude. The phasor diagram in Figure 16-42(c) illustrates that the two equal reactive currents (regardless of their exact values) completely cancel each other (i.e., net reactive current is zero). This leaves only resistive current to form I_T. We note the following characteristics of a parallel *RLC* circuit operating at its resonant frequency:

1. Inductive and capacitive reactances are equal.
2. Inductive and capacitive currents are equal.
3. Total current is equal to the resistive branch current.
4. Total current is minimum (only resistive current).
5. Circuit impedance is maximum (sometimes called a resonant rise of impedance).
6. Phase angle between total current and total voltage is zero (i.e., circuit acts resistive).
7. Power factor is one.

The reactive currents are determined by the reactance and the applied voltage as with any *RLC* circuit. However, at resonance, they may be substantially higher than the line current (I_T). The smaller line current only has to supply losses in the circuit (e.g., power dissipation in the resistor). This is like the small amount of energy that must be supplied to a mechanically resonant circuit to keep it oscillating. As with a mechanical system, the external energy must be timed correctly. In the case of an *RLC* circuit, the external energy must be added at the resonant frequency. If we shift the frequency, either higher or lower, we start to minimize the resonance effect as illustrated in Figures 16-42(a) and (b).

Series Resonance

Figure 16-43 shows a series *RLC* circuit being operated at three frequencies. In Figure 16-43(a), the circuit is operated below the resonant frequency. At low frequencies, X_L, which decreases with decreasing frequencies, will be lower than X_C, which increases with decreasing frequencies. Since both *L* and *C* have the same current (i.e., they are in series), we know that V_C will be greater than V_L (Ohm's Law). The phasor diagram in Figure 16-43(a) illustrates the relationships in the circuit.

In Figure 16-43(b), the circuit is operated above the resonant frequency. At high frequencies, X_L, which increases with increasing frequencies, will be higher than X_C, which decreases with increasing frequencies. Since both *L* and *C* have the same current (i.e., they are in series), we know that V_L will be greater than V_C (Ohm's Law). The phasor diagram in Figure 16-43(b) illustrates the relationships in the circuit.

Now let's consider the circuit conditions if the input frequency is adjusted such that the inductive reactance and the capacitive reactance are equal. This is the resonant frequency of the circuit. Figure 16-43(c) illustrates the *RLC* circuit at the resonant frequency. Because the inductive and capacitive reactances are equal, we know the two voltage drops will be equal. The phasor diagram in Figure 16-43(c) illustrates that the two equal reactive voltages (regardless of their exact values) completely cancel each other (i.e., net reactive voltage is zero). This means that the resistance will drop the entire applied voltage. We note the following characteristics of a series *RLC* circuit operating at its resonant frequency:

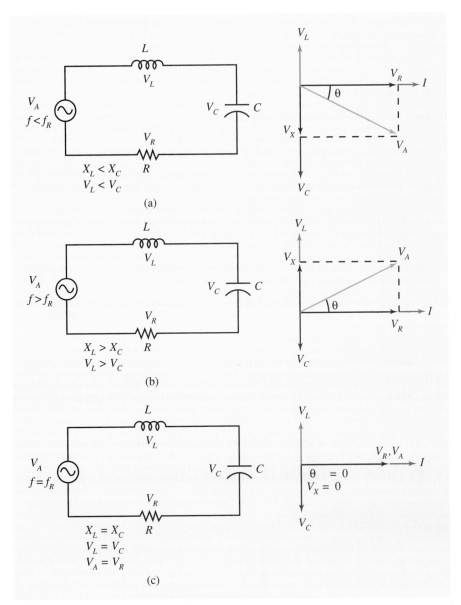

Figure 16-43. A series RLC circuit (a) below resonance, (b) above resonance, and (c) at resonance.

1. Inductive and capacitive reactances are equal.
2. Inductive and capacitive voltage drops are equal.
3. The resistive voltage drop is equal to the applied voltage.
4. Circuit impedance is minimum (reactances cancel).
5. Current is maximum (limited only by the resistance).
6. Phase angle between total current and total voltage is zero (i.e., circuit acts resistive).
7. Power factor is one.

⬤ KEY POINTS

At resonance, a series RLC circuit acts resistive, and the total impedance is equal to the resistance.

The reactive voltages are determined by the reactance and the current as with any *RLC* circuit. However, at resonance, they may be substantially higher than the applied voltage (V_A). This increase in voltage is often called a resonant rise in voltage.

Q of a Resonant Circuit

The degree to which the resonance effects are evident in an *RLC* circuit is determined by the quality, or *Q*, of the circuit. *Q* has no units and is a ratio between the resistance and the inductive reactance in the circuit at the resonant frequency. The higher the value of *Q*, the more marked the effects of resonance. Figure 16-44 shows a graph of current as a function of frequency for a series resonant circuit (recall that current is maximum at resonance in a series *RLC* circuit). There are three cases plotted on the graph representing low-, moderate-, and high-*Q RLC* circuits. The highest *Q* produces a very sharp response at resonance. The lower the *Q*, the less pronounced the resonance effect.

Values of *Q* range from less than 10 to over 100 for standard *RLC* circuits. In general, we classify circuits whose *Q* factor is less than 10 as low-*Q* circuits. By contrast, circuits with *Q* factors over 100 are considered to be high-*Q* circuits. Later in your electronics education, you will study circuits and devices that have *Q* factors greater than 1,000.

Figure 16-45 shows a graph of current as a function of frequency for a parallel *RLC* circuit (recall that current is minimum at resonance in a parallel *RLC* circuit). There are three cases plotted on the graph representing low-, moderate-, and high-*Q RLC* circuits. As with series *RLC* circuits, the highest *Q* produces a very sharp response at resonance. The lower the *Q*, the less pronounced the resonance effect.

Increased resistance in a series circuit decreases the *Q* of the circuit. In a parallel circuit, higher parallel resistance increases *Q*. We can compute the *Q* of a series or parallel *RLC* circuit with Equations 16-10 and 16-11, respectively. These equations are

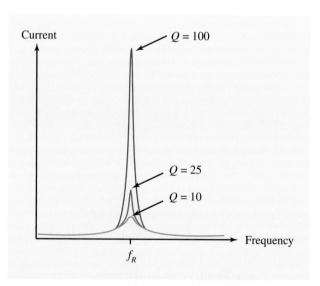

Figure 16-44. Current as a function of frequency for a series *RLC* circuit. The sharpness of the resonance response is determined by the *Q* of the circuit.

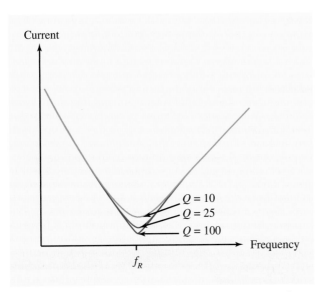

Figure 16-45. Current as a function of frequency for a parallel *RLC* circuit. The sharpness of the resonance response is determined by the *Q* of the circuit.

$$Q = \frac{X_L}{R_S} \qquad (16\text{-}10)$$

where R_S is the series resistance and X_L is the inductive reactance at the resonant frequency, and

$$Q = \frac{R_P}{X_L} \qquad (16\text{-}11)$$

where R_P is the resistance of the parallel resistive branch and X_L is the inductive reactance at the resonant frequency.

EXAMPLE SOLUTION

What is the Q of a series circuit with a resonant frequency of 3.11 MHz and a 1.75-μH coil, a 1,500-pF capacitor, and a 2.2-Ω resistor?

EXAMPLE SOLUTION

First, we compute the inductive reactance at the resonant frequency.

$X_L = 2\pi f L = 6.28 \times 3.11 \text{ MHz} \times 1.75 \text{ μH} \approx 34.18 \text{ Ω}$

Now we can apply Equation 16-10 to determine the Q of the circuit.

$$Q = \frac{X_L}{R_S} = \frac{34.18 \text{ Ω}}{2.2 \text{ Ω}} = 15.5$$

KEY POINTS

When a series *RLC* circuit reaches resonance, the voltage drop across each of the reactances will be larger than the applied voltage by the factor Q. Similarly, in a parallel *RLC* circuit, at resonance the impedance of the circuit is larger than the inductive reactance by the factor Q.

EXAMPLE SOLUTION

Find the Q of a parallel circuit that resonates at 1.37 MHz and has a 200-µH coil, a 68-pF capacitor, and a 68-kΩ resistor.

EXAMPLE **SOLUTION**

The inductive reactance at the resonant frequency must be computed first.

$$X_L = 2\pi f L = 6.28 \times 1.37 \text{ MHz} \times 200 \text{ µH} = 1.72 \text{ k}\Omega$$

Applying Equation 16-11 to determine Q gives us the following result.

$$Q = \frac{R_P}{X_L} = \frac{68 \text{ k}\Omega}{1.72 \text{ k}\Omega} = 39.5$$

Practice Problems

1. If a series RLC circuit consisting of a 10-µH coil, a 100-pF capacitor, and a 7-Ω resistor resonates at 5 MHz, what is the Q of the circuit?
2. What is the Q of a parallel RLC circuit that consists of a 250-µH coil, a 150-pF capacitor, and a 24-kΩ resistor, and has a resonant frequency of 822.3 kHz?

Answers to Practice Problems

1. 44.9   2.   18.6

Selectivity, Bandwidth, and Bandpass of a Resonant Circuit
SELECTIVITY

Selectivity of an RLC circuit describes how sharply the circuit distinguishes between the resonant frequency and frequencies on either side of the resonant frequency. Frequencies that cause a response that is 70.7% of the maximum (voltage, current, or impedance) response or greater are considered to be passed or selected by the circuit. The frequencies that cause a response of less than 70.7% of the maximum (voltage, current, or impedance) response are considered to be rejected (i.e., not selected) by the circuit.

Figure 16-46 shows the response of three RLC circuits with different Q factors. Their 100% responses have been adjusted to be equal. Regardless of the Q factor, all frequencies with responses above 70.7% of the maximum voltage, current, or impedance are considered to be selected or passed by the circuit. Clearly, there is a band or range of frequencies that are above the 70.7% level. The greater the selectivity (i.e., the higher the Q) of the circuit, the narrower the range of frequencies that produces a response of greater than 70.7% of the maximum.

In the case of parallel RLC circuits, we measure the response in terms of impedance. That is, the impedance is maximum at the resonant frequency (f_R) and decreases on either side of f_R. In the case of series RLC circuits, it is common to use resistive voltage, power, or circuit current as the indicator. All of these are maximum at the resonant frequency and decrease for frequencies on either side of f_R. In the case of current or voltage responses, we use 70.7% as the dividing line between selection and rejection. Since power is equal to the

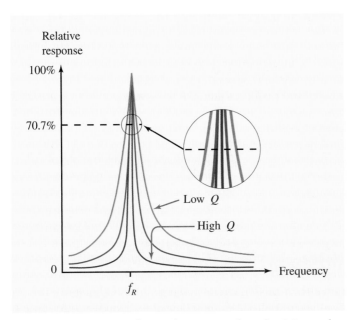

Figure 16-46. The Q factor determines the selectivity and bandwidth of an RLC circuit.

product of voltage and current, the dividing line for a power response is $70.7\% \times 70.7\%$ or 50%. For this reason, the points at which the response falls below 70.7% of the maximum voltage or current in a series RLC circuit are called the **half-power points.** Power is often measured in decibels, so these same points are also called 3-dB points (technically −3-dB points), since a −3-dB power loss corresponds to a 50% power reduction.

BANDPASS

The band of frequencies that causes a response that is above the half-power points is called the **bandpass** of the RLC circuit. Bandpass (also called passband) is expressed as a range of frequencies. The lowest frequency that still produces a response above the half-power point is called the lower **cutoff frequency** (f_L). The upper cutoff frequency (f_H) is the highest frequency that still produces a response above the half-power point. The bandpass of an RLC circuit is the range of frequencies between the lower and upper cutoff frequencies.

EXAMPLE SOLUTION

What is the bandpass of an RLC circuit that has 70.7% or higher of the maximum response for frequencies higher than 150 kHz and lower than 250 kHz?

EXAMPLE SOLUTION

The bandpass is simply the range of frequencies that cause a 70.7% or more of the maximum response. In the present example, the bandpass would be stated as the range of 150 kHz to 250 kHz.

BANDWIDTH

The width (measured in hertz) of the bandpass of an RLC circuit is called the **bandwidth** of the circuit. The bandwidth is not a measure of specific frequencies like the bandpass,

● KEY POINTS

Frequencies that produce 70.7% or more of the maximum circuit response (e.g., maximum current or component voltage drop in a series circuit, maximum impedance in a parallel circuit) in an RLC circuit are considered to be in the bandpass (or passband) of the circuit.

● KEY POINTS

The frequencies on the lower and upper edges of the passband are called the lower and upper cutoff frequencies.

but rather it is a measure of the width of the bandpass. We can compute bandwidth with Equation 16-12.

$$BW = f_H - f_L \qquad (16\text{-}12)$$

where f_L is the lower cutoff frequency and f_H is the upper cutoff frequency of the RLC circuit.

EXAMPLE SOLUTION

What is the bandwidth of an RLC circuit that has 70.7% or higher of the maximum response for frequencies higher than 150 kHz and lower than 250 kHz?

EXAMPLE **SOLUTION**

We apply Equation 16-12.

$$BW = f_H - f_L$$
$$= 250 \text{ kHz} - 150 \text{ kHz} = 100 \text{ kHz}$$

Clearly, the bandwidth of an RLC circuit is related to the Q factor of the circuit. For a given resonant frequency, higher Q factors produce narrower bandwidths. Conversely, lower Q factors produce wider bandwidths for a given center frequency (f_R). We can express the relationship of Q, f_R, and bandwidth with Equation 16-13.

$$BW = \frac{f_R}{Q} \qquad (16\text{-}13)$$

EXAMPLE SOLUTION

What is the bandwidth of an RLC circuit that has a resonant frequency of 2.5 MHz and a Q of 20?

EXAMPLE **SOLUTION**

We apply Equation 16-13.

$$BW = \frac{f_R}{Q} = \frac{2.5 \text{ MHz}}{20} = 125 \text{ kHz}$$

Figure 16-47 further clarifies the definitions and relationships affecting the bandwidth of an RLC circuit.

Calculations for a Resonant RLC Circuit

Now that we have an overall view of the characteristics of series and parallel RLC circuits, we can analyze them. It is very important for you to realize that literally all of the techniques you used to analyze series and parallel RLC circuits also apply to resonant RLC circuits. The resonant frequency is just a specific frequency where X_L and X_C are equal. We will learn a few new equations used by technicians to simplify calculations.

CALCULATION OF THE RESONANT FREQUENCY

We know that resonance occurs when the inductive and capacitive reactances are equal. It is very useful to be able to quickly and directly determine this frequency for a particular RLC circuit. We can evolve such an equation by setting the two reactance equations equal to each other and solving for frequency as follows:

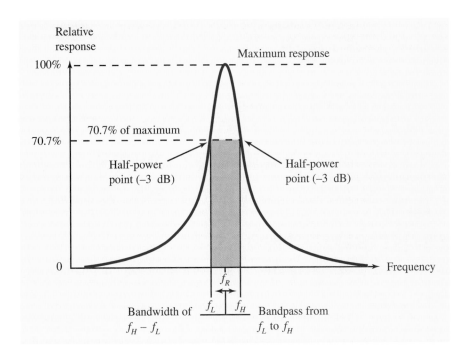

Figure 16-47. The relationship of bandpass and bandwidth to the 70.7% response level.

$$X_L = X_C$$

$$2\pi f L = \frac{1}{2\pi f C}$$

$$4\pi^2 f^2 LC = 1$$

$$f^2 = \frac{1}{4\pi^2 LC}$$

Finally, taking the square root of both sides gives us Equation 16-14.

$$f_R = \frac{1}{2\pi\sqrt{LC}} \qquad (16\text{-}14)$$

This equation applies to either series or parallel *RLC* circuits.

EXAMPLE SOLUTION

What is the resonant frequency of a series circuit consisting of a 200-μH inductor, a 1,000-pF capacitor, and a 5-Ω resistor?

EXAMPLE **SOLUTION**

We apply Equation 16-14.

$$f_R = \frac{1}{2\pi\sqrt{LC}}$$

$$= \frac{1}{6.28 \times \sqrt{200\ \mu H \times 1{,}000\ pF}}$$

$$= 356.1\ \text{kHz}$$

1. If a 10-µH coil, a 250-pF capacitor, and a 1.2-MΩ resistor are in parallel, what is the resonant frequency of the circuit?
2. What is the resonant frequency of a series circuit with a 1.5-mH coil, a 2,200-pF capacitor, and a 7.5-Ω resistor?
3. What value capacitor is needed in parallel with a 200-µH inductor to produce resonance at 1.13 MHz?
4. What value of inductor is needed in series with a 270-pF capacitor to resonate at 8.85 MHz?

Answers to Practice Problems

1. 3.18 MHz   2. 87.66 kHz   3. 99.3 pF   4. 1.2 µH

SERIES RESONANCE CALCULATIONS

Most of the calculations for a series resonant RLC circuit are based on Ohm's and Kirchhoff's Laws and basic series circuit rules. We will introduce other more specific equations as they are needed.

EXAMPLE SOLUTION

Analyze the series RLC circuit shown in Figure 16-48, and complete the solution matrix shown in Table 16-7. Assume that the circuit is operating at the resonant frequency.

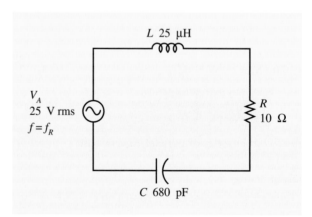

Figure 16-48. A series resonant RLC circuit.

EXAMPLE SOLUTION

First, we need to compute the resonant frequency with Equation 16-14.

$$f_R = \frac{1}{2\pi\sqrt{LC}} = \frac{1}{6.28 \times \sqrt{25\ \mu H \times 680\ pF}} = 1.22\ \text{MHz}$$

The two reactances can now be computed at the resonant frequency.

CIRCUIT QUANTITY	RESISTOR	INDUCTOR	CAPACITOR	TOTAL
Current (rms)				
Voltage (rms)				25 V
Resistance/Reactance/Impedance	10 Ω			
Resonant Frequency				
Q				
Bandwidth				
Phase Angle				
Power Factor				

Table 16-7. Solution Matrix for the Circuit in Figure 16-48

$$X_L = 2\pi fL = 6.28 \times 1.22 \text{ MHz} \times 25 \text{ μH} \approx 192 \text{ Ω}$$

$$X_C = \frac{1}{2\pi fC} = \frac{1}{6.28 \times 1.22 \text{ MHz} \times 680 \text{ pF}} \approx 192 \text{ Ω}$$

We need only compute one of the reactances, since we know they are equal at the resonant frequency. Computation of both values, however, is a convenient way to check your work.

The impedance of a series *RLC* circuit at the resonant frequency is equal to the resistance in the circuit (i.e., the reactances cancel). Nevertheless, we can compute the impedance of a series *RLC* circuit at any frequency with Equation 16-3.

$$Z = \sqrt{(X_L - X_C)^2 + R^2}$$
$$= \sqrt{(192 \text{ Ω} - 192 \text{ Ω})^2 + (10 \text{ Ω})^2}$$
$$= \sqrt{0 + 100} = \sqrt{100} = 10 \text{ Ω}$$

Ohm's Law can now be used to find the total current in the circuit.

$$I_T = \frac{V_A}{Z} = \frac{25 \text{ V}}{10 \text{ Ω}} = 2.5 \text{ A}$$

Again, we can apply Ohm's Law to determine the individual component voltage drops.

$$V_R = I_R R = 2.5 \text{ A} \times 10 \text{ Ω} = 25 \text{ V}$$
$$V_L = I_L X_L = 2.5 \text{ A} \times 192 \text{ Ω} = 480 \text{ V}$$
$$V_C = I_C X_C = 2.5 \text{ A} \times 192 \text{ Ω} = 480 \text{ V}$$

Note that the two reactive voltages are greater than the applied voltage and are equal in magnitude. You will recall that they are 180° out of phase with each other and effectively cancel each other. This is also confirmed by the observation that the resistive voltage drop is equal to the applied voltage.

Since this is a series *RLC* circuit, we can use Equation 16-10 to determine the *Q* of the circuit.

$$Q = \frac{X_L}{R_S} = \frac{192 \text{ Ω}}{10 \text{ Ω}} = 19.2$$

You will recall that the voltages across the reactances in a series resonant circuit exceed the value of applied voltage. We computed the exact values with Ohm's Law. This voltage increase as the circuit nears resonance is called a resonant rise or Q rise in voltage. It can also be computed with Equation 16-15.

$$V_L = V_C = QV_A \qquad (16\text{-}15)$$

This equation only applies to a series RLC circuit at the resonant frequency. It is also useful in a transposed version for computing the Q of the circuit when the component voltages are known (or measured in the laboratory).

We can use Equation 16-13 to determine the bandwidth of the circuit.

$$BW = \frac{f_R}{Q} = \frac{1.22 \text{ MHz}}{19.2} = 63.5 \text{ kHz}$$

The phase angle and power factor for a resonant RLC circuit are always zero and one, respectively, since the circuit acts resistive. If you forget this relationship, you can always compute the values as with any other RLC circuit.

This completes the analysis of the series resonant circuit shown in Figure 16-48. The completed solution matrix is provided in Table 16-8.

CIRCUIT QUANTITY	RESISTOR	INDUCTOR	CAPACITOR	TOTAL
Current (rms)	2.5 A			
Voltage (rms)	25 V	480 V	480 V	25 V
Resistance/Reactance/Impedance	10 Ω	192 Ω	192 Ω	10 Ω
Resonant Frequency	1.22 MHz			
Q	19.2			
Bandwidth	63.5 kHz			
Phase Angle	0°			
Power Factor	Unity (1)			

Table 16-8. Completed Solution Matrix for the Circuit in Figure 16-48

Practice Problems

1. Complete a solution matrix similar to Table 16-7 for the circuit shown in Figure 16-49.

2. A series RLC circuit has the following values: 2.5 mH, 82 pF, 68 Ω, and an 18-V rms voltage source. Complete a solution matrix similar to Table 16-7 for this circuit.

3. Complete a solution matrix similar to Table 16-7 for a series RLC circuit having the following values: 27 μH, 270 pF, 10 Ω, and a 5-V rms voltage source.

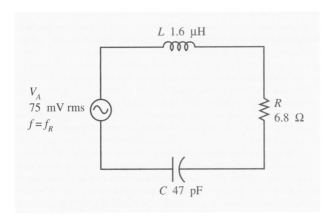

Figure 16-49. A series *RLC* circuit.

Answers to Practice Problems

1.

CIRCUIT QUANTITY	RESISTOR	INDUCTOR	CAPACITOR	TOTAL
Current (rms)	11.03 mA			
Voltage (rms)	75 mV	2.04 V	2.04 V	75 mV
Resistance/Reactance/Impedance	6.8 Ω	184.5 Ω	184.5 Ω	6.8 Ω
Resonant Frequency	18.36 MHz			
Q	27.13			
Bandwidth	676.7 kHz			
Phase Angle	0°			
Power Factor	Unity (1)			

2.

CIRCUIT QUANTITY	RESISTOR	INDUCTOR	CAPACITOR	TOTAL
Current (rms)	264.7 mA			
Voltage (rms)	18 V	1.46 kV	1.46 kV	18 V
Resistance/Reactance/Impedance	68 Ω	5.52 kΩ	5.52 kΩ	68 Ω
Resonant Frequency	351.7 kHz			
Q	81.18			
Bandwidth	4.33 kHz			
Phase Angle	0°			
Power Factor	Unity (1)			

3.

CIRCUIT QUANTITY	RESISTOR	INDUCTOR	CAPACITOR	TOTAL
Current (rms)	500 mA			
Voltage (rms)	5 V	158.1 V	158.1 V	5 V
Resistance/Reactance/Impedance	10 Ω	316.2 Ω	316.2 Ω	10 Ω
Resonant Frequency	1.87 MHz			
Q	31.62			
Bandwidth	59.14 kHz			
Phase Angle	0°			
Power Factor	Unity (1)			

PARALLEL RESONANCE CALCULATIONS

For the most part, analysis of parallel *RLC* circuits at resonance is no different than the analysis of the same circuit at some other frequency. We will, however, introduce alternate equations that can streamline your work in some cases.

EXAMPLE SOLUTION

Analyze the parallel *RLC* circuit shown in Figure 16-50, and complete the solution matrix given in Table 16-9. Assume that the circuit is operating at the resonant frequency.

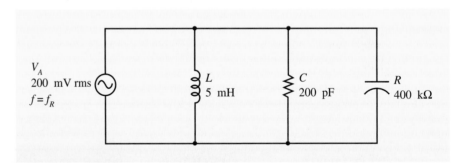

Figure 16-50. Analyze this parallel *RLC* circuit.

EXAMPLE SOLUTION

Finding the resonant frequency is a good place to start solving a problem like this. We apply Equation 16-14.

$$f_R = \frac{1}{2\pi\sqrt{LC}} = \frac{1}{6.28 \times \sqrt{(5 \text{ mH}) \times (200 \text{ pF})}} = 159.2 \text{ kHz}$$

Now, we can compute the reactance of the inductive and capacitive branches.

$$X_L = 2\pi fL = 6.28 \times 159.2 \text{ kHz} \times 5 \text{ mH} = 5 \text{ k}\Omega$$

$$X_C = \frac{1}{2\pi fC} = \frac{1}{6.28 \times 159.2 \text{ kHz} \times 200 \text{ pF}} = 5 \text{ k}\Omega$$

CIRCUIT QUANTITY	RESISTOR	INDUCTOR	CAPACITOR	TOTAL
Voltage (rms)	200 mV			
Current (rms)				
Resistance/Reactance/Impedance	400 kΩ			
Resonant Frequency				
Q				
Bandwidth				
Phase Angle				
Power Factor				

Table 16-9. Solution Matrix for the Circuit in Figure 16-50

The second calculation is not essential, since we know that the inductive and capacitive reactances are always equal at the resonant frequency.

Application of Ohm's Law will provide the value of current through each of the parallel branches.

$$I_R = \frac{V_R}{R} = \frac{200 \text{ mV}}{400 \text{ k}\Omega} = 500 \text{ nA}$$

$$I_L = \frac{V_L}{X_L} = \frac{200 \text{ mV}}{5 \text{ k}\Omega} = 40 \text{ }\mu\text{A}$$

$$I_C = \frac{V_C}{X_C} = \frac{200 \text{ mV}}{5 \text{ k}\Omega} = 40 \text{ }\mu\text{A}$$

Kirchhoff's Current Law can be used to determine the total current. We will sum the branch currents with phasor addition (Equation 16-8), since they are out of phase.

$$I_T = \sqrt{(I_L - I_C)^2 + I_R^2}$$
$$= \sqrt{(40 \text{ }\mu\text{A} - 40 \text{ }\mu\text{A})^2 + (500 \text{ nA})^2} = 500 \text{ nA}$$

We can now apply Ohm's Law to calculate the total impedance of the circuit.

$$Z = \frac{V_A}{I_T} = \frac{200 \text{ mV}}{500 \text{ nA}} = 400 \text{ k}\Omega$$

This calculation could have been avoided, since we know that the total impedance of a parallel *RLC* circuit at the resonant frequency is equal to the value of the resistive branch.

The Q of the circuit can be found with Equation 16-11.

$$Q = \frac{R_P}{X_L} = \frac{400 \text{ k}\Omega}{5 \text{ k}\Omega} = 80$$

You will recall that the impedance of a parallel resonant circuit is maximum at the resonant frequency. We computed the exact value with Ohm's Law. This impedance increase as the circuit nears resonance is called a resonant rise or Q rise in impedance. It can also be computed with Equation 16-16.

$$Z = QX_L \qquad\qquad (16\text{-}16)$$

This equation only applies to a parallel RLC circuit at the resonant frequency. It is simply a transposed version of Equation 16-11.

We can now use the computed value of Q and the previously computed resonant frequency to find the bandwidth of the circuit (Equation 16-13).

$$BW = \frac{f_R}{Q} = \frac{159.2 \text{ kHz}}{80} = 2 \text{ kHz}$$

We can draw a phasor diagram and compute the values of phase angle and power factor as with any other parallel RLC circuit, or we can simply remember that the phase angle is always zero and the power factor is always unity at the resonant frequency. The completed solution matrix for this analysis problem is given in Table 16-10.

CIRCUIT QUANTITY	RESISTOR	INDUCTOR	CAPACITOR	TOTAL
Voltage (rms)	200 mV			
Current (rms)	500 nA	40 µA	40 µA	500 nA
Resistance/Reactance/Impedance	400 kΩ	5 kΩ	5 kΩ	400 kΩ
Resonant Frequency	159.2 kHz			
Q	80			
Bandwidth	2 kHz			
Phase Angle	0°			
Power Factor	Unity (1)			

Table 16-10. Completed Solution Matrix for the Circuit in Figure 16-50

Practice Problems

1. Complete a solution matrix similar to Table 16-9 for the circuit shown in Figure 16-51.

2. A parallel RLC circuit has the following values: 2.5 µH, 120 pF, 5.0 kΩ, and a 12-V rms voltage source. Complete a solution matrix similar to Table 16-9 for this circuit.

3. Complete a solution matrix similar to Table 16-9 for a parallel RLC circuit that has the following values: 270 µH, 27 pF, 200 kΩ, and a 500-mV rms voltage source.

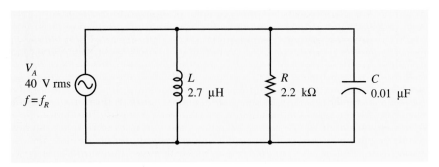

Figure 16-51. A parallel *RLC* circuit.

Answers to Practice Problems

1.

CIRCUIT QUANTITY	RESISTOR	INDUCTOR	CAPACITOR	TOTAL
Voltage (rms)	40 V			
Current (rms)	18.18 mA	2.44 A	2.44 A	18.18 mA
Resistance/Reactance/Impedance	2.2 kΩ	16.43 Ω	16.43 Ω	2.2 kΩ
Resonant Frequency	969.1 kHz			
Q	133.9			
Bandwidth	7.24 kHz			
Phase Angle	0°			
Power Factor	Unity (1)			

2.

CIRCUIT QUANTITY	RESISTOR	INDUCTOR	CAPACITOR	TOTAL
Voltage (rms)	12 V			
Current (rms)	2.4 mA	83.16 mA	83.16 mA	2.4 mA
Resistance/Reactance/Impedance	5 kΩ	144.3 Ω	144.3 Ω	5 kΩ
Resonant Frequency	9.19 MHz			
Q	34.65			
Bandwidth	265.2 kHz			
Phase Angle	0°			
Power Factor	Unity (1)			

3.

CIRCUIT QUANTITY	RESISTOR	INDUCTOR	CAPACITOR	TOTAL
Voltage (rms)	500 mV			
Current (rms)	2.5 µA	158.2 µA	158.2 µA	2.5 µA
Resistance/Reactance/Impedance	200 kΩ	3.16 kΩ	3.16 kΩ	200 kΩ
Resonant Frequency	1.865 MHz			
Q	63.29			
Bandwidth	29.47 kHz			
Phase Angle	0°			
Power Factor	Unity (1)			

Self-Resonance of Inductors and Capacitors

For many applications and analyses, we can assume that an inductor is purely inductive and that a capacitor is purely capacitive. In other cases (high-frequency circuits in particular), a technician must consider the nonideal effects of a practical inductor and a practical capacitor.

Figure 16-52 illustrates how a physical coil that is designed to have inductance also possesses capacitance. Recall that a capacitor is simply two conductors separated by an insulator. As shown in Figure 16-52(a), a practical coil has stray or parasitic capacitance that appears between adjacent turns on the coil. The stray capacitance appears electrically in parallel with the inductance. Additionally, the wire used to wind the coil has resistance. The wire resistance appears electrically in series with the inductance. Figure 16-52(b) shows the equivalent circuit for a physical coil.

For low frequencies, the stray capacitance can generally be ignored (i.e., the capacitive reactance is very high). At higher frequencies, however, the capacitance can play a significant role. In fact, since the coil begins to act like an *RLC* circuit at high frequencies, we can infer that it will also exhibit the characteristics of resonance at some frequency. This frequency is called the **self-resonant frequency** and is provided in the manufacturer's specification sheet for a given coil.

Practical capacitors also exhibit nonideal properties. They have inductance in the leads and resistive losses caused by the ESR of the capacitor (explained in Chapter 14). Figure 16-53 shows the equivalent circuit for a practical capacitor. The rated capacitance and the parasitic resistance and inductance all appear in series. Thus, the capacitor alone becomes a series resonant circuit at some high frequency.

At frequencies in the low kilohertz, the self-resonance of coils and capacitors can be safely ignored for most applications. As the frequency increases, the technician must be increasingly aware of the nonideal effects of capacitors and inductors in order to understand and explain the behavior of the circuits.

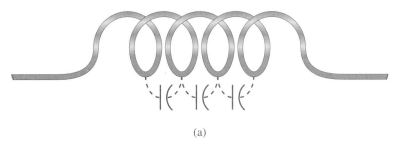

(a)

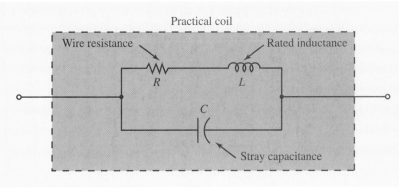

(b)

Figure 16-52. A physical coil also has resistance and capacitance.

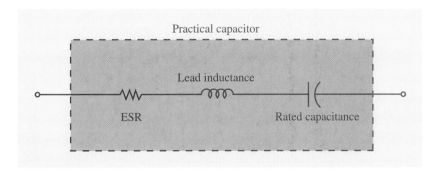

Figure 16-53. A practical capacitor acts like a series *RLC* circuit at high frequencies.

Exercise Problems 16.3

1. Describe two examples of mechanical resonance.

2. At frequencies above resonance in an *RLC* circuit, inductive reactance is (*greater than, less than, equal to*) capacitive reactance.

3. At resonance, the inductive reactance in an *RLC* circuit is (*greater than, less than, equal to*) the capacitive reactance.

4. The current in a series RLC circuit is maximum at resonance. (True or False)

5. The current through the resistive branch of a parallel RLC circuit is maximum at resonance and decreases at other frequencies. (True or False)

6. What is the phase angle between total voltage and total current in a parallel resonant circuit?

7. What is the power factor of a series resonant circuit?

8. What is meant by the phrase Q *rise in voltage* for a series RLC circuit?

9. Analyze the series RLC circuit shown in Figure 16-54, and complete Table 16-11. Assume that the circuit is operating at its resonant frequency.

10. Analyze the parallel RLC circuit shown in Figure 16-55, and complete Table 16-12. Assume that the circuit is operating at the resonant frequency.

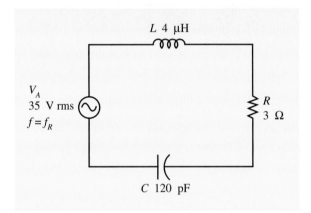

Figure 16-54. Analyze this circuit.

CIRCUIT QUANTITY	RESISTOR	INDUCTOR	CAPACITOR	TOTAL
Current (rms)				
Voltage (rms)				35 V
Resistance/Reactance/Impedance	$3\ \Omega$			
Resonant Frequency				
Q				
Bandwidth				
Phase Angle				
Power Factor				

Table 16-11. Solution Matrix for the Circuit in Figure 16-54

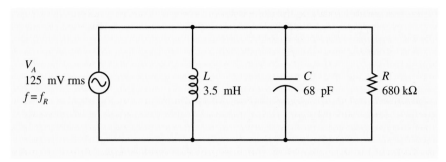

Figure 16-55. Analyze this circuit.

CIRCUIT QUANTITY	RESISTOR	INDUCTOR	CAPACITOR	TOTAL
Voltage (rms)	125 mV			
Current (rms)				
Resistance/Reactance/Impedance	680 kΩ			
Resonant Frequency				
Q				
Bandwidth				
Phase Angle				
Power Factor				

Table 16-12. Solution Matrix for the Circuit in Figure 16-55

16.4 Troubleshooting *RLC* Circuits

Malfunctions in *RLC* circuits can be categorized into two general classes: defective components and misalignment. We will consider these individually.

Defective Components in *RLC* Circuits

The list of possible defects in an *RLC* circuit is essentially a combination of the defects that may occur in *RL* and *RC* circuits. You will recall the following possibilities:

RESISTOR DEFECTS

- open
- shorted
- wrong value

CAPACITOR DEFECTS

- open
- shorted
- leaky (generally limited to electrolytics)
- wrong value

INDUCTOR DEFECTS

- open
- shorted turns
- coil-to-core short

The actual method for locating the defective component in an *RLC* circuit is identical to that previously discussed with reference to *RL* and *RC* circuits and will not be repeated here.

RLC Circuit Alignment

Most applications of *RLC* circuits require that the circuit be operated at its resonant frequency. In many cases, either the inductor or the capacitor is made variable, so the circuit can be tuned to a specific resonant frequency. If the circuit is tuned to the wrong resonant frequency, we say it is misaligned or out of alignment. A mistuned or misaligned *RLC* circuit will not perform as expected. It must be adjusted by a technician to obtain the desired performance.

TUNING PROCEDURES FOR A SERIES *RLC* CIRCUIT

Figure 16-56 shows a method that can be used to tune a series *RLC* circuit to the correct resonant frequency. The circuit is connected to a source voltage (typically a signal generator, which can be found in most electronic laboratories) that is operating at the desired resonant frequency. An oscilloscope is connected across the resistor and calibrated for a convenient display. The circuit is then tuned to resonance by adjusting *L* and/or *C* while monitoring the oscilloscope. As you know, series resonance will produce the maximum voltage across *R*. This, of course, is indicated by maximum amplitude on the oscilloscope display.

As a technician, you must be careful of multiple ground connections in circuits that you are testing. In the case of Figure 16-56, if one end of the resistor is connected to ground, then the ground side of the oscilloscope should be connected to this same point. If neither side of the resistor is connected to ground, then you can use a dual-channel oscilloscope to measure the voltage across the resistor. In this case, Channel A of the scope is connected to one end of the resistor and Channel B is connected to the other. The scope is set to measure the difference between the two channels (e.g., A–B on the vertical mode selector).

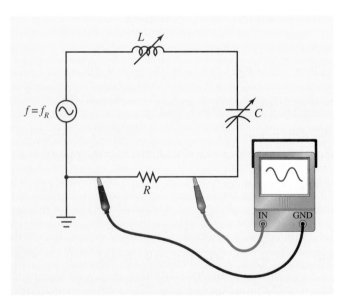

Figure 16-56. How to tune a series resonant circuit.

You may be tempted to insert a current meter in series with the *RLC* circuit and tune the circuit to obtain maximum current. While this is a valid approach in theory and in some practical cases, there is a potential problem; the current meter must be capable of measuring currents in the frequency range of the resonant circuit. Most common current meters are low-frequency measuring devices.

TUNING PROCEDURES FOR A PARALLEL *RLC* CIRCUIT

Figure 16-57 illustrates a procedure for tuning parallel *RLC* circuits to the correct frequency. Here again, a signal generator is used to provide a voltage at the correct frequency. A 1-Ω resistor (R_S) is connected in series with the *RLC* circuit. The oscilloscope measures the voltage across the series resistor. The circuit is tuned by adjusting the value of *L* and/or *C* and monitoring the scope display. At resonance, the *RLC* circuit will have maximum impedance and minimum current. Since the total line current flows through the series resistor, the voltage across it (and the scope display) will also be minimum when the circuit is tuned to the input frequency.

There are many other ways to calibrate an *RLC* circuit, but most require more specialized equipment that is not always available to the technician. Frequency counters, resonant dip meters, high-frequency voltmeters, high-frequency current meters, and absorption wave meters can also be used.

○ KEY POINTS

Resonance in a parallel circuit is indicated by a dipping of voltage across a small series resistor.

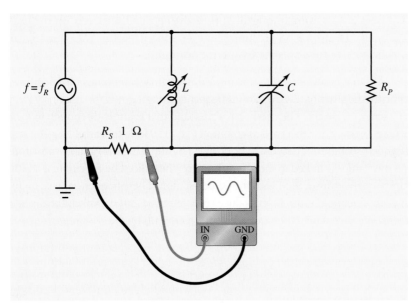

Figure 16-57. Procedure for tuning a parallel *RLC* circuit.

Exercise Problems 16.4

1. When a series *RLC* circuit is tuned to the same frequency as the source voltage, the voltage across the resistor will (*peak, dip*).

2. When a parallel *RLC* circuit is tuned to the same frequency as the source voltage, the line current will (*peak, dip*).

3. Based on your knowledge of series resonant circuits, think of a way to align it to a specific resonant frequency that was not described in this text.

4. Based on your knowledge of parallel resonant circuits, think of a way to align it to a specific resonant frequency that was not described in this text.

16.5 Applied Technology: *RLC* Circuits

RLC (often called *LC*) circuits have many uses in electronic circuits. In nearly all cases, their performance in a given application utilizes the characteristics of resonance.

Radio Frequency Selection

At any given point on earth (or anywhere else), there is a constant bombardment of electromagnetic waves of many different frequencies. For example, there are electromagnetic waves from many AM and FM radio stations, TV stations, police radios, business radios, cellular telephones, paging systems, portable telephones, satellite communications, microwave links, and a host of other sources such as motors, fluorescent lights, and computers, that unintentionally emit electromagnetic waves. Unintentional emissions are called interference or noise when detected by a radio receiver. In order to select a particular frequency or group of frequencies from among all of the existing signals, we rely on the use of resonant *RLC* circuits. Figure 16-58 illustrates how a resonant *LC* (or *RLC*) circuit can be used to tune a radio receiver to a specific frequency or station.

In Figure 16-58, an antenna is used to intercept all electromagnetic waves traveling through space in the vicinity of the antenna. As the various frequencies travel down the antenna wire, they encounter a series *LC* circuit. The circuit has a passband that is just wide enough to include all of the frequencies associated with a particular station (e.g., 150 kHz for each FM radio station). The resonant frequency of the *LC* circuit can be changed by adjusting the variable capacitor. The capacitor is mechanically linked to the tuning knob on the radio. The *LC* circuit is adjusted so that it passes the frequencies associated with the desired station. These frequencies pass on to the radio circuits and amplifiers and ultimately produce voice or music in the speaker. Frequencies from all other stations are rejected (i.e., they are outside the passband) by the *LC* circuit.

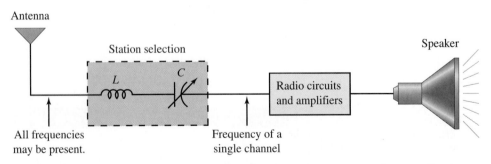

Figure 16-58. An *LC* circuit can select a particular radio station and reject all other frequencies.

Rejection of Interference

Figure 16-59 shows how an *RLC* circuit can be used to separate a desired frequency from another frequency, which would otherwise cause interference. As shown in Figure 16-59, the circuit is essentially a two-part voltage divider. The first part is the series resistance (*R*). The second part of the voltage divider is the series *LC* circuit. The *LC* circuit is tuned to the frequency of the interference.

Recall that a series *LC* circuit has minimum impedance at resonance (i.e., the net reactance is zero ohms). In the case of the circuit shown in Figure 16-59, there will be a zero-ohm (theoretically) impedance through the *LC* portion of the circuit to ground at the frequency of the interference. This means all of the interference voltage will be dropped across the series resistance and will not pass on to subsequent circuitry.

The desired frequency, on the other hand, is either higher or lower than the resonant frequency of the *LC* circuit. Recall that a series *LC* combination can have substantial reactance at frequencies on either side of the resonant frequency. In the case shown in Figure 16-59, the impedance of the *LC* circuit at the desired frequency is much greater than the value of series resistance. This means that nearly all of the desired signal voltage will be felt across the *LC* portion of the circuit and will be available for use in subsequent circuitry connected in parallel with the *LC* circuit.

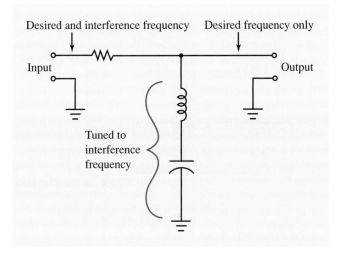

Figure 16-59. An *RLC* circuit can be used to eliminate interference.

Frequency Determination for Oscillators

An oscillator is an electronic circuit that can generate sine waves. It may be viewed as an amplifier that provides its own input. You have probably heard the shrill squeal that occurs when a microphone in a public address system is held too close to the speaker for the same system. Feedback occurs and you hear a loud squeal or oscillation. The output from the speaker provides its own input to the microphone.

Figure 16-60 illustrates how an *LC* circuit can be used to determine the frequency of an oscillator. All electronic circuits generate a wide range of electrical frequencies at very low levels (like the noise between stations on a television). In the case of the circuit illustrated in Figure 16-60, the random low-level noise from the oscillator is passed through a series LC circuit to the input of the amplifier. The LC circuit will offer opposition to off-resonance frequencies, but will offer low (ideally zero) opposition to signals at the resonant frequency. When the low-level signal at the resonant frequency returns to the input of the amplifier, it is amplified and appears even larger at the output. Now this stronger signal can make the feedback trip to the input of the amplifier and appear even larger at the output. This process continues until the circuit reaches its maximum amplitude. At that time, the circuit becomes stable, and the constant frequency, constant amplitude signal is available at the output of the amplifier. Since oscillation (self-generation of a signal) depends on feedback, and since the LC circuit only allows feedback at the resonant frequency, oscillation will occur at the frequency set by the LC network. Either L or C can be made variable to alter the frequency of oscillation.

When a portion of an amplifier's output is returned to its own input, the returned energy is called feedback. The phase of feedback relative to the original input can either aid (as described in the preceding discussion) or cancel. Feedback that aids the original input is called **positive (or regenerative) feedback.** If the feedback is out of phase with the input signal, it is called **negative (or degenerative) feedback.**

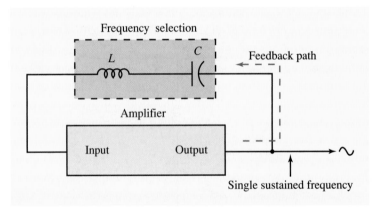

Figure 16-60. An oscillator consists of an amplifier whose output is returned to its own input.

Exercise Problems 16.5

1. Refer to Figure 16-58. With the circuit connected as shown, explain why a series LC circuit was used instead of a parallel LC circuit.

2. Refer to Figure 16-59. Explain what would happen to circuit operation if the series LC circuit were replaced with a parallel LC circuit.

3. Refer to Figure 16-60. If the value of C were made smaller, what would happen to the frequency of oscillation?

Chapter Summary

- This chapter introduced the characteristics of series and parallel *RLC* circuits. Many of the characteristics are similar to previously studied series and parallel circuits. For example, current is the same throughout a series circuit, the sum (phasor sum for reactive circuits) of the voltage drops in a series circuit always equals the supply voltage, the voltage is the same across all components in a parallel circuit, and the sum (phasor sum for reactive circuits) of the branch currents in a parallel circuit always equals the total supply current.

- *RLC* circuits also exhibit some interesting phenomena. For example, the voltage drop across the reactances in a series *RLC* circuit can individually exceed the value of supply voltage, the reactive branch currents in a parallel *RLC* circuit can exceed the supply current, and the total impedance of either a series or a parallel *RLC* circuit at resonance is equal to the value of the resistance in the circuit.

- Complete analysis of series, parallel, and series-parallel *RLC* circuits was presented. Analysis of complex *RLC* circuits was not formally presented, but is similar to the analytical procedures presented for other complex circuit configurations.

- All *RLC* circuits have a single frequency where the inductive and capacitive reactances are equal. We call this the resonant frequency of the circuit. The resonant frequency is strictly determined by the values of inductance and capacitance in the circuit. The characteristics of an *RLC* circuit at resonance are very important to a technician. In short, a series *RLC* circuit at resonance has minimum impedance, maximum current, and maximum component voltages. The power factor is unity, the phase angle is zero and the circuit acts purely resistive. In the case of a parallel *RLC* circuit at resonance, the circuit has maximum impedance, minimum line current, a unity power factor, and a phase angle of zero. It also acts purely resistive.

- On either side of the resonant frequency, a series or parallel circuit acts reactive. Series *RLC* circuits above and below resonance act inductive and capacitive, respectively. Parallel *RLC* circuits act capacitive and inductive, respectively, for frequencies above and below the resonant frequency.

- The *Q* factor of an *RLC* circuit determines how pronounced the effects of resonance will be in the vicinity of the resonant frequency. At frequencies on either side of the resonant frequency, circuit response diminishes. Frequencies on either side of the resonant frequency that cause a 70.7% or higher response relative to the maximum (100%) response at the resonant frequency are called the passband or bandpass of the circuit. The width of the passband is called the bandwidth of the circuit. Bandwidth and *Q* factor are inversely related. Higher *Q*s result in narrower bandwidths. The frequencies on the lower and upper edges of the passband are called the lower cutoff frequency and the upper cutoff frequency, respectively. In the case of series *RLC* circuits, these same points are also called the half-power or 3-dB points.

- The increase in reactive voltage (series *RLC* circuits) or impedance (parallel *RLC* circuits) is related to the *Q* factor. The voltage across the inductance or capacitance in a series *RLC* circuit at resonance is higher than the supply voltage by a factor of *Q*.

This is called a Q rise in voltage. Similarly, a Q rise in impedance occurs in a parallel *RLC* circuit and causes the total impedance to be higher than either reactance by a factor of Q.

• Analysis of *RLC* circuits at the resonant frequency is largely identical to the analysis at any other frequency. Computation of the resonant frequency and calculations involving Q and bandwidth are most easily accomplished using specialized equations.

• Troubleshooting an *RLC* circuit requires the same basic methods that are used to diagnose *RL* or *RC* circuits. In the case of an *RLC* circuit that is used for its resonance effects (the usual case), the circuit must be tuned for the correct resonant frequency. Either *L* or *C* may be made variable for the purpose of aligning or tuning the circuit.

Review Questions

Section 16.1: Introduction to *RLC* Circuits

1. The current is the same through every component in a series *RLC* circuit. (True or False)

2. The voltage drop is the same across every component in a series *RLC* circuit. (True or False)

3. What is the phase relationship between total current and inductive voltage in a series *RLC* circuit?

4. What is the phase relationship between total current and capacitive voltage in a series *RLC* circuit?

5. What is the phase relationship between inductive current and capacitive current in a series *RLC* circuit?

6. What is the phase relationship between inductive voltage and capacitive voltage in a series *RLC* circuit?

7. If a series *RLC* circuit has an inductive voltage drop of 120 mV, a capacitive voltage drop of 80 mV, and a resistive voltage drop of 40 mV, what is the value of applied voltage?

8. What is the impedance of a series *RLC* circuit with the following values: $X_L = 10\ \text{k}\Omega$, $X_C = 4\ \text{k}\Omega$, and $R = 8.2\ \text{k}\Omega$?

9. How much current flows in a series *RLC* circuit with 10 V rms applied and consisting of the following values: $X_L = 7.5\ \text{k}\Omega$, $X_C = 22\ \text{k}\Omega$, and $R = 10\ \text{k}\Omega$?

10. The current is the same through every component in a parallel *RLC* circuit. (True or False)

11. The voltage drop is the same across every component in a parallel *RLC* circuit. (True or False)

12. What is the phase relationship between total voltage and inductive current in a parallel *RLC* circuit?

13. What is the phase relationship between total voltage and capacitive current in a parallel *RLC* circuit?

14. What is the phase relationship between inductive current and capacitive current in a parallel *RLC* circuit?

15. What is the phase relationship between inductive voltage and capacitive voltage in a parallel *RLC* circuit?

16. If a parallel *RLC* circuit has an inductive branch current of 3.5 A, a capacitive branch current of 5 A, and a resistive branch current of 1.2 A, what is the value of source current?

17. If an *RLC* circuit has reactive powers of 20 VAR (P_{R_L}) and 15 VAR (P_{R_C}), and has a true power dissipation of 10 W, determine the value of apparent power in the circuit.

Section 16.2: *RLC* Circuit Calculations

Refer to Figure 16-61 for Questions 18–30.

18. What is the value of inductive reactance?

19. What is the value of capacitive reactance?

20. What is the value of impedance in the circuit?

21. How much current flows through L?

22. How much current flows through R?

23. What is the voltage drop across C?

24. What is the voltage drop across L?

25. What is the voltage drop across R?

26. What is the phase angle between total voltage and total current in the circuit?

27. What is the power factor of the circuit?

28. Compute the value of true power.

29. Compute the value of apparent power.

30. What is the value of net reactive power?

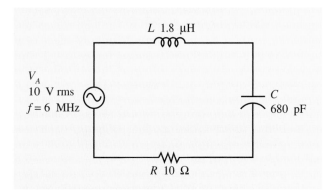

Figure 16-61. A series *RLC* circuit.

Refer to Figure 16-62 for Questions 31–45.

31. What is the value of inductive reactance?

32. What is the value of capacitive reactance?

33. How much current flows through the inductor?

34. How much current flows through the capacitor?

35. How much current flows through the resistor?

36. What is the value of source current?

37. What is the impedance of the circuit?

38. What is the admittance of the circuit?

39. What is the phase angle between total current and total voltage?

40. What is the power factor?

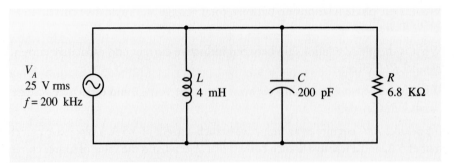

Figure 16-62. A parallel *RLC* circuit.

41. How much reactive power is due to the inductor?
42. How much reactive power is due to the capacitor?
43. What is the net reactive power?
44. What is the value of true power?
45. What is the apparent power in the circuit?

Refer to Figure 16-63 for Questions 46–60.

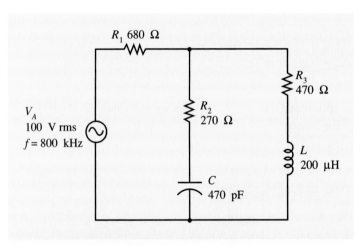

Figure 16-63. A series-parallel *RLC* circuit.

46. What is the capacitive reactance of *C*?
47. What is the inductive reactance of *L*?
48. Write the impedance of the R_3, *L* branch in polar and rectangular notation.
49. Write the impedance of the R_2, *C* branch in polar and rectangular notation.
50. What is the combined impedance of the R_2, *C* and the R_3, *L* branches?
51. What is the total impedance of the circuit?
52. What is the total current in the circuit?
53. How much current flows through R_1?

54. How much current flows through *L*?

55. What is the voltage drop across *C*?

56. What is the voltage drop across R_1?

57. What is the power factor of the circuit?

58. What is the value of apparent power for the circuit?

59. What is the phase angle between total voltage and total current in the circuit?

60. How much power is dissipated by R_3?

Section 16.3: Resonance

61. Only certain kinds of *RLC* circuits have a resonant frequency. (True or False)

62. What are the relative values of inductive and capacitive reactance in an *RLC* circuit that is operating at its resonant frequency?

63. A series *RLC* circuit that is operating below its resonant frequency will act (*resistive, inductive, capacitive*).

64. A series *RLC* circuit that is operating above its resonant frequency will act (*resistive, inductive, capacitive*).

65. A series *RLC* circuit that is operating at its resonant frequency will act (*resistive, inductive, capacitive*).

66. A parallel *RLC* circuit that is operating above its resonant frequency will act (*resistive, inductive, capacitive*).

67. A parallel *RLC* circuit that is operating below its resonant frequency will act (*resistive, inductive, capacitive*).

68. A parallel *RLC* circuit that is operating at its resonant frequency will act (*resistive, inductive, capacitive*).

69. The reactive voltage drops in a series resonant circuit are always equal. (True or False)

70. The impedance of a series *RLC* circuit at resonance is maximum. (True or False)

71. The current through a series *RLC* circuit at resonance is maximum. (True or False)

72. What is the phase angle between total current and total voltage in an *RLC* circuit that is operating at its resonant frequency?

73. The impedance of a parallel *RLC* circuit at resonance is maximum. (True or False)

74. The supply current for a parallel *RLC* circuit at resonance is minimum. (True or False)

75. An *RLC* circuit with a *Q* of greater than _____ is considered to have a high *Q*.

76. An *RLC* circuit with a *Q* of less than _____ is considered to have a low *Q*.

77. The higher the resistance in a series *RLC* circuit, the higher the *Q*. (True or False)

78. The *Q* of a parallel *RLC* circuit is directly proportional to the value of the resistive branch. (True or False)

79. For a given resonant frequency, high-*Q* circuits have wider bandwidths than low-*Q* circuits. (True or False)

80. What is the name given to the range of frequencies that extends between the lower cutoff frequency and the upper cutoff frequency of an RLC circuit?

81. Explain why the half-power point frequency in a series RLC circuit is also called the 3-dB frequency.

82. If the resonant frequency of an RLC circuit is 25 MHz, the lower cutoff frequency is 23.75 MHz, and the bandwidth is 2.5 MHz, what is the pass band of the circuit?

83. Refer to Question 82. What is the Q of the circuit?

84. Refer to Question 82. What is the upper cutoff frequency?

85. If the current in a series RLC circuit is 250 mA at resonance, what is the value of current at the lower cutoff frequency?

86. If the resistor voltage in a series RLC circuit is 5 V at the upper 3-dB frequency, what is the value of resistor voltage at the lower cutoff frequency?

87. Refer to Question 86. What is the value of resistor voltage at the resonant frequency?

88. What is the bandwidth of an RLC circuit that has a resonant frequency of 15 MHz and a Q of 35?

89. Compute the resonant frequency of a series RLC circuit that has a 175-μH inductor, an 860-pF capacitor, and a 3.5-Ω resistor.

90. Compute the resonant frequency of a parallel RLC circuit that has a 2-mH inductor, a 2,000-pF capacitor, and a 27-kΩ resistor.

91. Analyze the series resonant circuit shown in Figure 16-64, and complete the solution matrix provided in Table 16-13.

92. Analyze the parallel resonant circuit shown in Figure 16-65, and complete the solution matrix provided in Table 16-14.

93. Due to the nonideal effects of stray (interwinding) capacitance, a coil can exhibit the characteristics of a resonant circuit. The frequency at which this occurs is called the _____ frequency of the coil.

94. Due to the nonideal effects of lead inductance and resistance, a capacitor becomes a (*series, parallel*) resonant circuit at a frequency called the _____ frequency of the capacitor.

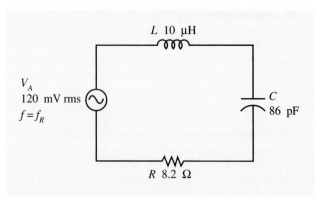

Figure 16-64. Analyze this circuit.

CIRCUIT QUANTITY	RESISTOR	INDUCTOR	CAPACITOR	TOTAL
Current (rms)				
Voltage (rms)				120 mV
Resistance/Reactance/Impedance	8.2 Ω			
Resonant Frequency				
Q				
Bandwidth				
Phase Angle				
Power Factor				

Table 16-13. Solution Matrix for the Circuit in Figure 16-64

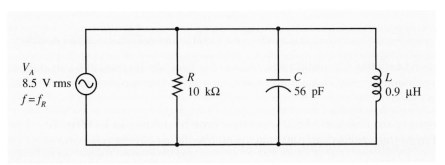

Figure 16-65. Analyze this circuit.

CIRCUIT QUANTITY	RESISTOR	INDUCTOR	CAPACITOR	TOTAL
Voltage (rms)		8.5 V		
Current (rms)				
Resistance/Reactance/Impedance	10 kΩ			
Resonant Frequency				
Q				
Bandwidth				
Phase Angle				
Power Factor				

Table 16-14. Solution Matrix for the Circuit in Figure 16-65

Section 16.4: Troubleshooting *RLC* Circuits

95. Malfunctions in *RLC* circuits can be caused by defective components or _____ of the *LC* circuit.

96. List six possible defects that may occur in an *RLC* circuit.

97. If you monitor the line current in a parallel *RLC* circuit as you tune it to resonance, it will peak as you tune through the resonant frequency. (True or False)

98. If you monitor the resistor voltage in a series *RLC* circuit as you tune it to resonance, it will peak as you tune through the resonant frequency. (True or False)

Section 16.5: Applied Technology: *RLC* Circuits

99. Refer to Figure 16-58. If *L* were smaller, you would be tuning in a radio station that has a higher channel frequency. (True or False)

100. Refer to Figure 16-59. If the frequency of interference were higher than the resonant frequency of the *LC* circuit, how could you alter the *LC* circuit to eliminate the interference?

TECHNICIAN CHALLENGE

You have been asked to design a resonant filter that can be used in a communications receiver manufactured by your company. The basic design parameters are listed in Figure 16-66.

The input to the filter will be 5 Vrms in the range of 100 kHz to 10 MHz. The filter must reject most frequencies and pass only those frequencies in the passband of 475 kHz to

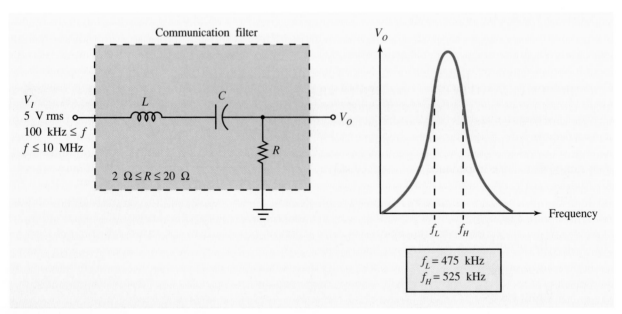

Figure 16-66. The basic design requirements for a communications filter.

525 kHz. For compatibility with existing circuitry, the resistance shown in Figure 16-66 must be in the range of 2 to 20 Ω.

Determine an acceptable set of values for L and C. After you complete your design, completely analyze the circuit to verify that the original design requirements have been satisfied. If you have access to the required materials, then measure the performance of your design in the laboratory.

Equation List

$(16\text{-}1)$ $V_{REACTIVE} = V_L + V_C$

$(16\text{-}2)$ $V_A = \sqrt{(V_L - V_C)^2 + V_R^2}$, or

$V_A = \sqrt{V_R^2 + (V_L - V_C)^2}$

$(16\text{-}3)$ $Z = \sqrt{(X_L - X_C)^2 + R^2}$, or

$Z = \sqrt{R^2 + (X_L - X_C)^2}$

$(16\text{-}4)$ $X_{NET} = X_L + X_C$

$(16\text{-}5)$ $P_{R_{NET}} = P_{R_L} + P_{R_C}$

$(16\text{-}6)$ $P_A = \sqrt{(P_{R_L} - R_{R_C})^2 + P_T^2}$, or

$P_A = \sqrt{P_T^2 + (P_{R_L} - P_{R_C})^2}$

$(16\text{-}7)$ $I_X = I_L + I_C$

$(16\text{-}8)$ $I_T = \sqrt{(I_L - I_C)^2 + I_R^2}$, or

$I_T = \sqrt{I_R^2 + (I_L - I_C)^2}$

$(16\text{-}9)$ $R_P = ESR \times Q^2$

$(16\text{-}10)$ $Q = \dfrac{X_L}{R_S}$

$(16\text{-}11)$ $Q = \dfrac{R_P}{X_L}$

$(16\text{-}12)$ $BW = f_H - f_L$

$(16\text{-}13)$ $BW = \dfrac{f_R}{Q}$

$(16\text{-}14)$ $f_R = \dfrac{1}{2\pi\sqrt{LC}}$

$(16\text{-}15)$ $V_L = V_C = QV_A$

$(16\text{-}16)$ $Z = QX_L$

key terms

coefficient of coupling   magnetizing current   step-down transformer
dot notation   mutual inductance   step-up transformer
Faraday shield   neutral   turns ratio
hot wire   primary
leakage flux   reflected impedance
leakage inductance   secondary

objectives

After completing this chapter, you should be able to:

1. Compute the total inductance of series- or parallel-connected coils that are magnetically coupled.

2. Analyze transformer circuits by applying the relationships between turns ratio, voltage ratio, current ratio, and impedance ratio between the primary and secondary windings.

3. Name and describe three types of transformer cores.

4. List and describe at least three types of transformer losses.

5. Solve problems relating to the efficiency of a transformer.

6. Test a transformer to locate defects.

7. Describe at least three applications for transformers.

8. Explain the terms step-up and step-down as applied to transformers.

9. Explain the operation of each of the following types of transformers: autotransformers, tapped transformers, and multiwinding transformers.

10. Describe the basic power distribution scheme used by commercial power companies.

11. Identify delta and wye transformer connections and state the relationship between leg and line currents and voltages.

Mutual Inductance and Transformers

Transformers are basic circuit components like resistors, capacitors, and inductors. The operation of transformers may seem natural to you, since their operation is based on the same magnetic principles discussed with reference to inductors. Transformers are used in nearly every electronic system that operates from the ac power line and in most battery-operated circuits as well. A technician must understand the theory of operation of a transformer, know how to analyze transformer circuits, and be able to troubleshoot transformers to locate defects.

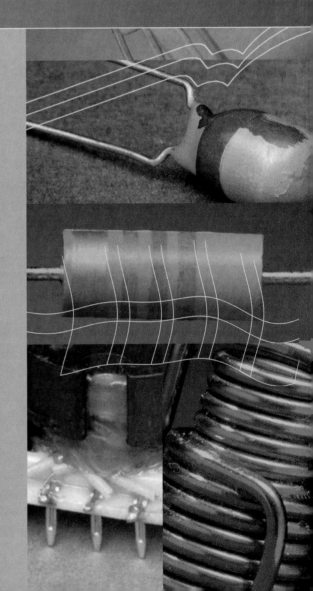

17.1 Mutual Inductance and Coupling

You already know that when current passes through an inductor, a magnetic field is produced around the inductor. The strength of the field is proportional to current and the physical properties of the coil (e.g., number of turns, type of core, and so on). You will also recall that a changing magnetic field can induce a voltage into the windings of an inductor. The amount of voltage induced is proportional to such things as the relative rate of movement of the field, strength of the field, and physical properties of the coil.

Figure 17-1 shows two coils that are physically close to each other. One coil is connected to an alternating voltage source, the other coil is connected to an ac voltmeter. The alternating current in L_1 will cause the surrounding flux (shown in Figure 17-1 as dashed lines) to continuously expand and contract. Because L_2 is nearby, some of the flux produced by L_1 cuts the windings of L_2 and induces a voltage into the L_2 coil. This is a measurable, useful voltage as indicated by the voltmeter in Figure 17-1. As indicated in Figure 17-1(a), not all of the flux lines produced by L_1 pass through L_2.

Now, if the two coils are brought closer together as shown in Figure 17-1(b), a greater percentage of the flux is shared by the two coils. Because more flux lines are cutting the L_2 turns, a higher voltage will be produced in L_2 as indicated by the voltmeter in Figure 17-1(b).

The percentage of flux from one coil that passes through a second coil is called the **coefficient of coupling.** The coefficient of coupling can range from zero (i.e., the two coils share no flux lines) to 100 (i.e., the two coils share 100% of the flux lines). Generally, the coefficient of coupling is expressed as a decimal, so the range is from zero to one. It is represented by the letter k.

When two coils are magnetically linked as represented in Figure 17-1, we say they have **mutual inductance.** You will recall that 1 H is the measure of inductance that causes 1 V to be induced when the current through it changes at the rate of 1 A/s. In the case of magnetically coupled coils, if the current through one coil changes at the rate of 1 A/s and a voltage of 1V is produced in the second coil, we say they have a mutual inductance of 1 H.

The value of mutual inductance (L_M) between two coils depends on the inductance of each winding (L_1 and L_2) and how closely they are coupled (k). We can express this formally with Equation 17-1.

$$L_M = k\sqrt{L_1 L_2} \qquad (17\text{-}1)$$

where k is the coefficient of coupling, and L_M, L_1, and L_2 are the inductance values in henrys of the mutual and coil inductances, respectively.

EXAMPLE SOLUTION

If two coils have a coefficient of coupling of 0.75 and have individual inductances of 100 mH and 250 mH, what is the value of mutual inductance?

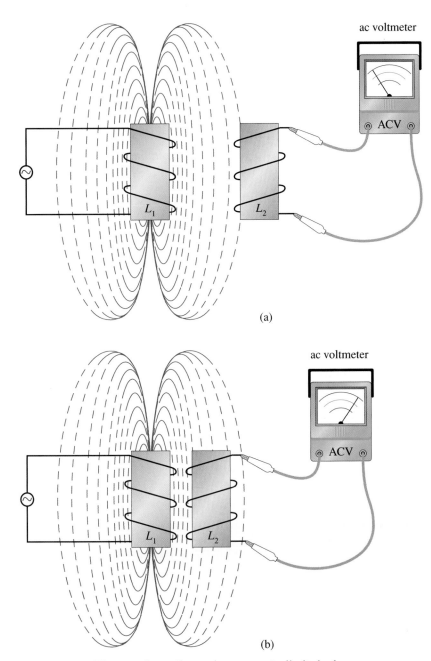

Figure 17-1. Two nearby coils can be magnetically linked.

EXAMPLE SOLUTION

We apply Equation 17-1.

$$L_M = k\sqrt{L_1 L_2}$$
$$= 0.75 \times \sqrt{(100 \text{ mH}) \times (250 \text{ mH})} = 118.6 \text{ mH}$$

Practice Problems

1. Compute the value of mutual inductance between two 50-mH coils that have a coefficient of coupling of 0.8.

2. What is the mutual inductance between a 50-µH coil and a 175-µH coil if they have a coefficient of coupling of 0.5?

3. When a 2.5-mH coil has a 0.9 coefficient of coupling to another coil, the mutual inductance is 4 mH. What is the inductance of the second coil?

Answers to Practice Problems

1. 40 mH   2. 46.77 µH   3. 7.9 mH

Effect of Mutual Inductance on Series Coils

KEY POINTS

When magnetically linked coils are connected in series, the total inductance is affected by the amount of mutual inductance in the coils.

You will recall that inductors in series combine like resistors in series (i.e., simple addition). You may also recall that this simple relationship requires that there be no magnetic flux linkage between the coils. We are now in a position to consider the effects of magnetic linkage between two coils (i.e., mutual inductance). Figure 17-2 shows two series coils that have some shared flux. That is, a portion of the flux from one inductor cuts the winding of the second inductor.

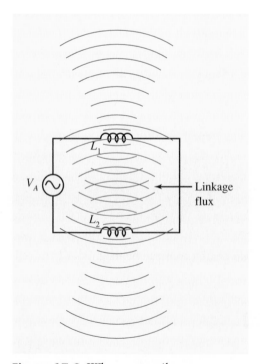

Figure 17-2. When two coils are magnetically linked, they have mutual inductance.

The total inductance of the two series coils is affected by the mutual inductance. Additionally, the mutual inductance may increase or decrease the total inductance, depending on the relative polarities of the two coils. Recall that the polarity of induced voltage is determined by the direction of magnetic flux motion and the direction of the coil winding (left-hand rule). If we change the direction of the winding for one of the coils, then we alter the additive or subtractive effects of the mutual inductance. This is illustrated in Figure 17-3.

In Figure 17-3(a), the magnetically linked coils are wound such that their magnetic fields are additive. This increases the total inductance in the circuit. If we change the direction of one of the coil windings, as shown in Figure 17-3(b), then the magnetic fields of the two coils oppose each other. This lessens the effective self-induced voltage and results in a lower total inductance.

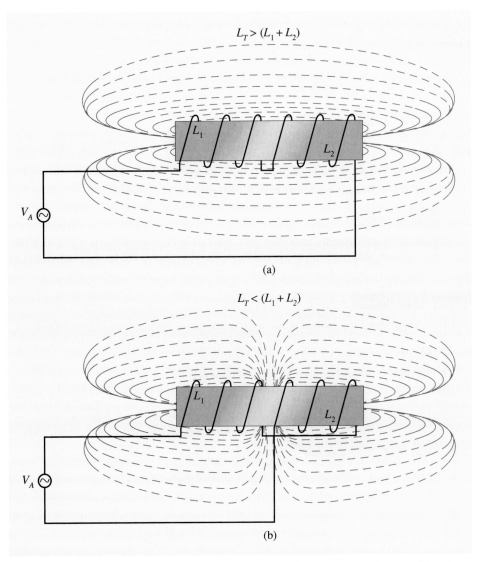

Figure 17-3. Series coils with magnetic coupling can be (a) series-aiding or (b) series-opposing.

The increase or decrease in total inductance as a result of aiding or opposing fields is expressed more formally in Equation 17-2, which computes the total inductance of two series coils.

$$L_T = L_1 + L_2 \pm 2L_M \qquad (17\text{-}2)$$

where L_1 and L_2 are the individual coil inductances, and L_M is the mutual inductance. The plus or minus ($\pm$) operator accounts for the two possible polarities of the coils. If the coil polarities aid, as shown in Figure 17-3(a), then the equation becomes

$$L_T = L_1 + L_2 + 2L_M$$

If, by contrast, the coil polarities are in opposition as illustrated in Figure 17-3(b), then Equation 17-2 becomes

$$L_T = L_1 + L_2 - 2L_M$$

EXAMPLE SOLUTION

Two 100-µH coils are connected in series so their magnetic fields aid one another. Their mutual inductance is 50 µH. What is the total inductance of the series combination?

EXAMPLE **SOLUTION**

We apply Equation 17-2. Since the fields are aiding, we add the mutual inductance factor.

$$L_T = L_1 + L_2 + 2L_M$$
$$= 100 \ \mu\text{H} + 100 \ \mu\text{H} + (2 \times 50 \ \mu\text{H}) = 300 \ \mu\text{H}$$

EXAMPLE SOLUTION

Two coils are connected in series such that their fields are opposing. One coil is 250 mH, the other coil is 400 mH, and the mutual inductance is 75 mH. What is the total inductance of the series circuit?

EXAMPLE **SOLUTION**

In this case, we subtract the mutual inductance factor, since the fields are in opposition.

$$L_T = L_1 + L_2 - 2L_M$$
$$= 250 \ \text{mH} + 400 \ \text{mH} - (2 \times 75 \ \text{mH}) = 500 \ \text{mH}$$

Practice Problems

1. If a 200-µH inductor and a 175-µH inductor are connected to be series-aiding with a mutual inductance of 50 µH, what is the value of total inductance?

2. What is the total inductance if a 5-mH coil is connected in series-opposition with a 2-mH coil with a mutual inductance of 1.5 mH?

3. A 25-mH coil is series-aiding with a 35-mH coil. They have a coefficient of coupling of 0.8. What is the total inductance of the series combination?

Answers to Practice Problems

1. 475 µH   **2.** 4 mH   **3.** 107.3 mH

Effect of Mutual Inductance on Parallel Coils

Magnetic flux linkage between parallel-connected inductors also affects the value of their combined inductance. Computation of total inductance is more involved, but it can be approximated with Equation 17-3.

$$L_T = \frac{L_1 L_2 - L_M{}^2}{L_1 + L_2 \pm 2L_M} \qquad (17\text{-}3)$$

The mutual inductance factor in the denominator $(2L_M)$ is positive when the magnetic fields of the two coils aid and negative when the fields oppose.

EXAMPLE SOLUTION

What is the approximate value of total inductance when two parallel coils (with aiding magnetic fields) have inductances of 400 µH and 300 µH, and a mutual inductance of 100 µH?

EXAMPLE SOLUTION

We apply Equation 17-3 directly.

$$L_T = \frac{L_1 L_2 - L_M{}^2}{L_1 + L_2 \pm 2L_M}$$

$$= \frac{(300\ \mu H \times 400\ \mu H) - (100\ \mu H)^2}{300\ \mu H + 400\ \mu H + (2 \times 100\ \mu H)}$$

$$= 122.2\ \mu H$$

Practice Problems

1. Calculate the approximate total inductance if two 150-mH coils are connected in parallel with their magnetic fields in opposition. The mutual inductance is 25 mH.

2. Two parallel coils have fields that are magnetically aiding. One coil is 6 mH, the other is 10 mH, and the mutual inductance is 0.8 mH. What is the approximate inductance of the parallel combination?

3. The magnetic fields of two parallel-connected coils are magnetically aiding and produce a mutual inductance of 5 µH. If each coil has an inductance of 35 µH, what is the total inductance of the parallel combination?

Answers to Practice Problems

1. 87.5 mH   **2.** 3.37 mH   **3.** 15 µH

Exercise Problems 17.1

1. If 100% of the flux lines generated by one coil also cut the windings of another coil, what is the coefficient of coupling between the two coils?

2. If 35% of the flux lines generated by one coil also cut the windings of another coil, what is the coefficient of coupling between the two coils?

3. If two coils have a coefficient of coupling of 0.85 and have individual inductances of 235 mH and 150 mH, what is the value of mutual inductance?

4. What is the mutual inductance between a 150-μH coil and a 75-μH coil if the coefficient of coupling is 0.6?

5. The presence of mutual inductance between two coils always increases the value of total inductance when the two coils are connected in series. (True or False)

6. Two 35-mH coils are connected in series, so their magnetic fields are additive. If their mutual inductance is 10 mH, what is the total inductance of the series combination?

7. Two coils are connected in series such that their fields are opposing. One coil is 200 μH, the other is 375 μH, and the mutual inductance is 50 μH. What is the total inductance of the series circuit?

8. If two 25-mH coils are connected in parallel with magnetically aiding fields, calculate the total inductance if the mutual inductance is 5 mH.

9. If connections to one of the coils described in Problem 8 were interchanged, what would be the value of total inductance?

17.2 Transformers

A transformer is a basic component whose operation depends on the magnetic linkage (mutual inductance) between two or more coils. The coils in the transformer are wound on a common core to increase the amount of flux linkage. In many cases, the coils are wound on overlapping layers to obtain the highest degree of coupling.

Basic Transformer Action

Figure 17-4 shows two coils wound on a common magnetic core. A sinusoidal voltage is applied to one of the windings. This winding is called the **primary** of the transformer. The inductive reactance of the primary winding limits the primary current. The magnetic flux in the core varies as the current in the primary varies. Because the core is permeable, much of the flux created by the primary passes through the secondary winding. The changing flux induces a voltage in the secondary winding. As illustrated in Figure 17-4, the secondary voltage is a measurable, usable voltage. As you might expect, the greater the coefficient of coupling, the higher the value of secondary voltage for a given primary voltage.

The polarity of the voltage in the secondary depends on the direction in which the secondary winding is wrapped. In certain applications, the primary and secondary windings must be connected into the circuit with proper phase relationships. Schematic

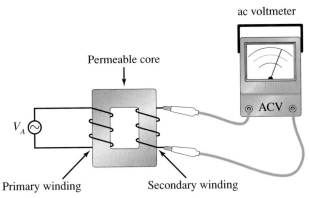

Figure 17-4. Basic transformer action.

KEY POINTS

The phase relationships between primary and secondary voltages are indicated on schematic diagrams with dot notation.

diagrams often indicate the phase relationships in the transformer with **dot notation.** Figure 17-5 shows the schematic symbol for one type of transformer and illustrates the use of dot notation.

As shown in Figure 17-5, dots are added to both primary and secondary windings. The location of the dots indicates similar instantaneous polarity. As indicated by the sine wave outputs in Figure 17-5, the input/output phase relationship is indicated by the relative positioning of the primary and secondary phase dots.

KEY POINTS

The phase dots on a transformer indicate points with similar instantaneous polarities.

Transformers can be constructed to provide secondary voltages that are different than the primary voltage. If the secondary voltage of a transformer is higher than its primary voltage, we call it a **step-up transformer.** Similarly, if the secondary voltage is lower than the primary voltage, we call it a **step-down transformer.** Figure 17-6 illustrates step-up and step-down transformers.

Transformer Types

KEY POINTS

Transformers can step-up (increase), step-down (decrease), or merely couple the supply voltage.

There are several ways to categorize transformers. They can be classified according to their intended application, the type of core material, or the way their windings are configured.

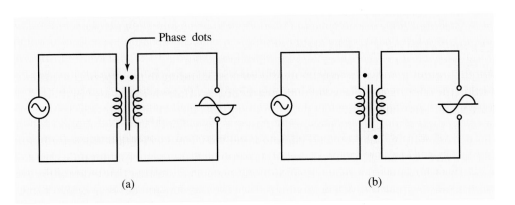

(a) (b)

Figure 17-5. Schematic symbol for a basic transformer including the use of dots to indicate transformer phase.

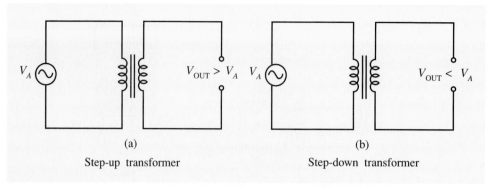

Figure 17-6. Illustration of step-up and step-down transformers.

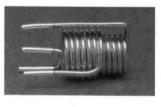

(a)

Figure 17-7.
A representative air-core transformer and its schematic symbol.

CLASSIFICATION BY APPLICATION

Transformer applications will be discussed later in this chapter, but let us briefly examine how transformers can be categorized based on their primary application. Some transformers are designed for connection to the power line and are used to provide power to electronic circuits. This type of transformer is generally called a power transformer. Power transformers can be used to increase or decrease the voltage applied to the primary. If the transformer does not increase or decrease the voltage, but is used strictly to provide isolation between the primary and secondary circuits, it is often called an isolation transformer.

Some transformers are specifically designed for operation throughout the audio range. They are, for example, used to connect audio amplifiers to speakers. In any case, these transformers are commonly called audio transformers.

Transformers that are constructed for operation at high radio frequencies are classified as r.f. (radio frequency) transformers. The internal circuits of a radio receiver utilize signals that are called intermediate frequencies. The transformers used in these circuits are called i.f. (intermediate frequency) transformers.

Many pulse (e.g., rectangular waveform) circuits utilize transformers for coupling signals from one point to another. These transformers are called pulse transformers and are optimized for coupling nonsinusoidal waveforms.

CLASSIFICATION BY CORE MATERIAL

Another common way to categorize transformers is according to the type of core material used in the transformer. There are three primary types of core material used in transformers: air, iron, and ferrite.

Air-core transformers, like air-core coils, are wound around a nonmagnetic coil form. The form may be plastic, cardboard, or any other material with a relative permeability near unity. Since the core material has such a low permeability, it follows that much of the flux escapes the core. The flux that is external to the core and does not cut both primary and secondary windings is called **leakage flux.** Leakage flux causes additional inductance to appear in series with the primary and secondary windings. This apparent inductance is called **leakage inductance.** Figure 17-7 shows an air-core transformer and its generic schematic symbol. These devices are generally used at high radio frequencies.

The primary and secondary windings in an iron-core transformer are wound on a common, high-permeability core. Because of the high permeability of the core, the coefficient of coupling approaches unity (i.e., there is minimal leakage flux). Iron-core transformers experience the same core losses as previously discussed for iron-core inductors. These losses increase with frequency. Iron-core transformers are generally limited to frequencies in the audio range. One of the most common applications for iron-core transformers is power transformers used to couple power from the ac power line. Figure 17-8 shows a typical iron-core transformer and its generic schematic symbol.

The third category of transformer cores is ferrite. Ferrite is a high-permeability ceramic material that is relatively brittle. Because the ferrite core has a fairly high dc resistance, it has less core losses at high frequencies than an equivalent iron-core transformer. Ferrite-core transformers are used for applications extending from audio well into the high-megahertz range.

The cores used in ferrite-core transformers come in many different shapes and sizes. Figure 17-9 shows some core shapes. The torroid core has a very low level of leakage flux; nearly all of the flux is contained within the continuous core material. Figure 17-10 shows a ferrite-core transformer and the generic schematic symbol.

(a)

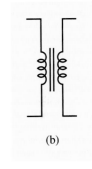

(b)

Figure 17-8.
A typical iron-core transformer and its schematic symbol.

CLASSIFICATION BY WINDING CONNECTION

Figure 17-11 shows a pictorial sketch and the schematic diagram for a transformer with multiple secondaries. The changing (i.e., sinusoidal) primary current causes corresponding changes in the magnetic flux. The high-permeability core material causes most of this changing flux to pass through the two secondary windings. Each of the secondary windings will have an induced voltage that can be used by subsequent circuits or devices. As you will see later in this chapter, the two (or more) secondary windings are essentially independent of each other.

(a)

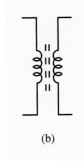

(b)

Figure 17-10.
A representative ferrite-core transformer and its schematic symbol.

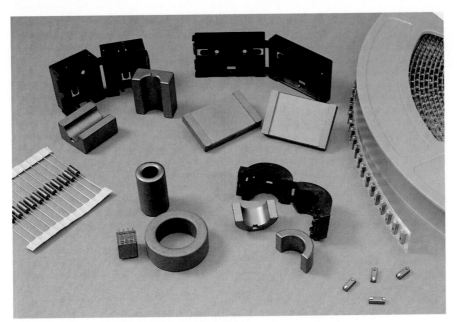

Figure 17-9. Ferrite cores are available in a variety of shapes and sizes. (Courtesy of Fair-Rite Products Corporation)

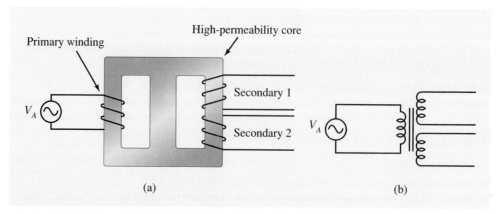

Figure 17-11. A transformer can have multiple secondary windings. Each secondary acts like a separate transformer.

Figure 17-12 shows another way to alter the connection of transformer windings. In the case shown in Figure 17-12(a), the secondary winding is tapped in the center. During manufacture, we can position the tap at any convenient point along the length of the winding. Although the winding is continuous, the tap allows us to access two different voltages (V_1 and V_2). As you would expect, the sum of V_1 and V_2 is equal to the voltage of the entire secondary. Figure 17-12(b) shows a secondary with multiple taps. We can obtain a useful voltage between any two of the connection points. Although not specifically pictured in Figure 17-12, we can also provide taps on the primary of the transformer. This option is often used when constructing transformers that can be operated from two different voltages (e.g., 120 Vac and 240 Vac). The voltage selection can be done by soldering to a different tap on the primary or, more commonly, by using a switch to change the primary connection.

Figure 17-13 shows yet another way that the windings of a transformer can be constructed. Here a single, tapped winding serves as both the primary and the secondary of the transformer. One end of the single coil serves as a common line for both primary and secondary windings. This configuration is called an *autotransformer.*

If the input voltage is applied between the two ends of the transformer winding, as illustrated in Figure 17-13(a), then the output voltage will be lower than the input voltage. By contrast, a step-up autotransformer can be constructed as shown in Figure 17-13(b) by connecting the primary voltage between the common and the tap. In either case, the amount of increase or

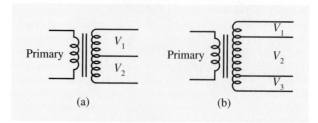

Figure 17-12. Tapped transformers provide multiple access points to the transformer winding.

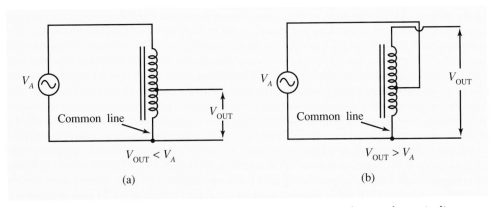

Figure 17-13. An autotransformer has a common primary and secondary winding.

decrease in primary voltage is determined by the relative location of the tap. Some manufacturers make autotransformers with the tap connected to a movable slider. The slider is connected to an external knob that can be used to adjust the position of the tap and, therefore, the amount of step-up or step-down provided by the transformer. Such a device is shown in Figure 17-14. Typically, autotransformers such as the one shown in Figure 17-14 can be used to adjust the 120-Vac power line voltage from 0 to about 150 V.

Autotransformers are also frequently used in high-voltage power supplies. A common example is the power supply in your television set, which generates voltages as high as 27,000 V.

Transformer Losses

A practical transformer has several nonideal losses that result in power dissipation. Some of the losses occur in the transformer windings, while others occur in the core of the transformer. Let's examine each of them individually.

Figure 17-14. A typical autotransformer that can be used to adjust the voltage provided by the power line. (Courtesy of Technipower)

LOSS DUE TO WIRE RESISTANCE

The copper wire used for the primary and secondary windings has a definite amount of ohmic resistance. The resistance of a wire is directly proportional to its length and inversely proportional to its cross-sectional area. Loss due to the resistance of the wire is called copper loss or I^2R loss. This is identical to the copper loss discussed with reference to inductors.

LOSS DUE TO SKIN EFFECT

When alternating current flows through a wire, each moving electron has an associated changing magnetic field. The fields from the various electrons interact with neighboring electrons and affect their movement. The result is that the electrons tend to flow near the surface of the wire. This phenomenon is called the skin effect and was introduced in Chapter 12. Skin-effect losses increase with frequency.

HYSTERESIS LOSS

You will recall from our study of basic magnetic theory that a magnetic material has domains that align with external magnetic fields. In the case of a magnetic-core transformer, the domains in the core must realign themselves every half-cycle. This requires energy, and the energy must come from the ac source. As the frequency is increased, the hysteresis loss also increases, since the domain-switching energy is being consumed more often.

EDDY-CURRENT LOSS

When a changing magnetic field intersects a conductor, it induces a voltage in the conductor. The induced voltage can cause current to flow. We know from our basic power formula ($P = VI$) that there must be an associated power dissipation.

When the changing magnetic fields of a transformer cut through its magnetic core, a voltage is induced in the core if the core is conductive. Many materials that are highly magnetic are also conductive. Since the core is relatively large, different potentials are induced into different regions of the core at any instant in time. This causes current to flow from one point in the core to another. However, since the flux pattern continually changes, so do the specific current paths within the core. Currents induced into the core of a transformer are called eddy currents. They are circulating currents whose paths are dynamic. They do draw energy from the power source and, therefore, represent a power loss. Eddy currents increase with higher frequencies, since the voltage induced into the core material increases.

We can reduce the power dissipation due to eddy currents by using a high-resistance core. This is one advantage of a ferrite-core or an air-core transformer over an iron-core transformer. An air-core transformer has no eddy-current loss. A ferrite-core transformer has a high ohmic resistance, which limits the magnitude of the eddy currents. Iron-core transformers use laminated cores to increase the ohmic resistance while maintaining a low permeability. The core consists of thin sheets of iron (actually silicon steel) separated by an insulative coating. This allows the core to provide the high permeability needed to obtain high values of inductance, but the insulative barriers keep the eddy-current losses to a minimum.

Faraday Shielding

Another nonideal characteristic of transformers is primary-to-secondary capacitance. Since the primary and secondary windings are conductors that are electrically separated, a capacitance is formed between them. Even in the case of an iron-core transformer, there is a capacitive path that extends from the primary winding to the core and from the core to the secondary winding. Although the value of primary-to-secondary capacitance may be only a few picofarads, it can cause problems at high frequencies, where the capacitive reactance becomes relatively low.

We can minimize the primary-to-secondary capacitance by separating the primary and secondary windings with a grounded, conductive barrier. This barrier is called a **Faraday shield** and is illustrated in Figure 17-15. Any energy that passes through the winding-to-shield capacitance is passed directly to ground and does not couple to the other winding.

Faraday shields are commonly used in high-frequency transformers to prevent the unwanted coupling of signals via the primary-to-secondary capacitance. Similar shields are sometimes used in communications line transformers to protect the electronic circuits from damage due to lightning, motor transients, and so on.

KEY POINTS

Primary-to-secondary winding capacitance is another nonideal characteristic. It can be effectively eliminated by including a Faraday shield between the primary and secondary windings.

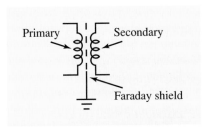

Figure 17-15. A Faraday shield reduces the primary-to-secondary capacitance.

Exercise Problems 17.2

1. Power is applied to the _____ winding of the transformer.
2. Power is supplied to external devices or circuits by the _____ winding of a transformer.
3. The relative phase relationship between primary and secondary voltages is indicated on a schematic diagram with _____ _____.
4. List three ways that transformers can be classified.
5. What is the name given to the magnetic flux that does not link the primary and secondary windings in a transformer?
6. List three types of material used for the cores in transformers.
7. What type of core material is a ceramic product?

8. If 24 Vac and 200 Vac were both required by a particular application, a transformer with multiple secondaries might be used. (True or False)

9. What is the name of the transformer that has a single winding that is common to both primary and secondary?

10. Explain the terms step-up and step-down as applied to transformers.

11. An autotransformer can be used to step-up or step-down the voltage applied to the primary. (True or False)

17.3 Analysis of Transformer Circuits

We shall limit our discussion to the analysis of iron-core transformer circuits. We will begin our analysis by assuming an ideal transformer that has unity coupling. This not only simplifies calculations, but it provides adequate accuracy for many applications encountered by a technician.

Turns Ratio

KEY POINTS

A transformer's turns ratio is an important link to all other transformer ratios. It is defined as the ratio of primary turns to secondary turns.

A given transformer has a certain number of turns in the primary and a certain number of turns in the secondary. The ratio of primary turns to secondary turns is called the **turns ratio** of the transformer. This ratio is the heart of many transformer calculations. We can express it formally with Equation 17-4.

$$k = \frac{N_P}{N_S} \qquad (17\text{-}4)$$

It should be noted that some textbooks define turns ratio as N_S/N_P. This is acceptable, provided the reciprocal of all other transformer ratios are used.

EXAMPLE SOLUTION

What is the turns ratio for a transformer that has 400 turns in the primary winding and 100 turns in the secondary winding?

EXAMPLE SOLUTION

The given transformer is illustrated in Figure 17-16. We compute the turns ratio by dividing the primary turns by the secondary turns (Equation 17-4). Rather than express the turns ratio as a single number (4 in this case), it is customary to express it as a reduced fraction (4/1 or simply 4:1 in this case).

EXAMPLE SOLUTION

What is the turns ratio of a transformer that has 225 turns on the primary and 300 turns on the secondary?

EXAMPLE SOLUTION

$$k = \frac{N_P}{N_S} = \frac{225 \text{ T}}{300 \text{ T}} = \frac{3}{4}, \text{ or } 3{:}4$$

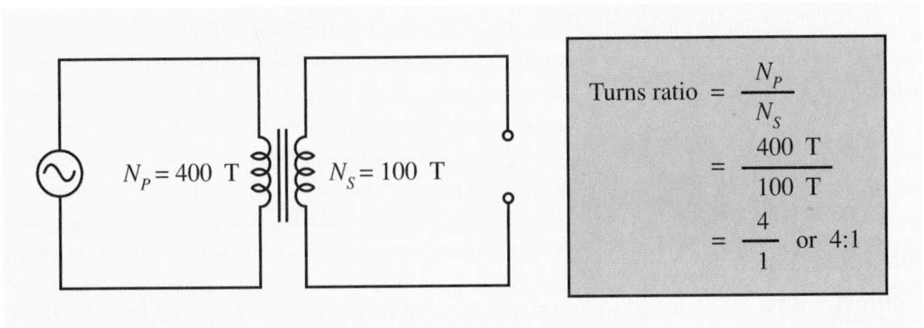

Figure 17-16. A transformer turns ratio example.

Practice Problems

1. If a transformer has 50 turns on the primary and 250 turns on the secondary, what is its turns ratio?
2. What is the turns ratio of a transformer with 800 turns on its primary and 80 turns on its secondary?
3. If a transformer with a 5:2 turns ratio has 100 turns on its primary, how many turns are on the secondary?

Answers to Practice Problems

1. 1:5   2. 10:1   3. 40

Voltage Ratio

For purposes of this discussion, we are assuming unity coupling. This means that exactly the same flux cuts both primary and secondary windings. Whatever voltage is induced into a single turn of the primary will also be induced into a single turn in the secondary. That is, the volts per turn are the same for both primary and secondary windings. This should seem reasonable, since the coils are wound on a common core and are cut by the same flux lines.

Now, we know from inspection (or Kirchhoff's Voltage Law) that the sum of the voltages induced into the various turns in the primary winding must equal the applied voltage. The voltage in the secondary, however, depends on the number of turns in the secondary (remember the volts per turn are the same as the primary). Let's examine this mathematically. First, we set the volts per turn ratios of the primary and secondary windings equal to each other, and then transpose the equation as follows:

$$\frac{V_P}{N_P} = \frac{V_S}{N_S}$$

$$V_P N_S = V_S N_P$$

$$\frac{N_P}{N_S} = \frac{V_P}{V_S}$$

The result shows that the ratio between primary and secondary voltages is equal to the turns ratio of the transformer. This is an important relationship and is formally stated as Equation 17-5.

$$\frac{N_P}{N_S} = \frac{V_P}{V_S} \qquad (17\text{-}5)$$

EXAMPLE SOLUTION

A transformer with a 4:3 turns ratio is connected to a 120-Vac source. Calculate the secondary voltage.

EXAMPLE **SOLUTION**

We apply Equation 17-5 as follows.

$$\frac{N_P}{N_S} = \frac{V_P}{V_S} \text{ or}$$

$$N_P V_S = N_S V_P \text{ or}$$

$$V_S = \frac{N_S V_P}{N_P}$$

$$V_S = \frac{3 \text{ T} \times 120 \text{ V}}{4 \text{ T}} = 90 \text{ V}$$

Practice Problems

1. A transformer has a 5:1 turns ratio and delivers a secondary voltage of 10 V. What is the value of the primary source voltage?
2. A transformer has a primary voltage of 120 V and a secondary voltage of 12 V. What is the turns ratio of the transformer?
3. If a transformer has 80 turns in its primary winding and provides 20 V in the secondary when its primary is connected to 5 V, how many turns are on the secondary winding?

Answers to Practice Problems

1. 50 V   2. 10:1   3. 320

Power Ratio

The power ratio for an ideal transformer is unity. That is, with 100% coupling and no transformer losses, the output power must be the same as the input power. In particular, power in the primary (i.e., power taken from the source) must always be equal to the power in the secondary (i.e., power delivered to devices or circuits connected to the secondary winding).

Practical transformers have losses and less than 100% coupling. Nevertheless, the ideal assumptions will satisfy our immediate needs.

Current Ratio

We know that power is the product of current and voltage (i.e., $P = VI$). We also know that the primary power must equal the secondary power. It follows, therefore, that if the secondary voltage is higher than the primary voltage, then the current in the secondary must be correspondingly smaller, so that the power will remain the same. This means that if the voltage is stepped up in a transformer, then current will always be stepped down.

Let's examine this relationship mathematically. We begin with the knowledge that primary and secondary powers are equal.

$$P_P = P_S; \text{ therefore}$$

$$V_P I_P = V_S I_S, \text{ or}$$

$$\frac{V_P}{V_S} = \frac{I_S}{I_P}$$

This shows us that the voltage ratio is equal to the inverse of the current ratio. We know the voltage ratio is equal to the turns ratio (Equation 17-5). It is common to express the current ratio in terms of the turns ratio as shown in Equation 17-6.

$$\frac{N_P}{N_S} = \frac{I_S}{I_P} \tag{17-6}$$

EXAMPLE SOLUTION

If the turns ratio of a transformer is 5:1 and the primary current is 100 mA, what is the value of secondary current?

EXAMPLE **SOLUTION**

We apply Equation 17-6 as follows:

$$\frac{N_P}{N_S} = \frac{I_S}{I_P}, \text{ or}$$

$$I_S = \frac{N_P I_P}{N_S}$$

$$I_S = \frac{5 \text{ T} \times 100 \text{ mA}}{1 \text{ T}} = 500 \text{ mA}$$

EXAMPLE SOLUTION

A transformer is connected to a 120-Vac source and supplies 12 Vac to a load. The load draws 1.5 A. What is the value of primary current?

EXAMPLE **SOLUTION**

We know that the voltage and turns ratios are equal (Equation 17-5). We use this relationship along with Equation 17-6 to solve for primary current.

$$\frac{V_P}{V_S} = \frac{N_P}{N_S} = \frac{I_S}{I_P}, \text{ so}$$

> **KEY POINTS**
>
> The current ratio in a transformer is the inverse of the turns ratio or voltage ratio. This implies that a transformer that steps up the voltage between primary and secondary will also step down the current.

$$I_P = \frac{V_S I_S}{V_P}$$

$$I_P = \frac{12 \text{ V} \times 1.5 \text{ A}}{120 \text{ V}} = 150 \text{ mA}$$

Practice Problems

1. Calculate the secondary current in a transformer with a 4:1 turns ratio and a primary current of 2.75 A.
2. What is the turns ratio of a transformer that has a secondary current of 200 mA, with a primary current of 50 mA?
3. If a transformer has a primary voltage of 220 V, a secondary voltage of 25 V, and a primary current of 100 mA, what is the value of secondary current?

Answers to Practice Problems

  1. 11 A   **2.** 4:1   **3.** 880 mA

Impedance Ratio

KEY POINTS

Any impedance connected to the secondary of a transformer will be reflected through the transformer and appear as an impedance to the source.

We know that a transformer can alter both voltage and current levels between the primary and secondary. It should seem reasonable, then, that a transformer can also alter the impedance between primary and secondary.

REFLECTED IMPEDANCE

Consider the circuit shown in Figure 17-17. In Figure 17-17(a), the secondary is connected to a high-resistance load. The flux produced by the secondary tends to cancel (i.e., it is moving in the opposite relative direction) some of the induced voltage in the primary. This allows more current to flow in the primary.

Now, if the resistance in the secondary is made smaller as in Figure 17-17(b), then more secondary current flows. This causes more secondary flux to induce an opposite polarity voltage in the primary, which again allows even more primary current to flow. Finally, Figure 17-17(c) shows that if the secondary load resistance is made even smaller, more secondary current and, therefore, more primary current will flow.

Since changes in the resistance or, more universally the impedance of the secondary, cause current changes in the primary (with the source voltage held constant), it follows that the impedance changes in the secondary are transferred to the primary. The impedance in the primary that results from the secondary load is called the **reflected impedance.** Changes in secondary impedance are reflected to the primary.

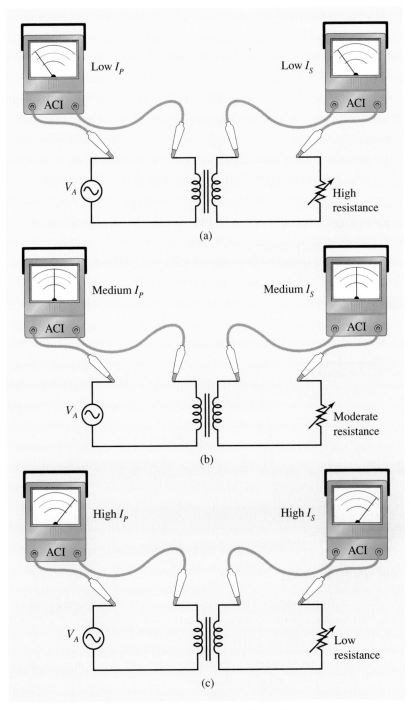

Figure 17-17. The impedance in the secondary of a transformer is reflected into the primary.

Impedance Calculations

Now, let us consider the magnitude of the reflected impedance. We shall make use of Ohm's Law, Equations 17-5 and 17-6, and relevant substitutions.

$$\frac{Z_P}{Z_S} = \frac{\dfrac{V_P}{I_P}}{\dfrac{V_S}{I_S}} = \frac{V_P I_S}{V_S I_P} \text{, or}$$

$$\frac{Z_P}{Z_S} = \left(\frac{N_P}{N_S}\right)\left(\frac{N_P}{N_S}\right) = \left(\frac{N_P}{N_S}\right)^2$$

This tells us that the impedance ratio (Z_P over Z_S) is equal to the square of the turns ratio. To be more consistent with the form of Equations 17-5 and 17-6, let's express this relationship as Equation 17-7.

$$\frac{N_P}{N_S} = \sqrt{\frac{Z_P}{Z_S}} \qquad\qquad (17\text{-}7)$$

EXAMPLE SOLUTION

The secondary of a transformer with a 5:1 turns ratio is connected to a 100-Ω resistor. What is the value of impedance reflected into the primary?

EXAMPLE **SOLUTION**

We apply Equation 17-7 as follows.

$$\frac{N_P}{N_S} = \sqrt{\frac{Z_P}{Z_S}} \text{, or}$$

$$\left(\frac{N_P}{N_S}\right)^2 = \frac{Z_P}{Z_S} \text{, so}$$

$$Z_P = Z_S\left(\frac{N_P}{N_S}\right)^2$$

$$Z_P = 100\ \Omega \times \left(\frac{5\ \text{T}}{1\ \text{T}}\right)^2$$

$$Z_P = 100\ \Omega \times 25 = 2.5\ \text{k}\Omega$$

This means that the source sees the 100-Ω resistance reflected back as a 2.5-kΩ resistance. That is, the current drawn from the primary source will be the same as if a 2.5-kΩ resistor were connected directly across the source.

The ability of a transformer to change impedances is an important property and is the sole purpose of the transformer in many applications.

Practice Problems

1. A transformer with a 2:1 turns ratio has a 1,000-Ω resistor connected across the secondary. What impedance is reflected into the primary?

2. A transformer with 120 V across the primary produces 24 V in the secondary. If the secondary load is 10 Ω, what is the reflected impedance in the primary?

3. If a transformer has 200 turns in the primary, 300 turns in the secondary, and has a 50-Ω resistor connected across the secondary, what is the impedance reflected into the primary?

4. What is the impedance ratio of a transformer that has 100 turns in the primary and 250 turns in the secondary?

Answers to Practice Problems

1. 4,000 Ω   2. 250   3. 22.22   4. 4:25

Transformer Efficiency

The transformers referenced in the preceding calculations had an assumed efficiency of 100%. This means that no power was lost in the transformer. Practical transformers do have winding and core losses that dissipate power in the form of heat. The efficiency (η) of a transformer is computed the same as for any other system (output power/input power). In the case of a transformer it is expressed as

$$\eta = \frac{P_{OUT}}{P_{IN}} = \frac{P_S}{P_P}$$

where P_S and P_P are the power dissipations in the secondary and primary, respectively. Primary power dissipation consists of three parts: power delivered to the secondary (P_S), power lost in the windings (copper and skin-effect losses), and power lost in the core (hysteresis and eddy currents). The expression for transformer efficiency is stated more completely by Equation 17-8.

$$\eta = \frac{P_S}{P_S + P_{WINDING} + P_{CORE}} \qquad (17\text{-}8)$$

EXAMPLE SOLUTION

The secondary of a transformer provides 25 W to a resistive load. The combined core losses are 0.8 W. The copper and skin-effect losses total 1.5 W. What is the efficiency of the transformer?

EXAMPLE **SOLUTION**

We apply Equation 17-8.

$$\eta = \frac{P_S}{P_S + P_{WINDING} + P_{CORE}}$$

$$= \frac{25\ W}{25\ W + 1.5\ W + 0.8\ W} = 91.6\ \%$$

Practice Problems

1. The secondary power of a transformer is 150 W. The core and winding losses are 3.75 W and 6.5 W, respectively. What is the efficiency of the transformer?

2. A transformer draws 200 mA from a 120-V source. The secondary maintains 20 V across a 17.5-Ω resistor. What is the efficiency of the transformer?

3. If the primary winding of a transformer with 92% efficiency is connected to a 120-V source and the secondary supplies 12 V with a current of 2.5 A, what is the value of primary current?

Answers to Practice Problems

1. 93.6%   **2.** 95.2%   **3.** 271.7 mA

Practical Transformer Considerations

The calculations in the preceding section focused on the analysis of ideal iron-core transformers with limited discussion of nonideal characteristics (e.g., efficiency). We are now in a position to examine a practical iron-core transformer and all of its nonideal characteristics.

To avoid getting lost in details, it is important for you to realize that for a great many applications likely to be encountered by a technician, there are only minor differences between an ideal iron-core transformer and a practical one. The following discussion, however, will provide you with additional insights regarding transformer operation.

PRACTICAL TRANSFORMER EQUIVALENT

When analyzing a new or complex device, it is often helpful to represent the characteristics of the device in a more obvious way. In the case of the iron-core transformer, we can represent the effect of its nonideal characteristics with a simple equivalent circuit. Figure 17-18 shows the equivalent circuit that we will be using. This equivalent circuit omits the parasitic capacitances between the windings, between the windings and core, and between the turns on a winding. These capacitances have only minimal effect except at high frequencies. We will discuss each component in the equivalent circuit in the following paragraphs.

The model consists of an ideal transformer (T_1) surrounded by series and parallel resistances and inductances that represent nonideal quantities. Resistances R_{W_P} and R_{W_S} represent the winding losses (copper and skin-effect) for the primary and secondary, respectively. Similarly, R_{CORE} represents the core losses (hysteresis and eddy currents). All

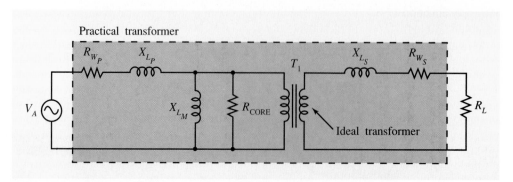

Figure 17-18. An equivalent circuit for a practical iron-core transformer.

of these resistances take power from the source that is never delivered to the load (R_L). They all reduce the efficiency to less than 100%.

Reactances X_{L_P} and X_{L_S} are equivalent reactances that represent the effects of leakage flux in the primary and secondary, respectively. The series reactances have the same effect on the overall circuit as flux that expands into the surrounding space without linking the primary and secondary. Finally, reactance X_{L_M} represents the reactance associated with the mutual inductance. The source "sees" this reactance as a parallel path.

UNLOADED SECONDARY

When the secondary of the transformer has no load (i.e., it is open-circuited), then none of the secondary components in Figure 17-18 are seen by the source. The primary current is largely limited by the inductive reactance of the primary winding. Only enough current flows to produce a self-induced voltage across the primary that is equal to the applied voltage. This small current is called the **magnetizing current.** Because the magnetizing current is small, the winding losses are also small ($P = I^2 R_{W_P}$ in Figure 17-18). The primary current then is composed of two components: magnetizing current (inductive) and current due to core losses (resistive current through R_{CORE} in Figure 17-18). Under no-load conditions, the core losses are low and the primary current is essentially inductive.

We can also express the transformer relationships with the aid of a phasor diagram. Figure 17-19 shows a phasor representation of the primary circuit under no-load conditions. I_M is the magnetizing current which, under no-load conditions, makes up the entire primary current. Note that θ is essentially 90° and the circuit appears to the source as a rather large inductor (i.e., voltage leads current by 90°).

FULLY LOADED SECONDARY

Now let's see what happens when a heavy load (i.e., small load resistance) is connected across the secondary. As you would expect, secondary current begins to flow. The increased secondary current creates corresponding magnetic flux, which cuts the primary winding. The voltage induced in the primary as a result of the secondary current is of a polarity to cancel a portion of the primary's self-induced voltage. Recall that it was the self-induced voltage in the primary that was limiting primary current. So, with some of the self-induced voltage cancelled, the primary current increases until the self-induced voltage is restored. Figure 17-20 shows a phasor representation of the primary circuit under full-load conditions.

In Figure 17-20, I'_P is the portion of the primary current that results from the flow of secondary current. Note that it is in phase with V_A, since the reflected load is resistive. I_M in Figure 17-20 is the magnetizing current. Finally, I_P is the phasor sum of the two primary current components. The resulting phase angle (measured between total voltage and total current) is a very small angle. If the load on a practical iron-core transformer is resistive and heavy, the primary current and primary voltage can be within a few degrees of each other. In summary, the phase relationship between primary current and voltage can range from near 90° to near 0° as the load varies from no load (infinite load resistance) to a heavy load (small load resistance). Reactive loads on the secondary reflect into the primary as reactive loads and cause corresponding phase shifts.

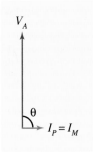

Figure 17-19.
Under no-load conditions, the primary current is inductive and limited by the reactance of the primary.

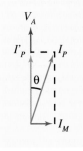

Figure 17-20.
Under full-load conditions (resistive load), the primary current is nearly resistive.

The reactance X_{L_s} and resistance R_{W_s} in Figure 17-18 introduce another important, but nonideal, behavior. We can see by inspection of Figure 17-18 that an increased secondary current will cause an increased voltage drop across the series elements X_{L_s} and R_{W_s}. This causes the output voltage (voltage across R_L) to decrease as secondary current is increased. The decrease in voltage depends on the design of the transformer and the extent of the loading, but it can be substantial. Transformers are generally designed such that their rated voltage is available when the secondary is supplying its rated current. At lesser currents, the secondary voltage may be higher.

Exercise Problems 17.3

1. What is the turns ratio of a transformer that has 665 turns in the primary and 95 turns in the secondary?

2. What is the turns ratio of a transformer with 500 turns in the primary and 625 turns in the secondary?

3. If a transformer with 200 turns in its primary has a 1:5 turns ratio, how many turns are in the secondary?

4. The primary of a transformer with a 5:2 turns ratio is connected across a 120-V source. What is the value of secondary voltage?

5. A transformer has a 4:1 turns ratio and delivers a secondary voltage of 25 V. What is the primary voltage?

6. If a transformer has 120 turns in its primary winding and provides 12 V in the secondary when its primary is connected to 10 V, how many turns are on the secondary winding?

7. If the turns ratio of a transformer is 3:1 and the primary current is 600 mA, what is the value of secondary current?

8. A transformer is connected to a 120-Vac source and supplies 50 V to a secondary load. The load draws 275 mA. What is the value of primary current?

9. Calculate the secondary current in a transformer that has a 5:2 turns ratio and a primary current of 150 mA.

10. What is the turns ratio of a transformer that has a secondary current of 2 A and a primary current of 25 mA?

11. The secondary of a transformer with a 2:1 turns ratio is connected to a 51-Ω resistor. What is the value of impedance (resistance in this case) reflected into the primary?

12. A transformer with a 1:4 turns ratio has a 1,500-Ω resistor connected across the secondary. What impedance is reflected into the primary?

13. A transformer has 200 turns in the primary and 50 turns in the secondary. The secondary load is reflected into the primary as a 1.2-kΩ resistance. What is the actual value of secondary load resistance?

14. The secondary power of a transformer is 100 W. The core and winding losses are 1.2 W and 4.5 W, respectively. What is the efficiency of the transformer?

15. A transformer draws 25 mA from a 120-V source. The secondary maintains 40 V across a 560-Ω resistor. What is the efficiency of the transformer?

16. With no load on the secondary of a transformer, the primary current is very low. (True or False)

17. If the secondary of a transformer is heavily loaded with a resistive load, the primary current will be nearly in phase with the primary voltage. (True or False)

18. A technician measures the secondary voltage of an unloaded transformer as 8.7 V. When the transformer is loaded, the technician measures the secondary voltage as 6.3 V. Explain the reason for these different measurements.

17.4 Applied Technology: Transformers

If one tries to think of all the electronic products in the world, it is far easier to name products that include a transformer than products without a transformer. We will briefly examine five specific applications that rely on the use of transformers.

Commercial Power Distribution

Power companies use many different kinds of transformers to effectively transfer electrical power from the generating plant to your home. Figure 17-21 illustrates a representative power distribution system. The exact voltage levels and sequence of voltage transformations vary greatly in different installations.

In the example shown in Figure 17-21, the power company alternators generate the initial power at 22 to 26 kV. The alternators actually generate three-phase power. That is, they generate three 60-Hz sine waves that are exactly 120° out of phase with each other. Many factories, farms, and businesses that have large motors or other high-power electrical equipment use all three phases. As indicated by Figure 17-21, only one of the three phases is generally brought to your door.

Immediately outside the generating plant, the voltage goes to a transformer which steps up the voltage to as high as 765 kV. Recall that when a transformer steps up the voltage, the current is reduced by a similar amount. The power companies use such high voltages to minimize the copper loss ($P = I^2 R$) and voltage drop ($V = IR$) in the hundreds or thousands of miles of wire needed to reach your home. The wires for these extremely high voltages are supported by huge steel towers. Generally, you can see three large-diameter (≈2 inches) wires and one smaller one. The three larger wires are the three phases and the smaller one is the neutral or ground wire. Less current flows in the neutral, so it can be smaller.

Some very large factories have step-down transformers that connect directly to the high-voltage transmission lines. Somewhere near your community, the high transmission voltage goes to another transformer, which steps down the voltage to the range of 12.5 to 138 kV. These are the transformers that are often referred to as substations. They are generally surrounded by high steel fences. The voltage is now routed on high wooden poles to smaller substations in your neighborhood. The voltage is still transferred as three-phase power. Some industrial plants have step-down transformers connected directly to the 12.5 to 138 kV lines.

KEY POINTS

The distribution of power throughout the country relies heavily on the use of transformers.

KEY POINTS

Transformers are used to step up the generated voltage for reduced loss during transmission.

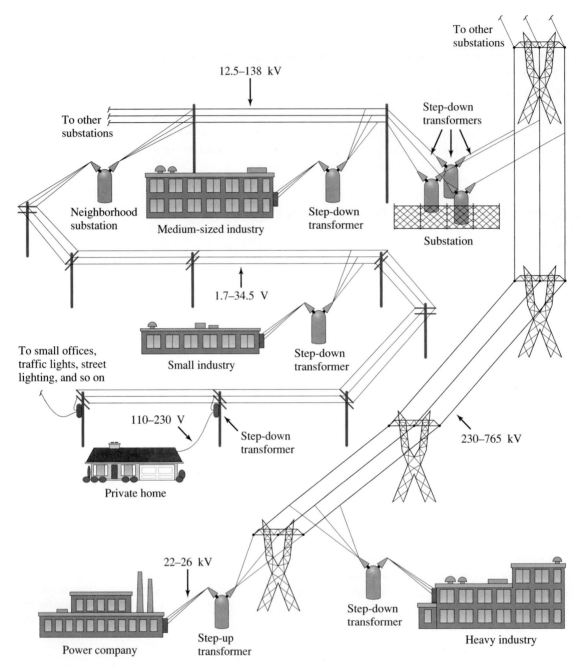

Figure 17-21. A representative power distribution scheme.

In the local substations, the voltage is reduced to a range of 1.7 to 34.5 kV by passing through another step-down transformer. All three phases of this lower-level voltage are connected to the step-down transformers of some smaller factories. Individual phases are used for streetlights and local businesses, each with its own step-down transformer. One of the phases of this power is routed to the power pole outside your home. Here, it passes through a step-down transformer with a center-tapped secondary. All three of the secondary wires are brought into the main electrical panel in your home. The transformer delivers a full secondary voltage of 220 to 240 V. Certain appliances like air conditioners and electric clothes dryers require 220 to 240 V and connect directly across the secondary of the transformer (via fuses or breakers). The lights and other smaller appliances in your home require 110 to 120 V. Some of them are connected between one side of the transformer secondary and the center tap. The rest are connected between the other side of the secondary and the center tap. The center tap of the secondary is physically connected to earth ground. Additional detail is provided later in this section.

Three-Phase Transformers

As previously mentioned, commercial power is generated as three-phase power. That is, three separate sets of coils are mounted on the alternator shaft. The coils are physically positioned 120° apart. Therefore, the sine waves of voltage that are produced in the coils are 120° out of phase with each other.

The primary and secondary windings of three-phase transformers can be connected in either of two ways: wye (tee) and delta (pi). In each of these cases, there are two measurements that are commonly used: leg and line. Figure 17-22 shows the two basic connections and definitions of leg and line voltage and current. Leg voltage and leg current are associated with a particular winding. Line voltage is measured between any two lines, and line current is the current flowing in a given line.

In a wye connection, we can see that line current and leg current are equal, since they are essentially the same current (see Figure 17-22a). This is expressed by Equation 17-9.

$$I_{line} = I_{leg} \text{ (wye connection)} \qquad \textbf{(17-9)}$$

Line voltage in a wye connection is actually the sum (according to Kirchhoff's Law) of two windings. However, since the voltages are 120° out of phase, the summation must account for the phase difference. Equation 17-10 describes this relationship:

$$V_{line} = \sqrt{3}V_{leg} \text{ (wye connection), or}$$
$$V_{line} = 1.732V_{leg} \text{ (wye connection)} \qquad \textbf{(17-10)}$$

By inspection of Figure 17-22(b), we can see that line current is actually the summation of two leg currents (Kirchhoff's Current Law). Again, we must remember that these currents are 120° out of phase. Equation 17-11 expresses the correct relationship.

$$I_{line} = \sqrt{3}I_{leg} \text{ (delta connection), or}$$
$$I_{line} = 1.732I_{leg} \text{ (delta connection)} \qquad \textbf{(17-11)}$$

Finally, we can see from Figure 17-22(b) that leg and line voltages in delta connections are essentially the same. We can express this relationship with Equation 17-12.

$$V_{line} = V_{leg} \text{ (delta connection)} \qquad \textbf{(17-12)}$$

Since both primary and secondary windings of a three-phase transformer can be either wye or delta connected, this leads to four possible combinations of primary/secondary configurations. Figure 17-23 illustrates all four connections.

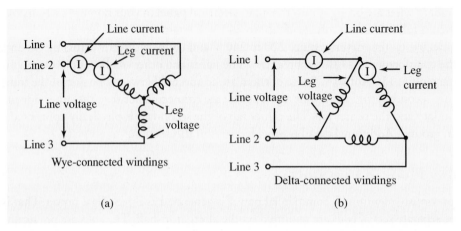

Figure 17-22. Three-phase transformer windings can be either (a) wye or (b) delta connected. In either case, both leg and line voltages and currents are useful measurements.

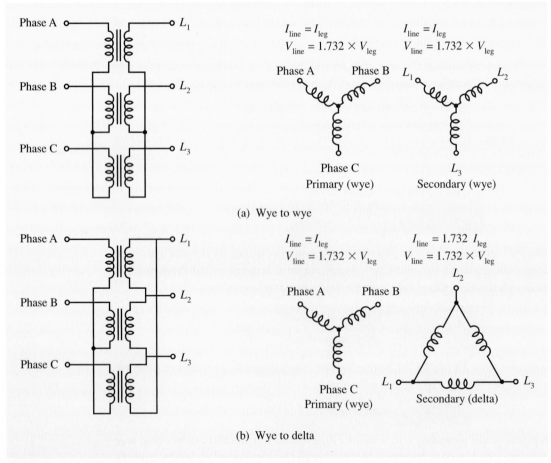

Figure 17-23. There are four combinations of wye and delta primary/secondary configurations.

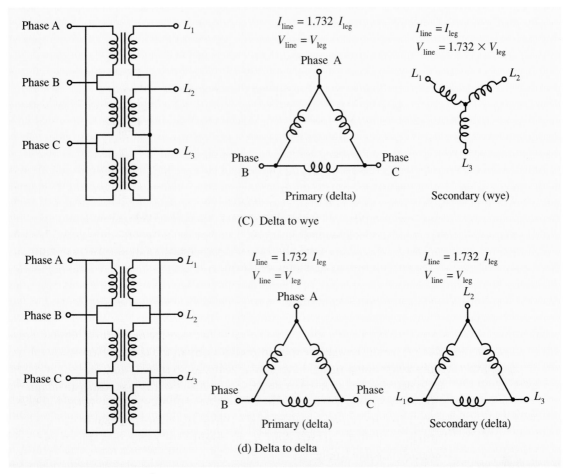

$I_{line} = 1.732\ I_{leg}$
$V_{line} = V_{leg}$

$I_{line} = I_{leg}$
$V_{line} = 1.732 \times V_{leg}$

Primary (delta)

Secondary (wye)

(C) Delta to wye

$I_{line} = 1.732\ I_{leg}$
$V_{line} = V_{leg}$

$I_{line} = 1.732\ I_{leg}$
$V_{line} = V_{leg}$

Primary (delta)

Secondary (delta)

(d) Delta to delta

Figure 17-23. *(Continued)*

POWER DISTRIBUTION TERMINOLOGY

Figure 17-24 illustrates a portion of the power distribution system used to deliver 60-Hz power to your home. The transformer located on the pole near your house has its primary connected to one of the three phases provided by your power company. The secondary of the pole transformer is center tapped. The voltage across the entire secondary is nominally 240 Vac. From the center tap to either side is 120 Vac. Common electrician jargon labels the center tap as the **neutral.** The other two wires are often called the **hot** wires, since they are both "hot" (i.e., 120 Vac) with respect to the neutral.

As indicated in Figure 17-24, the neutral is connected to earth ground. For this reason, it is also referred to as ground by some electricians. By inspection of Figure 17-24, we can see that if we wanted to power a 120-Vac device (e.g., lamps, fans, electric blankets, and so on) in your home, we would connect it between the neutral and one of the hot wires. In practice, part of your 120-Vac house wiring is supplied from one hot line and the rest of your house is supplied by the other hot line. The electric clothes dryer or other appliance that requires 240 Vac would be connected between the two hot lines (the neutral also goes to the appliance). Electricians often say, "Pick up two hots to get 240 volts."

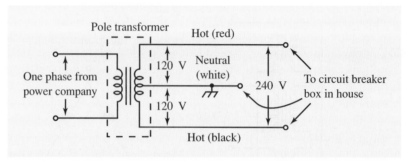

Figure 17-24. Power distribution terminology.

Electronic technicians should take note of the color codes indicated in Figure 17-24. For most house-wiring applications, the white wire is considered to be ground and the black wire is one of the hot wires. This can be confusing, since it is common practice to use black as the ground wire in electronic systems.

Isolation Transformers

Certain types of electronic equipment have no transformer in their internal power supply. One side of the power line is connected directly to the chassis. Now in use, this type of equipment is supposed to have a polarized plug, so there is only one way it can be plugged in. In practice, however, plugs do get inserted backwards (e.g., using an adapter to convert a three-prong plug to a two-prong plug). When this occurs, the chassis (and any exterior metal parts) becomes connected to the side of the power line that is 120 V with respect to earth ground. If you are standing on the ground and touch the cabinet, you will be shocked; possibly fatally.

Figure 17-25(a) illustrates the same basic problem, but in a situation likely to be experienced by a technician. Two electrical items (Equipment 1 and Equipment 2 in Figure 17-25) are being used or tested by a technician in a shop environment. One may be an oscilloscope and one may be a transformerless computer power supply, for example. In any case, there is a dangerous shock hazard for the technician. Additionally, if the technician attempts to connect the ground of Equipment 1 (e.g., the ground lead for an oscilloscope) to the chassis of Equipment 2 (e.g., the computer power supply), there will be fireworks. More specifically, the technician will essentially be placing a direct short across the 120-V power line. Trace the current path in Figure 17-25(a). At a minimum, the 120-V circuit breaker will blow, the equipment and test lead will have burn marks, and the technician will be worried about future employment opportunities.

Figure 17-25(b) shows how an isolation transformer can eliminate the problem. The figure shows a 1:1 transformer associated with each of the two equipment items. In practice, only one transformer is required and it can be on either power cord. The 1:1 isolation transformer has no effect on the voltage being supplied to the test equipment or the equipment being repaired, but it does remove or isolate the reference to earth ground. Now, either equipment can be plugged in either polarity without creating a difference in potential between the two cabinets. If it is your job to repair equipment that is internally referenced to earth ground, then be sure your shop is equipped with an isolation transformer.

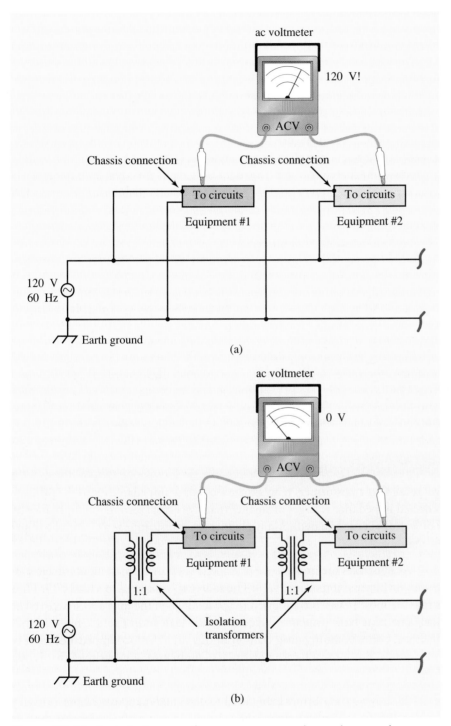

Figure 17-25. An isolation transformer can remove the earth-ground reference and eliminate a shock hazard for technicians.

dc Power Supplies

Most electronic products that operate from the standard power line have internal power supply circuits that convert the 120 Vac into lower dc voltages required by the electronic circuitry. Transformers are widely used to isolate the earth-ground reference (described previously) and, at the same time, transform the 120 Vac to a lower voltage. Figure 17-26 illustrates a typical power supply application.

KEY POINTS

Most equipment that operate from the ac power line utilize internal transformers to reduce the voltage from 120 V to lower voltages, which are then converted to dc as required by typical electronic circuits.

In the example illustrated in Figure 17-26, the transformer steps the 120-V power line voltage down to 12 Vac in the secondary. This lower ac voltage is then routed to the power supply circuits, which convert the ac into dc. The dc voltage is then used to power the electronic circuitry in the product (e.g., television, computer, satellite receiver, and so forth).

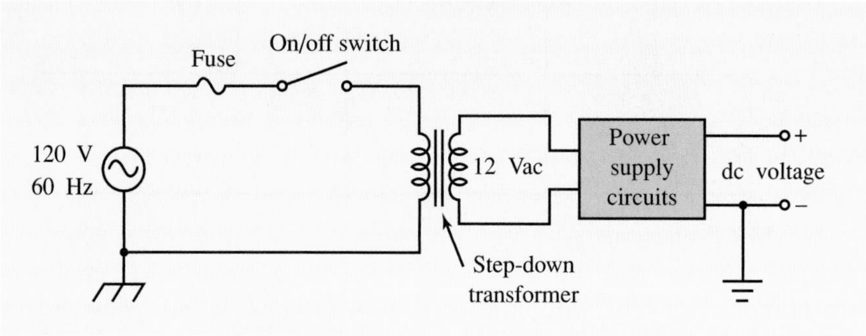

Figure 17-26. A transformer is used in dc power supplies to isolate earth ground and to transform the 120-V power to lower voltages.

Impedance Matching

KEY POINTS

Maximum power can be transferred from one circuit to another when the impedance of the source is matched to the impedance of the load.

You will recall that transformers reflect their secondary impedances into the primary. The reflected impedance is larger or smaller than the actual secondary impedance by a factor of k^2 (turns ratio squared). Many circuits use transformers to obtain the impedance transformation.

Figure 17-27 illustrates the importance of impedance transformation. A voltage source is shown with an internal impedance (R_S). The source is connected to a load (R_L). The graph shows that the most power is transferred to the load when the load and source resistances are equal. The same basic principle is true when the load and/or source has reactance associated with it. For maximum power transfer in this case, the load impedance must be the same magnitude, but it must have a polarity (i.e., inductive or capacitive) that is opposite the source.

KEY POINTS

Transformers are used to match impedances by selecting the appropriate turns ratio.

Figure 17-28 shows a transformer being used to drive an 8-Ω speaker from a signal generator that has a 50-Ω output resistance. The 5:2 step-down transformer makes the 8-Ω load appear to be a 50-Ω resistance when it is reflected into the primary. The source acts as if it were connected to a 50-Ω load, and the load acts as if it were connected to an 8-Ω source. Maximum power is transferred because the source and load resistances are equal. It should be noted that a practical speaker will appear somewhat inductive, but the concept of impedance matching is still the same.

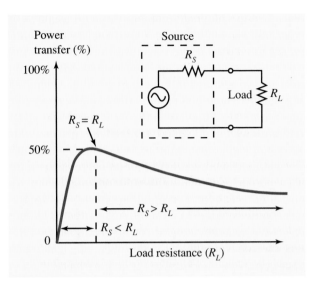

Figure 17-27. Maximum power transfer occurs when the source resistance is equal to the load resistance.

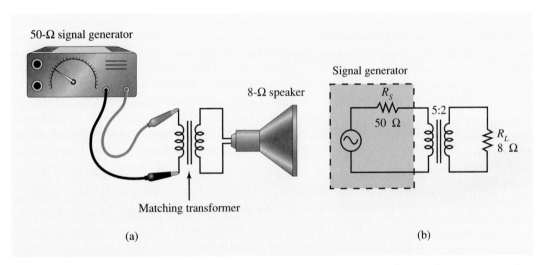

(a)   (b)

Figure 17-28. A transformer can match the 50-Ω output resistance of a signal generator to the 8-Ω resistance of a speaker.

Exercise Problems 17.4

1. Briefly explain why power companies use high voltages for transmission of power even though your home only requires 120 V.
2. Most power company substations utilize step-up transformers. (True or False)
3. What is the turns ratio of a typical isolation transformer?

4. Explain why a technician would probably not need to use an isolation transformer when working on equipment whose power supply utilizes an internal step-down transformer.

5. Determine the turns ratio needed for a transformer to match a preamplifier output of 250 Ω to an amplifier input of 25 kΩ.

17.5 Troubleshooting Transformers

In general, transformers can develop all of the same defects discussed with reference to inductors. Following is a list of common transformer defects:

- open primary
- open secondary(s)
- shorted turns or shorted winding
- winding shorted to the core
- short between primary and secondary

Most defective transformers can be detected through effective use of three things: technician observation, voltmeter checks, and ohmmeter checks.

Observation

When a transformer develops a short circuit, it generally results in increased current flow and increased internal power dissipation. If the primary of the transformer is fused, then a shorted winding will likely cause it to blow as soon as power is applied. If you disconnect one side of the secondary (one side of each secondary in the case of multiple windings), and the fuse still blows, then you can be sure the transformer is shorted. Always use your powers of observation before you begin an extensive troubleshooting process.

Iron-core transformers typically use varnish or shellac as insulation on the wires and between adjacent laminations in the core. If some of the turns in one winding of the transformer become shorted (e.g., the insulation breaks down), then the excessive heat will cause the shellac or varnish to emit an odor and possibly visible smoke. A similar symptom occurs if the transformer is not properly fused and is subjected to an excessive load in its secondary (e.g., a shorted component). In any case, the smell of an overheated transformer is very distinct. Once a technician has smelled the odor, it will be remembered forever. Be alert when troubleshooting. By detecting this aroma, you may be able to quickly locate a defective transformer and save a lot of troubleshooting time.

There may also be visible evidence indicating that a transformer has been overheated. The visible paper insulation may be discolored or the visible varnish may have a bubbled appearance. Use your powers of observation to help you quickly locate transformer defects.

Voltmeter Tests

If a winding on a transformer is open (either primary or secondary), there will be no voltage developed across the secondary winding. In the case of multiple secondaries, an open primary results in no voltage in any secondaries. An open secondary, on the other hand, results in no voltage across the defective secondary, but relatively normal voltages across all other secondaries.

Ohmmeter Tests

An ohmmeter can be used to absolutely confirm that a particular winding is open. Desolder one end of the suspected winding, and measure its resistance. An open winding will read infinite ohms. In many cases, the normal resistance of the transformer windings is so much lower than any parallel sneak paths that you can test for an open winding without desoldering one lead first. If it checks open, then it definitely is. If it checks good, then check carefully to be sure you are not measuring the resistance of a parallel sneak path.

An ohmmeter is also useful for detecting a winding-to-core short or a primary-to-secondary short. In either of these cases, the normal resistance is infinite ohms. If the measured resistance is very low or even zero, then you have located a short circuit.

Ohmmeters do not usually provide adequate resolution to reliably detect shorted turns in a transformer winding. That is, if the resistance of a normal winding were 2.8 Ω, then a few shorted turns might cause it to have a resistance of 2.75 Ω. Not only is this small change difficult to detect reliably, but it is well within the normal variation of different transformers (of the same type) and different ohmmeters (i.e., measurement error).

KEY POINTS

An open primary will prevent the generation of voltage across any secondaries. An open secondary will have zero volts across it, but normal voltage across any other secondaries.

KEY POINTS

Final confirmation of an open winding is best done with an ohmmeter.

Exercise Problems 17.5

1. While troubleshooting a transformer circuit, you measure infinite ohms across one of the secondary windings. What does this indicate?

2. If the primary winding measures infinite resistance, what would a voltmeter across the secondary measure with the transformer connected to power?

3. What is a normal resistance reading between the primary and secondary windings?

4. What is a normal resistance reading between one of the secondary windings and the core of the transformer?

5. Explain why observation can play an important role in the troubleshooting of transformer circuits.

Chapter Summary

• When two or more inductors have flux in common, we say they are magnetically linked. The portion of the total flux that does not link the coils is called leakage flux and gives rise to leakage inductance. The portion of the total flux that is common to the coils produces mutual inductance. The amount of mutual inductance is determined by the value of the inductances and by the coefficient of coupling (i.e., the percentage of total flux that links the coils).

- When magnetically linked coils are connected in series, the total inductance may be higher or lower (by a factor of two times the mutual inductance) than if the coils had no common flux. Whether total inductance is higher or lower depends on whether the magnetic polarities of the two coils aid or oppose. Similarly, when parallel-connected coils have mutual inductance, the total inductance may be higher or lower than the same two coils without mutual inductance.

- A transformer consists of two or more coils that are magnetically coupled. The coil to which power is applied is called the primary winding. All other coils are called secondary windings. The phase relationship between the primary and any given secondary is indicated on a schematic diagram with dot notation. The phase dots indicate points of similar instantaneous polarity.

- Transformers may be classified many ways, including by their application, by the type of core material, or by how their windings are constructed. There may be a single or multiple secondaries, and the primary and/or the secondary(s) may be tapped. An autotransformer has a special winding configuration, where the primary and secondary share a common, tapped winding.

- Practical transformers have winding losses and core losses. Winding losses include copper loss and losses due to skin effect. Core losses include hysteresis loss and loss due to eddy currents. Practical transformers have parasitic capacitance between adjacent turns and between the primary and secondary windings. This latter capacitance can be eliminated by using a Faraday shield between the primary and secondary windings.

- The ratio of the number of turns in the primary to the number of turns in the secondary is called the transformer turns ratio. The primary-to-secondary voltage ratio is equal to the turns ratio, because the volts per turn must be the same in both primary and secondary. The current ratio is equal to the inverse of the turns ratio. Primary and secondary power are the same with the exception of transformer losses. The ratio of secondary power to primary power is called the transformer efficiency. Iron-core transformers can have efficiencies on the order of 95%.

- Any impedance connected to the secondary of a transformer is reflected into the primary. The value of the reflected impedance is related to the actual secondary impedance by the square of the turns ratio.

- The primary of an unloaded transformer has a very small magnetizing current that is nearly 90° behind the primary voltage. The transformer acts inductive. Under conditions of heavy resistive loads, the primary current is largely resistive and the primary current and voltage are nearly in phase. The secondary voltage of a transformer generally decreases as more secondary current is drawn. This is due to the voltage drop across the nonideal secondary impedance consisting of the secondary leakage inductance and the resistance of the secondary winding.

- Transformers play a key role in the distribution of commercial power. They are used to step up (increase) and step down (decrease) the voltage at various points in the distribution network. Other transformer applications include isolation transformers, step-down transformers for low-voltage dc power supplies, and transformers used for impedance matching to attain maximum transfer of power between two circuits.

• Transformer defects include open or shorted windings and windings that are shorted to the core or to each other. Many defective transformers can be detected by smelling burned insulation or by visually locating indications of overheating. A voltmeter can be used to detect open windings, but results should be verified with an ohmmeter.

Review Questions

Section 17.1: Mutual Inductance and Coupling

1. If two magnetically linked coils are moved closer to each other, the coefficient of coupling will increase. (True or False)

2. If all of the flux produced by one coil cut through the windings of a second coil, what would be the coefficient of coupling for the two coils?

3. Mutual inductance can only exist if two or more coils have common flux. (True or False)

4. If two coils have a coefficient of coupling of 0.85 and have individual inductances of 2.5 μH and 4.8 μH, what is the value of mutual inductance?

5. Calculate the mutual inductance for two 175-mH coils that have a 0.5 coefficient of coupling.

6. What is the mutual inductance between a 75-mH coil and a 225-mH coil if they have a coefficient of coupling of 0.6?

7. When a 200-μH coil has a 0.9 coefficient of coupling to another coil, the mutual inductance is 350 μH. What is the inductance of the second coil?

8. When a 1.75-mH coil has a 0.75 coefficient of coupling with another coil, the mutual inductance is 750 μH. What is the inductance of the second coil?

9. When two magnetically linked coils are connected in series, the total inductance is always higher than if the coils were not magnetically linked. (True or False)

10. Compute the total inductance of two series coils having values of 25 mH and 75 mH if they have a mutual inductance of 10 mH and are connected to be series-aiding.

11. Two 225-mH coils are connected in series so that their magnetic fields are in opposition. The mutual inductance is 75 mH. What is the total inductance of the series circuit?

12. What is the total inductance if a 35-μH coil is connected series-aiding with a 10-μH coil with a mutual inductance of 8 μH?

13. What is the approximate value of total inductance when two parallel coils (with opposing magnetic fields) have inductances of 175 mH and 250 mH and a mutual inductance of 100 mH?

14. Two parallel coils have fields that are magnetically aiding. One coil is 650 μH, the other is 470 μH, and the mutual inductance is 450 μH. What is the approximate inductance of the parallel combination?

Section 17.2: Transformers

15. Transformers always have mutual inductance. (True or False)

16. The winding of a transformer where input power is connected is called the _____ winding.

17. How is the phase relationship between primary voltage and secondary voltage indicated on a schematic diagram?

18. A transformer whose secondary voltage is higher than its primary voltage is called a step-_____ transformer.

19. The secondary voltage in a step-down transformer is (*higher, lower*) than the primary voltage.

20. Transformers used in power supplies are often classified as _____ transformers.

21. Transformers intended for use in the audio frequency range are classified as _____ transformers.

22. Name the three types of material commonly used for cores in transformers.

23. What type of core material has the lowest permeability?

24. What type of core material would be used in a transformer designed for use at 60 Hz?

25. What type of material is generally used as the core material in a toroid-core transformer?

26. If the full secondary voltage of a center-tapped transformer is 30 V, how much voltage will be measured between either end of the transformer and the center tap?

27. When a transformer has multiple secondaries, each secondary can have its own turns ratio. (True or False)

28. How many windings are there in an autotransformer?

29. Is it possible to connect an autotransformer as a step-up transformer?

30. Can an autotransformer be used as a step-down transformer?

31. The power loss due to the resistance of the transformer windings is called _____.

32. What is the name of the phenomenon that causes high-frequency currents to flow only near the surface of a conductor?

33. Hysteresis loss can be classified as a loss due to winding resistance. (True or False)

34. The core of an iron-core transformer is laminated to reduce the losses due to hysteresis. (True or False)

35. What is done in the construction of an iron-core transformer to reduce losses due to eddy currents?

36. An ideal transformer core would have very high resistance and very high permeability. (True or False)

37. What is the primary purpose of a Faraday shield?

38. To be effective, the material used in a Faraday shield must have a high resistance. (True or False)

Section 17.3: Analysis of Transformer Circuits

39. What is the turns ratio of a transformer that has 350 turns in the primary and 1,750 turns in the secondary?

40. What is the turns ratio of a transformer that has 500 turns in the primary and 750 turns in the secondary?

41. What is the turns ratio of a transformer that has 700 turns in the primary and 525 turns in the secondary?

42. What is the turns ratio of a transformer that has 250 turns in the primary and 500 turns in the secondary?

43. If a transformer with a 2:1 turns ratio has 275 turns on its primary, how many turns are on the secondary?

44. If a transformer with a 4:3 turns ratio has 900 turns on its secondary, how many turns are on the primary?

45. If the volts per turn on the primary of a transformer is 1.75 V/T, what is the volts per turn on the secondary of the transformer?

46. A transformer with a 2:1 turns ratio has a secondary voltage of 6.3 V. What is the primary voltage?

47. A transformer with a 4:3 turns ratio has a primary voltage of 220 V. What is the secondary voltage?

48. A transformer has a turns ratio of 5:1 and delivers a secondary voltage of 100 V. What is the primary voltage?

49. A transformer has a 1:5 turns ratio and delivers a secondary voltage of 100 V. What is the primary voltage?

50. If the turns ratio of a transformer is 3:1 and the primary current is 250 mA, what is the value of secondary current?

51. A transformer is connected to a 120-V source and supplies 24 V to a load that draws 800 mA. What is the value of primary current?

52. If a transformer has a primary voltage of 100 V, a secondary voltage of 20 V, and a primary current of 1.2 A, what is the value of secondary current?

53. When the secondary current in a transformer increases due to a lower resistance load, the primary current will (*increase, decrease*).

54. In your own words, explain what is meant by the term *reflected impedance.*

55. Reflected impedance is always higher than the actual secondary impedance. (True or False)

56. The secondary of a transformer with a turns ratio of 4:1 is connected to a 250-Ω resistor. What is the value of impedance reflected into the primary?

57. A transformer with 220 V across the primary produces 10 V in the secondary. If the secondary load is 1.25 Ω, what is the current in the primary?

58. A transformer produces 12 V in the secondary when 120 V are connected to the primary. If the secondary has 100 mA of current, what is the impedance seen by the source?

59. The secondary of a transformer provides 100 W to a resistive load. The combined core losses are 3 W and the winding losses are 5 W. What is the efficiency of the transformer?

60. A transformer draws 2 A from a 120-V source. The secondary supplies 10 V to a 0.5-Ω resistance. What is the efficiency of the transformer?

61. If the secondary of a transformer is unloaded, the primary current is very small. (True or False)

62. Under no-load conditions, the primary current and voltage of a transformer are nearly 90° out of phase. (True or False)

63. When the secondary of a transformer is fully loaded with a resistive load, the primary acts almost purely inductive. (True or False)

64. What causes the secondary voltage of a transformer to decrease as more current is drawn from the secondary?

Section 17.4: Applied Technology: Transformers

65. The transformers that are located between a power generating plant and a high-voltage transmission line are step-up transformers. (True or False)

66. The transformer whose secondary supplies power to your home is a step-up transformer. (True or False)

67. What is the name of a transformer with a 1:1 turns ratio?

68. An autotransformer is used to eliminate the earth-ground reference from the power line. (True or False)

69. Explain why impedance matching is such an important consideration in electronic circuits.

Section 17.5: Troubleshooting Transformers

70. An open secondary will develop a higher-than-normal voltage. (True or False)

71. An open primary will cause the voltage across all secondaries to increase. (True or False)

72. Name two ways that some defective transformers can be identified without making any electrical tests.

73. What is the normal resistance between the primary and secondary windings of a transformer?

74. If a center-tapped secondary measures 10 V from the tap to one end, but measures 0 V from the tap to the other end, what is the most probable trouble?

75. One indication of an open transformer winding is a blown fuse in the primary. (True or False)

TECHNICIAN CHALLENGE

You are working as a computer technician and have just located a defective transformer in a video monitor. The monitor is old, and the manufacturer has gone out of business. You decide to have a local transformer manufacturer wind a new transformer for you, but you must provide the needed specifications. Here is what you know:

The transformer operates from 120 Vac. It has three secondary windings with voltages of 6.3 V, 12.6 V, and 48 V. The 6.3-V, 12.6-V, and 48-V secondaries have 2.5-A, 1.0-A, and 0.125-A fuses, respectively. The primary is also fused, but the fuse was physically burned so badly it could not be interpreted. You don't know what size fuse to use in the primary.

Take the technician challenge and determine the following information:

- required turns ratios
- required power ratings for each secondary
- required fuse size for the primary (assume 100% efficiency)

Equation List

$$(17\text{-}1) \quad L_M = k\sqrt{L_1 L_2}$$

$$(17\text{-}2) \quad L_T = L_1 + L_2 \pm 2L_M$$

$$(17\text{-}3) \quad L_T = \frac{L_1 L_2 - L_M^2}{L_1 + L_2 \pm 2L_M}$$

$$(17\text{-}4) \quad k = \frac{N_P}{N_S}$$

$$(17\text{-}5) \quad \frac{N_P}{N_S} = \frac{V_P}{V_S}$$

$$(17\text{-}6) \quad \frac{N_P}{N_S} = \frac{I_S}{I_P}$$

$$(17\text{-}7) \quad \frac{N_P}{N_S} = \sqrt{\frac{Z_P}{Z_S}}$$

$$(17\text{-}8) \quad \eta = \frac{P_S}{P_S + P_{\text{WINDING}} + P_{\text{CORE}}}$$

$$(17\text{-}9) \quad I_{\text{line}} = I_{\text{leg}} \text{ (wye connection)}$$

$$(17\text{-}10) \quad \begin{aligned} V_{\text{line}} &= \sqrt{3}V_{\text{leg}} \text{ (wye connection), or} \\ V_{\text{line}} &= 1.732V_{\text{leg}} \text{ (wye connection)} \end{aligned}$$

$$(17\text{-}11) \quad \begin{aligned} I_{\text{line}} &= \sqrt{3}I_{\text{leg}} \text{ (delta connection), or} \\ I_{\text{line}} &= 1.732I_{\text{leg}} \text{ (delta connection)} \end{aligned}$$

$$(17\text{-}12) \quad V_{\text{line}} = V_{\text{leg}} \text{ (delta connection)}$$

objectives

After completing this chapter, you should be able to:

1. Name the five general classes of passive filters.

2. Sketch the generic frequency response curve for each of the following filter types:
   a. low-pass
   b. high-pass
   c. bandpass
   d. bandstop
   e. notch

3. Compute the cutoff frequency for a given low-pass or high-pass RC or RL filter.

4. Distinguish between coupling and decoupling circuits.

5. Troubleshoot passive filter circuits.

18

Passive Filter Circuits

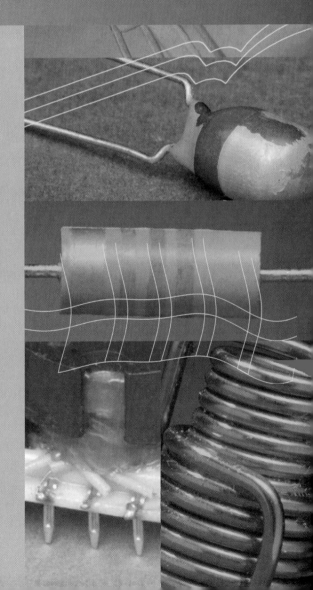

Filter circuits are used extensively in many types of electronic circuits. A filter, be it an air filter, a lint filter, an oil filter, or a passive electronic filter, accepts a broad range of inputs but only allows certain of these inputs to pass through to the output. Some filters are designed to let the "good stuff" pass through the filter while the filter catches or stops the "bad stuff." A lint filter in a washing machine and an air filter in a car are examples of this type of filter action. A filter can also be designed to catch the "good stuff" and let the "bad stuff" pass through the filter. A gold prospector's sieve is an example of this type of filter action.

*All of the filters given as examples in the preceding discussion are mechanical filters and discriminate between "good" and "bad" on the basis of physical size (e.g., size of a dust particle relative to an air molecule for an air filter). In the case of the passive filters described in this chapter, "good" and "bad" signals will be classified on the basis of their frequencies. The input to the filter circuits can be a broad range of frequencies, but the filter circuit will only allow a certain range of these frequencies to pass. All other frequencies will be rejected or attenuated (reduced in amplitude). The frequency (or frequencies) that separates the range of allowed frequencies from the range of rejected frequencies is called the **cutoff frequency** of the filter. As discussed in Chapter 16, signal voltages at the cutoff frequency are attenuated to 70.7% of the input voltage. This point is also called the **half-power point** as detailed in Chapter 16. You will also recall that the 70.7% response also corresponds to the –3-dB point. That is,*

$$dB = 20 \log \frac{V_{OUT}}{V_{IN}}$$
$$= 20 \log 0.707 = -3 \text{ dB}$$

We will examine the performance of five basic filter classes: low-pass, high-pass, bandpass, bandstop, and notch. It is important for a technician to be able to recognize and identify the schematic diagram for each of these circuits and to be able to determine the **frequency response** *of each circuit (i.e., determine which frequencies are passed and which ones are rejected).*

18.1 Low-Pass Filters

Low-pass filters, as the name implies, are designed to pass low frequencies and to reject higher frequencies. Figure 18-1 shows the basic frequency response of a low-pass filter.

There are several ways to obtain a frequency response similar to that shown in Figure 18-1. Figure 18-2 illustrates the fundamental requirements for low-pass filter action. As shown in Figure 18-2(a), low-frequency signals (i.e., frequencies below the cutoff frequency) must see a low series impedance and a high shunt impedance. Under these conditions,

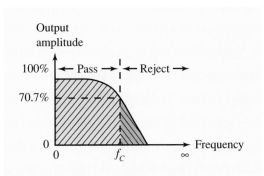

Figure 18-1. A low-pass filter allows frequencies below the cutoff frequency (f_C) to pass, but rejects or attenuates all higher frequencies.

basic voltage divider action will cause the output voltage (V_O) to be nearly the same amplitude as the input voltage (V_I). Figure 18-2(b) shows the condition that must exist for high-frequency signals (i.e., frequencies higher than the cutoff frequency). The high frequencies must see a high series impedance and a low shunt impedance. Voltage divider action, in this case, causes a severe attenuation of the input signals. That is, the high-frequency signals are not passed through the filter.

There are numerous ways to achieve the requirements outlined in Figure 18-2. We shall examine two common methods using *RC* and *RL* circuits.

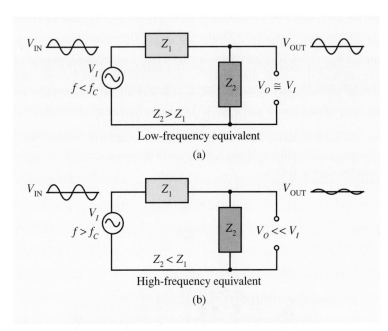

Figure 18-2. Basic requirements for a low-pass filter circuit.

RC Filters

Figure 18-3 shows a basic *RC* low-pass filter circuit. The range of input frequencies is shown to be from dc to infinity (i.e., all frequencies). The output frequency range, by contrast, will only consist of those frequencies that are below the cutoff frequency.

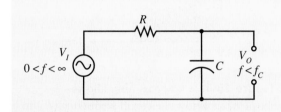

Figure 18-3. A basic *RC* low-pass filter circuit.

Figure 18-4 illustrates the relative behavior of the *RC* low-pass filter at three key frequencies: below the cutoff frequency, at the cutoff frequency, and above the cutoff frequency. In Figure 18-4(a), the input frequency is below the cutoff frequency. You will recall that capacitive reactance is inversely proportional to frequency. So as frequency goes down, capacitive reactance goes up. As shown in Figure 18-4(a), the voltage divider action caused by a capacitive reactance that is greater than the series resistance results in an output voltage (i.e., the voltage across the capacitor) that is nearly the same as the input voltage. That is, the frequencies below the cutoff frequency are passed through the filter and appear in the output with amplitudes that are at least 70.7% of the input amplitude (i.e., $V_O \cong V_I$).

As shown in Figure 18-4(b), the capacitive reactance in the circuit is equal to the resistance at the cutoff frequency. Voltage divider action (detailed in Chapter 15) causes equal voltage drops across *R* and *C*. Due to the phase shift in the circuit (45°), the output voltage will have an amplitude that is 70.7% of the input amplitude.

Finally, Figure 18-4(c) shows the equivalent circuit seen by frequencies above the cutoff frequency. Throughout this range, the capacitive reactance is less than the series resistance. Therefore, voltage divider action attenuates the signals such that the output voltage is less than 70.7% of the input voltage (i.e., $V_O \ll V_I$). The higher the frequency, the greater the attenuation.

As stated previously, the capacitive reactance is equal to the series resistance at the cutoff frequency. We can use this fact and the basic capacitive reactance equation to develop an important relationship as follows:

$$R = X_C$$

$$R = \frac{1}{2\pi f_C C}$$

$$f_C = \frac{1}{2\pi RC}$$

This result is expressed formally as Equation 18-1.

$$f_C = \frac{1}{2\pi RC} \qquad (18\text{-}1)$$

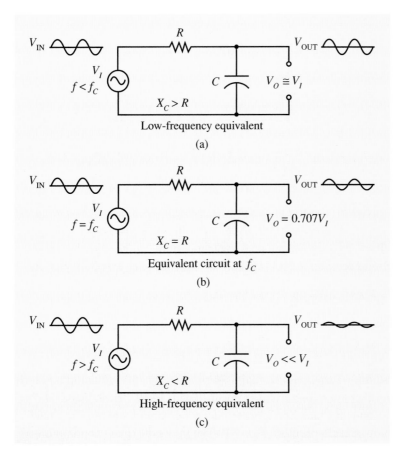

Figure 18-4. The equivalent circuits for an RC low-pass filter at three key frequencies.

EXAMPLE SOLUTION

Determine the cutoff frequency for the filter circuit shown in Figure 18-5.

EXAMPLE SOLUTION

We apply Equation 18-1.

$$f_C = \frac{1}{2\pi RC}$$

$$= \frac{1}{6.28 \times 10 \times 10^3 \times 0.01 \times 10^{-6}}$$

$$= 1.59 \text{ kHz}$$

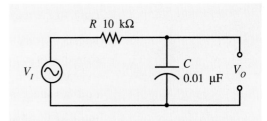

Figure 18-5. Determine the cutoff frequency for this filter circuit.

Practice Problems

1. Calculate the cutoff frequency for the circuit shown in Figure 18-6.

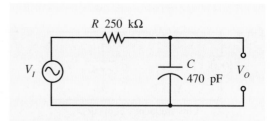

Figure 18-6. Find the cutoff frequency for this circuit.

2. A circuit similar to the one shown in Figure 18-6 has the following values: $R = 250\ \Omega$ and $C = 1{,}000$ pF. What is the cutoff frequency?

3. What value of series resistance is required in a circuit like that shown in Figure 18-6 to produce a 250-kHz cutoff frequency with a 500-pF capacitor?

Answers to Practice Problems

1. 1.36 kHz   **2.** 639.9 kHz   **3.** 1.27 kΩ

RL Filters

The low-pass filter response can also be obtained with an *RL* circuit. Figure 18-7 shows an *RL* low-pass filter and illustrates its response under conditions of three key frequencies: below cutoff, at cutoff, and above cutoff. Figure 18-7(a) shows the circuit conditions with an input frequency lower than the cutoff frequency (i.e., within the passband of the filter). The reactance of the inductor is less than the value of resistance. This means that voltage divider action will cause most of the voltage to be dropped across *R*. Since the output voltage is taken across *R*, then V_O will nearly equal V_I. That is, the lower frequencies are allowed to pass through the filter.

Figure 18-7(b) illustrates the *RL* low-pass filter circuit at cutoff. The inductive reactance is equal to the resistance. The two components drop equal voltages (70.7% of V_I).

Frequencies higher than the cutoff frequency cause the inductive reactance to be larger than the resistance as reflected in Figure 18-7(c). Voltage divider action results in most of the voltage being dropped across the inductive reactance and very little across *R*. Since the output voltage is taken across *R*, the output is greatly attenuated.

Because $X_L = R$ at the cutoff frequency, we can substitute the inductive reactance equation and obtain an expression for the cutoff frequency.

$$X_L = R$$
$$2\pi f_c L = R$$
$$f_c = \frac{R}{2\pi L}$$

This latter expression is given as Equation 18-2.

$$f_c = \frac{R}{2\pi L} \qquad\qquad (18\text{-}2)$$

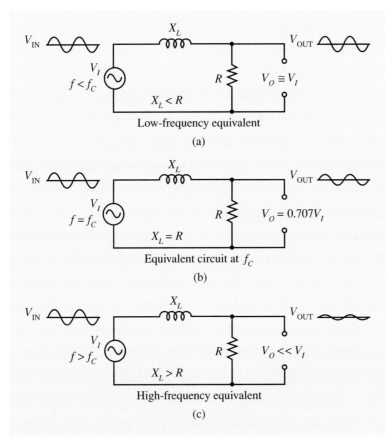

Figure 18-7. An *RL* low-pass filter at three key frequencies.

EXAMPLE SOLUTION

Determine the cutoff frequency for the low-pass filter circuit shown in Figure 18-8.

EXAMPLE SOLUTION

We apply Equation 18-2 as follows.

$$f_C = \frac{R}{2\pi L} = \frac{27 \text{ k}\Omega}{6.28 \times 200 \text{ μH}} = 21.5 \text{ MHz}$$

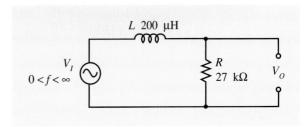

Figure 18-8. Find the cutoff frequency for this circuit.

Practice Problems

1. Compute the cutoff frequency for a circuit like the one shown in Figure 18-8 if it has the following component values: $R = 33 \text{ k}\Omega$ and $L = 375 \text{ μH}$.
2. A circuit similar to the one shown in Figure 18-8 has the following values: $R = 250 \text{ }\Omega$ and $L = 1.75 \text{ mH}$. What is the cutoff frequency?

3. What value of resistance (R) is required in a circuit like that shown in Figure 18-8 to produce a 750-kHz cutoff frequency with a 25-mH inductor?

Answers to Practice Problems

1. 14.01 MHz   **2.** 22.75 kHz   **3.** 117.8 kΩ

Describing Filter Response

There are many different ways to implement a low-pass filter function, but not all work equally well. It becomes important to describe the response of a given filter, so that comparisons and evaluations can be made. Figure 18-9 shows the frequency response of an ideal low-pass filter. Here, all frequencies below the cutoff frequency are passed with virtually no attenuation. All frequencies above the cutoff frequency are totally eliminated. While this performance represents an ideal low-pass filter, it does not represent the response of a practical filter. A practical low-pass filter must make a more gradual transition between the passband and stopband regions of the curve. Figure 18-10 shows several practical filter responses.

Clearly, some of the filter responses in Figure 18-10 are more ideal than others. Several factors can be considered when comparing filters. We will consider three major ones. The first and generally the most significant consideration is the actual cutoff frequency. This is given as a certain number of hertz.

The second consideration that is important for most filter applications is the steepness of the slope in the stopband region. As illustrated in Figures 18-9 and 18-10, the steeper the slope, the more ideal the filter. The steepness of the slope (also called **roll-off**) is generally specified in units of decibels (dB) per decade or decibels per octave. A decade represents a frequency change factor of ten. For example, 100 Hz, 1,000 Hz, and 10 kHz represent three frequencies that are one decade apart. If a filter were specified as having a –60 dB/decade slope between 1.0 kHz and 10.0 kHz, then the output signal would be 60 dB lower at 10.0 kHz than it is at 1.0 kHz.

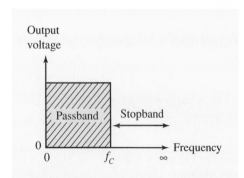

Figure 18-9. The frequency response of an ideal low-pass filter.

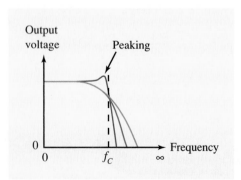

Figure 18-10. The frequency response of practical low-pass filters makes a more gradual transition between the passband and stopband regions.

Filter roll-off is sometimes stated in terms of dB/octave. An octave represents a frequency change of two (or one-half). Thus, 6.0 kHz is one octave higher than 3.0 kHz. If a filter has a –12 dB/octave slope between 3.0 kHz and 6.0 kHz, then the output voltage will be 12 dB lower at 6.0 kHz than at 3.0 kHz.

All of the *RC* and *RL* filters presented in this section have theoretical slopes of 20 dB/decade or 6 dB/octave. Steeper slopes can be obtained by cascading (connecting the output of one filter to the input of a second filter) filter sections. Even greater slopes can be realized by adding other circuitry (e.g., amplifiers with feedback).

The third factor that is important when contrasting the performance of filters is **peaking**. As illustrated in Figure 18-10, the response of some filter circuits peaks just before the edge of the passband. The degree of peaking for a low-pass filter is indicated by its **Q factor**. A *Q* of 1 has only slight peaking. *Q* values less than 1 have no peaking, and *Q* values more than 1 have more pronounced peaking. In most cases, there is a tradeoff between peaking (generally undesired) and steepness (generally desired) of the slope. None of the passive low-pass filters presented in this chapter have any degree of peaking.

KEY POINTS

A steep slope with minimal peaking is generally a desired response since it more closely approximates the response of an ideal low-pass filter.

Exercise Problems 18.1

1. Which of the circuits shown in Figure 18-11 will function as low-pass filters?
2. Which of the frequency response curves shown in Figure 18-12 represent the performance of a low-pass filter?
3. Calculate the cutoff frequency for the low-pass filter shown in Figure 18-13.
4. If the values in Figure 18-13 are changed to $R = 39\ k\Omega$ and $C = 1{,}800\ pF$, what is the new cutoff frequency?
5. What value of C must be used in a circuit like Figure 18-13 to provide a 100-kHz cutoff frequency when the resistor value is 200 kΩ?

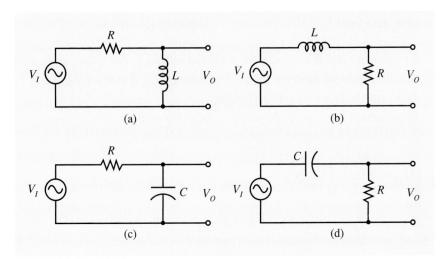

Figure 18-11. Which of these circuits are low-pass filters?

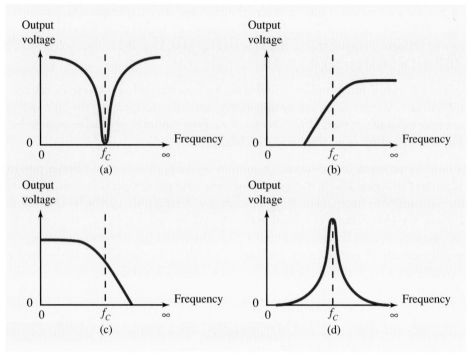

Figure 18-12. Which of these graphs represent the response of a low-pass filter?

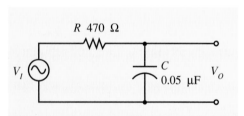

Figure 18-13. What is the cutoff frequency for this circuit?

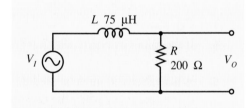

Figure 18-14. Calculate the cutoff frequency for this circuit.

6. Compute the cutoff frequency for the low-pass filter circuit shown in Figure 18-14.
7. If the component values in Figure 18-14 are changed to $R = 4.7$ kΩ and $L = 6.8$ mH, what is the new cutoff frequency?
8. What value of resistance (R) is needed in a circuit like Figure 18-14 to provide a 5.8-kHz cutoff frequency if the inductance (L) is 2.5 mH?

18.2 High-Pass Filters

A high-pass filter circuit rejects or attenuates all frequencies below the cutoff frequency. Frequencies above the cutoff frequency pass through the filter with little or no attenuation. Figure 18-15 illustrates the frequency response for an ideal high-pass filter circuit. Practical filters have a more gradual roll-off below the cutoff frequency.

● KEY POINTS

Filter circuits that pass frequencies above a certain frequency—called the cutoff frequency—with little or no loss but attenuate frequencies lower than the cutoff frequency are called high-pass filters.

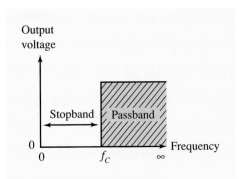

Figure 18-15. The frequency response of an ideal high-pass filter makes a step transition at the cutoff frequency.

KEY POINTS

High-pass filters can be constructed as *RC* or *RL* circuits where:

• The reactance (inductive or capacitive) is equal to the resistance at the cutoff frequency.

• The series element has a lower resistance or reactance than the shunt element at frequencies above the cutoff frequency.

• The series element has a higher resistance or reactance than the shunt element at frequencies below the cutoff frequency.

There are many ways to achieve a high-pass filter response, but the illustrations in Figure 18-16 show the basic requirements. Figure 18-16(a) shows that for frequencies below the cutoff frequency, the series impedance (Z_1) must be greater than the shunt impedance

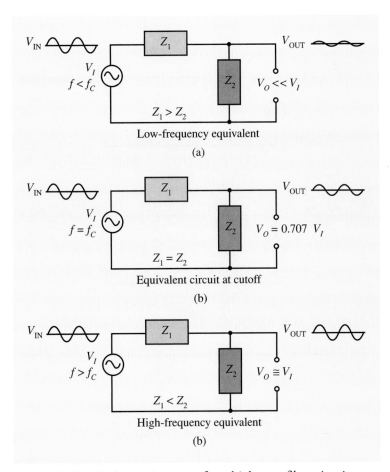

Low-frequency equivalent

(a)

Equivalent circuit at cutoff

(b)

High-frequency equivalent

(b)

Figure 18-16. Basic requirements for a high-pass filter circuit.

(Z_2). Under these conditions, voltage divider action will cause most of the voltage to be dropped across Z_1, leaving very little for the output voltage (voltage across Z_2). Figure 18-16(b) shows that the series and shunt impedances are equal at the cutoff frequency. The two impedances will have equal voltage drops (70.7% of the input voltage). Finally, as shown in Figure 18-16(c), frequencies above the cutoff frequency must see Z_1 as a lower impedance than Z_2. This causes most of the input voltage to be dropped across Z_2. Since the output voltage is taken across Z_2, we know that the output voltage will nearly equal the input voltage under these conditions.

RC Filters

Figure 18-17 shows how to construct a high-pass *RC* filter circuit. Figure 18-17(a) shows that the capacitive reactance is higher than the resistance for frequencies below the cutoff frequency. Voltage divider action will drop most of the input voltage across *C*, leaving very little for the output voltage (voltage across *R*). At the cutoff frequency, shown in Figure 18-17(b), the capacitive reactance and the resistance values are equal and drop equal voltages (70.7% of the input voltage). For frequencies higher than the cutoff frequency, the capacitive reactance is lower than the shunt resistance as shown in Figure 18-17(c). Voltage divider action results in most of the input voltage being dropped across the resistance and

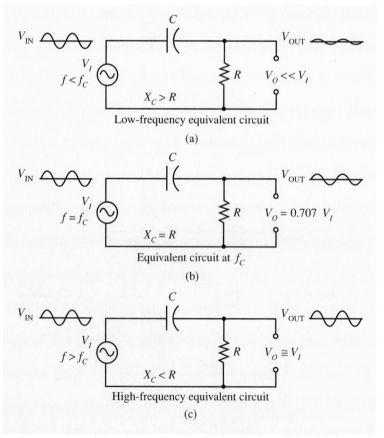

Figure 18-17. An *RC* circuit can provide a high-pass filter response.

appearing as output voltage. That is, frequencies above the cutoff frequency are passed through with minimal attenuation.

Since the reactance and resistance values for a high-pass filter are equal at the cutoff frequency, as they were in a low-pass filter, we can use Equation 18-1 to compute the cutoff frequency.

EXAMPLE SOLUTION

Determine the cutoff frequency for the high-pass filter circuit shown in Figure 18-18.

EXAMPLE **SOLUTION**

We apply Equation 18-1 as follows:

$$f_C = \frac{1}{2\pi RC}$$

$$= \frac{1}{6.28 \times 10 \text{ k}\Omega \times 0.05 \text{ μF}} = 318.5 \text{ Hz}$$

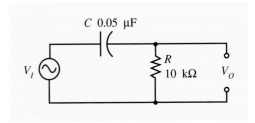

Figure 18-18. What is the cutoff frequency for this circuit?

Practice Problems

1. Determine the cutoff frequency for the *RC* high-pass filter circuit shown in Figure 18-19.
2. If the capacitor in Figure 18-19 were changed to 1.5 μF and the resistor was left as 18 kΩ, what would be the value of cutoff frequency?
3. What value of resistor would be required for a circuit like that shown in Figure 18-19 to produce a cutoff frequency of 250 Hz if the capacitor is 0.05 μF?

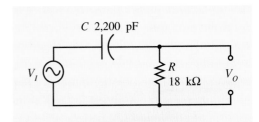

Figure 18-19. Calculate the cutoff frequency for this circuit.

Answers to Practice Problems

1. 4.02 kHz   2. 5.9 Hz   3. 12.74 kΩ

RL Filters

Figure 18-20 shows how an *RL* circuit can provide the response of a high-pass filter. Figure 18-20(a) shows the circuit's response to frequencies below the cutoff frequency (stopband frequencies). In this range, the inductive reactance is less than the resistance, so most of the input voltage is dropped across the resistance; very little voltage appears in the output (across X_L). Figure 18-20(b) shows the circuit response at the cutoff frequency. Since the inductive reactance and the resistance are equal, they will have equal voltage drops (70.7% of the input voltage). Frequencies higher than the cutoff frequency cause the inductive reactance to be higher than the resistance as shown in Figure 18-20(c). In this range (the passband), most of the input voltage is dropped across the inductive reactance, which means that the output voltage nearly equals the input voltage.

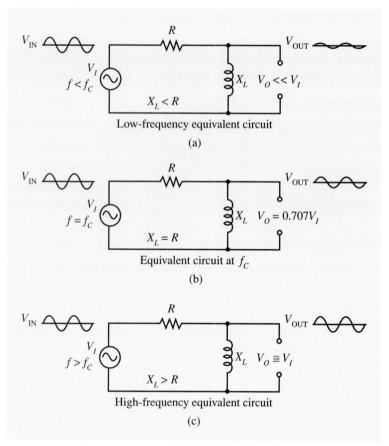

Figure 18-20. An *RL* high-pass filter passes all frequencies above the cutoff frequency.

As with *RL* low-pass filters, the value of inductive reactance is equal to the resistance at the cutoff frequency. This means that we can find the cutoff frequency for an *RL* high-pass filter by applying Equation 18-2.

EXAMPLE SOLUTION

Calculate the cutoff frequency for the *RL* high-pass filter shown in Figure 18-21.

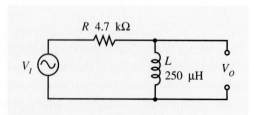

EXAMPLE **SOLUTION**

We apply Equation 18-2 as follows:

$$f_C = \frac{R}{2\pi L} = \frac{4.7 \text{ k}\Omega}{6.28 \times 250 \text{ μH}} \approx 3 \text{ MHz}$$

Figure 18-21. Calculate the cutoff frequency for this circuit.

Practice Problems

1. Find the cutoff frequency for the high-pass circuit shown in Figure 18-22.
2. If the inductor in Figure 18-22 was changed to 50 μH, what would be the new cutoff frequency?
3. What value of resistance is needed with a 5-mH coil to obtain a 100-kHz cutoff frequency in a filter circuit like the one shown in Figure 18-22?

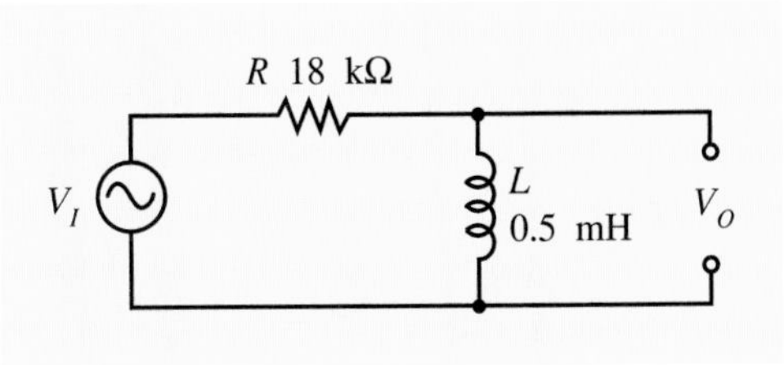

Answers to Practice Problems

1. 5.73 MHz   2. 57.3 MHz   3. 3.14 kΩ

Figure 18-22. Calculate the cutoff frequency for this circuit.

Exercise Problems 18.2

1. Which of the circuits shown in Figure 18-23 will function as high-pass filters?
2. Which of the frequency response curves shown in Figure 18-24 represent the performance of a high-pass filter?
3. Calculate the cutoff frequency for the high-pass filter shown in Figure 18-25.
4. If the values in Figure 18-25 are changed to $R = 47$ kΩ and $C = 1{,}500$ pF, what is the new cutoff frequency?
5. What value of C must be used in a circuit like Figure 18-25 to provide a 500-kHz cutoff frequency when the resistor value is 270 kΩ?
6. Compute the cutoff frequency for the high-pass filter circuit shown in Figure 18-26.

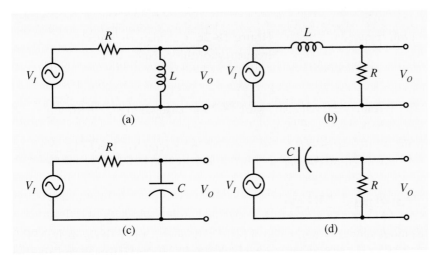

Figure 18-23. Which of these circuits are high-pass filters?

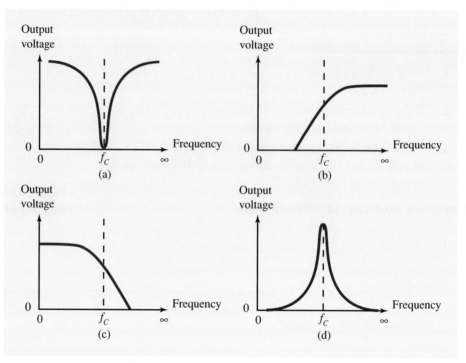

Figure 18-24. Which of these graphs represent the response of a high-pass filter?

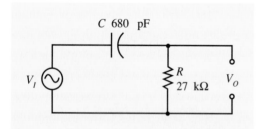

Figure 18-25. What is the cutoff frequency for this circuit?

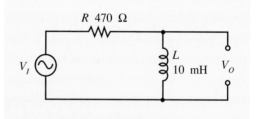

Figure 18-26. Calculate the cutoff frequency for this circuit.

KEY POINTS

Bandpass filter circuits allow a range of frequencies—called the **passband**—to go through the filter with minimal attenuation. Frequencies either lower or higher than the passband—called **stopband** frequencies—are greatly attenuated.

7. If the component values in Figure 18-26 are changed to $R = 6.8$ kΩ and $L = 250$ µH, what is the new cutoff frequency?

8. What value of resistance (R) is needed in a circuit like Figure 18-26 to provide a 12.5-kHz cutoff frequency, if the inductance (L) is 25 mH?

18.3 Bandpass Filters

Bandpass filters allow a certain range or band of frequencies to pass through the filter with minimal attenuation. Frequencies either higher or lower than the passband are attenuated by the filter. Figure 18-27 shows the frequency response curves for an ideal and a practical

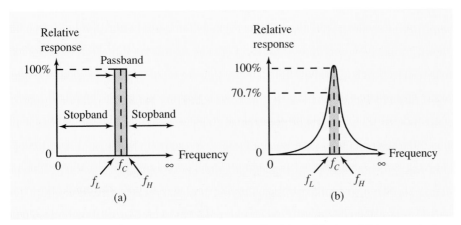

Figure 18-27. The frequency response curves for (a) an ideal and (b) a practical bandpass filter.

bandpass filter. Figure 18-27(a) shows the ideal response to have two cutoff frequencies (f_L and f_H). Input voltages at frequencies lower than f_L are completely eliminated (i.e., they do not appear in the output). Similarly, all frequencies higher than f_H are attenuated and do not appear in the output. By contrast, frequencies within the passband (i.e., higher than f_L and lower than f_H) are passed through the filter with no attenuation.

Figure 18-27(b) shows the frequency response of a practical bandpass filter. Again, there are two cutoff frequencies that set the limits for the passband. However, there is a more gradual transition between the stopband and the passband regions. The edges of the passband are defined as the points where the relative response has fallen to 70.7% of the maximum response. You will recall from Chapter 16 that these points are also called half-power points.

Figure 18-28 illustrates the basic requirements of a bandpass filter circuit. As reflected in Figures 18-28(a) and (c), the series impedance (Z_1) must be greater than the shunt impedance (Z_2) for frequencies outside the passband of the filter. Under these conditions, voltage divider action causes most of the input voltage to be dropped across Z_1. This leaves very little voltage across Z_2 (V_O).

Figure 18-28(b) shows the circuit conditions for frequencies that are within the passband of the filter. Here, the series impedance (Z_1) must be lower than the shunt impedance (Z_2). Voltage divider action causes most of the input voltage to be dropped across Z_2, where it appears as V_O. Ideally, there is no attenuation of input voltages for frequencies in the passband. In practice, there will always be a small voltage drop across Z_1 even in the passband.

Resonant Filters

A bandpass filter can be realized with a series- or parallel-resonant RLC circuit. A more complete mathematical analysis of series-resonant RLC circuits was presented in Chapter 16 and will not be repeated here. Rather, let us concentrate on the general behavior of RLC circuits as it applies to filter applications.

KEY POINTS

Ideally, the transitions between stopband and passband regions are abrupt. Practical filters, however, have more gradual transitions.

KEY POINTS

Bandpass filters can be made with series or parallel RLC circuits.

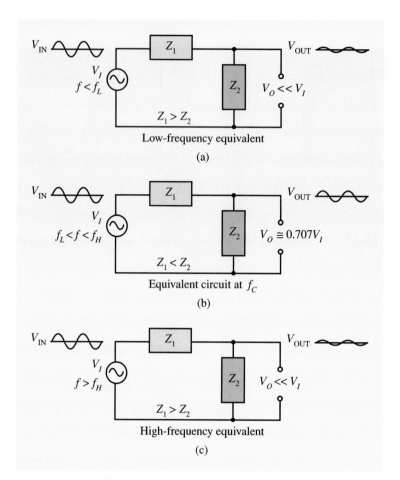

Figure 18-28. The basic requirements for a bandpass filter circuit.

SERIES-RESONANT BANDPASS FILTER

Figure 18-29 shows a series-resonant circuit being used as a bandpass filter. Recall the following important characteristics about a series LC circuit:

- At resonance, the inductive and capacitive reactances are equal and essentially cancel each other. Current is limited only by the resistance in the circuit. The LC portion of the circuit acts like a short circuit.

- Below resonance, the reactance of the capacitor is the dominant factor in determining impedance. As the operating frequency approaches dc, the capacitive reactance approaches infinity. The LC portion of the circuit acts like an open circuit.

- Above resonance, the reactance of the inductor is the dominant factor in determining impedance. As the operating frequency increases, so does the inductive reactance. At sufficiently high frequencies, the inductor and, therefore, the LC portion of the circuit begin to act like an open circuit.

Figure 18-30 illustrates the relative impedances in the RLC circuit at three key frequencies: below resonance, at resonance, and above resonance. At frequencies far below the lower

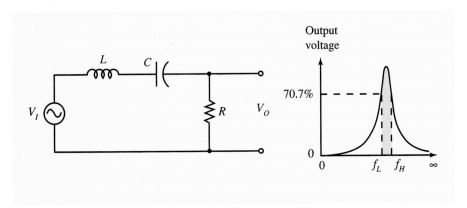

Figure 18-29. A bandpass filter using a series-resonant circuit.

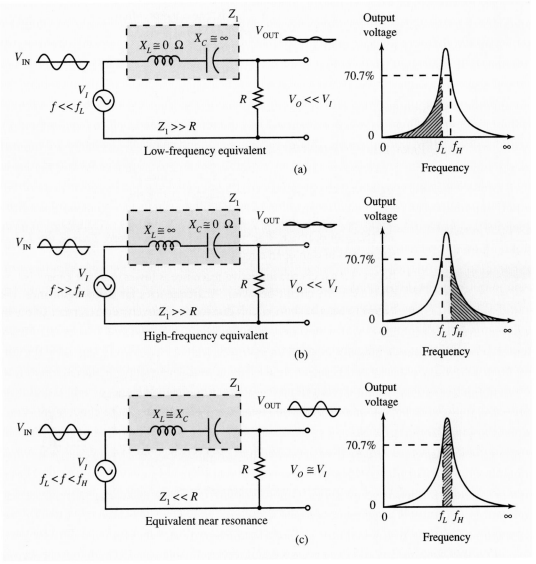

Figure 18-30. Circuit conditions in an *RLC* bandpass filter at three key frequencies.

cutoff frequency, X_C is so high (relative to R) that the net reactance of the LC portion of the circuit is essentially open (i.e., it drops nearly all of the applied voltage). This condition is represented in Figure 18-30(a). A similar condition occurs when the operating frequency is far above the upper cutoff frequency. In this case, as shown in Figure 18-30(b), X_L is so high (relative to R) that the net reactance of the LC portion of the circuit is essentially open and drops nearly all of the input voltage. So, for frequencies outside of the passband, the net reactance of the LC portion of the circuit (Z_1) is high relative to R. Voltage divider action causes most of the input voltage to be dropped across the LC network, leaving near zero to be dropped across R (output voltage).

Figure 18-30(c) shows the circuit conditions for frequencies in the passband of the filter (i.e., above the lower cutoff frequency and below the upper cutoff frequency). In this range, X_L is nearly the same value as X_C. These two reactances effectively cancel each other, leaving a near-zero net reactance for Z_1. Voltage divider action will cause most of the input voltage to be dropped across R. That is, the output voltage is nearly the same as the input voltage for frequencies in the passband. More specifically, as discussed in Chapter 16, the output voltage will be at least 70.7% of the input voltage for frequencies in the passband.

All of the calculations presented in Chapter 16 for series RLC circuits apply to filter circuits like the one shown in Figure 18-29. The output voltage is simply taken across the resistive portion of the circuit.

PARALLEL-RESONANT BANDPASS FILTER

Figure 18-31 shows how a parallel-resonant circuit can be used to implement a bandpass filter. Recall the following important characteristics about a parallel LC network:

- At resonance, the inductive and capacitive currents are equal in magnitude, but they are 180° out of phase. The net reactive current is zero. That is, the LC portion of the circuit acts like an open circuit.

- Below resonance, the inductive reactance is lower than the capacitive reactance and dominates the circuit behavior. At frequencies far below resonance, the LC circuit begins to act as a short circuit due to the relatively low reactance of the inductor.

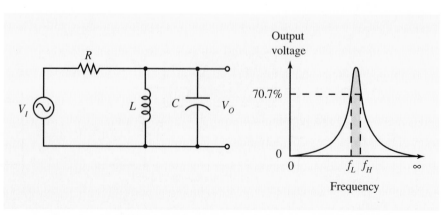

Figure 18-31. A parallel resonant circuit can serve as a bandpass filter.

- Above resonance, the capacitive reactance is lower than the inductive reactance and dominates the circuit characteristics. At frequencies far above resonance, the *LC* circuit begins to act like a short circuit due to the relatively low reactance of the capacitor.

Figure 18-32 shows a bandpass filter circuit at three key frequencies: below resonance, above resonance, and at resonance. At frequencies well below the lower cutoff frequency as shown in Figure 18-32(a), the low value of inductive reactance makes the *LC* portion of the circuit act as a short (i.e., low impedance relative to *R*). Voltage divider action causes most of the input voltage to be dropped across *R*, leaving very little across the *LC* circuit (V_O).

Figure 18-32(b) shows the circuit conditions at frequencies far above the upper cutoff frequency. In this range, the low value of capacitive reactance makes the *LC* portion of the circuit act as a short circuit (relative to *R*). Again, voltage divider action drops most of the input voltage across *R*, and produces an output voltage (V_O) that is greatly attenuated.

Figure 18-32(c) shows the response of the circuit for frequencies within the passband of the filter. In this case, X_L is nearly equal to X_C, and the two reactive currents effectively

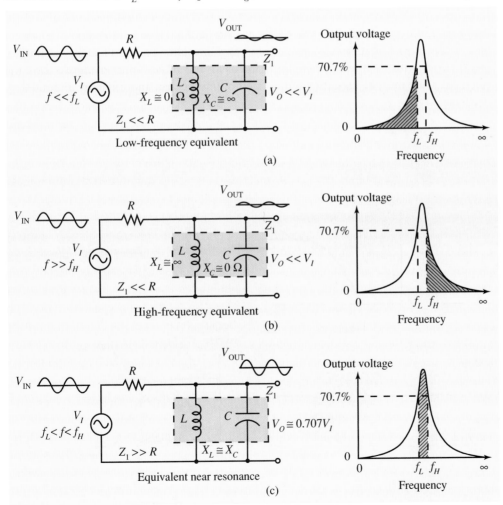

Figure 18-32. Circuit conditions in an RLC circuit at three key frequencies.

cancel each other. This means the LC portion of the circuit is acting like an open circuit (i.e., a high impedance relative to R). Voltage divider action causes most of the input voltage to be dropped across the high-impedance LC circuit and very little across R. The output voltage is very near the same as the input voltage. That is, frequencies in the pass-band receive minimal attenuation. By definition, all passband frequencies pass through the filter with an amplitude of at least 70.7% of the input voltage.

RC Bandpass Filters

Since a low-pass filter passes all frequencies below its cutoff frequency and a high-pass filter passes all frequencies above its cutoff frequency, we can cascade a low-pass and a high-pass filter to obtain the frequency response of a bandpass filter. This technique is illustrated in Figure 18-33. The cutoff frequency for the low-pass filter section establishes the upper cutoff frequency (f_H) for the overall bandpass filter. Likewise, the cutoff frequency for the high-pass filter section establishes the lower cutoff frequency (f_L) for the overall filter. Clearly, if the cutoff frequency of the low-pass section is lower than the cutoff frequency of the high-pass section, then all frequencies will be attenuated and the filter will be useless.

Although it is more common to use RC filters instead of RL filters due to cost, size, and weight, a bandpass filter can also be made by cascading a low-pass and a high-pass filter section made with RL filters. If a combination of RL and RC sections are used in the same filter, great care must be used due to the effects of unexpected, and generally undesired, resonances.

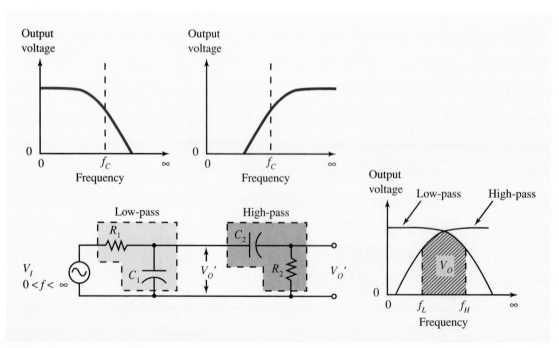

Figure 18-33. A bandpass filter formed by cascading low-pass and high-pass filter sections.

Exercise Problems 18.3

1. Which of the frequency response curves shown in Figure 18-34 represents the performance of a bandpass filter?

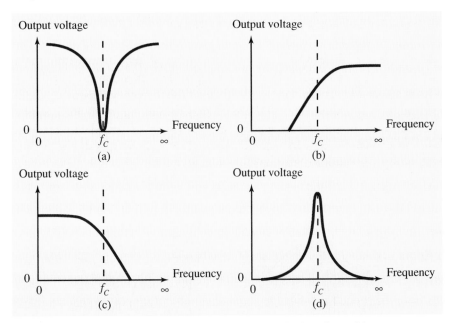

Figure 18-34. Which is the frequency response of a bandpass filter?

2. Which of the circuits shown in Figure 18-35 are bandpass filters?

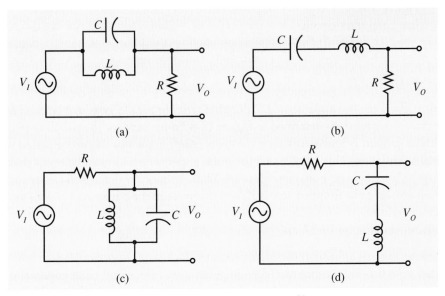

Figure 18-35. Which of these circuits are bandpass filters?

3. If high-pass and low-pass filter sections are cascaded to form a bandpass filter, the cutoff frequency of the high-pass section must be lower than the cutoff frequency of the low-pass section. (True or False)

4. All input frequencies within the passband of a bandpass filter have an amplitude that is at least _____% of the maximum response.

18.4 Bandstop or Band-Reject Filters

A bandstop—also called a band-reject—filter attenuates all frequencies within the stop-band of the filter. Frequencies that are lower or higher than the stopband pass through the filter with minimal attenuation. Figure 18-36 shows the frequency responses of both ideal and practical bandstop filters. The ideal filter, represented in Figure 18-36(a), makes an abrupt transition at the cutoff frequencies. Passband signals are passed with no attenuation, whereas stopband signals are reduced to zero. Practical filters, as shown in Figure 18-36(b), have more gradual transitions.

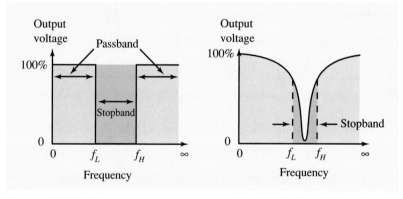

Figure 18-36. Frequency response curves for (a) ideal and (b) practical bandstop filter circuits.

Resonant Filters

Bandstop filters can be implemented with either series or parallel resonant circuits. In either case, the conditions described in Figure 18-37 must exist for the circuit to perform as a bandstop filter. Frequencies below the lower cutoff frequency should be passed with minimal attenuation. This implies that the series impedance (Z_1) must be lower than the shunt impedance (Z_2) as shown in Figure 18-37(a). Voltage divider action causes most of the input voltage to be dropped across Z_2 and to appear in the output as V_O. This same set of circuit conditions must also exist for frequencies higher than the upper cutoff frequency as represented in Figure 18-37(c).

Frequencies in the stopband of the filter must see a high series impedance and a low shunt impedance, so they will be attenuated. Voltage divider action under these conditions causes most of the input voltage to be dropped across Z_1, leaving very little to be dropped across Z_2. The stopband conditions are illustrated in Figure 18-37(b).

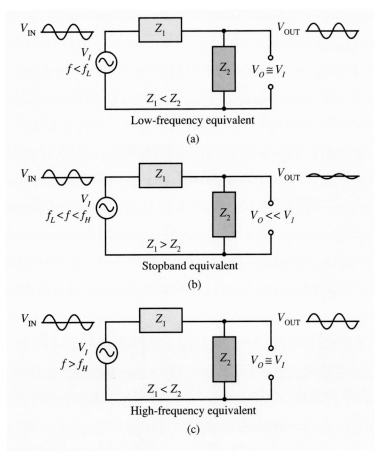

Figure 18-37. Conditions required for a bandstop filter circuit.

SERIES-RESONANT BANDSTOP FILTERS

Figure 18-38 shows a series-resonant circuit configured as a bandstop filter. You will recall that a series LC circuit has a low impedance at resonance and a higher impedance on either side of resonance. This characteristic satisfies the requirements of the shunt impedance (Z_2) in Figure 18-37.

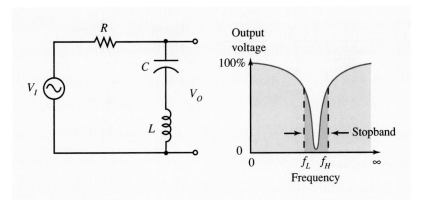

Figure 18-38. A series-resonant circuit can provide a bandstop filter response.

Figure 18-39 shows the circuit conditions for a bandstop filter at three key frequencies: below the lower cutoff frequency, above the upper cutoff frequency, and within the stopband. The response to frequencies below the stopband is illustrated in Figure 18-39(a). The relatively high capacitive reactance causes the *LC* portion of the circuit to have a high impedance relative to the series resistance. Voltage divider action will cause most of the input voltage to be dropped across the *LC* network. Since V_O is taken across the *LC* network, V_O will be nearly equal to the input voltage.

A nearly identical situation exists for frequencies higher than the stopband. Figure 18-39(b) shows that in this case, it is the high value of inductive reactance that causes the *LC* network (Z_2) to have a high impedance. Voltage divider action again causes the output voltage to be nearly equal to the input voltage.

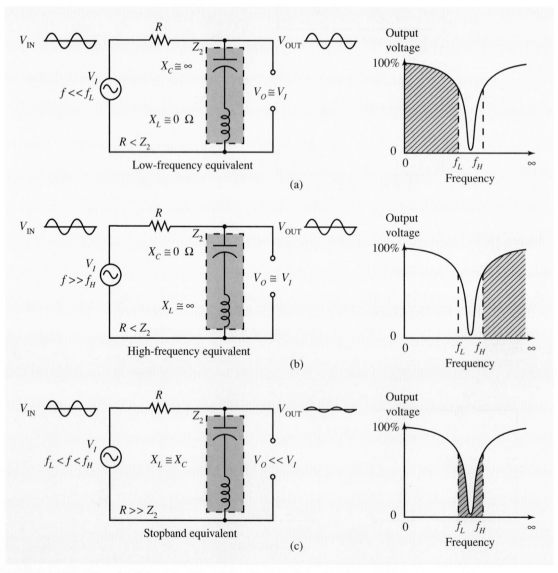

Figure 18-39. The conditions in a bandstop filter at three key frequencies.

Frequencies within the stopband cause nearly equal values for inductive and capacitive reactance. These out-of-phase components tend to cancel each other, which produces an overall low impedance for the LC network (Z_2). The LC network approximates a short circuit. This condition is represented in Figure 18-39(c). Voltage divider action in this case causes most of the input voltage to be dropped across the series resistance, which greatly attenuates the output voltage. Output voltages within the stopband will be reduced to 70.7% or less of the maximum response.

All of the calculations presented in Chapter 16 for series RLC circuits apply to filter circuits like the one shown in Figure 18-38. The output voltage is simply taken across the LC portion of the circuit.

PARALLEL-RESONANT BANDSTOP FILTERS

Figure 18-40 shows how a parallel-resonant circuit can be used to implement a bandstop filter. You will recall that a parallel LC circuit has a high impedance at resonance and a lower impedance on either side of resonance. This characteristic satisfies the requirements of the series impedance (Z_1) in Figure 18-37.

Figure 18-41 shows the circuit conditions for a bandstop filter at three key frequencies: below the lower cutoff frequency, above the upper cutoff frequency, and within the stopband. The response to frequencies below the stopband is illustrated in Figure 18-41(a). The relatively low inductive reactance causes the LC portion of the circuit to have a low impedance relative to the resistance (R). Voltage divider action will cause most of the input voltage to be dropped across the resistance. Since V_O is taken across the resistor, V_O will be nearly equal to the input voltage.

A nearly identical situation exists for frequencies higher than the stopband. Figure 18-41(b) shows that in this case, it is the low value of capacitive reactance that causes the LC network to have a low impedance. Voltage divider action again causes the output voltage to be nearly equal to the input voltage.

Frequencies within the stopband cause nearly equal values for inductive and capacitive reactance. The currents through these two reactances tend to cancel each other, which

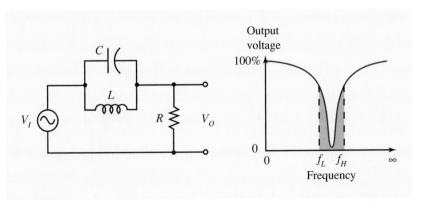

Figure 18-40. A parallel-resonant circuit can provide a bandstop filter response.

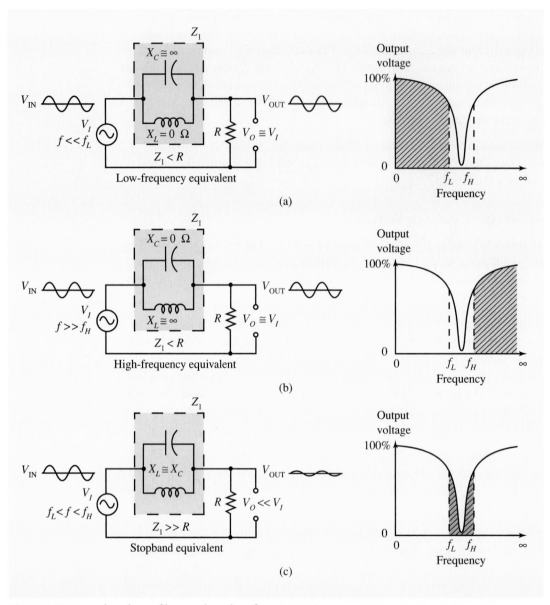

Figure 18-41. A bandstop filter at three key frequencies.

means the LC network acts much like an open circuit. This condition is represented in Figure 18-41(c). Voltage divider action in this case causes most of the input voltage to be dropped across the LC network. Very little voltage is available as V_O. Output voltages within the stopband will be reduced to 70.7% or less of the maximum response.

All of the calculations presented in Chapter 16 for series-parallel RLC circuits apply to filter circuits like the one shown in Figure 18-40. The output voltage is simply taken across the resistive portion of the circuit.

RC Bandstop Filters

In a previous section, we saw that an *RC* bandpass filter could be constructed by cascading low-pass and high-pass filter sections. In a similar manner, an *RC* bandstop filter can be constructed by connecting low-pass and high-pass filter sections in parallel. In this way, frequencies lower or higher than the stopband can pass easily through the low-pass or high-pass filter section, respectively. Those frequencies that are stopped by both filter sections will not appear in the output. Figure 18-42 shows the schematic diagram of an *RC* bandstop filter. This particular configuration is called a twin-tee network. It gets its name from the two *RC* tee networks. A similar filter function can be realized with *RL* circuits, but it is not normally done due to cost, weight, and physical size.

● **KEY POINTS**

A twin-tee filter is an *RC* circuit that provides a notch filter frequency response.

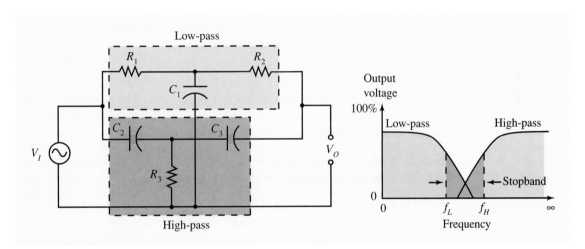

Figure 18-42. A notch filter attenuates a very narrow band of frequencies.

In Figure 18-42, the network consisting of R_1, R_2, and C_1 is, by itself, a low-pass filter. C_1 has a high reactance to low frequencies and allows the signal to pass. High-frequencies, by contrast, are shorted to ground by C_1.

The C_2, C_3, and R_3 network is, by itself, a high-pass filter. The low reactances of C_2 and C_3 pass high frequencies through to the output with minimal attenuation. These same capacitors are effective open circuits for low frequencies and prevent the passage of low-frequency signals.

The actual operation of the circuit is somewhat more complex and nearly achieves total attenuation of one particular frequency. Because the circuit is so selective, it is sometimes called a **notch filter.** The frequency at the center of the notch receives the most attenuation.

At the notch frequency, the reactance of C_1 is equal to the resistance of R_3, and the reactances of C_2 and C_3 are equal to the resistances of R_1 and R_2. This means that the high-pass and low-pass networks will provide equal attenuation to input voltages at the notch frequency. Additionally, notch-frequency input voltage will receive equal but opposite phase shifts from the two networks. The net result is that notch-frequency input voltages are almost completely cancelled at the output of the filter. For frequencies on either side of the notch frequency, attenuation is less. Figure 18-43 shows the frequency response curve

● **KEY POINTS**

A notch filter is essentially a bandstop filter, but it has a very narrow bandwidth (high Q).

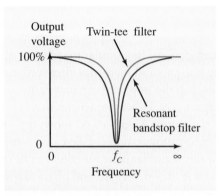

Figure 18-43. The frequency response of a twin-tee notch filter has a very narrow bandwidth.

for a practical twin-tee notch filter as compared to the response of a typical resonant band-stop filter. Both filters have the same center frequency, but the notch filter clearly has a narrower bandwidth. You will recall from Chapter 16 that the narrow bandwidth of the twin-tee filter corresponds to a relatively high Q ($Q = f_R/BW$).

Figure 18-44 shows the schematic diagram of a twin-tee notch filter. For proper operation, the following relationships must exist:

- $R_1 = R_2$
- $R_1 = 2R_3$
- $C_2 = C_3$
- $C_1 = 2C_2$

With these circuit relationships, the center frequency of the filter may be computed with Equation 18-3.

$$f_N = \frac{1}{2\pi R_1 C_2} \qquad \text{(18-3)}$$

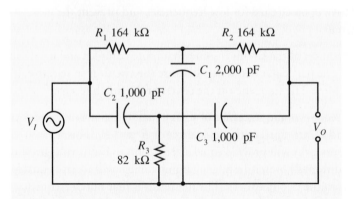

Figure 18-44. The schematic diagram of a twin-tee notch filter.

EXAMPLE SOLUTION

Determine the notch frequency for the twin-tee filter shown in Figure 18-44.

EXAMPLE **SOLUTION**

Since the component relationships satisfy the previously stated requirements, we can apply Equation 18-3 to compute the notch frequency.

$$f_N = \frac{1}{2\pi R_1 C_2}$$

$$= \frac{1}{6.28 \times 164 \text{ k}\Omega \times 1{,}000 \text{ pF}} = 971 \text{ Hz}$$

Practice Problems

1. Compute the notch frequency for a circuit like the one shown in Figure 18-44, if the circuit has the following component values: $R_1 = R_2 = 300$ kΩ, $R_3 = 150$ kΩ, $C_1 = 470$ pF, $C_2 = C_3 = 235$ pF.
2. If a twin-tee filter like the one shown in Figure 18-44 has a value of 75 Ω for R_3, what is the value of R_1?
3. What is the notch frequency of a circuit like that shown in Figure 18-44 if it has the following component values: $R_1 = R_2 = 1.5$ kΩ, $R_3 = 750$ Ω, $C_1 = 2.0$ μF, $C_2 = C_3 = 1.0$ μF?

Answers to Practice Problems

1. 2.26 kHz   2. 150 Ω 3. 106.2 Hz

Exercise Problems 18.4

1. Which of the frequency response curves shown in Figure 18-45 represents the performance of a bandstop filter?
2. Which of the circuits shown in Figure 18-46 are bandstop filters?
3. All input frequencies within the stopband of a bandstop filter have an amplitude that is at least 70.7% of the maximum response. (True or False)
4. A notch filter is generally characterized by a wide bandwidth. (True or False)
5. A twin-tee notch filter passes a narrow band of frequencies, but attenuates frequencies either higher or lower than the passband. (True or False)

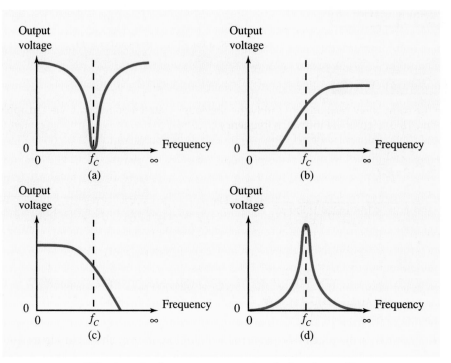

Figure 18-45. Which is the frequency response of a bandstop filter?

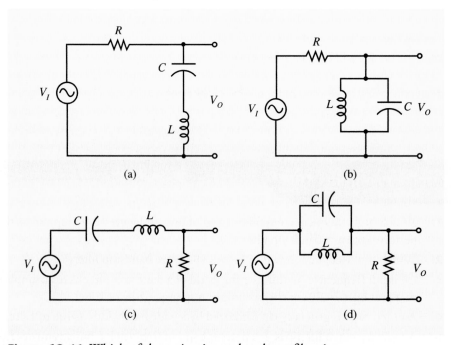

Figure 18-46. Which of these circuits are bandstop filters?

18.5 Applied Technology: Coupling and Decoupling Circuits

For proper operation of an electronic system, signal voltages must be routed to specified parts of the circuit and excluded from other parts. The routing of the signal voltages is often based on the frequency of the signal. In these cases, the routing can be done with filter circuits. We will now examine two common forms of signal routing: coupling and decoupling. Both of these techniques are used in nearly all nontrivial electronic systems.

Coupling Circuits

Coupling circuits provide a path for signal voltages to go between two points in the circuit, *neither of which is ground*. Figure 18-47 illustrates the basic function of a coupling circuit.

● **KEY POINTS**

Coupling circuits provide a low-impedance path between two points, neither of which is grounded.

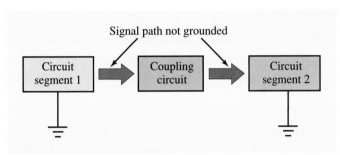

Figure 18-47. A coupling circuit provides passage for signal voltages between two nonground points.

CAPACITIVE COUPLING

One of the most common type of coupling circuit is capacitive coupling. It consists simply of connecting a capacitor between the two points to be coupled. Figure 18-48 illustrates capacitive coupling.

Examination of Figure 18-48 will show that signal voltages will be readily coupled from circuit segment 1 to circuit segment 2 as long as the capacitor has a low reactance

● **KEY POINTS**

Capacitive coupling circuits are essentially high-pass filters.

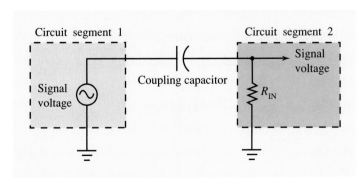

Figure 18-48. Capacitive coupling provides a signal path between two ungrounded points in a circuit.

compared to the resistance value of R_{IN}. R_{IN} is the effective resistance of circuit segment 2. Under these conditions, voltage divider action will cause most of the signal voltage to be developed across R_{IN}. You will recognize the *RC* portion of Figure 18-48 as a high-pass filter. That is, as the signal frequency decreases, the reactance of the coupling capacitor will increase. When the capacitive reactance is larger than the resistance (R_{IN}), then the circuit no longer serves as a good coupling circuit. Most of the signal voltage will be dropped across the coupling capacitor. Proper operation of a capacitive coupling circuit is assured by selecting a value of capacitor that will effectively couple the lowest frequency that is expected. If it couples adequately at the lowest frequency, then it will couple even better at higher frequencies where the capacitive reactance is even lower.

Transformer Coupling

Another popular coupling circuit is illustrated in Figure 18-49. Here, a transformer is used to provide the signal path between two points in the circuit. The transformer may have a 1:1 turns ratio and simply couple the signal, or it may have some other turns ratio, so that it not only couples the signal voltages, but also matches the impedance of circuit segment 1 to the impedance of circuit segment 2. Recall that maximum power is transferred between two points when the impedances are equal. Figure 18-49 shows an iron-core transformer. The same method is also used with other transformer types. The best transformer choice for a particular application is based on frequency.

KEY POINTS

Transformers are also used as coupling circuits. For many applications, transformers can be viewed as high-pass filters.

Regardless of the specific transformer type selected, the coupling circuit will be frequency selective (i.e., it acts like a filter circuit). The most obvious frequency selectivity of a transformer occurs as the frequency decreases. Recall that transformer operation is based on magnetic coupling. The voltage in the secondary exists because the winding is being cut by the magnetic flux lines of the primary. If the frequency is too low, then the voltage in the secondary falls off. At 0 Hz (i.e., dc), there is no relative movement of the magnetic flux and virtually no coupling between the primary and secondary. In this regard, the transformer acts like a high-pass filter.

If we study the circuit even more closely, we will recall that stray capacitance exists between the primary and secondary windings and between the turns on a given winding. The primary-to-secondary capacitance tends to bypass the transformer at high frequencies. The capacitance between adjacent turns on a given winding can cause the transformer to

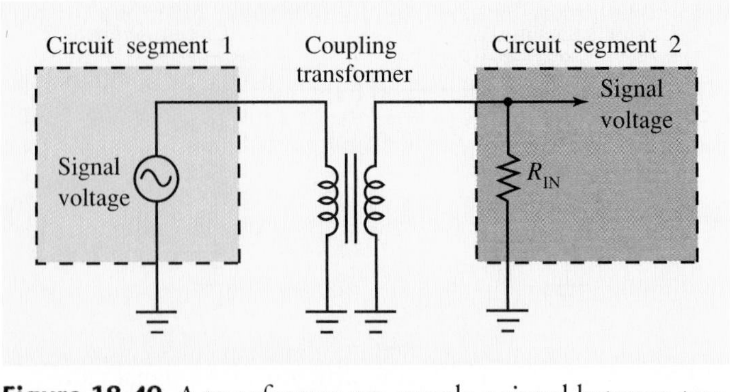

Figure 18-49. A transformer can couple a signal between two ungrounded points in a circuit.

become resonant at some frequency. The effects of parasitic capacitance cause the frequency response of the transformer to deviate from a normal low-pass filter curve. The exact response is highly dependent on the type of transformer. In noncritical applications, the effects of winding capacitance can often be ignored.

Decoupling Circuits

A **decoupling circuit** provides a low-impedance path between an ungrounded point and a grounded point in a circuit. Figure 18-50 illustrates a capacitive decoupling circuit. You will recognize the circuit in Figure 18-50 as a low-pass filter. As long as the signal frequencies are higher than the cutoff frequency of the filter, then minimal voltage (ideally zero) will be dropped across the decoupling capacitor. Since the voltage for circuit segment 2 is taken across the decoupling capacitor (i.e., parallel networks have equal voltage drops), there will be minimal signal voltage available at circuit segment 2. For proper operation, the decoupling capacitor must be selected such that it provides a low-impedance path to the lowest frequency that is to be decoupled. Higher frequencies will be attenuated even more.

The preceding discussion is based on the behavior of an ideal capacitor. You will recall that a practical capacitor has parasitic inductance (ESL) and resistance (ESR) associated with it. These parasitic components appear to be in series with the capacitor. Thus, the capacitor actually behaves like a series RLC circuit and exhibits resonance at some specific frequency. Above this self-resonant frequency, the capacitor behaves more like an inductor than a capacitor. It loses its value as a decoupling circuit for frequencies higher than the self-resonant frequency. For this reason, it is important to use the shortest possible leads on decoupling capacitors, since less lead inductance makes the resonant frequency higher.

Many digital circuits contain literally hundreds of integrated circuit packages. It is common design practice to provide a decoupling capacitor immediately adjacent to the power lead of each integrated circuit. The capacitor permits the dc supply voltage (i.e., low frequency) to pass with no attenuation. High-frequency switching noise generated within the integrated circuit, however, finds a low-reactance path through the decoupling capacitor to ground. This keeps the high-frequency noise from traveling to other portions of the circuit and interfering with proper operation.

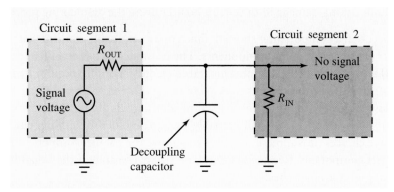

Figure 18-50. A decoupling circuit provides a low-impedance connection between an ungrounded and a ground point in a circuit.

Exercise Problems 18.5

1. Coupling circuits provide a low-impedance path between two ungrounded points in a circuit. (True or False)

2. Decoupling circuits provide a low-impedance path between two grounded points in a circuit. (True or False)

3. A coupling capacitor is selected to have a low reactance at the (*lowest, highest*) expected frequency. This ensures effective coupling at all other signal frequencies.

4. A capacitive coupling circuit is essentially a _____-pass filter.

5. A capacitive decoupling circuit is essentially a _____-pass filter.

18.6 Troubleshooting Passive Filter Circuits

Filter circuits are simply applications of *RLC*, *RL*, and *RC* circuits. Therefore, troubleshooting filter circuits is identical to troubleshooting any other comparable *RLC*, *RL*, or *RC* circuit. Filter-circuit defects include the following possibilities:

- open coil, capacitor, or resistor
- shorted coil, capacitor, or resistor
- wrong value coil, capacitor, or resistor
- intermittent or deteriorated component (e.g., leaky capacitor, poor solder joint).

Diagnosis of each of these defects has been discussed in preceding chapters and will not be repeated here. Rather, we shall provide you some technician troubleshooting tips that can speed your troubleshooting procedure.

⬤ **KEY POINTS**

Confirmation of suspected opens can be readily accomplished without desoldering by bridging the suspected component with a good component.

First, if you suspect an open component, then simply bridge across the suspected component with a known good one *without removing the "bad" component*. If the suspected device is actually defective, then the bridging will correct the problem. You can then spend the time necessary for desoldering and resoldering the components. If the problem still exists with the bridged component in place, then the suspected component is good (or there are multiple defects in the circuit). Bridging provides a quick way to diagnose this type of defect without risking damage to the circuit board during the desoldering process.

If you suspect that a component is shorted, then you can probably confirm your suspicions without having to desolder any components. The resistance of a shorted component will read near 0 Ω in or out of the circuit. This is true regardless of the specific circuit being tested. Occasionally, a component suspected of being shorted will be in parallel with a normally low impedance component (e.g., a transformer winding), which causes a sneak current path. In these cases, you will probably have to desolder something, but remember that you have choices. It is sometimes easier to remove the parallel component than the actual suspect component. Removing either one, however, achieves the isolation needed for a meaningful resistance check.

⬤ **KEY POINTS**

Shorted components can generally be detected in the circuit without desoldering.

Intermittent connections or poor solder joints are sometimes very difficult to locate in any circuit. They are often susceptible to mechanical shock or movement. Your goal is to move or vibrate individual joints while keeping all others still. This can be physically hard to do. One method used by many technicians is to use a pencil with a rubber eraser. By gently tapping or pressing with the eraser on various connections, you can stress one joint or area of the circuit board without stressing or jarring nearby areas. You must be particularly careful not to use a conductive probe to press on the board. It can slip, cause a short, and ultimately damage the circuit even more. Even a pencil may have a metal band near the eraser that demands care in use to avoid unexpected shorts.

Exercise Problems 18.6

1. Diagnosis of filter circuit defects requires specialized test equipment. (True or False)

2. Which of the following filter defects can be located by component bridging?
   - **a.** open coil
   - **b.** shorted capacitor
   - **c.** incorrect resistor value
   - **d.** open resistor
   - **e.** leaky capacitor
   - **f.** open capacitor
   - **g.** primary-to-secondary transformer short

3. When stressing printed-circuit-board connections to reveal a loose or intermittent connection, why is it important to use localized force to stress the connection?

Chapter Summary

- Filter circuits are used to selectively pass (passband) or reject (stopband) signal voltages based on their relative frequencies. Frequencies in the passband of a filter produce an output response that is at least 70.7% of the maximum response (ideally 100%). Frequencies in the stopband, by contrast, produce output responses less than 70.7% of the maximum response (ideally zero). The point between the passband(s) and stopband(s) is called the half-power point. It occurs at the cutoff frequency of the filter. The cutoff frequency produces an output response that is 70.7% of the maximum response.

- Low-pass filter circuits pass all frequencies below the cutoff frequency of the filter. Frequencies higher than the cutoff frequency are attenuated. The stopband for high-pass filters is defined as all frequencies below the cutoff frequency. The passband for high-pass filters consists of all frequencies above the cutoff frequency. High- and low-pass filters can be constructed from either RC or RL circuits.

- Bandpass filters have two stopbands separated by a passband. This type of filter passes those frequencies that lie between the upper and lower cutoff frequency, but attenuates frequencies outside of this passband. Bandstop filters reject (i.e., attenuate) all frequencies that lie between the upper and lower cutoff frequency. Frequencies lower than the lower cutoff frequency or higher than the upper cutoff frequency are passed with minimal attenuation. Bandstop filters can utilize series- or parallel-resonant circuits. They can also be realized with RC or RL networks.

- A special subclass of bandstop filters called notch filters have very narrow bandwidths. A narrow band of frequencies receives maximum attenuation, but frequencies on either side of this stopband have minimal attenuation.

• Coupling and decoupling of signal voltages are two important filter applications. A coupling circuit provides a low-impedance path between two ungrounded points in a circuit. A decoupling circuit provides a low-impedance path between a grounded point and an ungrounded point in a circuit. Capacitive coupling circuits are high-pass filters. Capacitive decoupling circuits are low-pass filters.

• Troubleshooting of passive filter circuits can be accomplished using the same techniques used for other *RC*, *RL*, and *RLC* circuits. Component bridging can help confirm an open component without desoldering. Localized board flexing can be used to identify loose connections or bad solder joints.

Review Questions

Section 18.1: Low-Pass Filters

1. Low-pass filters have two cutoff frequencies. (True or False)

2. The output response is 70.7% or more of the maximum response for frequencies higher than the cutoff frequency in a low-pass filter. (True or False)

3. The passband of a low-pass filter ideally extends to dc. (True or False)

4. The stopband of a low-pass filter ideally extends to an infinite frequency. (True or False)

5. The output response of a low-pass filter is 3 dB lower than the maximum response at the cutoff frequency. (True or False)

6. In most cases, a steeper slope on the frequency response curve is desired for a low-pass filter. (True or False)

7. In order for the circuit represented in Figure 18-51 to function as a low-pass filter, Z_1 must be (*less than, greater than, equal to*) Z_2 for all frequencies above the cutoff frequency.

8. In order for the circuit represented in Figure 18-51 to function as a low-pass filter, Z_1 must be (*less than, greater than, equal to*) Z_2 for all frequencies below the cutoff frequency.

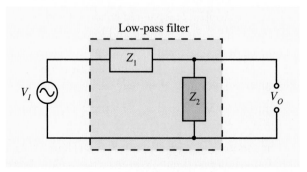

Figure 18-51. A low-pass filter diagram.

9. In order for the circuit represented in Figure 18-51 to function as a low-pass filter, Z_1 must be (*less than, greater than, equal to*) Z_2 at the cutoff frequency.

10. Figure 18-52 shows the frequency response curve for a low-pass filter. Match the following terms to the correct labeled point in Figure 18-52:
    Cutoff frequency_____
    Stopband_____
    Passband_____
    Half-power point_____

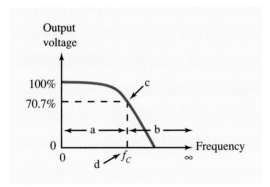

Figure 18-52. Identify the labeled points.

11. The circuit shown in Figure 18-53 is a low-pass filter. (True or False)

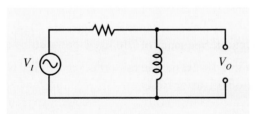

Figure 18-53. Is this a low-pass filter circuit?

12. Calculate the cutoff frequency for the circuit shown in Figure 18-54.

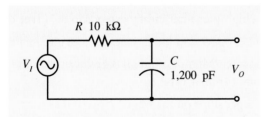

Figure 18-54. What is the cutoff frequency for this circuit?

13. If the component values in Figure 18-54 were changed to $R = 270$ kΩ and $C = 470$ pF, what would be the new cutoff frequency?

14. If the component values shown in Figure 18-54 were both doubled, the cutoff frequency would (*increase, decrease, remain the same*).

15. At frequencies above the cutoff frequency, the resistance in Figure 18-54 is greater than the capacitive reactance. (True or False)

16. Calculate the cutoff frequency for the circuit shown in Figure 18-55.

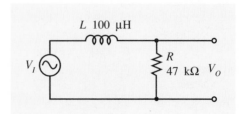

Figure 18-55. What is the cutoff frequency for this circuit?

17. If the component values in Figure 18-55 were changed to $R = 680\ \Omega$ and $L = 25$ mH, what would be the new cutoff frequency?

18. If the component values shown in Figure 18-55 were both doubled, the cutoff frequency would (*increase, decrease, remain the same*).

19. At frequencies below the cutoff frequency, the reactance of the coil in Figure 18-55 is less than the value of the resistor. (True or False)

20. Refer to Figure 18-55. If the resistance is changed to 12 kΩ, what value inductance is required to have a cutoff frequency of 50 kHz?

21. Refer to Figure 18-54. If the capacitance is changed to 470 pF, what value resistance is required to have a cutoff frequency of 780 kHz?

22. A low-pass filter with a 10-dB/decade roll-off is a more ideal filter than one that has a 40-dB/decade roll-off. (True or False)

Section 18.2: High-Pass Filters

23. High-pass filters have two cutoff frequencies. (True or False)

24. The output response is 70.7% or more of the maximum response for frequencies higher than the cutoff frequency in a high-pass filter. (True or False)

25. The passband of a high-pass filter ideally extends to dc. (True or False)

26. The stopband of a high-pass filter ideally extends to an infinite frequency. (True or False)

27. The output response of a high-pass filter is 3 dB lower than the maximum response at the cutoff frequency. (True or False)

28. In most cases, a steeper slope on the frequency response curve is desired for a high-pass filter. (True or False)

29. In order for the circuit represented in Figure 18-56 to function as a high-pass filter, Z_1 must be (*less than, greater than, equal to*) Z_2 for all frequencies above the cutoff frequency.

30. In order for the circuit represented in Figure 18-56 to function as a high-pass filter, Z_1 must be (*less than, greater than, equal to*) Z_2 for all frequencies below the cutoff frequency.

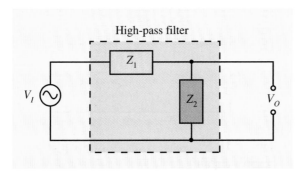

Figure 18-56. A high-pass filter diagram.

31. In order for the circuit represented in Figure 18-56 to function as a high-pass filter, Z_1 must be (*less than, greater than, equal to*) Z_2 at the cutoff frequency.

32. Figure 18-57 shows the frequency response curve for a high-pass filter. Match the following terms to the correct labeled point in Figure 18-57:

Cutoff frequency _____

Stopband _____

Passband _____

Half-power point _____

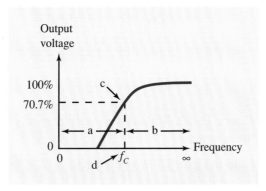

Figure 18-57. Identify the labeled points.

33. The circuit shown in Figure 18-58 is a high-pass filter. (True or False)

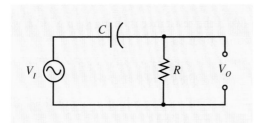

Figure 18-58. Is this a high-pass filter circuit?

34. Calculate the cutoff frequency for the circuit shown in Figure 18-59.

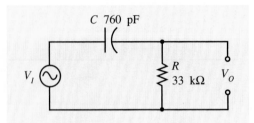

Figure 18-59. What is the cutoff frequency for this circuit?

35. If the component values in Figure 18-59 were changed to $R = 470$ kΩ and $C = 1,000$ pF, what would be the new cutoff frequency?

36. If the component values shown in Figure 18-59 were both doubled, the cutoff frequency would (*increase, decrease, remain the same*).

37. At frequencies above the cutoff frequency, the resistance in Figure 18-59 is greater than the capacitive reactance. (True or False)

38. Calculate the cutoff frequency for the circuit shown in Figure 18-60.

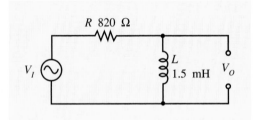

Figure 18-60. What is the cutoff frequency for this circuit?

39. If the component values in Figure 18-60 were changed to $R = 200$ kΩ and $L = 100$ mH, what would be the new cutoff frequency?

40. If the component values shown in Figure 18-60 were both doubled, the cutoff frequency would (*increase, decrease, remain the same*).

41. At frequencies below the cutoff frequency, the reactance of the coil in Figure 18-60 is less than the value of the resistor. (True or False)

42. Refer to Figure 18-60. If the resistance is changed to 47 kΩ, what value inductance is required to have a cutoff frequency of 75 kHz?

43. Refer to Figure 18-59. If the capacitance is changed to 470 pF, what value resistance is required to have a cutoff frequency of 2.5 kHz?

44. A high-pass filter with a 12-dB/octave roll-off is a more ideal filter than one that has a 6-dB/octave roll-off. (True or False)

Section 18.3: Bandpass Filters

45. Bandpass filters have two cutoff frequencies. (True or False)

46. Bandpass filters attenuate all frequencies above the upper cutoff frequency. (True or False)

47. Frequencies lower than the lower cutoff frequency pass through a bandpass filter with minimal attenuation. (True or False)

48. Bandpass filters provide minimal attenuation for frequencies within the passband. (True or False)

49. If the circuit in Figure 18-61 is a bandpass filter, then Z_1 must be greater than Z_2 for frequencies within the passband. (True or False)

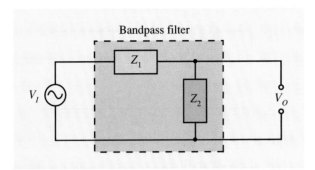

Figure 18-61. A bandpass filter diagram.

50. Z_1 must be greater than Z_2 in Figure 18-61 for frequencies in the stopband. (True or False)

51. If Z_1 and Z_2 in Figure 18-61 are equal, what can be said about the input frequency?

52. The circuit shown in Figure 18-62 is a bandpass filter. (True or False)

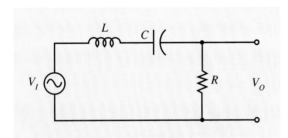

Figure 18-62. Is this a bandpass filter?

53. For frequencies below the lower cutoff frequency, the capacitive reactance in Figure 18-63 is greater than the inductive reactance. (True or False)

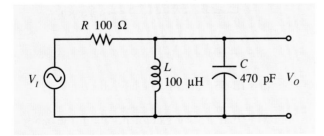

Figure 18-63. A bandpass filter.

54. The capacitive reactance in Figure 18-63 is equal to the inductive reactance at the center frequency of the passband. (True or False)

55. Compute the center frequency of the passband for the circuit shown in Figure 18-63.

56. If the series resistance (R) is doubled in value for the circuit shown in Figure 18-63, does the center frequency of the filter change?

57. Compute the center frequency for the circuit shown in Figure 18-64.

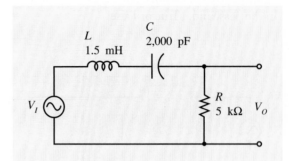

Figure 18-64. A filter circuit.

58. At frequencies in the passband of the circuit in Figure 18-64, the value of resistance is greater than the net reactance. (True or False)

59. At frequencies higher than the upper cutoff frequency in Figure 18-64, the combined reactance of L and C is greater than the value of resistance (R). (True or False)

60. Figure 18-65 shows the frequency response curve for a bandpass filter. Match the following terms to the correct labeled point in Figure 18-65:

Cutoff frequencies_____
Stopband_____
Passband_____
Half-power points_____

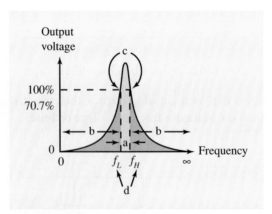

Figure 18-65. Identify the labeled points

61. Circle all of the following that are true statements about the bandpass filter shown in Figure 18-66.

a. Resistor R_1 must be greater than resistor R_2

b. Capacitor C_1 must be the same size as capacitor C_2.

c. The cutoff frequency of the R_1C_1 network must be the same as the cutoff frequency of the R_2C_2 network.

d. The cutoff frequency of the R_1C_1 network must be higher than the cutoff frequency of the R_2C_2 network.

e. This circuit cannot be used as a bandpass filter.

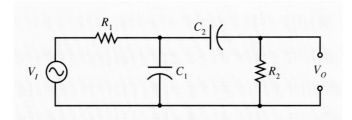

Figure 18-66. A possible filter circuit.

Section 18.4: Bandstop or Band-Reject Filters

62. Frequencies within the stopband of a bandstop filter receive minimal attenuation. (True or False)

63. Frequencies above the upper cutoff frequency receive minimal attenuation in a bandstop filter. (True or False)

64. Frequencies below the lower cutoff frequency receive minimal attenuation in a bandstop filter. (True or False)

65. A notch filter has a wider bandwidth than most bandstop filters.

66. An ideal bandstop filter would have a roll-off of 3 dB/decade. (True or False)

67. In order for the circuit shown in Figure 18-67 to function as a bandstop filter, Z_1 must be greater than Z_2 in the passband of the filter. (True or False)

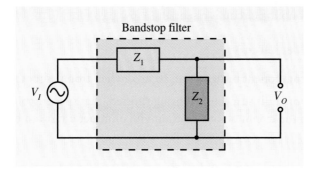

Figure 18-67. A bandstop filter.

68. Z_1 in Figure 18-67 must be greater than Z_2 for all frequencies higher than the lower cutoff frequency. (True or False)

69. Z_1 in Figure 18-67 must be greater than Z_2 for all frequencies higher than the upper cutoff frequency. (True or False)

70. If Z_1 has the same impedance as Z_2 in Figure 18-67, then the circuit is operating at a cutoff frequency. (True or False)

71. What is the name of the circuit configuration shown in Figure 18-68?

72. Calculate the center frequency of the circuit shown in Figure 18-68.

73. What value of resistance is required for R_2 in Figure 18-68 if the cutoff frequency is changed to 12 kHz? Assume the capacitor values remain unchanged.

74. The circuit shown in Figure 18-68 is characterized by a wide bandwidth. (True or False)

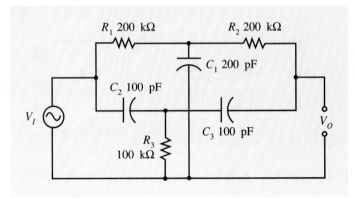

Figure 18-68. A filter circuit.

Section 18.5: Applied Technology: Coupling and Decoupling Circuits

75. A circuit that provides a low-impedance path between a grounded point and an ungrounded point in a circuit is called a coupling circuit. (True or False)

76. A typical decoupling circuit consists of a capacitor connected between ground and the point to be decoupled. (True or False)

77. Name an advantage of transformer coupling circuits over capacitive coupling circuits.

78. What is meant by the term *self-resonant frequency* with reference to capacitors?

79. Decoupling capacitors should always have the shortest possible leads. (True or False)

80. A capacitive coupling circuit is essentially a low-pass filter. (True or False)

Section 18.6: Troubleshooting Passive Filter Circuits

81. Discuss how component bridging can be used when troubleshooting a filter circuit.

82. When a component is suspected of being shorted, that specific component must be removed from the circuit before a meaningful ohmmeter test can be done. (True or False)

TECHNICIAN CHALLENGE

Following is a brief description of several filter applications. Based on the descriptions and your knowledge of filter circuits, select the type of circuit that should be used for each application. You may choose from any of the following filter types:

- low-pass
- high-pass
- bandpass
- bandstop
- notch

APPLICATION 1   The electrical system on an automobile is generating high-frequency noise that is interfering with radio reception. You have determined that the noise is getting to the radio circuit via the +12-Vdc power wire. What type of filter circuit can you put in the dc power wire to attenuate the interference?

APPLICATION 2   Individual stages in a multistage transistor audio-frequency amplifier circuit require dc voltages to operate, but the dc voltages in one stage are often different from the dc voltages in a subsequent stage. What type of filter could be used between successive stages to pass the audio signal but block the dc voltages?

APPLICATION 3   A composite television signal has both picture and sound information, but they are at different frequencies. The sound and picture information in the received signal must be split apart and processed by separate circuits. If the sound information is allowed to travel to the picture circuits, then picture distortion results. The sound information is contained in a well-defined band of frequencies. What kind of filter could be placed at the input of the picture circuits that would eliminate the sound information, but would allow information at all other frequencies to pass?

APPLICATION 4   The audio amplifier circuitry for a magnetic tape player is very sensitive, since it must respond to very small signals. Unfortunately, this high sensitivity can lead to problems if the 60-Hz power-line noise is allowed to get into the input amplifier circuitry. Normal audio signals range from 20 Hz to at least 15 kHz. Unfortunately, 60 Hz is in this band of desired frequencies. What type of filter could be used to attenuate 60-Hz noise and still pass most of the desired signals with minimal attenuation?

APPLICATION 5   You have an inexpensive transistor radio that has a "tinny" sound. You believe the radio would sound less annoying if the higher audio frequencies were prevented from reaching the speaker. What type of filter could be connected in line with the speaker to solve this problem?

Equation List

$$(18\text{-}1) \quad f_C = \frac{1}{2\pi RC}$$

$$(18\text{-}2) \quad f_C = \frac{R}{2\pi L}$$

$$(18\text{-}3) \quad f_N = \frac{1}{2\pi R_1 C_2}$$

appendix A

PShooter Measurement Values

1. 54.4 k	33. 51.7 k	65. 4.7 k	97. 4.7 k
2. 50	34. 0.5 μ	66. 120	98. low
3. 245 m	35. 10	67. zero	99. ∞
4. 470	36. ∞	68. 0.5 μ	100. 4.7 k
5. ∞	37. 7.56	69. 16.41	101. 8.2 k
6. 8.65	38. −25	70. 25	102. −12.8
7. 8.6	39. 6.8 k	71. 7.39	103. 25
8. 39 k	40. 5.48	72. zero	104. 10
9. 3.5	41. zero	73. 39 k	105. zero
10. 1.2	42. 2.99 k	74. normal	106. 17.3
11. 150	43. 1.0 k	75. 59.4	107. 49 k
12. 4.7 k	44. 1.69	76. −12.8	108. 2.99 k
13. 1.8 k	45. 27 k	77. 8.6 k	109. 11.9
14. 3.6 k	46. 24.5	78. 25	110. −16.1
15. 10 k	47. 3.8	79. indeterminate	111. 50
16. 4.7 k	48. 11.3 k	80. ∞	112. 470
17. 8.2 k	49. 10	81. 9.26 m	113. ∞
18. 2.2	50. 943 μ	82. 0 → ∞	114. 27
19. 50	51. 25	83. 3.29	115. 1.5 p
20. 7.56	52. 38	84. 2.7 k	116. 6.8 k
21. 25	53. 4.7 k	85. 16.41	117. 22.9
22. zero	54. 6.6	86. 39 k	118. 0.5 μ
23. 2.99 k	55. 47 k	87. normal	119. ∞
24. 10 k	56. 100	88. zero	120. −12.8
25. 5.48	57. zero	89. 39 k	121. 470
26. 4.7 k	58. ∞	90. 50	122. 42.2
27. zero	59. 1.8 k	91. 2.7 k	123. 4.5 k
28. 75	60. 3.8	92. 24.5	124. 38.4
29. 5.6	61. zero	93. 57 k	125. 59.4
30. 2.2 μ	62. 470	94. 0.7	126. 6.6
31. 61.2 k	63. 50	95. −25	127. 25
32. 0 → ∞	64. 590	96. 8.2 k	128. 15

129. 23.83	161. normal	193. 24.5	225. 20
130. −50	162. 11.3 k	194. 2.2 μ	226. 2 k
131. 67 k	163. 25	195. 1,000	227. 200
132. zero	164. 8.2 k	196. ∞	228. 20
133. 50	165. 20.8	197. 10	229. 500
134. 470	166. 3.9 k	198. 9.07 m	230. 20
135. 18 k	167. 6.28 m	199. 0.6	231. −18.54
136. 75	168. 7.56	200. zero	232. 31.46
137. zero	169. zero	201. 275	233. 560
138. ∞	170. normal	202. high	234. 330
139. 31.7 m	171. 2.99 k	203. 0.65	235. −50
140. 0 → ∞	172. −12.6	204. ∞	236. −18.54
141. zero	173. −6	205. 76 k	237. 330
142. 13.9 m	174. 24.5	206. −6	238. 890
143. 8.2 k	175. 275	207. normal	239. 330
144. zero	176. zero	208. 10	240. −18.54
145. high	177. ∞	209. 16.41	241. 9
146. 10	178. 7.56	210. −12	242. 11
147. 20.8	179. 5.5	211. 8.6 k	243. 2.89 k
148. 8.65	180. 25	212. 24.5	244. 11
149. 39 k	181. ∞	213. 1.15	245. 2.89 k
150. 120	182. 13.6 m	214. 2.7 k	246. 820
151. zero	183. 274 m	215. normal	247. 1.07 k
152. 16.41	184. 10	216. 18.86	248. 9
153. 3.9 k	185. 2.99 k	217. 25	249. 11
154. 25	186. 4.6 k	218. 200	250. 684
155. 5.48	187. 470	219. 50	251. 1.8 k
156. 44.4	188. zero	220. 8.88	252. 26
157. 7.58 m	189. 22.9	221. 20	253. 1.53 k
158. 50	190. 25	222. 900	254. 4.41 k
159. 8.2 k	191. 0 → ∞	223. 20	255. 37.6
160. 50	192. 171 k	224. 100	256. 2.21 k

appendix

A

257. 26	277. 3.99	297. 1.77	317. 520
258. 1.53 k	278. 680	298. 5.6 k	318. 0.6
259. 20.4	279. 1.2 k	299. 2.2 k	319. 23.3
260. 1.6 k	280. 5.6	300. 150	320. 150
261. 37.5	281. 3.13	301. 389	321. 35
262. 75	282. 680	302. 2.14 k	322. 150
263. 12 k	283. 9.1 k	303. 910	323. 1.2
264. 1,200	284. 13.5	304. 150	324. 5.97
265. 3.13	285. 1.39 k	305. 1.68	325. 879.7
266. 7.39 k	286. 150	306. 6.53	326. 29
267. 1.53 k	287. 23.3	307. 6.53	327. 52
268. 17.5	288. 12 k	308. 1.49 k	328. 150
269. 11.6	289. 1.39 k	309. 150	329. 5.97
270. 1,200	290. 330	310. 16.8	330. 1 k
271. 20.4	291. 150	311. 389	331. 35
272. 2.2 k	292. 13.5	312. 3.9 k	332. 100
273. 19.3	293. 1,200	313. 910	333. 35
274. 5.74	294. 150	314. 1,000	334. zero
275. 6.71 k	295. 0.86	315. 22 k	335. 879.7
276. 9.59 k	296. 6.53	316. 29	

Table of Standard Resistor Values

Note: All values are available in 5% tolerance. Values set in **bold-face** type are available in 10% tolerance.

1.0 Ω	**15 Ω**	**220 Ω**	**3.3 kΩ**
1.1 Ω	16 Ω	240 Ω	3.6 kΩ
1.2 Ω	**18 Ω**	**270 Ω**	**3.9 kΩ**
1.3 Ω	20 Ω	300 Ω	4.3 kΩ
1.5 Ω	**22 Ω**	**330 Ω**	**4.7 kΩ**
1.6 Ω	24 Ω	360 Ω	5.1 kΩ
1.8 Ω	**27 Ω**	**390 Ω**	**5.6 kΩ**
2.0 Ω	30 Ω	430 Ω	6.2 kΩ
2.2 Ω	**33 Ω**	**470 Ω**	**6.8 kΩ**
2.4 Ω	36 Ω	510 Ω	7.5 kΩ
2.7 Ω	**39 Ω**	**560 Ω**	**8.2 kΩ**
3.0 Ω	43 Ω	620 Ω	9.1 kΩ
3.3 Ω	**47 Ω**	**680 Ω**	**10 kΩ**
3.6 Ω	51 Ω	750 Ω	11 kΩ
3.9 Ω	**56 Ω**	**820 Ω**	**12 kΩ**
4.3 Ω	62 Ω	910 Ω	13 kΩ
4.7 Ω	**68 Ω**	**1.0 kΩ**	**15 kΩ**
5.1 Ω	75 Ω	1.1 kΩ	16 kΩ
5.6 Ω	**82 Ω**	**1.2 kΩ**	**18 kΩ**
6.2 Ω	91 Ω	1.3 kΩ	20 kΩ
6.8 Ω	**100 Ω**	**1.5 kΩ**	**22 kΩ**
7.5 Ω	110 Ω	1.6 kΩ	24 kΩ
8.2 Ω	**120 Ω**	**1.8 kΩ**	**27 kΩ**
9.1 Ω	130 Ω	2.0 kΩ	30 kΩ
10 Ω	**150 Ω**	**2.2 kΩ**	**33 kΩ**
11 Ω	160 Ω	2.4 kΩ	36 kΩ
12 Ω	**180 Ω**	**2.7 kΩ**	**39 kΩ**
13 Ω	200 Ω	3.0 kΩ	43 kΩ

appendix
B

47 kΩ	240 kΩ	**1.2 MΩ**	6.2 MΩ
51 kΩ	**270 kΩ**	1.3 MΩ	**6.8 MΩ**
56 kΩ	300 kΩ	**1.5 MΩ**	7.5 MΩ
62 kΩ	**330 kΩ**	1.6 MΩ	**8.2 MΩ**
68 kΩ	360 kΩ	**1.8 MΩ**	9.1 MΩ
75 kΩ	**390 kΩ**	2.0 MΩ	**10 MΩ**
82 kΩ	430 kΩ	**2.2 MΩ**	11 MΩ
91 kΩ	**470 kΩ**	2.4 MΩ	**12 MΩ**
100 kΩ	510 kΩ	**2.7 MΩ**	13 MΩ
110 kΩ	**560 kΩ**	3.0 MΩ	**15 MΩ**
120 kΩ	620 kΩ	**3.3 MΩ**	16 MΩ
130 kΩ	**680 kΩ**	3.6 MΩ	**18 MΩ**
150 kΩ	750 kΩ	**3.9 MΩ**	20 MΩ
160 kΩ	**820 kΩ**	4.3 MΩ	**22 MΩ**
180 kΩ	910 kΩ	**4.7 MΩ**	
200 kΩ	**1.0 MΩ**	5.1 MΩ	
220 kΩ	1.1 MΩ	**5.6 MΩ**	

appendix
C

Conversion Table for Magnetic Units

QUANTITY	SYMBOL OR ABBREVIATION	TO CONVERT FROM	TO	MULTIPLY BY
Flux	Φ	webers	maxwells	1×10^8
		maxwells	webers	1×10^{-8}
Flux density	B	tesla	gauss	1×10^4
		gauss	tesla	1×10^{-4}
Magnetizing force	H	ampere-turns per meter	oersteds	0.01256
		oersteds	ampere-turns per meter	79.577
Magnetomotive force	$\mathcal{F}$	ampere-turns	gilberts	1.257
		gilberts	ampere-turns	0.796
Absolute permeability	μ	webers per ampere-turn-meter	gauss per oersted	795.7×10^3
		gauss per oersted	webers per ampere-turn-meter	1.257×10^{-6}
Relative permeability		No units		
Permeance	$\mathcal{P}$	webers per ampere-turn	maxwells per gilbert	79.6×10^6
		maxwells per gilbert	webers per ampere-turn	1.257×10^{-8}
Reluctance	$\mathfrak{R}$	ampere-turns per weber	gilberts per maxwell	1.257×10^{-8}
		gilberts per maxwell	ampere-turns per weber	79.6×10^6

Shaded entries identify SI (MKS) units. Unshaded entries are CGS units.

glossary

ac resistance The effective resistance of a component or device to alternating current. Also called dynamic resistance.

Accuracy A meter specification that indicates the largest difference between actual and indicated values.

Admittance (Y) The reciprocal of impedance. The ability of RC, RL, and RLC circuits to pass alternating current. Measured in siemens.

Air gap Any portion of a magnetic circuit where the magnetic flux passes through air.

Alphanumeric Consisting of both digits (0–9) and letters (A–Z), and often including special characters (e.g., !, ~, &, %, and so on).

Alternating current (ac) Current that periodically changes direction.

Alternator A device that converts mechanical energy into alternating voltage and current.

Amalgamation Used in the manufacture of carbon-zinc cells to reduce local action by coating the zinc electrode with mercury.

Ammeter A device used to measure current.

Ampere (A) The unit of measure for current.

Ampere-hour (A · h) The unit of measure for the capacity of a cell or battery.

Ampere-turn (A · t) The unit of measure for magnetomotive force.

Amplification The process of converting a small changing signal into a larger, but corresponding, signal.

Amplitude The maximum (unless otherwise stated) value of a waveform.

Analog A class of devices or systems that utilize a continuous range of values.

Anode The negative terminal of a cell. Also the more positive terminal of a conducting diode.

Apparent power (P_A) The phasor sum of true power and reactive power.

Armature The movable part of a motor or relay. Also refers to the winding of an alternator or generator where the voltage is induced.

Asymmetrical Not symmetrical. Generally refers to a waveform with dissimilar half-cycles.

Atom The smallest part of an element that still has the properties of the element.

Attenuation A reduction or loss of signal amplitude.

Autotransformer A transformer with a single tapped winding that serves as both primary and secondary.

Average value The sum of several instantaneous values of an electrical quantity divided by the number of values. In the case of a sinusoidal waveform, the average value is zero for a full cycle and $0.636 \, V_{PEAK}$ for a half-cycle.

AWG An abbreviation for American wire gage.

Ayrton shunt *See* Universal Shunt.

Backoff scale An ohmmeter scale that reads from right to left. The nonlinear scale has zero ohms on the far right and infinite ohms on the far left of the scale.

Balanced bridge A condition in a bridge circuit where the two legs have equal resistance ratios and the voltage across the middle of the bridge is zero.

Band-reject filter A filter circuit that attenuates a band of frequencies, but allows higher and lower frequencies to pass with minimal attenuation. Also called a bandstop filter.

Bandstop filter *See* Band-reject Filter.

Bandpass The band of frequencies that pass through a circuit with at least 70.7% of the maximum amplitude.

Bandpass filter A filter circuit that passes a band of frequencies with minimal attenuation, but attenuates all frequencies lower or higher than the passband.

Bandwidth The width (in hertz) of the range between the half-power points of a resonant circuit or a filter circuit.

Battery A series and/or parallel connection of two or more cells.

BH curve A graph showing the relationship between flux density (B) and magnetizing force (H) for a given material.

Bimetallic strip A structure consisting of two dissimilar metals that bends in response to temperature changes due to the unequal coefficients of expansion.

Bleeder current The current in a voltage divider that flows with or without a load connected.

Block diagram A generalized sketch that represents the functional components of a system with graphic boxes. The boxes are interconnected with lines to show the flow (generally signal flow in block diagrams of electronic systems).

Branch A current path in a circuit that is in parallel with one or more other paths.

Break frequency The frequency where the output of a circuit falls to 70.7% of its maximum voltage or current response. This also corresponds to the half-power (–3 dB) point(s) on the frequency response curve.

Bridge circuit A special series-parallel circuit consisting of two parallel branches with two series resistors in each branch. The output is taken between the midpoints of the two parallel branches. Bridge circuits are widely used in instrumentation circuits.

Brush A graphite block that provides sliding contact to a slip ring (alternator or ac motor) or commutator (dc motor or generator).

Cable Two or more individually insulated wires bound in a common sheath.

Capacitance (C) The electrical property that allows a capacitor to store energy in an electrostatic field. Measured in farads.

Capacitive reactance (X_C) The opposition to alternating current flow offered by capacitance. It is measured in ohms.

Capacitor An electrical component consisting of two or more conductive plates separated by insulation and used to store energy in an electrostatic field.

Cathode The positive terminal of a cell. Also the more negative terminal of a conducting diode.

Cell A single, stand-alone source of electrical energy (e.g., electrochemical cell, fuel cell, solar cell).

Cemf The voltage induced in a conductor or inductor by a changing current. The

glossary

induced voltage opposes the changing current.

Center tap A coil or transformer winding that has a connection in the middle of the winding.

CGS An older system of measurement based on the centimeter, gram, and second.

Charge A term used to describe the condition that exists when a material or region has unequal numbers of electrons and protons.

Chassis The metal frame or cabinet used to house an electronic system. The chassis is frequently, but not always, connected to ground.

Chattering The audible noise that results when a normally closed contact of a relay is connected in series with its own coil.

Chip Refers to the physical semiconductor material used to make integrated circuits and other solid-state devices.

Choke A name given to an inductor when its primary purpose is to attenuate high frequencies.

Circuit Any configuration of electrical and/or electronic devices interconnected with conductors.

Circuit breaker A protective device that mechanically opens a circuit if the current through the circuit exceeds the trip current of the breaker.

Circular mil (cmil) A unit of measure for wire area. One circular mil is the cross-sectional area of a round wire that has a diameter of one mil (1/1,000th of an inch).

Closed circuit A circuit that has a complete path for current.

Coaxial cable A shielded, two-conductor cable where the braided outer conductor completely surrounds the insulated inner conductor.

Coefficient of coupling (k) A dimensionless number that represents the percentage of total flux that is common between two coils or circuits.

Coercive force The amount of magnetizing force needed to overcome or cancel the residual magnetism in a magnetic material.

Coil An inductor or other device formed by winding multiple turns of wire on a core (even an air core).

Coil resistance The dc resistance (measured with an ohmmeter) of a coil.

Color code A standardized scheme that uses colored markings to identify component values or part numbers. Also used to identify leads on transformers.

Commutator A segmented conductor that is mounted on the rotor of a dc generator or motor and makes sliding contact with the brushes. Allows power to be applied (or removed) from a rotating coil.

Complex circuit A circuit configuration that cannot be simplified by replacing sets of series and parallel components with equivalent components.

Complex number A number composed of both real (resistive) and imaginary (reactive) parts used to describe a phasor.

Complex voltage source A circuit with both series-aiding and series-opposing cells or voltage sources.

Compound A material that consists of two or more elements that are chemically bonded.

Conductance (*G*) The reciprocal of resistance. It describes the ability of a circuit to allow current flow. It is measured in siemens.

Conductor A material with a very low resistance that readily permits current flow.

Constant current A source of current that is constant at all times and is unaffected by changes in circuit resistance.

Constant voltage A source of voltage that is constant at all times and is unaffected by changes in circuit resistance.

Contact bounce The series of momentary opens and closures that occurs when mechanical switch or relay contacts are moved.

Continuity A term used to indicate the presence of a complete path for current.

Conventional current A convention that conceptualizes current as flowing in a direction opposite that of electron flow. Conventional current moves from positive to negative through a complete circuit.

Copper loss The energy converted to heat by the resistance in the windings of a transformer or other electromagnetic device.

Core The material used as the central part of an inductor. It concentrates the magnetic flux and physically supports the coil windings.

Core loss The energy converted to heat in the core material of an electromagnetic device. Core loss consists primarily of losses due to hysteresis and eddy currents.

Cosine A trigonometric function of an angle in a right triangle that is equal to the length of the adjacent side divided by the length of the hypotenuse.

Coulomb (C) The SI unit of measure for charge. One coulomb is the amount of charge of 6.25×10^{18} electrons.

Coulomb's Law The law that relates the force between two charged bodies as a function of the strength of the two charges and the distance between them.

Counter emf *See* Cemf.

CRT An abbreviation for cathode-ray tube. It is the display tube used in most oscilloscopes, televisions, and radars.

Cryogenics The study of the behavior of materials as they approach absolute zero (−273.2°C).

Crystal A material whose atoms form a consistent lattice pattern throughout the bulk of the material.

Current The movement of charged particles. Generally used to describe the movement of electrons.

Current divider A circuit that consists of two or more parallel branches. It is used to divide the total current into two or more components.

Current probe An ammeter attachment that can sense the value of current flow in a circuit *without* breaking the circuit.

Current source *See* Constant Current.

Cutoff frequency *See* Break Frequency.

Cycle One complete repetition of a periodic waveform consisting of two alternations.

glossary

D'Arsonval movement A common meter movement consisting of a moving coil attached to a pointer and suspended in the field of a permanent magnet.

Damping A technique used to prevent dramatic overswings in pointer movement on analog meters.

dc resistance The ohmic resistance as measured by an ohmmeter.

Decade A tenfold (i.e., 10:1 or 1:10) change in the value of a quantity.

Decibel (dB) The logarithmic unit of measure for a ratio.

Degauss Demagnetize.

Degenerative feedback *See* Negative Feedback.

Degree An angular measure equal to 1/360th of a full circle.

Delta configuration A configuration where three components are connected in a loop with connection made at each node. Also called a pi configuration. One of two common connections for three-phase transformer windings.

Diamagnetic A material with a relative permeability of less than one.

Dielectric An insulator.

Dielectric constant (*k*) A measure of the ability of a material (relative to air) to concentrate an electric field.

Dielectric strength A measure of a dielectric material's ability to withstand high voltage.

Digital A device or system that utilizes discrete (noncontinuous) values.

Diode A two-terminal electronic component that permits current flow in only one direction.

Direct current (dc) Current that flows in only one direction and generally has a constant value.

Directly proportional Two quantities are directly proportional when increases in one quantity cause corresponding increases in the second.

DMM Digital multimeter. The digital equivalent of an analog VOM.

Domain A region within a magnetic material that behaves like a small bar magnet. Alignment of the domains in a material results in the material being magnetized.

Dot notation A method used on schematics to identify in-phase points on a multiple-winding transformer.

Drop-out current The value of coil current that permits the contacts of a relay to return to their normal state after the relay has been energized.

Drop-out voltage The value of coil voltage that permits the contacts of a relay to return to their normal state after the relay has been energized.

Dropping resistor A name given to a resistor that is placed in series with another circuit and whose primary purpose is to drop a portion of the total voltage.

DVM Digital voltmeter. A digital meter that measures voltage.

Dynamic resistance *See* ac Resistance.

Eddy current A circulating current in the core of an electromagnetic device that results from the conductive core material being cut by changing magnetic flux. Eddy currents produce a heat loss in the core.

Effective value *See* Root-mean-square.

Efficiency The ratio of output power to input power. Ideally 100%, but always less in practice.

Electrode An electrical terminal of a cell or other electrical device that permits current to enter/exit the device.

Electrolyte A liquid or paste with mobile ions that reacts with the electrodes in an electrochemical cell and permits current flow between the anode and cathode of the cell.

Electromagnet A magnet produced by the flow of current through a coil. The magnetic field is present only so long as the current is flowing.

Electromagnetic field A magnetic field produced by current flow through a conductor.

Electromagnetic induction The process that causes a voltage to be produced in a wire as it is intercepted by moving magnetic flux.

Electromechanical A device whose operation is both electrical and mechanical.

Electromotive force (emf) A voltage source or potential difference that is sustained as charges are moved through the circuit. The electrical pressure that causes sustained current to flow in an electrical circuit. Measured in volts.

Electron A negatively charged atomic particle that orbits the nucleus of all atoms.

Electron current flow A convention that conceptualizes current as the movement of electrons. Electron current flows from negative to positive through a complete circuit.

Electronic Components or devices that control or regulate the flow of electrons with active devices (e.g., transistors).

Electrostatic field The region around a charged body where another charged body would experience a force of attraction or repulsion.

Element A material whose atoms are all identical. An element cannot be subdivided by chemical means.

Emf *See* Electromotive Force.

Energy The ability to do work.

Energy level Used to describe the energy content of an orbiting electron. Higher energy levels correspond to orbits that are farther from the nucleus.

Equivalent circuit A simplified circuit that provides similar performance to a more complex circuit under a given set of conditions.

Equivalent series inductance (ESL) The parasitic inductance of a capacitor caused primarily by the capacitor lead wires.

Equivalent series resistance (ESR) The effective resistance (i.e., equivalent power loss) of a capacitor caused primarily by dielectric losses.

EVM Electronic voltmeter. An analog VOM that has electronic circuitry, which makes the meter appear more ideal.

Farad (F) The unit of measure for capacitance.

Faraday shield A grounded conductive shield used to separate the primary and

glossary

secondary windings of a transformer to reduce coupling via the interwinding capacitance.

Fast-blo A type of fuse that is designed to open quickly.

Feedback The portion of a signal in an electronic system that is returned to a prior stage of the same system.

Ferrite A magnetic ceramic material used as a core material for transformers and coils. Also used in the form of a bead or clamp to attenuate high frequencies.

Ferromagnetic A material with a relative permeability much greater than one.

Field winding A coil in a motor, generator, or alternator that is used to create a steady magnetic field.

Filter A circuit designed to pass certain frequencies and reject (i.e., attenuate) other frequencies.

Firing voltage The minimum voltage required to ionize the gas in a neon lamp.

Flux Magnetic or electric lines of force

Flux density (B) The number of magnetic lines of force per unit area.

Flux leakage Magnetic flux that is intended to but does not link two circuits (e.g., primary and secondary of a transformer).

Free electron An electron that has been disassociated from its parent atom and is no longer in orbit.

Frequency (F) The number of complete cycles of a periodic waveform per unit time.

Fringing A phenomenon that causes magnetic flux lines to "bulge" when they pass between two high-permeability materials that are separated by a low-permeability material (e.g., an air gap in a magnetic core).

Fuel cell A special type of electrochemical cell that must have a continuous supply of external fuel (e.g., hydrogen and oxygen).

Full-scale current The amount of current required to fully deflect an analog meter movement.

Full-scale voltage The amount of voltage required to fully deflect an analog meter movement.

Fully-specified circuit A circuit with all component values given on the schematic diagram.

Function generator An electronic instrument that generates periodic waveforms such as sine waves, rectangular waves, and triangular waves.

Fuse A protective device that burns open to protect a circuit if the current through the circuit exceeds the rating of the fuse.

Ganged The mechanical connecting together of two or more adjustable components (e.g., switches or potentiometers) such that they operate simultaneously.

Gauss (G) The CGS unit of measure for magnetic flux density.

Generator An electromechanical device that converts mechanical energy into electrical energy in the form of direct voltage and current.

Giga (G) A prefix used to represent 10^9.

Gilbert The CGS unit of measure for magnetomotive force.

glossary

Graticule A grid on the face of an oscilloscope CRT used to increase reading accuracy.

Ground The point in a circuit to which all voltage measurements are referenced. Sometimes used to refer to earth ground.

Ground plane A metallic plane having zero potential with reference to circuit ground.

Half-digit A digital indicator capable of displaying ±1.

Half-power frequency *See* Break Frequency.

Harmonics Any whole multiple of a base frequency called the fundamental frequency.

Henry (H) The SI unit of measure for inductance.

Hertz (Hz) The SI unit of measure for frequency.

High-pass filter A filter circuit that passes all frequencies above the cutoff frequency with minimal attenuation, but attenuates all frequencies below the cutoff frequency.

Horsepower A unit of mechanical power. One horsepower is equivalent to 746 watts.

Hot Electrician's jargon for any wire that is 120 Vac with respect to earth ground.

Hydrometer A device used to measure the specific gravity of a liquid.

Hypotenuse The longest side of a right triangle and opposite the right angle.

Hysteresis The characteristic that causes the flux changes in a magnetic material to lag behind the changes in magnetizing force.

Hysteresis loss A heat loss in a magnetic material caused by the rapid switching of the magnetic domains.

Impedance (Z) Total opposition to current flow in a circuit containing both resistance and reactance. Measured in ohms.

Induced voltage A voltage that is created in a conductor as a result of being intercepted by a moving magnetic field.

Inductance (L) The electrical property that allows a coil to store energy in a magnetic field. It is a characteristic or property that tends to oppose any change in current. Measured in henries.

Induction The process of producing an induced voltage.

Inductive reactance (X_L) The opposition to alternating current flow offered by an inductance. It is measured in ohms.

Inductor A component (coil) that is specifically designed to have inductance.

Input The voltage, current, or power applied to an electrical circuit.

Inrush current A transient current that is higher than normal, but occurs only when power is initially applied to a device or circuit.

Instantaneous value The value of a changing electrical quantity at a specific point in time.

Insulator A material that has a very high resistance and allows no practical current flow.

Integrated circuit A semiconductor device that consists of thousands of transis-

glossary

tors, resistors, and so on integrated into a single wafer of silicon.

Internal resistance An effective resistance that appears to be internal to a device or component. Although it is not generally a physical resistance, it behaves like a series resistance.

Interpolate A process used to estimate the indicated value when the pointer of an analog meter comes to rest between scale markings.

Inversely proportional Two quantities are inversely proportional when increases in one quantity produce corresponding decreases in the second.

Ion A charged particle. An atom that has unequal numbers of electrons and protons.

Ionization The process of creating ions by adding or removing electrons from an otherwise neutral atom.

Ionization voltage The voltage required to ionize the gas in a neon bulb and cause the indicator to emit light.

Isolation transformer A 1:1 transformer used to isolate circuit grounds from earth grounds.

J operator A mathematical prefix or operation that indicates a 90° phase shift.

Joule The SI unit of measure for energy or work.

Junction A point in the circuit where two or more components are joined.

Kilo (k) A prefix representing 10^3.

Kilowatt-hour (kWh) The unit of energy commonly used by electrical power companies.

Kirchhoff's Current Law A fundamental law that says the current entering any point in a circuit must be equal to the current leaving that same point.

Kirchhoff's Voltage Law A fundamental law that states the sum of the voltage drops and voltage sources in any closed loop must be equal to zero.

L/R time constant The time required in an RL circuit for the current to increase or decrease by 63% of the total possible change. Measured in seconds.

Lag To be behind. Generally refers to one sine wave that occurs later in time (out of phase) than a second sine wave.

Laminated Built up from several layers. Material is usually different on adjacent layers.

Lead To be ahead of. Generally refers to one sine wave that occurs earlier in time (out of phase) than a second sine wave.

Leakage current A small (ideally zero) current that flows through an insulator.

Leakage flux *See* Flux Leakage.

Leakage inductance An apparent inductance caused by leakage flux.

LED *See* Light-Emitting Diode.

Lenz's Law The law that states the polarity of an induced voltage will oppose the current change that caused it.

Light-emitting diode (LED) A semiconductor device that emits light. LEDs are used for indicators and digital displays.

Linear Describes a relationship between two quantities that are directly proportional to each other.

Lines of flux Imaginary lines that indicate strength and direction of a magnetic field.

Lissajous pattern A pattern formed on an oscilloscope by applying a sinusoidal waveform to both vertical and horizontal channels. Lissajous patterns can be used to measure frequency and phase.

Load A device or component that draws current from a circuit.

Load current The current that flows through a load connected to a circuit.

Load resistor A resistor connected across the output of a circuit (e.g., a loaded voltage divider).

Loaded voltage divider A voltage divider that supplies power to other circuits or devices.

Loading The changing of a circuit quantity by the connection of another component, circuit, or device.

Local action A localized chemical activity in an electrochemical cell that destroys the electrodes and eventually ruins the cell. Local action can lessen the shelf life of a cell.

Long time constant A time constant (RC or L/R) that is at least ten times greater than the period of the input pulse waveform.

Loop Any closed path for current.

Low-pass filter A filter circuit that passes all frequencies below the cutoff frequency with minimal attenuation, but attenuates all frequencies above the cutoff frequency.

Magnet A material that possesses magnetism.

Magnetic domain *See* Domain.

Magnetic field The region around a magnet where another magnet would experience a force of attraction or repulsion.

Magnetic field intensity *See* Magnetizing Force.

Magnetic flux *See* Lines of Flux.

Magnetic polarity The relative direction of a magnet's field with respect to the earth's magnetic field. The north (actually north-seeking) pole of a magnet is attracted by the earth's north pole.

Magnetizing current The current that flows in the primary of a transformer with no load on the secondary. Power supplied by the magnetizing current is consumed as transformer core and winding losses.

Magnetizing force (H) The magnetomotive force per unit length.

Magnetomotive force ($\mathscr{F}$) The magnetic force produced by current flowing through a coil.

Magnitude The value of a circuit quantity without consideration of phase angle.

Make before break A type of switching contact where the movable part of the contact connects to the new position before contact is broken with the original position.

Matter Anything that occupies space and has weight.

Maximum power transfer theorem A theorem that states that maximum power (50%) can be transferred between two circuits when their resistances are equal.

Maxwell (Mx) A CGS unit of measure for magnetic flux.

Mega (M) A prefix representing 10^6.

glossary

Memory effect A characteristic of certain electrochemical cells (e.g., NiCad cells) that causes them to lose their ability to supply large amounts of power after an extended period of time with only modest loads.

Mesh analysis A technique used to analyze complex circuit configurations.

Meter shunt A resistance connected in parallel with a meter movement that bypasses part of the total current, thus extending the effective range of the movement to a value greater than full-scale current value of the meter movement itself.

Micro (μ) A prefix representing 10^{-6}.

Milli (m) A prefix representing 10^{-3}.

Millman's theorem A simplification technique that is particularly well-suited to circuits having several parallel (nonideal) voltage sources.

Mixture A material composed of more than one element or compound, but whose dissimilar atoms or molecules are not chemically bound together.

MKS *See* SI.

mmf *See* Magnetomotive Force.

Molecule The smallest part of a compound that still exhibits the properties of the compound.

Momentary-contact switch A switch whose contacts change from their normal position when activated (e.g., a button is pressed) but return to their normal position as soon as the mechanical activation force is removed.

Motor A device that converts electrical energy into rotating mechanical energy.

Multimeter A measuring instrument (e.g., VOM, EVM, DVM) capable of measuring resistance, voltage, and current.

Multiplier resistor A resistance connected in series with a meter movement to increase the effective value of voltage required for full-scale deflection.

Mutual inductance Inductance that is common to two magnetically linked coils.

Negative (–) The polarity of charge represented by an excess of electrons.

Negative feedback Feedback that is out of phase with the original signal and causes a decrease in amplitude. Also called degenerative feedback.

Negative ion An atom with fewer protons than electrons.

Network A configuration of electrical components; a circuit.

Neutral Electrician's terminology for the side of the 120-Vac power that is also connected to earth ground.

Neutron An atomic particle located in the nucleus of an atom and having no charge. Its mass is equivalent to that of a proton.

Nodal analysis A circuit analysis technique applicable to complex circuits.

Node Any point in a circuit where the current divides.

Nominal The ideal or expected value of a component or circuit quantity.

Nonlinear Describes a relationship between two quantities that are not directly proportional to each other. Uniform changes in one quantity cause corresponding changes in the second, but the resulting changes are not uniform.

glossary

Normally closed The contacts of a switch or relay that are closed without activating the switch or relay.

Normally open The contacts of a switch or relay that are open without activating the switch or relay.

Norton's theorem A simplification technique that converts a network into an equivalent circuit consisting of a current source and a parallel resistance.

Octave An eightfold (i.e., 8:1 or 1:8) change in the value of a quantity.

Oersted (Oe) The CGS unit of measure for magnetizing force.

Ohm (Ω) The unit of measure for resistance.

Ohm's Law A fundamental law that describes the relationship between current, voltage, and resistance in an electrical circuit.

Ohms per volt *See* Voltmeter Sensitivity.

Ohmic resistance *See* dc Resistance.

Ohmmeter An instrument used to measure resistance.

Open circuit A circuit that has no complete path for current flow; no continuity.

Oscillator A circuit that produces alternating voltage waveforms when a dc voltage is applied.

Oscilloscope An instrument used to display a graph of instantaneous circuit quantities as functions of time.

Output The voltage, current, or power taken from a circuit.

Parallax error An interpretation error on an analog meter movement caused by viewing the pointer and scale marks at an angle.

Parallel A method of connecting circuit components such that all components connect between the same two points.

Paramagnetic Materials with a relative permeability slightly greater than one.

Parameter Any electrical quantity or circuit characteristic.

Partially-specified circuit A circuit where one or more component values are unknown.

Passband The band of frequencies that are passed with minimal attenuation by a filter circuit.

Passive component An electrical component that cannot amplify or rectify. Resistors, inductors, and capacitors are passive components, whereas transistors and other solid-state devices are active components.

Peak value The maximum instantaneous value of a waveform.

Peak-to-peak value The measure of an alternating circuit quantity that describes the difference between the minimum and maximum levels.

Period (T) The time required for one complete cycle of a periodic waveform.

Periodic A waveform that repeats at regular intervals.

Permanent magnet A magnet that does not lose its magnetism when the magnetizing force is removed.

glossary

Permeability (μ) The ability of a material to concentrate magnetic flux. Often expressed as a dimensionless ratio of the permeability of the material relative to air.

Permeance (𝒫) A measure of the ease with which a magnetic field may be established in a material. Analogous to conductance in an electrical circuit.

Permittivity (∈) The ability of a dielectric material to concentrate an electric field.

Phase The timing of a waveform relative to another waveform that has an identical frequency.

Phasor A graphical or numerical representation of the magnitude and phase of an electrical quantity.

Pi configuration *See* Delta Configuration.

Pick-up voltage The coil voltage required to energize a relay.

Pico (p) A prefix representing 10^{-12}.

Piezoelectric effect A phenomenon that causes certain materials to generate a voltage when mechanically stressed.

Point of simplification The point in a circuit that separates components to be replaced by an equivalent circuit from those that will remain.

Polar notation A method of expressing a complex circuit quantity that includes both magnitude and phase information (e.g., $50 \angle 45°$)

Polarity A term used to describe the direction of the relative potential between two points in a circuit. Also used to describe the direction of a magnetic field.

Polarization A buildup of hydrogen gas on the positive electrode of an electro-

chemical cell that degrades the cell's operation.

Pole The movable contact of a relay or switch; also the area of maximum flux density in a magnet.

Positive (+) The polarity of charge represented by a deficiency of electrons.

Positive feedback Feedback that is in phase with the original signal and causes an increased amplitude. Also called regenerative feedback.

Positive ion An atom with fewer electrons than protons.

Potential The ability of an electrical charge to do work by moving electrons.

Potential difference Voltage.

Potentiometer A three-terminal variable resistor. The end-to-end resistance remains constant, but the resistance from either end to the wiper varies as the wiper is moved.

Power (P) The rate of energy consumption; the rate of doing work.

Power factor (pf) The ratio of true power to apparent power in a circuit.

Power supply A device or circuit used to supply electrical energy. In most cases, the power supply delivers direct current and voltage.

Powers of ten A method used to express large or small numbers as a modest number times a power of ten (i.e., ten raised to some exponent).

Primary The winding on a transformer where power is applied.

Primary cell An electrochemical cell that is not designed to be recharged. The chemical processes are not generally reversible (practically).

Proton A positively charged particle in the nucleus of an atom.

Pull-in current The coil current required to energize a relay.

Pythagorean theorem A fundamental theorem that describes the relationship between the sides of a right triangle ($c^2 = a^2 + b^2$).

Quality factor (Q) A dimensionless figure of merit for inductors, capacitors, and resonant circuits. Higher Q values correspond to less losses and more ideal performance.

Radian (rad) An angular measure equal to 57.3°.

RC time constant The time required for the capacitor voltage in an RC circuit to change by 63% of the total possible change.

Reactive power (P_R) A measure of the power that is taken from and subsequently returned to a circuit by the inductance and capacitance in a circuit.

Real number Any rational or irrational number.

Real power *See* True Power.

Rectangular notation A method of representing complex circuit quantities that includes both real (resistive) and imaginary (reactive) portions (e.g., $25 + j10$).

Recurrent sweep A type of sweep circuit in an oscilloscope that causes the hori-

zontal sweep to occur automatically (i.e., without a trigger).

Reflected impedance The apparent impedance in the primary of a transformer that results from a load in the secondary.

Regenerative feedback *See* Positive Feedback.

Relative permeability (μ_r) A dimensionless measurement of the permeability of a material relative to the permeability of air (or vacuum).

Relay An electromechanical device that consists of an electromagnet and switching contacts. The switch contacts are activated by energizing the electromagnet.

Reluctance ($\mathfrak{R}$) A measure of a material's opposition (magnetic resistance) to magnetic flux.

Residual magnetism The level of magnetism remaining in a material after the magnetizing force has been removed.

Resistance (R) Opposition to current flow in a circuit; measured in ohms.

Resistivity (ρ) The resistance of a given volume (e.g., one cubic meter) of a material.

Resistor An electrical component designed to provide a given opposition to current flow.

Resistor tolerance The maximum amount of deviation due to manufacturing tolerances between the nominal and actual values of a resistor.

Resolution The smallest change in a measured value that can be resolved or displayed by a measuring instrument.

glossary

Resonance A condition in an *LC* or *RLC* circuit where $X_L = X_C$ and the impedance of the network is maximum (parallel circuit) or minimum (series circuit).

Resonant circuit An *LC* or *RLC* circuit that is operating at its resonant frequency.

Resonant frequency The frequency in an *LC* or *RLC* circuit that causes the inductive reactance to be equal to the capacitive reactance.

Response time The time required for a fuse to open in response to an overcurrent condition.

Retentivity The property of a magnetic material that causes it to have residual magnetism.

Rheostat A two-terminal variable resistor.

Right angle A 90° angle; an angle of $\pi/2$ radians.

Roll-off The slope of the frequency response curve of a filter circuit beyond the cutoff frequency.

Root-mean-square (rms) A unit of measure for alternating voltage or current that is an amount that produces the same heating effect in a resistance as a similar value of dc. It is numerically equal to 70.7% of the peak value in a sinusoidal circuit.

Rotor The rotating coil of a motor, generator, or alternator.

Schematic A diagram that depicts the various components and interconnections in an electrical circuit.

Secondary A winding on a transformer from which power is removed.

Secondary cell An electrochemical cell that is designed to be recharged. The chemical processes are reversible.

Selectivity The ability of a circuit to pass certain frequencies and reject others.

Self-holding contacts A relay circuit configuration where a normally open set of contacts is in parallel with the switch used to activate the relay. Once energized, the relay remains energized or latched.

Self-inductance The property of a conductor or coil that opposes any change in current.

Self-resonant frequency The frequency at which the parasitic components in a coil or capacitor resonate with the intended quantity.

Sensitivity *See* Voltmeter Sensitivity.

Series A circuit configuration that results in a single path for current flow.

Series-aiding Voltage sources connected in series with similar polarities are series-aiding. The net voltage is the sum of the individual sources.

Series-opposing Voltage sources connected in series with opposite polarities are series-opposing. The net voltage is equal to the difference between the two source voltages.

Series-parallel A circuit configuration that consists of groups of parallel and series components.

Seven-segment display A digital display consisting of seven illuminated segments. Selective illumination of the segments allows display of the numbers 0–9.

Shelf life The length of time that an electrochemical cell can remain inactive and still be expected to deliver its rated characteristics.

Shells Orbital levels of electrons in an atom. Each shell is a discrete region.

Shield A conductive or permeable sheet or enclosure that provides electromagnetic isolation between two circuits.

Short circuit A low-resistance path that essentially bypasses the shorted component(s).

Short time constant An RC or L/R time constant that is less than one-tenth the period of the input pulse.

Shorting switch *See* Make Before Break.

Shunt Parallel; also a resistor used to extend the current range of an ammeter.

SI Système International. A system of measurement based on the meter, kilogram, and second. Sometimes called the modern MKS system.

Siemens (S) The SI unit of measure for conductance, admittance, and susceptance.

Signal A current or voltage that is present in a circuit. Frequently refers to an ac waveform.

Sine A trigonometric function of an angle in a right triangle that is equal to the length of the opposite side divided by the length of the hypotenuse.

Sine wave A periodic waveform whose amplitude-versus-time graph is the same shape as a graph of the trigonometric sine function versus degrees.

Sinusoidal Any waveform (regardless of phase) that is generally shaped like a sine wave.

Skin effect A phenomenon that causes high-frequency currents to flow near the surface of a conductor, which reduces the effective cross-sectional area of the wire and increases the effective (ac) resistance of the wire.

Slip rings Metal rings on the rotor of an alternator or ac motor that make sliding contact with the brushes and provide a means of connecting or removing power from a rotating coil.

Slo-Blo A type of fuse that can withstand currents greater than its rating as long as the excessive current is only momentary.

Solar cell A solid-state device that converts light energy into electricity.

Solder A tin and lead alloy with a low melting point that is used to permanently bond electrical connections.

Solenoid In general, any coil wound on a long coil form. Also refers to an electromagnetic device that converts electrical energy into a pushing, pulling, or twisting motion.

Specific gravity The ratio of the weights of equal quantities of some liquid and water.

Spiraling A manufacturing process used to trim film resistors to the correct value.

Stator The stationary or nonrotating part of a motor, alternator, or generator.

Step-down transformer A transformer whose secondary voltage is lower than its primary voltage.

glossary

Step-up transformer A transformer whose secondary voltage is higher than its primary voltage.

Substrate The insulating base material used as the core of film resistors. Also the base semiconductor layer in a transistor or integrated circuit.

Superconductivity A characteristic of certain materials that causes their effective resistance to drop to zero as their temperature approaches absolute zero.

Superposition A simplification procedure for multiple-source linear circuits.

Surge A short-duration transient or current burst.

Susceptance (B) The ease with which current flows through a reactive component; the reciprocal of reactance; measured in siemens.

Switch An electrical component used to open and close a current path in a circuit.

Symmetrical A waveform whose positive and negative alternations are equal in amplitude, duration, and shape.

Tangent A trigonometric function of an angle in a right triangle that is equal to the length of the opposite side divided by the length of the adjacent side.

Tank circuit A parallel LC circuit.

Taper Describes the relationship (linear or logarithmic) between the resistance and the angle of rotation of a variable resistor.

Tee configuration A circuit configuration where one end of each of three components connects to a common point. Also called a wye configuration. One of two

common connections for three-phase transformer windings.

Temperature coefficient Describes the relationship between a circuit parameter and its temperature. May be positive, negative, or zero.

Temporary magnets A magnetic material with minimal residual magnetism.

Tesla The SI unit of measure for flux density.

Thermistor An electrical device whose resistance varies with temperature in a specific way.

Thermocouple A bimetallic junction that converts heat energy into electrical energy.

Thermopile A series connection of several thermocouples.

Thevenin's theorem A simplification technique that converts a network into an equivalent circuit consisting of a voltage source and a series resistance.

Throw Identifies the number of circuits that are opened or closed by each pole of a switch.

Time constant A fixed time interval determined by the RC or RL values in an RC or RL circuit that specifies the time required for the voltage and current to change by 63% of the total possible change.

Tolerance The maximum amount of deviation due to manufacturing tolerances between the nominal and actual values of a component.

Toroid A donut-shaped object. Usually refers to a toroidal-shaped core used for a transformer or coil.

Transformer A device that couples electrical energy from one circuit (primary winding) to another (secondary winding) via magnetic flux linkage.

Transient A momentary, short-duration voltage or current surge.

Triggered sweep A type of sweep circuit used in oscilloscopes that produces no horizontal sweep until a specific set of conditions exists (such as the input voltage is at a given level).

Trimmer A variable component used to adjust a circuit parameter to a precise value. Generally provides a very limited range of adjustment (fine tuning).

Tripped A term used to describe the condition of a circuit breaker after it has opened the circuit.

Troubleshooting The process of locating the defective components in an electrical circuit.

True power (P_T) The power, measured in watts, that is dissipated by the resistance in a circuit.

Turns ratio The ratio of primary-to-secondary turns in a transformer.

Universal shunt A method of connecting and switching an ammeter shunt to provide multiple ranges.

Unloaded voltage divider A voltage divider that does not provide power to other circuits or devices.

Valence electrons Electrons in the outermost orbit of an atom.

VAR Abbreviation for volt-ampere-reactive. The unit of measure for reactive power.

Vector A graphical representation of the magnitude and phase of an electrical quantity or the magnitude and direction of a mechanical force.

Volt (V) The SI unit of measure for electromotive force.

Voltage (V) The amount of potential difference between two points that can be used to cause current flow.

Voltage breakdown The condition that exists when an insulator material is subjected to a sufficiently high voltage to destroy the high-resistance characteristics of the insulator. Breakdown results in a high current that may or may not permanently damage the insulator.

Voltage divider A series circuit used to reduce a supply voltage to one or more lower voltages.

Voltage drop (V) The voltage developed across a component as a result of the current flowing through it.

Voltage regulation The process of maintaining a relatively constant output from a voltage source even when the load current or power-line voltage varies.

Voltage source A circuit or device that provides a relatively constant voltage to other circuits or devices.

Voltmeter An instrument used to measure voltage.

Voltmeter loading Occurs when the internal resistance of a voltmeter causes the measured voltage to be less than the actual (unloaded) voltage in the circuit. The higher the internal resistance of a voltmeter, the less it disrupts or loads the circuit.

glossary

Voltmeter sensitivity The ohms-per-volt rating of a voltmeter.

VOM Volt-ohm-milliammeter. An analog meter that measures voltage, current, and resistance.

Watt (W) The SI unit of measure for power.

Watt's Law A fundamental law that describes the relationships between power, current, voltage, and resistance.

Wattmeter An instrument used to measure true power.

Waveform A graph of voltage, current, or other circuit parameter versus time.

Wavelength The physical distance that an acoustical or electromagnetic wave travels in the time required for one complete cycle of the waveform.

Weber (Wb) The SI unit of measure for magnetic flux.

Wheatstone bridge *See* Bridge Circuit.

Winding The turns of wire on an electromagnetic device such as a coil, transformer, relay, or motor.

Wiper A movable, sliding contact; the center terminal of a potentiometer.

Work The expenditure of energy.

Wye configuration *See* Tee Configuration.

Z-axis Refers to intensity control of the CRT display in an oscilloscope.

Zeroed The condition of an ohmmeter that has been calibrated such that when the leads are shorted together, the pointer indicates zero ohms.

index

)